Optimization and Flow Characteristics in Advanced Fluid Machinery

Optimization and Flow Characteristics in Advanced Fluid Machinery

Editors

Chuan Wang

Li Cheng

Qiaorui Si

Bo Hu

MDPI • Basel • Beijing • Wuhan • Barcelona • Belgrade • Manchester • Tokyo • Cluj • Tianjin

Editors
Chuan Wang
College of Hydraulic Science and Engineering
Yangzhou University
Yangzhou
China

Li Cheng
College of Hydraulic Science and Engineering
Yangzhou University
Yangzhou
China

Qiaorui Si
National Research Center of Pumps
Jiangsu University
Zhenjiang
China

Bo Hu
Department of Energy and Power Engineering
Tsinghua University
Beijing
China

Editorial Office
MDPI
St. Alban-Anlage 66
4052 Basel, Switzerland

This is a reprint of articles from the Special Issue published online in the open access journal *Machines* (ISSN 2075-1702) (available at: www.mdpi.com/journal/machines/special_issues/Advanced_Fluid_Machinery).

For citation purposes, cite each article independently as indicated on the article page online and as indicated below:

LastName, A.A.; LastName, B.B.; LastName, C.C. Article Title. *Journal Name* **Year**, *Volume Number*, Page Range.

ISBN 978-3-0365-8293-1 (Hbk)
ISBN 978-3-0365-8292-4 (PDF)

Contents

About the Editors

Chuan Wang

Chuan Wang is currently a Full Professor of Hydraulic Engineering at Yangzhou University, China. He received his PhD degree in Fluid Machinery and Engineering from Jiangsu University in December 2015 and stayed on as a member of the faculty before transferring to Yangzhou University in 2019. At present, his main research interests include internal flow theory and the optimal design of pumps, the interference mechanism between a jet and rotating fluid, etc. As the leader, he has completed two National Natural Science Foundation of China projects and two Postdoctoral Science Foundation of China projects. As the first author or corresponding author, he has published more than 30 SCI-indexed papers, including 6 papers that are included in the CAS Top Journals, 12 papers that are included in the ESI Highly Cited Papers (top 1%), and 6 papers that are included in the ESI Hot Papers (top 0.1%). Moreover, he received the First Prize of the Science and Technology Award of the China Business Federation (ranked 1st), First Prize of the Innovation Achievement Award of the China Industry-University-Research Cooperation (ranked 3rd), First Prize of the Science and Technology Progress Award of the China Machinery Industry (ranked 4th), Second Prize of the Science and Technology Progress Award of Jiangsu Province (ranked 5th), and Third Prize of the Science and Technology Progress Award of Shandong Province (ranked 3rd). He was invited to be a correspondence reviewer for the National Natural Science Foundation of China several times. His research results were recognized as a "Key Scientific Article" by the Canadian Advances in Engineering website. His supervised work, "Research on the Key Technology of High-Efficiency Self-priming Centrifugal Pumps for Rapid Disaster Relief and Rescue", won the Grand Prize at the 16th Challenge Cup Works Competition of China.

Li Cheng

Li Cheng is a Full Professor at the College of Hydraulic Science and Engineering, Yangzhou University, China. Currently, he is the director of the China Water Resources Society, and the deputy director of the Pumps and Pumping Stations Specialized Committee of the China Water Resources Society.

Qiaorui Si

Qiaorui Si is a Full Professor at the Research Center of Fluid Machinery Engineering and Technology, Jiangsu University, China. He received his PhD degree in Fluid Machinery and Engineering from Jiangsu University in December 2014 and stayed on as a member of the faculty.

Bo Hu

Bo Hu is a Full Assistant Professor in the Department of Energy and Power Engineering, Tsinghua University, China. Currently, he has published more than 20 manuscripts in the past 10 years, and most of them have been indexed by SCIE.

machines

MDPI

Editorial

Special Issue on Optimization and Flow Characteristics in Advanced Fluid Machinery

Chuan Wang [1,2]

[1] International Shipping Research Institute, GongQing Institute of Science and Technology, Jiujiang 332020, China; wangchuan198710@126.com

[2] College of Hydraulic Science and Engineering, Yangzhou University, Yangzhou 225009, China

Abstract: This editorial discusses the importance of Advanced Fluid Machinery in the sustainable development of energy. Fluid machinery is crucial in many engineering applications, including aerospace, civil, mechanical, and chemical engineering. This Special Issue, entitled "Optimization and Flow Characteristics in Advanced Fluid Machinery", features several research articles exploring flow characteristics and optimization in fluid mechanics. The authors present innovative ideas, methodologies, and techniques to advance the field of fluid mechanics. The papers cover a wide range of topics, including computational fluid dynamics (CFD), turbulence modeling, heat transfer, multiphase flow, and fluid–structure interactions. The articles featured in this Special Issue also investigate the relevant hydrodynamic attributes of turbomachinery, high-pressure jets, marine propulsion systems, and internal combustion engines to a considerable extent, significantly expanding the scope of research within the Special Issue.

Keywords: fluid machinery; optimization; flow characteristics; CFD

Citation: Wang, C. Special Issue on Optimization and Flow Characteristics in Advanced Fluid Machinery. *Machines* **2023**, *11*, 718. https://doi.org/10.3390/machines11070718

Received: 28 June 2023
Accepted: 3 July 2023
Published: 6 July 2023

Advanced Fluid Machinery is the key component for the sustainable development of energy and water resources, including various transport processes for liquids. Where fluid flows, fluid machinery works. Therefore, fluid machinery occupies an important position in the social economy. This Special Issue, entitled "Optimization and Flow Characteristics in Advanced Fluid Machinery", will promote a platform for the sharing of knowledge among researchers in the field of fluid machinery, including theoretical analysis, numerical simulation, and experimental study.

Fluid mechanics is a field that is fundamental to numerous engineering applications such as aerospace, civil, mechanical, and chemical engineering. Understanding the behavior of fluids and their flow characteristics is crucial in designing effective and efficient systems. Optimization of these systems is equally important to enhance their performance and reduce energy consumption. Therefore, the Special Issue focuses on the optimization of fluid mechanics systems to improve their efficiency and performance.

The Special Issue comprises several research articles that explore various aspects of flow characteristics and optimization in fluid mechanics. The papers in this Special Issue cover a wide range of topics, including computational fluid dynamics, turbulence modeling, heat transfer, multiphase flow, and fluid–structure interactions. The authors have presented innovative ideas, methodologies, and techniques that have the potential to advance the field of fluid mechanics.

The ability to move fluids is essential for many industrial processes, and pumps are often the most efficient and effective way to achieve this. By providing a steady flow of fluid, pumps help to maintain consistent pressure, temperature, and other important parameters, which is critical for ensuring that processes operate correctly and safely. In [1], a volute reference scheme with passive rotation speed is proposed, which provides new ideas about rotor–rotor interference in dishwasher innovation. Another study [2] proposes impellers with different hub radii for self-priming pumps and analyzes the internal flow characteristics, blade surface load, pressure pulsation characteristics, and radial force distribution of each scheme. The optimization of diffuser blades in multistage submersible pumps is discussed in [3]. The study used numerical simulation to optimize the lean and sweep of the diffuser blades to improve efficiency and pressure recovery coefficient while reducing non-uniformity. Similarly, the influence of blade structure on the performance of a vortex pump is investigated in [4]. In [5], an automated optimization platform is presented for improving the operating efficiency of mixed flow pumps driven by hydraulic motors. The platform optimizes both the meridional shape and blade profiles of the impeller and diffuser simultaneously using CFD and genetic algorithms. In [6], the internal flow characteristics of a slurry pump are investigated, and the effects of different particle volume concentrations, particle sizes, and pump speeds on the impeller's wear characteristics and hydraulic performance are studied. In [7], the cavitation performance of a liquefied natural gas (LNG) submerged pump and pressure pulsation characteristics are analyzed. In [8], the correlation between internal flow patterns and blade load distributions in centrifugal impellers is investigated using particle image velocimetry (PIV). Another study [9] investigates the formation mechanism of the secondary tip leakage vortex (S-TLV) in cavitating flows using numerical simulations. Tip clearance is another important parameter affecting pump performance, as investigated in [10]. The article studies the influence of tip clearance on the performance and energy dissipation of an axial flow pump, as well as an axial flow pump operating as a turbine. The hydrodynamic characteristics of bidirectional axial flow pumps are investigated in [11], including pressure pulsations, under positive and negative operation. Article [12] uses CFD, one-dimensional theory, and response surface methodology to enhance the performance of a diving tubular pump, an important part of underground water conveyance equipment. In [13], a numerical simulation study of a vertical axial flow pump is presented. Article [14] investigates the impact of non-uniform suction flow on the transient characteristics of a vertical axial flow pump through unsteady calculations and external characteristic tests. The effects of different coolants and working temperatures on the hydraulic and cavitation performances of two impellers with different types of blades in a centrifugal pump for aircraft cooling systems are investigated in [15]. Article [16] presents an optimization system based on CFD, optimized Latin hypercube sampling (OLHS), machine learning (ML), and multi-island genetic algorithm (MIGA) to improve the efficiency of high-specific-speed axial flow pumps under non-design conditions. The Gaussian process regression (GPR) model was found to have the highest prediction accuracy for impeller head and weighted efficiency. The authors of [17] investigated the effect of blade number on the internal flow condition of a high-specific-speed centrifugal pump. The start-up characteristics of large axial flow pump systems and the challenges they face during the start-up process are investigated in [18].

The articles featured in this Special Issue also investigate the relevant hydrodynamic attributes of turbomachinery, high-pressure jets, marine propulsion systems, and internal combustion engines to a considerable extent, thereby significantly expanding the scope of research within the Special Issue. A new method for precise position control of hydraulic cylinders in hydraulic support used in fully mechanized mining faces is proposed in [19]. Article [20] investigates the cavitation behavior of submerged high-pressure jets with organ pipe nozzles of different outlet shapes. Article [21] optimizes the structural parameters of a coal-breaking and punching nozzle for high-pressure water jet technology in low-permeability coal seams to improve jet performance and increase coal seam permeability. Article [22] focuses on the static characteristics of aerostatic journal bearings with porous bushing, and a flow model is established based on the Reynolds lubrication equation, Darcy equation for porous material, and continuity equation. A new framework for the application of the Harmonic Balance Method (HBM) in the open-source software OpenFOAM is introduced in [23], which is commonly used for turbomachinery calculations. Article [24] evaluates the impact of the Miller cycle on the internal aerodynamics of a spark ignition engine and recommends the need to intensify the internal aerodynamics to prevent significant turbulence degradation while still gaining efficiency in the Miller cycle. In addition, the cavitation resistance of small ship propellers in the coastal waters of South Korea is investigated through a demonstration test in [25]. Finally, Article [26] examines the flow characteristics within the pre-swirl system of a marine gas turbine at low rotational speed and shows the effects of nozzle pressure on radial velocity, core swirl ratio, and pressure.

Overall, this Special Issue presents an excellent opportunity for researchers to explore new ideas and approaches to optimize fluid mechanics systems. The papers in this issue highlight the importance of optimization in enhancing the performance and efficiency of fluid mechanics applications. It is hoped that this Special Issue will inspire further research in this area, leading to new breakthroughs and innovations in the field of fluid mechanics.

Acknowledgments: Thanks to all the authors and peer reviewers for their valuable contributions to the Special Issue "Optimization and Flow Characteristics in Advanced Fluid Machinery".

Conflicts of Interest: The author declares no conflict of interest.

References

1. Zhu, Y.; Zhang, J.; Li, Y.; Huang, P.; Xu, H.; Zheng, F. Experimental Investigation of Unsteady Pressure Pulsation in New Type Dishwasher Pump with Special Double-Tongue Volute. *Machines* **2021**, *9*, 288.
2. Chang, H.; Hong, S.; Wang, C.; Peng, G.; Fan, F.; Ye, D. Investigation on the Transient Characteristics of Self-Priming Pumps with Different Hub Radii. *Machines* **2021**, *9*, 311. [CrossRef]
3. Ning, C.; Cao, P.; Gong, X.; Zhu, R. Optimization of Sweep and Blade Lean for Diffuser to Suppress Hub Corner Vortex in Multistage Pump. *Machines* **2021**, *9*, 316. [CrossRef]
4. Quan, H.; Li, Y.; Kang, L.; Yu, X.; Song, K.; Wu, Y. Influence of Blade Type on the Flow Structure of a Vortex Pump for Solid-Liquid Two-Phase Flow. *Machines* **2021**, *9*, 353. [CrossRef]
5. Lu, R.; Yuan, J.; Wei, G.; Zhang, Y.; Lei, X.; Si, Q. Optimization Design of Energy-Saving Mixed Flow Pump Based on MIGA-RBF Algorithm. *Machines* **2021**, *9*, 365.
6. Peng, G.; Tian, L.; Chang, H.; Hong, S.; Ye, D.; You, B. Numerical and Experimental Study of Hydraulic Performance and Wear Characteristics of a Slurry Pump. *Machines* **2021**, *9*, 373. [CrossRef]
7. Li, W.; Li, S.; Ji, L.; Zhao, X.; Shi, W.; Agarwal, R.K.; Awais, M.; Yang, Y. A Study on the Cavitation and Pressure Pulsation Characteristics in the Impeller of an LNG Submerged Pump. *Machines* **2022**, *10*, 14. [CrossRef]
8. Chen, B.; Chen, X.; Zhu, Z.; Li, X. Correlation between the Internal Flow Pattern and the Blade Load Distribution of the Centrifugal Impeller. *Machines* **2022**, *10*, 40. [CrossRef]
9. Zhang, H.; Zang, J.; Zhang, D.; Shi, W.; Shen, J. Analysis of the Formation Mechanism of Secondary Tip Leakage Vortex (S-TLV) in an Axial Flow Pump. *Machines* **2022**, *10*, 41. [CrossRef]
10. Li, Y.; Lin, Q.; Meng, F.; Zheng, Y.; Xu, X. Research on the Influence of Tip Clearance of Axial-Flow Pump on Energy Characteristics under Pump and Turbine Conditions. *Machines* **2022**, *10*, 56. [CrossRef]
11. Zhang, X.; Tang, F.; Chen, Y.; Huang, C.; Chen, Y.; Wang, L.; Shi, L. Experimental Study on the Internal Pressure Pulsation Characteristics of a Bidirectional Axial Flow Pump Operating in Forward and Reverse Directions. *Machines* **2022**, *10*, 167. [CrossRef]
12. Yang, X.; Tian, D.; Si, Q.; Liao, M.; He, J.; He, X.; Liu, Z. Flow Loss Analysis and Optimal Design of a Diving Tubular Pump. *Machines* **2022**, *10*, 175. [CrossRef]
13. Liu, X.; Xu, F.; Cheng, L.; Pan, W.; Jiao, W. Stress Characteristics Analysis of Vertical Bi-Directional Flow Channel Axial Pump Blades Based on Fluid–Structure Coupling. *Machines* **2022**, *10*, 368. [CrossRef]
14. Meng, F.; Qin, Z.; Li, Y.; Chen, J. Investigation of Transient Characteristics of a Vertical Axial-Flow Pump with Non-Uniform Suction Flow. *Machines* **2022**, *10*, 855. [CrossRef]
15. Wu, A.; Zhao, R.; Wang, F.; Zhang, D.; Wang, X. Effects of Coolant and Working Temperature on the Cavitation in an Aeronautic Cooling Pump with High Rotation Speed. *Machines* **2022**, *10*, 904. [CrossRef]
16. Sun, Z.; Tang, F.; Shi, L.; Liu, H. Multi-Conditional Optimization of a High-Specific-Speed Axial Flow Pump Impeller Based on Machine Learning. *Machines* **2022**, *10*, 1037. [CrossRef]
17. Wang, C.; Chen, X.; Ge, J.; Cao, W.; Zhang, Q.; Zhu, Y.; Chang, H. Internal Flow Characteristics of High-Specific-Speed Centrifugal Pumps with Different Number of Impeller Blades under Large Flow Conditions. *Machines* **2023**, *11*, 138.
18. Zhang, X.; Jiang, Y.; Tang, F.; Song, X.; Liu, Y.; Yang, F.; Shi, L. Investigation on Start-Up Characteristics of Large Axial Flow Pump Systems Considering the Influence of Auxiliary Safety Facilities. *Machines* **2023**, *11*, 182.
19. Zhou, R.; Meng, L.; Yuan, X.; Qiao, Z. Research and Experimental Analysis of Hydraulic Cylinder Position Control Mechanism Based on Pressure Detection. *Machines* **2022**, *10*, 1.
20. Wang, G.; Yang, Y.; Wang, C.; Shi, W.; Li, W.; Pan, B. Effect of Nozzle Outlet Shape on Cavitation Behavior of Submerged High-Pressure Jet. *Machines* **2022**, *10*, 4.

21. Chen, L.; Cheng, M.; Cai, Y.; Guo, L.; Gao, D. Design and Optimization of High-Pressure Water Jet for Coal Breaking and Punching Nozzle Considering Structural Parameter Interaction. *Machines* **2022**, *10*, 60. [CrossRef]
22. Gu, Y.; Cheng, J.; Xie, C.; Li, L.; Zheng, C. Theoretical and Numerical Investigations on Static Characteristics of Aerostatic Porous Journal Bearings. *Machines* **2022**, *10*, 171. [CrossRef]
23. Oliani, S.; Casari, N.; Carnevale, M. A New Framework for the Harmonic Balance Method in OpenFOAM. *Machines* **2022**, *10*, 279. [CrossRef]
24. Perceau, M.; Guibert, P.; Clenci, A.; Iorga-Simăn, V.; Niculae, M.; Guilain, S. Investigation of the Aerodynamic Performance of the Miller Cycle from Transparent Engine Experiments and CFD Simulations. *Machines* **2022**, *10*, 467. [CrossRef]
25. Ju, H.-J.; Choi, J.-S. Experimental Study of Cavitation Damage to Marine Propellers Based on the Rotational Speed in the Coastal Waters. *Machines* **2022**, *10*, 793. [CrossRef]
26. Hu, B.; Yao, Y.; Wang, M.; Wang, C.; Liu, Y. Flow and Performance of the Disk Cavity of a Marine Gas Turbine at Varying Nozzle Pressure and Low Rotation Speeds: A Numerical Investigation. *Machines* **2023**, *11*, 68.

machines

MDPI

Article

Investigation on Start-Up Characteristics of Large Axial Flow Pump Systems Considering the Influence of Auxiliary Safety Facilities

Xiaowen Zhang, Yuhang Jiang, Fangping Tang *, Xijie Song, Yuxi Liu, Fan Yang and Lijian Shi

College of Hydraulic Science and Engineering, Yangzhou University, Yangzhou 225009, China
* Correspondence: tangfp_yzu@126.com; Tel.: +86-1395-1059-552

Abstract: A large number of operation practices show that the reliability and stability of large axial flow pump systems will face significant challenges during the start-up process. If the start-up control mode or safety auxiliary facilities of large axial flow pump stations are unreasonable, start-up failure will easily follow. In order to find a scientific control strategy for the start-up of large axial flow pump stations, the start-up characteristics of large axial flow pump stations must be fully understood first. In this paper, based on the secondary development of Flowmaster software, a simulation study of the start-up process of a large axial flow pump system equipped with different safety aids is carried out. It is found that it is a very dangerous start-up control mode to delay the opening of the rapid-drop gate to reduce the maximum reflux value and reflux duration when the pump system is initially started. When the rapid-drop gate opens with a delay of 4 s, the power overload coefficient reaches 23.49, indicating that the possibility of start-up failure of the large axial flow pump system increases sharply the longer the gate delay is opened. The method of adding a flap valve to the rapid-drop gate can significantly weaken the instantaneous impact power of the unit and prevent the unit from overload. When safety auxiliary facilities with an additional disc valve on the fast descending gate are adopted, the backflow coefficient is within 0.2, the impact head coefficient is within 2, and the power overload coefficient is less than 0. The research results will provide an important reference value for comprehensively understanding the start-up characteristics of large axial flow pump stations and finding scientific and safe start-up control strategies.

Keywords: large axial flow pump system; start-up characteristics; safety auxiliary facilities; numerical simulation; model experiment

Citation: Zhang, X.; Jiang, Y.; Tang, F.; Song, X.; Liu, Y.; Yang, F.; Shi, L. Investigation on Start-Up Characteristics of Large Axial Flow Pump Systems Considering the Influence of Auxiliary Safety Facilities. *Machines* **2023**, *11*, 182. https://doi.org/10.3390/machines11020182

Academic Editor: Kim Tiow Ooi

Received: 27 December 2022
Revised: 23 January 2023
Accepted: 26 January 2023
Published: 28 January 2023

1. Introduction

In order to meet the needs of drainage, irrigation, and inter-basin water transfer, the large axial flow pump station system (LAPS) is widely used in many countries in the world. At present, the optimization and improvement of the energy characteristics of the LAPS have reached a design bottleneck, and the reliability and stability of the LAPS have become the main problems in the development of the LAPS [1–3]. A number of operation practices show that the reliability and stability of the LAPS will face the greatest challenge during the start-up process, and operating parameters such as flow, head, shaft power and impeller torque change sharply during the start-up process of the LAPS. The inertia of fluid flow and the inertia of machine motion cause great dynamic additional load, and most of the accidents involving a LAPS occur during the start-up of units. Exploring a reasonable start-up control mode and finding effective auxiliary facilities to improve the safety and reliability of the unit in the start-up process has become an urgent task for the development of the LAPS [4–6].

Many experts and scholars have carried out research on the transition process of hydraulic machinery and its systems, and the research on the transition process of hydropower stations or pump stations is more mature [7,8]. Its theoretical basis is mainly rigid water hammer theory and elastic water hammer theory, and the analysis method is mainly the characteristic line method in computational fluid dynamics. The characteristic line method in computational fluid dynamics has been tested by engineering field tests, and has good adaptability and accuracy in dealing with the transition process of hydraulic machinery. In recent years, some scholars have tried to use the method of three-dimensional numerical simulation to study the evolution law of the internal flow field in the transition process of hydraulic machinery. However, application of three-dimensional numerical simulation in hydraulic machinery systems, especially in the LAPS, is very time-consuming. It often requires a huge calculation cost to calculate a single case, and the accuracy is difficult to guarantee. It is difficult to compare and discuss a large number of different cases. At the same time, hydraulic machinery systems such as the LAPS are often equipped with many kinds of safety auxiliary facilities [9–11], such as overflow holes, flap valves, rapid-drop gates and so on. Three-dimensional numerical simulation will face multiple exit boundaries such as overflow holes, flap valves, rapid-drop gates and so on, and the outlet boundary conditions of the flap valve and the rapid-drop gate change in real time, so the convergence of three-dimensional numerical simulation is extremely difficult [12,13].

At present, the progress of research on different hydraulic machinery systems is very uneven, and most of the research results are concentrated on hydropower stations, pumped storage power stations and centrifugal pump stations. The research on the transition process of the LAPS is almost non-existent [14–17]. On the one hand, the related research of the LAPS started relatively late, on the other hand, the problem of the transition process of the LAPS has its particularity [18,19]. Firstly, the inlet system and outlet system of the LAPS are distorted and irregular short pipes, and most of them are cast-in-place reinforced concrete structures, so water compressibility, wave propagation and pipeline elasticity cannot be the decisive factors of the transition process. Secondly, the core content of the dynamic process research of the LAPS is not to study the abnormal water hammer pressure in the system, but to establish the relationship between the characteristic parameters of the pump unit to prevent overload and lack of control [20–22].

According to the field operation experience of the LAPS, if the start-up control mode or safety auxiliary facilities of the LAPS are unreasonable, its start-up will easily fail. There are two most common main reasons for start-up failure of the LAPS [23–25]. Firstly, the LAPS falls into the saddle zone of the axial flow pump during the start-up process, which leads to the instability or even failure of the start-up of the LAPS. Secondly, during the start-up of the LAPS, the instantaneous impact power exceeds the maximum power limit of the unit, which leads to the overload of the unit and the failure of the start-up. Therefore, whether the LAPS falls into the saddle zone of the axial flow pump during the start-up process, and whether the instantaneous impact power is overloaded, are the core factors in reviewing whether the transition process of the LAPS is safe [26–28]. At the same time, because the backflow in the LAPS will cause impact and damage to the pump blade at the initial stage of the start-up of the LAPS, parameters such as the maximum backflow flow (MBF) and backflow duration of the LAPS during the start-up process are also worthy of attention.

In this paper, the start-up characteristics of the LAPS are studied by combining one-dimensional modeling, three-dimensional modeling and a model experiment. Firstly, three-dimensional modeling of the LAPS is carried out, and the pressure drop curves of the inlet channel, outlet channel and flap valve under different flow conditions are established by the CFD method. Secondly, the secondary development of Flowmaster V7 software is carried out, and the pump experiment performance curve obtained from the model experiment is stored in the database of Flowmaster in the form of the Suter curve, and the pressure drop characteristics of inlet and outlet channels predicted by CFD are added to the self-developed flow resistance element. Thirdly, an energy characteristic experiment and power-off runaway experiment are carried out on the pump system model, and the experimental results are compared with the software simulation results to verify the accuracy of the simulation calculation. Finally, the start-up characteristics of LAPS units with different auxiliary facilities and different start-up control modes are studied, and the relationship and variation law of characteristic parameters in the start-up process are revealed.

2. Numerical Simulation and Experimental Setup

2.1. Physical Model

This paper took a typical LAPS in China as its research subject. A typical LAPS consists of an inlet channel, axial flow pump, outlet elbow, outlet channel and safety auxiliary facilities. Among them, the safety auxiliary facilities include a rapid-drop gate, flap valve and overflow hole. Figure 1 shows the diagram and main parameters of the LAPS. Since the start-up under the maximum net head H_n is the most unfavorable condition in the start-up process of the LAPS, the start-up process research in this paper is based on the maximum net head as the calculation boundary. The specific speed of the pump is 806, the impeller inlet diameter is 1.87 m, and the impeller outlet diameter is 1.96 m. The inertia moment of the LAPS I_p is 425.80 kg·m^2, the motor rotation inertia I_m is 3350 kg·m^2, the upper limit of motor power P_u is 1000 kW, and the area of the rapid-drop gate A_r is 13.25 m^2. It should be noted that the rapid-drop gate is opened or closed at a constant speed.

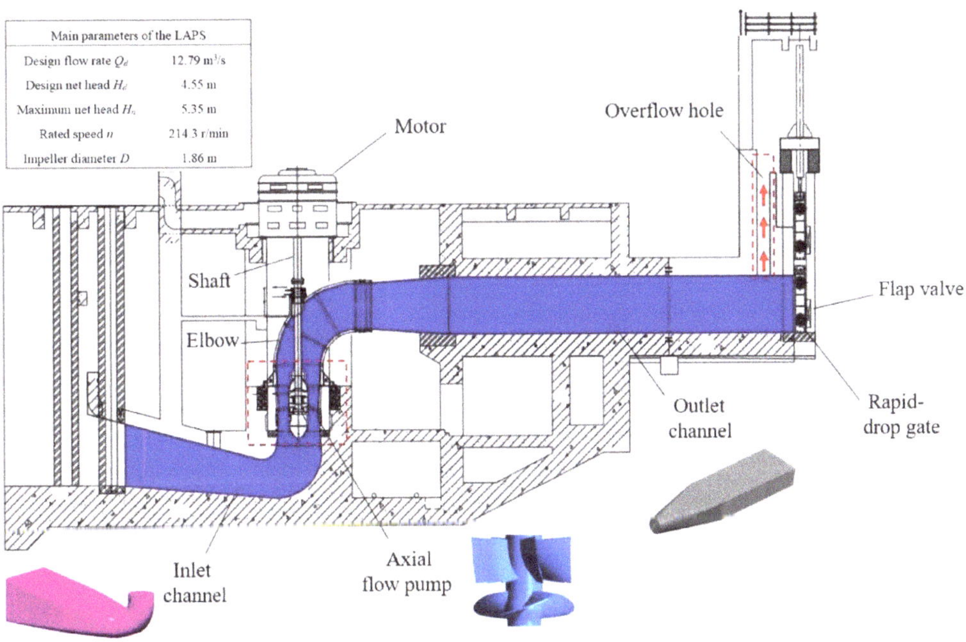

Main parameters of the LAPS	
Design flow rate Q_d	12.79 m^3/s
Design net head H_d	4.55 m
Maximum net head H_n	5.35 m
Rated speed n	214.3 r/min
Impeller diameter D	1.86 m

Figure 1. Diagram and main parameters of the large axial flow pump system.

2.2. Simulation Process and Mathematical Model

Before the simulation based on Flowmaster was carried out, the pressure drop characteristics of the inlet and outlet channels were calculated by CFD method (based on ANSYS CFX 17.2 software). The grid diagram of the core components of LAPS is shown in Figure 2, which has been densified at the blade shroud. The impeller and guide vane are divided by structured grids, and the other parts are divided by unstructured grids. Figure 3 shows the grid independence analysis of the head and efficiency of LAPS under the design flow condition during the steady-state condition. It can be seen that when the total number of grids increases to 5.16 million, the head and efficiency are basically stable and meet the requirements of grid independence. In order to save calculation time and cost, the scheme in this paper is simulated numerically with the number of grids set at 5.16 million. Table 1 shows the number of grids and nodes of each flow passage component. It can be seen that the number of grids and nodes of the inlet channel is 1,083,798 and 191,421 respectively, and the number of grids and nodes of the outlet channel is 910,224 and 158,193. The CFD calculation is based on the Reynolds time-averaged N-S equation, which is discretized by the finite volume method and the SIMPLEC algorithm [29–31]. The pressure term is in the standard format, and the momentum, turbulent kinetic energy and dissipation rate terms are in the second-order upwind scheme. For the setting of boundary conditions, velocity inlet conditions and free outflow conditions are used as inlet and outlet boundary conditions, respectively.

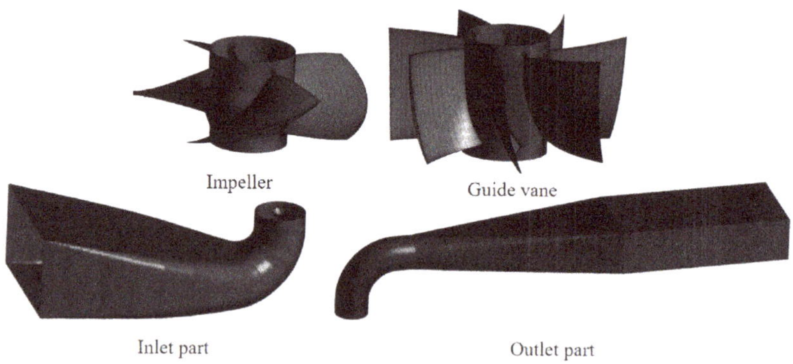

Figure 2. Grid diagram of each flow passage components.

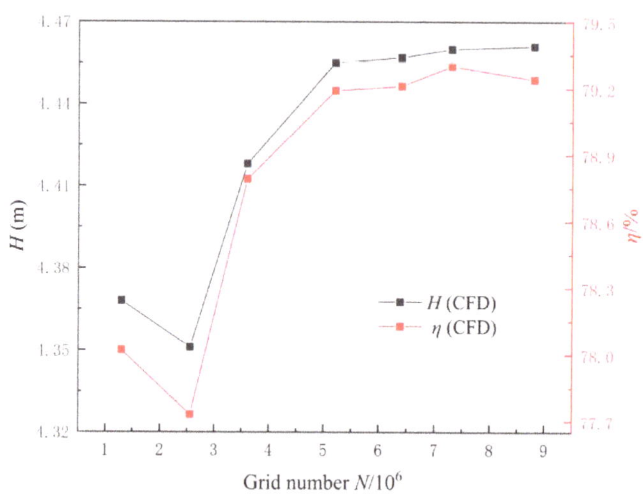

Figure 3. Grid independence analysis of head and efficiency of the large axial flow pump system.

Table 1. Number of grids and nodes of each flow passage component.

Flow Passage Component	Number of Grids	Number of Nodes
Inlet channel	1,083,798	191,421
Impeller	1,296,256	1,373,248
Guide vane	1,438,255	1,538,250
Outlet elbow	432,533	78,380
Outlet channel	910,224	158,193
Total	5,161,066	3,339,492

The simulation of the start-up transition process was realized by Flowmaster software and its self-defined secondary development platform [32]. Flowmaster regards the fluid pipeline system as the object, and regards the fluid pipeline system as a series of fluid pipeline elements. The pipeline is directly connected by nodes, and the continuity equation and momentum equation are used to linearly describe the fluidity of each element in the system. The simplified fluid network solution equation is as follows [33]:

$$\overline{v}\frac{\partial h}{\partial \mathrm{x}} + \frac{\partial h}{\partial t} - \overline{v}\sin\alpha + \frac{a^2}{\mathrm{g}}\frac{\partial \overline{v}}{x} = 0 \quad (1)$$

$$g\frac{\partial h}{\partial x} + \overline{v}\frac{\partial \overline{v}}{\partial x} + \frac{\partial \overline{v}}{\partial t} + \frac{f\overline{v}|\overline{v}|}{2\mathrm{D}} = 0 \quad (2)$$

where h is the head along the way, in the pipeline expressed as the sum of pressure energy and potential energy and in the free surface reservoir expressed as the water level; v is the average velocity of the fluid on the cross-section; g is the gravitational acceleration; f is the friction factor; α is the angle between the pipe centerline and the horizontal line; D is the pipe diameter; a is the wave velocity.

In the simulation, the overall linear equations of the whole fluid network were solved first to obtain the flow and head parameters of each element in the steady state, and then the transient process was calculated. The model of each element in the fluid network was mainly based on the pressure-flow relationship, and the linear equations of the fluid network were derived according to the following derivation.

Taking the pipeline as an example, the linear equation of the element was derived. It was assumed that Q_{m1} and P_1 were the fluid mass flow and pressure at the inlet of the pipeline, and Q_{m2} and P_2 were the fluid mass flow and pressure at the outlet of the pipeline, respectively. Assuming that the cross-sectional area along the pipeline is constant, the relationship between the pressure difference and the flow can be expressed as follows [34]:

$$\Delta P = k\frac{\rho v^2}{2} = \frac{kQ_m^2}{2\rho A^2} \quad (3)$$

where Q_m is the mass flow; ΔP is the pressure difference between the inlet and outlet of the pipeline, $\Delta P = P_2 - P_1$; A is the cross-section area of the element; ρ is the liquid density; k is the pressure difference coefficient.

Considering the direction of the mass flow of the fluid, the formula can be changed to [35]:

$$\Delta P = \frac{kQ_m|Q_m|}{2\rho A^2} \quad (4)$$

Thus, the linear equation of pipeline inlet and outlet can be deduced [33]:

$$Q_{m1} = -\frac{2\rho A^2}{k|Q_{m1}|}P_1 + \frac{2\rho A^2}{k|Q_{m1}|}P_2 \quad (5)$$

$$Q_{m2} = \frac{2\rho A^2}{k|Q_{m2}|}P_1 - \frac{2\rho A^2}{k|Q_{m2}|}P_2 \quad (6)$$

Because of the continuity of the fluid $Q_{m1} = Q_{m2}$, the linear equation can be obtained [34]:

$$Q_{m1} = A_1\rho_1 + A_2\rho_2 + B_1 \tag{7}$$

$$Q_{m2} = A_3\rho_3 + A_4\rho_4 + B_2 \tag{8}$$

where $A_1 = -\frac{2\rho A^2}{k|Q_{m1}|}$, $A_2 = -A_1$, $B_2 = 0$, $A_3 = -A_1$, $A_4 = A_1$, $B_2 = 0$

Since any fluid node in the pipeline system must satisfy the continuity equation, the overall equation set of the whole pipeline system can be obtained by simultaneously establishing the continuity equations of all nodes. For any fluid node m on the pipeline system [35]:

$$\sum_{i=1}^{M} Q_{in} = q_m \tag{9}$$

where Q_{in} is the node flow of unit i connected to node m; the left side of the formula is the sum of node traffic of all M units contributing to the node; and q_m is the total traffic entering the node.

Before the simulation based on the Flowmaster platform, the pump experiment and the three-dimensional modeling of the LAPS were carried out, and the performance curve of the pump test and the pressure drop curve of the flow channel were obtained by the pump experiment and CFD method [36–38]. Then, the secondary development of Flowmaster software was carried out, and the pump test performance curve obtained from the model test was stored in the database of Flowmaster in the form of the Suter curve, and the pressure drop characteristics of inlet and outlet channel predicted by CFD were added to the self-developed flow resistance element. Figure 4 shows the numerical strategy block diagram of this numerical calculation.

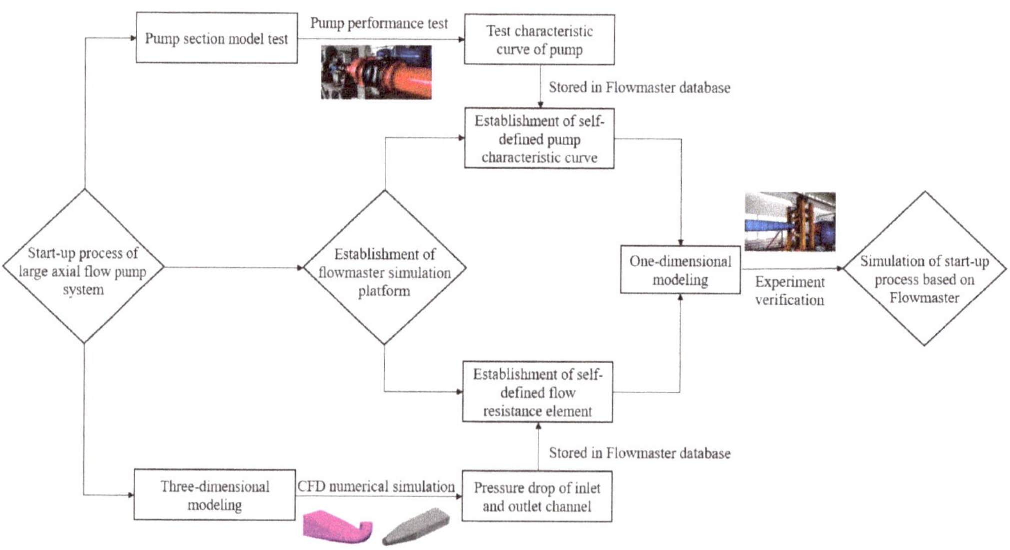

Figure 4. Flow chart of start-up process simulation strategy for the large axial flow pump system.

2.3. Simulation Model Construction and Boundary Conditions

A simulation model based on the Flowmaster platform was built after secondary development. Figure 5 shows the simulation model platform involving all components such as the overflow hole, flap valve and rapid-drop gate. Among them, element 1 and element 3 represent the inlet and outlet channels, which were realized in the form of self-defined flow resistance elements and rigid pipes. Element 2 represents the hydraulic model of the LAPS, and the pump speed control was realized by the pump speed controller. Element 4 represents the overflow hole, which was realized in the form of a rigid pipe and check valve. Element 5 represents the rapid-drop gate, and the gate control was realized by the gate controller. Element 6 represents the flap valve attached to the rapid-drop gate. It should be noted that in the study method used this paper, the simulation models of different schemes can delete the elements not involved, according to need.

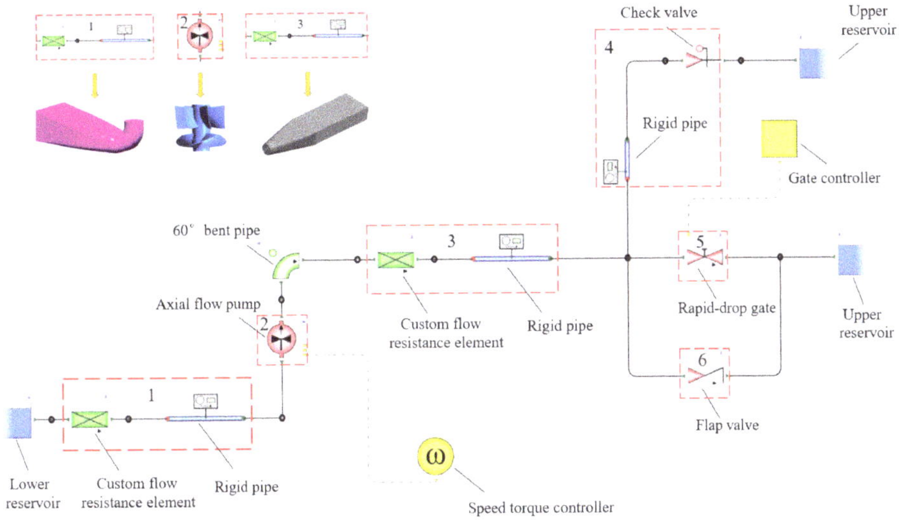

Figure 5. Schematic diagram of simulation model of the large axial flow pump system.

The control of the pump start-up speed in the Flowmaster platform was based on the field measured data of the LAPS. The pump speed rose linearly from 0, taking 6 s reach to the rated speed of 214.3 r/min. The length of the pipeline in element 1 in the Flowmaster platform took the length of the centerline of the inlet channel, which was 10.62 m. Twenty-five sections were uniformly selected from the inlet to the outlet of the inlet channel, and the weighted average of the hydraulic diameter of these 25 sections was 2.82 m, the diameter of the pipe in element 1. The length of the pipeline in element 3 in the Flowmaster platform took the length of the centerline of the outlet channel, which was 23.87 m. Twenty-five sections were uniformly selected from the inlet to the outlet of the outlet channel, and the weighted average of the hydraulic diameter of these 25 sections was 2.87 m, the diameter of the pipe in element 3. After the independence test of the calculation time step, 0.0025 s was selected as the time step for transient calculation. The total calculation time of transient simulation was 250 s.

2.4. Scheme and Its Specific Description

In order to better describe the influence of adding different auxiliary measures on the characteristic parameters of LAPS, this section uses Table 2 to describe in detail the hydraulic components added in different schemes and the changes of control methods when using different auxiliary measures.

The Flowmaster platform and the British fluid Society have carried out a number of experiments on the resistance coefficient of the flap valve, and the loss coefficient of different types of flap valves can be selected in the Flowmaster database. The different loss coefficients ξ caused by the change of the opening angle of the flap valve in this simulation are shown in Table 3.

Table 2. Scheme and its specific hydraulic components description.

Start-Up Control Simulation	Safety Auxiliary Facility Components	Time for Pump to Reach Rated Speed	Time Required for Gate Opening	Relation between Motor Starting and Gate Opening	Scheme Design				
Influence of opening speed of rapid-drop gate (Section 4.1)	Rapid-drop gate	6 s		Synchronization	Time required for gate opening (**TRRO**)				
					20 s	45 s	70 s	95 s	120 s
Influence of delaying opening of rapid-drop gate (Section 4.2)	Rapid-drop gate	6 s	120 s	Asynchronization	Time of gate opening delay (**TOOD**)				
					1 s	2 s	3 s	4 s	
Influence of adding the flap valve (Section 4.3)	Rapid-drop gate and flap valve	6 s	120 s	Synchronization	Area of flap valve (AOF)				
					2.0 m^2	3.5 m^2	5.0 m^2		
Influence of adding the overflow hole (Section 4.4)	Rapid-drop gate and overflow hole	6 s	120 s	Synchronization	Elevation of overflow hole (**EOOH**)				
					5.65 m	6.15 m	6.65 m		
Influence of adding both the flap valve and overflow hole (Section 4.5)	Rapid-drop gate, flap valve and overflow hole	6 s	120 s	Synchronization	Elevation of overflow hole (AOF =3.5 m^2)				
					5.65 m	5.90 m	6.15 m		

Table 3. Hydraulic loss coefficient of the flap valve.

Flap valve opening angle α	20	30	40	50	60
Loss coefficients ζ	6.3	4	3.2	2.8	2.5

2.5. Experimental Setup

The experimental test in this paper was carried out on the high-precision hydraulic machinery test bench of Jiangsu Key Laboratory of Hydraulic and Power Engineering. The system comprehensive uncertainty of the experimental test was related to the precision of the measuring instrument, and its calculation formula is as follows [39]:

$$E_{\eta,s} = \pm\sqrt{E_{q,s}^2 + E_{H,s}^2 + E_{n,s}^2 + E_{M,s}^2} \tag{10}$$

where $E_{\eta,s}$ is the system comprehensive uncertainty %; $E_{q,s}$ is the flow test system uncertainty %; $E_{H,s}$ is the head test system uncertainty %; $E_{n,s}$ is the speed test system uncertainty %; $E_{M,s}$ is the torque test system uncertainty %.

The system comprehensive uncertainty of the test bench was $\pm$0.39%. Figure 6 shows the schematic diagram of the high-precision hydraulic machinery test bench. The parameters of the main measuring instruments of the test bench are shown in Table 4. Figure 7 shows the physical diagram of axial flow pump.

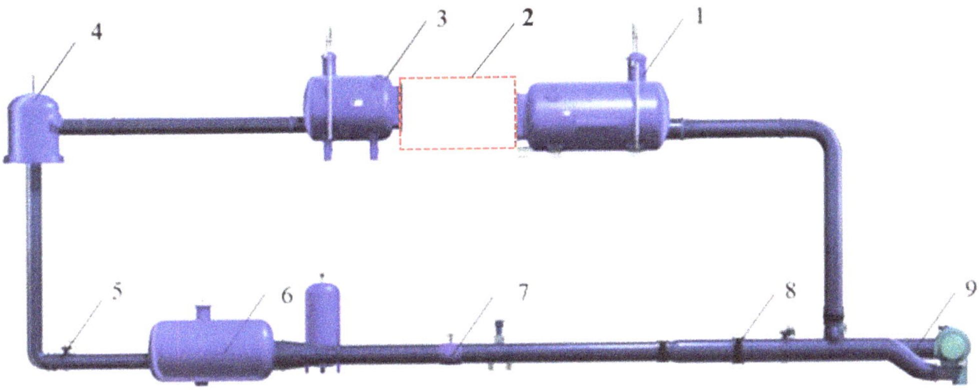

Figure 6. Schematic diagram of the high-precision hydraulic machinery test bench. 1. Intake tank. 2. Tested pump unit and drive motor. 3. Pressure outlet tank. 4. Bifurcation tank. 5. Condition regulating gate valve. 6. Voltage regulating rectifier. 7. Electromagnetic flowmeter. 8. System forward and reverse operation control gate valve. 9. Auxiliary pump unit.

Table 4. Parameters of main measuring instruments of the test bench.

Measuring Items	Instrument Name	Instrument Types	Instrument Range	Calibration Accuracy
Head	Difference pressure transmitter	EJA 110A	0~200 kPa	±0.1%
Flow	Electromagnetic flowmeter	E-mag type	DN400 mm	±0.20%
Torque	Speed and torque sensor	ZJ	200 N·m	±0.15%

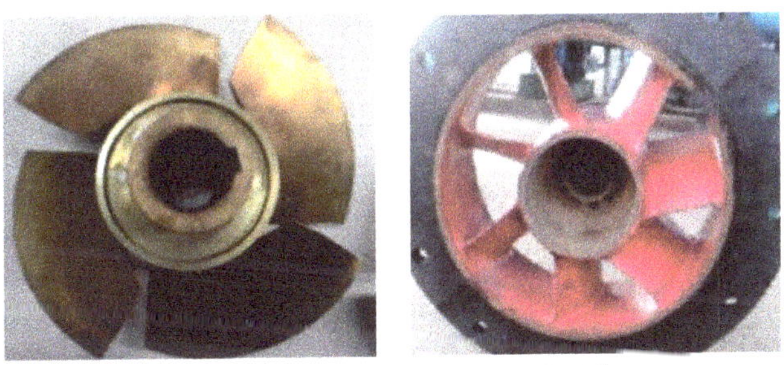

(**a**) Impeller (**b**) Guide vane

Figure 7. Physical diagram of axial flow pump.

3. Model Validation

In order to verify the feasibility of the numerical strategy and the accuracy of the simulation results, for this paper we machined the LAPS model and conducted an energy characteristic experiment and power-off runaway experiment on the LAPS model. The model experiment diagram of the LAPS is shown in Figure 8. The energy characteristic experiment carried out was to verify the steady simulation model, and the power-off runaway experiment was to verify the transient simulation model. The actual test process was completely based on the acceptance test specifications of the pump model and device model (SL 140-2006). The energy characteristics of the LAPS under different flow conditions and the runaway characteristics of the pump under different lifts were tested.

Figure 8. Model experiment of the large axial flow pump system.

Figure 9 shows the comparison between the experimental results and the simulation results [40,41]. Figure 9a shows the energy characteristics, and Figure 9b shows the runaway speed of the pump after power-off. From Figure 9a, it can be seen that the experiment head and shaft power were very close to the simulation head and shaft power, and the maximum error was less than 5%, indicating that the steady simulation of the LAPS based on the numerical model in this paper has high accuracy. From Figure 9b, it can be seen that the runaway speed of the pump after the power-off of the LAPS under the different heads was very close to that of the simulation, the variation law was highly consistent and the maximum error was less than 3%, indicating that the transient simulation of the LAPS based on the numerical model in this paper also has high accuracy [42,43].

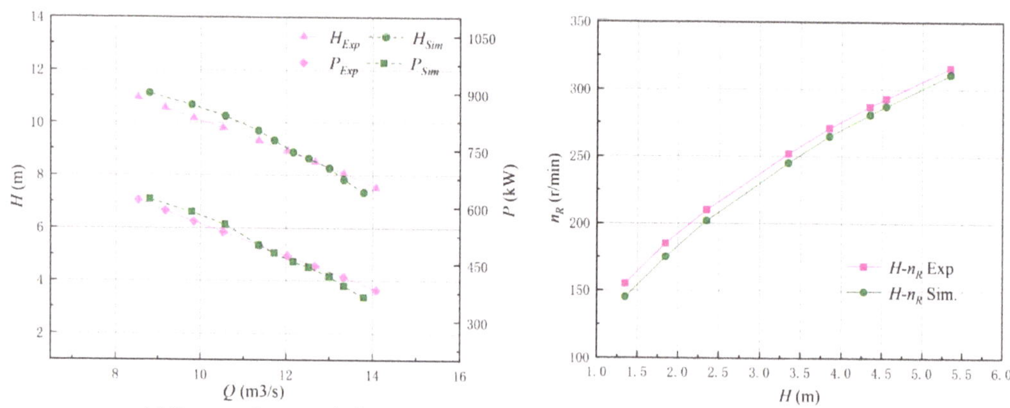

(**a**) Energy characteristics (**b**) Runaway speed of pump after power-off

Figure 9. Comparison between experimental results and simulation results.

Figure 10 shows the error analysis of numerical simulation. From the error of head and shaft power in Figure 10a, it can be seen that with the increase of flow, the error value of head and shaft power obtained by numerical simulation gradually decreased. At the design flow, the error of head was about 3.0%, and the error of power was about 0.5%. From Figure 10b, it can be seen that the error value of runaway speed obtained by numerical simulation gradually decreased with the increase of head. At the design head, the error value of runaway speed was about 2.0%.

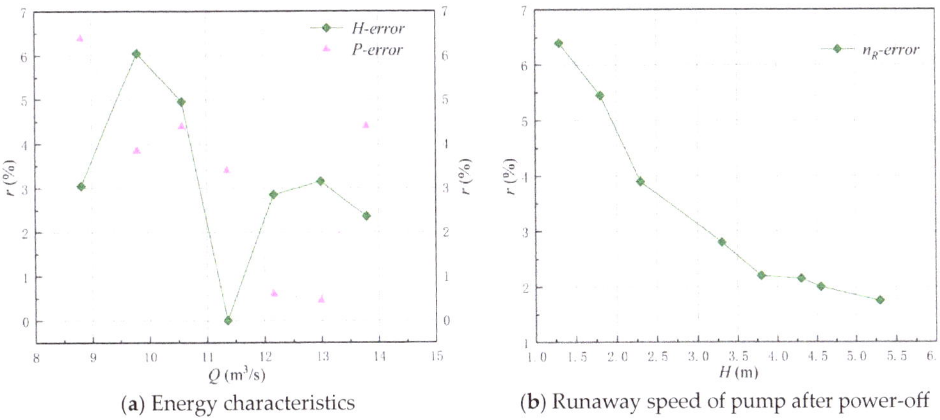

(**a**) Energy characteristics (**b**) Runaway speed of pump after power-off

Figure 10. Error analysis of numerical simulation.

4. Results and Discussion

4.1. Influence of Opening Speed of Rapid-Drop Gate

In order to explore the influence of rapid-drop gate opening speed on the start-up process characteristics of the LAPS, it was necessary to transiently simulate the start-up process of the LAPS with different gate opening speeds (gate opening and unit start-up synchronization). Figure 11 shows the variation law of the key characteristic parameters of the LAPS with time; the time required for gate opening was 120 s. It can be seen from Figure 11 that the fluid in the pump will be backflow at the initial stage of the start-up because the speed of the impeller has not been raised to the rated speed. With the gradual increase of the speed of the impeller, the phenomenon of backflow in the pump gradually disappears, and the pump head increases steadily. When t = 6 s, the unit reaches the rated speed, and the instantaneous impact head and instantaneous impact power (IIP) of the pump also reach the maximum. When the unit reaches the rated speed, the LAPS gradually transitions to the conventional steady operation, and the fluctuation of the key characteristic parameters tends to be smooth.

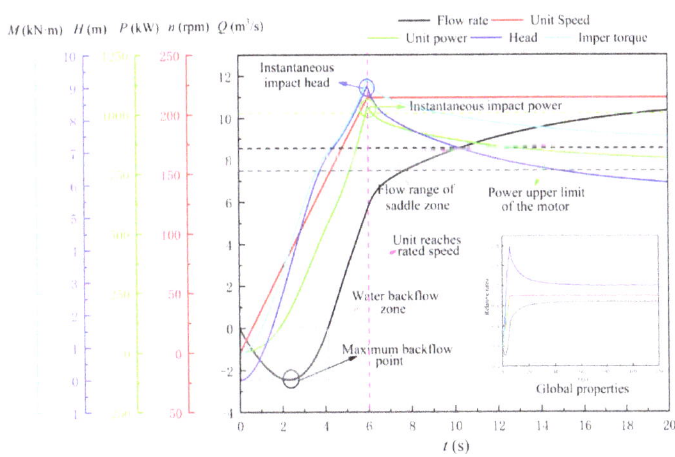

Figure 11. Variation law of the key characteristic parameters of the large axial flow pump system (time required for gate opening is 120 s).

According to the actual operation experience of the LAPS, as long as the corresponding instantaneous flow rate does not fall within the flow range of the saddle area when the LAPS reaches the rated speed, it can basically ensure that the start-up of the LAPS will not lead to instability of the LAPS due to the influence of the saddle zone. Through the performance experiment of the pump, it was found that the flow rate range of the saddle zone was 7.5 m^3/s~8.5 m^3/s. When t = 6 s, the instantaneous impact head of the pump was 9.06 m, and the corresponding flow was 5.76 m^3/s, which was not within the flow range of the saddle zone, indicating that when the LAPS starts with this scheme, it cannot be unstable due to the influence of the saddle zone. However, when t = 6 s, the IIP of the unit is 1035.88 kW, which exceeds the upper limit of the maximum power of the motor. Therefore, it is likely to cause the LAPS to fail on start-up.

Figure 12 shows the variation law of flow rate and head under different times required for gate opening (TRRO). From Figure 12, the following conclusions can be drawn. Firstly, no matter which TRRO is adopted, the corresponding instantaneous flow rate is not within the flow range of the saddle area when the LAPS reaches the rated speed, so the start-up of the LAPS cannot be unstable due to the influence of the saddle zone. Secondly, no matter which TRRO is adopted, the instantaneous impact head in the LAPS reaches the maximum when the unit reaches the rated speed, but the slower the gate opening speed is, the faster the increasing speed of the pump head is, and the greater the maximum instantaneous impact head is. Thirdly, the slower the gate opening speed is, the more favorable it is for reducing the backflow and the duration of backflow at the initial stage of the start-up of the LAPS. It can be observed that with the increase in the TRRO, the maximum backflow flow (MBF) in the start-up process of the LAPS gradually decreases, and the duration of the backflow state in the LAPS is also gradually shortened.

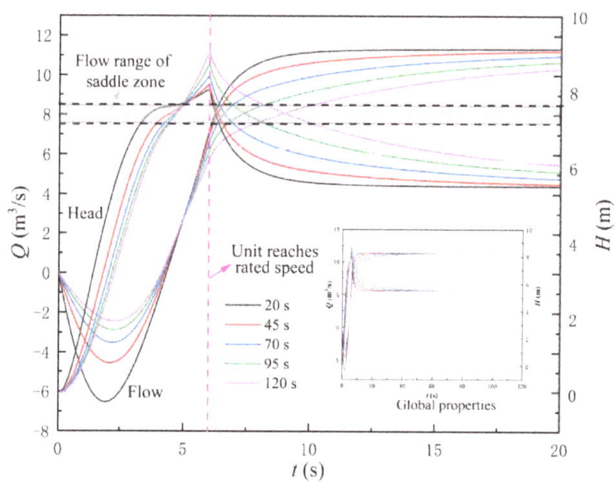

Figure 12. Variation law of flow rate and head with time under different times required for gate opening.

Figure 13 shows the MBF and IIP of the LAPS under the different TRRO. The following conclusions can be drawn from Figure 13. Firstly, with the gradual increase in the TRRO, the MBF decreases gradually. Under the TRRO = 20 s scheme, the MBF in the LAPS is 6.53 m^3/s, and under the TRRO = 120 s scheme, the MBF in the LAPS has been reduced to 2.44 m^3/s. Secondly, the decline rate of the MBF in the LAPS is not linear, but slows down with the gradual increase of the TRRO. Thirdly, the larger the TRRO, the greater the IIP of the unit is. Under the TRRO = 20 s scheme, the IIP of the unit reaches 1009.36 kW, and under the TRRO = 120 s scheme, the IIP of the unit reaches 1035.88 kW. It is obvious that if the gate starts too slowly, the IIP will exceed the upper limit of the maximum power of the motor, which will lead to the overload of the unit and finally lead to start-up failure.

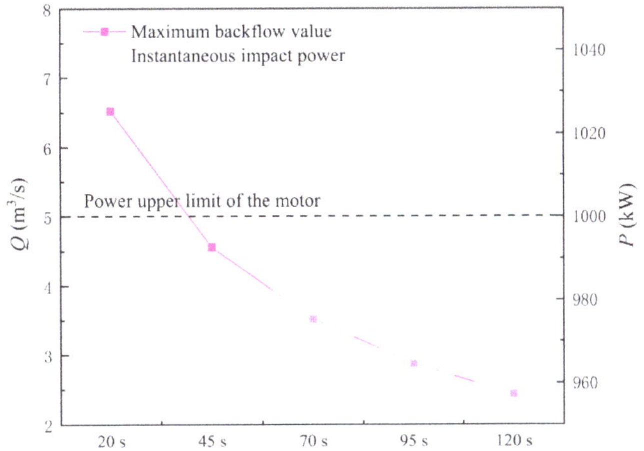

Figure 13. Maximum backflow flow and instantaneous impact power of the large axial flow pump system with different times required for gate opening.

4.2. Influence of Delaying Opening Time of Rapid-Drop Gate

From Section 4.1, it can be found that in the initial stage of the start-up of the LAPS, if the rapid-drop gate and the motor are opened synchronously, backflow will occur in the LAPS, which will cause impact and damage to the pump blade. This section studies the influence of delaying opening time of the rapid-drop gate in the start-up process characteristics of the LAPS.

Figure 14 shows the variation law of the flow rate and head of the LAPS with different times of gate opening delay (TOOD). The following conclusions can be drawn from Figure 14. Firstly, when adopting the four different TOOD schemes, the flow rate of the LAPS corresponding to the synchronous speed of the unit is not within the flow range of the saddle zone, and the start-up of the LAPS cannot be unstable due to the influence of the saddle zone. Secondly, delaying the opening of the gate will result in a surge of instantaneous impact head, but the instantaneous impact head in the LAPS will still reach the maximum when the unit reaches the rated speed, and the longer the gate lag is, the greater the instantaneous impact head of the pump. Finally, delaying the opening of the gate can significantly reduce the MBF and the duration of backflow at the initial start-up of the LAPS. It can be observed that the longer the TOOD, the smaller the MBF during the start-up of the LAPS is, and the shorter the duration of the fluid backflow state in the LAPS is.

Figure 15 shows the MBF and backflow duration of the LAPS with the different TOOD. The following conclusions can be drawn from Figure 15. Firstly, with the delay of the rapid-drop gate opening, the MBF of the LAPS gradually decreases, and the backflow duration of the LAPS continues to shorten. When the TOOD = 1 s, the MBF in the LAPS is 1.60 m^3/s, and the duration of backflow is 3.08 s. When the TOOD = 4 s, the MBF and the duration of backflow in the LAPS are 0, indicating that the backflow in the LAPS disappears completely. Secondly, when the TOOD = 2 s, the decreasing rate of the MBF and backflow duration in the LAPS tend to be linear, and when the TOOD is more than 2 s, the decline rate of the MBF and backflow duration in the LAPS slows down gradually.

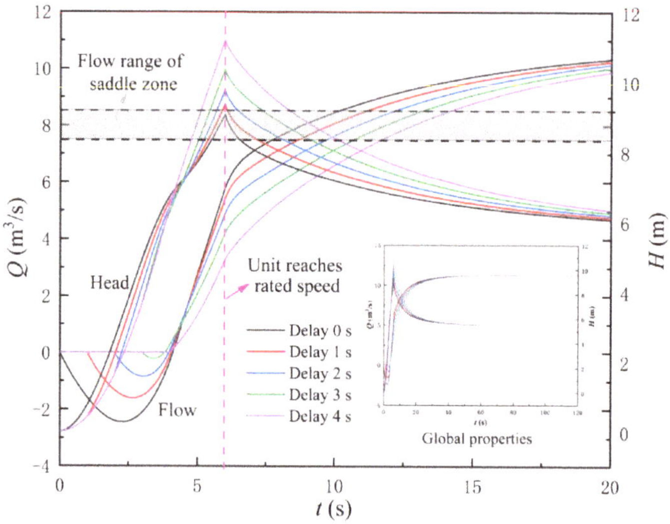

Figure 14. Variation law of the flow rate and head of the large axial flow pump system with different times of gate opening delay.

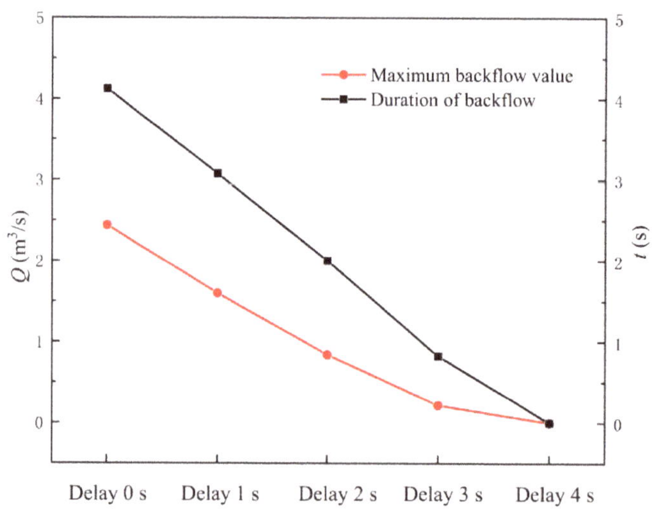

Figure 15. Maximum backflow flow of the large axial flow pump system with different times of gate opening delay.

Figure 16 shows the IIP and the growth rate of the LAPS with different TOOD. The following conclusions can be drawn from Figure 16. Firstly, the opening of the rapid-drop gate lags the opening of the motor, which will bring great risk to the start-up of the LAPS. With the delayed opening of the rapid-drop gate, the IIP of the unit increases rapidly. The IIP of the unit is 1061.4 kW when the TOOD = 1 s, and the IIP of the unit has reached 1234.89 kW when the TOOD = 4 s, which far exceeds the upper limit of the motor power. Secondly, the longer the opening of the rapid-drop gate lags the opening of the motor, the faster the IIP growth rate of the unit will rise. When the TOOD = 1 s, the growth rate of the IIP of the unit is 2.46%. When the TOOD = 4 s, the growth rate of the IIP of the unit has jumped to 7.15%.

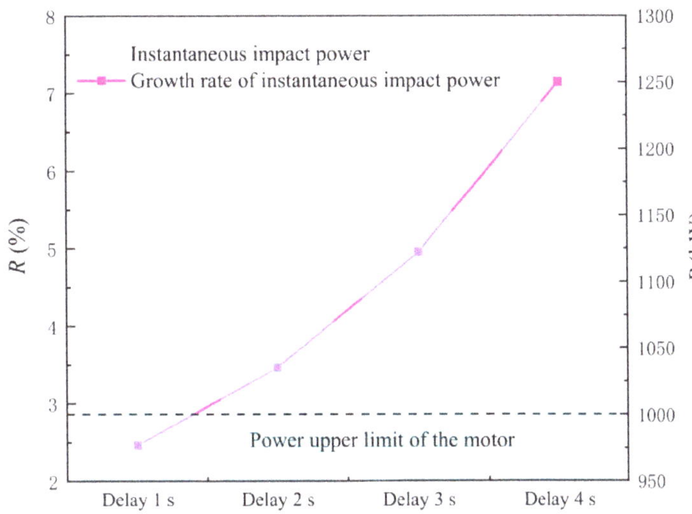

Figure 16. Instantaneous impact power of the large axial flow pump system with different times of gate opening delay.

4.3. Influence of Adding the Flap Valve

From Section 4.1, it can be found that if the gate opening speed is too slow, the head and power of the LAPS will increase sharply in a short time, which will cause the motor to overpower and the pump unit start-up will fail. This section studies the influence of adding the flap valve to the start-up process characteristics of the LAPS.

Figure 17 shows the variation law of the flow rate and head of the LAPS after adding the flap valve. Figure 18 shows the flow rate at the flap valve. The following conclusions can be drawn by combining Figures 17 and 18. Firstly, when adopting the three different areas of flap valve (AOF), the flow rate corresponding to the synchronous speed of the LAPS is not within the flow range of the saddle zone, and the start-up of the LAPS cannot be unstable due to the influence of the saddle zone. Secondly, the addition of the flap valve to the rapid-drop gate will significantly reduce the maximum instantaneous impact head, and with the increase of the AOF, the maximum instantaneous impact head will gradually decrease. Thirdly, the scheme of adding the flap valve is not helpful to shorten the backflow state at the initial stage of the LAPS start-up, and the flow rate variation law of the LAPS with different AOF is basically the same as that without the flap valve. Finally, with the increase of the AOF, the shunt effect of the flap valve in the start-up process is gradually enhanced.

Figure 19 shows the MBF and IIP of the LAPS after adding the flap valve. The following conclusions can be drawn from Figure 19. Firstly, the addition of the flap valve to the rapid-drop gate will weaken the IIP of the unit, prevent the overload of the unit and significantly improve the reliability of the LAPS start-up. When AOF = 2.0 m^2 (15% of the area of the gate), the IIP of the unit has been reduced to 974.26 kW, which reduces the 61.62 kW compared with the start-up scheme without a flap valve. With the gradual increase of the AOF, the IIP of the unit decreases gradually. Secondly, the method of adding the flap valve is not helpful to reduce the MBF at the initial stage of the LAPS start-up. No matter how big the AOF is, the MBF of the LAPS is unchanged, consistent with MBF start-up without adding the flap valve, which is 2.44 m^3/s.

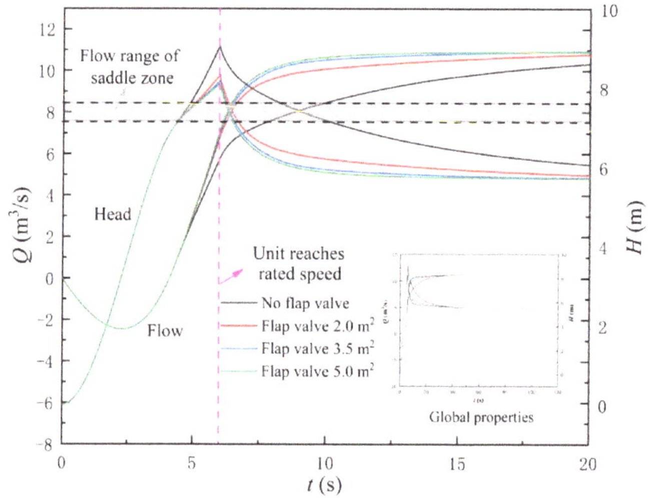

Figure 17. Variation law of the flow rate and head of the large axial flow pump system with different areas of flap valve.

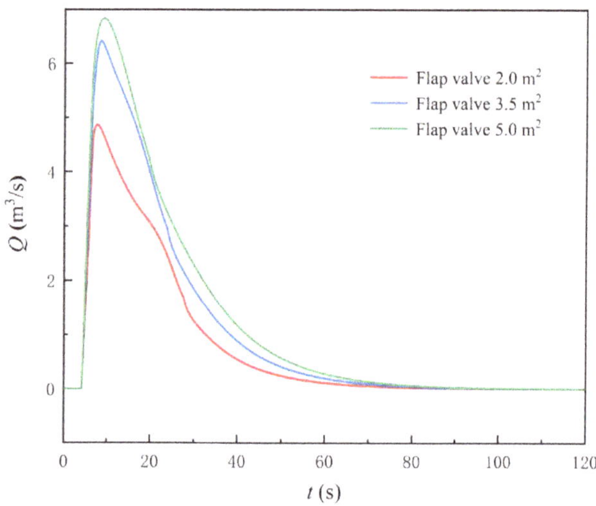

Figure 18. Flow rate at the flap valve.

4.4. *Influence of Adding the Overflow Hole*

This section studies the influence of adding an overflow hole in the start-up process characteristics of the LAPS. Figure 20 shows the variation law of the flow rate and head of the LAPS after adding the overflow hole. Figure 21 shows the flow rate at the overflow hole. Combining Figures 20 and 21, the following conclusions can be drawn. Firstly, with different elevation of overflow hole (EOOH), the corresponding flow rate of the LAPS is not within the flow range of the saddle zone when the unit reaches the synchronous speed, and the start-up of the LAPS cannot be unstable due to the influence of the saddle zone. Secondly, after adding the overflow hole, the instantaneous impact head of the start-up process is reduced, but the extent of the reduction is not as obvious as that of the flap valve. When the overflow hole is not added, the instantaneous impact head is 9.06 m, and when the EOOH = 5.65 m (1.06 times of the maximum net head), the instantaneous impact head still reaches 8.45 m. Thirdly, adding overflow holes does not help to shorten the backflow state at the initial stage of the LAPS start-up, and the backflow variation law of the LAPS with different overflow hole elevations is basically the same as that of the LAPS without an overflow hole. Finally, with the increase of the EOOH, the shunt effect in the start-up process of the overflow hole is gradually weakened.

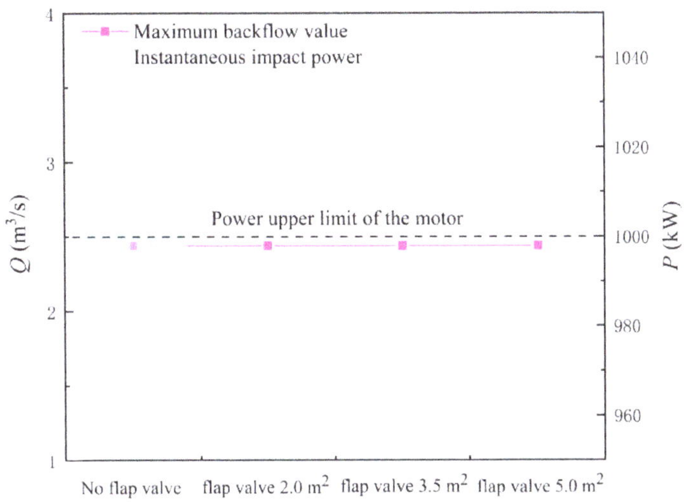

Figure 19. Maximum backflow flow and instantaneous impact power of the large axial flow pump system with different areas of flap valve.

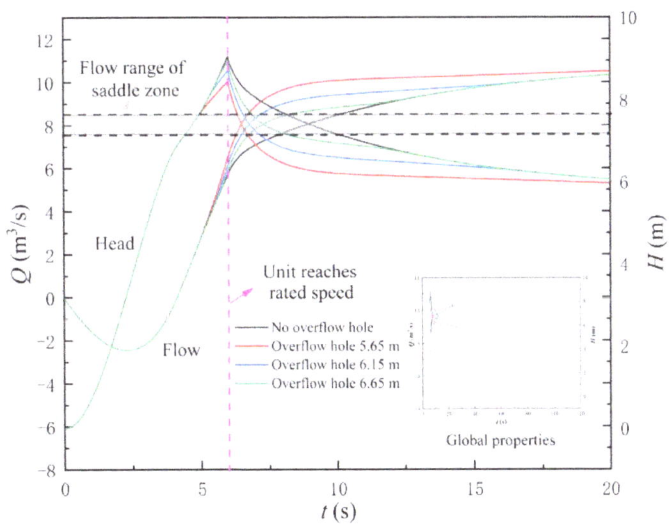

Figure 20. Variation law of the flow rate and head of the large axial flow pump system with different elevations of overflow hole.

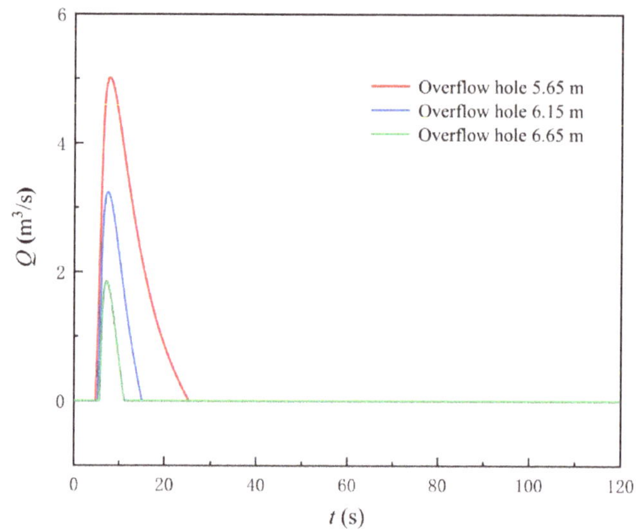

Figure 21. Flow rate at the overflow hole.

Figure 22 shows the MBF and IIP of the LAPS after adding the overflow hole. The following conclusions can be drawn from Figure 22. Firstly, adding the overflow hole can appropriately reduce the IIP of the unit, but cannot significantly improve the start-up reliability of the LAPS. With different overflow holes, only the IIP of the EOOH = 5.65 m (1.06 times of the maximum net head) is lower than the maximum power limit of the motor. The IIP of EOOH = 6.15 m (1.15 times of the maximum net head) and EOOH = 6.65 m (1.24 times of the maximum net head) is still as high as 1007.13 kW and 1024.89 kW, which cannot meet the start-up load requirements. Secondly, the method of adding the overflow hole is not helpful to reduce the MBF at the initial stage of the LAPS start-up. No matter how higher the EOOH is, the MBF of the LAPS is unchanged, consistent with the MBF start-up without adding the flap valve, which is 2.44 m^3/s.

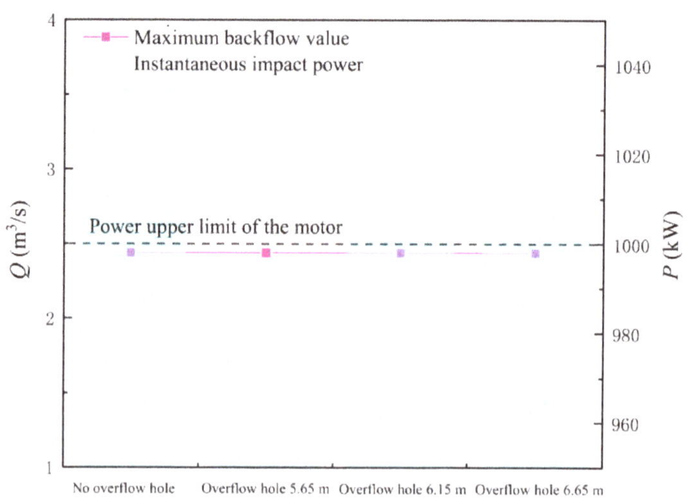

Figure 22. Maximum backflow flow and instantaneous impact power of the large axial flow pump system with different elevations of overflow hole.

4.5. Influence of Adding Both the Flap Valve and Overflow Hole

According to the analysis results of Sections 4.3 and 4.4, adding the flap valve is more effective than adding the overflow hole to improve reliability in the start-up process. This section further studies the influence of adding the overflow hole on the start-up process characteristics of the LAPS on the basis of adding a flap valve to the rapid-drop gate.

Figure 23 shows the variation law of the flow rate and head of the LAPS when the overflow hole is further added. The following conclusions can be drawn from Figure 23. Firstly, after adding overflow holes with the three different elevations, the corresponding flow rate of the LAPS is not within the flow range of the saddle zone when the unit reaches the synchronous speed, and the start-up of the LAPS cannot be unstable due to the influence of the saddle zone. Secondly, adding the flap valve on the basis of the rapid-drop gate and then adding the overflow hole is not helpful in further reducing the instantaneous impact head in the start-up process. The instantaneous impact head is 8.18 m when only adding the flap valve and not adding the overflow hole, the pump head is still 8.18 m when a different elevation overflow hole is added. Finally, on the basis of adding the flap valve, adding an overflow hole is still not helpful to shorten the backflow at the initial stage of the LAPS start-up, and the flow rate variation law of the LAPS with time has not changed after adding overflow holes of different elevations.

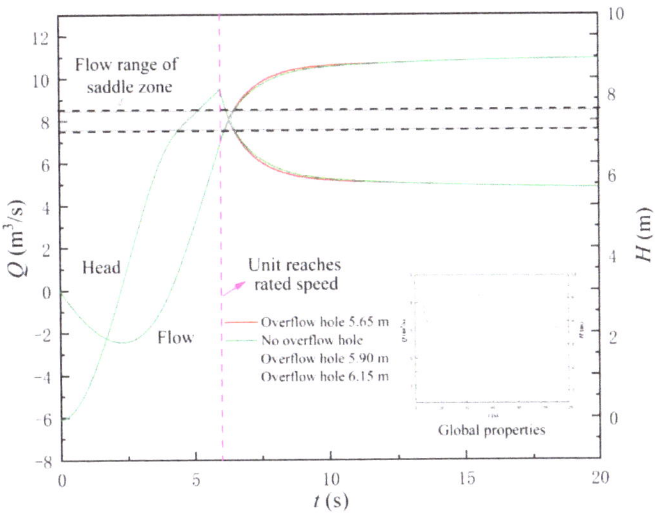

Figure 23. Variation law of the flow rate and head of the large axial flow pump system with different elevations of overflow hole (area of flap valve is 3.5 m^2).

Figure 24 shows the flow rate at the flap valve and overflow hole. Figure 25 shows the MBF and IIP of the LAPS when an overflow hole is further added. The following conclusions can be drawn by combining Figures 24 and 25. Firstly, the flap valve diversion plays an absolutely dominant role in the initial start-up of the LAPS, and further setting the overflow hole on the basis of adding the flap valve (26% of the rapid-drop gate area) cannot play an obvious role in the further diversion in the initial start-up of the LAPS. Only the overflow hole with an elevation of 5.65 m could divert the maximum flow rate of 0.81 m^3/s at the initial stage of the start-up of the LAPS. Secondly, on the basis of adding the flap valve (26% of the area of the rapid-drop gate), the addition of the overflow hole has no significant influence on the MBF and IIP of the LAPS. On the basis of adding the flap valve (26% of the area of the rapid-drop gate), it can be considered whether or not to adjust the overflow hole further.

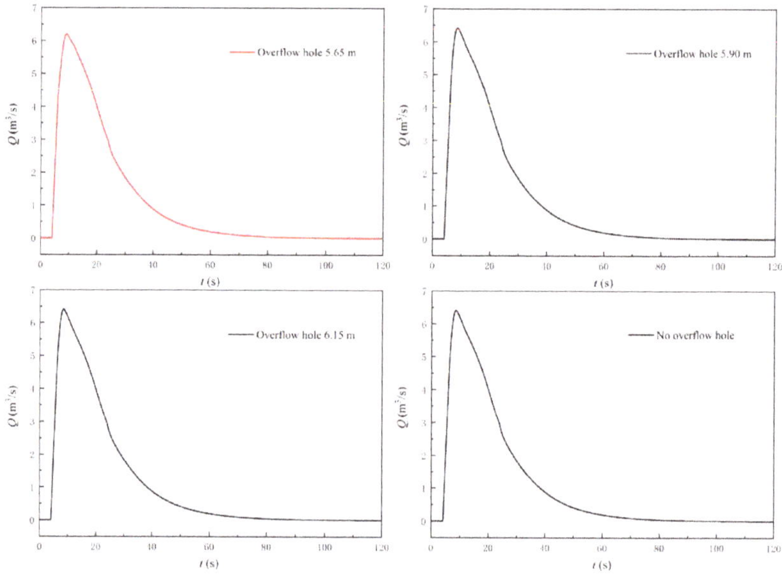

(**a**) Flap valve

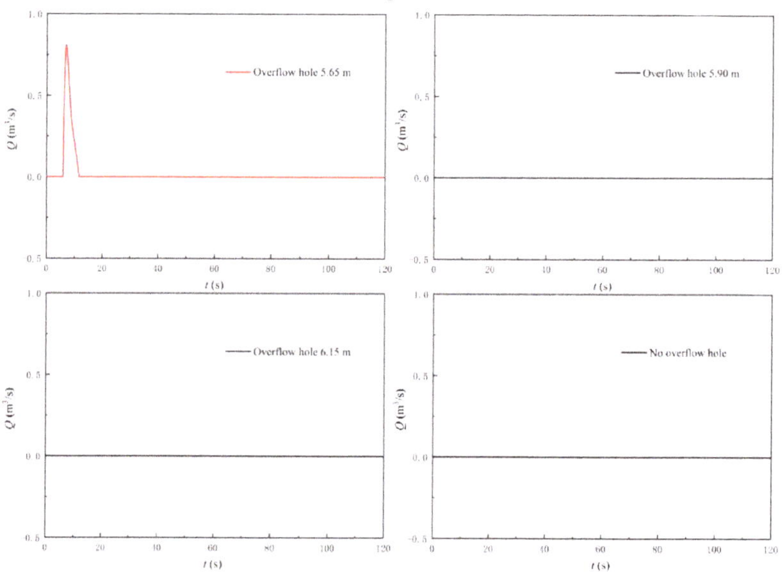

(**b**) Overflow hole

Figure 24. Flow rate at flap valve and overflow hole.

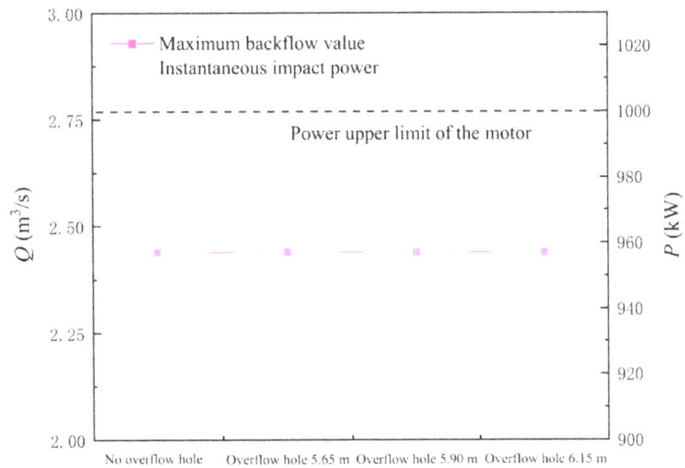

Figure 25. Maximum backflow flow and instantaneous impact power of the large axial flow pump system with different elevation of overflow hole (area of flap valve is 3.5 m^2).

4.6. Comparison of the Influence of Different Safety Auxiliary Measures

In order to study the influence of different safety auxiliary measures on the start-up characteristics of the LAPS, this section extracts the key characteristic indexes in the start-up process of the LAPS and carries out the dimensionless treatment. The dimensionless coefficients include backflow coefficient K_Q, impact head coefficient K_H and power overload coefficient K_P. The dimensionless method is as follows [44]:

$$K_H = \frac{H_i}{H_d} \tag{11}$$

$$K_Q = \frac{Q_b}{Q_d} \times 10 \tag{12}$$

$$K_P = \frac{P_i - P_u}{P_u} \times 100 \tag{13}$$

where H_i is the maximum instantaneous impact head in the start-up process; H_d is the design head of the LAPS; K_H is the impact head coefficient; Q_b is the maximum backflow during start-up; Q_d is the design flow of the LAPS; K_Q is the backflow coefficient; P_i is the maximum IIP in the start-up process; P_u is the upper limit of motor power; K_P is the power overload coefficient.

Figure 26 shows the key characteristic indexes (dimensionless) during the start-up of the LAPS after taking different safety auxiliary measures. The following conclusions can be drawn from Figure 26. Firstly, when no other facilities are used, only the opening speed of the rapid-drop gate is adjusted, so it is difficult to take into account the backflow coefficient index and the impact head coefficient index at the same time. Once the setting of the rapid-drop gate opening law is unreasonable, it will often lead to a sharp increase in the backflow or a sharp increase in the instantaneous impact head. When the total opening time of the rapid-drop gate is 20 s, the backflow coefficient reaches 0.51, which will cause a certain impact and damage to the pump blade. Secondly, the delayed opening of the rapid-drop gate has a very great risk, although the backflow coefficient index has been optimized to a certain extent, the power overload coefficient is more than 5. When the rapid-drop gate opens with a delay of 4 s, the power overload coefficient reaches 23.49, indicating that the possibility of start-up failure of the LAPS increases sharply the longer the gate opening is delayed. Thirdly, the improvement effect of adding a flap valve or overflow hole on the quality of the start-up process is similar, but the indexes of the start-up process of the LAPS with the addition of a flap valve show it is better. When the safety auxiliary facilities of an additional flap valve on the rapid-drop gate are adopted, the backflow coefficient is within 0.2, the impact head coefficient is within 2, and the power overload coefficient is less than 0. Combining the three key characteristic indexes of backflow coefficient, impact head coefficient and power overload coefficient to evaluate them, the method of adding a flap valve on the rapid-drop gate can obtain a higher quality of the start-up process. Finally, on the basis of the existing flap valve, further adding overflow holes has little influence on improving the quality of the start-up process, and the three key characteristic indexes have little change. Only an overflow hole with a very low elevation can further slightly improve the start-up process quality of the LAPS.

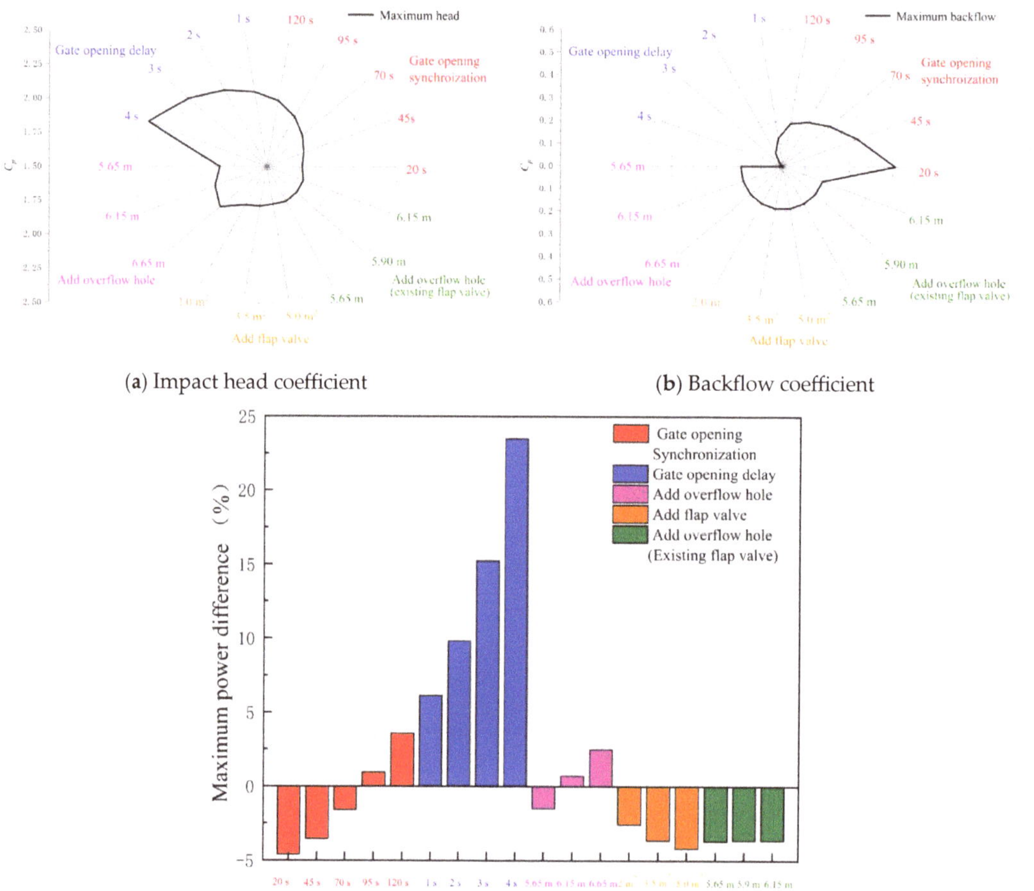

(**a**) Impact head coefficient

(**b**) Backflow coefficient

(**c**) Power overload coefficient

Figure 26. Key characteristic indexes (dimensionless) in the start-up process of the large axial flow pump system.

In order to further comprehensively show the changes in the start-up characteristics of the LAPS before and after the adoption of safety ancillary measures, this paper selects the one with the highest quality of all the start-up transition process calculation schemes to display. Figure 27 shows variation law of the key characteristic parameters of the LAPS. In Section 4.1, Figure 11 shows the variation law of key characteristic parameters of the LAPS (TRRO = 120 s). Compared with Figure 11, the curve of the key characteristic parameters in Figure 27 is relatively smooth, the fluctuation of the parameters is relatively smooth at the initial stage of the LAPS start-up, the coverage of the backflow area is small and the IIP is within the allowable range. After the unit reaches the rated speed, the characteristics of the LAPS quickly transition to a stable operation state.

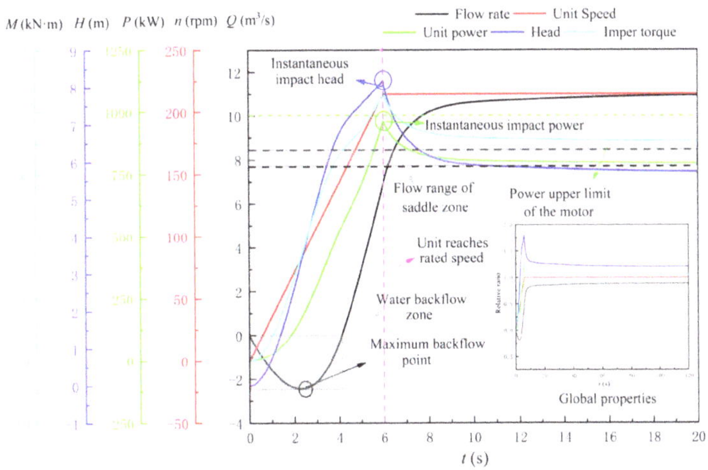

Figure 27. Variation law of the key characteristic parameters of the large axial flow pump system: time required for gate opening is 120 s, time of gate opening delay is 0 s, area of flap valve is 3.5 m^2, elevation of overflow hole is 5.65 m).

5. Conclusions

In this paper, considering the defects of theoretical derivation method and CFD method in obtaining a large number of different pump start-up schemes, a new simulation strategy is proposed. Based on the secondary development of Flowmaster, the start-up process of a LAPS with various safety auxiliary facilities such as a pat door and overflow hole is numerically simulated, and the start-up characteristics of a large axial flow pump station are comprehensively explored. This paper fills the gap in the research on the influence of safety auxiliary facilities on the LAPS start-up process, and can provide important reference value for LAPS systems to find a scientific and safe start-up control strategy. The main conclusions are as follows:

(1) During the start-up transition process, if only the opening speed of the rapid-drop gate is adjusted without any safety auxiliary facilities, it will be difficult to give consideration to the index of backflow coefficient and the index of impact head coefficient. Once the gate opening speed is too fast, it will cause a sharp increase in backflow, while if the gate opening speed is too slow, it will cause a sharp increase in the instantaneous impact head and the instantaneous impact power.

(2) It is a very dangerous start-up mode to delay the opening of the rapid-drop gate to reduce the maximum backflow value and the duration of backflow at the initial start-up of the pump system. When the opening of the rapid-drop gate lags behind the start of the motor, the instantaneous impact power of the unit will increase rapidly, and the possibility of start-up failure of the pump station will be greatly increased.

(3) The method of adding the flap valve to the rapid-drop gate has no obvious benefit in shortening the backflow state at the initial stage of the pump start-up, but it will significantly weaken the instantaneous impact power of the unit and prevent overload of the unit. When the safety auxiliary facilities of an additional flap valve on the rapid-drop gate are adopted, the backflow coefficient is within 0.2, the impact head coefficient is within 2, and the power overload coefficient is less than 0. Synthesizing the three key characteristic indexes in the start-up process evaluation, the method of

adding a flap valve on the rapid-drop gate can obtain a higher quality of the start-up transition process.

The influence of the overflow hole on the start-up process characteristics of the pump station is similar to that of the flap valve, but the effect of improving the start-up process quality of the pump station is not as good as that of the flap valve. Moreover, it should be pointed out that, on the basis of the existing flap valve, the further addition of an overflow hole has little effect on improving the quality of the start-up transition process, and the three key characteristic indexes show little change. In the case of the existing flap valve, the start-up transition process quality of the pump station can be further slightly improved only if an overflow hole with extremely low elevation is used. In this paper, a one-dimensional simulation of the start-up process of a large axial flow pump system equipped with safety auxiliary facilities is carried out. The main purpose of the research is to reveal the variation of key characteristic parameters of a LAPS equipped with different safety auxiliary facilities during the start-up process, and to evaluate the influence of different safety auxiliary facilities on the safety of LAPS during the start-up process. However, the three-dimensional internal flow field of the LAPS has not been studied in detail. In the future research, the three-dimensional numerical simulation of LAPS will be carried out based on CFD method to reveal the hydraulic transient flow characteristics during LAPS start-up.

Author Contributions: X.Z.: conceptualization, software, methodology, writing—original draft preparation, and supervision; Y.J.: model experiment, data curation, and visualization; F.T.: writing—reviewing, supervision and funding acquisition; X.S.: supervision and visualization; Y.L.: model experiment and visualization; F.Y.: data curation and visualization; L.S.: funding acquisition. All authors have read and agreed to the published version of the manuscript.

Funding: This research was supported by the National Natural Science Foundation of China (No. 52209116); the Scientific and Technological Research and Development Program of South-to-North Water Transfer in Jiangsu Province (No. JSNSBD202201); the Jiangsu Water Conservancy Science and Technology Project (No. 2021012).

Data Availability Statement: The data presented in this study are available on request from the corresponding author.

Conflicts of Interest: The authors declare no conflict of interest.

Nomenclature

g	Local acceleration of gravity (m/s^2)
H	Head (m)
M	The impeller torque (kN·m)
n	Rated speed (r/min)
H	the maximum net head (m)
D	Impeller diameter
I_p	The inertia moment of the LAPS ($kg \cdot m^2$)
I_m	The motor rotation inertia (m^3/s)
A_r	The area of the rapid-drop gate (m^2)
t	Time (s)
ρ	The density of flow (kg/m^3)
ω	The angular velocity of the impeller (rad/s)
η	Efficiency (%)
η_{exp}	Experimental efficiency (%)
η_{sim}	Simulated efficiency (%)
Q_b	The maximum backflow
Q_d	Designed flow
H_{exp}	Experimental head (m)
H_{sim}	Simulated head (m)
H_i	Maximum instantaneous impact head
H_d	The design head of the pump system

K_H	The impact head coefficient
K_Q	The backflow coefficient
K_p	The power overload coefficient
P_i	The maximum IIP
P_u	The upper limit of motor power

Abbreviations

CFD	Computational fluid dynamics
LAPS	Large axial flow pump station system
IIP	Instantaneous impact power
TRRO	Time required for gate opening
TOOD	Time of gate opening delay
AOF	Area of flap valve
EOOH	Elevation of overflow hole
MBF	Maximum backflow flow

References

1. Ge, X.F.; Zhang, J.; Zhang, J.; Liu, D.; Zheng, Y.; Chen, H.X. Review of Research on the Three-Dimensional Transition Process of Large-Scale Low-Lift Pump. *Energies* **2022**, *15*, 8338. [CrossRef]
2. Xu, H.; Gong, Y.; Cao, C.J.; Chen, H.X.; Kan, K.; Feng, J.G. Research on the three-dimensional transition process of the full discharge system of the axial extension tubular pump station. *J. Agric. Mach.* **2021**, *52*, 143–151.
3. Zhang, X.X.; Cheng, Y.G.; Yang, J.D.; Xia, L.S.; Lai, X. Simulation of the load rejection transient process of a francis turbine by using a 1-D-3-D coupling approach. *J. Hydrodyn.* **2014**, *26*, 715–724. [CrossRef]
4. Thanapandi, P.; Prasad, R. Centrifugal pump transient characteristics and analysis using the method of characteristics. *Int. J. Mech. Sci.* **1995**, *37*, 77–89. [CrossRef]
5. Zhang, J.P. *The Study on Characteristics of Starting Transient Process of Shaft Tubular Pumping Station*; Yangzhou University: Yangzhou, China, 2015.
6. Fu, S.F.; Zheng, Y.; Kan, K.; Chen, H.X.; Han, X.X.; Liang, X.L.; Liu, H.W.; Tian, X.Q. Numerical simulation and experimental study of transient characteristics in an axial flow pump during start-up. *Renew. Energy* **2020**, *146*, 1879–1887. [CrossRef]
7. Long, Y.; Lin, B.; Fang, J.; Zhu, R.S.; Fu, Q. Research on the Transient Hydraulic Characteristics of Multistage Centrifugal Pump During Start-Up Process. *Front. Energy Res.* **2020**, *8*, 76. [CrossRef]
8. Li, T.X.; Li, Y.C.; He, Z.S.; Luo, W. Application of the Energy-saving Flap Valve in Pumping Station Rebuild. *Fluid Mach* **2008**, *8*, 48–49, 65.
9. Liang, X.; Zhang, J.K.; Li, Z.H.; Xiao, B.C.; Chen, S. Optimizing protection scheme for accident shutdown of axial flow pumping station under super hump conditions. *J. Drain. Irrig. Mach. Eng.* **2020**, *38*, 1010–1015.
10. Xu, Z.; Zheng, Y.; Kan, K.; Chen, H.X.; Lu, Y.; Dai, J. Accelerated starting transition process of rapid gate in large tubular pumping station. *J. Drain. Irrig. Mach. Eng.* **2020**, *38*, 1016–1022.
11. Li, Q.; Ma, X.; Wu, P.; Yang, S.; Huang, B.; Wu, D.Z. Study on the Transient Characteristics of the Centrifugal Pump during the Startup Period with Assisted Valve. *Processes* **2020**, *8*, 1241. [CrossRef]
12. Lu, W. *Studies on the Mechanical Characteristics of Flap Valve Cutoff Device in Pumping Stations*; Shanghai University: Shanghai, China, 2009.
13. Wang, Z.X.; Du, Y.T. Opening Angle and Impact Force Analysis of Different Types Flap Valve in Pumping Stations. *IOP Conf. Ser. Earth Environ. Sci.* **2019**, *371*, 022067. [CrossRef]
14. Tsukamoto, H.; Ohashi, H. Transient Characteristics of a Centrifugal Pump during Starting Period. *J. Fluids Eng.* **1982**, *104*, 6–13. [CrossRef]
15. Li, Z.F.; Wu, P.; Wu, D.Z.; Wang, L.Q. Experimental and numerical study of transient flow in a centrifugal pump during startup. *J. Mech. Sci. Technol.* **2011**, *25*, 749–757. [CrossRef]
16. Zhang, X.W.; Tang, F.P.; Liu, C.; Shi, L.J.; Liu, H.Y.; Sun, Z.Z.; Hu, W.Z. Numerical Simulation of Transient Characteristics of Start-Up Transition Process of Large Vertical Siphon Axial Flow Pump Station. *Front. Energy Res.* **2021**, *9*, 706975. [CrossRef]
17. Wang, Y.; Xie, L.; Chen, J.; Liu, H.L.; Luo, K.K.; Zhang, Z.L.; Cao, M.H. Experimental study on transient startup characteristics of a super low specific speed centrifugal pump. *J. Chem. Eng. Jpn.* **2019**, *52*, 743–750. [CrossRef]
18. Liu, Y.F.; Zhou, D.Q.; Zheng, Y.; Zhang, H.S.; Xu, J.Y. Numerical simulation of starting process of axial flow pump with quick-stop gate. *South North Water Transf. Water Sci. Technol.* **2017**, *15*, 167–172.
19. Ge, Q.E.; Duan, X.H.; Xu, L.L.; Ma, B.; Chen, S.S.; Yan, D.F. Experimental Study of the Transition Course on the Starts of Synchronous Motor in South-North Water Diversion Project. *Large Electric Machine and Hydraulic Turbine* **2006**, *5*, 1–5.
20. Sun, Y.M.; Yu, G.P.; Liu, C. Research on Closing Control of the Sluice Gate for Sudden Power Off of Large Pump. *Adv. Sci. Lett.* **2011**, *4*, 2316–2320. [CrossRef]

21. Zhang, C.Y.; Wang, W.J.; Pei, J.; Yuan, S.Q. Numerical simulation of unsteady flow characteristics in a centrifugal pump during runaway process caused by power failure. *J. Phys. Conf. Ser.* **2022**, *2217*, 0212047. [CrossRef]
22. Kan, K.; Zheng, Y.; Chen, H.X.; Zhou, D.Q.; Dai, J.; Binama, M.; Yu, A. Numerical simulation of transient flow in a shaft extension tubular pump unit during runaway process caused by power failure. *Renew. Energy* **2020**, *154*, 1153–1164. [CrossRef]
23. Kan, K.; Chen, H.X.; Zheng, Y.; Zhou, D.Q.; Binama, M.; Dai, J. Transient Characteristics during Power-Off Process in a Shaft Extension Tubular Pump by Using a Suitable Numerical Model. *Renew. Energy* **2021**, *164*, 109–121. [CrossRef]
24. Zhang, X.W.; Tang, F.P. Investigation on hydrodynamic characteristics of coastal axial flow pump system model under full working condition of forward rotation based on experiment and CFD method. *Ocean Eng.* **2022**, *253*, 0029–8018. [CrossRef]
25. Feng, J.J.; Ge, Z.G.; Zhang, Y.; Zhu, G.; Wu, G.J.; Lu, J.L.; Luo, X.Q. Numerical investigation on characteristics of transient process in centrifugal pumps during power failure. *Renew. Energy* **2021**, *170*, 267–276. [CrossRef]
26. Gou, D.M.; Guo, P.C.; Luo, X.Q.; Zhan, L.G.; Guo, Y.F. 3-D combined simulation of power-off runaway transient process of pumped storage power station under pump condition. *Chin. J. Hydrodyn.* **2018**, *33*, 28–39.
27. Yang, Z.Y.; Liu, Z.R.; Cheng, Y.G.; Zhang, X.X.; Liu, K.; Xia, L.S. Differences of Flow Patterns and Pressure Pulsations in Four Prototype Pump-Turbines during Runaway Transient Processes. *Energies* **2020**, *13*, 5269. [CrossRef]
28. Yang, Z.Y.; Zhang, X.X.; Liu, D.M.; Cheng, Y.G.; Zhao, M.D. CFD simulation of the pump trip runaway transient process of a pumped-storage power plant with head 700 m. *IOP Conf. Ser. Earth Environ. Sci.* **2021**, *774*, 012102. [CrossRef]
29. Song, X.J.; Liu, C. Experimental study of the floor-attached vortices in pump sump using V3V. *Renew. Energy* **2021**, *164*, 0960–1481.
30. Song, X.J.; Luo, Y.Y.; Wang, Z.W. Numerical prediction of the influence of free surface vortex air-entrainment on pump unit performance. *Ocean Eng.* **2022**, *256*, 111503. [CrossRef]
31. Yang, F.; Li, Z.B.; Yuan, Y.; Lin, Z.K.; Zhou, G.X.; Ji, Q.W. Study on vortex flow and pressure fluctuation in dustpan-shaped conduit of a low head axial-flow pump as turbine. *Renew. Energy* **2022**, *196*, 856–869. [CrossRef]
32. Rohani, M.; Afshar, M. Simulation of transient flow caused by pump failure: Point-Implicit Method of Characteristics. *Ann. Nucl. Energy* **2010**, *37*, 1742–1750. [CrossRef]
33. Wang, F.J.; Bai, M.M.; Xiao R., F. Application of Flowmaster in the analysis of pump station transition process. *J. Drain. Irrig. Mech. Eng.* **2010**, *28*, 144–148.
34. Hu, X.Y. *Research on Hydraulic Transition Process in Pump Station System Based on Flowmaster*; Xihua University: Chengdu, China, 2012.
35. Geng, T.H. *Numerical Simulation of Hydraulic Transient Process of Francis Turbine Generator Unit based on Flowmaster*; Xihua University: Chengdu, China, 2020.
36. Yang, F.; Li, Z.B.; Fu, J.G.; Lv, Y.T.; Ji, Q.W.; Jian, H.F. Numerical and experimental analysis of transient flow field and pressure pulsations of an axial-flow pump considering the pump-pipeline interaction. *J. Mar. Sci. Eng.* **2022**, *10*, 258. [CrossRef]
37. He, Z.; Bao, H.; Gao, C.; Shu, L. Research of runaway transition process in the full flow system of tubular shaft-extension pump. *Fluid Mach* **2021**, *49*, 57–64.
38. Yang, Z.Y.; Cheng, Y.G.; Xia, L.S.; Meng, W.W.; Liu, K.; Zhang, X.X. Evolutions of flow patterns and pressure fluctuations in a prototype pump-turbine during the runaway transient process after pump-trip. *Renew. Energy* **2020**, *152*, 1149–1159. [CrossRef]
39. Wang, L.; Tang, F.P.; Chen, Y.; Liu, H.Y. Evolution Characteristics of Suction-Side-Perpendicular Cavitating Vortex in Axial Flow Pump under Low Flow Condition. *J. Mar. Sci. Eng.* **2021**, *9*, 1058. [CrossRef]
40. Song, X.J.; Liu, C.; Wang, Z.W. Prediction on the pressure pulsation induced by the free surface vortex based on experimental investigation and Biot-Savart Law. *Ocean Eng.* **2022**, *250*, 0029–8018. [CrossRef]
41. Yang, F.; Li, Z.B.; Hu, W.Z.; Liu, C.; Jiang, D.J.; Liu, D.S.; Nasr, A. Analysis of flow loss characteristics of slanted axial-flow pump device based on entropy production theory. *R. Soc. Open Sci.* **2022**, *9*, 211208. [CrossRef] [PubMed]
42. Yang, Z.Y.; Cheng, Y.G.; Xia, L.S.; Meng, W.W.; Gao, L.W.; Dai, Y.X. Evolutions of guide vane moment of a pump-turbine during runaway transient process after pump trip. *IOP Conf. Ser. Earth Environ. Sci.* **2019**, *240*, 072033. [CrossRef]
43. Liu, J.T.; Li, Z.; Wang, L.; Jiao, L. Numerical Simulation of the Transient Flow in a Radial Flow Pump during Stopping Period. *J. Fluids Eng.* **2011**, *133*, 111101. [CrossRef]
44. Zhang, R.T.; Zhu, H.G.; Li C., X.; Yao, L.B. Shut-down Transient of Bulb Tubular Pumping Station with VFD. *Trans. Chin. Soc. Agric. Mach.* **2013**, *44*, 45–49.

machines

MDPI

Article

Internal Flow Characteristics of High-Specific-Speed Centrifugal Pumps with Different Number of Impeller Blades under Large Flow Conditions

Chuan Wang [1,2], Xionghuan Chen [2,*], Jie Ge [3], Weidong Cao [4], Qiqi Zhang [5], Yong Zhu [5] and Hao Chang [1]

1 International Shipping Research Institute, GongQing Institute of Science and Technology, Jiujiang 332020, China
2 College of Hydraulic Science and Engineering, Yangzhou University, Yangzhou 225009, China
3 SHIMGE Pump Co., Ltd., Taizhou 317525, China
4 Institute of Fluid Engineering Equipment, JITRI, Zhenjiang 212013, China
5 National Research Center of Pumps, Jiangsu University, Zhenjiang 212013, China
* Correspondence: chen_xionghuan@hotmail.com

Abstract: As compared with a conventional centrifugal pump, a high-specific-speed centrifugal pump mostly operates under large flow conditions. In this paper, a typical high-specific-speed centrifugal pump is examined, and the effect of the blade number on the internal flow condition is investigated numerically. The numerical predictions have been verified through measurement. It was found that the predictions and the measurements are in good agreement of discrepancy. Serious cavitation could be observed within the pump when the flow rate reached 1300 m^3/h. Meanwhile, the effect of the blade number on the cavitation intensity was extremely obvious. The cavitation area at the inlet edge of the blades significantly reduced when the blade number increased from three to six. In addition, the turbulent kinetic energy within the pump was more uniformly distributed. This demonstrates that the blade number can be reasonably chosen to improve the internal flow pattern within the pump, which could provide a theoretical basis for the practical application of high-specific-speed centrifugal pumps.

Keywords: high-specific-speed centrifugal pump; number of impeller blades; large flow condition; numerical simulation

Citation: Wang, C.; Chen, X.; Ge, J.; Cao, W.; Zhang, Q.; Zhu, Y.; Chang, H. Internal Flow Characteristics of High-Specific-Speed Centrifugal Pumps with Different Number of Impeller Blades under Large Flow Conditions. *Machines* **2023**, *11*, 138. https://doi.org/10.3390/machines11020138

Academic Editors: Francesco Castellani and Kim Tiow Ooi

Received: 27 November 2022
Revised: 11 January 2023
Accepted: 16 January 2023
Published: 19 January 2023

1. Introduction

The performance and operational stability of centrifugal pumps have been a popular research topic in the field of fluid mechanics. Based on the anti-design approach, Hoang et al. [1] adjusted the parameters of the impeller meridian to improve the pump performance. Shen et al. [2] studied the effect of the groove structure on the performance and the flow pattern. This showed that the appropriate arrangement of the groove structure could not only improve the performance of the centrifugal pump at large flow conditions, but could also extend the stable operating range of the centrifugal pump. Based on the above research, the optimization theory of centrifugal pumps has been gradually refined. In the 19th century, the concept of specific speed was first applied to centrifugal pumps and became the most important factor during the design process [3–5]. Currently, high-specific-speed centrifugal pumps are widely used in urban buildings for sewage treatment, firefighting, and water supply because of their high conveying flow [6–8].

The open literature shows that the combination of theoretical derivations, numerical simulations and experimental studies is performed to optimize the performance of high-specific-speed centrifugal pumps. Hao et al. [9] carried out a theoretical analysis on the current status and deficiencies of existing high-specific-speed centrifugal pumps, which could provide theoretical support for the application of high-specific-speed centrifugal pumps. Ding et al. [10] numerically investigated the effect of blade exit angle on the performance of a high-specific-speed centrifugal pump and found that an appropriate selection of the blade exit angle was important to improve pump performance. Cheng et al. [11] studied the effect of balance orifice area on the performance of high-specific-speed centrifugal pumps. It was found that the area of the balancing bore could change the axial force of the rotor. Huan et al. [12] experimentally investigated the evolution of cavitation within a high-specific-speed centrifugal pump, which provided the basis for improving the cavitation resistance of centrifugal pumps. Since cavitation is an unavoidable problem for centrifugal pumps under high flow conditions, many scholars have also conducted research on the thermal effects of cavitation. Ge et al. [13] analyzed the three unstable states of sheet cavitation, periodic single cloud cavitation and non-periodic multi-cloud cavitation through cavitation flow experiments. They found that the effect of hydrodynamic cavitation was reduced when the water temperature was in the range of 55–60 °C. Rajavamsi et al. [14] elaborated the cavitation phenomenon in the centrifugal pump in detail, predicted the existence of cavitation phenomenon using the acoustic spectrum, and provided an effective method to reduce the noise in the pump. Hu et al. [15] studied the flow characteristics within the pre-swirl system of a marine gas turbine at low rotational speed by varying the pressure at the pre-swirl nozzle. Ge et al. [16] studied the influence of temperature on the dynamics of hydraulic cavitation in a closed loop cavitation tunnel and derived the thermal effect parameters that can provide a reference for avoiding cavitation-induced vibrations. In the meantime, scholars have also conducted a lot of research on the operation of centrifugal pumps under high flow conditions. Zhang et al. [17] analyzed the pressure pulsation characteristics of a high-specific-speed centrifugal pump. The lower intensity pressure pulsation was observed at high flow conditions. Chao et al. [18] investigated cavitation in pumps under high flow conditions and found that the reduction in volumetric displacement was beneficial to improve the operational stability of the pump system. Wang et al. [19] investigated the weak compressibility effect (WCE) of pressure fluctuations in rotor–stator interactions (RSI), which could provide some reference for efficient operation of the high-specific-speed centrifugal pump. However, the operating performance of the pump is often influenced by the geometric parameters of the impeller. Therefore, the internal flow condition within a high-specific-speed centrifugal pump with different blade numbers are investigated in this paper, which may provide some theoretical basis for the design and operational stability improvement of high-specific-speed centrifugal pumps.

2. Model and Numerical Methods

2.1. Model Parameters

The impeller and the volute are the main hydraulic components of a centrifugal pump. The role of the impeller is to transfer the energy of the prime mover to the medium. Then, the performance of the pump, such as the head, flow, efficiency and cavitation, is closely related to the parameters of the impeller. The function of the volute is to collect and export the medium. The performance of the pump will be directly influenced by the performance of the volute itself and how well it is matched to the impeller. In this paper, the inlet diameter D_1 of the impeller is 305 mm, the outer diameter D_2 of the impeller is 415 mm, the outlet width b_2 of the impeller is 107 mm, and the blade exit angle β_2 is 30°. Four impellers with 3, 4, 5 and 6 blades are studied in this paper. The diameter of spiral case base circle D_3 = 425 mm, and the width of spiral case inlet b_3 = 220 mm. The hydraulic model of the impeller and volute is depicted in Figure 1.

The rated flow of the pump adopted is Q = 1000 m^3/h, the head H = 20 m, the rated speed n = 980 r/min, and the specific speed n_s = 200. Based on the 2D (two-dimensional) drawings, the main overflow components were modeled in 3D (three-dimensional) using Creo software for point lines and surfaces, which is shown in Figure 2.

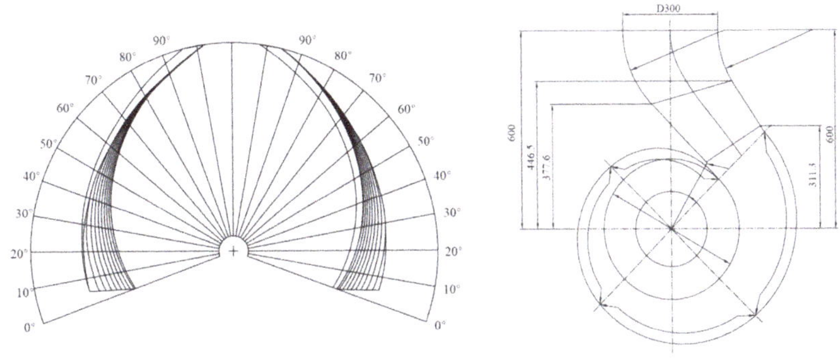

(**a**) (**b**)

Figure 1. Impeller and volute hydraulic model diagram: (**a**) impeller; (**b**) volute.

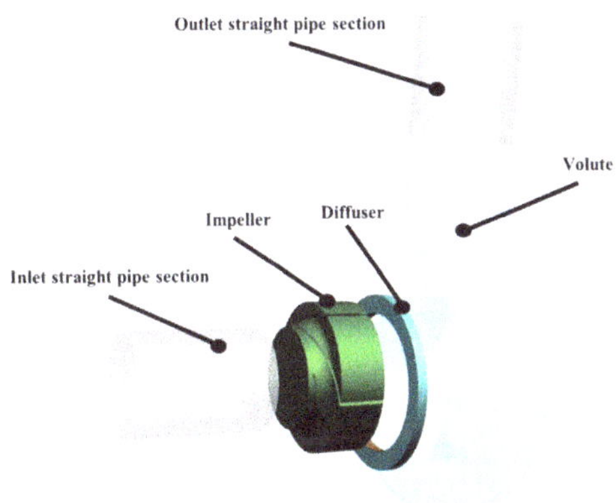

Figure 2. Three-dimensional diagram of high specific speed centrifugal pump.

2.2. Meshing and Irrelevance Verification

FLUENT MESHING was selected for mesh generation and optimization. Considering the twisted blades and wide flow path of the high-speed centrifugal pump, the hexahedral mesh is chosen for the impeller subdomain. For the other subdomains, the hexahedral structured mesh is adopted. The quality of all grids is above 0.31, which can meet the requirements of subsequent numerical calculations. The grids of the impeller and volute are shown in Figure 3.

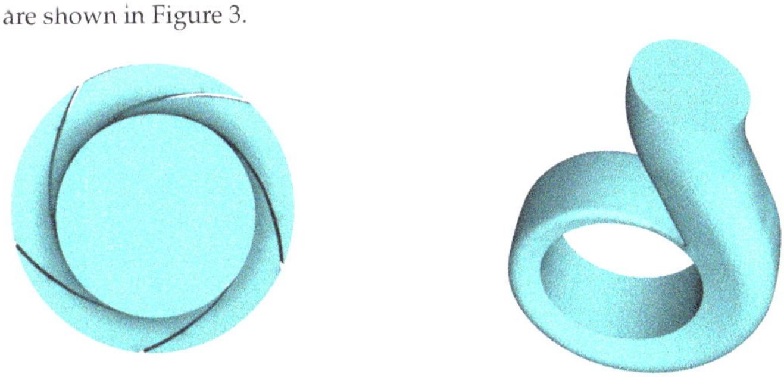

(**a**) (**b**)

Figure 3. Centrifugal pump grid diagram: (**a**) impeller; (**b**) volute.

In order to reduce the influence of the grid numbers on the numerical results, grid-independent analysis was conducted. The head of the centrifugal pump at a flow rate of 1000 m^3/h was used as an indicator. As depicted in Figure 4, the pump head no longer varies significantly when the total number of grids is greater than 1.2 million. Thus, the final total number of grids was determined to be 1.2 million. The grid numbers of impeller, volute, inlet straight pipe section and outlet straight pipe section were 469,873, 554,281, 1,250,810 and 67,000, respectively.

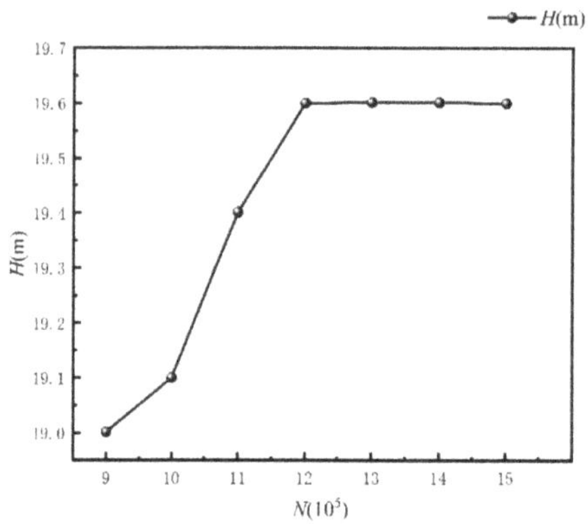

Figure 4. The grid independence analysis.

2.3. Boundary Condition Setting

The inlet and outlet, impeller, volute and diffuser are the five subdomains of the calculated domain. For the inlet, outlet and volute, the wall condition is set as a stationary wall, whereas the impeller is a rotor component. As a rotating domain, the impeller rotation speed is given as 980 r/min. The standard *k*-ε model was chosen for all the calculations. The total pressure is set as the inlet boundary condition, and the mass flow rate is set as the outlet boundary condition. For the cavitation calculation, the volume fraction of the gas phase is 1, the liquid phase is water at 25 °C, and the volume fraction is 0. The saturated steam pressure is set to 3574 Pa. The interfaces are employed to connect various sub-domains. For the steady numerical simulation, the coordinate system transformation of the interface between the inlet pipe and the impeller and between the impeller and the volute is set to frozen-rotor mode in the dynamic–static intersection. The number of solution steps is set to 4000, and the residual target is set to 10^{-4}.

In this paper, the Zwart model is used for cavitation prediction. The Zwart model is one of the widely used models for cavitation simulation. Its expressions of evaporation source term and condensation source term are the transport equations solved based on the Rayleigh Plesset cavitation growth equation, taking into account the changes in bubble volume during the generation and development of a single bubble. The expressions of evaporation source term and condensation source term are as follows.

$$R_e = F_{vap}\frac{3\alpha_{ruc}(1-\alpha)\rho_v}{R_B}\sqrt{\frac{2}{3}\frac{P_v-P}{\rho_l}};\ P < P_v \quad (1)$$

$$R_c = F_{cond}\frac{3\alpha_v\rho_v}{R_B}\sqrt{\frac{2}{3}\frac{P-P_v}{\rho_l}};\ P > P_v \quad (2)$$

where α_{ruc} is the volume fraction of the nucleation site; R_B is the cavitation radius, m; P is the flow field pressure, Pa; P_v is the vaporization pressure, and Pa; F_{vap} is the empirical correction coefficient of evaporation process, taking 50; The empirical correction coefficient of F_{cond} condensation process is 0.01.

3. Test Verification

3.1. Test Bench Construction

The performance tests with different impellers, which have three and six blades, were carried out on a comprehensive experimental platform with Class II accuracy. The test bench is depicted in Figure 5, with inlet and outlet valves, inlet and outlet pressure gauges, and pressure transducers being included. Measurement accuracy is 0.5% for electromagnetic flow meters and inlet and outlet pressure gauges. The data were collected and transferred to the computer, and the performance of the pump was calculated.

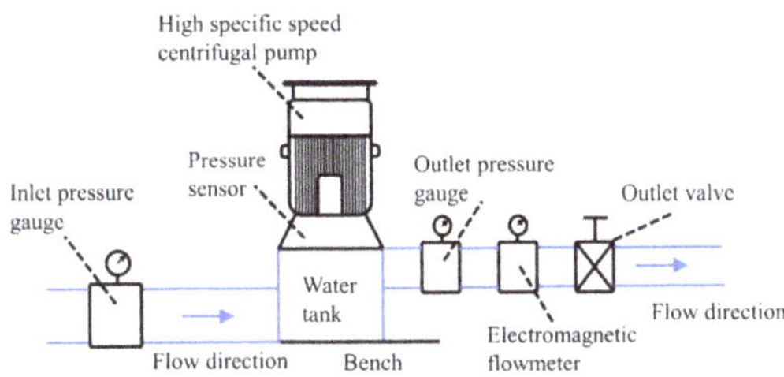

Figure 5. Schematic diagram of external characteristic test bench.

3.2. Comparison and Analysis of Test Results and Numerical Simulation Data

Figure 6 shows the numerical results and the experimental results with different blade numbers. The predicted values of head are slightly higher than the test values for most operating conditions, as is the efficiency. The higher predicted values of head at low flow conditions are due to uneven internal flow at low flow conditions. As the flow rate becomes larger, the relationship between predicted and tested values stabilizes. When the flow rate reaches 1300 m^3/h, the head and efficiency values show a large deviation from the test values. This may be the result of the inevitable cavitation that occurs with high flows. As a whole, there is a good agreement between the numerical results and the experimental results.

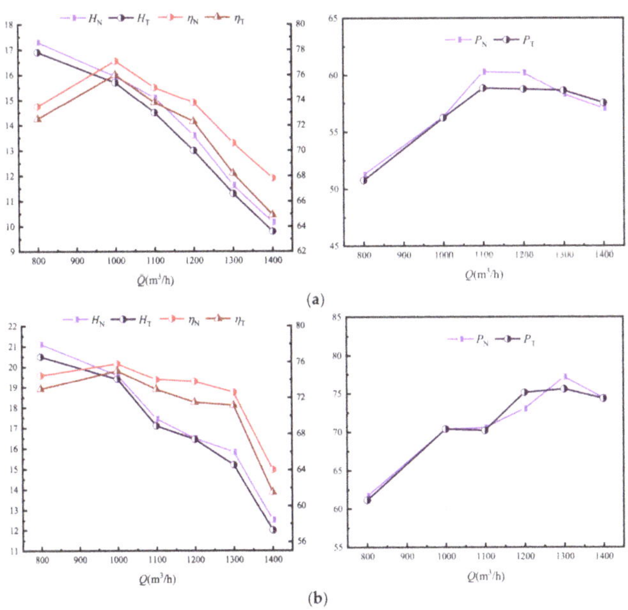

Figure 6. Comparison of performance between CFD prediction and experiment: (**a**) three-blade; (**b**) six-blade. Note: H, η and P represent head (m), efficiency (%) and power (kw), respectively. Subscripts N and T represent numerical calculated value and test value, respectively.

4. Results and Discussion

4.1. Flow Field Characteristics of Centrifugal Pumps under Different Operating Conditions

The velocity distribution from the inlet to the impeller section at different flow conditions is depicted in Figure 7. Smooth velocity changes at different large flow conditions and uniform velocity distribution from the inlet to the impeller shaft surface. The velocity from the fluid inlet channel to the axial surface of the impeller increases with the increase in flow rate, but the increase is only about 20%. At the inlet of the impeller, a small high-speed area could be observed. This is due to the curved fluid on the wall creating a vortex here. This indicates that the flow pattern in the inlet section is good under different flow conditions. However, under large flow condition, the flow structure at the impeller inlet is damaged due to the disturbance of rotating blades, which could generate a large amount of energy loss.

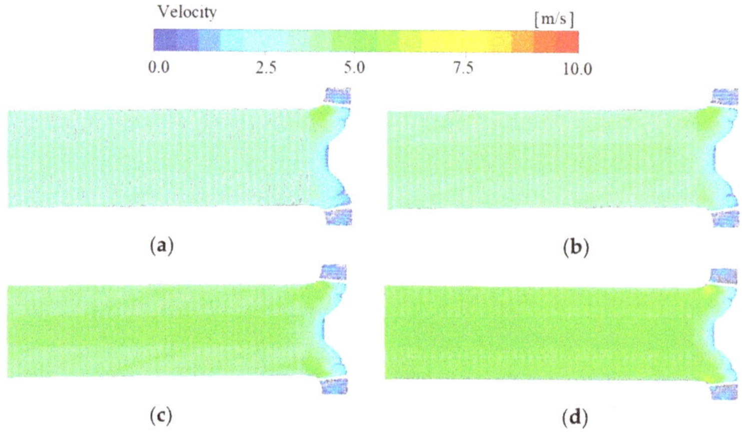

Figure 7. Relative velocity vectors in the axial section of the pump at different flow rates: (**a**) 800 m^3/h; (**b**) 1000 m^3/h; (**c**) 1100 m^3/h; (**d**) 1300 m^3/h.

The velocity and flow distribution of the mid-section in the volute under different operating conditions is depicted in Figure 8. There is a high-speed area in the impeller flow channel, and it can be seen that the speed of the blade suction surface in the impeller flow channel is greater than the pressure surface. When the flow rate was increased, it was found that the range of velocity variation around the impeller was not very large. Under the flow conditions of 1100 and 1300 m^3/h, the velocity distribution at the volute outlet is uneven. There is a low-speed zone at frame A, which expands from the right to the left with the increase in flow. The speed streamline in frame B is also gradually disordered. This shows that there is a trend of flow separation within the volute under large flow conditions, which may lead to further loss of energy.

Figure 9 depicts the turbulent kinetic energy distribution in the middle section of the volute. The results show that compared with the internal velocity distribution, the variation range of turbulent kinetic energy in the volute is large. When the flow rate is 800 m^3/h, the value of turbulent kinetic energy is large. Most areas in the volute are 1.5 m^2/s^2, of which the larger value is 4.0 m^2/s^2 outside the volute, as shown in box A. As the flow rate gradually increases, the turbulent energy values on the outside of the volute gradually decrease, while the turbulent energy values at the inlet gradually increase. When the flow reaches 1100 m^3/h, the turbulence intensity of the volute becomes uniform. This indicates that the flow is not complicated at this time, and the energy loss is small. However, the turbulence kinetic energy at the outlet is large, as depicted in box B. This is consistent with the velocity distribution under the large flow condition, which verifies that the internal flow at the inlet under the large flow condition is complex and requires in-depth study.

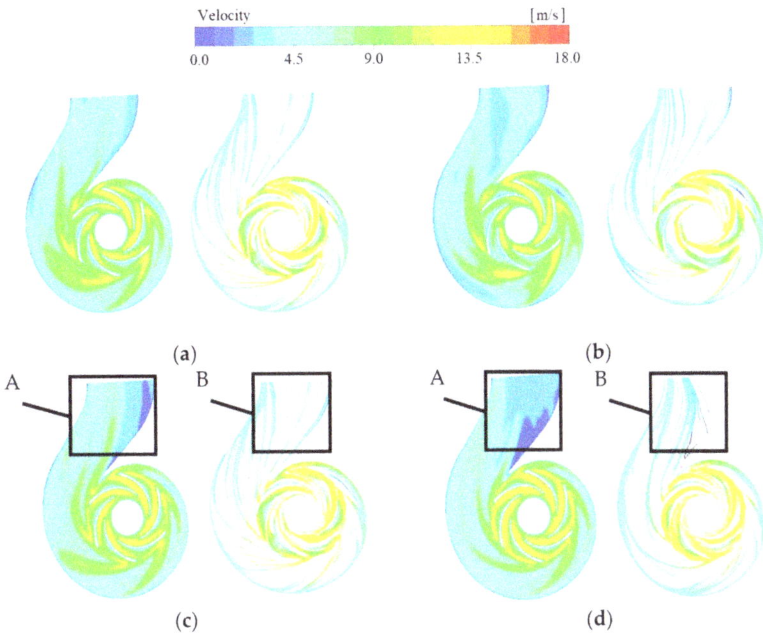

Figure 8. Velocity and streamline distribution of the volute at different flow rates: (**a**) 800 m^3/h; (**b**)1000 m^3/h; (**c**) 1100 m^3/h; (**d**) 1300 m^3/h.

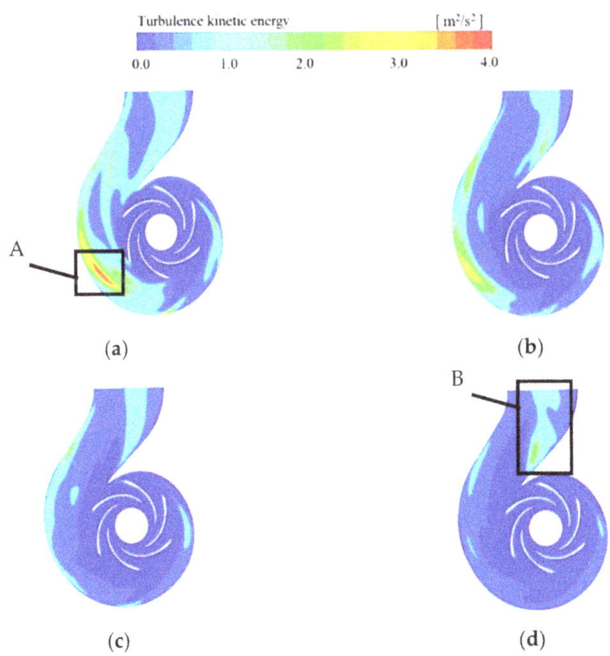

Figure 9. Turbulence kinetic energy in pump cross-section at different flow rates: (**a**) 800 m^3/h; (**b**) 1000 m^3/h; (**c**) 1100 m^3/h; (**d**) 1300 m^3/h.

The total pressure distribution in the impeller mid-section for different flow conditions is depicted in Figure 10. Obviously, within the impeller, the total pressure is evenly distributed, and there is no significant change in the size, while the total pressure distribution in the volute is uneven. When the flow rate is 800 and 1100 m^3/h, the relative high pressure appears at the volute near the impeller, and most of the high-pressure values reach 300 kPa. With the increase in flow rate, the pressure value inside the volute decreases, and the pressure value distribution is still uneven. The area with high pressure in the volute near the impeller decreases, and the area with low pressure in the impeller flow passage decreases relatively. Under the flow condition of 1100 m^3/h, a large range of high pressure appears at the volute near the impeller. A comparison of the total pressure diagrams for the four different flow conditions shows that there is no significant change in the impeller when the flow rate becomes larger. While the impeller low pressure range is reduced, the total pressure value in the volute is also reduced.

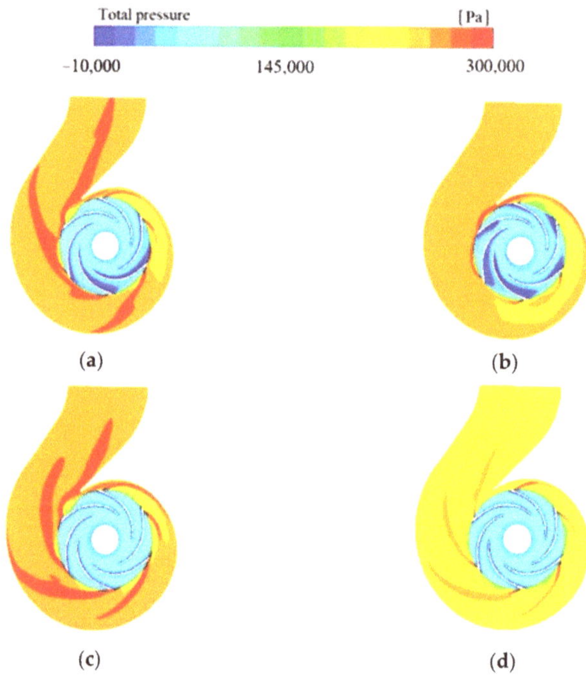

Figure 10. The total pressure in pump cross-section at different flow rates: (**a**) 800 m^3/h; (**b**) 1000 m^3/h; (**c**) 1100 m^3/h; (**d**) 1300 m^3/h.

The cavitation number is defined as:

$$\sigma = (p_{in} - p_v)/0.5\rho u^2 \quad (3)$$

where p_{in} is the pressure at the pump inlet (Pa); p_v is the saturated vapor pressure (Pa) corresponding to clean water at 25 °C.

The development of cavitation in the centrifugal pump could lead to hydraulic loss, which in turn leads to the deterioration of pump performance in two aspects. On the one hand, the cavitation structure blocks the flow channel, leading to changes in the flow pattern within the pump, which directly causes hydraulic losses. On the other hand, the cavitation structure induces the pressure fluctuation near the blade surface and the generation of vortex structure within the flow channel, which affects the work ability of the blade and leads to head loss. In order to analyze the cavitation morphology in the impeller of a high-specific-speed centrifugal pump under different working conditions, four different flows are selected in this paper to study the bubble distribution when the cavitation number is 0.55, as shown in Figure 11. According to the numerical results, when the flow rate is 500 m^3/h, the pump head drops by less than 1%. At this time, the centrifugal pump is in the initial stage of cavitation. It can be seen that only a small range of cavitation is generated at the leading edge (LE) of the blade inlet during the initial stage of cavitation. As shown by the red dotted line in the figure, the blade is used to compare the change of cavitation under different flow rates. It was found that with the increase in flow rates, a large range of cavitation attachment appears on the back of the blade, and the cavitation thickness gradually increases. When Q = 110 m^3/h, the cavitation volume occupies the edge of the blade, which affects the performance of the pump. This shows that cavitation can easily occur under large flow conditions, which may also be the reason for large deviation between the numerical results and the test results under large flow conditions.

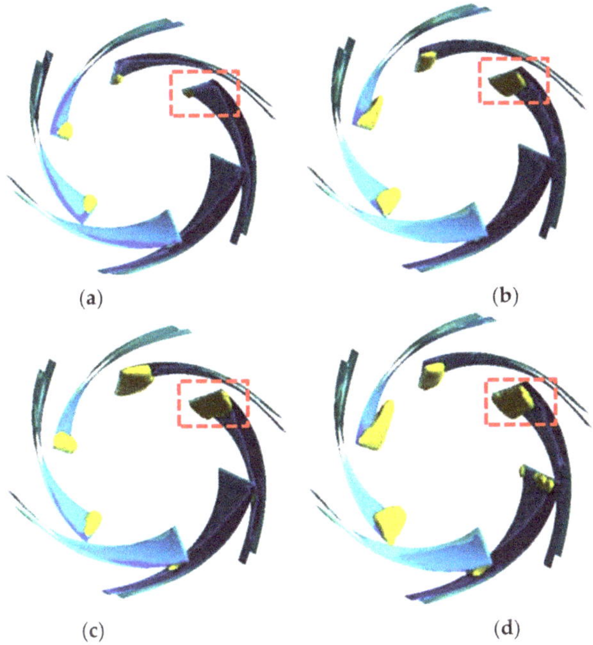

Figure 11. Cavitation distribution under different working conditions: (**a**) 500 m^3/h; (**b**) 800 m^3/h; (**c**) 1000 m^3/h; (**d**) 1100 m^3/h.

4.2. Flow Field Characteristics of Centrifugal Pumps with Different Blade Numbers under Large Flow Conditions

Figure 12 illustrates the distribution of the vapor pocket volume for the different number of impeller blades at a cavitation number of 0.55 and a flow rate of 1300 m^3/h. The cavitation around the blade inlet edge is clearly visible with three blades. The cavitation is most severe under these conditions, especially with the largest cavitation volume of A2. With the increase in blade numbers, the cavitation area of the blade decreases with the same cavitation number and flow rate. The main cavitation places are at the leading edge of the blade, where the curvature is relatively large, and it is easy to collide with the water flow. When the number of blades is five, the cavitation volume at A4 is 0, indicating that when the number of blades increases, the cavitation degree at the blade edge will decrease. The cavitation of three blades is relatively serious, which will have a great impact on the performance of the pump. Therefore, when considering cavitation under large flow conditions, the impact of blade numbers on cavitation should be considered.

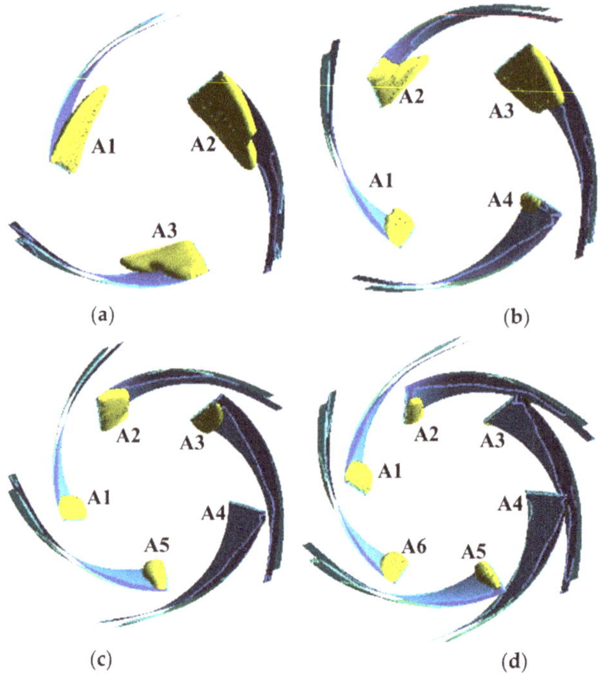

Figure 12. Cavitation volume distribution with different numbers of blades: (**a**) three-blade; (**b**) four-blade; (**c**) five-blade; (**d**) six-blade. Note: The setting angle of impeller blade inlet is 22°.

Figure 13 shows the velocity distribution of the sections with different impeller blade numbers at 1300 m^3/h. It is obvious from the figure that when the blade number is small, there is a low-speed area at the inlet of the volute, and the flow velocity near the impeller is larger. With the increase in the blade number, the low-speed area at the volute inlet is significantly reduced. At the same time, the speed of the impeller decreases slightly. Generally, the larger speed of the impeller is concentrated in the flow passage, which is generally about 13 m/s, and the speed in other areas is relatively uniform. Therefore, the increase in blade number is conducive to improving the flow pattern distribution near the volute inlet and impeller and to reducing the internal energy loss of the pump.

Figure 14 depicts the distribution of turbulent kinetic energy in the middle section of the centrifugal pump with different numbers of blades when the flow is 1300 m^3/h. When the number of blades is increased, the turbulent kinetic energy in the impeller is basically unchanged, while it changes regularly in the volute. Specifically, the turbulent kinetic energy on the outer side of the volute increases rapidly and unevenly, while the relative increase on the inner side is gentle. When the number of impeller blades is three and four, the turbulent kinetic energy in some areas increases sharply at the outlet of the volute. The maximum value of 4.0 J/kg indicates that this part of the flow is more complex and has a higher energy loss. When the number of impeller blades is four and five, most areas in the volute reach 1.5 J/kg. Under the number of six blades, the turbulent kinetic energy value is small, and the distribution of impeller and volute is relatively uniform. The turbulent energy changes of different blade numbers under large flow conditions further show that the reasonable increase in impeller blade numbers can improve the internal flow structure under large flow conditions and improve the operational stability of the pump.

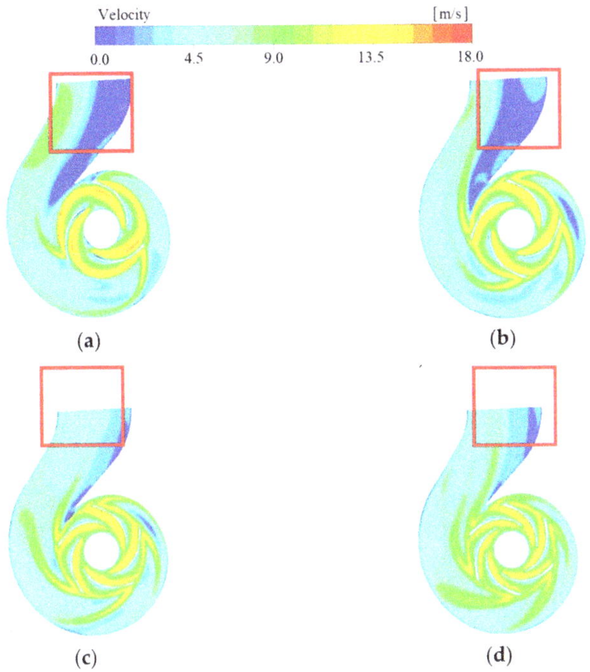

Figure 13. Relative velocity vectors in pump cross-section at different numbers of blades: (**a**) three-blade; (**b**) four-blade; (**c**) five-blade; (**d**) six-blade.

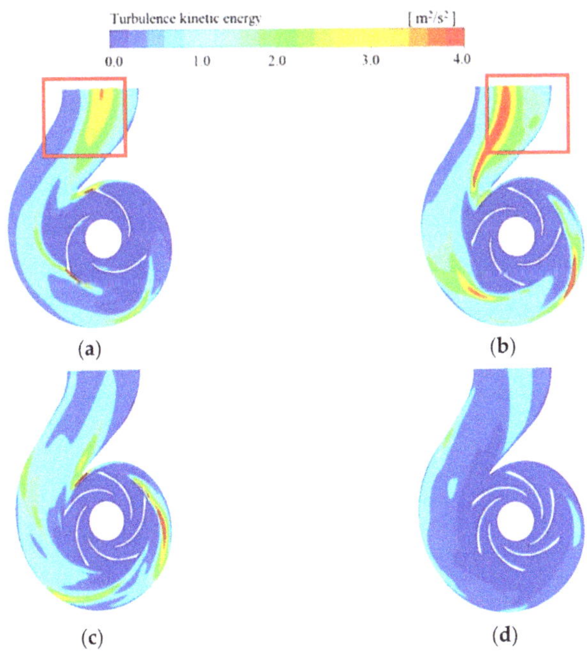

Figure 14. Turbulence kinetic energy in pump cross-section at different numbers of blades: (**a**) three-blade; (**b**) four-blade; (**c**) five-blade; (**d**) six-blade.

Figure 15 depicts the total pressure distribution in the middle section with different blade numbers when the flow is 1300 m^3/h. It is obviously observed that, from the impeller inlet to the whole impeller fluid domain, the fluid pressure basically has no change. However, lower pressures occur near the impeller blades. The total pressure distribution in the volute is not uniform and has a wide range of variation. The relatively high pressure appears near the volute near the impeller, and some areas reach 350 kPa. With the increase in the number of blades, the pressure value inside the volute increases continuously, and the pressure value distribution is still uneven. At this point, the area of lower pressure in the impeller runners is relatively reduced.

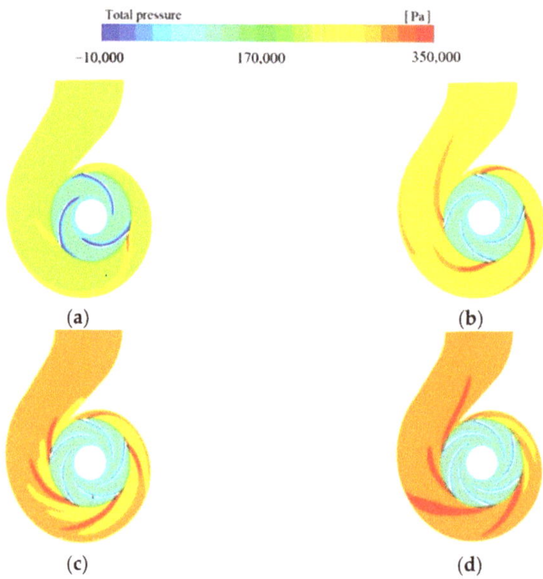

Figure 15. The total pressure in pump cross-section at different numbers of blades: (**a**) three-blade; (**b**) four-blade; (**c**) five-blade; (**d**) six-blade.

5. Conclusions

In this paper, the internal flow characteristics of a high-specific-speed centrifugal pump with different blade numbers are investigated under large flow conditions. Measurements have been conducted to verify the numerical predictions. The conclusions are as follows:

(1) The measurements and the prediction have good overall agreement and permissible errors.

(2) The results also show that when the flow rate increases, the inlet runner velocity increases. However, the speed of the impeller inlet is significantly reduced, and backflow occurs. The results show that the cavitation volume increases with the increase in flow rate, implying that significant cavitation occurs with higher flowrates. At high flow rate conditions, it is also observed that the velocity distribution at the outlet of the volute becomes unevenly chaotic, resulting in decreasing turbulent energy at the inlet of the volute. Hence, under a higher flowrate condition, this shows that the internal flow state is disordered and the operation is unstable.

(3) When the number of blades is six, the range of turbulent energy variations inside the impeller is low. The low-pressure range of the impeller runners is reduced, and the turbulent energy on the inside of the worm gear remains unchanged. At the same time, the degree of cavitation in the pump decreases with the increase in the blade numbers. Therefore, an appropriate selection of the blade number is important to improve the internal flow condition of the pump.

Author Contributions: Data curation, C.W. and X.C.; formal analysis, Q.Z. and W.C.; writing—original draft, Q.Z. and C.W.; writing—review and editing, J.G. and Y.Z.; software, H.C. All authors have read and agreed to the published version of the manuscript.

Funding: This research was supported by the National Natural Science Foundation of China (Grant No: 51979240 and 52009013), and the National Key R&D Program of China (Grant No: 2020YFC1512402).

Data Availability Statement: Not applicable.

Conflicts of Interest: The authors declare no conflict of interest.

References

1. Hoang, T.; Truong, V.; Shrestha, U.; Choi, Y. Optimization of the Meridional Plane Shape Design Parameters in a Screw Centrifugal Pump Impeller. *Proc. Korea Fluid Mach. Soc.* **2021**, *24*, 15–25. [CrossRef]
2. Shen, Z.; Chu, W. Influence of grooves arrangement in volute casing on internal flow characteristics of centrifugal pump. *J. Huazhong Univ. Sci. Technol.* **2019**, *47*, 37–42. [CrossRef]
3. Tang, S.; Zhu, Y.; Yuan, S. Intelligent fault identification of hydraulic pump using deep adaptive normalized CNN and synchrosqueezed wavelet transform. *Reliab. Eng. Syst. Saf.* **2022**, *224*, 108560. [CrossRef]
4. Yang, Y.; Zhou, L.; Zhou, H.; Lv, W.; Wang, J.; Shi, W.; He, Z. Optimal design of slit impeller for low specific speed centrifugal pump based on orthogonal test. *J. Mar. Sci. Eng.* **2021**, *9*, 121. [CrossRef]
5. Xi, B.; Wang, C.; Xi, W.; Liu, Y.; Wang, H.; Yang, Y. Experimental investigation on the water hammer characteristic of stalling fluid in eccentric casing-tubing annulus. *Energy* **2022**, *253*, 124113. [CrossRef]
6. Guo, X.; Zhu, L.; Zhu, Z.; Cui, B.; Li, Y. Numerical and experimental investigations on the cavitation characteristics of a high-speed centrifugal pump with a splitter-blade inducer. *J. Mech. Sci. Technol.* **2015**, *29*, 259–267. [CrossRef]
7. Skrzypacz, J.; Bieganowski, M. The influence of micro grooves on the parameters of the centrifugal pump impeller. *Int. J. Mech. Sci.* **2018**, *144*, 827–835. [CrossRef]
8. Xue, R.; Lin, X.; Zhang, B.; Zhou, H.; Lai, T.; Hou, Y. CFD and Energy Loss Model Analysis of High-Speed Centrifugal Pump with Low Specific Speed. *Appl. Sci.* **2022**, *12*, 7435. [CrossRef]
9. Hao, Y.; Hao, J.; Zhu, Z.; Su, X.; Lu, W.; Gruszczynski, M.; Ding, Q.; Panlong, G. Review of the Hydraulic and Structural Design of High-Speed Centrifugal Pumps. *Front. Energy Res.* **2022**, *10*. [CrossRef]
10. Ding, H.; Li, Z.; Gong, X.; Li, M. The influence of blade outlet angle on the performance of centrifugal pump with high specific speed. *Vacuum* **2019**, *159*, 239–246. [CrossRef]
11. Cheng, X.; Chang, Z.; Jiang, Y. Study on the influence of the specific area of balance hole on cavitation performance of high-speed centrifugal pump. *J. Mech. Sci. Technol.* **2020**, *34*, 3325–3334. [CrossRef]
12. Huan, Y.; Liu, Y.; Li, X.; Zhu, Z.; Qu, J.; Zhe, L.; Han, A. Experimental and numerical investigations of cavitation evolution in a high-speed centrifugal pump with inducer. *J. Hydrodyn.* **2021**, *33*, 140–149. [CrossRef]
13. Ge, M.; Manikkam, P.; Ghossein, J.; Subramanian, R.; Coutier-Delgosha, O.; Zhang, G. Dynamic mode decomposition to classify cavitating flow regimes induced by thermodynamic effects. *Energy* **2022**, *254*, 124426. [CrossRef]
14. Gangipamula, R.; Ranjan, P.; Patil, S. Flow-induced noise sources and reduction methods in centrifugal pumps: A literature review. *Phys. Fluids* **2022**, *34*, 081302. [CrossRef]
15. Hu, B.; Yao, Y.; Wang, M.; Wang, C.; Liu, Y. Flow and Performance of the Disk Cavity of a Marine Gas Turbine at Varying Nozzle Pressure and Low Rotation Speeds: A Numerical Investigation. *Machines* **2023**, *11*, 68. [CrossRef]
16. Ge, M.; Petkovšek, M.; Zhang, G.; Jacobs, D.; Coutier-Delgosha, O. Cavitation dynamics and thermodynamic effects at elevated temperatures in a small Venturi channel. *Int. J. Heat Mass Transf.* **2021**, *170*, 120970. [CrossRef]
17. Zhang, J.; Yang, H.; Liu, H.; Xu, L.; Lv, Y. Pressure Fluctuation Characteristics of High-Speed Centrifugal Pump with Enlarged Flow Design. *Processes* **2021**, *9*, 2261. [CrossRef]
18. Chao, Q.; Zhang, J.; Xu, B.; Huang, H.; Zhai, J. Centrifugal effects on cavitation in the cylinder chambers for high-speed axial piston pumps. *Meccanica* **2019**, *54*, 815–829. [CrossRef]
19. Wang, S.; Chen, X.; Li, X.; Cui, B.; Zhu, Z. Weak compressibility effects on the pressure fluctuation at RSI in a high speed centrifugal pump. *J. Mech. Sci. Technol.* **2022**, *36*, 5047–5057. [CrossRef]

machines
MDPI

Article

Flow and Performance of the Disk Cavity of a Marine Gas Turbine at Varying Nozzle Pressure and Low Rotation Speeds: A Numerical Investigation

Bo Hu [1], Yulong Yao [2], Minfeng Wang [3], Chuan Wang [2,*] and Yanming Liu [1,3,*]

1 Department of Energy and Power Engineering, Tsinghua University, Beijing 100084, China
2 College of Hydraulic Science and Engineering, Yangzhou University, Yangzhou 225009, China
3 Harbin Marine Boiler and Turbine Research Institute, Harbin 150078, China
* Correspondence: wangchuan198710@126.com (C.W.); hrbliuym@163.com (Y.L.)

Abstract: In marine gas turbines, variations in rotational speed occur all the time. To ensure adequate cooling effects on the turbine blades, the valves need to be adjusted to change the pressure upstream of the pre-swirl nozzle. Changing such pressure will have significant effects on the local or overall parameters, such as core swirl ratio, temperature, flow rate coefficient, moment coefficient, axial thrust coefficient, etc. In this paper, we studied the flow characteristics within the pre-swirl system of a marine gas turbine at low rotational speed by varying the pressure at the pre-swirl nozzle. The corresponding global Reynolds number ranged from $Re = 2.3793 \times 10^5$ to 9.5172×10^5. The flow in the rotor-stator cavities was analyzed to find the effects of nozzle pressure on the radial velocity, core swirl ratio, and pressure. According to the simulation results, we introduced a new leakage flow term into the formulary in the references to calculate the values of K between the inner seal and the pre-swirl nozzle. The matching characteristics between the pre-swirl nozzle and the inclined receiving hole was predicted. Performance of the pre-swirl system was also analyzed, such as the pressure drop, through-flow capacity, and cooling effects. After that, the moment coefficient and the axial thrust coefficient were predicted. This study provides some reference for designers to better design the pre-swirl system.

Keywords: pre-swirl rotor-stator cavity; nozzle pressure; core swirl ratio; flow rate coefficient; moment coefficient; axial thrust coefficient

Citation: Hu, B.; Yao, Y.; Wang, M.; Wang, C.; Liu, Y. Flow and Performance of the Disk Cavity of a Marine Gas Turbine at Varying Nozzle Pressure and Low Rotation Speeds: A Numerical Investigation. *Machines* **2023**, *11*, 68. https://doi.org/10.3390/machines11010068

Academic Editor: Davide Astolfi

Received: 30 November 2022
Revised: 20 December 2022
Accepted: 21 December 2022
Published: 5 January 2023

1. Introduction

With the continuous development of marine gas turbines, the demand for high performance and long service life are constantly growing. Effective cooling of the turbine blade and rotor without significantly increasing the amount of cooling air are important components that are always popular research subjects. They require insight into the flow in a pre-swirl disk cavity in the secondary air system.

The pre-swirl nozzle is the core component in the pre-swirl disk cavity. It provides the cooling air with large tangential velocity by twisting it. The jet flow is mixed with the cavity flow downstream of the nozzle. The difference in tangential velocity between the air and the disk then becomes small. Normally, the relative total temperature at the receiving hole inlet is used to estimate the cooling performance of delivered air on the turbine blade. With a relatively smaller tangential velocity difference, the cooling performance is improved. Meanwhile, the excessive amount of pre-swirl jet flow mixes with the cavity flow and may cause some problems, such as excessive pressure losses, axial thrust, disk frictional losses, etc. Over the past decades, a lot of related studies have been conducted to solve these problems. The scholars found three key parameters for the above parameters, namely through flow rate (C_w), circumferential Reynolds number (Re), and axial gap width (G).

In recent decades, a great number of studies have been conducted on rotor-stator cavities, in particular on the pre-swirl system. EL-Oun and Owen [1] developed a theoretical model of adiabatic efficiency, which represents the level of temperature reduction capacity, based on the Reynolds analogy. Chew et al. [2,3] developed a basic model to calculate the temperature of the blade cooling air through experiments and numerical simulations. Ciampoli and Chew [4] put forward an automatic design method for a pre-swirl nozzle with CFD (computational fluid dynamics).

To improve the performance of a pre-swirl nozzle, the impact of geometric parameters is studied, such as the radial position, swirl angle, inlet diameter, and channel shape [5–9]. In the following up studies, Lee et al. [10] optimized a vane-type pre-swirl nozzle to increase the through-flow rate and enhance the temperature reduction. The optimized model improved the span-wise uniformity of the flow path, achieving an increase in the discharged capacity by 2.57%. Yang et al. [11] illustrated the combined and separate influences of jet impingement and air entrainment on the flow and heat transfer within the rotor-stator cavity. Liu et al. [12–15] modified the nozzle geometry, lead angle, and aspect ratio to optimize the performance of a pre-swirl system. Hu et al. [16] carried out numerical simulations to study the flow characteristics, temperature drop, and specific entropy increment for the pre-swirl system with a different radial angle for the pre-swirl nozzle. Lin et al. [17] investigated the impact of flow rate and speed of rotation on the flow dynamics in the rotor-stator cavity with the pre-swirl jet flow at a low radius.

In terms of heat transfer, Cheng et al. [18] used numerical calculations to analyze the effects of three different inlet pre-swirl ratios and turbulence coefficients on the heat transfer characteristics in the pre-swirl disk cavity. Liu et al. [19,20] put forward a vane-hole shape pre-swirl nozzle. They demonstrated its advantage by comparing the pressure losses and temperature reduction with those of the vane nozzle. Zhao et al. [21], Wu et al. [22], and Wang et al. [23,24] studied the heat transfer in the rotor-stator cavity and revealed the mechanism of heat transfer of such a cavity.

The application of CFD during the optimal design of a turbomachine is the focus of numerical simulation nowadays [25]. In the numerical simulation of the rotor-stator cavity, many scholars tried different simulation methods to investigate the mechanism of flow and heat transfer. Damavandi et al. [26] studied the flow in a rotor–stator cavity with a curved rotor disk using the LES (large eddy simulation) technique. Liao et al. [27] used the conjugate CFD to investigate the flow and heat transfer in the rotor–stator cavity. They compared the results with steady-state simulation and experimental results. Jia et al. [28] proposed a numerical model for the ingestion of hot main flow into a rotor–stator cavity model, considering the rotor vibration. The results show that the sealing effectiveness at the outer rim seal increases due to the vibration of the rotor; meanwhile, the amplitudes of the pressure fluctuation decrease in the sealing gap. Schroeder et al. [29,30] presented an experimental investigation of centripetal leakage flow with and without pre-swirl in the rotor–stator cavities by analyzing the results from two test rigs. The influences of Re, G, and C_w on the radial pressure distribution, axial thrust, and frictional torque are presented. Onori et al. [31] used the proper orthogonal decomposition (POD) method to estimate the results from LES to identify a set of orthonormal basis functions for the Galerkin projection of the Navier–Stokes equations.

Besides the pre-swirl nozzle, there are a number of significant influences on the pre-swirl system. Lin et al. [32] revealed the heat transfer mechanism and energy conversion characteristics with the influences of many factors, such as the seal flow, friction, and work performed by either the rotating or the stationary parts. He et al. [33] and Shi et al. [34] focused on the effect of flow rate on cavity flow. Zhao et al. [35] numerically investigated the impact of blade fracture on cavity flow with a centrifugal superposed flow with the non-axis-symmetry boundary conditions. When the turbine blade fractures, the velocity obviously changes in the downstream region at a high radius, especially when the local Reynolds number, Re_φ, is large. Ren et al. [36] designed and performed an experimental investigation to determine the flow characteristics in a rotor–stator cavity with a large gap ratio and inlet at a low radius. The results indicate that the flow in the rotor–stator cavity is still dominated by circumferential motion. However, the velocity increases with the increase of radial position. In different axial positions, the velocity distribution is different. When the cavity flow approaches the rotor, it becomes uneven.

In a marine gas turbine, it is more and more important to widen its operating conditions, even below 20% of the rated speed of rotation. Therefore, the secondary air system is required to work at a small speed of rotation, even around 20% of the rated value. Therefore, it is very important to check whether the secondary air system is functional. In addition, in a marine gas turbine, there are some valves to adjust the pressure at the inlet of the pre-swirl nozzle, aiming to improve the performances at off-design conditions, especially at a small speed of rotation (below 50% maximum rotating speed). It is also possible that the performance of the pre-swirl system is improved if we bleed the cooling air from the later stage of the compressor.

Hence, this paper uses the CFD method to investigate the pressure (70% to 140% of the design values) at the pre-swirl nozzle on the flow characteristics and overall performance of the pre-swirl system. The ranges of the speed of rotation are from 12.5% to 50% of the maximum speed of rotation of the gas turbine. The flow characteristics, performance of the pre-swirl system, disk frictional losses, and axial thrust are predicted. The results can expand knowledge and provide a reference for designers to better consider the effects of pressure at the nozzle inlet on performance at a small speed of rotation.

2. Definitions of Main Parameters

Global circumferential and local circumferential Reynolds number *Re*:

$$Re = \frac{\omega \cdot b^2}{\nu} \tag{1}$$

where b stands for the outer radius of the disk, ω is the angular velocity of the disk, and ν represents the kinematic viscosity.

Local circumferential and local circumferential Reynolds number *Re*:

$$Re_\varphi = \frac{\omega \cdot r^2}{\nu} \tag{2}$$

where r is the radial coordinate.

Discharge coefficient of the receiving hole:

$$C_d = \frac{m}{m_i} \tag{3}$$

where m is the actual mass flow rate and m_i is the isentropic mass flow rate.

The ideal mass flow rate of a long orifice:

$$m_i = A \cdot \frac{p_{out}}{RT_1*} \cdot \left(\frac{p_{out}}{p_{in}*}\right)^{1/k} \cdot \sqrt{\frac{2k}{k-1} RT_1 * \left[1 - \left(\frac{p_{out}}{p_{in}*}\right)^{(k-1)/k}\right] + 2(U_2 V_{2\varphi} - U_1 V_{1\varphi}) - V_{2\varphi}^2} \tag{4}$$

where A is the nozzle outlet area, R is the air constant, and its value is 287. T_1* is the inlet total temperature. k is the adiabatic exponent of gas, and its value is 1.4. p_{out} is the static pressure at the outlet. $V_{1\varphi}$ is the circumferential velocity of the gas inlet. $V_{2\varphi}$ is the circumferential velocity of the gas outlet. U_1 is the circumferential velocity of the disc inlet. U_2 is the circumferential velocity of the disc outlet.

Local flow rate coefficient:

$$C_{qr} = \frac{Q \cdot Re_\varphi^{0.2}}{2\pi \cdot \omega r^3} \tag{5}$$

where Q is the volumetric through-flow rate.

Axial thrust coefficient:

$$C_F = \int_a^b \frac{2\pi \cdot (p_b - p) \cdot r}{\rho \omega^2 b^4} dr \tag{6}$$

where a is the hub radius, p represents the pressure, p_b stands for the pressure at $r = b$, and ρ is the density of air.

Moment coefficient:

$$C_M = \frac{2 \cdot |M|}{\rho \omega^2 b^5} \tag{7}$$

where M is the friction torque.

Pressure coefficient:

$$C_p = p*(x=1) - p*(x)\ ;\ x = \frac{r}{b}\ ;\ p* = \frac{p}{\rho \omega^2 b^2} \tag{8}$$

where x is the non-dimensional radial coordinate and p^* represents the non-dimensional pressure.

Non-dimensional temperature $T*$:

$$T* = \frac{T - T_{min}}{T_{max} - T_{min}} \tag{9}$$

where T is the temperature, T_{max} is the maximum temperature in the cavity, T_{min} is the minimum temperature in the cavity, c_p is the isobaric-specific heat capacity, and its value is 1.039. $T_{out}{}^*$ is the relative total temperature at the outlet of receiving hole. $T_{in}{}^*$ is the total temperature at the inlet of the pre-swirl nozzle.

Non-dimensional temperature drop

$$\triangle T* = \frac{c_p(T_{in}* - T_{out}*)}{0.5(\omega r_b)^2} \tag{10}$$

where c_p is the isobaric-specific heat capacity, and its value is 1.039. $T_{out}{}^*$ is the relative total temperature at the outlet of receiving hole. $T_{in}{}^*$ is the total temperature at the inlet of the pre-swirl nozzle.

3. Numerical Simulation Set-Ups

3.1. Domains for Numerical Simulation

The model in this study is shown in Figure 1. It includes three cavities, noted as C1 to C3 cavities. The C1 cavity is a typical rotor–stator cavity with centrifugal through-flow. The C2 cavity is a co-rotating cavity, which is a simplification of the flow path to the rotor blade. The C3 cavity is a pre-swirl cavity, which is the main research object. The main geometric parameters are shown in Table 1. The pre-swirl nozzle is on the stator, which is slightly below the receiving hole in the radial direction. The cooling air enters the C3 cavity in the form of pre-swirl jet flow and mixes with the gas in the disk cavity. Then, the essential flow enters the receiving hole, while the minor flow goes through the A1 labyrinth seal into the C1 cavity. It finally leaves the C1 and interacts with the hot main flow at the upper outlet.

Table 1. Main geometric parameters.

Parameter	Values
Hub radius a/b	0.736
Outer radius of the disk b/b	1
Axial gap of the front cavity s/b	0.044
Diameter of air inlet channel d_p/b	0.011
Intake angle θ (°)	20
Radius of inlet center r_p/b	0.096
Diameter of outlet channel d_b/b	0.015
Radius of outlet center r_b/b	0.115

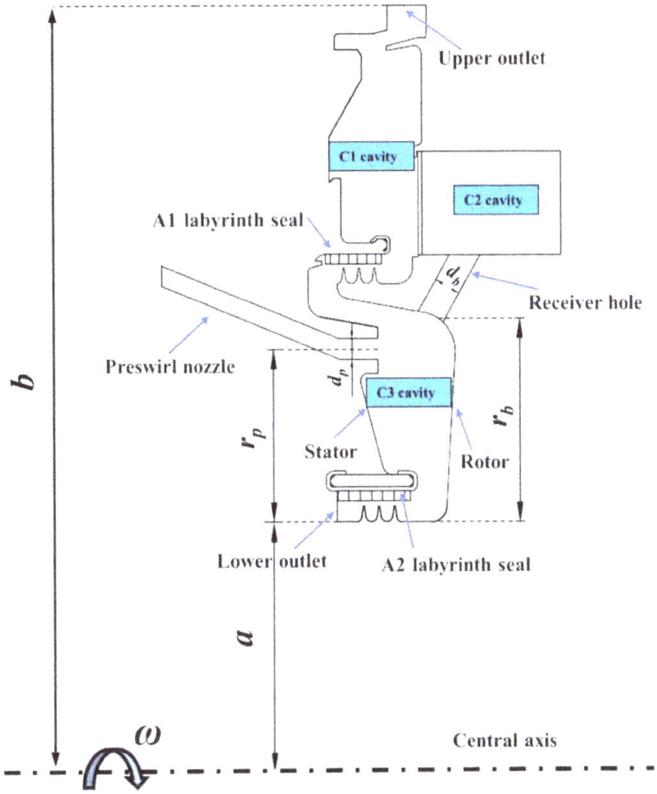

Figure 1. Sketch of pre-swirl rotor–stator cavity.

3.2. *Turbulence Model and Boundary Conditions*

Because of the axial symmetry of the C3 cavity, a segment of 4 degrees was modeled, shown in Figure 2. The ANSYS CFX 14.0 was used for numerical simulation. The simulation type was set as a steady state. Some researchers, such as Barabas et al. [37] and the authors' group [38,39], found that the simulation results from the shear stress transport SST k-ω turbulence model in combination with the scalable wall functions are in good agreement with the measured pressure in a rotor–stator cavity with air. The deviations of the pressure measurement are less than 1%. Hence, in this study, the same turbulence model and wall functions were used. The turbulent numeric was set as second-order upwind. The mesh was generated by NUMECA FINE OPEN, which is characterized by unstructured mesh dominated by hexahedrons. The thickness of the grids in the near wall regions decreases until the values of y^+ are less than 1, which satisfies the requirement from the SST k-ω turbulence model. The walls in contact with air were set as non-slip walls. The pressure and temperature at each branch and the temperature of the stator were set according to the values from the designers. The heat transfer coefficient of the rotor and the labyrinth seal was set according to the values from conjugate heat transfer. The flow was selected as air ideal gas, and the physical properties, such as specific heat, were from the database. The convergence criteria were set as 10^{-5} in max type. We conducted 40 simulations with different relative total pressure $p_{in,\ total}$ (made non-dimensional by dividing those at the design value for each speed of rotation) at the nozzle inlet and speed of rotation, shown in Table 2. The "−30% design pressure" is equal to $p_{in,\ total} = 0.7$, while "+40% design pressure" is equal to $p_{in,\ total} = 1.4$.

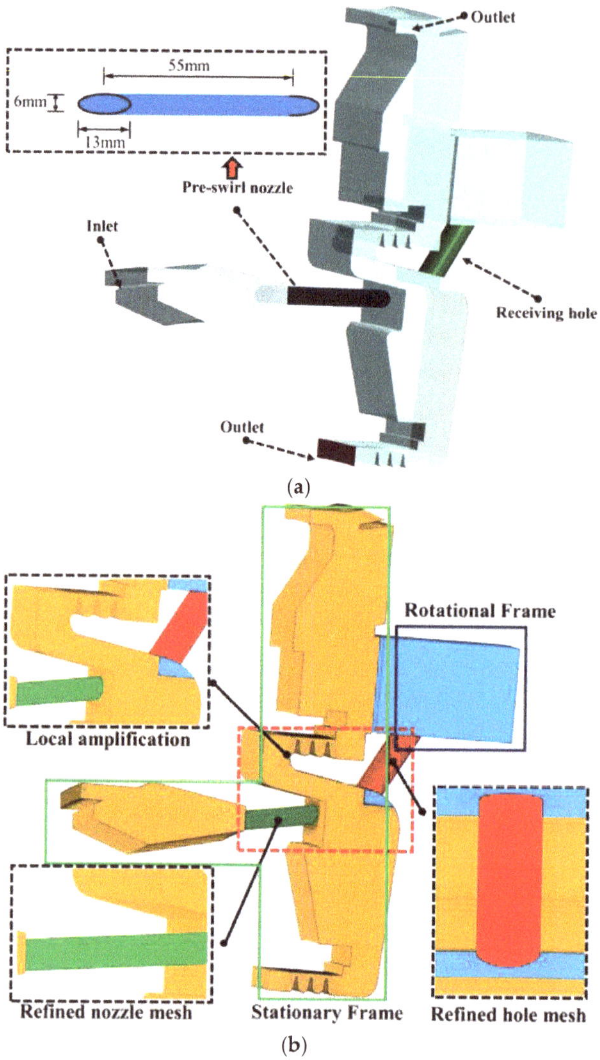

Figure 2. Domains and grids for numerical simulation. (**a**) Domains for numerical simulation; (**b**) Grids for numerical simulation.

Table 2. Main parameters of different simulations.

Inlet Pressure of the Pre-Swirl Nozzle	Rotational Speed/Maximum One
1. −30% design pressure	1. 12.5%
2. −20% design pressure	2. 25%
3. −10% design pressure	3. 37.5%
4. Design pressure	4. 45%
5. +10% design pressure	5. 50%
6. +20% design pressure	
7. +30% design pressure	
8. +40% design pressure	

3.3. Numerical Method Validation

In this paper, we measured the static pressure in the disk cavity of a marine gas turbine at four speeds of rotation, which are close to those in Table 2. The speed of rotation was modified to the same value during the validation of the numerical method. The mesh was modified until the maximum difference between simulation results and experimental data was less than 4.3%. The total grid count was 28.23 million. Hence, the numerical simulation setups are considered reasonable.

4. Results and Discussion

4.1. Flow Characteristics

4.1.1. Radial Velocity Distribution

The radial velocity is important in finding the circulation of air in a rotor–stator cavity. With different nozzle inlet pressures at $Re = 9.5172 \times 10^5$, the distribution of radial velocity on the meridional section is shown in Figure 3. In the C3 cavity, the flow in region 1 and region 4 (near the rotor) mainly moves radial outward. On the other hand, it flows toward the shaft below the nozzle in region 3. At the entrance of receiving hole, there is a vortex (see the region with different flow directions in region 2). In the receiving hole, we can see that there is a backflow area (region 5). Its scope enlarges with increasing $p_{in,\ total}$ at $p_{in,\ total} \geq 0.9$. The results indicate that the flow resistance is large in such a region. In the C1 cavity, there are two backflow areas, namely region 6 and region 7. The backflow in region 6 is a typical flow distribution near the stator, while the backflow in region 7 is related to the change of rotor shape and the existence of a rim seal. In the C2 cavity, there is a backflow area (region 8), although the C2 cavity is co-rotating. This is attributed to the formation of a vortex in this region due to the strong jet flow. The distributions of radial pressure are similar for other speeds of rotation. The results show that the pre-swirl nozzle and receiving hole do not match well. Some modifications have to be made to improve performance at low rotation speeds.

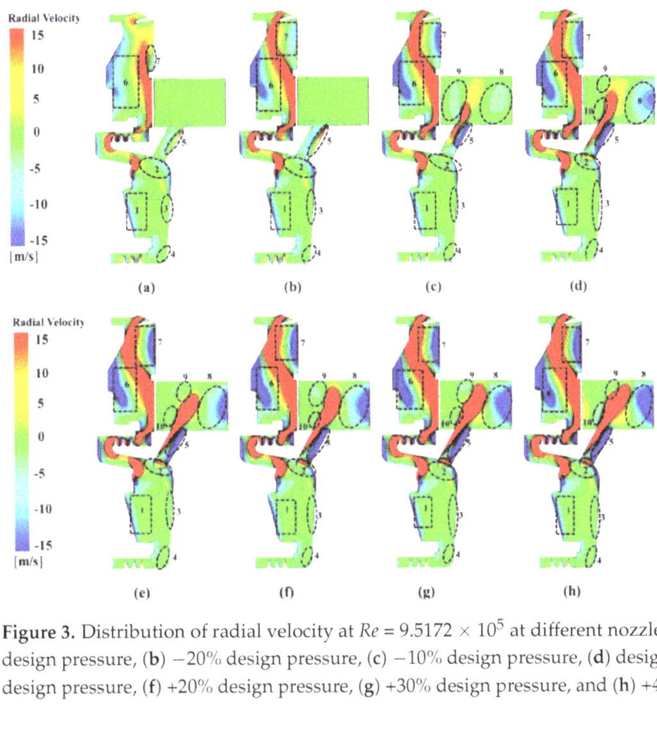

Figure 3. Distribution of radial velocity at $Re = 9.5172 \times 10^5$ at different nozzle pressure: (**a**) −30% design pressure, (**b**) −20% design pressure, (**c**) −10% design pressure, (**d**) design pressure, (**e**) +10% design pressure, (**f**) +20% design pressure, (**g**) +30% design pressure, and (**h**) +40% design pressure.

4.1.2. Core Swirl Ratio K

In order to predict the axial thrust, the pressure distribution along the radius of the disk should be estimated. In a typical rotor–stator cavity, the flow pattern changes as C_{qr} increases with either centrifugal or centripetal non-pre-swirl through-flow. The relationship between K and C_{qr} for different Re in the C1 cavity of this paper is shown in Figure 4. The values of K firstly decrease, then increase with increasing C_{qr}. The decreasing trend happens in the backflow area (mainly region 6 in Figure 3). With increasing $p_{in,\ total}$, the values of K in this area become small. The reason for this is that the leakage flow rate is larger with larger values of $p_{in,\ total}$. The work performed by the rotor on the volumetric flow drops. When x further increases, the values of K start to increase in most cases, except for cases where hot gas is ingested (see the radial velocity at the outer radius in Figure 3a) from the main flow passage through the rim seal.

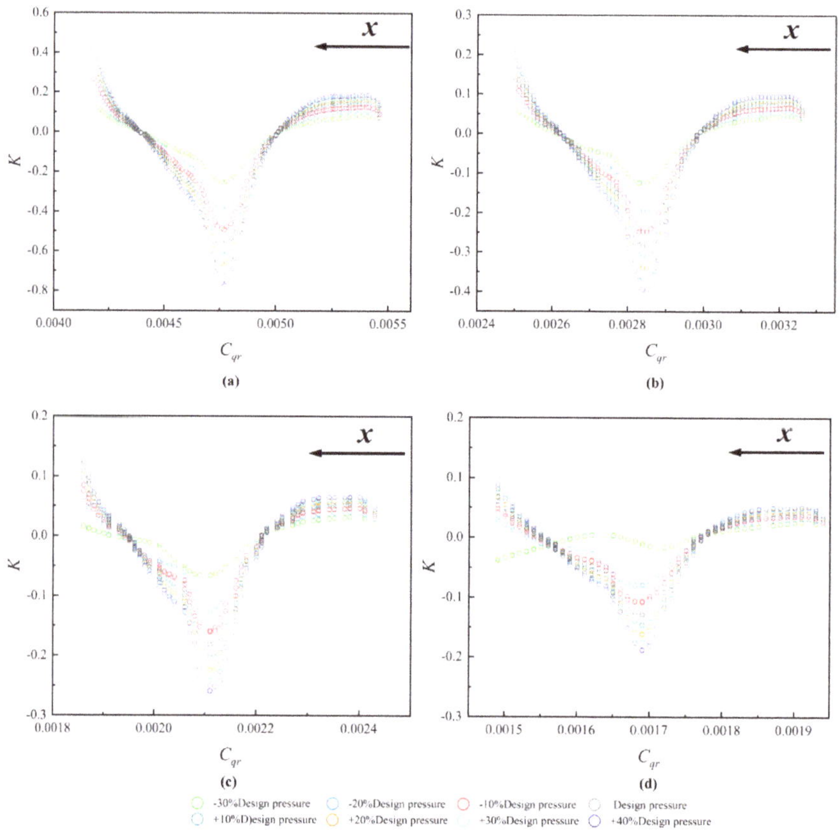

Figure 4. Radial distribution of core swirl ratio in C1 cavity at different Re: (**a**) $Re = 2.3793 \times 10^5$, (**b**) $Re = 4.7586 \times 10^5$, (**c**) $Re = 7.1379 \times 10^5$, and (**d**) $Re = 9.5172 \times 10^5$.

The distribution of K in the C3 cavity is shown in Figure 5. Above the receiving hole, the results fit relatively well with Equation (11). Below the pre-swirl nozzle, a relatively large difference occurs. Therefore, we determined a leakage flow term $(-44.44e^{-6.479C_{qr}} + 47.3)$ and extended the correlation in Ref. [40], which can be written in Equation (12). The results are compared in Figure 4. As C_{qr} increases, the values of K firstly significantly drop and almost remain the same. When C_{qr} further increases, the values of K gradually drop. The results from Equation (12) are in good agreement with those from numerical simulation with different $p_{in,\ total}$ and different *Re*, compared with those in Refs. [38–41].

$$K = \left(\frac{r_p}{r_i}\right)^2 K_p + \frac{M_r + M_s}{m r_i V_{\varphi,r=r_i}} \tag{11}$$

$$K = \left(\frac{r_p}{r_i}\right)^2 K_p + \frac{M_r + M_s}{m r_i V_{\varphi,r=r_i}} + \left(-44.44e^{-6.479C_{qr}} + 47.3\right) \tag{12}$$

where $2.3793 \times 10^5 \leq Re \leq 9.5172 \times 10^5$, 1.9191 MPa $\leq$ total inlet pressure $\leq$ 3.8381 MPa and $G = 0.044$.

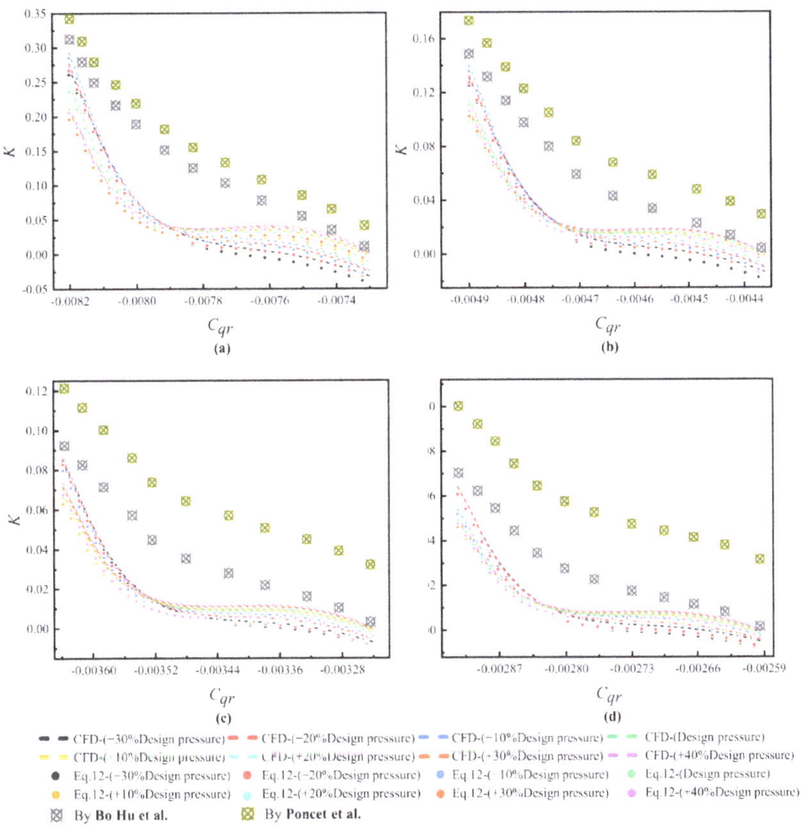

Figure 5. Comparison of core swirl ratio in C3 cavity at: (**a**) $Re = 2.3793 \times 10^5$, (**b**) $Re = 4.7586 \times 10^5$, (**c**) $Re = 7.1379 \times 10^5$, and (**d**) $Re = 9.5172 \times 10^5$ [38–41].

4.1.3. Pressure Coefficient C_p

The radial pressure distribution on the meridional section is depicted in Figure 6 for different inlet pressures at $Re = 9.5172 \times 10^5$ (at 0.5 times the maximum rotational speed). It can be seen that the pressure in the C3 cavity rises with increasing $p_{in,\ total}$. Near the inlet of receiving hole, there is a relatively high-pressure zone for $p_{in,\ total} \geq 1$. The rise of p_s also results in the increase of the pressure gradient from the pre-swirl nozzle to the outer labyrinth seal in C3. The high-pressure zone enlarges as $p_{in,\ total}$ increases. Conversely, the pressure at the outlet of the receiving hole minimally changes. In the downstream C1 cavity, there is a low-pressure zone downstream of the inner labyrinth seal. The scope of the zone increases with increasing $p_{in,\ total}$. With a larger pressure difference at either side of the labyrinth seal, the leakage flow rate is expected to increase.

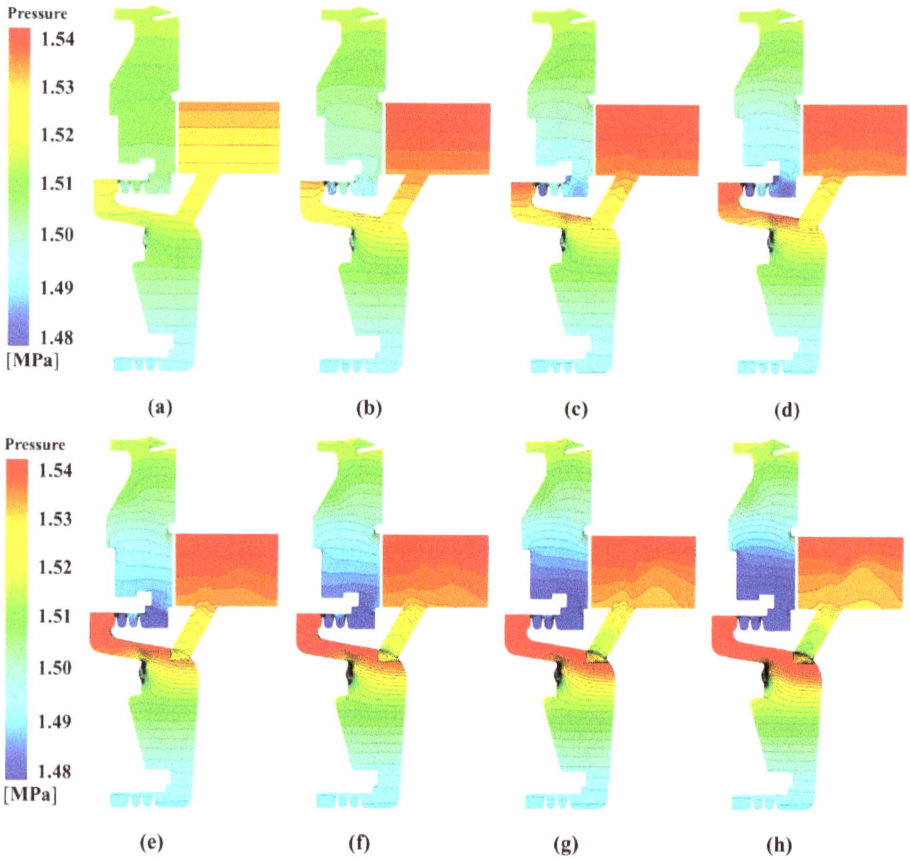

Figure 6. Pressure distribution of pre-swirl cavity at $Re = 9.5172 \times 10^5$ with different nozzle pressure: (**a**) −30% design pressure, (**b**) −20% design pressure, (**c**) −10% design pressure, (**d**) -design pressure, (**e**) +10% design pressure, (**f**) +20% design pressure, (**g**) +30% design pressure, and (**h**) +40% design pressure.

Since the axial thrust is an important parameter for the safe operation of a gas turbine, the radial distribution of pressure coefficient C_p was compared. The reference pressure was taken at the radius of labyrinth A1 from numerical simulation. Since the pressure drop is radial inwards, the values of the pressure coefficient C_p are positive, as depicted in Figure 7. In all the cases, the increasing trend of C_p towards the shaft are similar. The parameter C_p is positively correlated with p_s at each Re. When the Reynolds number is large, the C_p increases with the nozzle inlet pressure, but the increments are small, and the effect of increasing p_s on C_p is relatively weak. The flow below the pre-swirl nozzle is mainly controlled by the strong swirl flow in the C3 cavity and the pressure ratio is less influenced by $p_{in,\ total}$ at this point. The larger values of p_s cause the more severe pressure difference along the disk.

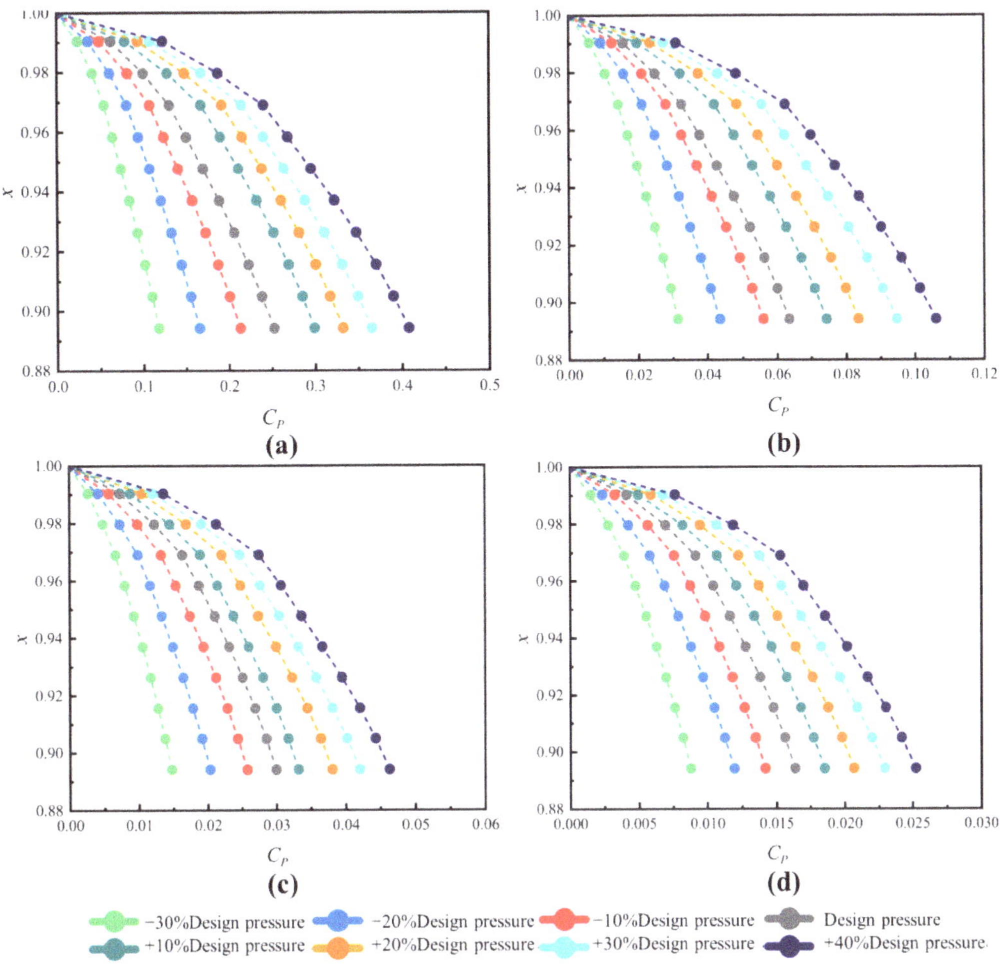

Figure 7. Radial distribution of pressure coefficients in C3 cavity at different Re: (**a**) $Re = 2.3793 \times 10^5$, (**b**) $Re = 4.7586 \times 10^5$, (**c**) $Re = 7.1379 \times 10^5$, and (**d**) $Re = 9.5172 \times 10^5$.

4.2. Performances of Pre-Swirl System

4.2.1. Non-Dimensional Pressure Drop

In this paper, we estimate the non-dimensional pressure drop ζ to predict the capacity of cooling air delivery. It is defined in Equation (13). The variation of ζ are plotted in Figure 8. The values of ζ increase with the total pressure $p_{in,\ total}$ at the nozzle inlet. With the increase of Re, the dominant rotation is strong in C3, and the values of ζ are less influenced by $p_{in,\ total}$. When Re reaches the maximum value (9.5172×10^5), the increment rate of ζ is less than 2%.

$$\zeta = \frac{P_{in}* - P_{out}*}{\frac{1}{2}\rho\omega^2 r_p^2} \tag{13}$$

where $P_{n,t}$ is the total pressure at the pre-swirl nozzle, P_r is the static pressure at the receiving hole, $P_{in}{}^*$ is the total pressure at the inlet, and $P_{out}{}^*$ is the total pressure at the outlet.

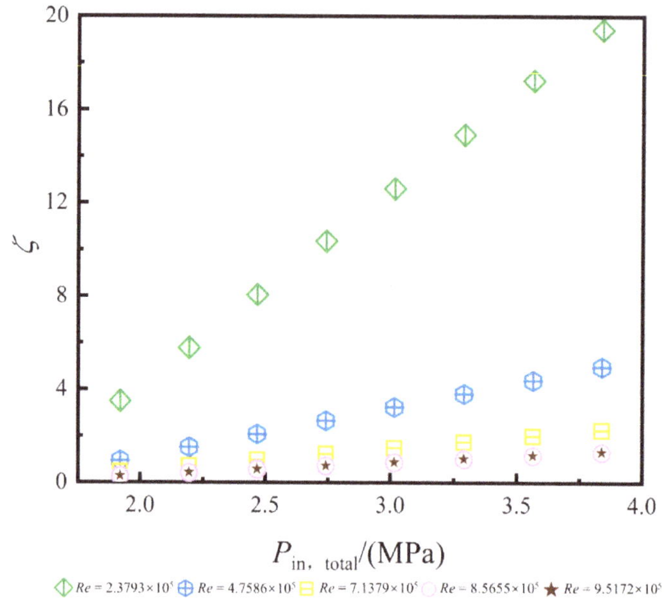

Figure 8. Distribution of ζ at different *Re*.

4.2.2. Through-Flow Capacity of Nozzle and Receiving Hole

The total pressure distribution in the mid-section of the pre-swirl nozzle at $Re = 9.5172 \times 10^5$ is depicted in Figure 9. It shows that as the $p_{in,total}$ continues to increase, there is a significant increase in the pressure, and the values of total pressure near the wall are larger than those near the center of the flow passage, which indicates that there is more flow resistance in the wall region.

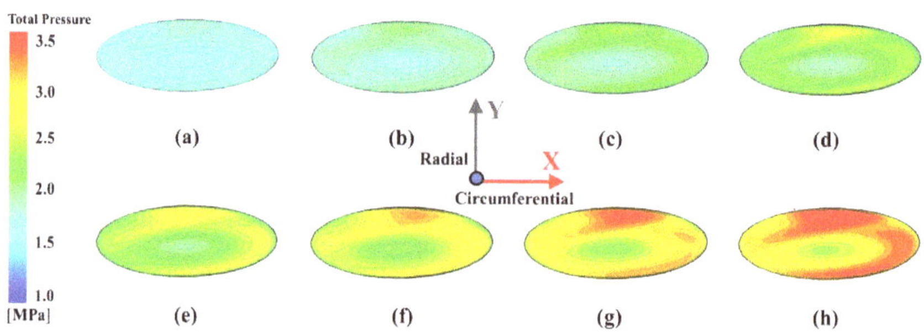

Figure 9. Distribution of total pressure at the middle section in the pre-swirl nozzle at $Re = 9.5172 \times 10^5$ with different nozzle pressure: (**a**) −30% design pressure, (**b**) −20% design pressure, (**c**) −10% design pressure, (**d**) -design pressure, (**e**) +10% design pressure, (**f**) +20% design pressure, (**g**) +30% design pressure, and (**h**) +40% design pressure.

The pressure distribution of each section at different axial positions of the pre-swirl nozzle with either maximum or minimum $p_{in,total}$ is shown in Figure 10. The circumferential velocity of airflow increases due to rotation. The pressure of the whole pre-swirl nozzle decreases along the flow direction. It is obvious that the through-flow capacity of the nozzle is good in the range of $p_{in,total}$. For other *Re*, the trends are similar.

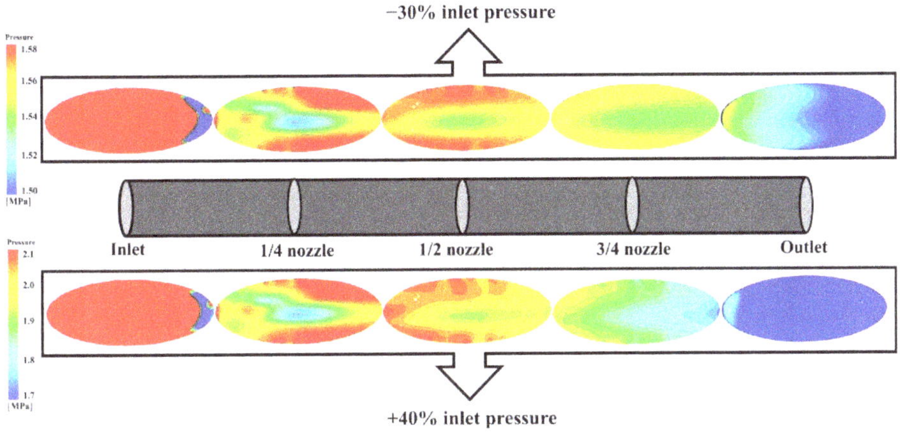

Figure 10. Pressure distribution in the pre-swirl nozzle at different axial positions ($Re = 9.5172 \times 10^5$).

Figure 11 reflects the variation of C_d versus $p_{in,\ total}$ for the pre-swirl nozzle at four different Reynolds numbers. These values show an increasing trend versus $p_{in,\ total}$. The maximum relative increment is up to 2% when Re increases from 2.39×10^5 to 9.5172×10^5 at a fixed $p_{in,\ total}$.

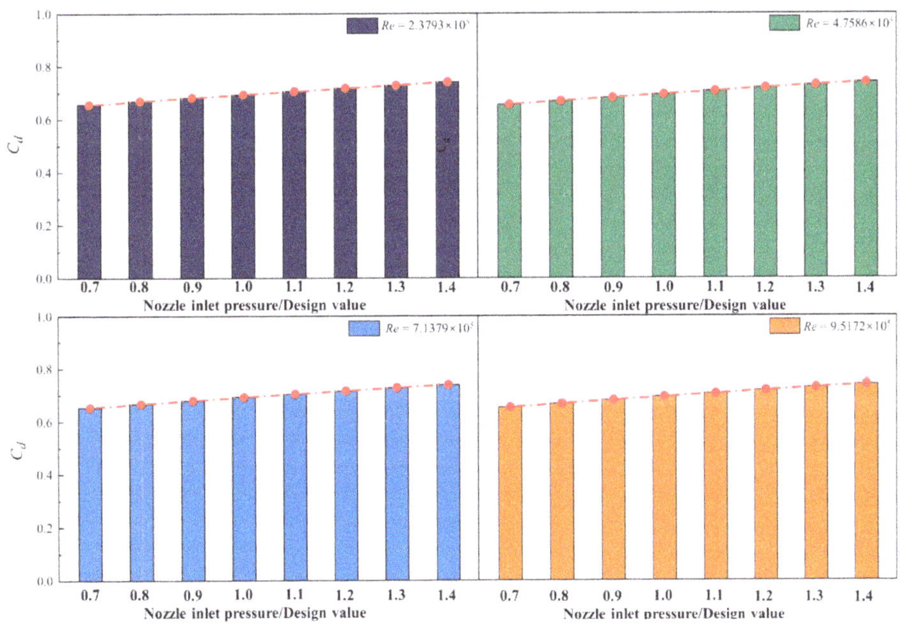

Figure 11. Distribution of flow rate coefficients of pre-swirl nozzle at different Re.

In Figure 12, the total pressure distribution in the mid-section of the receiving hole at $Re = 9.5172 \times 10^5$ is shown. The receiving hole is a long orifice since the gas reattaches to the wall after a flow separation near the inlet. In the separation area, the values of pressure are large, which is represented by the red color. With increasing $p_{in,\ total}$, the scope of the separation area enlarges. In the rest of the area, the values of pressure drop, indicating better through-flow capacity.

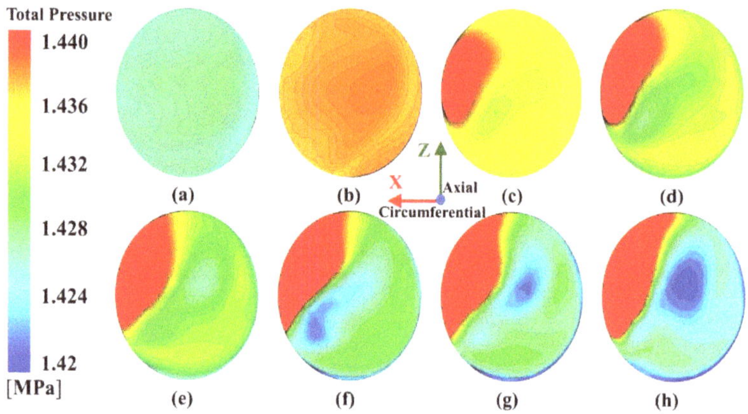

Figure 12. Distribution of total pressure on the middle section of receiving hole at $Re = 9.5172 \times 10^5$ with different nozzle pressure: (**a**) −30% design pressure, (**b**) −20% design pressure, (**c**) −10% design pressure, (**d**) design pressure, (**e**) +10% design pressure, (**f**) +20% design pressure, (**g**) +30% design pressure, and (**h**) +40% design pressure.

The static pressure distribution of each section at different axial positions of the receiving hole with either maximum or minimum $p_{in,total}$ is shown in Figure 13. The values of pressure increase along the passage of the receiving hole. At the same axial position, the pressure is large with an increase in $p_{in,total}$. At the minimum $p_{in,total}$, there is a region with large pressure. When the air flows through the passage of the receiving hole, the pressure violently fluctuates. At the outlet, the pressure reaches the maximum value, which indicates the most flow resistance.

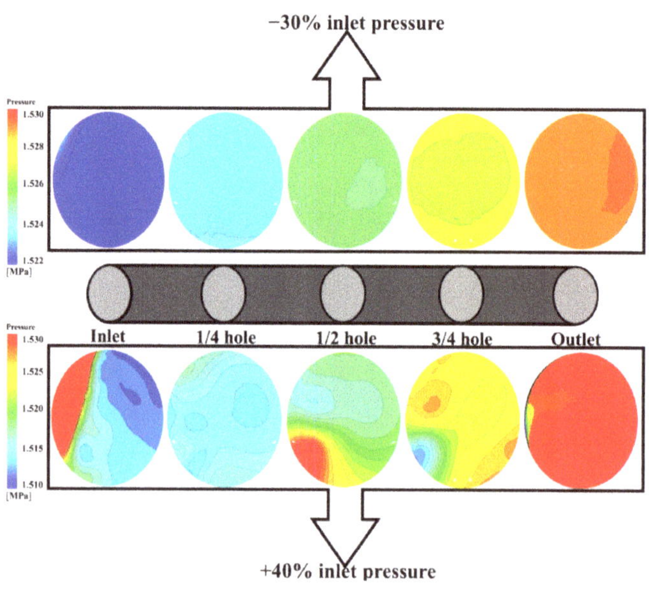

Figure 13. Pressure distribution in the receiving hole at different axial positions ($Re = 9.5172 \times 10^5$).

The variations of the receiving hole flow rate coefficient versus $p_{in,\ total}$ are depicted in Figure 14. The values of C_d ranges from 0.43 to 0.73. It increases first and then decreases with increasing $p_{in,\ total}$ at a fixed *Re*. There is an optimal value of $p_{in,\ total}$, where the best through-flow rate can be achieved. The values increase up to 13.5% with increasing *Re*.

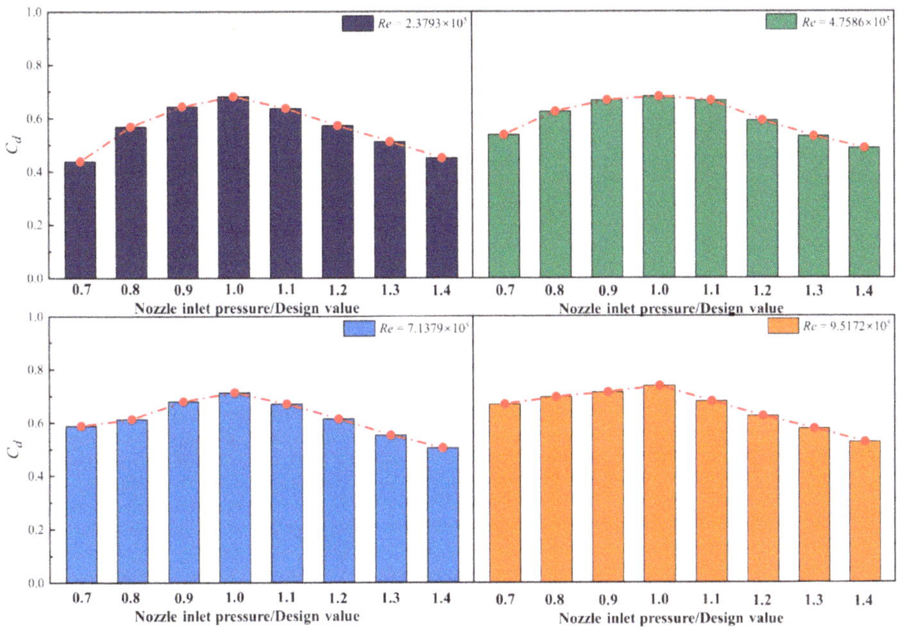

Figure 14. Distribution of C_d for receiving hole at different *Re*.

4.2.3. Cooling Effects of the Cavity Flow

The temperature distribution on the meridional surface at different inlet pressures for $Re = 9.5172 \times 10^5$ is depicted in Figure 15. From the figure, the whole temperature in the cavity decreases with increasing $p_{in,\ total}$. In the C1 cavity, the effects of $p_{in,\ total}$ on temperature becomes weaker when $p_{in,\ total}$ exceeds 1.1. In the C3 cavity, there is a low-temperature region between the nozzle, receiving hole, and A2 labyrinth seal. It is a comprehensive influence of impingement cooling and swirl flow cooling. The temperature at the outlet of receiving hole drops with increasing $p_{in,\ total}$ as well. This indicates that increasing $p_{in,\ total}$ enhances the cooling effect on the turbine blade.

In Figure 16, the radial distribution of T^* in the C1 cavity at different *Re* is depicted. It can be observed that T^* in the C1 cavity has a general tendency to fall and then rise along the radial direction, and there is a low-temperature zone in the middle section of the cavity. As the nozzle inlet pressure increases, T^* is negatively correlated at the high and low radius of the C1 cavity, but a positive correlation in the middle section, and the variation is greater in this area. When the speed increases, the relative velocity between the rotor and the gas decreases, this reduces the heat transfer effect of the gas on the rotor and causes an upward shift of the low-temperature zone. The temperature in the C3 cavity decreases in the radial direction, as shown in Figure 17. When the pre-swirl jet flow enters the C3 cavity, it is mixed with the cavity flow and then impinges on the rotor. With the increase of *Re*, the change of T^* is small. On the other hand, the values of T^* decrease up to 21.4% when the $p_{in,\ total}$ decreases. This can be explained by the more intense impingement cooling and swirl flow cooling.

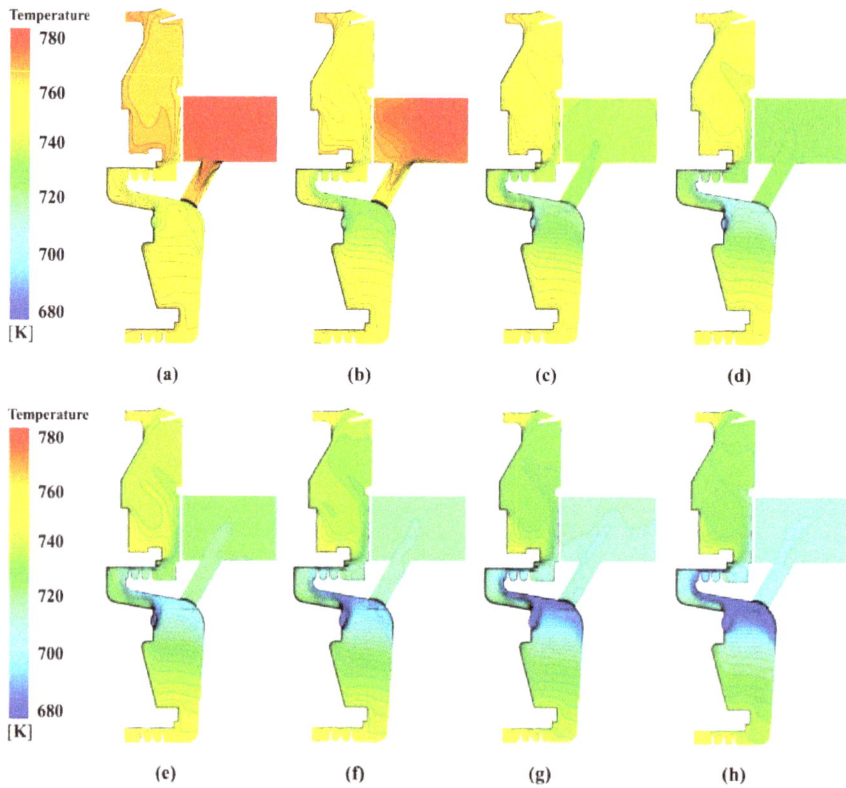

Figure 15. Temperature distribution of pre-swirl cavity at $Re = 9.5172 \times 10^5$ with different nozzle pressure: (**a**) −30% design pressure, (**b**) −20% design pressure, (**c**) −10% design pressure, (**d**) design pressure, (**e**) +10% design pressure, (**f**) +20% design pressure, (**g**) +30% design pressure, and (**h**) +40% design pressure.

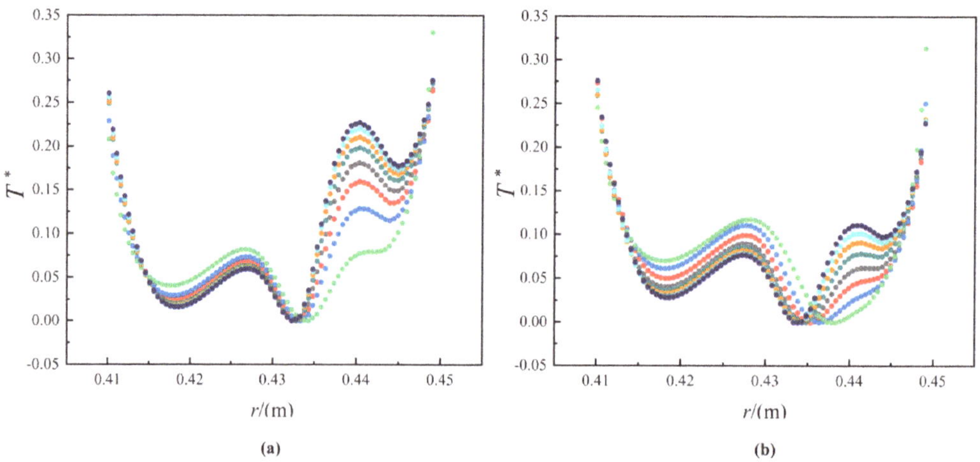

Figure 16. *Cont.*

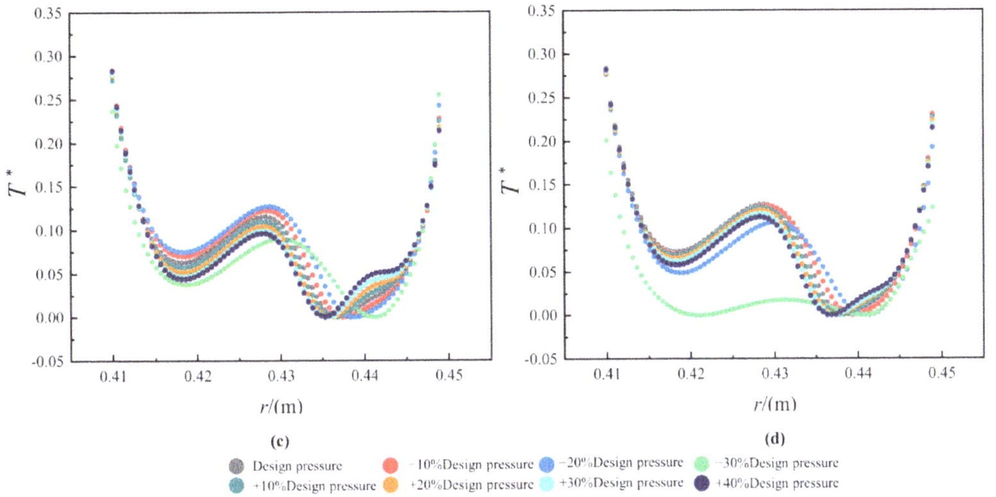

Figure 16. Radial distribution of non-dimensional temperature in C1 cavity for different *Re*: (**a**) *Re* = 2.3793 × 105, (**b**) *Re* = 4.7586 × 105, (**c**) *Re* = 7.1379 × 105, and (**d**) *Re* = 9.5172 × 105.

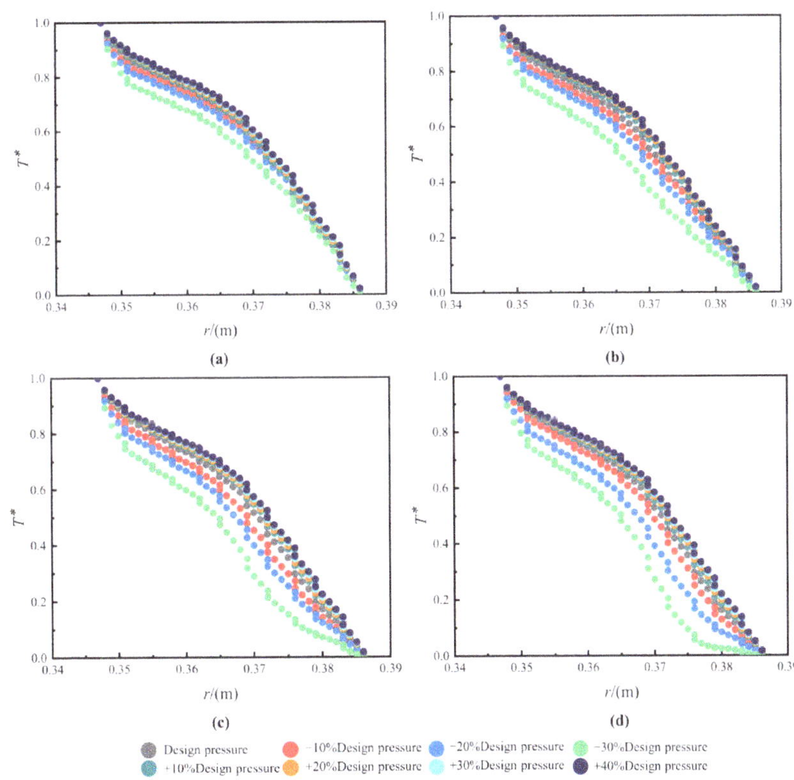

Figure 17. Radial distribution of non-dimensional temperature in C3 cavity for different *Re*: (**a**) *Re* = 2.3793 × 10^5, (**b**) *Re* = 4.7586 × 10^5, (**c**) *Re* = 7.1379 × 10^5, and (**d**) *Re* = 9.5172 × 10^5.

The change of ΔT^* is plotted versus $p_{in,\ total}$ in Figure 18. For different Re, the values of ΔT^* increase with increasing $p_{in,\ total}$, which indicates better cooling performances. When the nozzle inlet pressure is fixed, the values of ΔT^* decrease with the increase of Re.

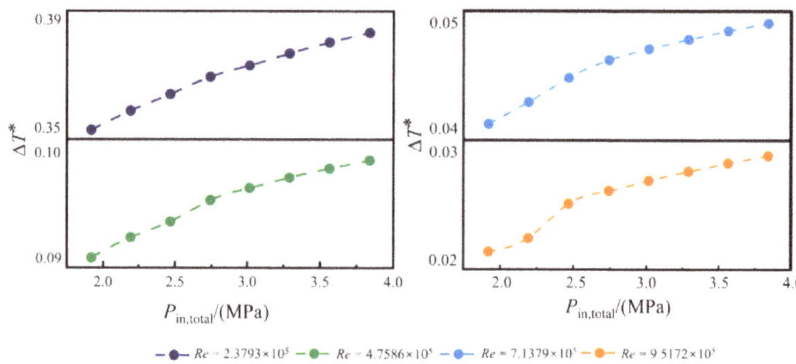

Figure 18. Distribution of ΔT^* at different Re.

4.2.4. Moment Coefficient C_M

The moment coefficients of the C1 cavity are compared in Figure 19 with increasing Re. The decreasing trend is similar to the experimental results by Hu et al. [38,39]. The results also show that the flow resistance on the stationary wall is larger than those on the rotor, which should not be neglected. Meanwhile, the variation of $p_{in,\ total}$ has a larger influence on the values of C_M on the stationary wall than those on the rotor.

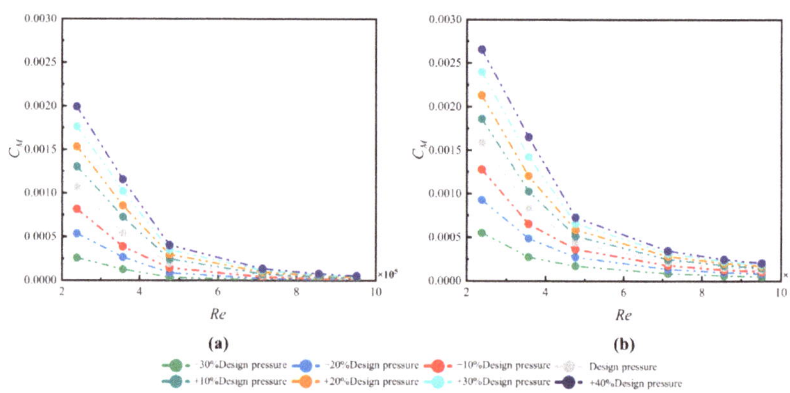

Figure 19. Variations of C_M in C1 cavity at different $p_{in,\ total}$ and Re on: (a) rotor and (b) stator.

Figure 20 shows the variation of different inlet pressures in the C3 cavity on the moment coefficients of the rotor and stator, respectively. It can be observed that the C_M values of the rotor decrease as Re increases. The trend is similar to the trend of the rotor–stator cavity with either centrifugal or centripetal through-flow [38,39]. Meanwhile, the moment coefficient of the stator is up to 75% larger than that of the rotor. With increasing $p_{in,\ total}$, the values of C_M obviously rise. This is because more work has to be performed on the cavity flow by the rotating disk with a more mass flow rate through the nozzle. With regard to the stator, the tangential velocity increases with increasing $p_{in,\ total}$, which contributes to larger wall shear stress. Unlike those on the stator, the values of C_M on the rotor are almost not influenced by $p_{in,\ total}$ at the maximum value of Re. The parameter of C_M on the stator are more sensitive to $p_{in,\ total}$ than those on the rotor.

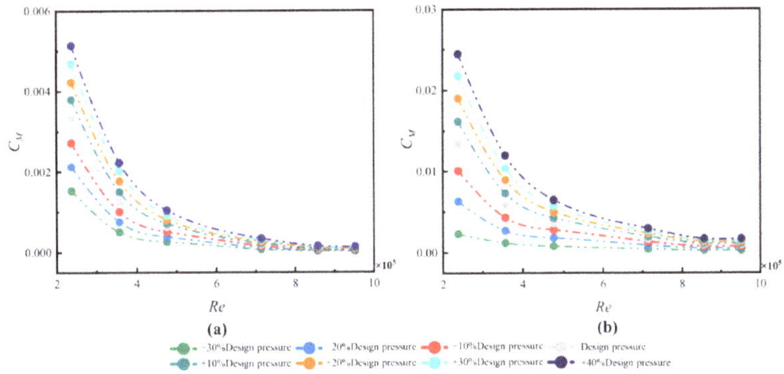

Figure 20. Variations of C_M in C3 cavity at different $p_{in,\ total}$ and Re on: (**a**) rotor and (**b**) stator.

4.2.5. Axial Thrust Coefficient C_F

The variations of C_F in the C1 cavity are depicted in Figure 21. The values drop with increasing Re or with increasing $p_{in,\ total}$. The variations of C_F in the C3 cavity for different Re and different $p_{in,\ total}$ are made in Figure 22. As for the rotor, the C_F decreases with increasing Re, with an average reduction of 73%; meanwhile, it increases with increasing $p_{in,\ total}$. Increasing $p_{in,\ total}$ results in the rise of C_F on both the rotor and the stator, although the influence is much weaker on the stator.

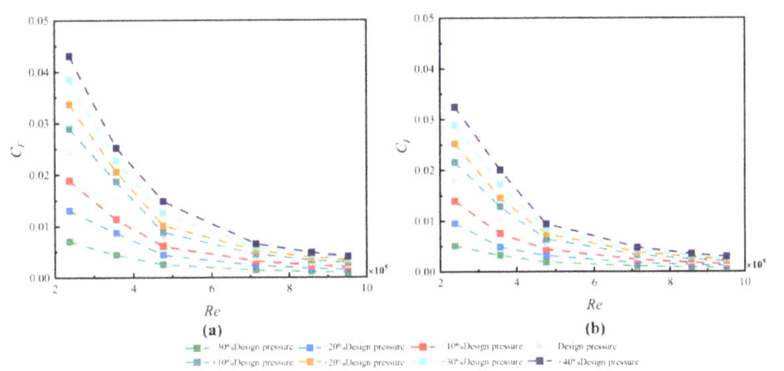

Figure 21. C_F distribution in the C1 cavity at different $p_{in,\ total}$ and Re on the (**a**) rotor and (**b**) stator.

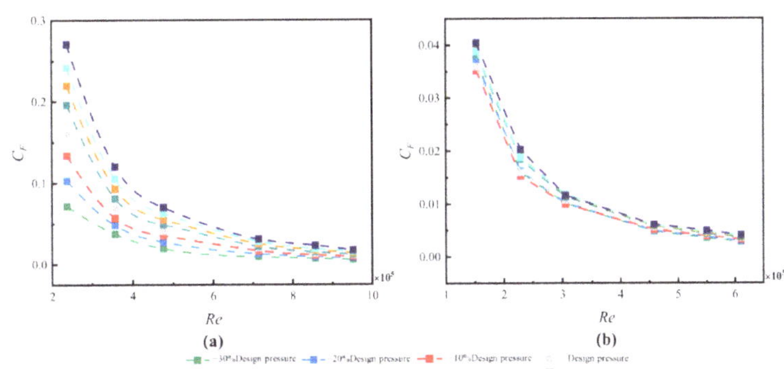

Figure 22. C_F distribution in C3 cavity at different $p_{in,\ total}$ and Re on the (**a**) rotor and (**b**) stator.

5. Conclusions

In this paper, the influence of $p_{in,\ total}$ (70−140% of design values) on the flow characteristics and performance of a disk cavity at $Re = 2.3793 \times 10^5$ to 9.5172×10^5 are numerically studied. Some conclusions are as follows:

1. To better predict the distribution of K between the inner seal and nozzle, the empirical formula for K in the literature was extended by adding a leakage flow term.
2. The variation of T^* in the pre-swirl rotor–stator cavity is mainly influenced by convective heat transfer and impingement cooling. As the nozzle inlet pressure increases, T^* is negatively correlated at both the high and low radii of the C1 cavity. The values of T^* in the C3 cavity decrease along the radial direction, with a relative increment of up to 21.4% for different inlet pressures.
3. The current disk cavity shows relatively good performance at low speeds of rotation. Firstly, the flow rate coefficient of the receiving hole is not positively correlated with $p_{in,\ total}$. Increasing $p_{in,\ total}$ may cause excessive flow resistance due to the backflow in the receiving hole. Secondly, the pre-swirl system has a good cooling effect on the C3 cavity at low rotation speeds. Thirdly, when the $p_{in,\ total}$ values exceed the design values, the performance of the pre-swirl system decreased, which lowered the cooling ability of the turbine blades.
4. The moment coefficient and axial thrust coefficient show decreasing trend as Re increases. Increasing $p_{in,\ total}$ causes the rise of C_M and C_F, which cause additional frictional losses and have a negative influence on the bearing systems.

Author Contributions: Conceptualization, B.H. and C.W.; methodology, Y.Y.; software, Y.Y.; validation, M.W., B.H. and Y.L.; formal analysis, Y.Y.; investigation, B.H.; resources, M.W.; data curation, Y.L.; writing—original draft preparation, Y.Y.; writing—review and editing, B.H.; visualization, M.W.; supervision, C.W.; project administration, Y.L.; funding acquisition, B.H. All authors have read and agreed to the published version of the manuscript.

Funding: This study is supported by the National Science and Technology Major Project (J2019-IV-0022-0090) and Natural Science Foundation of China No. 52206047.

Data Availability Statement: Data on the analysis and reporting results during the study can be obtained by contacting the authors.

Conflicts of Interest: The authors declare no conflict of interest.

Nomenclature

A	Nozzle outlet area, m^2
A_r	Rotor surface area, m^2
A_s	Stator surface area, m^2
A	Hub radius, m
B	Outer radius of the disk, m
C_d	Flow rate coefficient
C_F	Axial thrust coefficient
C_M	Moment coefficient
C_p	Pressure coefficient
c_p	Isobaric specific heat capacity
C_{qr}	Local flow rate coefficient
d_b	Receiving hole diameter, m
d_p	Nozzle hole diameter, m
K	Core swirl ratio
K_p	Core swirl ratio at nozzle
K	Adiabatic exponent of gas, $k = 1.4$
M	Friction torque, $N \cdot$m
M_r	Rotor surface torque
M_s	Stator surface torque

M	Actual mass flow rate, kg/s
m_i	Isentropic mass flow rate, kg/s
p	Pressure, Pa
p_b	Pressure at r = b, Pa
$p_{in}{}^*$	Total pressure at inlet, Pa
$p_{n,t}$	Total pressure at the pre-swirl nozzle, Pa
p_{out}	Static pressure at outlet, Pa
$p_{out}{}^*$	Total pressure at outlet, Pa
p_r	Static pressure at the receiving hole, Pa
p^*	Non-dimensional pressure
Q	Volumetric through-flow rate, m^3/s
R	Air constant, R = 287
Re	Global circumferential Reynolds number
Re_φ	Local circumferential Reynolds number
R	Radial coordinate, m
r_b	Receiving hole to position, m
r_p	Nozzle radial position, m
s	Axial gap of the front cavity, m
T	Temperature, K
T^*	Non-dimensional temperature
ΔT^*	Non-dimensional temperature drop
$T_1{}^*$	Inlet total temperature, K
$T_{in}{}^*$	Total temperature at the inlet of the pre-swirl nozzle, K
$T_{out}{}^*$	Relative total temperature at the main outlet of the system, K
T_{max}	Maximum temperature in the cavity, K
T_{min}	Minimum temperature in the cavity, K
$V_{1\varphi}$	Inlet tangential velocity, m/s
$V_{2\varphi}$	Outlet tangential velocity, m/s
x	Non-dimensional radial coordinate, $x = r/b$
$V_{\varphi,r=ri}$	Tangential velocity at $r = r_i$ m/s
Greek symbols	
ρ	Density of air, kg/m^3
μ	Aerodynamic viscosity, N·s/m^2
υ	Kinematic viscosity, m^2/s
ζ	Non-dimensional pressure drop
θ	Inlet angle of pre-swirl nozzle,°
ω	Angular velocity of the disk, rad/s
Subscripts	
x	circumferential
y	radial
z	axial

References

1. El-Oun, Z.; Owen, J. Pre-Swirl Blade-Cooling Effectiveness in an Adiabatic Rotor-Stator System. *J. Turbomach.* **1989**, *111*, 522–529. [CrossRef]
2. Chew, J.; Hills, N.; Khalatov, S.; Scanlon, T.; Turner, A. Measurement and Analysis of Flow in a Pre-Swirled Cooling Air Delivery System. In Proceedings of the Turbo Expo: Power for Land, Sea, and Air, Atlanta, GA, USA, 16–19 June 2003. [CrossRef]
3. Chew, J.; Ciampoli, F.; Hills, N.; Scanlon, T. Pre-Swirled Cooling Air Delivery System Performance. In Proceedings of the Turbo Expo: Power for Land, Sea, and Air, Reno-Tahoe, NV, USA, 6–9 June 2005. [CrossRef]
4. Ciampoli, F.; Chew, J.; Shahpar, S.; Willocq, E. Automatic Optimization of Pre-swirl Nozzle Design. *J. Eng. Gas Turbines Power* **2007**, *129*, 387–393. [CrossRef]
5. Lewis, P.; Wilson, M.; Lock, G.; Owen, J. Effect of Radial Location of Nozzles on Performance of Pre-Swirl Systems. In Proceedings of the Turbo Expo: Power for Land, Sea, and Air, Berlin, Germany, 9–13 June 2008. [CrossRef]
6. Javiya, U.; Chew, J.; Hills, N.; Scanlon, T. A Comparative Study of Cascade Vanes and Drilled Nozzle Designs for Pre-Swirl. In Proceedings of the Turbo Expo: Turbine Technical Conference and Exposition, Vancouver, BC, Canada, 6–10 June 2011. [CrossRef]
7. Zhang, F.; Wang, X.; Li, J. Numerical investigation of flow and heat transfer characteristics in radial pre-swirl system with different pre-swirl nozzle angles. *Int. J. Heat Mass Transf.* **2016**, *95*, 984–995. [CrossRef]

8. Zhang, F.; Wang, X.; Li, J. Numerical investigation on the flow and heat transfer characteristics in radial pre-swirl system with different fillet radius at the junction of inlet cavity and nozzle. *Appl. Therm. Eng.* **2016**, *106*, 1165–1175. [CrossRef]
9. Lee, H.; Lee, J.; Kim, D.; Cho, J. Optimization of pre-swirl nozzle shape and radial location to increase discharge coefficient and temperature drop. *J. Mech. Sci. Technol.* **2019**, *33*, 4855–4866. [CrossRef]
10. Lee, J.; Lee, H.; Park, H.; Cho, G.; Kim, D.; Cho, J. Design optimization of a vane type pre-swirl nozzle. *Eng. Appl. Comput. Fluid Mech.* **2021**, *15*, 164–179. [CrossRef]
11. Yang, X.; Ren, Z.; Li, X.; Ren, J.; Ligrani, P. Flow and heat transfer characteristics in a pre-swirl rotor-stator cavity. *Int. J. Therm. Sci.* **2022**, *172*, 107271. [CrossRef]
12. Liu, G.; Gong, W.; Wu, H.; Pang, L.; Lin, A. Theoretical and experimental evaluation of temperature drop and power consumption in a cover-plate pre-swirl system for gas turbine cooling. *Case Stud. Therm. Eng.* **2021**, *27*, 101221. [CrossRef]
13. Liu, G.; Zang, L.; Li, B.; Feng, Q. Investigation on Flow Characteristics and Temperature Drop of an Aerodynamic-Hole Typed Pre-Swirl Nozzle. *J. Propuls. Technol.* **2013**, *34*, 390–396. (In Chinese) [CrossRef]
14. Liu, G.; Li, B.; Jiang, Z.; Zang, L. Effect of Pre-Swirl Angle on Flow Characteristics of Pre-Swirl Nozzle. *J. Propuls. Technol.* **2012**, *33*, 7. [CrossRef]
15. Liu, G.; Zang, L.; Wu, W.; Feng, Q. Numerical Simulations on the Flow Characteristics of the Pre-Swirl Nozzles with Different Length-to-Diameter Ratios. *J. Propuls. Technol.* **2013**, *34*, 644–650. (In Chinese) [CrossRef]
16. Hu, W.; Wang, S.; Mao, S. Numerical Study on Influence of pre-swirl nozzle radial angles on pre-swirl Characteristic. *J. Aerosp. Power* **2019**, *34*, 8. (In Chinese) [CrossRef]
17. Lin, L.; Tan, Q.; Wu, K. Study of Flow Dynamics in Rotor-Stator Cavity with Low Radius Pre-swirl Inlet. *J. Propuls. Technol.* **2016**, *37*, 8. (In Chinese) [CrossRef]
18. Cheng, S.; Gao, T.; Li, J. Numerical Study on Flow and Heat Transfer Characteristics in Rotor-Stator Cavity of Circular Hole Pre-swirl Nozzle. *J. Eng. Therm. Energy Power* **2016**, *31*, 59–66. (In Chinese) [CrossRef]
19. Liu, Y.; Liu, G.; Wu, H.; Tian, S. Numerical Investigation on Flow characteristics of a Vane Shaped Hole Pre-swirl Nozzle. *J. Propuls. Technol.* **2016**, *37*, 7. (In Chinese) [CrossRef]
20. Liu, Y.; Liu, G.; Kong, X.; Wang, Y. Experimental Testing and Numerical Analysis on Flow Characteristics and Cooling Performance for Two Vane Pre-Swirl Nozzles. *J. Propuls. Technol.* **2019**, *10*. (In Chinese) [CrossRef]
21. Zhao, L.; Liu, G.; Wu, H.; Feng, Q.; Zheng, L. Experimental Investigation on the Temperature Drop of a Cover-plate Pre-swirl System. *J. Eng. Thermophys.* **2020**, *41*, 8. (In Chinese)
22. Wu, H.; Feng, Q.; Liu, G.; Wang, Y. Entropy analysis of a Cover-Plate Pre-Swirl System. *J. Propuls. Technol.* **2016**, *37*, 2048–2054. (In Chinese) [CrossRef]
23. Wang, S.; Zhu, Q.; Luan, H.; Zhang, Y. Experimental study on heat transfer in rotor-stator cavity with high-positioned pre-swirl inflow. *J. Aerosp. Power* **2007**, *22*, 1216–1221. (In Chinese) [CrossRef]
24. Wang, S.; Zhu, Q.; Zhang, Y.; Luan, H.; Huang, A. Numerical studies of heat transfer in a rotor-stator cavity with different radial positions of pre-swirl inlet. *J. Aerosp. Power* **2007**, *22*, 1227–1232. (In Chinese) [CrossRef]
25. Wang, C.N.; Yang, F.C.; Nguyen, V.T.T.; Vo, N.T. CFD analysis and optimum design for a centrifugal pump using an effectively artificial intelligent algorithm. *Micromachines* **2022**, *13*, 1208. [CrossRef]
26. Damavandi, M.; Nejat, A. Flow Characteristics of Curved Rotor Stator Systems Using Large Eddy Simulation. *Flow Turbul. Combust.* **2019**, *103*, 111–140. [CrossRef]
27. Liao, G.; Liu, L.; Zhang, F.; E, J.; Chen, J. A comparison of numerical investigations on the flow and heat transfer characteristics in the rotor-stator cavity. *Appl. Therm. Eng.* **2019**, *162*, 114231. [CrossRef]
28. Jia, X.; He, L.; Zhang, H. Effect of turbine rotor disc vibration on hot gas ingestion and rotor-stator cavity flow. *Aerosp. Sci. Technol.* **2020**, *98*, 105719. [CrossRef]
29. Schröder, T.; Dohmen, H.; Brillert, D.; Benra, F. Impact of Leakage Inlet Swirl Angle in a Rotor–Stator Cavity on Flow Pattern, Radial Pressure Distribution and Frictional Torque in a Wide Circumferential Reynolds Number Range. *Int. J. Turbomach. Propuls. Power* **2020**, *5*, 7. [CrossRef]
30. Schröder, T.; Schuster, S.; Brillert, D. Experimental Investigation of Centrifugal Flow in Rotor–Stator Cavities at High Reynolds Numbers > 10^8. *Int. J. Turbomach. Propuls. Power* **2021**, *6*, 13. [CrossRef]
31. Onori, M.; Hills, N. Reduced order modelling for a rotor-stator cavity using proper orthogonal decomposition. *Comput. Fluids* **2021**, *216*, 104818. [CrossRef]
32. Lin, A.; Liu, G.; Wang, X.; Feng, Q. Comprehensive evaluations on performance and energy consumption of pre-swirl rotor–stator system in gas turbine engines. *Energy Convers. Manag.* **2021**, *244*, 114440. [CrossRef]
33. He, Z.; Zhou, J.; Yang, C.; Li, B.; Qian, J. Unsteady numerical investigation on gas ingestion into the rotor-stator disk cavities of a 1.5-stage turbine. *Aeronaut. J.* **2022**, *126*, 1–18. [CrossRef]
34. Shi, J.; Zhao, Z.; Song, W.; Jin, Y.; Lu, J. Numerical simulation analysis of flow characteristics in the cavity of the rotor-stator system. *Eng. Appl. Comput. Fluid Mech.* **2022**, *16*, 501–513. [CrossRef]
35. Zhao, G.; Qiu, T.; Liu, P. Influence of Blade Fracture on the Flow of Rotor-Stator Systems with Centrifugal Superposed Flow. *Aerospace* **2022**, *9*, 106. [CrossRef]
36. Ren, H.; Zhang, J.; Cheng, H.; Chai, J.; Fu, S. Experimental Investigation on Flow Characteristics in a Turbine Rotor-Stator Cavity with Inlet at Low Radius. *J. Appl. Fluid Mech.* **2022**, *15*, 73–84. [CrossRef]

37. Barabas, B.; Clauss, S.; Schuster, S.; Benra, F.K.; Dohmen, H.J.; Brillert, D. Experimental and numerical determination of pressure and velocity distribution inside a rotor-stator cavity at very high circumferential Reynolds numbers. In Proceedings of the 11th European Conference on Turbomachinery Fluid Dynamics and Thermodynamics, Madrid, Spain, 23–27 March 2015.
38. Hu, B.; Brillert, D.; Dohmen, H.; Benra, F. Investigation on the flow in a rotor-stator cavity with centripetal through-flow. *Int. J. Turbomach. Propuls. Power* **2017**, *2*, 18. [CrossRef]
39. Hu, B.; Brillert, D.; Dohmen, H.; Benra, F. Investigation on thrust and moment coefficients of a centrifugal turbomachine. *Int. J. Turbomach. Propuls. Power* **2018**, *3*, 9. [CrossRef]
40. Poncet, S.; Chauve, M.; Legal, P. Turbulent rotating disk flow with inward throughflow. *J. Fluid Mech.* **2005**, *522*, 253–262. [CrossRef]
41. Poncet, S.; Chauve, M.; Schiestel, R. Batchelor versus Stewartson flow structures in a rotor-stator cavity with throughflow. *Phys. Fluids* **2005**, *17*, 075110. [CrossRef]

machines

Article

Multi-Conditional Optimization of a High-Specific-Speed Axial Flow Pump Impeller Based on Machine Learning

Zhuangzhuang Sun, Fangping Tang *, Lijian Shi and Haiyu Liu

College of Hydraulic Science and Engineering, Yangzhou University, Yangzhou 225000, China
* Correspondence: tangfp@yzu.edu.cn

Abstract: In order to widen the range of high-efficiency area of a high-specific-speed axial flow pump and to improve the operating efficiency under non-design conditions, the parameters of the axial flow pump blades were optimized. An optimization system based on computational fluid dynamics (CFD), optimized Latin hypercube sampling (OLHS), machine learning (ML), and multi-island genetic algorithm (MIGA) was established. The prediction effects of three machine learning models based on Bayesian optimization, support vector machine regression (SVR), Gaussian process regression (GPR), and fully connected neural network (FNN) on the performance of the axial flow pump were compared. The results show that the GPR model has the highest prediction accuracy for the impeller head and weighted efficiency. Compared to the original impeller, the optimized impeller is forward skewed and backward swept, and the weighted efficiency of the impeller increases by 1.31 percentage points. The efficiency of the pump section at $0.8Q_d$, $1.0Q_d$, and $1.2Q_d$ increases by about 1.1, 1.4, and 1.6 percentage points, respectively, which meets the optimization requirements. After optimization, the internal flow field of the impeller is more stable; the entropy production in the impeller reduces; the spanwise distribution of the total pressure coefficient and the axial velocity coefficient at the impeller outlet are more uniform; and the flow separation near the hub at the blade trailing edge is restrained. This research can provide a reference for the efficient operation of pumping stations and the optimal design of axial flow pumps under multiple working conditions.

Keywords: axial flow pump; multi-objective optimization; machine learning; skew and sweep

Citation: Sun, Z.; Tang, F.; Shi, L.; Liu, H. Multi-Conditional Optimization of a High-Specific-Speed Axial Flow Pump Impeller Based on Machine Learning. *Machines* **2022**, *10*, 1037. https://doi.org/10.3390/machines10111037

Academic Editor: Davide Astolfi

Received: 15 September 2022
Accepted: 2 November 2022
Published: 7 November 2022

1. Introduction

Axial flow pumps are widely used in various fields due to their large flow and low head. In some areas along rivers and lakes, the water level difference between the inlet and outlet rivers is small, and it is more suitable for the application of low head, high specific speed axial flow pump [1,2]. At present, there are few models of high-specific-speed axial flow pumps. When the pump deviates from the design flow rate, the hydraulic efficiency of the pump is relatively low, and there are obvious vortices and backflow in the impeller [3–5]. Therefore, it has become an urgent problem to optimize the geometric parameters of the impeller, improve its efficiency under non-design conditions, and widen the range of its high-efficiency area.

With the development of computer technologies, optimization design methods based on a combination of CFD and optimization algorithms have been widely used. The optimization methods for the impeller of axial flow pump are mainly divided into direct optimization and surrogate model-based optimization. Direct optimization [6–8] uses global or gradient optimization algorithms to directly optimize parameters, which requires a lot of computing resources and time for high-precision simulation operations under multi-objective problems. The surrogate model [9,10] usually builds an approximate functional relationship between the impeller parameters and the target value through machine learning, which can often improve the optimization efficiency. Ma et al. [11] used a combination of radial basis function (RBF) neural network and genetic algorithm to carry out a three-objective optimization of a double-blade sewage pump, and the RBF model predicted the results with high accuracy. Zhu et al. [12] also used this method to optimize the blade profile of a current energy turbine, which improved the hydrofoil lift–drag ratio at three angles of attack and suppressed the hydrofoil stall phenomenon. Wang et al. [13–15] used the method of coupling a response surface model, a multi-objective genetic algorithm, and an inverse problem design to carry out a multi-objective optimization of a mixed-flow pump impeller, which improved the hydraulic efficiency of the pump under multiple working conditions. Pei et al. [16] used an artificial neural network (ANN) and a particle swarm algorithm to optimize 11 shape parameters of the water inlet pipe of a vertical pipeline pump. The results showed that the ANN could accurately reflect the relationship between pump efficiency and design variables, and the maximum pump efficiency significantly improved after optimization. Miao et al. [17] optimized a hydraulic turbine blade profile based on a neural network–genetic algorithm, and improved the hydraulic efficiency of the hydraulic turbine under multiple operating conditions. In addition, Feng [18], Zhang [19], Wang [20–22] et al. adopted the method of combining machine learning and optimization algorithm to optimize hydraulic machinery.

In the field of hydraulic machinery, machine learning is also widely used in performance prediction and state recognition. He et al. [23] predicted the gas–liquid two-phase pressure rise of a centrifugal pump based on machine learning, and the prediction result had a true value error of less than 15%. Yang et al. [24] used the proper orthogonal decomposition and SVR model to achieve rapid prediction of the temperature distribution and the overflow water distribution on the outer surface of a hot gas anti-icing skin. Wang et al. [25] established a performance prediction model for semi-hermetic reciprocating compressors based on the BP network. Huang et al. [26] proposed a centrifugal pump energy performance prediction model based on a hybrid neural network. Compared to the experimental value, the mean square error (MSE) value was less than 0.02. Li et al. [27] proposed a new method to measure the gas–liquid flow rate of wet gas by combining a conical throttling device with machine learning techniques, and predicted the gas-phase and liquid-phase flow rates in the throttling device. Chen et al. [28] determined the identification of different leakage states of hydraulic pumps based on the wavelet decomposition and deep neural network, and the identification accuracy reached 99.3%. Panda et al. [29] used a support vector machine method to diagnose and classify centrifugal pump faults at different speeds, and showed high accuracy in multiple fault classifications. Bordoloi et al. [30] used a support vector machine (SVM) model for the diagnosis of clogging levels and cavitation degrees at different pump speeds. Rapur et al. [31] also used SVM to monitor the status of the blockage of the inlet pipe of a centrifugal pump under different flow rates and rotational speeds.

The geometric parameters of the impeller of an axial flow pump mainly include the solidity, the stagger angle of the hydrofoil, and the maximum camber of the hydrofoil. In addition to the above parameters, the stacking modes of the hydrofoils also have a great influence on the performance of the impeller. The stacking modes are divided into circumferential "skew" (the stacking point moves circumferentially) and axial "sweep" (the stacking point moves axially). It is usually defined that the stack point movement is "forward" when it is in the same direction of rotation or in the opposite direction to the mainstream, and "backward" when it is in the opposite direction of rotation or in the same direction as the mainstream [32,33]. Appropriate skew and sweep can realize the redistribution of the flow and load in the spanwise direction and reduce the loss of secondary flow. At present, the skew and sweep are often ignored or optimized separately during the impeller optimization of axial flow pumps, and the impeller parameters are not optimized as a whole [34,35]. Based on the above research findings, three different machine learning models are used in this paper to construct the approximate functional relationship between the impeller solidity, the stagger angle, the skew and sweep parameters, and the target value. At the same time, the prediction results of the machine learning models are evaluated, and the optimal model is selected as the proxy model. Finally, the MIGA is selected as the optimization algorithm to optimize multi-conditions, which provides a reference for the design of axial flow pumps in multi-conditions.

2. Research Object

Taking a large-scale, low-head pumping station as the object, the pumping station has a design head of 3.15 m, a single-machine flow rate of 64 m^3/s, an impeller diameter of 4.4 m, a pump speed of 85.7 r/min, and a specific speed of about 1060. It is a typical low-head pumping station. After being converted to a model pump with an impeller diameter of 300 mm according to the principle of equal nD values, the model pump design has a flow rate Q_d = 297.51 L/S and a speed n = 1256.93 r/min, and the design head of the pump has H_d = 3.32 m. According to the research and analysis of pump station selection, the hydraulic model is ZM25. The main design parameters of the ZM25 model are as follows: the number of blades is 3, the hub ratio is 0.35, the number of guide vanes is 5, and the solidity at the hub and the shroud of the impeller blade is 0.754 and 0.58, respectively. A three-dimensional model of the pump section is shown in Figure 1.

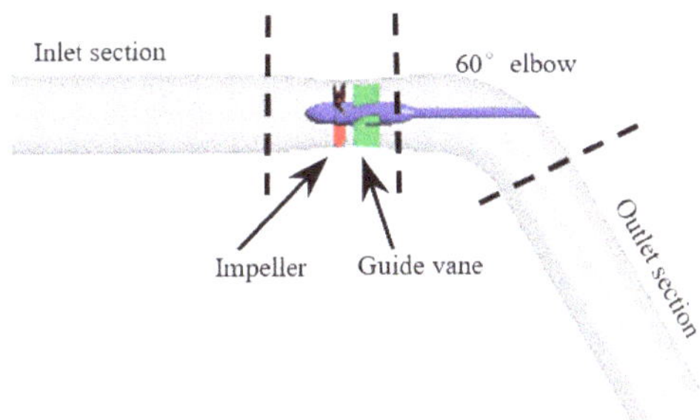

Figure 1. Three-dimensional model of the pump section.

3. Numerical Calculation Method

3.1. Turbulence Model and Boundary Conditions

The steady performance of an axial flow pump attracts more attention during the optimal design of the axial flow pump, so steady numerical calculation is adopted. The Reynolds-averaged N–S equations are solved using ANSYS CFX, and the equations are closed with the SST k-ω model [36,37]. The discrete format of the advection scheme and turbulence numerics is "High Resolution". The inlet adopts the mass flow inlet condition, and the outlet adopts the pressure outlet condition. The entire computational domain is divided into a rotating domain and a static domain, in which the impeller is a rotating domain and the rest are static domains. The "Stage" model (also known as the Mixing-Plane model) is used to deal with the parameter transfer of the flow in the dynamic and static couplings between the inlet and outlet of the impeller. The no-slip condition is applied to the solid wall, while the automatic wall treatment in CFX is used in the near-wall region to accommodate the turbulent flow model.

3.2. Meshing and Irrelevance Analysis

The hexahedral structured mesh is divided using the block strategy, and the mesh near the wall is refined. The y^+ of the main flow components, such as the impeller, is all within 100, which basically meets the requirements of the application of the SST k-ω turbulence model for the quality of near-wall mesh [38]. Among them, the inlet section, the outlet section, and the elbow section are meshed using ICEM, and the O-block topology is adopted. The impeller and guide vane section are meshed using TurboGrid and the J/O topology and H/O topology, respectively, while periodically arranging the flow channels. For areas with complex structures, appropriate mesh refinement is done. The tip gap size is 0.2 mm, and 15 nodes are arranged in the gap. We keep the same topology structure, modify the maximum mesh size to generate different numbers of meshes, and use the calculation result of the head under the design flow as the index to perform mesh irrelevance analysis. The results are shown in Figure 2. When the number of full machine grids increases to 4.11 million, continuing to increase the grid has little effect on the calculation results. Considering the calculation accuracy and calculation time, the number of grids is determined to be 4.11 million. The final grid is shown in Figure 3.

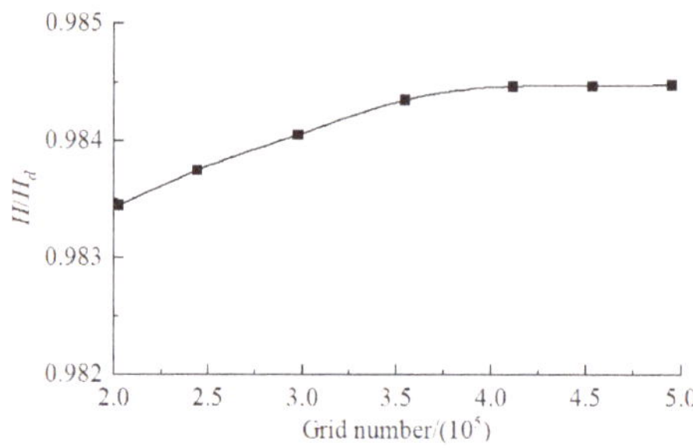

Figure 2. Grid independency test.

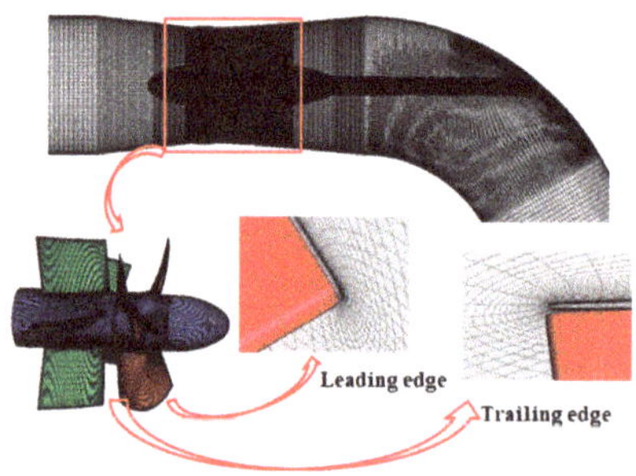

Figure 3. Meshing of Pump Sections.

3.3. *Verification of Numerical Calculation Results*

Figure 4 is a comparison diagram of the CFD results of the external characteristics of the pump section and the experimental (EXP) results. The numerical simulation results are basically consistent with the experimental results under different flow rates, and the head and efficiency errors near the design point are less than 3%, indicating that the numerical calculation method is reasonable and the results have high accuracy.

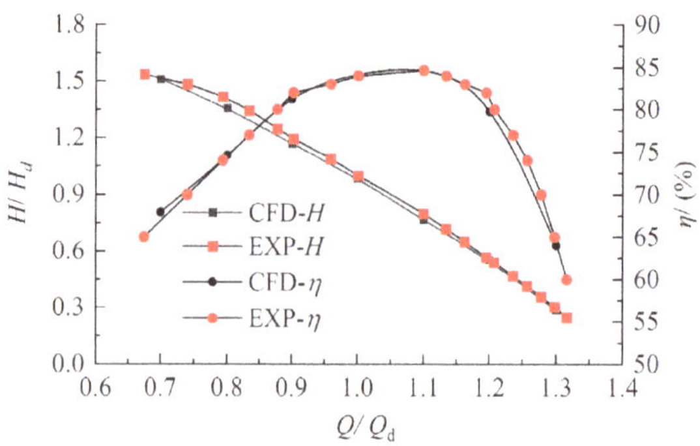

Figure 4. Comparison of the CFD results and the experiment results on hydraulic performance.

4. Machine Learning Models

4.1. Support Vector Machine Regression

Support vector machine regression (SVR) is a general machine learning method proposed by the statistician Vapnik [39,40]. Using the idea of support vector, it can non-linearly map low-dimensional data to a high-dimensional space, so that linear regression can be solved in the high-dimensional space. The prediction function has the following form:

$$\hat{Y} = \kappa^T \varphi(X) + b \tag{1}$$

where X is the input variable; $\hat{Y}$ is the predicted value of the variable; Y is the real value of the variable; $\varphi(X)$ is the nonlinear function that maps the sample to the high-dimensional space; and κ and b are the undetermined parameters, which are also the keys to the training of the SVR model.

SVR aims to find a hyperplane in a multi-dimensional space, so that all the data in a set are closest to the plane. Specifically, given the accuracy ε ($\varepsilon \geq 0$), the error between the predicted value and the actual value is not greater than ε; that is, the prediction can be considered correct, no loss is calculated, and only the loss of data points with an error greater than ε is calculated. Introducing the slack variables ξ and ξ^* ($\xi, \xi^* \geq 0$) and regularization parameter C ($C > 0$), κ and b training is transformed into an optimization problem:

$$\begin{aligned} &\min\left(\tfrac{1}{2}\kappa^T\kappa + C\sum_{i=1}^{N}\xi_i + C\sum_{i=1}^{N}\xi_i^*\right) \\ &\text{s.t. } \omega^T\varphi(X_i) + b - Y_i \leq \varepsilon + \xi_i^* \\ &\quad -\omega^T\varphi(X_i) - b + Y_i \leq \varepsilon + \xi_i \end{aligned} \tag{2}$$

where X_i is the i-th input variable.

This optimization problem can be transformed into a Lagrangian dual form, whereby the prediction function is equal to the following equation:

$$\hat{Y}(x) = \sum_{i=1}^{N}(\alpha_i - \alpha_i^*)K(X, X_i) + b \tag{3}$$

where α_i^* and α_i are the Lagrangian multipliers, and $K(X, X_i) = \varphi(X)\varphi^T(X_i)$ is the kernel function.

4.2. Gaussian Process Regression

A common application of Gaussian process (GP) in machine learning is the Gaussian process regression (GPR) [39,41], which has good applicability to dealing with complex problems, such as small samples, nonlinearity, and high dimensionality. The derivation of the Gaussian process regression can also start from a general linear regression, and a general linear regression model with noise has the following form:

$$\hat{Y} = X^T\kappa + \theta \tag{4}$$

where θ follows a Gaussian distribution with mean 0 and standard deviation σ, $\theta \sim N(0, \sigma^2)$, where any Gaussian distribution is completely determined by its first and second central moments (mean function and covariance function). The GPR introduces the functions $f(X)$ ($f(X)$ follows a Gaussian distribution with mean 0 and covariance function $k(X, X_i)$) and $h(X)$ (basis functions) to interpret this response and to project the functions to a high-dimensional space.

$$\hat{Y} = h(X)^T\kappa + f(X) \tag{5}$$

The GPR model is a probabilistic model, and each observation xi introduces a latent variable $f(X_i)$, making the GPR model nonparametric.

4.3. Fully Connected Neural Network

A complete fully connected neural network (FNN) [39,42] consists of an input, a fully connected layer, an activation function, and an output. The first fully connected layer of the neural network connects the input layer, and the fully connected layer consists of a weight matrix and a bias vector. Each fully connected layer multiplies the input by a weight matrix and then adds a bias vector. The activation function is the function that operates on the neurons of the neural network and is responsible for mapping the input of the neuron to the output. In machine learning, parameters, such as the weight matrices of the fully connected layers, are updated after each iteration. When the model encounters noisy samples, the parameters of the fully connected layer will also be synchronized with the noise (the weight matrix changes greatly), resulting in overfitting. Therefore, a regularization method is introduced into the model. The L_2 regularization method is a common regularization method to reduce overfitting, and a penalty coefficient is added to the weight matrix to avoid overfitting caused by too large parameters in the weight matrix.

4.4. Data Standardization and Evaluation Indicators

The data need to be preprocessed before training to improve the training effect. The Z-score standardization method is used to make the data conform to the Gaussian distribution with mean 0 and variance 1. The formula is as follows:

$$X^* = \frac{X - \mu}{\sigma^2} \tag{6}$$

where μ is the mean value.

In order to evaluate the prediction performance of the model, the correlation coefficient R^2, the mean square error (MSE), the mean absolute percentage error (MAPE), the relative absolute error (RAE), and the Willmott's Index of Agreement (WIA) are introduced as evaluation indicators, and the relevant definitions are as follows:

$$R^2 = 1 - \frac{\sum_{i=1}^{N} (Y_i - \hat{Y}_i)^2}{\sum_{i=1}^{n} (Y_i - \overline{Y})^2} \tag{7}$$

$$\text{MSE} = \frac{1}{N}\sum_{i=1}^{N} (Y_i - \hat{Y}_i)^2 \tag{8}$$

$$\text{MAPE} = \frac{100}{N}\sum_{i=1}^{N} \frac{|Y_i - \hat{Y}_i|}{Y_i} \tag{9}$$

$$\text{RAE} = \frac{\sum_{i=1}^{N} |Y_i - \hat{Y}_i|}{\sum_{i=1}^{n} |Y_i - \overline{Y}|} \tag{10}$$

$$\text{WIA} = 1 - \frac{\sum_{i=1}^{N} (Y_i - \hat{Y}_i)^2}{\sum_{i=1}^{N} (|Y_i - \overline{Y}| + |\hat{Y}_i - \overline{Y}|)^2} \tag{11}$$

Among the evaluation indicators, the smaller the RMSE, MAPE, and RAE are, and the closer R^2 and WIA are to 1, the more accurate the prediction results are.

4.5. Hyperparameter Optimization

Hyperparameters are parameters that cannot be obtained through learning in machine learning models (such as the kernel functions of SVR and GPR models, and the number of fully connected layers of FNN models), and their selection directly affects the training effect of the model. Hyperparameter optimization in machine learning aims to find the hyperparameter combination that makes the machine learning algorithm perform the best on the validation set. Manual parameter tuning requires a lot of experience and is time-consuming. Therefore, many automatic parameter tuning methods have been developed, such as grid search, random search, and Bayesian optimization. In grid search and random search, each hyperparameter is independent of each other, and the previous calculation result does not affect the latter calculation result, which usually takes more time. Bayesian optimization uses Bayes' theorem to estimate the posterior distribution of the objective function based on the data, and then selects the next sampled hyperparameter combination based on the distribution. It makes full use of the information of the previous sampling point, and its optimization works by learning the shape of the objective function and finding the parameters that can improve the result to the global maximum. Therefore, this paper adopts the Bayesian optimization method when adjusting the hyperparameters. In order to prevent overfitting during model training, a 5-fold cross-validation is used, and the principle of minimum mean square error (MSE) is used for 100 iterations. Table 1 shows the hyperparameters and the search range of the hyperparameters in each ML model.

Table 1. Hyperparameter and its search range in each ML model.

Model	Hyperparameter and Its Search Range
SVR	Kernel functions: Gaussian, linear, cubic, and quadratic Regularization factor: [0.001,1000] Kernel scale: [0.001,1000] ε: [0.001,100]/1.349·iqr(Y)
GPR	Basis functions: zero, constant, and linear. Kernel functions: Non-isotropic and isotropic exponential, quadratic rational, squared exponential, Matern 5/2, and Matern 3/2 Kernel scale: [0.001, 1] ·(max(X)–min(X)) Standard Deviation: [0.0001, max (0.001, 10×std(Y))]
FNN	Number of fully connected layers: [1,2,3] Size of each connection layer: [1,300] Activation function: Rectified Linear Unit (RELU), Tanh, None, and Sigmoid Regularization strength: [0,1250]

5. Optimization Method of Impeller

5.1. Optimization Objective

According to the needs of working on multiple operating conditions, the weighted efficiency η_d under the conditions of $0.8Q_d$, $1.0Q_d$, and $1.2Q_d$ is taken as the optimization target. In the optimization process, a large number of sample points need to be obtained, and the full flow channel calculation requires a lot of computing resources and time, so only single-channel calculation is used for the impeller. At the same time, the head change should not be too large under the design flow to ensure that the specific revolutions are consistent, so that the axial flow pump can meet the design requirements in engineering applications. The head change before and after the impeller optimization is restricted to be less than 0.1 m. The optimization model is defined as follows:

$$\begin{aligned} &\eta_d = k_1\eta_1 + k_2\eta_2 + k_3\eta_3 \\ &\text{s.t. } H_{ip2} = H_{ip1} \pm 0.1 \end{aligned} \tag{12}$$

where η_1, η_2, and η_3 are the impeller efficiencies under the conditions of $0.8Q_{\rm d}$, $1.0Q_d$, and $1.2Q_d$, respectively; k_1, k_2, and k_3 are the weighting factors and, based on the relevant literature [6], the values are 0.2786, 0.4059, and 0.315, respectively; H_{ip} is the impeller head under the design flow; and H_{ip1} and H_{ip2} are the impeller head under the design flow before and after optimization, in m.

5.2. Optimization Parameters

There are 11 sections from the hub to the shroud, and the section hydrofoil is NACA66 (mod). The spanwise dimensionless distance r^* is defined as follows:

$$r^* = \frac{r - r_h}{r_t - r_h} \tag{13}$$

where r is the radius of the section; r_t is the radius at the shroud of the blade; and r_h is the radius at the hub of the blade, in m.

The main design parameters of the blade include the following: the solidity c/t (where c is the chord length and t is the pitch, in m), the stagger angle β, and the maximum camber ratio a/c, as shown in Figure 5. By specifying the hydrofoil stagger angles β_1, β_2, and β_3, and maximum camber ratios $(a/c)_1$, $(a/c)_2$, and $(a/c)_3$ of section r^* = 0, 0.47 and 1, the remaining section parameters can be obtained using quadratic function interpolation. Given the solidity $(c/t)_1$ and $(c/t)_2$ at the hub and the shroud of the blade, the chord lengths of the remaining sections are obtained using linear interpolation.

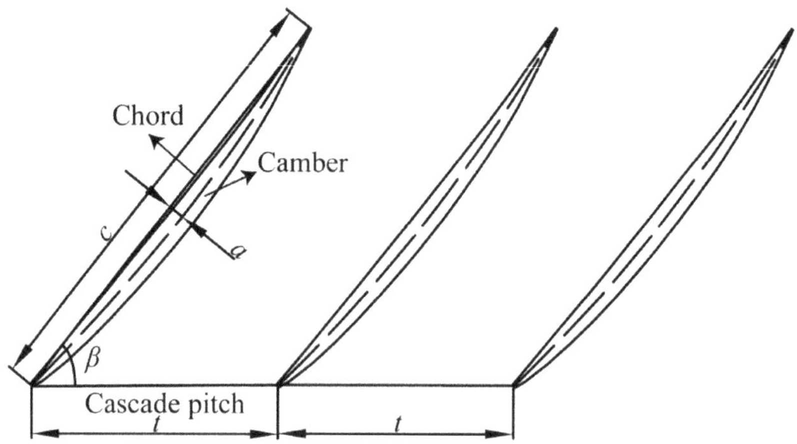

Figure 5. Schematic diagram of the blade parameters.

Taking the circumferential angle and the axial displacement at the hub as 0, the circumferential skew angles α_1 and α_2 and the axial sweep displacements γ_1 and γ_2 of the r^* = 0.47 and 1 sections are specified. The skew and sweep parameters of the remaining sections are also obtained using quadratic function interpolation. Among them, the circumferential skew angle and the axial sweep displacement are "+", which means backward skew and backward sweep, respectively. The schematic diagram of the blade stacking modes is shown in Figure 6.

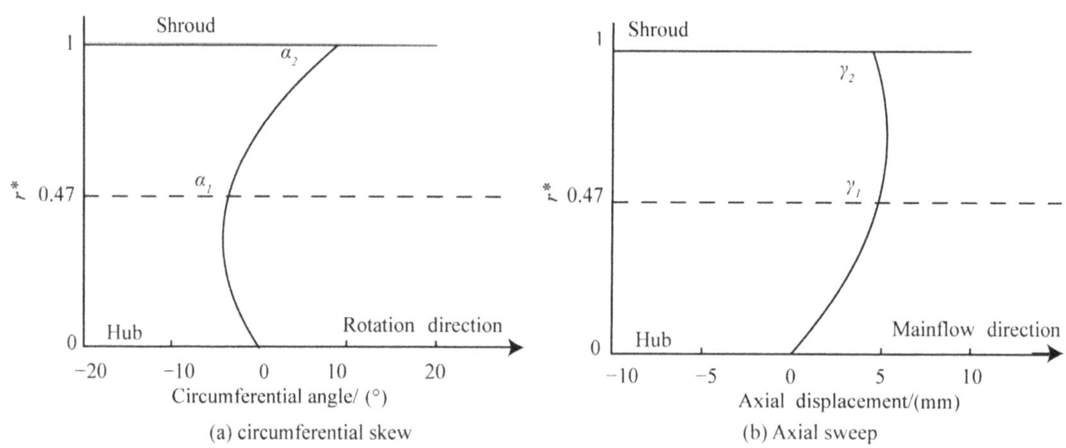

Figure 6. Schematic diagram of the blade stacking modes.

Based on the above analysis, the impeller has a total of 12 design parameters, and the value range of each parameter is shown in Table 2. Among them, the variation range of the solidity, hydrofoil stagger angle, and hydrofoil camber is ±10% of the original values.

Table 2. Range of design parameters.

Design Parameters	Low Level (−)	High Level (+)
$(c/t)_1/$-	0.679	0.829
$(c/t)_2/$-	0.522	0.638
$\beta_1/°$	39.435	48.199
$\beta_2/°$	24.816	30.330
$\beta_3/°$	15.817	19.331
$(a/c)_1/\%$	5.470	6.686
$(a/c)_2/\%$	3.154	3.854
$(a/c)_3/\%$	1.362	1.664
γ_1/mm	−5	10
γ_2/mm	−5	10
$\alpha_1/°$	−15	15
$\alpha_2/°$	−15	15

5.3. Optimization Progress

The optimization process is shown in Figure 7. First, an optimized Latin hypercube sampling (OLHS) is used to obtain sample data within the design range. An automatic numerical simulation platform is built through Isight to quickly obtain the optimal target value of the sample point, and the approximate relationship between the variables and the optimization target is fitted through machine learning. After comparing the training results of the SVR, GPR, and FNN models, the appropriate approximate model is selected. Finally, the approximate model is solved using the multi-island genetic algorithm (MIGA) to obtain the optimal parameter combination, which will be confirmed by CFD. The MIGA is essentially an improvement of the parallel distributed genetic algorithm, which has better global solving ability and computing efficiency than traditional genetic algorithms [43]. Therefore, the MIGA is selected for global optimization.

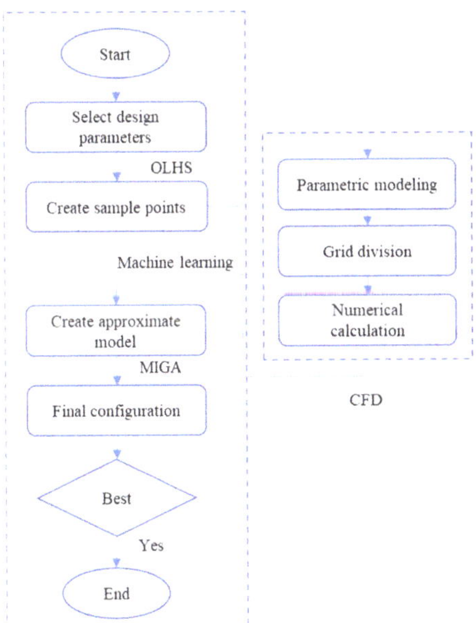

Figure 7. Optimization progress.

6. Results & Analysis

6.1. Data Set Partitioning

In the optimization process, the predicted variables are the weighted efficiency η_d and the impeller head H_{ip2}. The OHLS is used to generate 516 sets of samples, of which 85% are used for training and 15% are used for testing. Table 3 shows the results of the division of the sample set. The statistical characteristics of the training set and the test set are similar and highly representative.

Table 3. Data set partitioning.

Data Set		Sample Size	Max Value	Min Value	Mean Value	Standard Deviation
H_{ip2}	Training set	439	5.123	1.952	3.494	0.660
	Testing set	77	5.114	1.993	3.540	0.669
η_d	Training set	439	91.796	73.497	88.707	2.281
	Testing set	77	91.972	79.107	88.795	1.904

6.2. Comparison of Training Results

The hyperparameters after Bayesian optimization are used to establish the prediction model. Table 4 shows the comparison of the indicators of the prediction results of the three machine learning models. It can be seen from the table that the GPR model has the highest prediction accuracy for H_{ip2} and η_d, and the model has high generalization ability. On the training set of H_{ip2}, the evaluation indicators R^2, MSE, MAPE, RAE, and WIA of the GPR model are 0.997, 0.001, 0.772, 0.047, and 0.999, respectively. On the testing set of H_{ip2}, the evaluation indexes R^2, MSE, MAPE, RAE, and WIA of the GPR model are 0.998, 0.001, 0.634, 0.041, and 1.000, respectively. On the training set of η_d, the evaluation indicators R^2, MSE, MAPE, RAE, and WIA of the GPR model are 0.988, 0.063, 0.198, 0.101, and 0.997, respectively. On the test set of η_d, the evaluation indicators R^2, MSE, MAPE, RAE, and WIA of the GPR model are 0.981, 0.070, 0.195, 0.124, and 0.995, respectively.

Table 4. Comparison of the prediction results of the three models.

	Evaluation Indicators	SVR		GPR		FNN	
		Training Set	Testing Set	Training Set	Testing Set	Training Set	Testing Set
H_{ip2}	R^2	0.995	0.997	0.997	0.998	0.982	0.991
	MSE	0.002	0.001	0.001	0.001	0.008	0.004
	MAPE	1.016	0.945	0.772	0.634	2.082	1.416
	RAE	0.061	0.058	0.047	0.041	0.122	0.087
	WIA	0.999	0.999	0.999	1.000	0.995	0.998
η_d	R^2	0.937	0.898	0.988	0.986	0.971	0.954
	MSE	0.324	0.364	0.063	0.070	0.150	0.166
	MAPE	0.440	0.429	0.198	0.195	0.346	0.348
	RAE	0.224	0.270	0.101	0.124	0.179	0.222
	WIA	0.984	0.975	0.997	0.995	0.993	0.988

Figure 8 shows the probability distribution functions (PDF) of the predicted and observed H_{ip2} and η_d. From the shape of the PDF, there is a certain gap between the predicted values of the SVR, GPR, and FNN models and the observed values when the η_d is less than 80% in the prediction of efficiency. When the η_d is greater than 80%, the GPR model shows the best agreement with the observations, followed by the FNN and SVR models. In the prediction of H_{ip2}, all three models produce acceptable predictions. However, the GPR model shows the best agreement with the observed values, with the GPR model in the box plots predicting results that are closer to the median, the upper quartile, and the lower quartile. In addition, the Taylor diagram of the prediction results of the SVR, GPR, and FNN models for H_{ip2} and η_d is shown in Figure 9. For the prediction of H_{ip2} and η_d, the GPR model is closer to the position of the target point, and its prediction results are the most accurate.

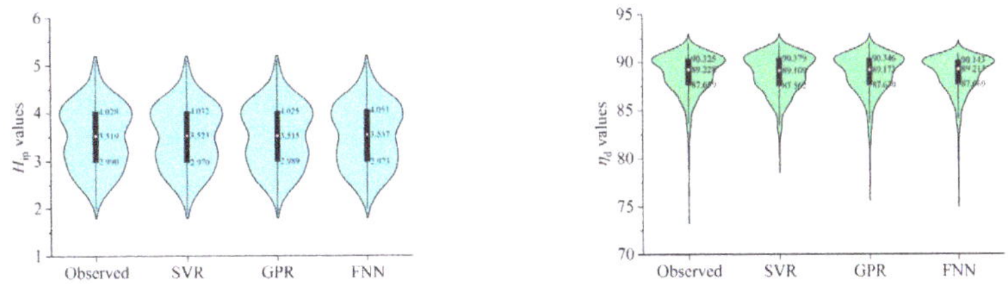

(a) (b)

Figure 8. PDF of predicted and observed values: (**a**) H_{ip2} and (**b**) η_d.

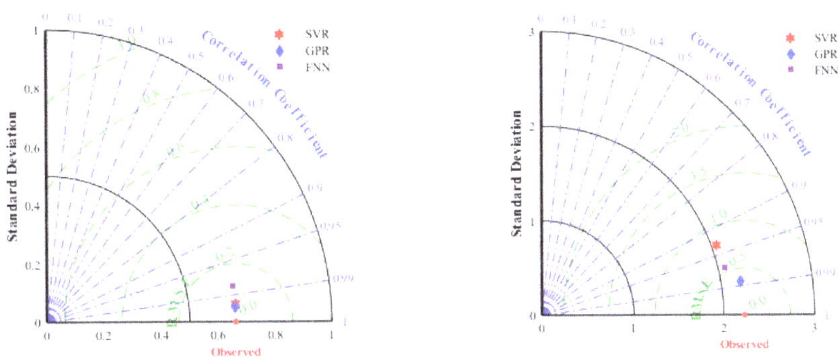

(a) (b)

Figure 9. The Taylor diagram of predicted and observed values: (**a**) H_{ip2} and (**b**) η_d.

Figure 10 shows the relative deviation (RD) of all models on the training and test data. The smaller the RD range of the model is, the more efficient the model results are. For the prediction results of H_{ip2}, the RD of the GPR model is in the range of [−4, 4] and [−2, 2] in the training and testing sets. Compared to the SVR model (training set [−4, 6] and testing set [−4, 4]) and the FNN model (training set [−15, 15] and testing set [−7, 10]), the margin of error is smaller. For the prediction results of η_d, the RD of the GPR model is mainly distributed in the range of [−1,1] on the training and testing sets. The range is smaller compared to the SVR ([−2, 2]) and FNN ([−1.5, 1]) models. When combining the cumulative frequency and the absolute relative deviation (ARD) percentages (see Figure 11), for the prediction results of H_{ip2}, the cumulative frequency of the GPR model is within 3% and the ARD is 98.64%, which is significantly higher than that of the SVR (96.71%) and FNN (78.87%) models. For the prediction of η_d, the cumulative frequency of the GPR model is within 1% and the ARD is 99.41%, which is also higher than that of the SVR (92.64%) and FNN (97.09%) models. Based on the above analyses, it can be seen that the GPR model has a higher prediction accuracy for the weighted efficiency and the impeller head, and the GPR model is finally determined as the proxy model for impeller optimization.

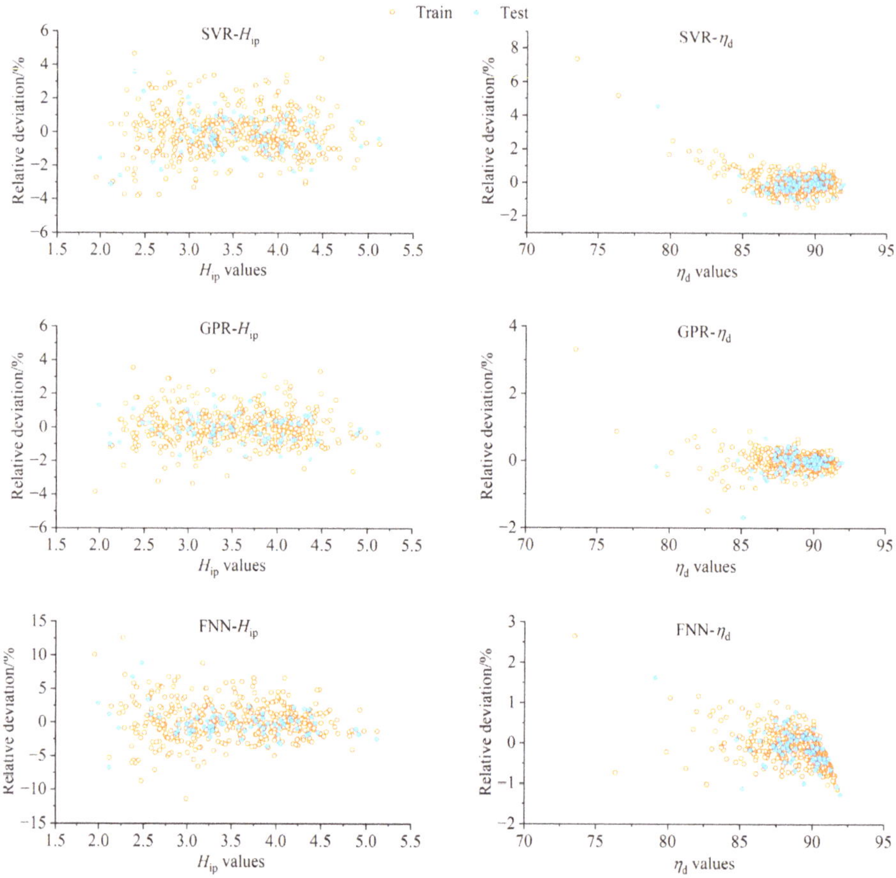

Figure 10. The relative deviation in H_{ip2} and η_d of the three models.

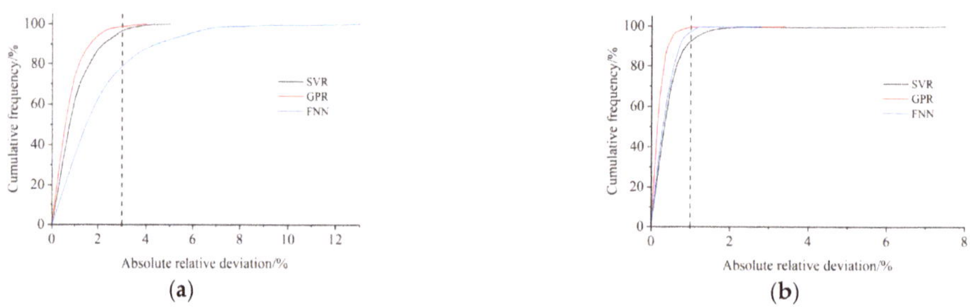

Figure 11. The cumulative frequency of absolute relative deviation: (**a**) H_{ip2} and (**b**) η_d.

6.3. Analysis of Optimization Results

The prediction model obtained through the GPR training is further optimized based on the multi-island genetic algorithm. The algorithm is set as follows: the number of subgroups is 20, the number of islands is 20, the crossover rate is one, the inter-island mobility and the mutation probability are both 0.01, and the migration interval is five generations [43]. According to the optimization results of the MIGA, the CFD calculation is re-calculated, and the main parameters of the optimized impeller are obtained (see

Table 5). The optimized parameters are within the allowable range. After optimization, the axial direction of the blade is backward and the circumferential direction is skewed forward. The shape of the impeller is shown in Figure 12. The hydrofoil stagger angle at the blade hub increases, while the hydrofoil stagger angle at the blade shroud decreases, indicating that the work at the blade hub increases and the work at the blade tip decreases, which balances the outlet head of the blade. The solidity at the hub and the shroud of the blade increases, and the chord length of the blade becomes longer as a whole. At the same time, the reduction of the maximum camber of the hydrofoil is also conducive to the improvement of the lift–drag ratio of the hydrofoil at a small angle of attack, which means the blade has a higher hydraulic efficiency at a large flow. The weighted efficiency and the head of the optimized impeller are 92.22% and 3.699 m, respectively, which is 1.31 percentage points higher than the original impeller's weighted efficiency. The change of the impeller head under the design flow is less than 0.04 m, which meets the optimization requirements. At the same time, the weighted efficiency and the impeller head predicted by the GPR model are 92.18% and 3.704 m, respectively, which are 0.04 percentage points and 0.005 m away from the actual values of the numerical simulation. This proves the high precision of the GPR model.

Table 5. Comparison of the parameters between the original and optimized models.

Design Parameters	Original	Optimized
$(c/t)_1$/-	0.754	0.819
$(c/t)_2$/-	0.580	0.594
$\beta_1/°$	43.817	47.792
$\beta_2/°$	27.573	28.106
$\beta_3/°$	17.574	16.847
$(a/c)_1$/%	6.078	5.571
$(a/c)_2$/%	3.503	3.845
$(a/c)_3$/%	1.513	1.655
γ_1/mm	0	6.531
γ_2/mm	0	6.981
$\alpha_1/°$	0	−3.93
$\alpha_2/°$	0	−7.98
η_d/%	90.91	92.22
H_{ip}/m	3.660	3.699

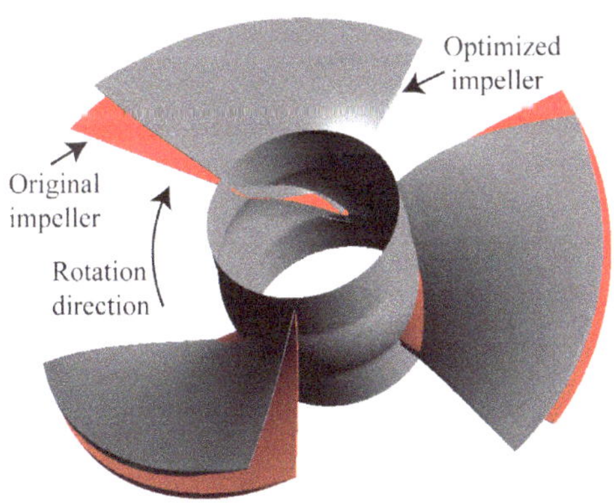

Figure 12. Comparison of the impeller shapes before and after optimization.

In order to further analyze the performance changes of the optimized axial flow pump under multiple working conditions, the optimized impeller plus the flow-passing components, such as guide vanes, are combined to form the axial flow pump section for numerical calculation. When comparing the performance curves obtained using CFD of the pump section before and after optimization (see Figure 13), the efficiency curve of the pump section after optimization is overall higher than the original design curve. Under the flow conditions of $0.8Q_d$, $1.0Q_d$, and $1.2Q_d$, the efficiencies are 75.71%, 85.53%, and 81.38%, and the efficiency increases by about 1.1, 1.4, and 1.6 percentage points, respectively. After optimization, the high-efficiency area of the axial flow pump is significantly wider.

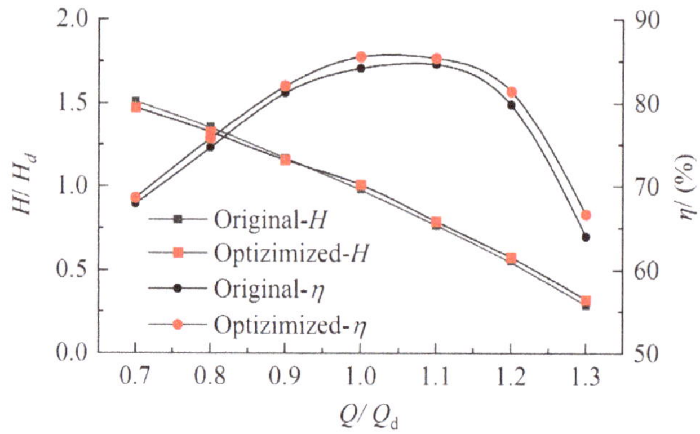

Figure 13. Performance comparison of the pump section before and after optimization.

6.4. Analysis of the Internal Flow Field of the Impeller before and after Optimization

To better compare the variation in the flow field parameters, the following dimensionless parameters are defined.

Velocity coefficient:

$$v^* = \frac{v}{u_t} \tag{14}$$

Total pressure rise coefficient:

$$P_t^* = \frac{P_t - P_{tin}}{0.5\rho {u_t}^2} \tag{15}$$

Static pressure coefficient:

$$P^* = \frac{P - P_{in}}{0.5\rho {u_t}^2} \tag{16}$$

where v is velocity; u_t is the circumferential velocity at the shroud, in m/s; P_t and P are the total pressure and static pressure, in Pa; P_{tin} and P_{in} are the inlet total pressure and static pressure, in Pa; and ρ is the density, in kg/m^3.

When comparing the axial velocity coefficient v_t^* distribution of the impeller outlet before and after optimization (see Figure 14a), it can be seen that, under different flow conditions, the axial velocity distribution at the blade outlet after optimization shows a trend of decreasing at the blade shroud and increasing at the blade hub, and the axial velocity distribution in the entire spanwise direction is more uniform. At the same time, under the condition of $0.8Q_d$, the axial velocity near the blade hub is negative, indicating that there is a small-scale recirculation zone there, which results in energy loss, and the scope of the recirculation zone is significantly reduced after optimization. The change of the total pressure rise coefficient P_t^* (see Figure 14b) at the impeller outlet is similar to the change of the axial velocity. The total pressure rise coefficient increases at the blade hub and decreases at the blade shroud. The total pressure rise coefficient of the original model has a negative value near the hub under the flow condition of $0.8Q_d$, which means that the impeller in this area does negative work. After optimization, this phenomenon is eliminated and the distribution of the total pressure at the impeller outlet is more uniform, indicating that the flow state at the impeller outlet has significantly improved.

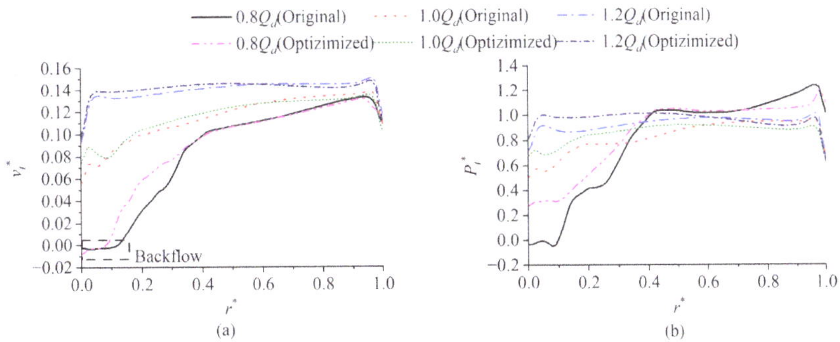

Figure 14. Comparison of the flow field at the impeller outlet: (**a**) axial velocity coefficient v_t^*, and (**b**) total pressure rise coefficient P_t^*.

Figure 15 shows the static pressure coefficient and the streamline distribution on the blade surface. As can be seen from Figure 15a, due to the influence of the hub (the hub is spherical), a small range of flow separation is formed near the outlet side of the pressure surface of the blade under the conditions of $1.0Q_d$ and $1.2Q_d$. Under the condition of $0.8Q_d$, the flow separation area of the pressure surface of the original model accounts for about 1/5 of the blade area. This is mainly due to the strong backflow intensity at the hub and the secondary flow formed by the interaction between the backflow and the mainstream, which causes the streamline near the hub at the outlet side to flow from the hub to the middle of the blade. At the same time, an obvious saddle point and a node are formed on the pressure surface of the blade, which results in obvious flow separation. After optimization, the backflow intensity weakens and the flow separation area significantly reduces. As can be seen from Figure 15b, a flow separation phenomenon near the outlet side hub of the suction surface of the blade is observed. It can be seen from the pressure distribution on the surface of the blade that the pressure distortion phenomenon occurs in the hub at the outlet of the suction surface of the blade, resulting in the pressure gradient direction being perpendicular to the mainstream direction, which increases the pressure difference between the shroud and the hub on the blade surface and leads to the secondary flow from the hub to the shroud. After optimization, the flow separation area under each flow condition significantly reduces; the pressure distribution on the surface of the blade is more reasonable; and the pressure distortion near the outlet side hub becomes suppressed.

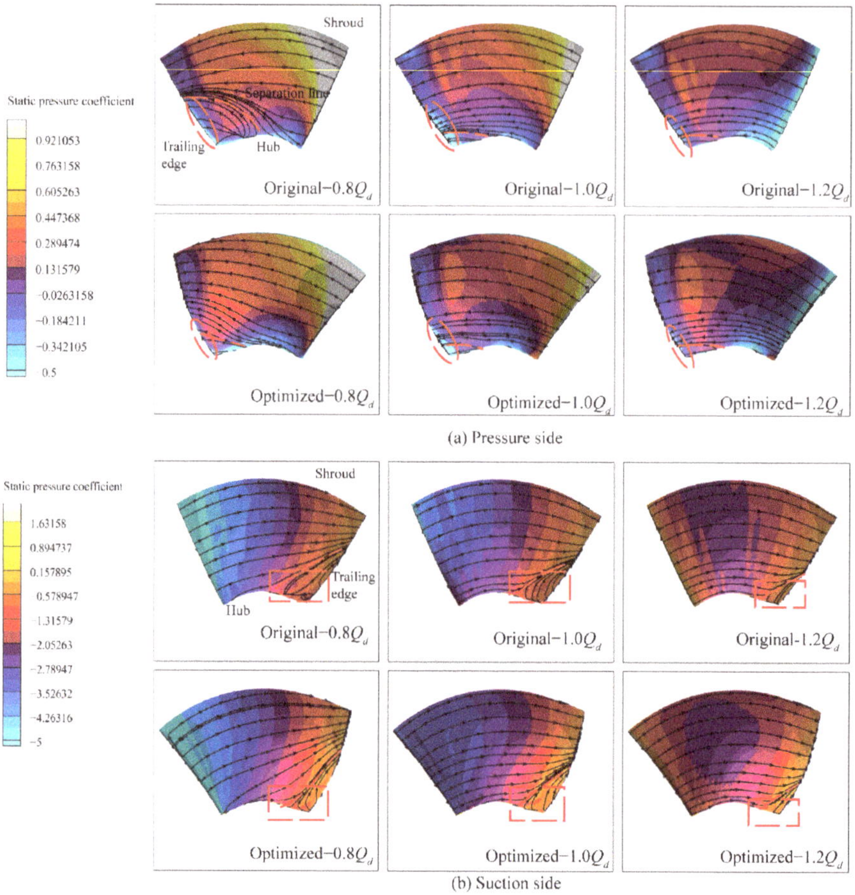

Figure 15. Surface pressure and streamline distribution of the blade.

When the fluid works inside the pump, due to the action of viscous force and internal friction, a part of the energy is converted into the internal energy of the system, and the entropy increases during the whole process. In recent years, the entropy production theory has been widely used in the analysis of the internal flow loss of hydraulic machinery [44,45]. Compared to the traditional differential pressure method, the entropy production theory can directly reflect the position of energy loss, which has guiding significance for the optimal design of hydraulic machinery. For turbulent flow, the entropy generation rate can be divided into two parts: one is caused by the average velocity, called the direct dissipation term, and the other is caused by the pulsating velocity, called the turbulent dissipation term. The entropy production rate (EPR) is defined as follows:

$$\dot{S}_D''' = \dot{S}_{\overline{D}}''' + \dot{S}_{D'}''' \tag{17}$$

where $\dot{S}_{D}'''$ is the EPR; $\dot{S}_{\overline{D}}'''$ is the EPR caused by the average speed; and $\dot{S}_{D'}'''$ is the EPR caused by the pulsating speed, in W $m^{-3}K^{-1}$. The entropy production caused by the average speed is defined as follows:

$$\begin{aligned}\dot{S}_{\overline{D}}''' = \frac{\mu}{T}\left[\left(\frac{\partial \overline{u}}{\partial y}+\frac{\partial \overline{v}}{\partial x}\right)^2+\left(\frac{\partial \overline{u}}{\partial z}+\frac{\partial \overline{w}}{\partial x}\right)^2+\left(\frac{\partial \overline{v}}{\partial z}+\frac{\partial \overline{w}}{\partial y}\right)^2\right] \\ +2\frac{\mu}{T}\left[\left(\frac{\partial \overline{u}}{\partial x}\right)^2+\left(\frac{\partial \overline{v}}{\partial y}\right)^2+\left(\frac{\partial \overline{w}}{\partial z}\right)^2\right]\end{aligned} \tag{18}$$

where $\overline{u}$, $\overline{v}$, and $\overline{w}$ are the components of the velocity in the x, y, and z directions, in m/s;μ is the dynamic viscosity, in Pa/s; and T is the temperature, in K.

The pulsation entropy production can be approximated using the following formula:

$$\dot{S}_{D'}''' = \psi \cdot \frac{\rho f k}{T} \tag{19}$$

where the coefficient ψ is 0.09; f is the turbulence eddy frequency, in s^{-1}; and k is the turbulent kinetic energy, in m^2/s^2. Figure 16 shows the comparison of the internal entropy production rate of the impeller before and after optimization under different flow conditions. Under the condition of $1.2Q_d$, the entropy production in the impeller is mainly concentrated on the blade surface and near the wake vortex. However, under the flow conditions of $0.8Q_d$ and $1.0Q_d$, there is a large area of high entropy production near the exit of the r^* = 0.02, 0.5 section, which is related to the backflow and the flow separation analyzed above. Compared to the original model, under the condition of $0.8Q_d$, the entropy production on the blade surface slightly increases, but the backflow and the flow separation near the outlet are restrained, and the entropy production near the impeller outlet significantly reduces. The entropy production on the blade surface decreases under $1.0Q_d$ and $1.2Q_d$, and the entropy production area near the outlet also decreases significantly under $1.0Q_d$.

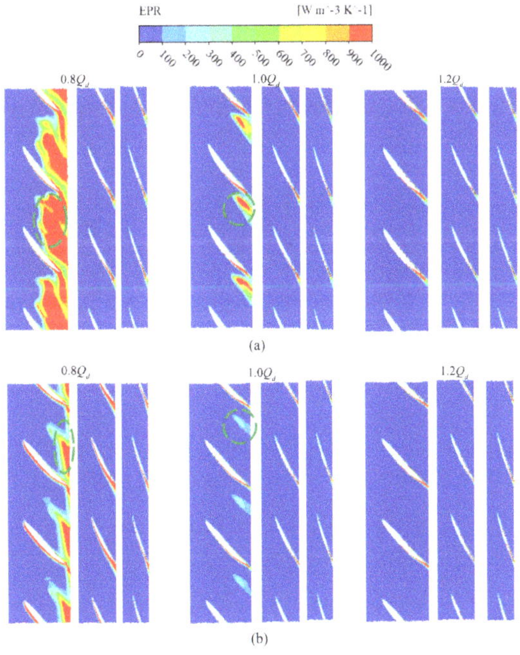

Figure 16. The distribution of EPR in the impeller:(**a**) Original impeller, and (**b**) Optimized impeller. (From left to right of each item is r^* = 0.02, 0.5, and 0.98).

7. Conclusions

In order to improve the hydraulic efficiency of a high-specific-speed axial flow pump impeller under multiple working conditions, the parameters of the blade, such as the solidity, the hydrofoil stagger angle, and the skew and sweep, are optimized. The specific conclusions are as follows:

1. An optimization system composed of the CFD, OLHS, ML, and MIGA is proposed, which provides a reference for the optimal design of axial flow pumps in the future.
2. Based on Bayesian optimization, the hyperparameters of the SVR, FNN, and GPR models are optimized, and the optimized hyperparameter combination is used to establish the prediction model of the weighted efficiency and the impeller head. Compared to the SVR and FNN models, the GPR model has better generalization and the highest prediction accuracy, and the GPR model is better adaptable to the nonlinear relationship between the fit optimization parameters and the target in the optimal design.
3. Compared to the original model, the weighted efficiency of the optimized impeller increases by 1.31 percentage points, and the efficiency of the pump section at $0.8Q_d$, $1.0Q_d$, and $1.2Q_d$ increases by about 1.1, 1.4, and 1.6 percentage points, respectively. The operating range of the high-efficiency area of the axial flow pump is improved.
4. The optimized impeller is forward skewed and backward swept, which is beneficial for reducing flow separation on the blade surface. After optimization, the flow field at the impeller outlet significantly improves; the total pressure and the axial velocity along the spanwise direction are more uniform; the flow separation at the trailing edge of the blade improves; and the entropy production in the impeller reduces.

Author Contributions: Data curation, Z.S.; Formal analysis, F.T.; Writing—original draft, Z.S.; Writing—review & editing, L.S.; Software, H.L. All authors have read and agreed to the published version of the manuscript.

Funding: This research work was supported by the National Natural Science Foundation of China (Grant No. 51376155), the National Natural Science Foundation of China (Grant No. 52209116), the Jiangsu Province Water Conservancy Science and Technology Project (Grant No. 2021012), and the Yangzhou Science and Technology Plan Project City-School Cooperation Special Project (Grant No. YZ2022178).

Institutional Review Board Statement: Not applicable.

Informed Consent Statement: Not applicable.

Data Availability Statement: The data presented in this study are available on request from the corresponding author.

Acknowledgments: This is a project funded by the Priority Academic Program Development (PAPD) of the Jiangsu Higher Education Institutions Support for construction.

Conflicts of Interest: The authors declare no conflict of interest.

References

1. Liu, C. Researches and Developments of Axial-flow Pump System. *Trans. Chin. Soc. Agric. Mach.* **2015**, *46*, 11.
2. Wang, Z.; Peng, G.; Zhou, L.; Hu, D. Hydraulic performance of a large slanted axial-flow pump. *Eng. Comput.* **2010**, *27*, 243–256. [CrossRef]
3. Yang, J.; Guan, X. Hydraulic Design of High Specific Speed Model Axial-flow Pump. *Trans. Chin. Soc. Agric. Mach.* **2008**, *39*, 4.
4. Shi, W. Design of Axial Flow Pump Hydraulic Model ZM931 On High Specific Speed. *Trans. Chin. Soc. Agric. Mach.* **1998**, *29*, 49–53.
5. Tang, F.; Wang, G.; Liu, C.; Zhou, J.; Cheng, L. Design and Numerical Analysis On An Axial-flow Pump With High Specific Speed. *J. Mech. Eng.* **2005**, *41*, 5. [CrossRef]
6. Yuan, J.; Fan, M. Orthogonal optimum design method for high specific-speed axial-flow pumps. *Trans. Chin. Soc. Agric. Eng.* **2018**, *37*, 22.
7. Tao, R.; Xiao, R.; Yang, W. Optimization design for axial flow pump based on genetic algorithm. *J. Drain. Irrig. Mach. Eng.* **2018**, *36*, 7.

8. Wang, L.; Wang, T.; Luo, Y. Improved NSGA- in Multi-Objective Optimization Studies of Wind Turbine Blades. *Appl. Math. Mech.* **2011**, *32*, 739–748. [CrossRef]
9. Mu, X.; Yao, W.; Yu, X.; Liu, K.; Xue, F. A survey of surrogate models used in MDO. *Chin. J. Comput. Mech.* **2005**, *22*, 5.
10. Gu, H.; Lin, Y.; Hu, Z.; Yu, J. Surrogate Models Based Optimization Methods for the Design of Underwater Glider Wing. *J. Mech. Eng.* **2009**, *45*, 8. [CrossRef]
11. Ma, S.B.; Kim, S.; Kim, J.H. Optimization design of a two-vane pump for wastewater treatment using machine-learning-based surrogate modeling. *Processes* **2020**, *8*, 1170. [CrossRef]
12. Zhu, G.; Feng, J.; Guo, P.; Luo, X. Optimization of hydrofoil for marine current turbine based on radial basis function neural network and genetic algorithm. *Trans. Chin. Soc. Agric. Eng.* **2014**, *30*, 9.
13. Wang, M.; Li, Y.; Yuan, J.; Yuan, S. Effects of different vortex designs on optimization results of mixed-flow pump. *Eng. Appl. Comput. Fluid Mech.* **2022**, *16*, 36–57. [CrossRef]
14. Wang, M.; Li, Y.; Yuan, J.; Meng, F.; Chen, J. Comprehensive Improvement of Mixed-Flow Pump Impeller Based on Multi-Objective Optimization. *Processes* **2020**, *8*, 905. [CrossRef]
15. Wang, M.; Li, Y.; Yuan, J.; Yuan, J.; Osman, F.K. Matching Optimization of a Mixed Flow Pump Impeller and Diffuser Based on the Inverse Design Method. *Processes* **2021**, *3*, 260. [CrossRef]
16. Pei, J.; Gan, X.; Wang, W.; Yuan, S.; Tang, Y. Hydraulic Optimization on Inlet Pipe of Vertical Inline Pump Based on Artificial Neural Network. *Trans. Chin. Soc. Agric. Mach.* **2018**, *49*, 8.
17. Miao, S.; Yang, J.; Wang, X.; Li, J.; Li, T. Blade pattern optimization of the hydraulic turbine based on neural network and genetic algorithm. *J. Aerosp. Power* **2015**, *30*, 8.
18. Feng, Y. Multi-Objective Optimum Design of Centrifugal Double-channel Pump Based on RBF Neural Network and Particle Swarm Optimization Algorithm. Ph.D. Thesis, Jiangsu University, Zhenjiang, China, 2018.
19. Zhang, X.; Hu, B.; Feng, Y.; Liu, K.; Wang, C. Multi-objective optimization design of screw centrifugal pump based on RBF neural network and differential evolution algorithm. *J. Drain. Irrig. Mach. Eng.* **2022**, *40*, 7.
20. Wang, C.; Ye, J.; Zeng, C.; Xia, Y.; Luo, B. Multi-objective optimum design of high specific speed mixed-flow pump based on NSGA-genetic algorithm. *Trans. Chin. Soc. Agric. Eng.* **2015**, *31*, 7.
21. Wang, C.; Feng, Y.; Ye, J.; Luo, B.; Liu, K. Multi-objective parameters optimization of centrifugal slurry pump based on RBF neural network and NSGA- genetic algorithm. *Trans. Chin. Soc. Agric. Eng.* **2017**, *33*, 109–115.
22. Wang, C.; Hu, B.; Feng, Y.; Liu, K. Multi-objective optimization of double vane pump based on radial basis neural network and particle swarm. *Trans. Chin. Soc. Agric. Eng.* **2019**, *35*, 25–32.
23. He, D.; Li, R.; Sun, S.; Guo, P. Prediction of gas-liquid two-phase pressure increment of a centrifugal pump based on machine learning. *Trans. Chin. Soc. Agric. Eng.* **2022**, *38*, 9.
24. Yang, Q.; Guo, X.; Li, Q.; Dong, W. Hot air anti-icing performance estimation method based on POD and surrogate model. *Acta Aeronautica et Astronautica Sinica* **2022**, *44*, 126992.
25. Wang, W.; Yao, Y.; Ma, Z. Model of Compressor Performance Prediction Based on Error Back-propagation Artificial Neural work. *Fluid Mach.* **2005**, *33*, 4.
26. Huang, R.; Zhang, Z.; Zhang, W.; Mou, J.; Zhou, P.; Wang, Y. Energy performance prediction of the centrifugal pumps by using a hybrid neural network. *Energy* **2020**, *213*, 119005. [CrossRef]
27. Li, S.; Zhao, F.; Zheng, X.; He, D.; Bai, B. Wet gas metering by cone throttle device with machine learning. *Measurement* **2020**, *164*, 108080. [CrossRef]
28. Chen, J.; Chen, L.; Wang, Z. The Recognition of Hydraulic Pump Leakage State Based on Wavelet Decomposition and Deep Learning. *Modul. Mach. Tool Autom. Manuf. Tech.* **2021**, *4*, 006.
29. Panda, A.; Rapur, J.; Tiwari, R. Prediction of flow blockages and impending cavitation in centrifugal pumps using Support Vector Machine (SVM) algorithms based on vibration measurements. *Measurement* **2018**, *130*, 44–56. [CrossRef]
30. Bordoloi, D.J.; Tiwari, R. Identification of suction flow blockages and casing cavitations in centrifugal pumps by optimal support vector machine techniques. *J. Braz. Soc. Mech. Sci. Eng.* **2017**, *39*, 2957–2968. [CrossRef]
31. Rapur, S.; Tiwari, R. Severity Assessment and Classification of Blockage of Centrifugal Pumps in Frequency Domain of Vibration Data Using Support Vector Machine Algorithms. In Proceedings of the Vibrations in Rotating Machinery (VIRM-11), Manchester, UK, 13–15 September 2016.
32. Zhang, S.; Tian, S.; Zhang, X.; Li, H.; Xi, D.; He, W.; Zhang, A. Research Progress of Skew and Sweep Aerodynamic Technology for Turbomachinery Blades. *J. Propuls. Technol.* **2021**, *42*, 15.
33. Mao, M.; Song, Y.; Wang, Z. Research on Design of Swept and Bowed Rotor in a Transonic Axial Flow Compressor. *J. Power Eng.* **2008**, *28*, 6.
34. Wei, B.; Wu, K. Review on the Research of high performance Skew-Swept Blades and Its Influence on Inner Flow. *Fluid Mach.* **2001**, *29*, 31–34.
35. Li, Y.; Ouyang, H.; Du, Z. Optimized Design Based on Skewed and Swept Blade Technology. *J. Eng. Therm. Energy Power* **2007**, *22*, 605–609.
36. Menter, F.; Kuntz, M.; Langtry, R. Ten years of industrial experience with the SST turbulence model. *Heat Mass Transf.* **2003**, *4*, 625–632.

37. Florianr, M. Zonal two equation kw turbulence models for aerodynamic flows. In Proceedings of the 23rd Fluid Dynamics, Plasmadynamics and Lasers Conference, Orlando, FL, USA, 6–9 July 1993; p. 2906.
38. Lee, K.S.; Ziaul, H.; Han, S.E. A study on the y+ effects on turbulence model of unstructured grid for CFD analysis of wind turbine. *J. Korean Assoc. Spat. Struct.* **2015**, *15*, 75–84. [CrossRef]
39. Zhou, Z.H. *Machine Learning*; Springer Nature: Berlin/Heidelberg, Germany, 2021.
40. Vapnik, V.; Golowich, S.; Smola, A. Support Vector Method for Function Approximation, Regression Estimation and Signal Processing. *Adv. Neural Inf. Process. Syst.* **1996**, *9*, 281–287.
41. Rasmussen, C.E.; Williams, C.K.I. *Gaussian Processes for Machine Learning*; The MIT Press: Cambridge, MA, USA, 2006.
42. Hsu, K.Y.; Li, H.Y.; Psaltis, D. Holographic implementation of a fully connected neural network. *Proc. IEEE* **1990**, *78*, 1637–1645. [CrossRef]
43. Gourisaria, M.K.; Mishra, B.; Dehuri, S. A Hybrid Parallel Multi-Objective Genetic Algorithm: HybJacIsCone Model. *Int. J. Comput. Appl.* **2013**, *66*, 1–6.
44. Sun, L.; Pan, Q.; Zhang, D.; Zhao, R.; van Esch, B. Numerical study of the energy loss in the bulb tubular pump system focusing on the off-design conditions based on combined energy analysis methods. *Energy* **2022**, *258*, 124794. [CrossRef]
45. Ji, L.; Li, W.; Shi, W.; Chang, H.; Yang, Z. Energy characteristics of mixed-flow pump under different tip clearances based on entropy production analysis. *Energy* **2020**, *199*, 117447. [CrossRef]

machines

Article

Effects of Coolant and Working Temperature on the Cavitation in an Aeronautic Cooling Pump with High Rotation Speed

Ao Wu [1], Ruijie Zhao [1,*], Fei Wang [2], Desheng Zhang [1] and Xikun Wang [1]

1 Research Center of Fluid Machinery Engineering and Technology, Jiangsu University, Zhenjiang 212013, China
2 Aviation Key Laboratory of Science and Technology on Aero-Electromechanical System Integration, Nanjing 211106, China
* Correspondence: rjzhao@ujs.edu.cn

Abstract: The centrifugal pump with high rotation speed is the key component in the cooling system of an aircraft. Because of the high rotation speed, the impeller inlet is very prone to cavitation. Two impellers with different types of blades (cylindrical and splitter) are designed, and the numerical models of the pumps are built. The authenticity of the numerical models is validated with the corresponding experiments in terms of both the hydraulic and cavitation characteristics. Then, the effects of different coolants and working temperatures on the hydraulic and cavitation performances of the prototype models are studied based on the numerical simulations. The results show that the head and efficiency of the pump for conveying water are higher than those for conveying ethylene glycol (EG) aqueous solution and propylene glycol (PG) aqueous solution (EGaq and PGaq are defined to represent the EG aqueous solution and the PG aqueous solution, respectively). The hydraulic performance in the EGaq is slightly better than that in the PGaq. The cavitation performance of water is far less than that of the EGaq and PGaq under high working temperature. The volume of cavitation in EGaq is smaller than that in PGaq, and the volume of cavitation in the splitter blades is slightly smaller than that in the cylindrical blades. It is suggested that EGaq be used as the first option. The splitter blades can improve the cavitation performance somehow while the improvement by using the splitter blades is very limited at high rotation speeds, and the design of the short blades should be careful in order to obtain a smooth internal flow field.

Keywords: aviation liquid cooling pump; pump performance; cavitation; numerical simulation; organic coolant

Citation: Wu, A.; Zhao, R.; Wang, F.; Zhang, D.; Wang, X. Effects of Coolant and Working Temperature on the Cavitation in an Aeronautic Cooling Pump with High Rotation Speed. *Machines* **2022**, *10*, 904. https://doi.org/10.3390/machines10100904

Academic Editor: Davide Astolfi

Received: 14 September 2022
Accepted: 5 October 2022
Published: 7 October 2022

1. Introduction

As the power of onboard electronic devices is continuously increased, the liquid cooling pump with high performance and reliability is critical to maintain the working environment for such devices. A small pump diameter and high rotation speed are the main characteristics of the centrifugal pump used in aircraft cooling systems because the aerial environment limits the size and weight of the cooling pump, and a high rotation speed can provide sufficient pressure head to circuit coolant in the cooling system [1]. Such pumps are also termed as the low-specific-speed pump, in which the flow passages of the impeller are long and narrow, and the fluid velocity at pump inlet is high [2,3]. Therefore, the flow is prone to cavitation, which will change the pattern of internal flow and severely deteriorate the pump performance.

Cavitation in pumps has been intensively studied during the last several decades. Rayleigh [4] took spherical symmetric cavitation as research object and put forward the famous Rayleigh equation. Plesset et al. [5,6] considered that the effects of gas, fluid viscosity and surface tension contained in the cavitation improved the cavitation dynamics theory and formulated the Rayleigh-Plesset equation. Brennen et al. [7] compared and discussed the dynamic transfer functions of two cavitating inducers with the same geometry but different dimensions from the experiments and further confirmed the validation of the theoretical model. Katz [8] studied the phenomenon of cavitation on four axisymmetric bodies whose boundary layers underwent a laminar separation and subsequent turbulent reattachment. The conditions for cavitation inception and desinence were determined by holography and the Schlieren flow visualization technique, and several holograms were recorded just prior to and at the onset of cavitation. Tanaka et al. [9] experimentally investigated the transient behavior of a cavitating centrifugal pump and found that the oscillating cavitation during the pump start-up and the separation of water column during the sudden pump shutdown are the main reasons for the fluctuations in the pump pressure and flow rate. Huang [10] evaluated Four cavitation models, including the Kubota model, Singhal model, Merkle model and Kunz model in the present study, which provides a theoretical and technical basis for the research of cavitation numerical simulation. Cheng et al. [11] summarized the research progress of cavitation, including cavitation characteristics, numerical methods, and the impact of cavitation on the flow field and proposed some frontier topics, which are of great significance for promoting cavitation research.

Computational Fluid Dynamics (CFD) technique has become popular during the last two decades for research on fluid problems in pumps. Besides accurate prediction of pressure and fluid velocity in a pump, turbulence should also be precisely predicted since occurrence and evolution of cavitation are strongly correlated with the local turbulence [12]. Much effort has been devoted to choosing a proper turbulence model for modeling the flow in pump [13–15]. Sun et al. [16] used the CFD method to simulate the internal flow in a pump used as turbine under different working conditions, and the comparison between the simulation result and the experiment demonstrated that the Detached Eddy Simulation (DES) model performed better than all models in the family of two-equation models. Feng et al. [17] performed both the steady-state and transient simulations of a centrifugal pump at design and off-design conditions by employing different turbulence models, and the results indicated that the choice of turbulence model has little effect on the prediction of pump head and energy efficiency. The standard k-ε model can obtain better results for the fluid velocity while the shear stress transport (SST) k-ω model was superior in the prediction of turbulent parameters. Zhou et al. [18] simulated a double-suction centrifugal pump with a specific-speed of 120 using both the S-A DES (one-equation DES) model and the SST k-ω model. The results show that both models can predict accurately the required power and pressure pulsation at different flow rates, and the performance of the former one is slightly better. Zhang et al. [19] employed the Delayed Detached Eddy Simulation (DDES) model to simulate the stalling effect in a centrifugal pump with low speed, and the simulation results reproduced well the transient flow behaviors and structures, especially for the jet wake at the outlet of the impeller blade.

The numerical simulation of cavitation in low-specific-speed pumps has been also received much attention by researchers. Wang et al. [20] modified turbulence models, such as the Re-Normalization Group (RNG) k-ε, SST k-ω, and Filter-based model (FBM) through the density correction, and used them for the prediction of cavitation performance of a low-specific-speed centrifugal pump. The results predicted by the modified turbulence models were in better agreement with the experiments. Phillip et al. [21] used the commercial CFD software CFX to simulate the cavitating flow in a low-specific-speed centrifugal pump under different working conditions and surface roughness. The simulation result was in good agreement with the corresponding experiment in the non-cavitating state. In addition, the inception of cavitation at the diaphragm tongue can be predicted more accurately by using the low-Reynolds-number methods. Gao et al. [22] analyzed the relationships between cavitation and vibration characteristics of a low-specific-speed centrifugal pump based on the experimental and simulation results, specifically focusing on the relationships between 10~500Hz low-frequency vibrations and cavitation evolution.

The effect of the impeller's structure on the behavior of cavitation in low-specific-speed pumps was also studied in some works. It was found that the cavitation can be effectively suppressed when the traditional impeller blade is replaced by a combination of long and short blades, or splitter blades [23]. Zhang et al. [19] numerically studied the effect of the curvature of short blades on the pressure fluctuation in a low-specific-speed centrifugal pump under different flow rates. It was found that if the outlet angle of the short blades was set as 12° to the pressure side of the blades, the streamlines in the flow channels were more in line with the blade shape, and both the overall pressure fluctuation at the pump outlet and the turbulence intensity were decreased. Hu et al. [24] carried out numerical simulation and hydraulic test for a low-specific-speed centrifugal pump under three attacking angles to study the effect of the inlet attacking angle of blade on the development of cavitation in the impeller, as well as the evolution of cavitating flow in the flow channels. Results show that as the cavitation number decreases, the cavity first occurs at the suction side of the leading edge of the blade and then expands to the outlet of impeller rapidly along the blade. The load of the blade near the tongue is heavier than that of other blades whether cavitation occurs or not.

From the above literature review, it is clear that the two-equation turbulence models are usually employed in modeling cavitation in pumps, probably because their predictions of the inception and evolution of cavitation are of reasonable accuracy. The more advanced turbulence models, such as DES, DDES, and LES, can predict more details of the flow structure in pumps while the required computational resources are also significantly increased. As for the cavitation in the aeronautic cooling pump, the material properties of the coolant are very different from those of pure water, and they are also more sensitive to the temperature [25,26]. The effects of cooling media and working temperature on the cavitation behavior in such pumps have yet to be revealed.

In this work, a pump model with two types of impeller's blades is firstly designed for an aircraft cooling system. The numerical model of the pump based on CFD is built and introduced in Section 2. The simulation procedures are validated with the corresponding experiments in terms of both the hydraulic and cavitation performances in Section 3. The effects of different coolants and working temperatures on the pattern of hydraulic performance and cavitation in the pumps are studied based on the simulation results in Section 4. The conclusions are summarized in the end.

2. Numerical Model

2.1. Physical Model

In this paper, a pump model with two types of impeller blades is designed based on a set of the design parameters of an aeronautic cooling pump. The two impellers are of cylindrical and splitter blades. To design a low-specific-speed pump, Yuan et al. [27] proposed enlarging the design flow rate and the specific-speed in order to improve the pump efficiency at the design point. The enlarged design flow rate and specific-speed are determined by Equations (1) and (2):

$$Q' = K_1^2 Q \tag{1}$$

$$n_s' = K_2 n_s \tag{2}$$

where Q and Q' denote the original design and modified design flow rates, respectively. n_s and n_s' indicated the original and modified design specific-speed, respectively. $K_1 = 1.2$, $K_2 = 1.17$ are the magnification index.

The design parameters of the prototype pump studied in this paper are presented in Table 1. The original dynamic specific-speed of the pump n_s' is calculated as:

$$n_s = \frac{3.65 n \sqrt{Q}}{H^{3/4}} \tag{3}$$

where H is the pressure head in the unit of m and n the rotation speed in the unit of r/min. According to Equations (1) and (2), the enlarged design flow rate is $Q_d' = 15.84\ \frac{\text{m}^3}{\text{h}}$, and the enlarged specific-speed is $n_s' = 73.04$. Then, a new design pressure head is calculated according to Equation (3). The modified design parameters are also shown in Table 1. The main parameters of the model pump are shown in Table 2. Two blade shapes are designed based on the modified design parameters, and the water bodies of the corresponding CFD model are shown in Figure 1a–c. It is noted that an extra pipe with a length of 8D is modeled in front of the pump inlet, and the outlet pipe is extended to 10D to ensure that the turbulent flow is fully developed.

Table 1. Design parameters of the aeronautic cooling pump.

	Flow Rate Q_d (m^3/h)	Pump Head H (m)	Speed n (r/min)
Original design parameters	13.2	132	11,000
Modified design parameters	15.84	120.91	11,000

Table 2. The model pump main parameters.

Parameters	Value
Rated speed n_d	11,000 r/min
Specific speed n_s	62
Impeller inlet diameter D_1	32 mm
Impeller outlet diameter D_2	94 mm
Impeller Outlet Width b_2	4 mm
Volute outlet diameter D_4	20 mm

(**a**) (**b**) (**c**)

Figure 1. Water bodies of the two impellers and volute of the CFD pump model: (**a**) cylindrical blade impeller; (**b**) splitter blade impeller; (**c**) pump model.

2.2. Governing Equations

The vapor volume fraction transport equation is added in the homogeneous multiphase flow model to simulate the formation and evolution of cavitation. The conservation equations of continuity and momentum of a fluid can be written in the following formula:

$$\frac{\partial \rho_m}{\partial t} + \frac{\partial (\rho_m u_i)}{\partial x_i} = 0 \quad (4)$$

$$\frac{\partial(\rho_m u_i)}{\partial t}+\frac{\partial(\rho_m u_i u_j)}{\partial x_j}=-\frac{\partial p}{\partial x_i}+\frac{\partial}{\partial x_i}\left[(\mu_m+\mu_t)\left(\frac{\partial u_i}{\partial x_j}+\frac{\partial u_j}{\partial x_i}-\frac{2}{3}\frac{\partial u_k}{\partial x_k}\delta_{ij}\right)\right] \tag{5}$$

where ρ_m and μ_m represent mixing density and hybrid viscosity, respectively, and they are expressed as

$$\rho_m=\rho_l\alpha_l+\rho_v\alpha_v \tag{6}$$

$$\mu_m=\mu_l\alpha_l+\mu_v\alpha_v \tag{7}$$

The subscripts l, v, and m represent liquid phase, gas phase, and mixed phase, respectively; p, u, and μ_t stand for pressure, velocity, and turbulent viscosity, respectively; i, j, and k, respectively, represent the three directions of the coordinate system, without considering the gravitational action term. Due to the little effect of the gravity in the pump of high pressure head, the gravitational force is usually ignored in the simulation [28,29].

The SST k-ω turbulence model is used in the numerical model [30,31]. It is a combination of the k-ω and k-ε models. It takes full advantage of the superior performance of the k-ω model in predicting the near-wall region and the k-ε model in predicting the far-field free flow. The performance of the inverse pressure gradient boundary layer model is significantly improved. The SST k-ω turbulence model is derived from the following formula:

$$\frac{\partial\rho k}{\partial t}+\frac{\partial}{\partial x_j}[\rho(u_j-V_j)k]=\widetilde{P}_k-D_k+\frac{\partial}{\partial x_j}\left[(\mu+\sigma_k\mu_t)\frac{\partial k}{\partial x_j}\right] \tag{8}$$

$$\frac{\partial\rho\omega}{\partial t}+\frac{\partial}{\partial x_j}[\rho(u_j-V_j)\omega]=P_\omega-D_\omega+\frac{\partial}{\partial x_j}\left[(\mu+\sigma_\omega\mu_t)\frac{\partial\omega}{\partial x_j}\right]+CD_{k\omega}(1-F_1) \tag{9}$$

$$CD_{k\omega}=2\rho\sigma_{\omega2}\frac{1}{\omega}\frac{\partial k}{\partial x_j}\frac{\partial\omega}{\partial x_j} \tag{10}$$

Turbulent viscosity is defined as follows:

$$\mu_t=\frac{\rho k a_1}{max(a_1\omega,|S_{ij}|F_2)} \tag{11}$$

where μ_t is the turbulent viscosity, and $|S_{ij}|=\sqrt{2S_{ij}S_{ij}}$ is the scalar of the strain rate tensor S_{ij}. F_1 and F_2 are mixing functions that ensure proper selection of the k-ω and k-ε regions [32]. P_k and P_ω are production terms; D_k and D_ω are dissipation terms. Their expressions are referred to in [31] for the sake of brevity.

The cavitation model based on the vapor volume fraction transport equation is used to solve the mass transfer of the vapor–liquid phase change caused by cavitation. The governing equation is expressed as follows:

$$\frac{\partial(\rho_v\alpha_v)}{\partial t}+\nabla(\rho_v\alpha_v u)=\dot{m}_{vap}-\dot{m}_{con} \tag{12}$$

where $\dot{m}_{vap}$ and $\dot{m}_{con}$ are the mass transfer rates of vaporization and condensation, proposed by Zwart et al. [33], and the formulas are as follows:

$$\dot{m}_{vap}=F_{vap}\frac{3\alpha_{nuc}(1-\alpha_v)\rho_v}{R_b}\sqrt{\frac{2}{3}\frac{\max(p_v-p,0)}{\rho_l}} \tag{13}$$

$$\dot{m}_{con}=F_{con}\frac{3\alpha\rho_v}{R_b}\sqrt{\frac{2}{3}\frac{\max(p-p_v)}{\rho_l}} \tag{14}$$

where F_{vap} and F_{con} are the empirical coefficients of vaporization and condensation rates [34], P_v is the vaporization pressure, R_b is the bubble radius, and α_{nuc} is the volume fraction of nucleation point. The reference values of the above parameters are $R_b = 1 \times 10^{-6}\ m$, $\alpha_{nuc} = 5 \times 10^{-4}$, $F_{vap} = 50$, and $F_{con} = 0.01$ by default. Through the above equations, the fluid modeling can be defined in the CFD software, and the simulation of the fluid transport in the pump model can be carried out by setting the boundary conditions and initial conditions.

2.3. Boundary Conditions and Settings

For the fluid boundary conditions, the total pressure of 1.0 atm is selected at the inlet, and the mass flow rate determined by the simulated flow rate is specified at the exit. During the cavitation modeling, the inlet pressure is reduced successively to achieve different cavitating conditions. In addition, as a rotating part, the domain of the impeller is set to the rated speed, and the blades and front and rear cover plates are set as rotating walls. The no-slip condition is defined on all walls. When setting the interfaces between different components, the interfaces between the impeller and the inlet section and between the impeller and the volute are set as the dynamic-static interface while the other interfaces are set as the static interface. Table 3 presents the material properties of three cooling media at 20 and 60 °C used in the simulations [35], wherein the ethylene glycol (EG) aqueous solution and propylene glycol (PG) aqueous solution are prepared with the organic solvent of 60% in volume fraction. EGaq and PGaq are defined to represent the EG aqueous solution and the PG aqueous solution, respectively.

Table 3. Material properties of three cooling medium at different temperatures.

	Temperature/*t* (°C)	**Saturation Pressure/P(bar)**	**Viscosity/$mPa*s$**	**Density/$\rho(kg/m^3)$**	**Specific Heat Capacity/ $c(J*kg^{-1}\ K^{-1})$**
EGaq	20	1582	5.38	1086.27	3084
	60	13,600	1.69	1063.69	3258
PGaq	20	1760	10.04	1048.25	3339
	60	15,435	2.22	1020.66	3515
Water	20	2330	1.005	998.2	4183
	60	1582	5.38	1086.27	3084

The governing equations are discretized by the finite volume method in space. In the steady-state calculation, the iteration steps are set as 1000. The second order discretization schemes are applied for the advection terms and the turbulent parameters. In the unsteady numerical simulation, the time step is defined as the impeller rotates the 2° in each time step. The transient term adopts the second-order backward Euler scheme. The basic settings of cavitation and non-cavitation conditions are the same except for the cavitation model. Firstly, the steady-state numerical simulation is carried out under the condition of non-cavitation. After the calculation converges, the cavitation model is activated. A steady-state simulation of 1000 steps takes 9 h with 8 cores of 2.5 GHz while a transient simulation of 0.054 s (10 rotating cycles) takes 51 h with 16 cores of 2.5GHz. For transient simulation, the number of iterations in each time step is set as 10.

The mesh independence of the simulation result is studied. Five sets of grids with different numbers of control volumes are created for each type of blade, and the dimensionless coefficients of pressure head as defined in Equation (15) are calculated

$$\varphi = \frac{H}{u_2^2/\ 2g} \tag{15}$$

where $u_2 = \pi D_2 n/60\ m/s$ is the circumferential speed, D_2 is the impeller outside diameter, H is the pressure head at the outlet and g the gravitational acceleration. The meshes in the flow channels of the CFD pump model are shown in Figure 2. Hybrid meshes are employed in the mesh of the pump model. Structured meshes are created in the components of impellers while unstructured meshes are formed in the components of volute. Several boundary layer meshes are defined on all wall surfaces. The thickness of the first boundary layer is 0.05 mm, and 10 boundary layers are grown with a growth rate of 1.1. y+ values are larger than 30 and less than 100 on the major wall surfaces. Figure 3 presents the results, and it is found that the pressure head coefficient reaches a saturated value as the fourth mesh is used for both the cylindrical blade impeller and the splitter blade impeller. Therefore, the fourth meshes are chosen for the following simulations.

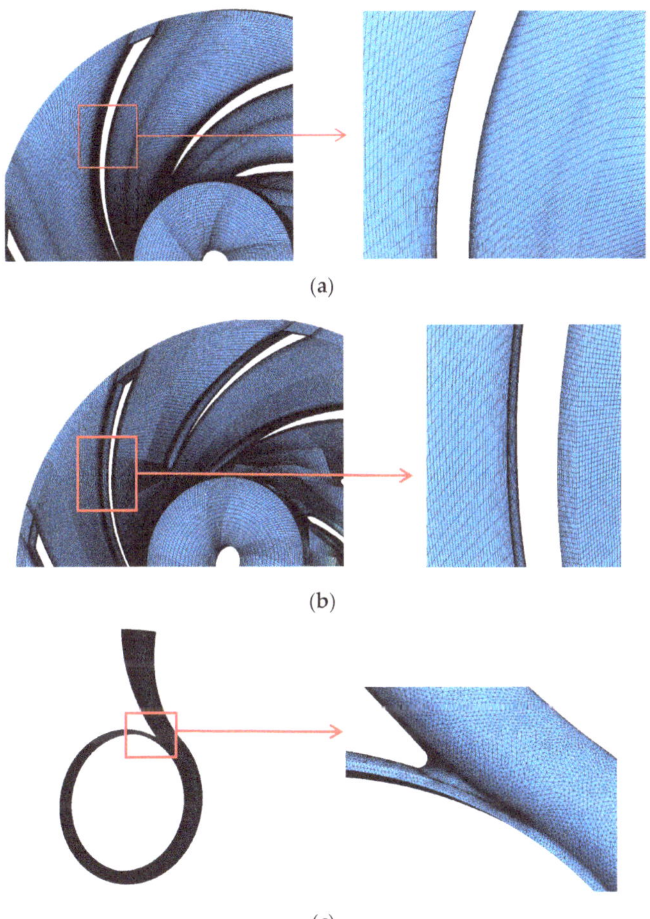

(a)

(b)

(c)

Figure 2. Meshes in the flow channels of the CFD pump model: (**a**) cylindrical blades impeller; (**b**) splitter blades; (**c**) volute.

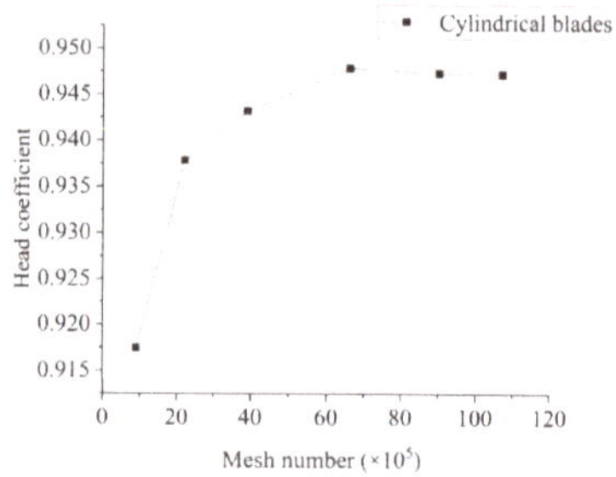

(a)

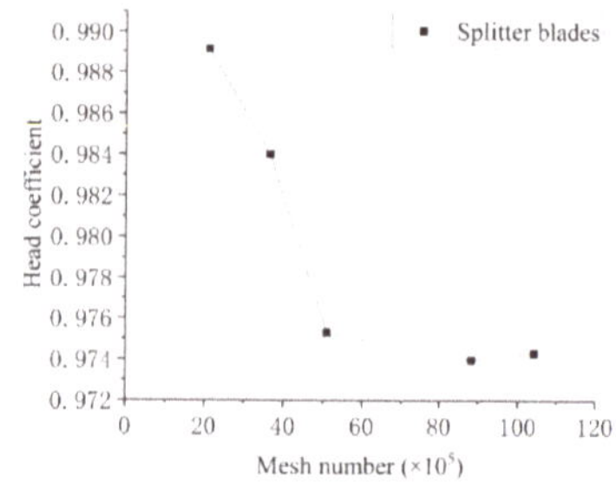

(b)

Figure 3. Verification of mesh independence: (**a**) cylindrical blades; (**b**) splitter blades.

3. Experimental Validation

3.1. Experimental Facilities

The authenticity of the numerical model is validated with the experiment performed in a tested pump model. The pump model is made of plexiglass for high-speed photography to study the cavitation characteristics in the pump. Due to the high rotation speed, the experiments are difficult to perform in the prototype pump. Therefore, a speed-reduced model based on the law of similarity of pump is created with the same specific-speed of the prototype pump. The conversions of the design parameters based on the law of similarity are expressed as:

$$\frac{Q_m}{Q} = \left(\frac{n_m}{n}\right)\left(\frac{D_{2m}}{D_2}\right)^3 \tag{16}$$

$$\frac{H_m}{H} = \left(\frac{n_m}{n}\right)^2\left(\frac{D_{2m}}{D_2}\right)^2 \tag{17}$$

$$\frac{P_m}{P} = \left(\frac{n_m}{n}\right)^3\left(\frac{D_{2m}}{D_2}\right)^5 \tag{18}$$

In the experiments, the impellers of both cylindrical and splitter blades and the associated volute are made of plexiglass for visualizing the cavitating flow in the pump model. Figure 4a,b show the impeller models of cylindrical and splitter blades, respectively. Figure 4c shows the on-site pump model. In order to validate the accuracy of the numerical procedures, the physical pump model was tested at 1450 r/min, and the hydraulic and cavitation performances were obtained from the experiments.

(a)

(b)

Figure 4. *Cont.*

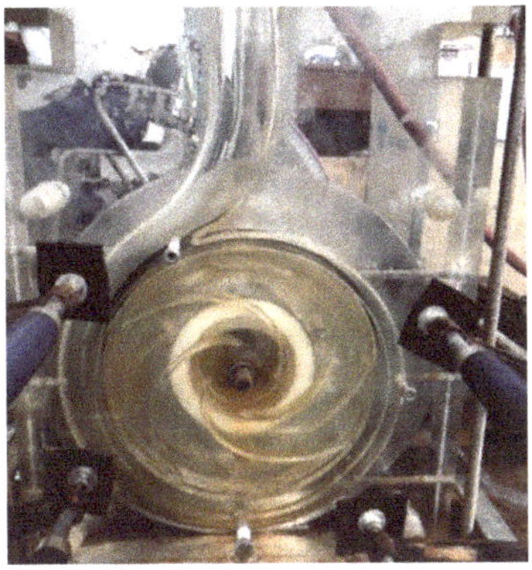

(**c**)

Figure 4. (**a**) Impeller mode of cylindrical blades; (**b**) impeller model of splitter blades; (**c**) on-site physical pump model.

A testing loop is built to measure the hydraulic and cavitation performances of the tested pump model and to photograph the cavitation in the pump model. As shown in Figure 5, the loop is mainly composed of a closed water tank with a vacuum pump (for cavitation test), inlet and outlet valves, the pump model, inlet and outlet pipelines, vortex flowmeter, inlet and outlet pressure transmitter, and electrical control cabinet. A high-speed camera as shown in Figure 6 is used to visualize the cavitation inside the pump. The adopted high-speed camera model is i-speed3 high-speed camera with a resolution of 1028 × 1024 at a maximum of 2000 frames per second.

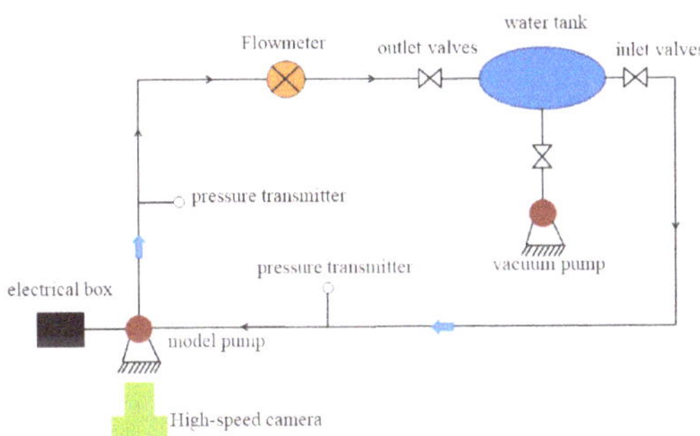

Figure 5. Schematic diagram of the testing loop.

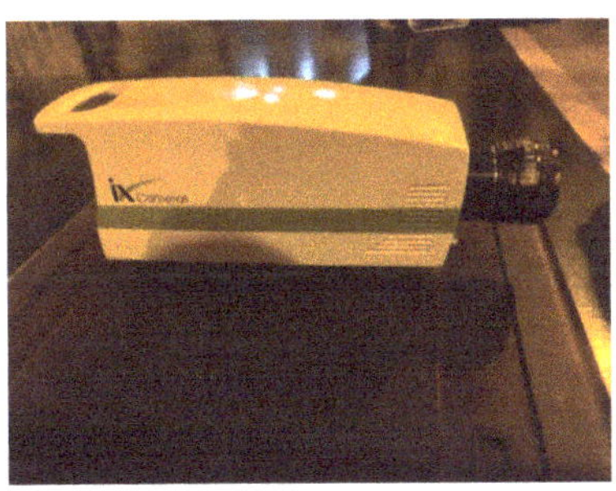

Figure 6. High-speed camera.

3.2. Validation of Hydraulic and Cavitation Performances

The measured parameters were normalized before presenting in the figures. The dimensionless coefficient of the pressure head is calculated in Equation (15). The dimensionless coefficients of flow rate and cavitation number are expressed as:

$$\psi = \frac{Q}{\pi D_2 b_2 u_2} \tag{19}$$

$$\sigma = \frac{P_{in} - p_v}{1/2\rho u_2^2} \tag{20}$$

where P_{in} is the pressure at inlet; P_v is the saturated pressure of the fluid, and u_2 is the circumferential speed of impellers. b_2 is the outlet width of the impeller.

In order to compare with the experiment, the numerical model of the tested pump model was built based on the converted design parameters. Both the cylindrical and splitter blades were simulated in the numerical model. The procedure of the numerical simulation is similar to that of creating the numerical model for the prototype except for defining the different geometries and boundary conditions. The information of the numerical procedure will not be repeated here for the sake of brevity.

By taking the simulation result of single-phase flow as the initial value, the cavitating flow in the tested pump was numerically simulated. Figure 7 shows the cavitation curves obtained from both the experiment and simulation. It can be seen that the characteristics of the cavitation curves between the simulation and experiment are in good agreement. The pressure head coefficients remain basically constant at first and then decrease with the decreasing cavitation number. When the cavitation number is large, the pressure heads of the two pumps are almost equal, indicating that the impeller of splitter blades can obtain the same pressure head as the cylindrical blades. However, when the cavitation number is reduced to the cavitating state, the decreasing rate of the pressure head coefficient of the impeller of splitter blades is smaller than that of the impeller of cylindrical blades. This phenomenon appears in both the simulation and experimental results, which demonstrates that the impeller of splitter blades possesses better cavitation performance. It is noted that the predicted pressure heads in the simulation are larger than those in the experiment. This discrepancy is attributed to the simplifications made in the geometry of the numerical model. When severe cavitation occurs under small cavitation numbers, the pressure head coefficients of the simulation are decreased significantly more than those in the experiment. From the photographs of the cavitating flow in the tested pump presented in Section 3.3, it is found that the numerical model over-predicts the cavitation zone and the resulting drop in the pressure head.

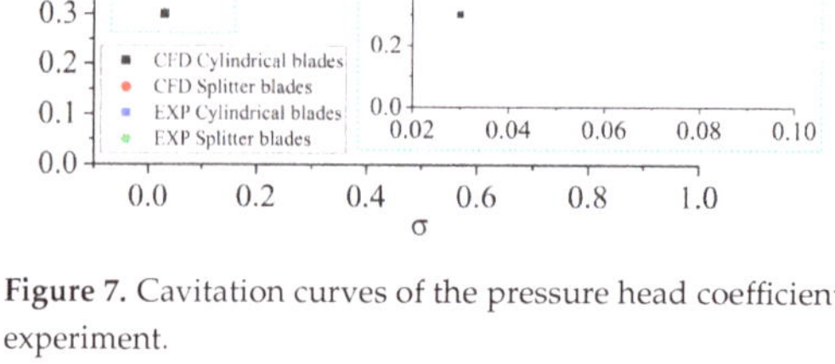

Figure 7. Cavitation curves of the pressure head coefficients of two impellers from simulation and experiment.

3.3. Validation of Cavitation Visualization

The cavitation zones in the impellers with two types of blades are shown in Figures 8 and 9 under the inlet pressures of 0.09 atm (σ = 0.058) and 0.07 atm (σ = 0.04), where different degrees of cavitating flows are observed in the photographs. The cavitation zone is represented by the iso-surface of vapor volume fraction α_v = 0.5 in the simulation results. It is seen that cavitation occurs slightly in the flow channels of both impellers at σ = 0.058. The original area for cavitation in both impellers is located at the suction side of blades near the inlet. Compared with the cylindrical blades, the cavitations in the splitter blades are distributed more uniformly and the cavitation intensity is smaller in each flow passage.

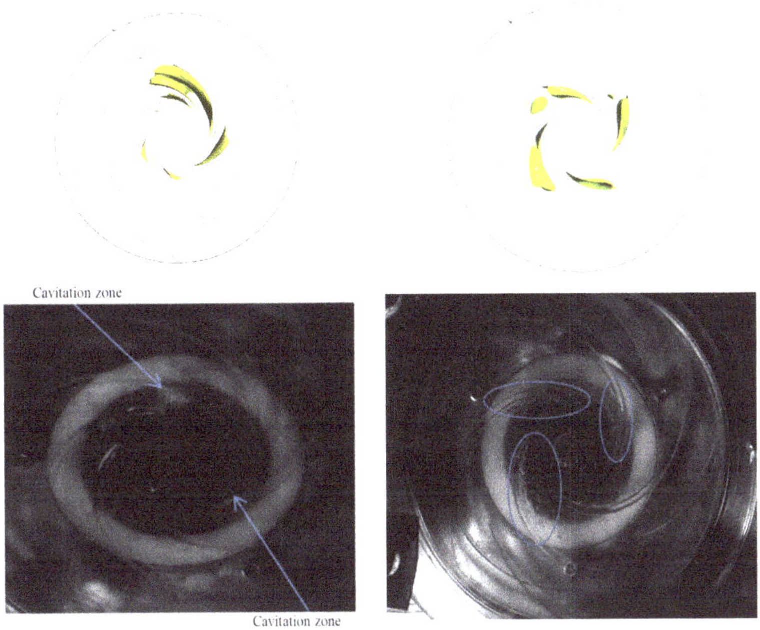

(a) (b)

Figure 8. Cavitation zones in both cylindrical and splitter blades at σ = 0.058. (**a**) Cylindrical blades; (**b**) Splitter blades.

When σ = 0.04, severe cavitation has occurred in both impellers, and the pressure head has also decreased greatly. However, in the splitter blades, the cavitation degree is significantly less than that in the cylindrical blades as shown in both the simulations and photographs, and the same feature is also applied to the pressure head coefficient. The cavitation performance of the impeller of splitter blades is better than that of the cylindrical blades. It is also noted that the depicted cavitation zones in the simulations are smooth and distinct from the water body while the cavitation bulbs in the high-speed photographs are fragmentized and mixed with the water body. This is attributed to the employed turbulence model in which the Reynolds-averaged velocity field is solved and the detailed turbulence field is missed. However, the simulation can predict the hydraulic and cavitation performances of the pump model with reasonable accuracy. Based on the above validations, the same procedure of the numerical simulation will be applied in the following simulations for the prototype.

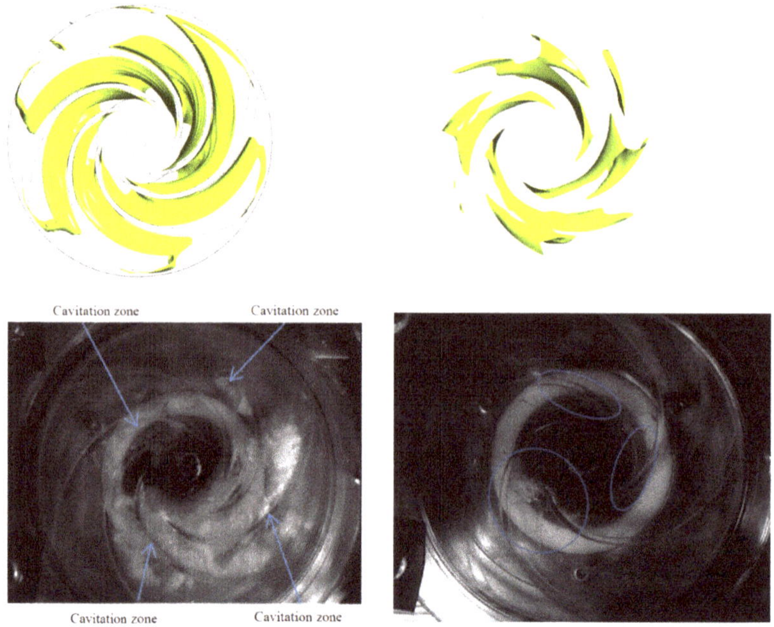

(a) (b)

Figure 9. Cavitation zones in both cylindrical and splitter blades at $\sigma = 0.04$. (**a**) Cylindrical blades; (**b**) Splitter blades.

4. Results and Discussion

In the preceding section, the authenticity of the numerical model was validated with the experiment in terms of both the hydraulic and cavitation performances. In this section, the numerical model of the prototype pump is employed to study its performance at high rotation speed. Based on the cylindrical and splitter impellers, the effects of coolant and working temperature on the characteristics of the hydraulic performance and cavitation are investigated in the pump of high rotation speed.

4.1. Characteristics of Pump Hydraulic Performance

The pump hydraulic performance was numerically simulated for conveying water, EGaq, and PGaq at different flow rates. The pressure head and pump efficiency of the pumps with cylindrical and splitter blades are shown in Figures 10 and 11, respectively. It is shown that the pressure head and pump efficiency for conveying water are higher than those for conveying EGaq and PGaq because the water viscosity is much smaller, resulting in less energy dissipation. A comparison between the two solutions shows that the performance of EGaq is slightly better than that of PGaq. The pressure head for conveying water reaches 132 m, and its efficiency is 62.9% at the nominal flow rate, which agrees well with the experimental date tested for the prototype. The relatively low efficiency is mainly caused by the energy dissipation of fluid friction occurring in the gap between the impeller of high rotation speed and the static back case. Due to the larger viscosity of the organic solutions, the pressure heads are only in 120~125 m, and the pump efficiencies are in 53~57% for conveying EGaq and PGaq. The point of the maximum pump efficiency agrees well with the nominal flow rate, indicating a good design for the hydraulic performance. It is also found the two organic solutions have a wider range of high efficiency, compared with that possessed by water. The type of impeller blade has little influence on pump performance, since the characteristics of the performance curves are very similar between the two impellers. Compared to the impeller of cylindrical blades, the impeller of splitter blades shows a slightly better performance for conveying EGaq than PGaq in terms of the predicted pressure head and efficiency.

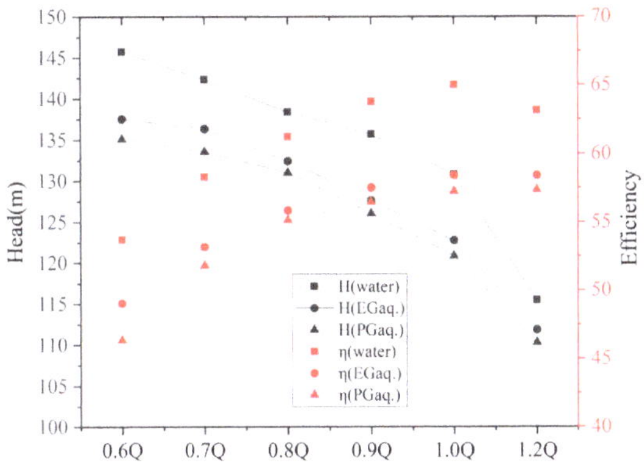

Figure 10. Characteristic curves of the pump with cylindrical blades.

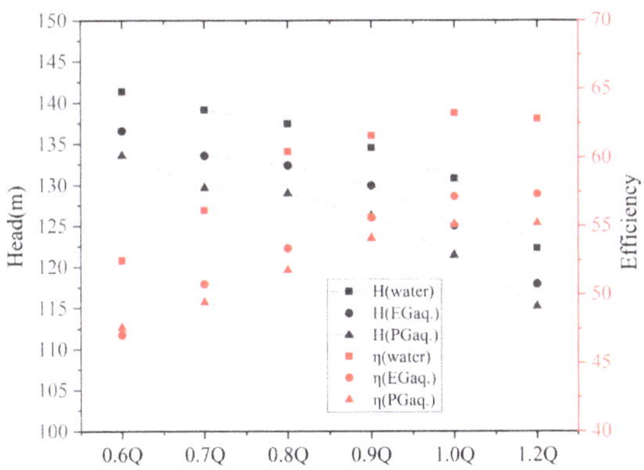

Figure 11. Characteristic curves of the pump with splitter blades.

4.2. Characteristics of Pump Internal Flow

The characteristics of pump internal flow for conveying water, EGaq, and PGaq at three typical flow rates are investigated. The streamlines and contours of fluid velocity in the impellers of cylindrical and splitter blades are shown in Figures 12 and 13. In general, fewer vortices and reserve flow are observed in the flow channels at the low, nominal, and high flow rates, indicating the designed impellers have good control for the internal flow. Specifically, the streamlines in the impeller of cylindrical blades for conveying EGaq and PGaq are smoother than those in water at $0.7Q_d$ while the streamlines at $1.0Q_d$ and $1.2Q_d$ are very similar among three media. The streamlines in the impeller of splitter blades are less smooth than those in the impeller of cylindrical blades. Small vortex occurs in water near the outlet of the impeller at $0.7Q_d$. EGaq and PGaq show better internal flow fields than water, and the internal flow in the impeller of splitter blades could be further improved by optimizing the shape and position of the short blades.

Figure 12. The velocity of the middle plane of the cylindrical blade impeller at different flow rates. (**a**) $0.7Q_d$; (**b**) $0.7Q_d$; (**c**) $0.7Q_d$.

Figure 13. The velocity of the middle plane of the splitter blade impeller at different flow rates. (**a**) $0.7Q_d$; (**b**) $0.7Q_d$; (**c**) $0.7Q_d$.

4.3. Characteristics of Pump Cavitation

The characteristics of pump cavitation for conveying water, EGaq, and PGaq are numerically investigated at different working temperatures, i.e., 20 and 60 °C. The correlations of cavitation number and inlet pressure for the impellers of cylindrical and splitter blades are presented in Figures 14–17. At 20 °C, the pressure head of the pump with cylindrical impeller blades as shown in Figure 14 can be retained until the inlet pressure is decreased to 0.5 atm. The pressure heads for conveying water, EGaq and PGaq are slightly decreased at 0.4 atm, indicating an inception of cavitation in the pump. The pressure heads are significantly decreased by 8.8%, 8.1%, and 9.5% at 0.3 atm, indicating severe cavitations in the pump. When the working temperature is at 60 °C, the pressure head as shown in Figure 15 can be retained until the inlet pressure is decreased to 0.6 atm. The pressure heads begin dropping at 0.5 atm, and they are decreased by 13.6%, 10.9%, and 11.8% at 0.4 atm, respectively, for conveying water, EGaq, and PGaq. The pressure head of the pump with splitter blades as shown in Figure 16 can be retained also at 0.5 atm, which is coincident with the pump with cylindrical blades at 20 °C. The pressure heads for conveying water, EGaq, and PGaq are decreased by 7.2%, 7.2%, and 7.3% at 0.3 atm, respectively. At 60 °C, the pressure heads begin dropping at 0.5 atm, and they are significantly decreased by 16.5%, 9.5%, and 10.1% at 0.4 atm, respectively.

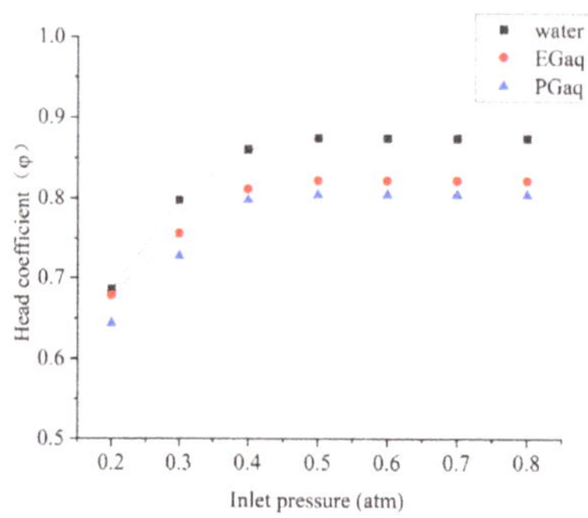

Figure 14. Cavitation curve of the pump with cylindrical blades at 20 °C.

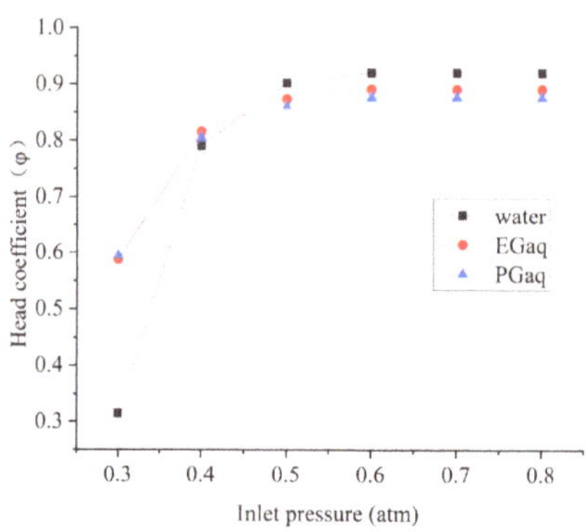

Figure 15. Cavitation curve of the pump with cylindrical blades at 60 °C.

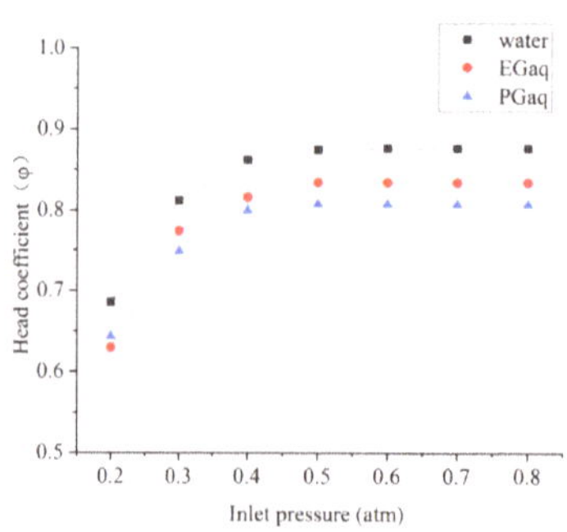

Figure 16. Cavitation curve of the pump with splitter blades at 20 °C.

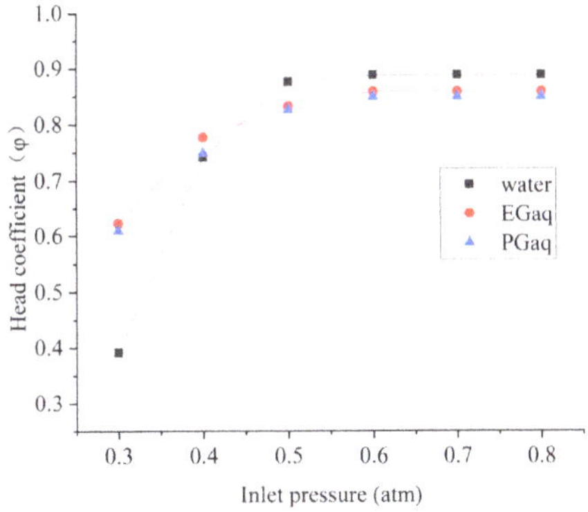

Figure 17. Cavitation curve of the pump with splitter blades at 60 °C.

It is found that the inception of cavitation occurs almost at the same inlet pressure between two types of impeller blades. The drops of pressure head in the pump with splitter blades are slightly smaller than those in the pump with cylindrical blades, indicating that the former impeller can improve the cavitation performance of the pump. The higher the working temperature is, the larger the critical inlet pressure for occurring cavitation in the pump is. At 60 °C, cavitation occurs at the pump inlet pressure below 0.6 atm, which can be easily encountered in aeronautic environment. Among different cooling media, the descending rate of the pressure head in water when occurring cavitation is much faster than that of the other two organic coolants, especially for the high working temperature. Compared between the two organic coolants, the pressure head in EGaq drops more slowly than that in PGaq under all simulated conditions, which demonstrates that the cavitation performance of EGaq is better than that of PGaq.

According to the above analyses, the cavitation zones in the different pump impellers are represented by depicting the iso-surface of the vapor volume fraction of 0.5 as shown in Figures 18 and 19. The cavitation zones are shown at 20 °C with the inlet pressure of 0.3 atm and 60 °C with the inlet pressure of 0.4 atm. In Figure 18, it is shown that the volumes of cavitation zones in the three coolants are very close at the low working temperature. They occur firstly at the suction side of each blade. However, the volume of the cavitation zone in water is significantly increased at the high working temperature while the increase is very limited in EGaq and PGaq. EGaq shows the smallest inflation in the cavitation volume, which agrees well with the results obtained from the analyses on the cavitation curves. The result indicates that the volume of cavitation is related not only to the saturated pressure but also to the viscosity of cooling medium. As the temperature rises, the influence of saturated pressure on cavitation becomes more significant. In the impeller of splitter blades, the cavitation only occurs at the suction side of the long blades, which are extend to the pump inlet. The length of the cavitation zone is much longer than the counterparts in the cylindrical blades. The correlations between the working temperature and the cavitation zone among different coolants are very similar with those in the cylindrical blades. EGaq shows the smallest inflation in the cavitation volume with the increasing temperature.

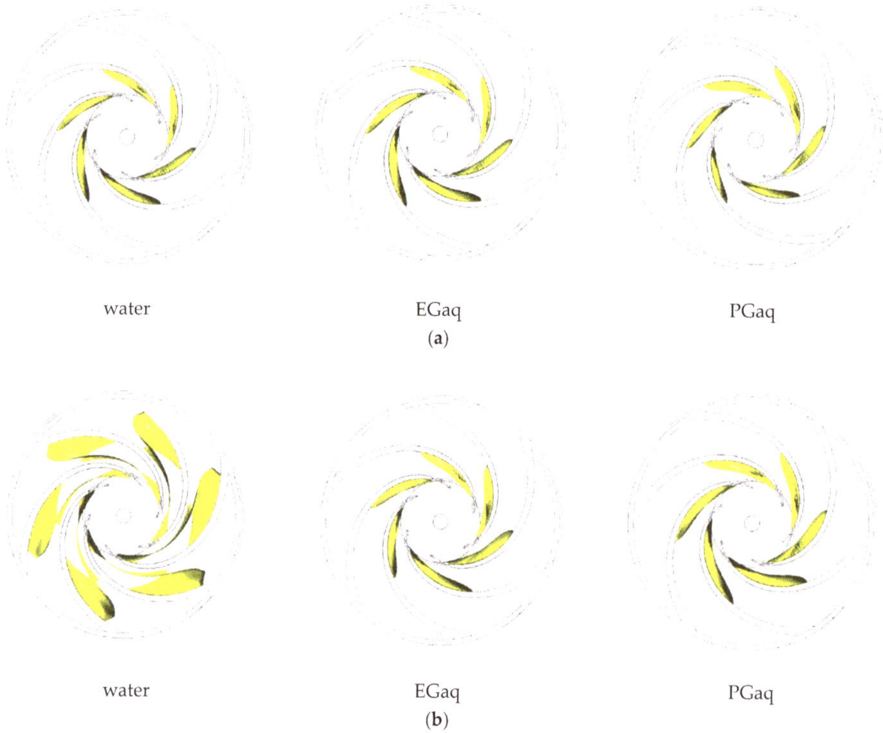

Figure 18. Iso-surface of cavitation distribution in the impeller of cylindrical blades. (**a**) 20 °C at 0.3 atm; (**b**) 60 °C at 0.3 atm.

In order to quantitatively estimate the cavitation performances in different conditions, the volumes of the cavitation zones are statistically presented in Figures 20 and 21. It is found that the cavitation performance of water is far less than those of EGaq and PGaq under high working temperature while the cavitation volume of EG is slightly smaller than that of PG. The increasing temperature has less effect on EGaq and PGaq. Between two types of impellers, the cavitation volume in the impeller of splitter blades is slightly smaller than that in the cylindrical blades (only one exception). It is found that the splitter blades can improve the cavitation in the experiments where the pump model works at a low rotation speed (1450 rpm). The finding is supported by the results shown in Figures 7 and 9. However, the cavitation in the pump model working at high rotation speed (11,000 rpm) behaves very differently. The splitter blades at such condition are not helpful in reducing the cavitation in the pump model compared with the cylindrical blades. This may be attributed to the factor that the circumferential speed of the blade is very high at the high rotation speed. Although the splitter blades reduce the number of blades at the inlet and increase the area of the flow channel, the positive effect on reducing the occurrence of cavitation is diminished by the high circumferential speed of the inlet blades. Therefore, it can be concluded that the effect of splitter blades on reducing the cavitation is prominent at a low rotation speed while this effect is marginal at a high rotation speed, such as 11,000 rpm.

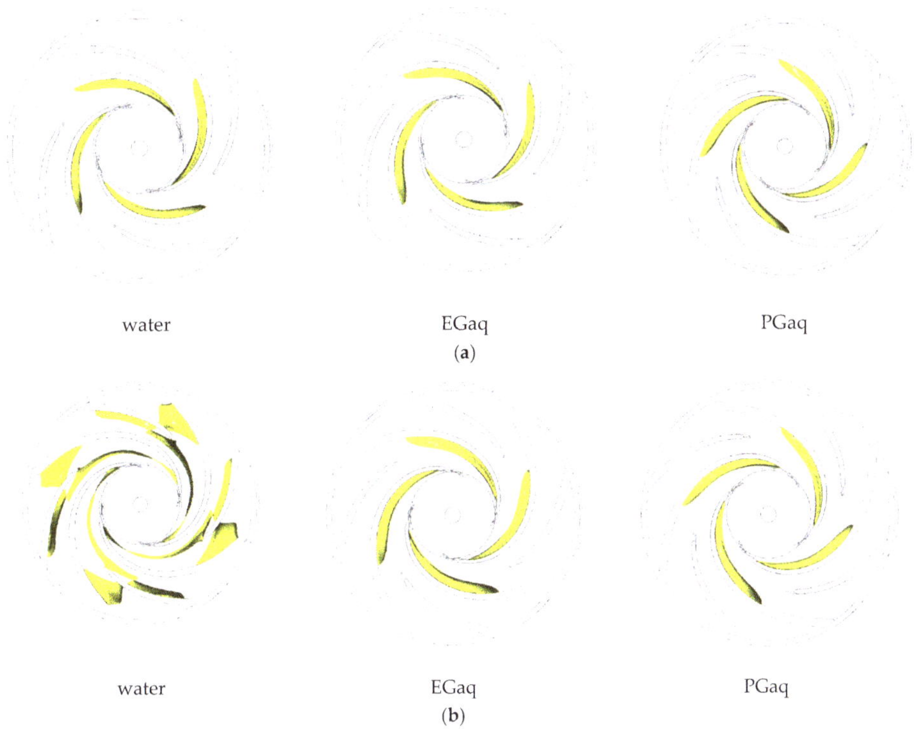

Figure 19. Iso-surface of cavitation distribution in the impeller of splitter blades. (**a**) 20 °C at 0.3 atm; (**b**) 60 °C at 0.3 atm.

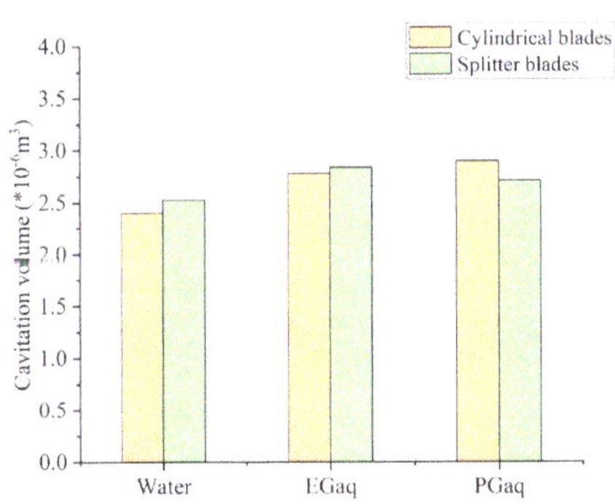

Figure 20. Volumes of cavitation zone in different impellers and coolants at 20 °C with an inlet pressure of 0.3 atm.

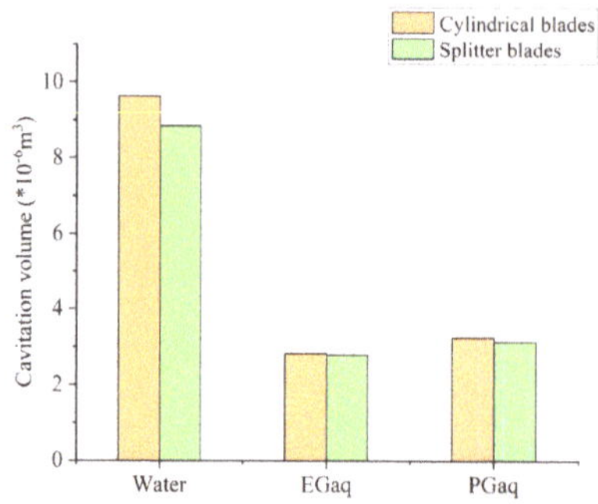

Figure 21. Volumes of cavitation zone in different impellers and coolants at 60 °C with an inlet pressure of 0.4 atm.

5. Conclusions

In this paper, a pump model with two types of impellers was designed based on the design parameters of an aeronautic cooling pump. A design method consisting in enlarging the design flow rate was used for designing the pump model, and the corresponding CFD model was built and simulated. Moreover, the accuracy of the numerical procedures was verified by comparing the simulation results with the performed experiments in terms of both hydraulic and cavitation performances.

Based on the verified numerical model, the effects of different types of impellers, cooling media, and working temperature on the hydraulic and cavitation performances of the aeronautic cooling pump were numerically investigated. It is found that the pressure head and pump efficiency for conveying water are higher than those for conveying EGaq and PGaq at 20 °C while the hydraulic performance of EGaq is slightly better than that of PG. In addition, two organic solutions have a wider range of high efficiency, compared with water. The impeller of splitter blades shows a slightly better performance for conveying EGaq than PGaq in terms of the predicted pressure head and pump efficiency. As for the characteristics of pump internal flow, the streamlines in the impeller of splitter blades are less smooth than those in the impeller of cylinder blades at low, nominal, and high flow rates. EGaq and PGaq show better internal flow fields than water, and the internal flow in the impeller of splitter blades could be further improved by optimizing the shape and position of the short blades. Comparing EGaq and PGaq, there is no significant difference in the internal flow field.

It is found that the cavitation in the pump is very similar at 20 °C for the three coolants. However, water has been severely cavitated at 60 °C with an inlet pressure of 0.4 atm. The cavitation performance of water is far less than that of EGaq and PGaq under high working temperature. The increasing temperature has less effect on EGaq and PGaq. The cavitation volume of EGaq is smaller than that of PGaq, and the cavitation volume in the splitter blades is slightly smaller than that in the cylindrical impeller. Since the coolant is usually working at high temperature, EGaq and PGaq are much better than water as the coolant. It is suggested to use EGaq as the first option. At 60 °C, which most often is the working condition of the cooling pump, the cavitation volume in the splitter blades is slightly less than that in the cylindrical blades in either water or any of the two aqueous solutions. However, the improvement by using the splitter blades is very limited. The splitter blades can improve the cavitation performance, but at the same time, they diminish the efficient performance of the pump. Therefore, the design of the short blades should be optimized in order to obtain a smooth internal flow field.

Author Contributions: Conceptualization, A.W. and R.Z.; formal analysis, A.W. and R.Z.; data curation, A.W.; writing—original draft preparation, A.W. and R.Z.; writing—review and editing, A.W., R.Z., F.W, D.Z. and X.W.; funding acquisition, F.W., D.Z. and X.W. All authors have read and agreed to the published version of the manuscript.

Funding: The authors are grateful for the financial supports from the National Natural Science Foundation of China (Grant No.: 52176038), Key R & D projects in Jiangsu Province (Grant No.: BE2021073), Aeronautical Science Fund (No. 201728R3001) and Primary Research & Development Plan of Shandong Province (No.: 2019TSLH0304).

Institutional Review Board Statement: Not applicable.

Informed Consent Statement: Not applicable.

Data Availability Statement: The data used to support the findings of this study are included within the article.

Conflicts of Interest: The authors declare no conflict of interest.

References

1. Guan, X.F. *Modern Pump Theory and Design*; China Aerospace Publishing House: Beijing, China, 2011. (In Chinese)
2. Jin, J.; Fan, Y.; Han, W.; Hu, J. Design and Analysis on Hydraulic Model of The Ultra -Low Specific-Speed Centrifugal Pump. *Procedia Eng.* **2012**, *31*, 110–114. [CrossRef]
3. Zhu, Z.C. Experimental study on high-speed centrifugal pumps with different impellers. *Chin. J. Mech. Eng. (Engl. Ed.)* **2002**, *15*, 372–375. [CrossRef]
4. Lindsay, R.B. On the Pressure Developed in a Liquid During the Collapse of a Spherical Cavity (1917). In *Men of Physics Lord Rayleigh—The Man & His Work*; Elsevier: Amsterdam, The Netherlands, 1970; pp. 221–226. [CrossRef]
5. Plesset, M.S.; Devine, R.E. Effect of Exposure Time on Cavitation Damage. *J. Basic Eng. Trans. ASME* **1966**, *88*, 31. [CrossRef]
6. Plessett, M.S.; Prosperetti, A. Bubble Dynamics and Cavitation. *Annu. Rev. Fluid Mech.* **1977**, *9*, 1945–1985. [CrossRef]
7. Brennen, C.E.; Meissner, C.; Lo, E.Y.; Hoffman, G.S. Scale Effects in the Dynamic Transfer Functions for Cavitating Inducers. *J. Fluids Eng.* **1982**, *104*, 428–433. [CrossRef]
8. Katz, J. Cavitation Inception in Separated Flows. Ph.D. Thesis, California Institute of Technology, Pasadena, CA, USA, 1982. [CrossRef]
9. Tanaka, T.; Tsukamoto, H. Transient Behavior of a Cavitating Centrifugal Pump at Rapid Change in Operating Condition (1st Report, Transient Phenomena at Opening/Closure of Discharge Valve). *Trans. Jpn. Soc. Mech. Eng. Ser. B* **1997**, *63*, 3976–3983. [CrossRef]
10. Huang, B.; Wang, G.Y.; Zhang, B.; Yu, Z.Y. Evaluation of the Cavitation Models on the Numerical Simulation of Cloud Cavitating Flows around a Hydrofoil. *Beijing Ligong Daxue Xuebao/Trans. Beijing Inst. Technol.* **2009**, *29*, 785–789.
11. Cheng, H.; Ji, B.; Long, X.; Huai, W.; Farhat, M. A Review of Cavitation in Tip-Leakage Flow and Its Control. *J. Hydrodyn.* **2021**, *33*, 226–242. [CrossRef]
12. Singhal, A.K.; Athavale, M.M.; Li, H.; Jiang, Y. Mathematical Basis and Validation of the Full Cavitation Model. In Proceedings of the ASME FEDSM'01, 2001 ASME Fluids Engineering Division Summer Meeting, New Orleans, LA, USA, 29 May–1 June 2001; 2003; Volume 1, pp. 379–406.
13. Shih, T.-H.; Liou, W.W.; Shabbir, A.; Yang, Z.; Zhu, J. A New K-ϵ Eddy Viscosity Model for High Reynolds Number Turbulent Flows. *Computers & Fluids* **1995**, *24*, 227–238. [CrossRef]
14. Wang, G.; Ostoja-Starzewski, M. Large Eddy Simulation of a Sheet/Cloud Cavitation on a NACA0015 Hydrofoil. *Applied Mathematical Modelling* **2007**, *31*, 417–447. [CrossRef]
15. Kral, L.D. Recent experience with different turbulence models applied to the calculation of flow over aircraft components. *Prog Aerosp. Sci.* **1998**, *34*, 481–541. [CrossRef]
16. Sun, H.; Xiao, R.; Wang, F.; Xiao, Y.; Liu, W. Analysis of the Pump-Turbine S Characteristics Using the Detached Eddy Simulation Method. *Chin. J. Mech. Eng. Engl. Ed.* **2015**, *28*, 115–122. [CrossRef]
17. Feng, J.; Benra, F.K.; Dohmen, H.J. Application of Different Turbulence Models in Unsteady Flow Simulations of a Radial Diffuser Pump. *Forsch. Ing.* **2010**, *74*, 123–133. [CrossRef]
18. Zhou, P.J.; Wang, F.J.; Yang, M. Internal Flow Numerical Simulation of Double-Suction Centrifugal Pump Using DES Model. *IOP Conf. Ser.: Earth Environ. Sci.* **2012**, *15*. [CrossRef]
19. Zhang, J.; Li, G.; Mao, J.; Yuan, S.; Qu, Y.; Jia, J. Numerical Investigation of the Effects of Splitter Blade Deflection on the Pressure Pulsation in a Low Specific Speed Centrifugal Pump. *Proc. Inst. Mech. Eng. Part A J. Power Energy* **2019**, *234*, 420–432. [CrossRef]
20. Wang, C.; Zhang, Y.X.; Zhang, J.Y.; Li, Z.W.; Xu, A. Comparative Research on Various Turbulence Closure Models for Unsteady Cavitation Flows in the Ultra-Low Specific Speed Centrifugal Pump. *IOP Conf. Ser.: Earth Environ. Sci.* **2019**, *240*, 062038. [CrossRef]
21. Limbach, P.; Skoda, R. Numerical and Experimental Analysis of Cavitating Flow in a Low Specific Speed Centrifugal Pump with Different Surface Roughness. *J. Fluids Eng. Trans. ASME* **2017**, *139*, 101201. [CrossRef]
22. Hindawi. Experimental Investigation on Cavitating Flow Induced Vibration Characteristics of a Low Specific Speed Centrifugal Pump. Available online: https://www.hindawi.com/journals/sv/2017/6568930/ (accessed on 11 September 2022).
23. Zhang, Y.L.; Yuan, S.Q.; Zhang, J.F.; Feng, Y.N.; Lu, J.X. Numerical Investigation of the Effects of Splitter Blades on the Cavitation Performance of a Centrifugal Pump. *IOP Conf. Ser. Earth Environ. Sci.* **2014**, *22*, 052003. [CrossRef]

24. Hu, Q.; Yang, Y.; Shi, W. Cavitation Simulation of Centrifugal Pump with Different Inlet Attack Angles. *Int. J. Simul. Model.* **2020**, *19*, 279–290. [CrossRef]
25. Yang, Q.W.; Zhang, W.; Li, X.J. Numerical simulation of cavitation and analysis of flow characteristic of thermo-sensitive fluid. *J. Drain. Irrig. Mach. Eng. (JDIME)* **2021**, *39*, 699–705. (In Chinese)
26. Sun, T.; Ma, X.; Wei, Y.; Wang, C. Computational modeling of cavitating flows in liquid nitrogen by an extended transport-based cavitation model. *Sci. China Technol. Sci.* **2016**, *59*, 337–346. [CrossRef]
27. Yuan, S.Q. *Theory and Design of Low Specific Speed Centrifugal Pump*; China Machine Press: Beijing, China, 1997. (In Chinese)
28. Zhang, D.; Liu, Y.; Kang, J.; Zhang, Y.; Meng, F. The Effect of Discharge Areas on the Operational Performance of a Liquid-Ring Vacuum Pump: Numerical Simulation and Experimental Verification. *Vacuum* **2022**, *206*, 111425. [CrossRef]
29. Li, W.; Li, S.; Ji, L.; Zhao, X.; Shi, W.; Agarwal, R.K.; Awais, M.; Yang, Y. A Study on the Cavitation and Pressure Pulsation Characteristics in the Impeller of an LNG Submerged Pump. *Machines* **2022**, *10*, 14. [CrossRef]
30. Menter, F.R. Two-Equation Eddy-Viscosity Turbulence Models for Engineering Applications. *AIAA J.* **1994**, *32*, 1598–1605. [CrossRef]
31. Menter, F.R.; Kuntz, M.; Langtry, R. Ten Years of Industrial Experience with the SST Turbulence Model. *Turbul. Heat Mass Transf.* **2003**, *4*, 625–632.
32. Menter, F.R. Zonal Two Equation Kappa-Omega Turbulence Models for Aerodynamic Flows. In Proceedings of the AIAA 24th Fluid Dynamic Conference, Orlando, FL, USA, 6–9 July 1993.
33. Zwart, P.J.; Gerber, A.G.; Belamri, T. A Two-Phase Flow Model for Predicting Cavitation Dynamics. In Proceedings of the 5th International Conference on Multiphase Flow, Yokohama, Japan, May 30–June 3 2004; Volume 152.
34. Zhang, D.; Shi, L.; Shi, W.; Zhao, R.; Wang, H.; Van Esch, B.B. Numerical Analysis of Unsteady Tip Leakage Vortex Cavitation Cloud and Unstable Suction-Side-Perpendicular Cavitating Vortices in an Axial Flow Pump. *Int. J. Multiph. Flow* **2015**, *77*, 244–259. [CrossRef]
35. *2017 ASHRAE Handbook—Fundamentals (I-P Edition)*; American Society of Heating, Refrigerating and Air-Conditioning Engineers: Atlanta, GA, USA, 2017; Volume 30, pp. 875–887.

machines

MDPI

Article

Investigation of Transient Characteristics of a Vertical Axial-Flow Pump with Non-Uniform Suction Flow

Fan Meng [1,2,*], Zhongjian Qin [3], Yanjun Li [1] and Jia Chen [1,4]

1 Research Center of Fluid Machinery Engineering and Technology, Jiangsu University, Zhenjiang 212013, China
2 Wenling Fluid Machinery Technology Institute of Jiangsu University, Wenling 317525, China
3 China Water Huaihe Planning Design and Research Co., Ltd., Hefei 230601, China
4 Crane Fengqiu (Zhejiang) Pump Co., Ltd., Zhuji 311800, China
* Correspondence: mf@ujs.edu.cn

Abstract: The aim of this paper is to study the influence of non-uniform suction flow on the transient characteristics of a vertical axial-flow pump device. The unsteady calculation is employed to forecast the unstable flow structure with three inlet deflection angles α, and the calculation accuracy under uniform inlet flow is verified by the external characteristic test. The results depict that a promotion in the α will increase the head and shaft power and thus improve the stress and fatigue failure risk of the impeller. At the impeller inlet, the pressure pulsation intensity (PPI) with $\alpha = 40°$ is lower than that with $\alpha = 0°$ caused by a decline in the axial velocity. The dominant frequency of the unsteady pressure signal is the blade-passing frequency (BPF), and the dominant frequency amplitude rises with the increase in α due to the improvement of the pre-rotation impact intensity. At the guide vanes inlet, the dominant frequency of the unsteady pressure signal at the guide vane inlet is also the blade-passing frequency. An improvement in α magnifies the angle between the trailing edge jet of the impeller and the leading edge of the guide vanes under $0.8Q_{des}$ and $1.0Q_{des}$, while it diminishes the angle under $1.2Q_{des}$. Thus, the PPI and dominant frequency amplitude with $\alpha = 40°$ are higher than that with $\alpha = 0°$ under $0.8Q_{des}$ and $1.0Q_{des}$, but these are lower than that with $\alpha = 0°$ under $1.2Q_{des}$.

Keywords: vertical axial-flow pump device; unsteady pressure signal; peak to peak; pressure pulsation intensity

Citation: Meng, F.; Qin, Z.; Li, Y.; Chen, J. Investigation of Transient Characteristics of a Vertical Axial-Flow Pump with Non-Uniform Suction Flow. *Machines* **2022**, *10*, 855. https://doi.org/10.3390/machines10100855

Academic Editor: Kim Tiow Ooi

Received: 25 August 2022
Accepted: 16 September 2022
Published: 26 September 2022

1. Introduction

A vertical axial-flow pump is a common piece of hydraulic machinery widely used in inter regional water transfer, agricultural irrigation [1,2], and flooding projects [3] due to its large flow rate and compact structure. However, the hydrodynamic pulsation characteristics of vertical axial-flow pumps have been being a key concern due to the fact that the curved elbow area of the inflow conduit is susceptible to the formation of a flow separation caused by the inverse pressure gradient [4]. Furthermore, the inlets of some vertical axial-flow pumps present a non-uniform velocity profile in actual applications due to a forebay with a side intake, which further increases the internal vortex scale, aggravates unit vibration, and directly threatens the safety of the operation of the vertical axial-flow pump [5]. The characteristic parameter of pressure pulsation is an important evaluation index of the operational stability of a vertical axial-flow pump [6,7]. Establishing a physical connection between the non-uniform suction flow and pressure pulsation characteristics can provide a theoretical reference to ensure that vertical axial-flow pumps operate safely and efficiently.

Recently, with the rapid growth of meshing techniques [8], numerical algorithms [9], and post-processing methods [10], the investigation of the internal pressure pulsation of axial-flow pumps based on computational fluid dynamics (CFD) has made remarkable progress [11,12]. Yang et al. [13] found that rotor–stator interference between the impeller and guide vanes is one of the most important inductors for strong pressure pulsation inside the axial-flow pump. As a result, the main frequency amplitude of pressure pulsation at the impeller outlet is high due to the turbulent flow regime, and the main frequency is the blade-passing frequency. Shi et al. [14] showed that blade angle deviations enhanced the rotor–stator interference effect resulting from the enhanced strength of the blade wake vortex, which increased the pulsation amplitude of the flow-field pressure and shaft power. Zhang [15] used a two-way fluid–structure coupling method to investigate the effect of hydrodynamic excitation forces induced by dynamic–static interference on the structural dynamics of an axial-flow pump. They reported that the main frequency of the equivalent force pulsation on the guide vanes is the same as the blade-passing frequency. In addition, the clearance leakage flow is another important inducer of strong pressure pulsation [16]. Shen et al. [17] studied the effect of the radius of tip clearance on the unsteady flow characteristic inside an axial-flow pump. The pressure difference at the blade surface drives the impeller outflow back to the impeller inlet along the clearance passage, which leads to tip leakage vortex formation and an improvement in the pressure pulsation amplitude in the tip clearance.

Previous studies of the pressure pulsation characteristics of axial pumps were mainly based on uniform inflow. Although non-uniform inflow leads to the formation of large-scale secondary vortices and a deteriorated outflow state inside the inflow conduit of vertical axial pumps [18], the impact of non-uniform inflow on the pressure pulsation characteristic inside the impeller and guide vanes of vertical axial pumps has hardly been explored in detail. The existed investigation of the unsteady characteristics of other types of pumps under non-uniform inflow can be used as a guideline. In a reactor coolant pump, the non-uniform inflow caused by the channel head significantly increased the pressure pulsation amplitude at the inlet and impeller radial force [19]. The main frequency of pressure pulsation at the inlet of the guide vanes is the blade-passing frequency and its harmonics [20]. The main frequency amplitude in the impeller and guide vanes are susceptible to rotational speed so a reduction in the rotational speed can significantly result in a drop in the pressure pulsation energy level and thus enhance the operational stability [21]. In a waterjet pump, the non-uniform inflow caused by the intake duct enhances the main frequency amplitude of pressure pulsation near the surfaces of the blades and guide vanes. The main frequency of the blade surface pressure pulsation is the shaft frequency and 11 times the shaft frequency, respectively, under non-uniform inflow and uniform inflow [22]. Under cavitation conditions with non-uniform inflow, the pressure pulsation characteristics were found to have a close relationship with the cavity volume near the blades' surfaces. In a time-domain analysis, the pressure pulsation curve on the monitor near the front of the impeller was similar to the cavity volume acceleration curve [23].

The inflow non-uniformity of vertical axial-flow pumps can be quantitatively described by the inflow deflection angle. Therefore, the main aim of this paper was to study the effect of the inflow deflection angle on the spatial distribution characteristics of the pressure pulsation intensity (PPI) and pressure signal transmission characteristics inside a vertical axial-flow pump. The external characteristic parameters of the vertical axial-flow pump under different inflow angles were predicted by solving the unsteady Reynolds time-averaged equation. The calculated results under uniform inflow were verified by an external characteristic test. The standard deviation of pressure pulsation was used as the evaluating indicator to spatially analyze the PPI inside the impeller and guide vanes. Time- and frequency-domain analyses of the unsteady pressure pulsation signals at the monitoring points near the inlet of the impeller and guide vanes were performed.

2. Numerical Simulation

2.1. Computational Model

A vertical axial-flow pump device with $n_s = 893.4$ was the subject of investigation in this study. The main hydraulic components are the elbow inlet channel, impeller, guide vanes, and elbow outlet channel, which are presented in Figure 1. In the calculation domain, the Z axis of the Cartesian coordinate system was the impeller rotation axis. Based on the engineering requirements, the design rotation speed and flow rate were 1340 r/min and 308 L/s, respectively. The design efficiency and head of the pump device were 78.3% and 4.4 m according to the unsteady numerical simulation. The impeller diameter, tip radius, and blade number were 300 mm, 0.3 mm, and 3, respectively. The number of guide vanes was finally determined to be 6 to better match the impeller.

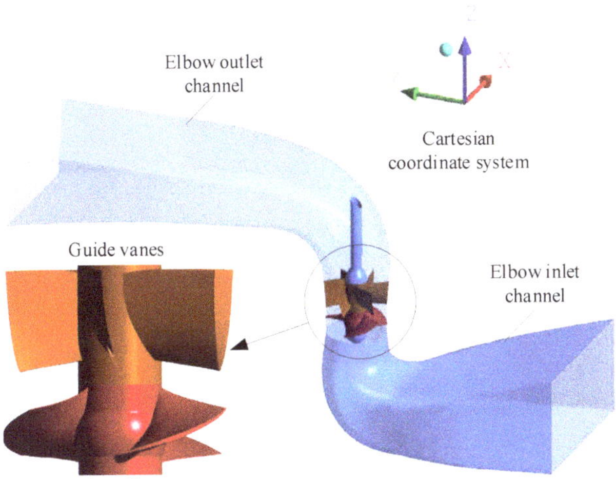

Figure 1. Geometrical structure of vertical axial-flow pump device.

2.2. Mesh and Boundary Conditions

To accurately predict the hydraulic performance and capture the unstable flow structure, all the calculation domains were divided into hexahedral grids, as shown in Figure 2a, and the near-wall grids were refined as shown in Figure 2b. The computational domains of the impeller, elbow inlet channel, and elbow outlet channel were meshed by using ICEM CFD and that of the guide vanes was meshed by applying TurboGrid.

The three-dimensional Navier–Stokes equations with the Reynolds-averaged method were solved using Ansys CFX to carry out the numerical simulation. The roughness values of the wall of the impeller and guide vanes were set at 0.0125 mm and those of the elbow inlet channel and elbow outlet channel were set at 0.05 mm. The inlet condition was set to "Normal Speed" or "Cart. Vel Components" depending on the inflow deviation angle. The outlet condition was determined as "Opening Pres. and Dirn". Furthermore, the results of the steady numerical simulation were taken as the initial values in the unsteady numerical simulation. In the steady numerical simulation, the interface condition between stators was "None" and that between the impeller and stator was "Stage". In the unsteady numerical simulation, the interface condition between stators was also "None", but that between the impeller and stator was "Transient Rotor Stator". In addition, the root mean square was selected as the convergence criterion and the residual target was set at 5×10^{-5}. The time step was 0.000373134 s, i.e., a rotation of 3° for each step, and the total time was 0.62686567 s, i.e., the number of rotations was 14.

(a) (b)

Figure 2. (**a**) Mesh of the vertical axial-flow pump device and (**b**) Mesh refinement of the impeller and guide vane wall.

The grid independence analysis was completed to accurately capture the internal vortex structure and turbulent characteristics of the vertical axial-flow pump with limited computational resources. The head and efficiency were calculated using Equations (1) and (2) [10].

$$H = \frac{p_{\text{out}} - p_{\text{in}}}{\rho g} \quad (1)$$

Here, H, p_{out}, and p_{in} stand for the head, outlet total pressure, and inlet total pressure, ρ and g are the water density and gravity acceleration.

The design head H_{des} of the last 120 time steps were recorded and then averaged as the quantitative evaluation criteria for the grid independence analysis due to the fluctuating nature of the unsteady calculation results. The number of grid nodes (N_g) for each hydraulic component and the H_{des} for the four cases are shown in Table 1. The error deviation can be calculated by $\frac{H_{\text{des},casei}}{H_{\text{des},case3}} \times 100\%$ (i = 1, 2, 3, 4). Grid 3 was finally selected as the optimal case for the next study due to the relative deviation of the H_{des} being only 0.28% compared with Grid 4. Moreover, Figure 3 shows the local distribution of Yplus on the wall of the impeller and guide vanes. The Yplus value is defined as a dimensionless variable that indicates the distance between the first layer mesh and the wall surface of the computational domain. It can be employed to detect whether the quality of the boundary layer mesh is consistent with the requirements of the turbulence model. In Grid C, the average Y+ values of the elbow inlet channel, impeller guide vanes, and elbow outlet channel are 4.97, 10.11, 10.79, and 15.95. These results can be considered as being within the constraint conditions of the turbulence model SST k-omega for Yplus.

Table 1. Mesh independence analysis.

Case	N_g of Inlet Channel	N_g of Impeller	N_g of Guide Vanes	N_g of Outlet Channel	Total N_g	H_{des}	Error Deviation
Grid 1	725,010	813,522	521,532	748,212	2,808,276	4.393 m	−2.16%
Grid 2	986,910	1,346,832	848,952	937,272	4,119,966	4.398 m	−1.58%
Grid 3	1,332,635	1,965,942	1,226,352	1,150,632	5,675,561	4.449 m	0.0%
Grid 4	1,613,460	2,462,532	1,773,486	1,379,392	7,228,870	4.454 m	0.28%

Figure 3. Yplus distribution on the wall of the impeller and guide vanes.

2.3. The Definition of Non-Uniform Inflow

In this work, the uniform area-averaged velocity and non-uniform area-averaged velocity at the inlet are defined as two-dimensional velocity vectors in the X-Y plane, which is perpendicular to the axis of rotation, and the Z axis is identical to the axis of rotation, as shown in Figure 4. $\vec{V_\alpha}$ can be divided into two parts: one is the velocity in the X-direction and the other is the one lone Y-direction in a Cartesian coordinate system, where the α is the angle between $\vec{V_\alpha}$ and $\vec{V_x}$. The uniform inflow conditions are represented by $\left|\vec{V_\alpha}\right| = \left|\vec{V_y}\right|$ and $\left|\vec{V_x}\right| = 0$, which means $\alpha = 0$. Under non-uniform inflow conditions (or $\alpha \neq 0$), two velocity components $\left|\vec{V_x}\right|$ and $\left|\vec{V_y}\right|$ can be calculated by $(-\cos\alpha) \times \left|\vec{V_\alpha}\right|$ and $\sin\alpha \times \left|\vec{V_\alpha}\right|$, respectively. It is worth noting that the inlet flow rate is equal to $\vec{V_\alpha}$ multiplied by the inlet area.

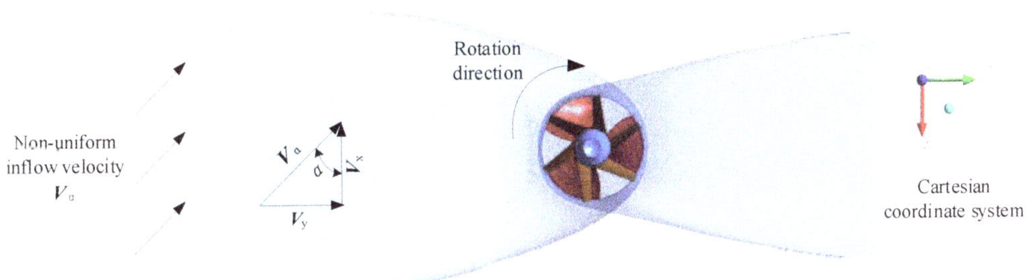

Figure 4. Schematic diagram of the inlet deflection angle.

3. Results and Discussion

3.1. Test Validation

An external characteristic test of the vertical axial-flow pump device was conducted to verify whether the mesh generation method and the boundary conditions were appropriate. Figure 5 depicts the specific structure of the test bench, which can be divided into an upper layer and a lower layer. The intelligent electromagnetic flowmeter was adopted for the measurement of the operating flow, which was arranged in the lower layer and the measurement uncertainty was 0.2%. An intelligent differential pressure transmitter with 0.1% measurement uncertainty was adopted for the head, and the monitoring points of the inlet and outlet pressures were arranged in the low-pressure tank and high-pressure tank, respectively, to relieve the interference of water pressure pulsation. Intelligent torque and speed sensors were used to measure the torque and the measurement uncertainty was 0.1%.

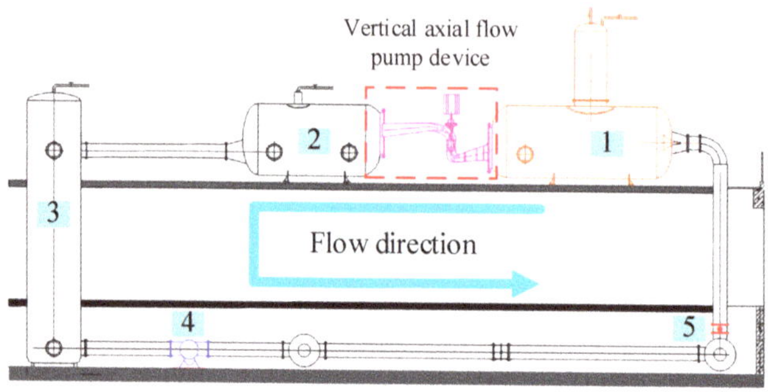

Figure 5. Test bench structure. (1 Low-Pressure Tank; 2 High-Pressure Tank 1; 3 High-Pressure Tank 2; 4 Circulating Pump; 5 Valve).

Figure 6 shows the performance results of the unsteady numerical simulation and test measurement data under uniform inflow conditions ($\alpha = 0$). The unsteady numerical simulation head and efficiency coefficients are the average value of the last 120 time-steps. As shown in the figure, the head and efficiency curves based on computational fluid dynamics (CFD) are similar to those of the test. The head of the CFD and test decreases with the increase in the flow rate, and the best efficiency point of the CFD and test can be found near $1.0Q_{des}$. Compared with the design head and efficiency of the test, the relative deviations of those of the CFD are 2.58% and 1.34%, respectively, which indicates that the calculation results are reliable.

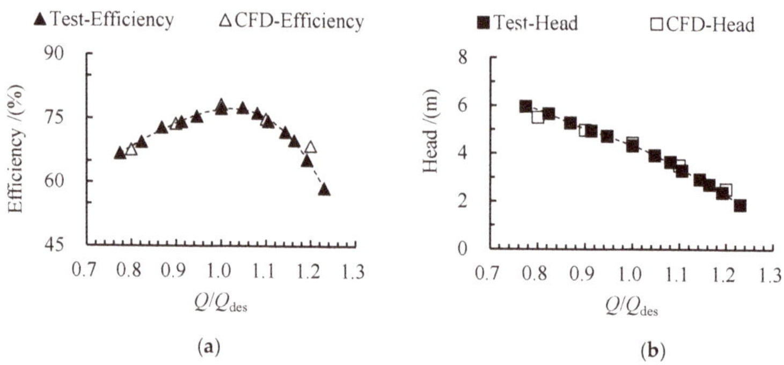

Figure 6. Comparison of (**a**) efficiency and (**b**) head between the CFD and the test under uniform inflow. ($\alpha = 0°$).

3.2. External Characteristic Parameters under Three Inlet Deflection Angles

Figure 7a–c denote the efficiency, head, and shaft power under three α values, respectively. The efficiency drops with the increase in α under $0.8Q_{des}$–$1.0Q_{des}$, but it ascends with the increase in α under $1.2Q_{des}$. This means that an improvement in α varies the impeller attack angle and thus causes the best efficiency point (BEP) to shift toward the larger flow rate. Moreover, the head and shaft power elevate with the increase in α under the same flow rate. The improvement of the force on the impeller due to the elevation of the shaft power not only magnifies the inner vibration energy but also raises the risk of fatigue failure. Therefore, the effect of α on the spatial distribution of the PPI and the transmission characteristics of the unsteady pressure signal need to be compared and analyzed to ensure the safe operation of the vertical axial-flow pump device.

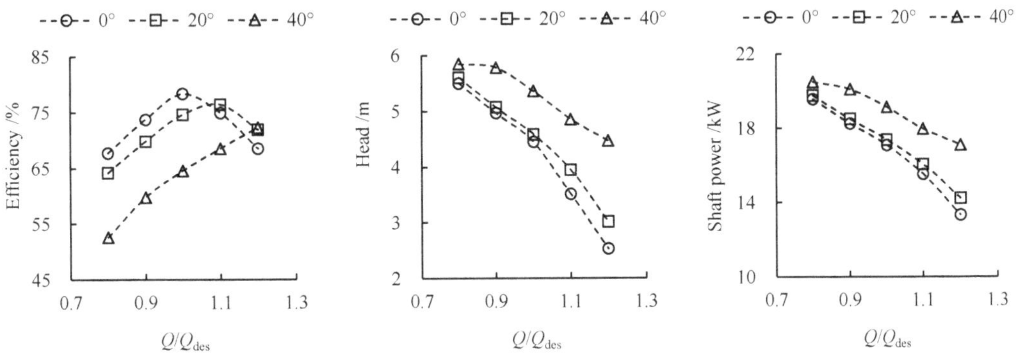

Figure 7. The (**a**) efficiency, (**b**) head, and (**c**) shaft power curves with flow rates under three α values.

3.3. Spatial Distribution of PPI under Different Inlet Deflection Angles

In this work, the standard deviation of pressure for the last 120 time-steps was employed as the evaluation parameter of PPI. The PPI and radial coefficient span can be calculated as follows:

$$\text{Span} = \frac{R - R_h}{R_s - R_h} \tag{2}$$

$$\text{PPI} = \sqrt{\frac{1}{m-1}\sum_{i=1}^{m}(p_{\text{i}} - \overline{p})^2} \tag{3}$$

$$\overline{p} = \frac{1}{m}\sum_{i=1}^{m} p_{\text{i}} \tag{4}$$

where R_{h}, R_{s}, and R depict the hub radius, shroud radius, and calculated radius, respectively. p_{i} and m are the instantaneous pressure at the ith time step and time step number, respectively.

Figures 8 and 9 show the radial distribution of the PPI in the impeller inlet and the axial velocity distribution in the vertical mid-section of the impeller, respectively. The flow field inside the impeller is a rotating domain and the high PPI at the impeller inlet is mainly caused by the impact effect of the axial inflow on the pre-rotation. Thus, firstly, the PPI increases with the increase in the flow rate owing to the improvement of the axial velocity. Secondly, the pre-rotation intensity at the impeller inlet gradually increases along the direction from the hub to the shroud, which leads to the PPI near the shroud being significantly higher than that near the hub. Thirdly, under a small flow rate, the influence of α on the PPI is not significant due to the fact that the impact effect of the axial inflow on pre-rotation is weak caused by the low axial velocity. However, under the design and large flow rate, the PPI decreases with the increase in α due to the decline in the axial velocity of the inflow.

Figure 10 shows the radial diction of the PPI at the inlet of the guide vanes under three α values. The impact effect of the trailing edge jet (TEJ) of the impeller blade on the guide vanes is the major effect of the high PPI at the inlet of the guide vanes. Under small design flow rates, the angle of the TEJ with $\alpha = 0°$ is more than the inlet angle of the guide vanes. An improvement of α will further increase the angle and impact effect of the TEJ, which leads to the elevation of the PPI. Under a large flow rate, the angle of the TEJ with $\alpha = 0°$ is less than the inlet angle of the guide vanes, and an increase in α can reduce the angle between the TEJ and the leading edge of the guide vanes. Consequently, the PPI with $\alpha = 0$ is higher than that with $\alpha = 20°$ and $40°$ due to the decreased impact effect.

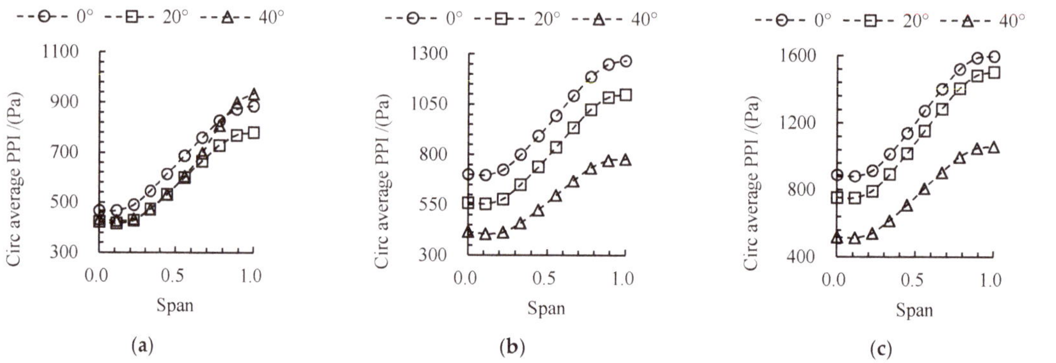

Figure 8. The PPI at the impeller inlet with three α values under (**a**) $0.8Q_{des}$, (**b**) $1.0Q_{des}$, and (**c**) $1.2Q_{des}$.

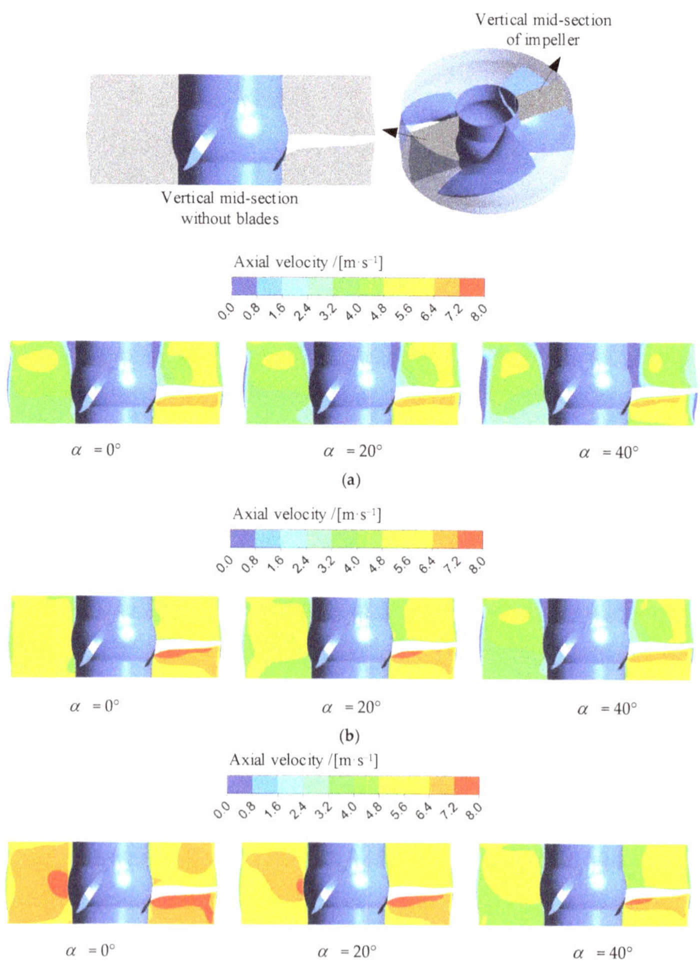

Figure 9. The axial velocity at the vertical mid-section of the impeller with three α values under (**a**) $0.8Q_{des}$, (**b**) $1.0Q_{des}$, and (**c**) $1.2Q_{des}$.

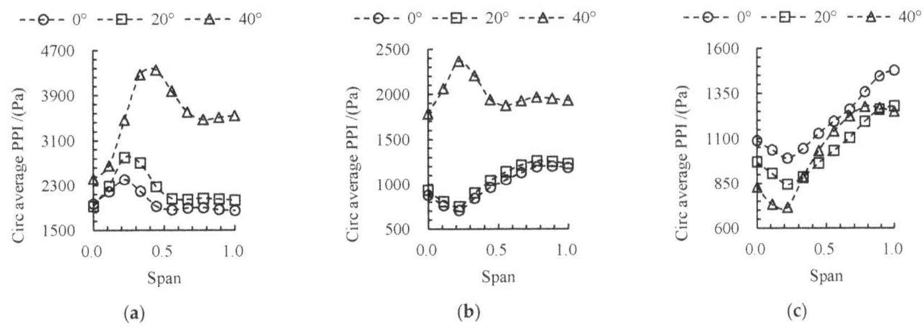

Figure 10. The PPI at the inlet of guide vanes with three α values under (**a**) $0.8Q_{des}$, (**b**) $1.0Q_{des}$, and (**c**) $1.2Q_{des}$.

Figure 11 is the radial distribution of the PPI at the outlet of the guide vanes under three α values. The PPI at the outlet of the guide vanes is significantly lower than that at the inlet of the impeller and guide vanes. This indicates that the guide vanes have converted most of the unstable rotational kinetic energy inside the impeller outflow into stable static energy. However, the PPI with $\alpha = 0°$ is higher than that with $\alpha = 20°$ and 40° under $0.8Q_{des}$, but the PPI with $\alpha = 40°$ is higher than that with $\alpha = 0°$ and 20° under $1.0Q_{des}$ and $1.2Q_{des}$. To explain this phenomenon, the velocity distributions at the vertical mid-sections of the guide vanes and outlet channels are depicted in Figure 12. Under a small flow rate, the outflow angle of the guide vanes is too large and a large area of backflow occurs at the outlet of the guide vane, which collides with the main flow and generates a strong pressure pulsation. With the increase in α, the impact position of the main flow and the backflow gradually moves away from the guide vanes outlet, and thus the impact effect on guide vanes outlet decreases. Under the design flow rate, the impact effect of the backflow at the outlet of the guide vane improves as the α increases due to the enlargement of the impact area between the backflow and the main flow. Under a large flow rate, the flow pattern at the outlet of the guide vane is relatively stable under $\alpha = 0°$. However, the large-area backflow occurs near guide vanes outlet under $\alpha = 40°$.

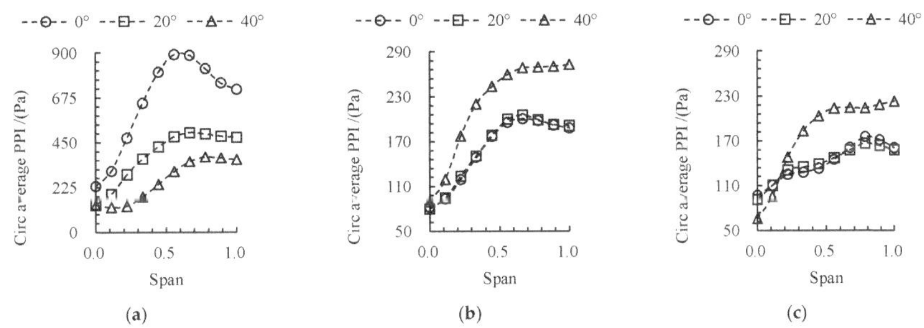

Figure 11. The PPI at the outlet of guide vanes with three α values under (**a**) $0.8Q_{des}$, (**b**) $1.0Q_{des}$, and (**c**) $1.2Q_{des}$.

Figure 13 describes the distribution of the PPI in the passage of the impeller and guide vanes with three α. Under a small flow rate, the large area of the high PPI is mainly located near the inlet and the suction side of the guide vanes due to the impact of the main flow on the flow separation vortex boundary. As the α increases, the flow separation area at the suction side increases and forms a barrier effect on the inlet and the area of the high PPI near the inlet rises with the increase in α in the guide vanes. For the design flow rate, an area with a high PPI is found near the impeller outlet from the mixing between the TEJ and main flow and there is no large area of high PPI in the guide vane caused by the stable flow regime. However, as the α increases, the TEJ angle will also change naturally, and a flow separation can be found near the suction side in the guide vane. Therefore, the high

PPI area in the vicinity of the suction side of the guide vanes rises with the increase in α. Under a large flow rate, the axial-flow rate is high and a stable flow pattern can clearly be observed inside the guide vane. Thus, the high PPI is found to be mainly located at the outlet and suction side of the impeller. When α increased to 40°, the area of high PPI in the vicinity of the outlet and suction side of the blade declined sharply due to the drop in the axial velocity, and some high PPIs in the vicinity of the suction surface of the guide vanes occurred induced by flow separation.

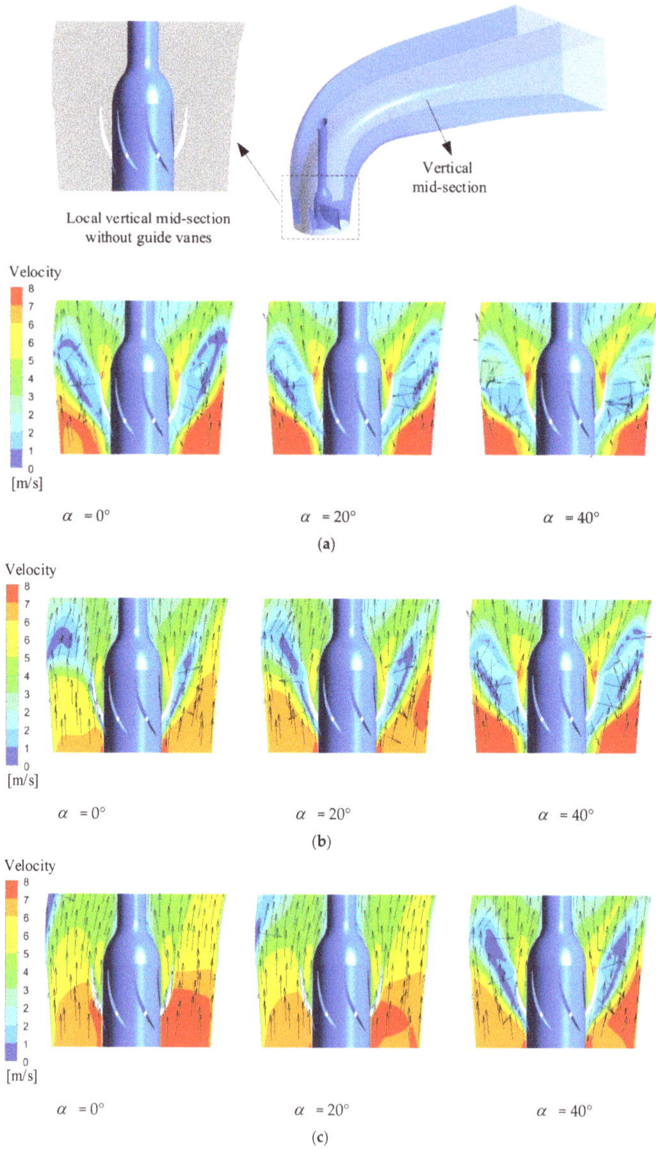

Figure 12. The velocity at the vertical mid-section of the guide vanes and elbow outflow channel with three α values under (**a**) $0.8Q_{des}$, (**b**) $1.0Q_{des}$, and (**c**) $1.2Q_{des}$.

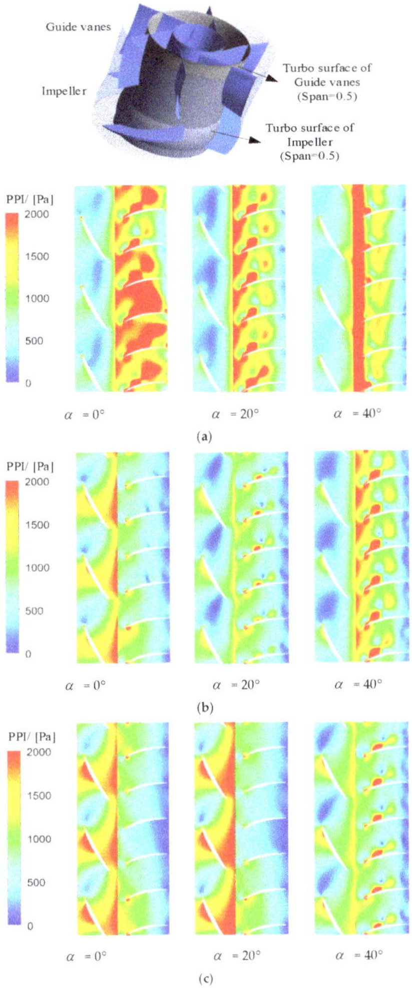

Figure 13. The PPI distribution in the passage of the impeller and guide vanes with three α values under (**a**) $0.8Q_{des}$, (**b**) $1.0Q_{des}$, and (**c**) $1.2Q_{des}$.

3.4. Transmission Characteristics of Unsteady Pressure Signal under Three Inlet Deflection Angles

According to Figures 8, 10 and 11, the PPI at the rotor–stator interface (inlet channel–impeller and impeller–guide vanes interface) is more obvious than at the stator interface (guide vanes–outlet channel). To analyze the transmission characteristics of the pressure fields upstream and downstream in the rotating domain, six monitoring points were arranged at the inlet of the impeller and guide vane interface as presented in Figure 14. The results of the time-domain analysis of the unsteady pressure signal at A1–A3 with three α are shown in Figure 15. For the last rotation period, there are three peak points and three valley points (three pulsation periods). In addition, peak-to-peak (PTP) of the pressure pulsation equals the difference between the maximum and the minimum in the last pulsation period. Since the impact of axial inflow on the pre-rotation is key to the pressure pulsation amplitude near the impeller inlet, the PTP in the vicinity of the shroud seems higher than that near the hub due to the fact that the pre-rotation intensity increases along the radial coefficient. The improvement of α can aggravate the pre-rotation caused by the increase in the circumferential velocity, and thus the PTP with $\alpha = 40°$ is higher than that with $\alpha = 0°$. The relative growth rates of PTP at A3 are 63.3%, 35.3%, and 21.7% under $0.8Q_{des}$, $1.0Q_{des}$, and $1.2Q_{des}$, respectively.

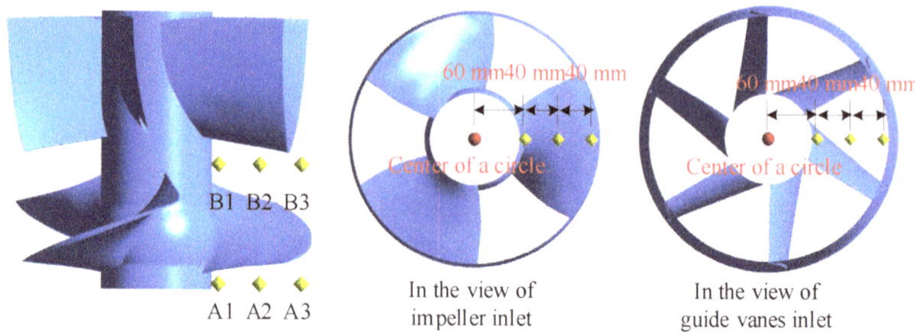

Figure 14. The position of monitor points near the inlet of the impeller and guide vanes.

Figure 16 presents the time-domain analysis of the unsteady pressure signal at B1–B3 with three α. Similar to the unsteady pressure signal at A1–A3, three peak points and three valley points can be found in the last rotating period. The PTP of the pressure pulsation in the vicinity of the shroud seems also higher than that near the hub. Under $0.8Q_{des}$ and $1.0Q_{des}$, the angle of the TEJ with $\alpha = 0°$ is more than the inlet angle of the guide vanes, and the improvement of α will further increase the jet angle and impeller–guide vanes' interference effect. Compared with $\alpha = 0°$, the growth rate of $\alpha = 40°$at A2 is 132.8% under $0.8Q_{des}$ and that at A1 is 174.8% under $1.0Q_{des}$. Under $1.2Q_{des}$, the angle of the TEJ with $\alpha = 0°$ is less than the inlet angle of the guide vanes, and the improvement of α decreases the angle between the jet angle and inlet angle, which leads to the reduction in the impeller–guide vane interference effect. Compared with $\alpha = 0°$, the decrease in $\alpha = 40°$ at A1 is 46.9%.

In this work, the fast Fourier function with a rectangular window was used to transform the pressure signal from the time-domain to the frequency-domain. Figure 17 shows the frequency-domain analysis of the unsteady pressure signal at A1–A3 with three α. The dominant frequency of the pressure signal at A1–A3 means the blade-passing frequency (BPF), and the dominant frequency amplitude of A3 is found to be the highest. The frequency with high amplitude is mainly the harmonic frequency of the BPF and it is within 0–400 Hz. In addition, an improvement in the α cannot change the dominant frequency of the pressure signal but can promote the dominant frequency amplitude of the pressure signal. This indicates that the frequency characteristics of the flow field inside the impeller for upstream transmission are mainly determined by the pre-rotation caused by the blade rotation, and the intensity of the pre-rotation is promoted with an increase in α. Under $0.8Q_{des}$ and $1.2Q_{des}$, compared with $\alpha = 0°$, the relative growth rates in $\alpha = 40°$ at A3 are 43.9% and 34.8%, respectively. Under $1.0Q_{des}$, the increase at A1 is 28.2%.

Figure 18 depicts the frequency-domain analysis of the unsteady pressure signal at B1–B3 with three α. The BPF is still the main frequency of the pressure signal on B1–B3, which is not influenced by the inflow deflection angle. The high amplitude frequencies can be found within 0–400 Hz and are mainly the harmonic frequencies of the BPF. Moreover, the increase in α has a significant effect on the dominant frequency amplitude of the pressure signal on B1–B3 due to the change in the TEJ angle of the blade. Under $0.8Q_{des}$ and $1.0Q_{des}$, the dominant frequency amplitude with $\alpha = 40°$ is more than $\alpha = 0°$ resulting from the increased angle between the TEJ and inlet angle of the guide vanes caused by the improvement of α. The maximum relative growth rates are 126.9% (at B2) and 136.4% (at B3) under $0.8Q_{des}$ and $1.0Q_{des}$, respectively. However, the dominant frequency amplitude with $\alpha = 40°$ is less than $\alpha = 0°$ due to the fact that the angle between the TEJ and inlet angle of the guide vanes decreased with the promotion of α. The maximum relative drop rate can be obtained at B1 and it is 43.8%.

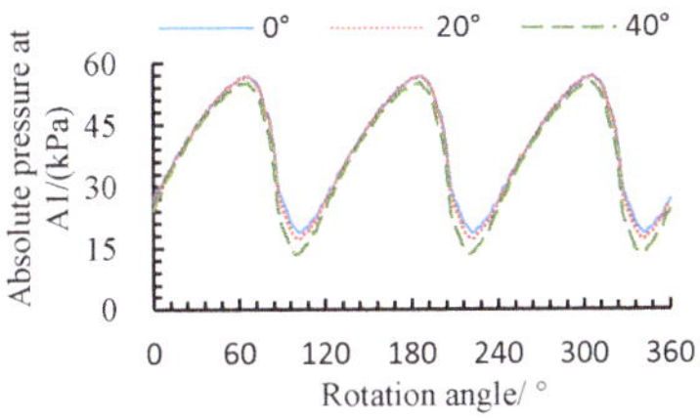

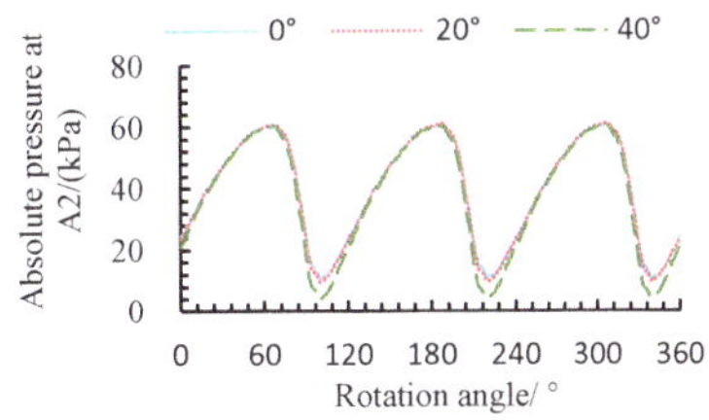

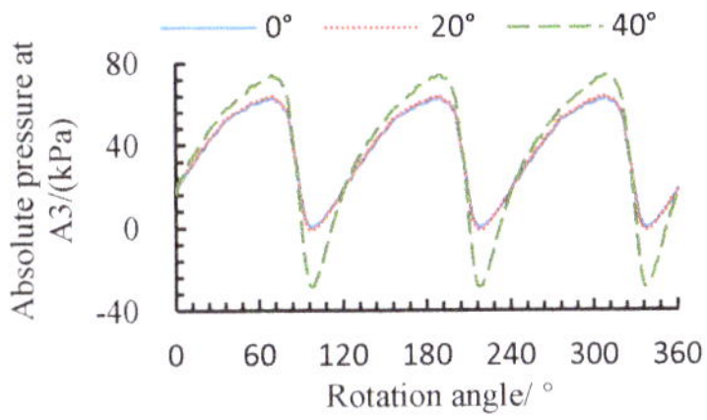

PTP distribution at monitor points

α	0°	20°	40°
PTP at A1 (kPa)	38.2	39.6	41.5
PTP at A2 (kPa)	49.6	51.5	56.1
PTP at A3 (kPa)	62.4	64.9	101.9

(**a**)

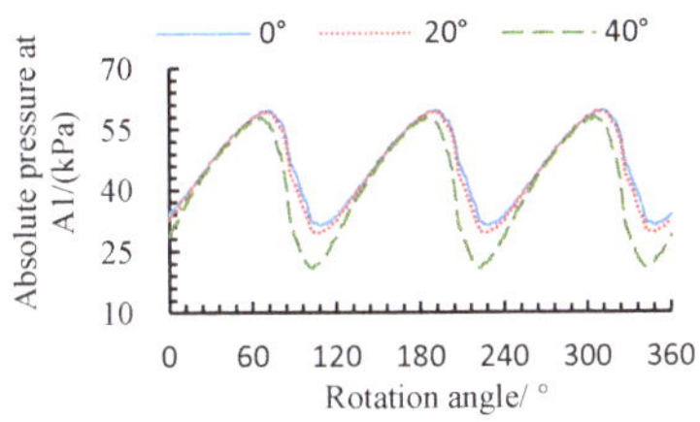

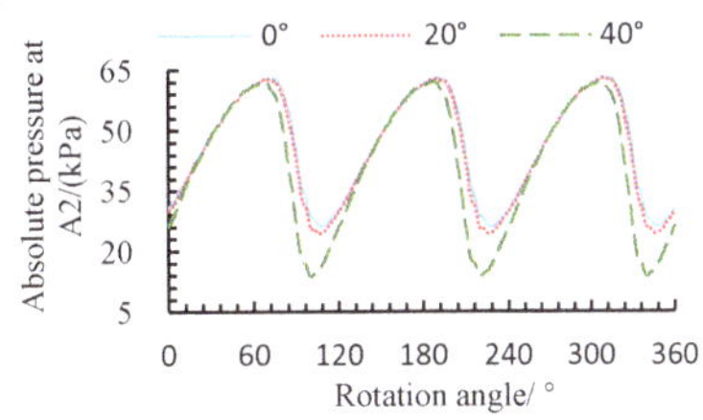

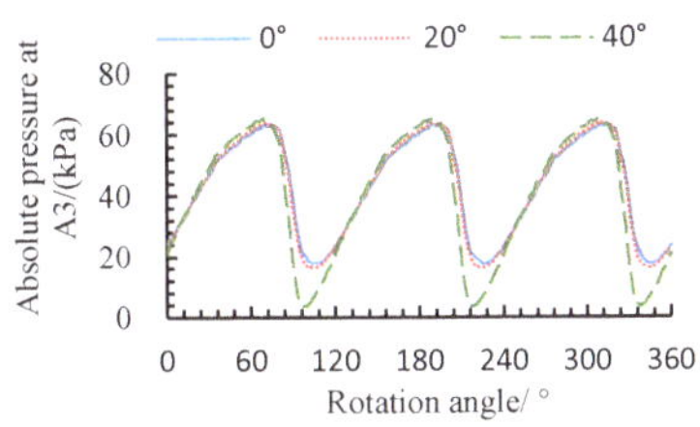

PTP distribution at monitor points

α	0°	20°	40°
PTP at A1 (kPa)	28.2	29.6	37.0
PTP at A2 (kPa)	36.9	38.5	48.2
PTP at A3 (kPa)	45.4	47.7	61.4

(**b**)

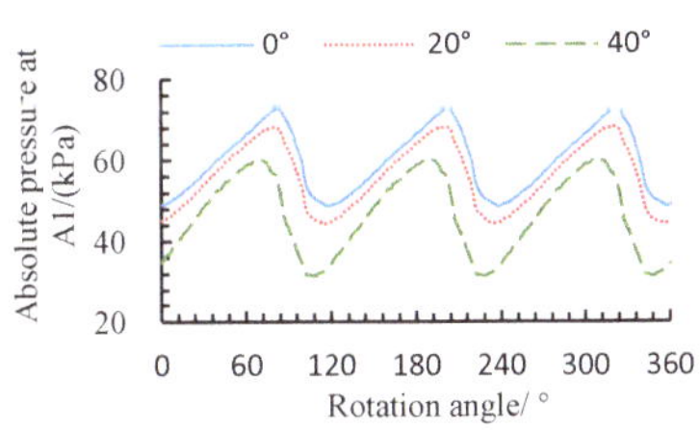

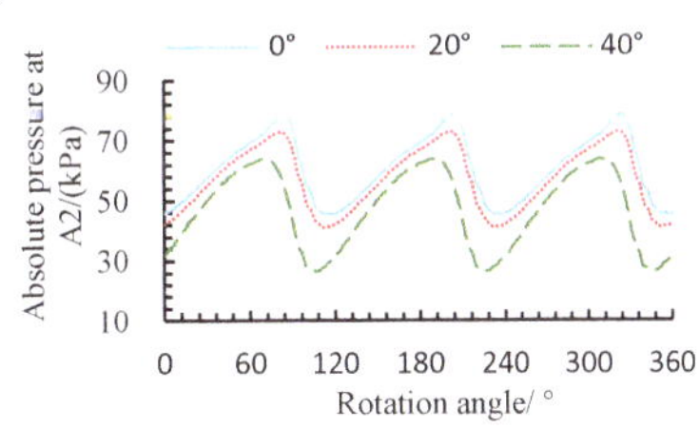

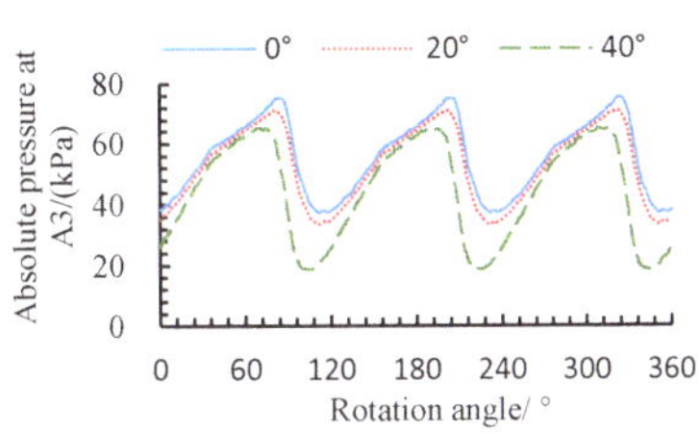

PTP distribution at monitor points

α	0°	20°	40°
PTP at A1 (kPa)	23.9	23.8	28.6
PTP at A2 (kPa)	33.3	31.9	37.1
PTP at A3 (kPa)	38.2	37.9	46.5

(**c**)

Figure 15. Time-domain analysis of different monitor points (A1, A2 and A3) with three α values near the impeller inlet under (**a**) $0.8Q_{des}$, (**b**) $1.0Q_{des}$, and (**c**) $1.2Q_{des}$.

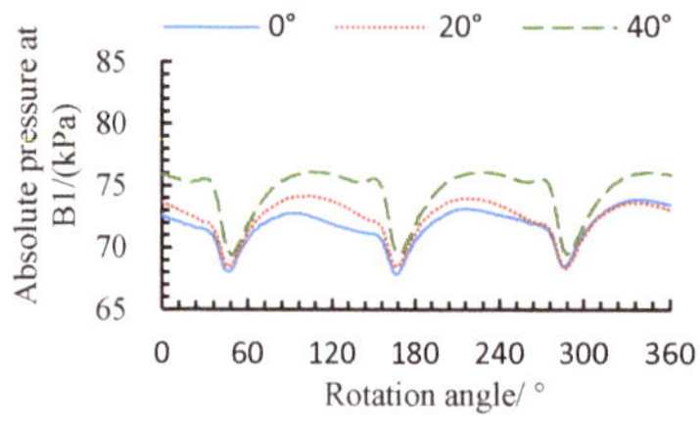

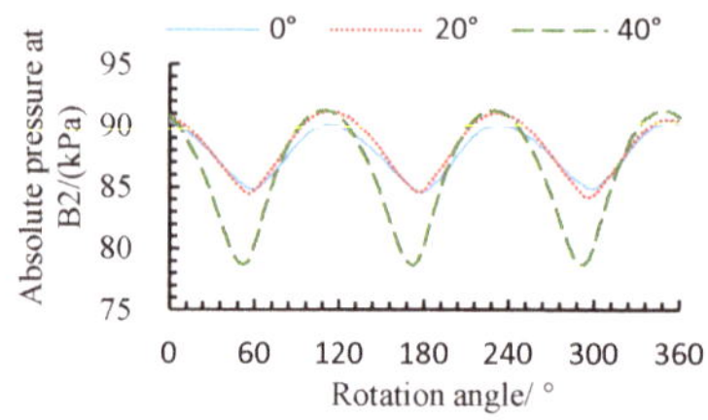

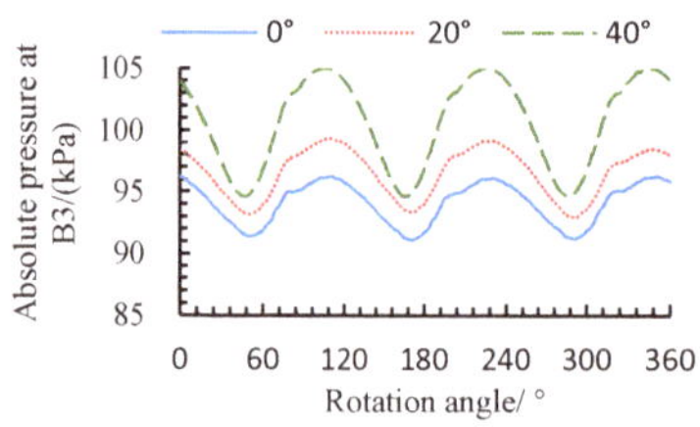

PTP distribution at monitor points

α	0°	20°	40°
PTP at B1 (kPa)	5.4	5.3	6.6
PTP at B2 (kPa)	5.4	6.5	12.5
PTP at B3 (kPa)	5.0	5.5	10.4

(**a**)

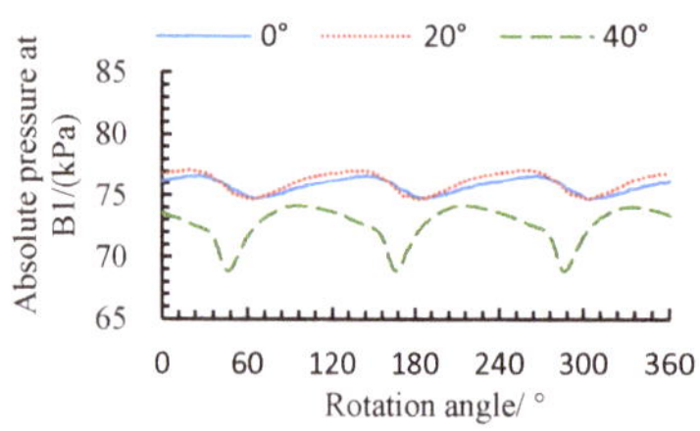

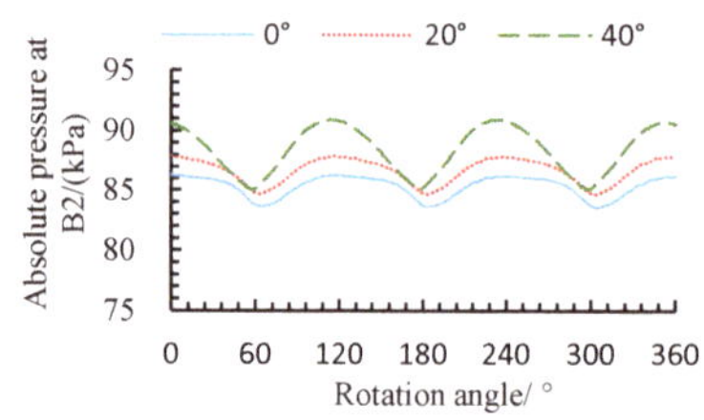

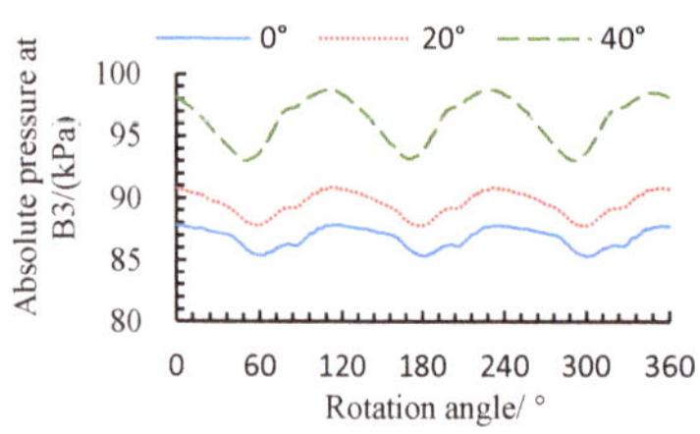

PTP distribution at monitor points

α	0°	20°	40°
PTP at B1 (kPa)	1.9	2.3	5.1
PTP at B2 (kPa)	2.6	3.2	5.6
PTP at B3 (kPa)	2.4	3.0	5.4

(**b**)

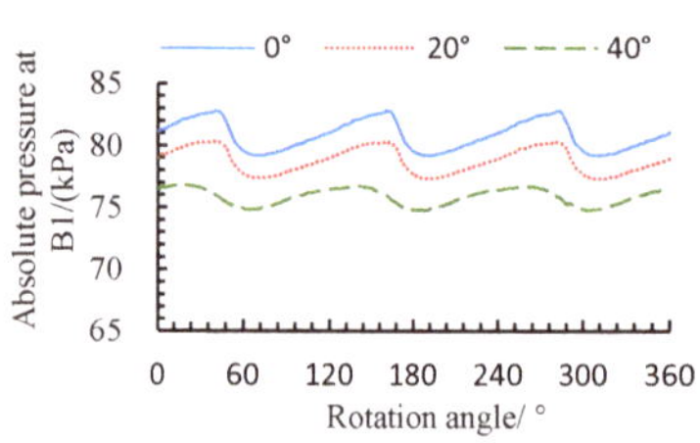

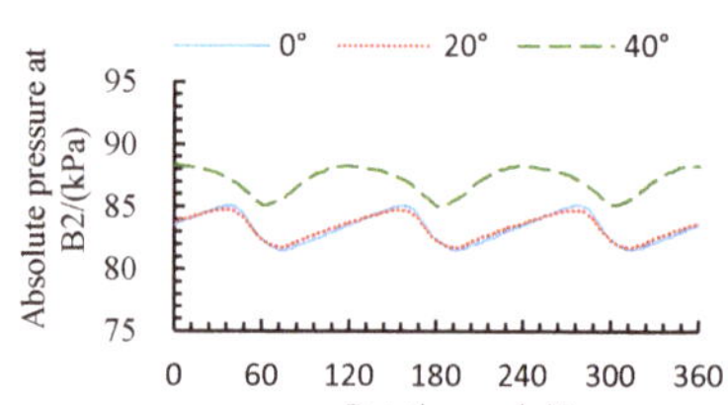

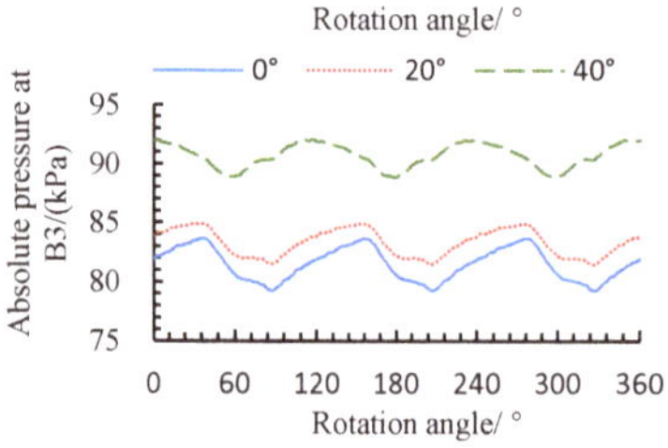

PTP distribution at monitor points

α	0°	20°	40°
PTP at B1 (kPa)	3.6	2.9	1.9
PTP at B2 (kPa)	3.6	3.0	3.1
PTP at B3 (kPa)	4.4	3.4	3.1

(**c**)

Figure 16. Time-domain analysis of different monitor points (B1, B2 and B3) with three α values near the inlet of the guide vanes under (**a**) $0.8Q_{des}$, (**b**) $1.0Q_{des}$, and (**c**) $1.2Q_{des}$.

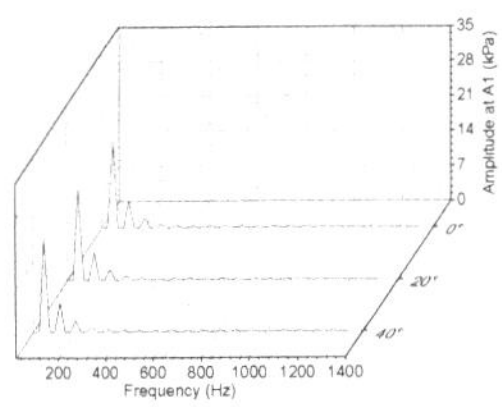

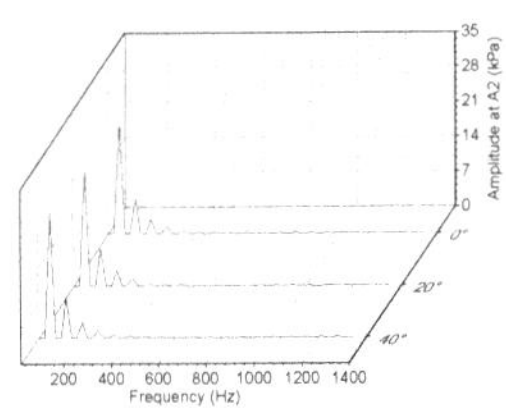

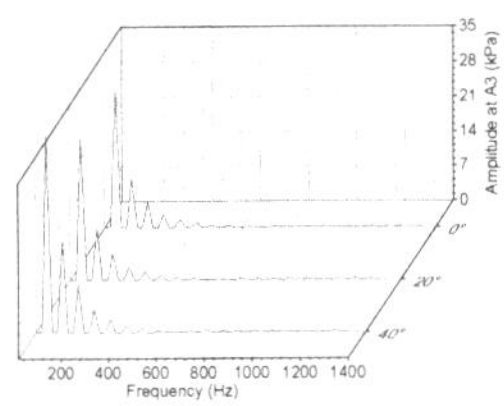

Amplitude of dominant frequency at monitor points			
α	0°	20°	40°
Amplitude of A1 (kPa)	17.3	18.0	18.8
Amplitude of A2 (kPa)	22.6	23.3	25.3
Amplitude of A3 (kPa)	27.7	28.6	39.6

(a)

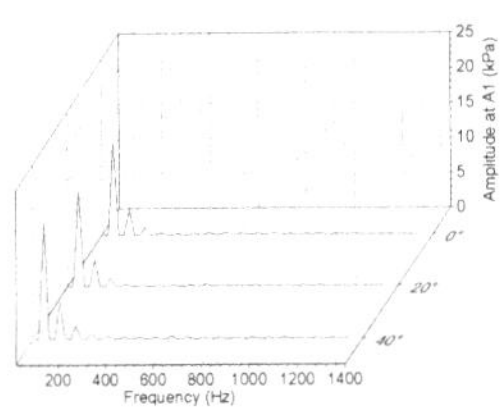

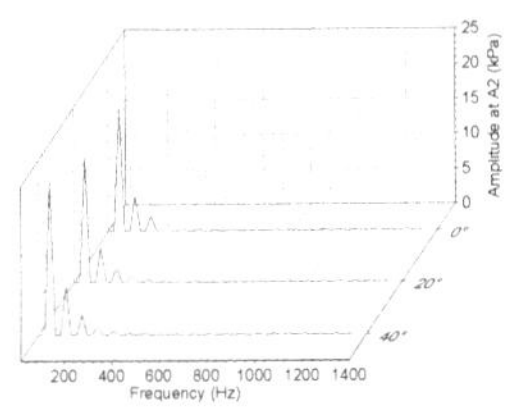

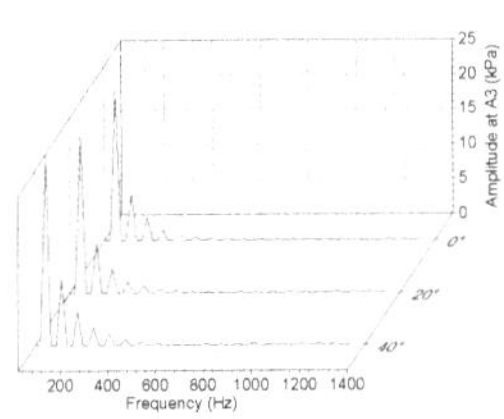

Amplitude of dominant frequency at monitor points			
α	0°	20°	40°
Amplitude of A1 (kPa)	13.1	13.8	16.8
Amplitude of A2 (kPa)	17.3	18.0	21.9
Amplitude of A3 (kPa)	21.3	22.3	27.2

(b)

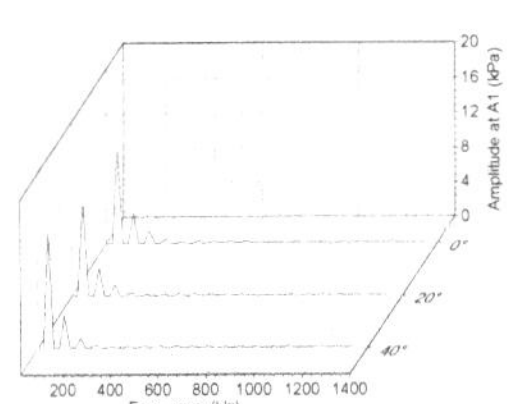

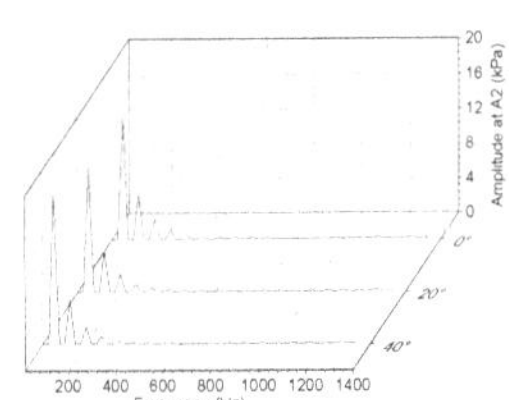

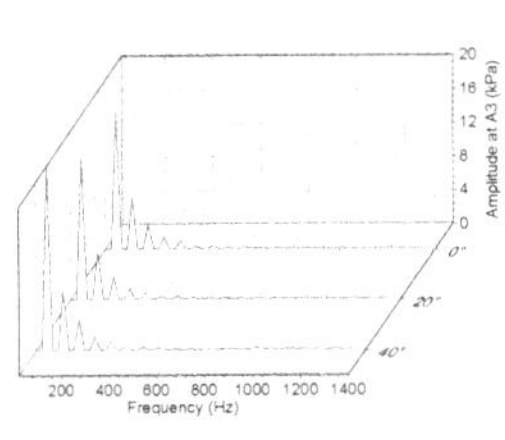

Amplitude of dominant frequency at monitor points			
α	0°	20°	40°
Amplitude of A1 (kPa)	10.5	10.8	13.3
Amplitude of A2 (kPa)	14.0	14.1	17.5
Amplitude of A3 (kPa)	16.1	16.6	21.7

(c)

Figure 17. Frequency-domain analysis of different monitor points (A1, A2 and A3) with three α values near the impeller inlet under (**a**) $0.8Q_{des}$, (**b**) $1.0Q_{des}$, and (**c**) $1.2Q_{des}$.

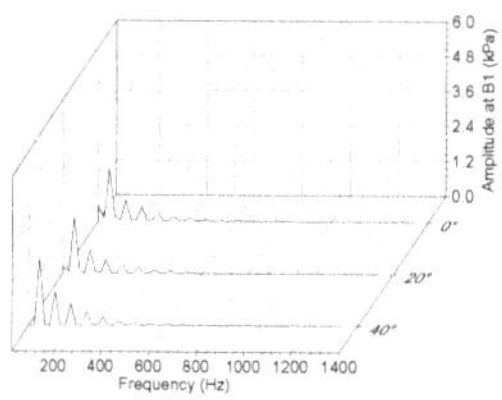

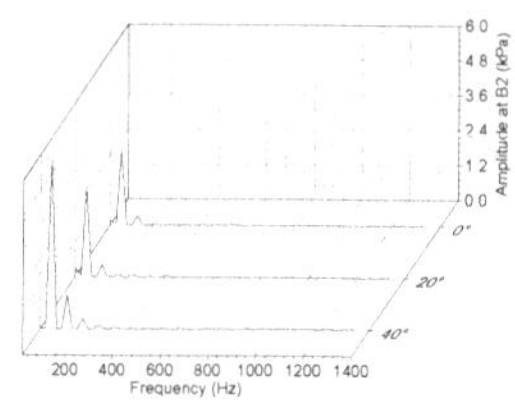

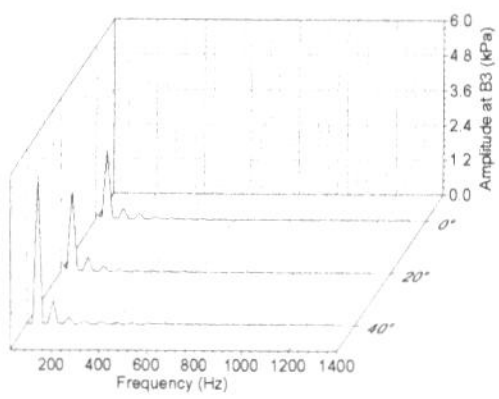

Amplitude of dominant frequency at monitor points

α	0°	20°	40°
Amplitude of B1 (kPa)	1.8	2.0	2.3
Amplitude of B2 (kPa)	2.6	3.1	5.9
Amplitude of B3 (kPa)	2.3	2.8	5.1

(a)

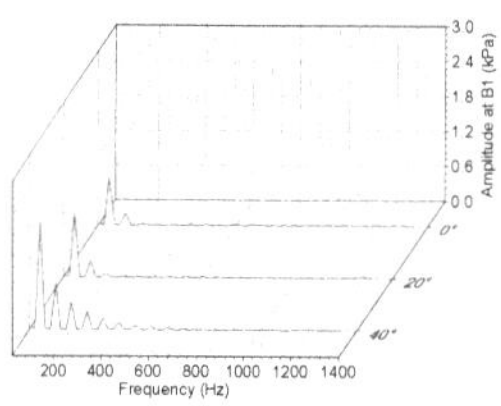

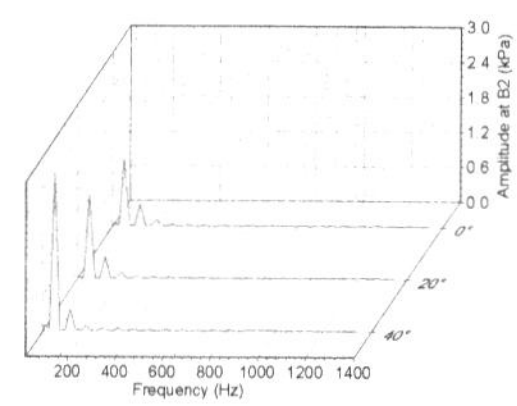

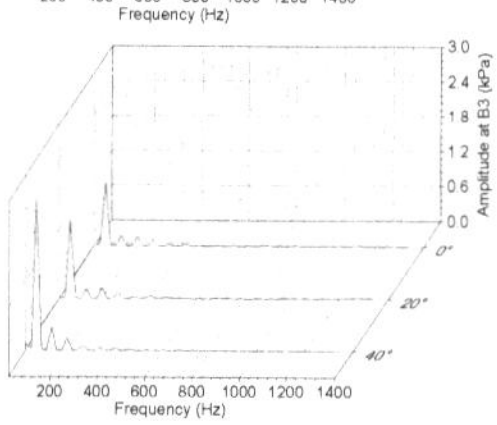

Amplitude of dominant frequency at monitor points

α	0°	20°	40°
Amplitude of B1 (kPa)	0.8	1.1	1.8
Amplitude of B2 (kPa)	1.2	1.4	2.7
Amplitude of B3 (kPa)	1.1	1.4	2.6

(b)

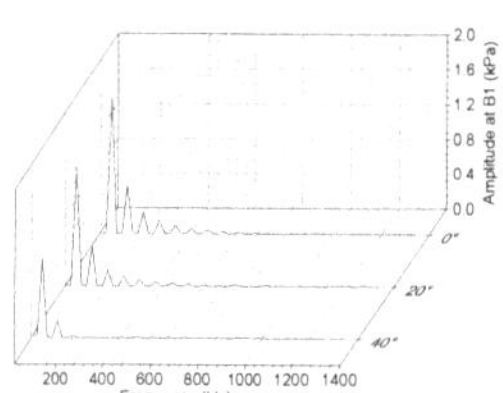

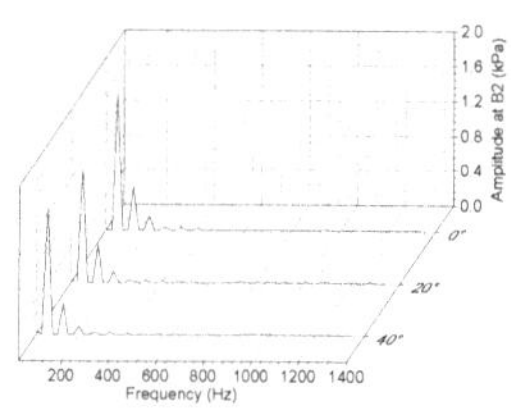

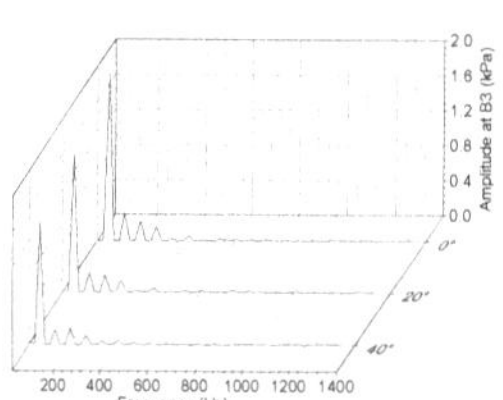

Amplitude of dominant frequency at monitor points

α	0°	20°	40°
Amplitude of B1 (kPa)	1.6	1.4	0.9
Amplitude of B2 (kPa)	1.6	1.3	1.4
Amplitude of B3 (kPa)	1.9	1.6	1.4

(c)

Figure 18. Frequency-domain analysis of different monitor points (B1, B2 and B3) with three α values near the inlet of the guide vanes under (**a**) $0.8Q_{des}$, (**b**) $1.0Q_{des}$, and (**c**) $1.2Q_{des}$.

4. Conclusions

In this work, the unsteady calculation in a vertical axial-flow pump device with three α was employed, and the standard deviation of the pressure pulsation was applied as an evaluation parameter to describe the spatial distribution of the PPI. To investigate the transmission features of the pressure signal, the PTP and fast Fourier function were applied to perform the time- and frequency-domain analyses. The main conclusions are as follows:

(1) As the α increases, the BEP of the vertical axial-flow pump device shifts to a large flow and the head and shaft power gradually increase. This indicates that an increase in α will increase the force on the impeller and further hazard fatigue failure.

(2) The PPI at the impeller inlet with $\alpha = 0°$ is higher than $\alpha = 40°$, resulting from the increased α that reduces the impact effects and axial velocity of the impeller inflow. In addition, an increase in α magnifies the angle between the TEJ and inlet angle of the guide vanes under $0.8Q_{des}$ and $1.0Q_{des}$, but it diminishes the angle between the TEJ and inlet angle of the guide vanes under $1.2Q_{des}$. Thus, the PPI at the inlet of the guide vanes with $\alpha = 40°$ is higher than that with $\alpha = 0°$ under $Q = 0.8Q_{des}$ and $1.0Q_{des}$, while the PPT with $\alpha = 40°$ is lower than that with $\alpha = 0°$ under $Q = 1.2Q_{des}$.

(3) Under $0.8Q_{des}$ and $1.0Q_{des}$, the PTP near the inlet of the impeller and guide vanes increases with the improvement of α. Under $1.2Q_{des}$, the PTP with $\alpha = 40°$ is lower and higher than that with $\alpha = 0°$ near the inlet of guide vanes and impeller, respectively. Furthermore, the dominant frequency is the blade-passing frequency under three α and the amplitude of the dominant frequency rises with an increase in α under $0.8Q_{des}$ and $1.0Q_{des}$.

This work can provide a theoretical reference for fault diagnosis and the optimization design of a vertical axial-flow pump device.

Author Contributions: Conceptualization, F.M. and Z.Q.; Data curation, F.M. and Y.L.; Methodology, F.M.; Formal analysis, F.M. and Z.Q.; Funding acquisition, F.M.; Resources, F.M. and Y.L.; Investigation, F.M.; Software, F.M. and J.C.; Project administration, Y.L.; Supervision, Z.Q.; Validation, F.M. and Z.Q.; Visualization, F.M.; Writing—original draft, F.M.; Writing—review and editing, J.C. and Z.Q. All authors have read and agreed to the published version of the manuscript.

Funding: This research was funded by ranking the top of the list for science and technology projects in Yunnan Province (Grant No. 202204BW050001).

Data Availability Statement: Not applicable.

Conflicts of Interest: The authors declare that there are no conflict of interest.

Nomenclature

H (m)	Head of vertical axial-flow pump device
P_{out} (Pa)	Total pressure in the outlet
P_{in} (Pa)	Total pressure in the inlet
ρ (kg/m^3)	Water density
g (m/s^2)	Gravity acceleration
α (°)	Inlet deflection angle
Span	Radial coefficient
R_s (mm)	Shroud radius
R_h (mm)	Hub radius
CFD	Computational Fluid Dynamics
PPI	Pressure pulsation intensity
PTP	Peak to peak
BPF	Blade-passing frequency
BEP	Best efficiency point
TEJ	Trailing edge jet

References

1. Zhang, X.; Tang, F.; Liu, C.; Shi, L.; Liu, H.; Sun, Z.; Hu, W. Numerical Simulation of Transient Characteristics of Start-Up Transition Process of Large Vertical Siphon Axial Flow Pump Station. *Front. Energy Res.* **2021**, *9*, 382. [CrossRef]
2. Yang, F.; Li, Z.; Hu, W.; Liu, C.; Jiang, D.; Liu, D.; Nasr, A. Analysis of flow loss characteristics of slanted axial-flow pump device based on entropy production theory. *R. Soc. Open Sci.* **2022**, *9*, 211208. [CrossRef] [PubMed]
3. Liu, C. Technical Innovation and Development Analysis of Axial Flow Pump System. *Trans. Chin. Soc. Agric. Mach.* **2015**, *46*, 49–59.
4. Liu, X.; Xu, F.; Cheng, L.; Pan, W.; Jiao, W. Stress Characteristics Analysis of Vertical Bi-Directional Flow Channel Axial Pump Blades Based on Fluid–Structure Coupling. *Machines* **2022**, *10*, 368. [CrossRef]
5. Duan, X.; Tang, F.; Duan, W.; Zhou, W.; Shi, L. Experimental investigation on the correlation of pressure pulsation and vibration of axial flow pump. *Adv. Mech. Eng.* **2019**, *11*, 1687814019889473. [CrossRef]
6. Li, Y.; Lu, R.; Zhang, H.; Deng, F.; Yuan, J. Improvement of intake structures in a two-way pumping station with experimental analysis. *Appl. Sci.* **2020**, *10*, 6842. [CrossRef]
7. Yang, F.; Zhao, H.; Liu, C.; He, J.; Tang, F. Experiment and Analysis on Outlet Flow Pattern and Pressure Fluctuation in Inlet Conduit of Vertical Axial-flow Pumping System. *Trans. Chin. Soc. Agric. Mach.* **2017**, *48*, 141–146.
8. Gao, B.; Zhang, N.; Li, Z.; Ni, D.; Yang, M. Influence of the blade trailing edge profile on the performance and unsteady pressure pulsations in a low specific speed centrifugal pump. *J. Fluids Eng.* **2016**, *138*, 051106. [CrossRef]
9. Cui, B.; Li, W.; Zhang, C. Effect of blade trailing edge cutting angle on unstable flow and vibration in a centrifugal pump. *J. Fluids Eng.* **2020**, *142*, 101203. [CrossRef]
10. Zhang, H.; Zang, J.; Zhang, D.; Shi, W.; Shen, J. Analysis of the Formation Mechanism of Secondary Tip Leakage Vortex (S-TLV) in an Axial Flow Pump. *Machines* **2022**, *10*, 41. [CrossRef]
11. Yang, F.; Chang, P.; Hu, W.; Mao, B.; Liu, C.; Li, Z. Numerical study on pressure pulsation in a slanted axial-flow pump device under partial loads. *Processes* **2021**, *9*, 1404. [CrossRef]
12. Yang, F.; Chang, P.; Li, C.; Shen, Q.; Qian, J.; Li, J. Numerical analysis of pressure pulsation in vertical submersible axial flow pump device under bidirectional operation. *AIP Adv.* **2022**, *12*, 025107. [CrossRef]
13. Yang, F.; Chang, P.; Yuan, Y.; Li, N.; Xie, R.; Zhang, X.; Lin, Z. Analysis of Timing Effect on Flow Field and Pulsation in Vertical Axial Flow Pump. *J. Mar. Sci. Eng.* **2021**, *9*, 1429. [CrossRef]
14. Shi, L.; Zhu, J.; Yuan, Y.; Tang, F.; Huang, P.; Zhang, W.; Liu, H.; Zhang, X. Numerical simulation and experiment of the effects of blade angle deviation on the hydraulic characteristics and pressure pulsation of an axial-flow pump. *Shock Vib.* **2021**, *2021*, 6673002. [CrossRef]
15. Zhang, L.; Wang, S.; Yin, G.; Guan, C. Fluid–structure interaction analysis of fluid pressure pulsation and structural vibration features in a vertical axial pump. *Adv. Mech. Eng.* **2019**, *11*, 1687814019828585. [CrossRef]
16. Shi, L.; Yuan, Y.; Jiao, H.; Tang, F.; Cheng, L.; Yang, F.; Jin, Y.; Zhu, J. Numerical investigation and experiment on pressure pulsation characteristics in a full tubular pump. *Renew. Energy* **2021**, *163*, 987–1000. [CrossRef]
17. Shen, J.; Xu, F.; Cheng, L.; Pan, W.; Ge, Y.; Li, J.; Zhang, J. Simulation of Internal Flow Characteristics of an Axial Flow Pump with Variable Tip Clearance. *Water* **2022**, *14*, 1652. [CrossRef]
18. Yang, F.; Li, Z.; Yuan, Y.; Liu, C.; Zhang, Y.; Jin, Y. Numerical and experimental investigation of internal flow characteristics and pressure fluctuation in inlet passage of axial flow pump under deflection flow conditions. *Energies* **2021**, *14*, 5245. [CrossRef]
19. Long, Y.; Wang, D.; Yin, J.; Hu, Y.; Ran, H. Numerical investigation on the unsteady characteristics of reactor coolant pumps with non-uniform inflow. *Nucl. Eng. Des.* **2017**, *320*, 65–76.
20. Zhou, Q.; Li, H.; Pei, L.; Zhong, Z. Research on non-uniform pressure pulsation of the diffuser in a nuclear reactor coolant pump. *Nucl. Eng. Technol.* **2021**, *53*, 1020–1028. [CrossRef]
21. Xu, R.; Long, Y.; Wang, D. Effects of rotating speed on the unsteady pressure pulsation of reactor coolant pumps with steam-generator simulator. *Nucl. Eng. Des.* **2018**, *333*, 25–44. [CrossRef]
22. Luo, X.; Ye, W.; Huang, R.; Wang, Y.; Du, T.; Huang, C. Numerical investigations of the energy performance and pressure fluctuations for a waterjet pump in a non-uniform inflow. *Renew. Energy* **2020**, *153*, 1042–1052. [CrossRef]
23. Huang, R.; Wang, Y.; Du, T.; Luo, X.; Zhang, W.; Dai, Y. Mechanism analyses of the unsteady vortical cavitation behaviors for a waterjet pump in a non-uniform inflow. *Ocean. Eng.* **2021**, *233*, 108798. [CrossRef]

machines

Article

Experimental Study of Cavitation Damage to Marine Propellers Based on the Rotational Speed in the Coastal Waters

Hae-ji Ju [1] and Jung-sik Choi [2,*]

1 Department of Mechanical Engineering, University of Science and Technology, Daejeon 34113, Korea
2 Division of Coast Guard, Mokpo National Maritime University, Mokpo 58628, Korea
* Correspondence: jungsikchoi@mmu.ac.kr

Abstract: Cavitation in a propeller causes erosion, accelerating corrosion, and tearing off blades in severe cases. Despite the maintenance requirements caused by cavitation, few studies have investigated the cavitation resistance of small ship propellers. Therefore, in this study, these characteristics were investigated through a demonstration test in the coastal waters of South Korea. Cavitation erosion characteristics were analyzed according to the low- and high-rotational speed of the propeller, and the weight was measured every 10 h for 100 h of sailing. The erosion pits were qualitatively compared through liquid penetrant testing (PT) and quantitatively compared by image processing with color edge detection. The results showed that propeller weight loss at high speed was double that at low speed. The cavitation erosion exhibited one cycle each of incubation, acceleration, deceleration, and steady state at low speed, while the acceleration and deceleration periods repeated at high speed. According to PT and color edge detection, the concentration of pits in the low- and high-speed conditions increased towards the trailing and leading edges, respectively. Further, in the radial direction, the trend was similar in both conditions, and the largest number of pits were detected in the region of 0.7–0.9R, where R is the propeller radius.

Keywords: marine propeller; cavitation; erosion; liquid penetrant testing; color edge detection

Citation: Ju, H.-j.; Choi, J.-s. Experimental Study of Cavitation Damage to Marine Propellers Based on the Rotational Speed in the Coastal Waters. *Machines* **2022**, *10*, 793. https://doi.org/10.3390/machines10090793

Academic Editors: Chuan Wang, Li Cheng, Qiaorui Si and Bo Hu

Received: 12 August 2022
Accepted: 7 September 2022
Published: 9 September 2022

1. Introduction

Cavitation is a phenomenon in which the static pressure of a liquid reduces to below its vapor pressure, leading to phase change and causing the vapor bubbles (cavities) to occur, grow, and collapse. The cavitation can occur in a rotating body, such as a propeller, impeller, and hydraulic torque converter, and the cavitation intensity increases as the rotational speed increases [1]. When a marine propeller rotates in water, cavitation occurs primarily on the suction side of the propeller blade. The cavities generated on the suction side grow while traveling along the fluid flow. Then, when the pressure of the fluid increases beyond the saturated vapor pressure, the cavities undergo a rapid transition to liquid, i.e., collapse [2]. The pressure of the collapse impacts the blade surface, causing erosion, noise, and vibration, thereby reducing the propulsion efficiency of the propeller. In severe cases, a blade can tear off due to fatigue failure [3,4]. Therefore, the government mandates propeller inspections through regular and midterm surveys, and the propeller will undergo maintenance or replacement of damaged parts. Despite the risk of cavitation, small marine propellers such as fishing boats are currently manufactured at a basic design level that only considers the average pitch without performing detailed design through model testing or theoretical analysis; in other words, cavitation resistance is not considered [5]. Small marine propellers are known to cause excessive cavitation, and consequently, economic loss occurs due to excessive fuel consumption or the requirements of regular maintenance [5].

The cavitation resistance characteristics of ship propellers have been mainly investigated for large ships, and the problem can be approached from the perspective of hydrodynamic design or materials [6]. The hydrodynamic design approach is useful because it can also achieve the goal of improving the propulsion efficiency of the propeller. However, there are disadvantages in that the application scope for the target vessel is limited, and it is difficult to standardize or generalize the findings. In contrast, the material approach has the advantage of a broader application scope for the target vessel to improve both cavitation and corrosion resistance. In particular, in the case of small ships with varying gross tonnage, the latter approach is considered more suitable in terms of research efficiency. The commonly used materials for ship propellers include aluminum, stainless steel, and alloys of manganese bronze and nickel–aluminum–brass [7]. These materials are applied to outboard motors, outboard motors or small ships, and medium-to-large sized ships, respectively. Aluminum is light and economical, but its tensile strength is approximately half that of stainless steel, resulting in weaker durability. Stainless steel has the highest durability, but its repair cost is approximately twice that of the other materials. Therefore, for medium-to-large sized ships, manganese bronze or nickel-aluminum-brass are applied, which have relatively reasonable costs of repair [7]. Further, composite materials have been studied [8,9], which have many advantages, such as light weight, excellent corrosion resistance, and low noise and vibration induced by cavitation. However, they are sensitive to environmental uncertainties leading to material variation such as matrix cracking and fiber fracture [10]. On the other hand, in the case of the AISI 1045 material, the tensile strength is similar to that of AISI 316 (stainless steel) (565 and 580 MPa, respectively), and the Vickers hardness is higher (240 and 180 HV, respectively), and it is also more economical; however, it has limitations concerning cavitation erosion and corrosion resistance [11,12].

Most studies on the cavitation erosion characteristics of propellers have been performed using material specimens with an ultrasonic system based on the ASTM-32 standard test method [13–15]. To the best of our knowledge, few studies have demonstrated cavitation characteristics using real propellers in sea trials. Demonstration tests of propellers are often replaced by cavitation tunnel or open water tests using model propellers [16]. The reason for this may be due to the difficulties associated with ship operation and measurement. However, model testing focuses on cavitation inception, vibration, and noise measurement for propellers of medium and large ships [17–19]. Only a limited number of studies have reported on cavitation erosion using the painting method [20–22]. In the painting method, paint is applied on the blade surface, and the degree of the peeling of the paint due to cavitation erosion is observed; however, the results of this method may be dependent on the experience of the researchers and the method also has drawbacks such as difficulty in predicting the incubation period and erosion rate of cavitation erosion [23]. Recently, numerical methods [24,25] have been utilized to investigate cavitation in terms of cavity inception, noise, and vibration, but it is also difficult to discuss cavitation erosion.

Demonstration tests were performed with 100 h of sea trial in coastal waters using a leisure boat with twin outboard motors to investigate the cavitation erosion characteristics according to the rotational speed of a small ship propeller. Two propellers were fabricated using AISI 1045, considering its low cost; one of the propellers was surface treated using chromizing and nitriding techniques to enhance erosion and corrosion resistance. The cavitation erosion characteristics of the propeller were analyzed based on its rotational speed through weight measurement every 10 h of sailing. In addition, for comparative analysis of the characteristics of the erosion pits formed on the blades, the liquid penetrant testing method, a nondestructive testing method, was used, and the color edge detection method was applied to compare the concentration of the pits quantitatively. Section 2 describes the experimental and measurement methods of the demonstration test, and Section 3 presents the results and related discussion.

2. Experimental Methods

2.1. Propellers

The four propellers were fabricated based on a drawing obtained from a 3D scan of a standard propeller (maker: Solas Science & Engineering Co., Ltd., Taichung, Taiwan, model number: 3331-114-12) used in outboard engines. The propellers had three blades with dimensions of 28.89 cm and 30.5 cm in diameter and pitch, respectively. A 3D scanner (maker: Leica Geosystems, St. Gallen, Switzerland, model: P50) and a 5-axis machining center (maker: Hwacheon, Gwangju, Korea, model: M4-5AX) were used to fabricate the propellers with AISI 1045 as the material. The scanner had a measuring accuracy of 3 mm, and the machining center had a high-precision rotary encoder controlled to the extent of 0.0001°. The surface of two propellers was treated with plasma nano-nitriding [26], followed by pack-chromizing coating [27]. Further details of the surface treatment techniques are not described as they are beyond the scope of this study.

2.2. *Sea Trial in Coastal Waters*

The demonstration test was conducted in adjacent waters of the marina of Mokpo National Maritime University in South Korea. The propeller and vessel used in the test are shown in Figure 1a,b, respectively. The ship (registration number: MB-01-8663) had a total tonnage of 0.6 tons and a total length of 5.99 m, and it was equipped with two 37 kW outboard motors (maker: Honda, Tokyo, Japan, model: BF50DK4). The sea trials for the demonstration test were conducted for 100 h each at two different average rotational speeds of 720 and 1440 rpm, during which the average ship speeds were approximately 3.5 and 5.4 m/s, respectively. A set of propellers, surface-treated and untreated propellers, were applied to each test. The advance coefficient ($J = V/nD$, where V is the ship speed, n is the rotational speed, and D is the diameter of the propeller), which is also called the apparent advance coefficient [28], was approximately 1.00 (for 720 rpm) and 0.78 (for 1440 rpm). To minimize the corrosion effect, the propeller was removed and stored on land while berthing. The weight of the propeller was measured every 10 h of sailing, and nondestructive testing was performed after the sea trial. Although the seawater conditions (such as density) in the coastal waters were not strictly controlled in the demonstration test, the effect is considered insignificant since the sea trial was conducted on the same route every time.

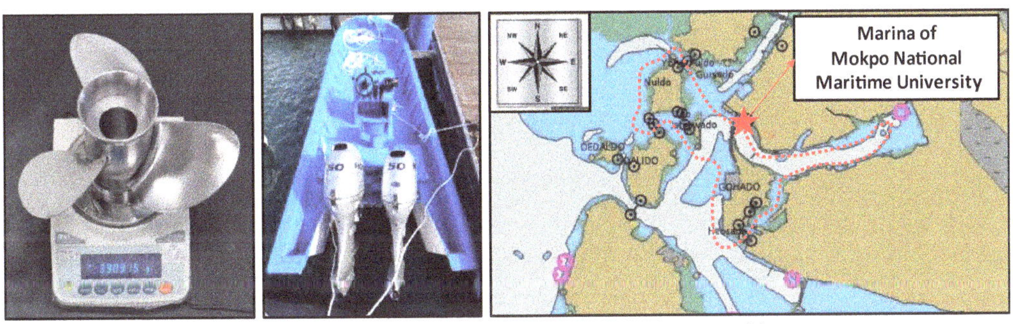

(a) (b) (c)

Figure 1. (**a**) Shape of the propeller and weight measurement, (**b**) a ship used for the demonstration test, and (**c**) the sailing route near (the red dot line) the Marina of Mokpo National Maritime University.

2.2.1. Weight Measurement of the Propeller

Since weight loss of the propeller occurs from the separation of materials due to erosion, measurement of the weight loss of the specimen is used to analyze the cavitation erosion characteristics, generally using ultrasonic systems [29]. In the demonstration test, it is impossible to detach the propeller in water; thus, it is not practically possible to track the weight change of the propeller. However, in this study, an outboard motor was used to enable the detachment of the propellers. The weight of the propeller was measured every 10 h of sailing using an electronic balance (maker: A&D Korea Ltd., Seoul, Korea, model: FX-5000i,) (see Figure 1a), which has a weight capacity of 5200 g and a resolution of 0.01 g. Korea Marine Equipment Research Institute, a test accreditation organization for marine equipment, calibrated the electronic balance and issued the test certificate (Certification no. KOMERI-0601-21T4176) for the weight measurement. The certificate included the mean weight value with three measurements. Before weighing, to remove corrosion products and other attachments, the propeller was cleaned using Clark's solution according to the method in ASTM G1-03 [30].

2.2.2. Nondestructive Testing (NDT)

The nondestructive testing (NDT) method is useful for the inspection of cracks or defects of an object without destroying its material. For the inspection of propellers in industrial applications, NDT is mandatory in regular or midterm surveys during the dry-dock period [31]. In this study, among the NDT methods, liquid penetrant testing (PT) defined in ISO 3452 [32] was used for the nondestructive examination of propeller erosion damage. The PT was performed by CMS Service Corp., which was approved by Korean Register [33] (Service Provider Approval Certificate No.: MKPNT-0001-21), a company that specializes in NDT. PT is an NDT method for inspecting discontinuities (that is, cracks) on the surface of a specimen regardless of the material. If a penetrant is applied to the surface of a specimen, the part of the penetrant that has penetrated a discontinuity remains within the discontinuity even after the penetrant is wiped off the surface. After the penetrant is wiped from the surface, if a developer such as a dye is applied to the surface, the penetrant in the discontinuity is extracted, enabling observation of the presence of defects and their shapes.

2.2.3. Image Processing

For quantitative analysis of the erosion status according to the local region in the propeller blade, color edge detection [34] was performed using the image obtained through the PT in the lighting environment using a LED light (maker: KOMI, Suwon, Korea, model: Cyclops I). In the images obtained by the PT method, the background color of the blade becomes uniform due to the dye application. Therefore, it is a suitable environment for the application of the RGB color edge detection technique using the color change of the blade surface. The basic concept of edge detecting is to detect a point in an image with a large intensity change (that is, an intensity gradient). Although image data are discontinuous, since they are arranged at regular intervals, the gradient is detected by differentiating adjacent pixels. Well-known edge detector methods [34–36] include Sobel, Prewitt, Canny, and Laplacian. However, they are sensitive to gradients, and for complex images, these methods have the drawback of extracting too many edges. Figure 2 shows the edge detection results using the PT image at low-speed condition. Figure 2b shows the result of applying the Canny detector, and it can be seen that small pits are detected with a thick distribution of too many edges. Based on this example, the above detection methods that use only intensity changes in 2D space are unsuitable for detecting fine pits, as required in this study. Therefore, the color edge detection method using the gradient in 3D space was applied using the RGB values of the image in the mask of Sobel 3 × 3. Assuming that the r, g, and b of each pixel in the image are unit vectors in the R, G, and B axes of the RGB color space, the vectors in each direction are defined in Equation (1), and the direction and magnitude of the maximum rate of change are defined in Equations (2) and (3), respectively [34,37]. The image obtained by color edge detection is saved in grayscale, and the image shown in Figure 2c is obtained accordingly.

$$u = \frac{\partial R}{\partial x}r + \frac{\partial G}{\partial x}g + \frac{\partial B}{\partial x}b,\ v = \frac{\partial R}{\partial y}r + \frac{\partial G}{\partial y}g + \frac{\partial B}{\partial y}b \tag{1}$$

$$\theta(x, y) = \frac{1}{2}\tan^{-1}\left[\frac{2g_{xy}}{g_{xx} - g_{yy}}\right] \tag{2}$$

$$F_{\theta}(x, y) = \left\{1/2\left[\left(g_{xx} - g_{yy}\right) + \left(g_{xx} - g_{yy}\right)\cos 2\theta(x, y) + 2g_{xy}\sin 2\theta(x, y)\right]\right\}^{1/2} \tag{3}$$

where, g_{xx}, g_{yy}, and g_{xy} are defined as the dot product of vectors, u·u, v·v, and u·v, respectively.

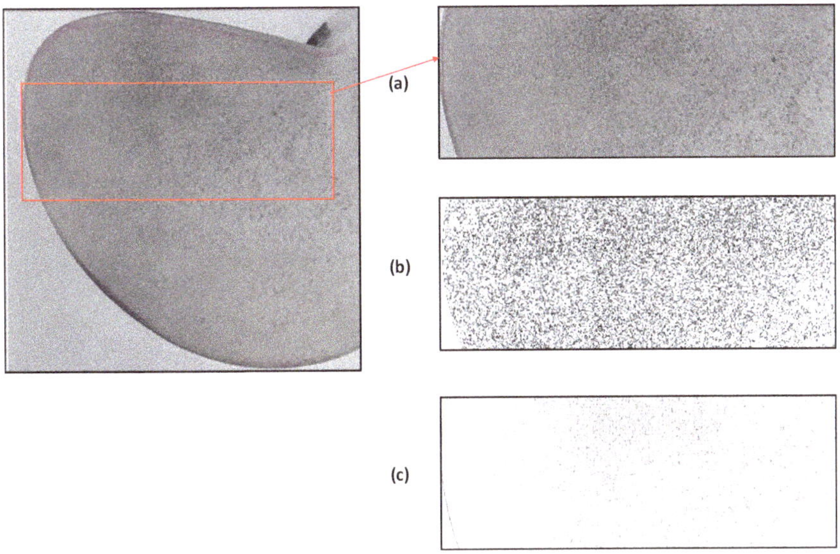

Figure 2. Edge detection images: (**a**) a photo from the liquid penetrant testing of the blade suction side after 100 h of low-speed condition, (**b**) an image processed with Canny edge detection, and (**c**) an image processed with color edge detection.

The light intensity was tried to maintain the same in the PT images of low- and high-speed conditions, but the ambient lighting could be slightly different because the pictures were captured at different times due to the extended experimental time. The light intensity can affect the gradients of RGB values in the region that includes the background of the picture near the blade tip of the propeller. Nevertheless, the gradient values except for the region mentioned above cannot be significantly affected by the light intensity when the light intensity variation is insignificant because the value is a gradient of nearby RGB values in the mask with the same light intensity.

3. Results and Discussion

3.1. Weight Change of the Propeller

Figure 3 illustrates the weight loss (%) of the propeller under the different rotational speeds, and the standard deviation of the measured values from the mean was less than 0.0095, and the error was less than 0.0054 in all cases. It can be seen that the treated propeller did not show any change in weight related to the rotational speed, which indicates that the cavitation resistance of the propeller made of AISI 1045 was improved through surface treatment. However, the propeller without surface treatment showed a decrease in weight of approximately 0.58% (22.5 g) and 0.28% (10.7 g) in the high and low-speed conditions, respectively, for 100 h of sailing. The high-speed condition showed approximately twice the weight loss compared to the low-speed condition. This is because, as shown in Equation (4), the number of cavitations (σ_n) decreased as the rotational speed of the propeller increased, and the lower this value, the higher the possibility of cavitation [38].

$$\sigma_n = \frac{P - P_v}{0.5\rho(nD)^2} \tag{4}$$

where P is static pressure, P_v is vapor pressure, ρ is liquid density, n denotes rotational speed, and D represents diameter. When the static pressure is assumed by $P_A + \rho g h$, where P_A is atmospheric pressure and h is propeller shaft immersion [39], the cavitation number was estimated at 1.95 for low speed and 0.5 for high speed in the propeller local radius of 0.7.

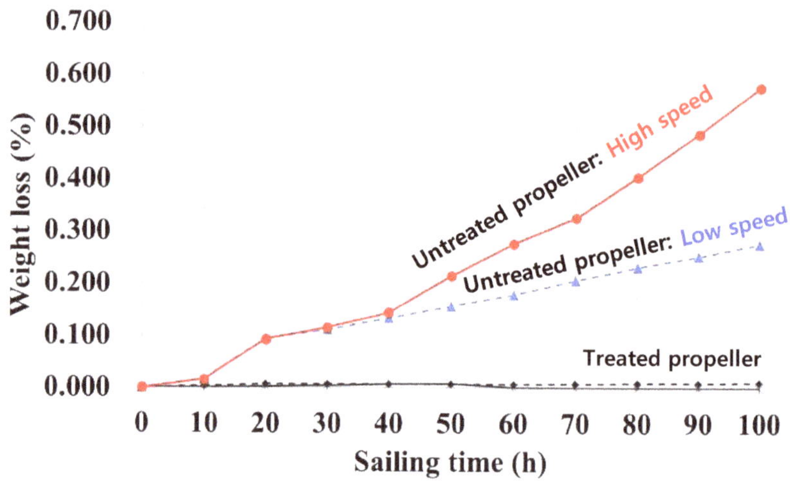

Figure 3. Weight loss as a function of sailing time for the untreated and treated propellers at high- and low-speed conditions.

Figure 4 illustrates the weight loss measured every 10 h in terms of the rate of change per unit of time. Since the treated propeller showed no weight loss, only the untreated propeller is presented in the figure. According to [40], cavitation erosion characteristics are classified into four periods: incubation, acceleration, deceleration, and steady state. Similar trends were observed in low-speed condition. The incubation period refers to the period with minimal weight change, which is related to the indentation of the material due to the collapse energy [21]. From Figure 4, it is considered that this period occurred prior to the 10 h mark. After this period, when the erosion pits began to overlap, the acceleration period showed a rapid increase in weight loss [21]. Up to the end of the acceleration period (that is, up to 20 h), the rotational speed of the propeller was not affected. In the deceleration period, the rate of weight loss began to reduce because, due to a change in the surface geometry from overlapping pits, the fluid filled the pits and dampened the cavitation impact [40]. The damping effect caused the cavitation curve to enter the deceleration period under both rotational speed conditions, but the effect was more prominent in the low-speed condition. As the rotational speed of the propeller increases, the centrifugal force increases, resulting in a reduction of the volume of fluid that causes the damping effect in the pits. After the deceleration period, the low-speed condition entered the steady state and remained in this state relatively well, whereas the high-speed condition repeated the increase/decrease after a temporary period of steady state and eventually (after 80 h) showed a larger value of erosion rate than that originally shown in the acceleration period. Unlike in the low-speed condition, where the distribution of cavitation erosion pits was concentrated on the trailing edge side, as shown in Figure 2, the high-speed condition showed the erosion pits distributed across the area of the blade, as shown in Figure 5 which shows the PT images before and after sailing. This is because the threshold leading to weight loss differs depending on the blade region. The weight loss during the first 40 h was similar for the two speeds, resulting from the similar erosion pits near the trailing edge. After 40 h, the damping effect due to water filling in the pits resulted in a steady rate of weight loss in the low-speed condition. However, in the high-speed condition, the trend of rate of weight loss could have been affected by two possibilities: the formation of a new erosion region and the increased erosion in the existing pits despite the damping effect. We consider that the formation of a new erosion region is the reason for the repetition during

40–70 h. The impact of the weight loss of the erosion pits on the leading-edge side appeared relatively later in the experiment; therefore, it is considered that a second cycle occurred. In other words, the threshold value would be larger on the leading-edge side, considering that the pit distribution on the leading-edge side is mainly observed in high-speed condition. After 70 h, the trend of rate of weight loss could have resulted from the increased erosion in the existing pits. The relative amount of erosion pits in each blade region is described in detail in Section 3.2.

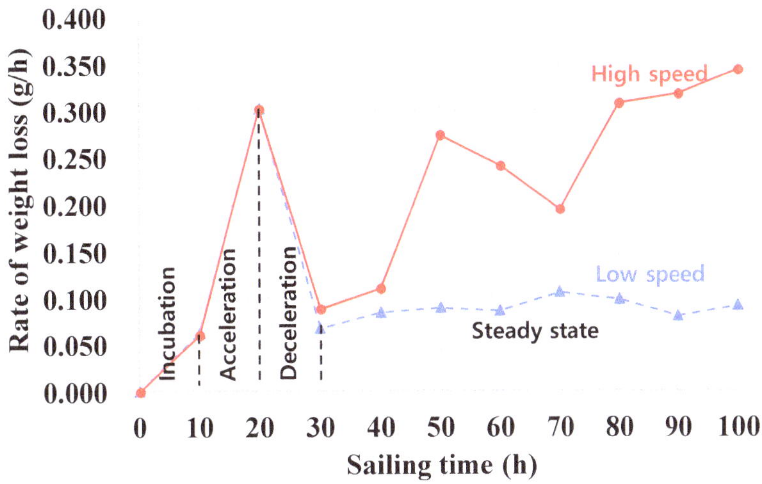

Figure 4. Rate of weight loss as a function of sailing time for the untreated propellers at high- and low-speed conditions.

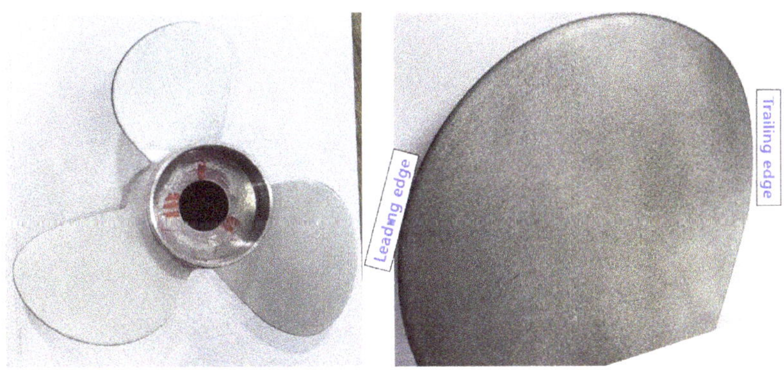

Figure 5. Photo from the liquid penetrant testing of the untreated propeller blade: (**left**) the suction side before sea trial and (**right**) one of the suction side blades after 100 h of sea trial with high-speed condition.

3.2. Erosion Pit Detection

As described in Section 2.2.3, since the PT image has a uniform background color and facilitates the identification of pits, a quantitative comparison of the pits in each blade area according to rotational speed was performed through color imaging detection, as shown in Figure 6. Figure 6 is a grayscale image, a result of the color edge detection, and it has a resolution (i.e., matrix of size, M × N) the same as that of the PT image. The value with grayscale means a gradient of RGB in the mask of Sobel (3 × 3), and the stronger pigment (closer to black) indicates the increased presence of cavitation erosion. R denotes the radius of the propeller. A more precise relative comparison is possible with the processed image compared to the original photograph, and the area with the largest pit depth (indicated by the red circle in Figure 6) can be identified.

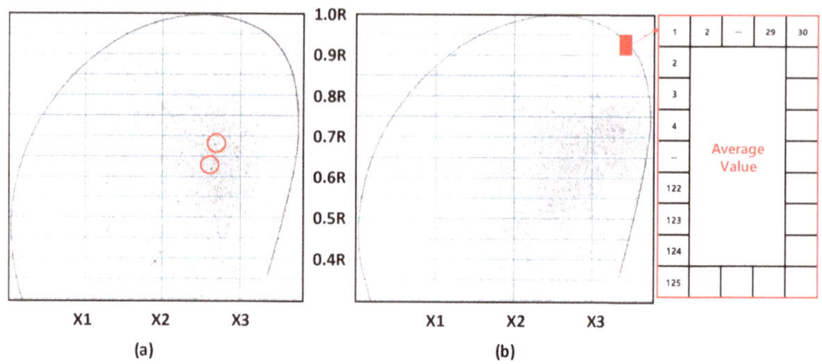

Figure 6. Images from the color edge detection for the PT photo after 100 h of sailing: (**a**) for high-speed condition and (**b**) for low-speed condition.

For a more quantitative comparison, Figure 7 shows the difference of the values averaging with a matrix of size 125 × 30 between the images with low- and high-speed conditions, i.e., the size of the x direction is the original matrix size M divided by 30, and that of the y direction is the original matrix size N divided by 125 as shown in the box in Figure 6. Each graph in Figure 7 represents for the local region of the propeller in the radial direction from 0.4R to R, and the y-axis represents the difference in grayscale values of low- and high-speed conditions. The x-axis represents the blade position in the horizontal direction, and X1–X3 are the same as the positions in the x-direction indicated in Figure 6. The 0 level refers to the same pit concentration in the same region between images with low- and high-speed conditions. A red area greater than 0 indicates that the pit concentration of the high-speed condition is large, and a blue area less than 0 indicates that it of the low-speed condition is large.

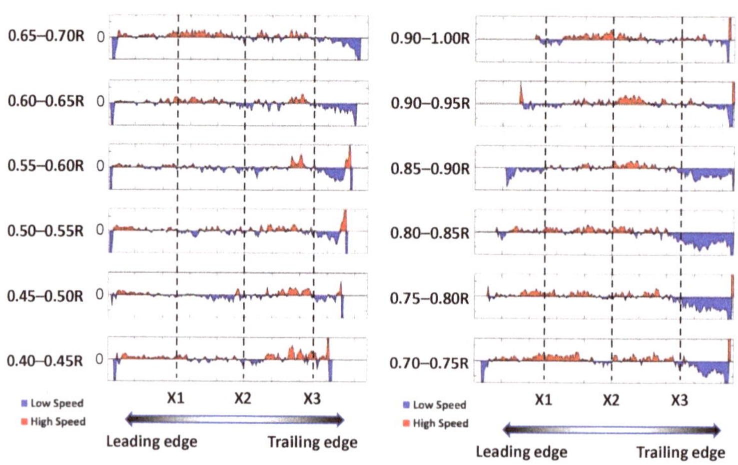

Figure 7. Comparison of grayscale values for each area within the blade according to the rotation speed condition: red is high speed, and blue is low speed.

As predicted in Section 3.1, in the high-speed condition, the erosion pits are more concentrated before X3 than in the low-speed condition. In particular, in the region between X2 and X3 of Figure 6, although the two conditions appear to show similar results, it can be seen in Figure 7 that the pit area of the high-speed condition is larger. After X3 (trailing-edge side), the distribution of erosion pits was concentrated in the low-speed condition. Figure 8 presents the variation in the radial direction obtained by normalizing the grayscale average value with the largest value, an average value in the 0.8–0.85R region at the low-speed condition, in the region of each radius. In all regions of radial direction except 0.75–0.9R, more erosion pits were detected at the high-speed condition. It can also be seen that the value of the tip side was larger than that of the hub side of the propeller in both conditions.

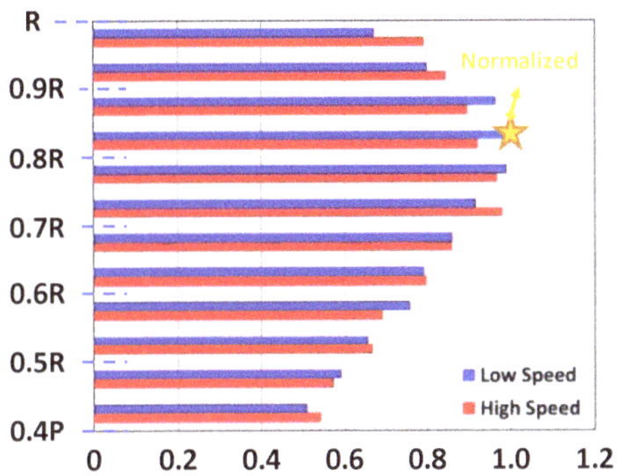

Figure 8. Changes in erosion pit concentration in the radial direction of the propeller.

4. Conclusions

In this study, cavitation characteristics of small ship propellers were investigated with different rotational speed conditions through a demonstration test with sea trials in the coastal waters of South Korea. As a result of sailing for 100 h at each propeller rotation condition of 720 and 1440 rpm using a small boat equipped with twin outboard motors, the propeller made of AISI 1045 showed twice as much weight loss in the high-speed condition as that in the low-speed condition. This indicates that the cavitation erosion increased with an increase in the rotational speed of the propeller. As a result of comparing the trend based on the values of the propeller weight change measured every 10 h of sailing, four periods, i.e., incubation, acceleration, deceleration, and a steady state, were observed in a single cycle at the low speed condition as reported in previous studies. At the high-speed condition, after the deceleration period, there was a repetition of acceleration and deceleration, and then the acceleration section appeared again. It is considered that depending on the position in the blade, both the cavitation impact and the threshold value leading to weight loss are different.

In the photo obtained from PT and the images obtained from color edge detection, it can be seen that in the low-speed condition, erosion pits were mainly observed on the trailing edge side, whereas in the high-speed condition, erosion pits were distributed across the blade area. As a result of performing color edge detection, a comparison of the erosion pits under high-speed and low-speed conditions was possible according to different positions in the blade. With reference to the 3/4 point in the horizontal direction, the concentration of pits in the low-speed condition increased toward the trailing edge, and the concentration of pits in the high-speed condition increased toward the leading edge. In addition, similar trends were observed in the radial direction in both conditions, and the largest concentration of pits was detected in the region of 0.7–0.9R.

Author Contributions: Conceptualization, H.-j.J. and J.-s.C.; methodology, H.-j.J.; software, H.-j.J.; validation, H.-j.J. and J.-s.C.; formal analysis, H.-j.J.; investigation, H.-j.J. and J.-s.C.; resources,

H.-j.J.; data curation, H.-j.J. and J.-s.C.; writing—original draft preparation, H.-j.J.; writing—review and editing, H.-j.J. and J.-s.C.; visualization, H.-j.J.; supervision, J.-s.C.; project administration, J.-s.C.; funding acquisition, J.-s.C.; All authors have read and agreed to the published version of the manuscript.

Funding: This work was supported by project for Collabo R&D between Industry, Academy, and Research Institute funded by Korea Ministry of SMEs and Startups in 2020 (Project No. S2908407).

Institutional Review Board Statement: Not applicable.

Informed Consent Statement: Not applicable.

Data Availability Statement: Data associated with this research are available and can be obtained by contacting the corresponding authors.

Acknowledgments: The authors kindly acknowledge the Korea Ministry of SMEs and Startups for their support of R&D project No. S2908407. Furthermore, special thanks to CEO Kim, Geun-myun of CMS Service Corp. for his contribution to Non-Destructive Testing (NDT).

Conflicts of Interest: The authors declare that they have no conflict of interest.

References

1. Guo, M.; Liu, C.; Yan, Q.; Wei, W.; Khoo, B.C. The Effect of Rotating Speeds on the Cavitation Characteristics in Hydraulic Torque Converter. *Machines* **2022**, *10*, 80. [CrossRef]
2. Fivel, M.C.; Franc, J.P. Cavitation Erosion. In *ASM Handbook, Volume 18: Friction, Lubrication, and Wear Technology*; Totten, G.E., Ed.; ASM International: Novelty, OH, USA, 2018; pp. 290–301. Available online: https://hal.archives-ouvertes.fr/hal-01697361 (accessed on 12 August 2022).
3. Helal, M.M.; Ahmed, T.M.; Banawan, A.A.; Kotb, M.A. Numerical prediction of sheet cavitation on marine propellers using CFD simulation with transition-sensitive turbulence model. *Alex. Eng. J.* **2018**, *57*, 3805–3815. [CrossRef]
4. Carlton, J. Chapter 27—Propeller Maintenance and Repair. In *Marine Propellers and Propulsion*, 4th ed.; Butterworth-Heinemann: Oxford, UK, 2019.
5. Lee, W.J.; Kim, M.C.; Chun, J.H.; Jang, J.Y.; Mun, W.J.; Lee, C.S. Study on Propeller Design for Fishing Vessel's High Efficiency. *J. Korean Soc. Mar. Environ. Energy* **2011**, *14*, 73–80. [CrossRef]
6. Karimi, A.; Martin, J.L. Cavitation erosion of materials. *Int. Met. Rev.* **1986**, *31*, 1986. [CrossRef]
7. Koko, T.S.; Shahin, K.O.; Akpan, U.O.; Norwood, M.E. Review of Composite Propeller Developments and Strategy for Modeling Composite Propellers using PVAST. In *DRDC Atlantic CR-2011-156*; Defence R & D Canada: Ottawa, ON, Canada, 2012.
8. Kim, J.-H.; Lee, H.; Kim, S.-H.; Choi, H.-Y.; Hah, Z.-H.; Seol, H.-S. Performance Prediction of Composite Marine Propeller in Non-Cavitating and Cavitating Flow. *Appl. Sci.* **2022**, *12*, 5170. [CrossRef]
9. Islam, F.; Caldwell, R.; Phillips, A.W.; St John, N.A.; Prusty, B.G. A review of relevant impact behaviour for improved durability of marine composite propellers. *Compos. Part C Open Access* **2022**, *8*, 100251. [CrossRef]
10. Young, Y.L. Analysis and Design of Advanced Marine Propulsors. In Proceedings of the ASME 2010 29th International Conference on Ocean, Offshore and Arctic Engineering, Shanghai, China, 6–11 June 2010.
11. Hattori, S.; Ishikura, R. Revision of cavitation erosion database and analysis of stainless steel data. *Wear* **2010**, *268*, 109–116. [CrossRef]
12. Hattori, S.; Ishikura, R.; Zhang, Q. Construction of database on cavitation erosion and analyses of carbon steel data. *Wear* **2004**, *257*, 1022–1029. [CrossRef]
13. Basumatary, J.; Nie, M.; Wood, R.J.K. The synergistic effects of cavitation erosion—Corrosion in ship propeller materials. *J Bio-Tribo-Corros.* **2015**, *1*, 12. [CrossRef]
14. Tomlinson, W.J.; Talks, M.G. Erosion and corrosion of cast iron under cavitation conditions. *Tribol. Int.* **1991**, *24*, 67–75. [CrossRef]
15. Tian, Y.; Zhao, H.; Yang, R.; Liu, X.; Li, H.; Chen, X. Effects of Bacillus sp. adhesion on cavitation erosion behaviour of nickel aluminium bronze in artificial seawater. *Wear* **2022**, *498*, 204344. [CrossRef]
16. Bertram, V. Chapter 2—Propellers. In *Practical Ship Hydrodynamics*; Elsevier: Amsterdam, The Netherlands, 2011. [CrossRef]
17. Choi, G.H.; Chang, B.J.; Cho, D.S. Comparison of cavitation patterns between model scale observations using model and full-scale wakes and full scale observations for a propeller of crude oil carrier. *J. Nav. Archit. Mar. Eng.* **2011**, *48*, 15–22.
18. Sampson, R.; Turkmen, S.; Aktas, B.; Shi, W.; Fitzsimmons, P.; Atlar, M. On the full scale and model scale cavitation comparisons of a Deep-V catamaran research vessel. In Proceedings of the Fourth International Symposium on Marine Propulsors SMP'15, Austin, TX, USA, 31 May–4 June 2015.
19. Atlar, M.; Aktas, B.; Sampson, R.; Seo, K.C.; Viola, I.M.; Fitzsimmons, P.; Fetherstonhaug, C. A multi-purpose marine science and technology research vessel for full-scale observations and measurements. In Proceedings of the 3rd International Conference on Advanced Model Measurement Technologies for the Marine Industry, Gdansk, Poland, 17–18 September 2013.
20. ITTC 2005. The specialist committee on cavitation erosion on propellers and appendages on high powered/high speed ships. In Proceedings of the 24th ITTC–Volume II, Edinburgh, UK, 11–13 April 2005.

21. Bark, G.; Berchiche, N.; Grekula, M. Application of principles for observation and analysis of eroding cavitation. In *The EROCAV Observation Handbook*; Department of Shipping and Marine Technology, Chalmers University of Technology: Göteborg, Sweden, 2004.
22. Aktas, B.; Ustab, O.; Atlar, M. Systematic investigation of coating application methods and soft paint types to detect cavitation erosion on marine propellers. *Appl. Ocean. Res.* **2020**, *94*, 101868. [CrossRef]
23. Van Terwisga, T.J.C.; van Wijngaarden, H.C.J.; Bosschers, J.; Kuiper, G. Cavitation research on ship propellers—A review of achievements and challenges. In Proceedings of the Sixth International Symposium on Cavitation, CAV2006, Wageningen, The Netherlands, 11–15 September 2006. Report No. 1543-P.
24. Kimmerl, J.; Mertes, P.; Abdel-Maksoud, M. Application of Large Eddy Simulation to Predict Underwater Noise of Marine Propulsors. Part 1: Cavitation Dynamics. *J. Mar. Sci. Eng.* **2021**, *9*, 792. [CrossRef]
25. Köksal, Ç.S.; Usta, O.; Aktas, B.; Atlar, M.; Korkut, E. Numerical prediction of cavitation erosion to investigate the effect of wake on marine propellers. *Ocean Eng.* **2021**, *239*, 109820. [CrossRef]
26. Kim, S.G.; Kim, S.W. The effect of activated nitrogen species for diffusion rate during a plasma nitriding process. *J. Korean Soc. Heat Treat.* **2010**, *23*, 150–155.
27. Kim, S.G.; Yeo, K.H.; Lee, J.H.; Cho, Y.K.; Okumiya, M. Effect of Formation of Concentrated Stainless Steel Layer by Thermochemical Reaction and Addition of Hydrocarbon Gas on Sintered Part. *Adv. Mater. Phys. Chem.* **2018**, *8*, 227. [CrossRef]
28. ITTC Propulsion Committee. Model manufacture, propeller models terminology and nomenclature for propeller geometry. In Proceedings of the International Towing Tank Conference, Fukuoka, Japan,, 14–20 September 2008.
29. Lin, C.; Zhao, Q.; Zhao, X.; Yang, Y. Cavitation Erosion of Metallic Materials. *Int. J. Georesour. Environ.* **2018**, *4*, 1–8. [CrossRef]
30. ASTM G1-03. Standard Practice for Preparing, Cleaning, and Evaluating Corrosion Test Specimens. ASTM: West Conshohocken, PA, USA, 2011.
31. Korea Resister. Part 1—Classification and surveys. In *Rules and Guidance for the Classification of Steel Ships*; Bureau Veritas Marine & Offshore: Seoul, South Korea, 2020.
32. ISO 3452:2021. Non-Destructive Testing—Penetrant Testing. ISO: Geneva, Switzerland, 2021.
33. Korea Resister. Available online: https://www.krs.co.kr/kor/ (accessed on 12 August 2022).
34. Rafael, C.G.; Richard, E.W.; Steven, L.E. *Digital Image Processing using MATLAB*; McGraw-Hill: New York, NY, USA, 2011; ISBN 978-0-07108-478-9.
35. Gonzalez, R.; Woods, R. *Digital Image Processing*, 4th ed.; Pearson Education: New York, NY, USA, 2018; ISBN 978-1-29222-304-9.
36. Jain, A.; Gupta, M.; Tazi, S.N.; Deepika. Comparison of edge detectors. In Proceedings of the International Conference on Medical Imaging, m-Health and Emerging Communication Systems (MedCom), Greater Noida, India, 7–8 November 2014; pp. 289–294. [CrossRef]
37. Di Zenzo, S. A note on the gradient of a multi-image. *Comput. Vis. Graph. Image Processing* **1986**, *33*, 116–125. [CrossRef]
38. Lee, I.; Park, S.; Seok, W.; Rhee, S.H. A study on the cavitation model for the cavitating flow analysis around the marine propeller. *Math. Probl. Eng.* **2021**, *2021*, 2423784. [CrossRef]
39. Molland, A.F.; Turnock, S.R.; Hudson, D.A. Propeller Characteristics. In *Ship Resistance and Propulsion*; Cambridge University Press: Cambridge, UK, 2017; pp. 277–312. [CrossRef]
40. Zakrzewska, D.E.; Krella, A.K. Cavitation erosion resistance influence of material properties. *Adv. Mater. Sci.* **2019**, *19*, 18–34. [CrossRef]

machines MDPI

Article

Investigation of the Aerodynamic Performance of the Miller Cycle from Transparent Engine Experiments and CFD Simulations

Marcellin Perceau [1,2], Philippe Guibert [1], Adrian Clenci [3,*], Victor Iorga-Simăn [4], Mihai Niculae [3] and Stéphane Guilain [2]

1 Institut Jean le Rond d'Alembert, Sorbonne Université, CNRS UMR 7190, 78210 Saint-Cyr-l'Ecole, France; marcellin.perceau@sorbonne-universite.fr (M.P.); philippe.guibert@sorbonne-universite.fr (P.G.)
2 Renault, 78280 Guyancourt, France; stephane.guilain@renault.com
3 Department Automobiles and Transport, University of Pitești, 1, Tg. din Vale Street, 110040 Pitești, Argeș, Romania; mihai.niculae@upit.ro
4 Regional R&D Center for Materials, Processes and Innovative Products, University of Pitești, 1, Tg. din Vale Street, 110040 Pitești, Argeș, Romania; victor.iorga@upit.ro
* Correspondence: adi.clenci@upit.ro

Abstract: This paper assesses the effect of the Miller cycle upon the internal aerodynamics of a motored transparent spark ignition engine via CFD simulation and particle image velocimetry. Since the transparent Miller engine does not allow for measurements in the roof of the combustion chamber, the extraction of information regarding the aerodynamic phenomena occurring here is based on CFD simulation, i.e., the results of the CFD simulation are used to allow for the extrapolation of the experimental data; thus, they are used to complete the picture regarding the aerodynamic phenomena occurring inside the whole cylinder. The results indicate that implementing the early intake valve closing strategy to obtain the Miller cycle has a negative impact on the mean kinetic energy, turbulent kinetic energy, and fluctuating velocity toward the end of the compression stroke, thus affecting, the combustion process. This supports the need to intensify the internal aerodynamics when applying the Miller cycle such that the turbulence degradation is not too big and, consequently, to still gain efficiency in the Miller cycle.

Keywords: Miller cycle; EIVC; PIV; CFD; RANS; turbulence

Citation: Perceau, M.; Guibert, P.; Clenci, A.; Iorga-Simăn, V.; Niculae, M.; Guilain, S. Investigation of the Aerodynamic Performance of the Miller Cycle from Transparent Engine Experiments and CFD Simulations. *Machines* **2022**, *10*, 467. https://doi.org/10.3390/machines10060467

Academic Editors: Davide Astolfi, Chuan Wang, Li Cheng, Qiaorui Si and Bo Hu

Received: 12 April 2022
Accepted: 6 June 2022
Published: 11 June 2022

1. Introduction

The knocking and the emission of NO_x in a spark-ignition (SI) engine depend on the pressure and temperature of the gases at the end of compression—in other words, on the volumetric compression ratio—while the thermodynamic efficiency of the SI engine depends in particular on the work provided during the expansion. Consequently, one way to improve the thermodynamic efficiency is to equip the engine with a volumetric expansion ratio that is higher than the volumetric compression ratio. A relatively simple method to achieve different compression and expansion ratios was proposed by Miller [1,2]. The "classic" configuration of the engine is retained, the compression stroke equals the expansion stroke, but the reduction in the volumetric compression ratio is obtained by either early or late intake valves closing (EIVC/LIVC). The valves close either before the arrival of the piston in the bottom dead center (BDC) position, which is followed by an expansion of the air over the rest of the intake stroke, or they remain open during the start of the compression phase and part of the gas admitted is rejected into the intake [2,3]. According to [4], compared with the conventionally throttled Otto cycle, the Miller over-expansion cycle revealed an efficiency improvement of 6.3% in the LIVC mode and of 7% in the EIVC mode. In fact, as stated in the same study, due to its particularities (see above), the Miller cycle can provide a decrease in NO_x emissions of up to 46%.

As shown before, the Miller cycle makes it possible to benefit from an additional rate of expansion, leading to a better thermodynamic efficiency with neither the inconvenience of the risk of knocking nor the aggravation of NO_x emissions. Both aspects are crucial for the future of the SI engine, as the first is directly related to the CO_2 emission, which must be lowered by any means so as to reduce the intensification of the greenhouse effect, whilst the second is associated with pollution and, consequently, with human health [5,6].

Moreover, in the context of the current and future European homologation standards (e.g., Euro6d-full and Euro7), the use of the Miller cycle on SI engines will be one of the preferred choices among the manufacturers [7–9]. After September 2017, as a result of the adoption of homologation in RDE (Real Driving Emissions) conditions [10], it has been imperative to control the engine at lambda 1 over the entire operating range in order to ensure that the best efficiency of the exhaust aftertreatment system is obtained (i.e., maximum depollution). However, operating at lambda 1 under high loads and speeds causes a significant increase in the temperature of the exhaust gases. Usually, before the application of the RDE procedure, the manufacturers enriched the mixture in order to reduce the temperature of the exhaust gases below the limit of the traditional materials used on the exhausts of the vehicles [11]. Obviously, this was done at the cost of pollutants and fuel consumption (i.e., CO_2 emissions). This is no longer possible in the current regulatory context; hence, there is a need to find other solutions to otherwise reduce the exhaust temperature. Due to over-expansion, the use of the Miller cycle also goes in this direction. Therefore, its application determines a significant improvement of the SI engine's efficiency, but it also allows for a significant drop in the temperature of the exhaust gases to the level where it is even possible to use a classic variable geometry turbine (VGT) with further positive effects on fuel efficiency and drivability [12,13]. For instance, on this matter, the research highlighted in paper [8] showed that the low intake charge associated with the EIVC strategy generates a reduced final compression temperature, which permits the compression ratio (i.e., expansion ratio) to be raised to 12.5:1 without encountering extra NO_x emissions or knocking issues and with maximum exhaust temperatures of only 880 °C. Thus, the synergy between the Miller EIVC and VGT was proved. Furthermore, the use of turbocharging is beneficial, as one drawback of the Miller cycle is the reduction of the intake charge/mass generated by the alteration of the valve timing (either the EIVC or LIVC), as shown in [4]. In other words, turbocharging may compensate for the inherent power reduction of the Miller engine compared with a standard SI engine based on the Otto cycle.

Another significant downside of the EIVC Miller cycle is that it strongly affects the internal aerodynamics, which are paramount for the air-fuel mixing and the flame propagation velocity. Indeed, considering the EIVC case, because of a reduced opening time, the tumble motion specific to the SI engine is disturbed, thus having a negative impact on the turbulent kinetic energy later on in the compression stroke [14]. The results of the experimental investigation presented in paper [3] also support the previous statement: the turbulent kinetic energy is reduced to almost zero toward the end of the compression stroke before producing the spark, finally leading to inefficient combustion. In fact, these studies [3,14], underline that the Miller cycle is capable of improvements in the matter of efficiency, but the EIVC strategy has a strong influence on the engine flow field [3], which may need further improvements, such as intake valve masking, intake valve seat masking, or specially-designed intake ducts for a higher tumble, as studied by Niculae et al. [15]. In other words, to gain efficiency, the turbulence degradation inherent to the Miller cycle must not be too high. However, as argued in [14], the EIVC drawbacks regarding the reduced turbulence intensity may be advantageous at high loads and at lower and medium engine speeds for the best fuel efficiency potential and CO_2 reduction potential.

Regarding the implementation of the Miller cycle, both versions can be found on the market. Some manufacturers prefer to implement the EIVC version, while some prefer the LIVC version and others implement both, depending on the engine operating point. For example, in [7], the EIVC strategy was used due to the best fuel efficiency results that can be achieved in the partial-load area with a combination between the early closing and the large overlap of the intake/exhaust valves. Additionally, the EIVC offers advantages concerning the knock conditions for the supercharged engines. Moreover, as presented in paper [16], the turbulence inside the cylinder at the end of the compression stroke under the EIVC strategy is higher than the turbulence achieved with the LIVC strategy.

The aim of this study is to assess the effect of the Miller cycle upon the internal aerodynamics of a motored/driven SI engine via particle image velocimetry (PIV) and 3D computational fluid dynamics (CFD) investigations. Relating to the PIV results, the transparent engine does not allow for measurements in the roof of the combustion chamber (i.e., this one is unobservable). Only the inside of the transparent cylinder is accessible. Consequently, the extraction of information regarding the aerodynamic phenomena occurring in the upper part of the cylinder when approaching the top dead center (TDC) position was impossible. Since this piece of information is crucial for the flame propagation occurring around the TDC position, to estimate the lacking information, CFD simulation was used as a complementary investigation tool. In other words, the results of the 3D CFD numerical simulation were used to allow for the extrapolation of the data gathered through the PIV investigation and, finally, to complete the picture regarding the aerodynamic phenomena occurring inside the cylinder for a Miller SI engine.

Thus, after this introduction outlining the study in the current context, the remaining part of the paper is organized into three main sections: Section 2 presents the PIV investigation focused on gathering data about the internal aerodynamics of the engine during the intake and the compression strokes, Section 3 shows the CFD approach as a tool to allow for the estimation of aerodynamic phenomena in the region where the PIV is no longer able to provide information, and Section 4 deals with the results and a discussion. Finally, the conclusions of this study are summarized.

2. Experimental Setup

As already introduced, the internal aerodynamics of a Miller cycle SI engine based on the EIVC strategy was first studied experimentally via the PIV technique, which is presented below.

2.1. Model Engine

The model engine was representative of a 0.9 L three-cylinder SI engine for passenger cars and is used solely for non-combustion cycle studies. It was made of a single transparent cylinder offering wide optical access to study the aerodynamic movements inside the combustion chamber. The cylinder head contains four valves—two for the intake and two for the exhaust—in a pent-roof geometry. The engine has a 12.3 compression ratio, an 81.2 mm stroke, and a 72.2 mm bore. A flat piston head was used because a real piston head shape can induce distortions in the laser sheet that could lead to inaccurate positioning and reflections. The cylinder head–piston distance has been changed to maintain the reference compression ratio value, as explained in [17]. Obviously, since the real piston is not used, the internal aerodynamics of the original engine are not fully replicated in the transparent engine. In other words, the effect of the real piston is not captured; yet, it may be assumed that it has a low influence.

All the measurements were made with the engine driven at 1200 rpm by an electric motor and at an intake pressure of 1 bar. Since the engine has a glass cylinder without lubrication, this relatively low speed was chosen in order to avoid damaging it.

The PIV experiments were performed in the tumble plane, which passes through the axis of the cylinder and the middle of the two intake and exhaust valves (Figure 1).

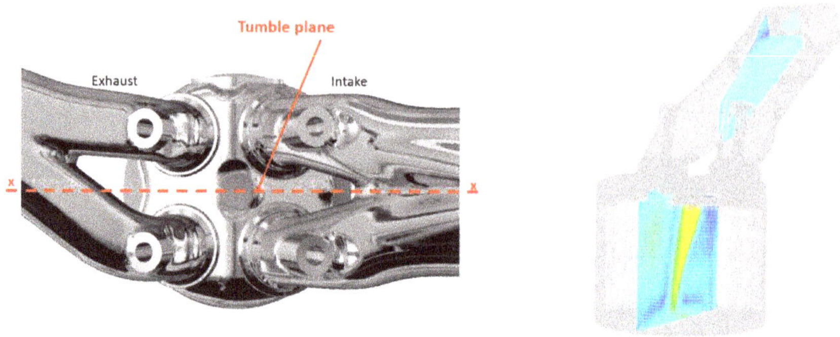

Figure 1. The tumble plane investigated with the particle image velocimetry (PIV).

Figure 2 comparatively illustrates the valve lift laws for the Miller case (the dashed curves) and for the Otto case (the continuous curves). Thus, it can be observed that the transparent engine followed an EIVC Miller cycle. In fact, to highlight the influence of an earlier intake valve closing, two Miller cases were used in this study: one characterized by a 30° CA offset at the intake camshaft (see the curve represented with long blue dashes) and the other by a 20° CA offset (see the curve represented with small blue dashes). These two intake valve lift laws were coupled with the following exhaust offsets: 0° CA and 30° CA (see the curves represented with small and long red dashes). Both cases were possible thanks to a variable valve timing (VVT) system. The intake valve opening duration measured for a reference lift of 0.7 mm is 140° CA in the Miller case and 160° CA in the Otto case. The original engine is meant to operate following the Otto cycle only in the full load condition.

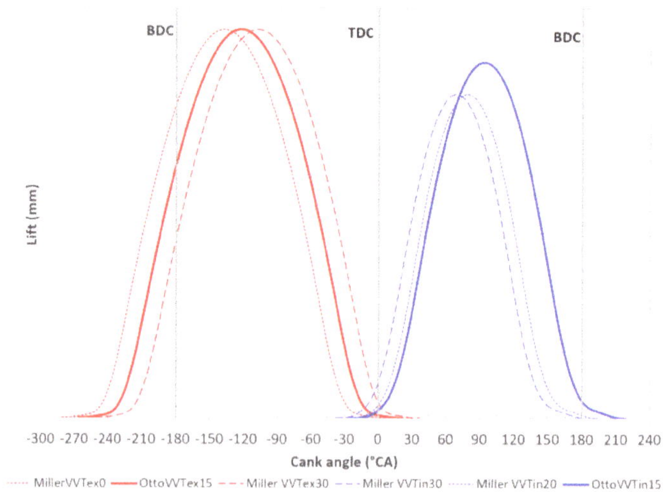

Figure 2. Exhaust and intake events: valve lift laws.

2.2. *Particle Image Velocimetry Measurements*

There are several types of PIV technologies that can be used on optically accessible engines. For example, the PIV system used by the Institute of Aerodynamics from RWTH Aachen University is a high-speed stereo PIV [18]. This system is used to study the internal aerodynamics of an externally driven (1500 rpm) optical Miller cycle engine. The laser light sheet is directed into the cylinder using a 45° mirror. The measurements are conducted at a temporal resolution of 0.1° CA.

Another example of a high-speed PIV system used with the Miller cycle was at the Institute of Technology of Tokyo [19]. This study was conducted on an optically accessible engine driven at 2000 rpm. The PIV was performed at 12 kHz with the following characteristics: a resolution of 1° CA; tracer particles with a diameter between 2–2.5 μm illuminated with a laser with a maximum power of 50 W and a wavelength of 532 nm; a laser pulse width of 200 ns; a laser sheet thickness of 313 μm at the bore center.

The PIV system, available at Darmstadt University and used for optical measurements, is a Tomographic PIV applied for a motored single-cylinder engine driven at 800 rpm [20]. In this case, the PIV system is equipped with four cameras in a "Scheimpflug" arrangement in a circular plane around the cylinder. The laser light sheet is also directed into the cylinder using a 45° mirror.

Another type available at the same institute is the Thermographic PIV, which uses luminous phosphor particles to trace the fluid motion, these particles also being used for the measurement of the particles' temperature [21].

Certainly there are other types of PIV systems such as micro PIV, holographic PIV, tomographic PIV [22], scanning PIV [23], laser speckle velocimetry [24], molecular tagging velocimetry [25], or laser doppler velocimetry [26], each one with advantages and drawbacks.

In our case, the in-cylinder flow was seeded with oil particles. An aerosol generator produced particles of approximately 1–10 μm at a concentration of more than 108 particles per cm^3. These particles were then illuminated using a light sheet with a thickness of less than 1 mm. Light was emitted from a double pulsed Nd-YAG laser (Specta Physics PIV 200) at a wavelength of 532 nm. The separation time between the two laser pulses was 15μs. The maximum output energy was around 200 mJ per pulse, with a pulse duration in the range of 8–10 ns. This laser provided a stable cycle-to-cycle illumination of the transparent cylinder at a $10^{\pm 2}$ Hz frequency (Figure 3), which means that one velocity field is recorded for every engine cycle at the same crank angle. After going through a Nikon AF Nikkor 50 mm f/1.8D, the scattered light was captured on a monochrome charge-coupled device camera (Jai CV-M2 CL) with a resolution of 1600 × 1200 pixels.

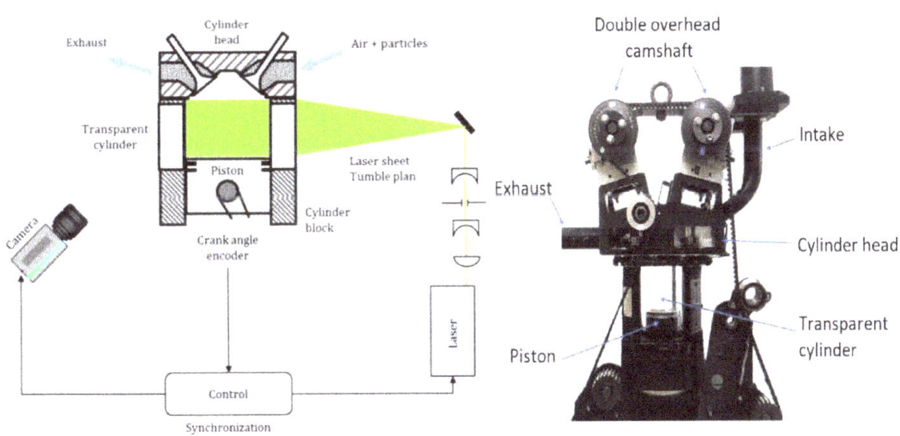

Figure 3. Particle image velocimetry (PIV) measuring system and transparent engine.

The synchronization of these devices was performed using a controller that took the signal of the engine crank angle encoder as the input. For a given crank angle, the measurement began after the rotation speed of the engine was stable. It then took 300 pairs of images corresponding to 300 engine cycles at the desired crank angle, which were used to obtain the average flow field, as shown in Section 4.1. This procedure was repeated for any other crank angle for which the flow fields are desired, meaning that the engine is switched off before recording another crank angle. An adaptive cross-correlation PIV algorithm from Dantec Dynamic Studio v6.8 software is used to process the pairs of images. The interrogation area went from 32 × 32 to 64 × 64 pixels according to the particle density, and a grid step size of 16 × 16 pixels was used. For each instantaneous velocity field, 76 × 61 instantaneous velocity vectors were finally provided, with a spatial resolution of about 2 mm.

3. Numerical Investigations

As already stated, this purpose of the CFD investigation is to provide data for the extrapolation of the PIV results toward the end of the compression. The CFD investigation is a 3D RANS-based (Reynold-average Navier–Stokes) cold simulation of the transparent single-cylinder Miller cycle engine motored at 1200 rpm. The simulation is termed cold because the engine is not fired. The CFD investigation was performed using the AVL Fire^{TM} 2022 R1 software (Graz, Austria), which is based on the finite volume approach. Additionally, the k-ζ-f turbulence model was used [27,28].

3.1. *Reynolod-Average Navier–Stokes Formalism and Turbulence Model*

The movement of the air inside the cylinder during the intake and the compression strokes or the internal aerodynamics of an engine are very important factors in the combustion process. Internal aerodynamics play a major role in volumetric efficiency, the heat exchange with cylinder walls, the homogenization of fresh mixture, and, finally, chemical reactions during combustion, which depend on the characteristics of the transport and turbulent diffusion.

Navier–Stokes equations are used to describe the fluid dynamics as laminar or turbulent. They are non-linear, second-order partial differential equations without exact solutions. From an engineering standpoint, sometimes only the averaged mean velocity fields are looked for (e.g., the fluid–structure interaction). Consequently, for these reasons, the CFD methods consider the statistically averaged equations derived from the principle of Reynolds decomposition, i.e., RANS. This approach is about decomposing the flow properties into an average value and a fluctuation component, which gives rise to a new, unknown Reynold-stress tensor that has to be modeled in order to solve the rest of the equations [29]. The RANS model offers a good balance between the duration of the simulation, the accuracy of the results, and cost effectiveness. The RANS model is usually used to predict the engine in-cylinder mean flow and turbulence distribution since it is good enough to capture the overall qualitative flow trends [30]. This is also supported by studies [31,32] that highlight the fact that CFD simulation using the RANS approach is an efficient and reliable tool for the fluid flow of the internal combustion (IC) engine. Nevertheless, the approach is limited for unsteady flows, while the hybrid models such as the large eddy simulation (LES) or unsteady Reynolds-averaged Navier–Stokes (URANS) provide a better prediction of smaller fluctuations. However, hybrid models require higher numerical schemes, smaller time steps, and higher resolutions for the computational grids. Therefore, the RANS approach can be used to save processing power (CPU) and reduce the simulation time, depending on the type of research.

According to [33], choosing a suitable turbulence model for IC engines is an unresolved issue in the scientific community. It is argued that most turbulence models are validated for specific well-defined test cases, but it is not certain that they are also valid for other type of flows generated by complex geometries where the flow might be different. The models used in [33] were: LES (Smagorinsky, WALE, Sigma), the hybrid model, and the second generation of unsteady Reynolds-averaged Navier–Stokes (URANS). The results indicated that LES with the Sigma turbulence model is suitable for IC engine flow and, compared with other hybrid models, highlights a higher number of resolved turbulent structures. Nevertheless, for most of the cases, the URANS approach offered reasonable results to describe the fluid flow for higher engine speeds and a higher Reynolds number [33]. On the other hand, as discussed in [34], most URANS models are poor in recreating the swirling flow in steady-flow configurations, while the LES model offers better results. Nonetheless, as mentioned in [35,36], LES applications are not straightforward due to the need for specific boundary conditions and severe grid resolutions/quality compared to RANS. In fact, as seen in the scientific literature, there are many methods and approaches already available, but none of them are suitable for every situation.

In terms of the turbulence models that can be used with the RANS approach, there are several such examples, each with its own advantages and disadvantages. For example, the k-ε model is simple to implement, leads to stable calculations, and is reasonably accurate for a variety of flows, but it is only valid for fully turbulent flows, predicts the swirling and rotating flows rather poorly, and comes down to a simplistic ε equation. The k-ω model solves two additional partial differential equations, but it suffers from similar drawbacks as the k-ε model. The Reynolds stress model (RSM) has a good accuracy in predicting complex flows such as swirling flows, rotating flows, or flows that involve separation, but it requires double or even triple the processing power in order to perform the numerical simulation [37].

In order to obtain a smoother scaling, especially near solid walls, the k-g model can be used, where g represents the square root of a characteristic turbulent time scale [38,39]:

$$g = \sqrt{\frac{k}{\varepsilon}}, \tag{1}$$

As shown in [38], this model reveals good and consistent results for steady intake jet flows in the IC engine, while for more complex reciprocating flows, the results are mixed.

Another turbulence model used, similar to the k-ε model, is k-ζ-f, which provides a good balance between computational cost and the accuracy of the results due to the particularities of the ζ term modeling [40]. According to [40], both RANS with RSM and the k-ζ-f model produce reasonable results with a good accuracy in resolving the dominant structures of the turbulent flow. In comparison with LES, the k-ζ-f model has a better advantage in managing computational cost efficiency, even compared with the RSM, which it is 20% faster than. Regarding the two RANS models, paper [40] showed that the k-ζ-f model provides a better prediction of flow separation and reattachment. The k-ζ-f model has been developed by Hanjalic, Popovac and Hadziabdic [27]. The authors propose a version of the eddy viscosity (ζ) model based on Durbin's elliptic relaxation concept, f [41]. As stated in [27,28], the aim was to improve the numerical stability of the original $v2 - f$ model, which had become increasingly popular as empirical damping functions were removed due to the employment of an additional velocity scale $v2$ derived by using an elliptic relaxation concept (f). As argued in [42], "the four-equation k-ζ-f model is very robust and more accurate than the simpler two equation eddy viscosity models. On average, the computing time is increased by up to 15% compared with the computing time needed for the k-ε model calculations". Nevertheless, as stated in [42], "the model is usable for a relatively coarse mesh next to the wall, but the cell next to the wall should reach a non-dimensional wall distance y^+ of 3 as a maximum". Moreover, as reported in the same publication [42], "the flow in an intake port shows good agreement between computed and measured discharge coefficient". Equally, one particular advantage is that it is sufficiently robust to be used for computations involving grids with moving boundaries and highly compressed flows, as is the case in IC engines.

In light of the above considerations, the k-ζ-f turbulence model with a maximum cell size of 1.0 mm was chosen for our CFD simulation.

3.2. Simulation Description

The 3D geometry of the engine can be seen in Figure 4, which also presents the positions of the sensors used to continuously monitor the inlet and outlet pressures, as well as the in-cylinder pressure. The first two signals (Figure 5) were used to impose the boundary conditions (BC), while the in-cylinder pressure signal served to provide a mean for comparison with the result of the CFD simulation.

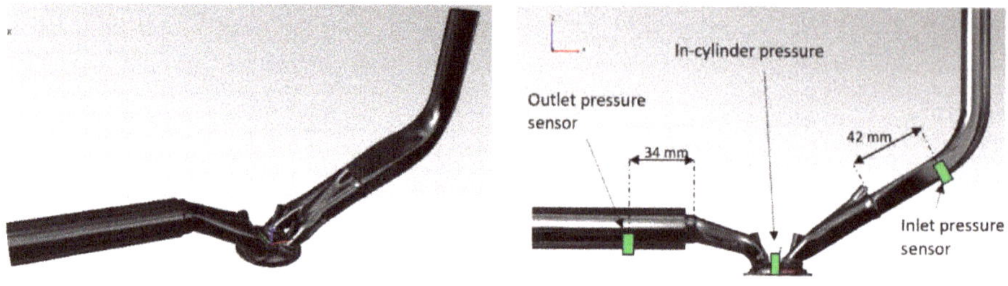

Figure 4. 3D engine geometry.

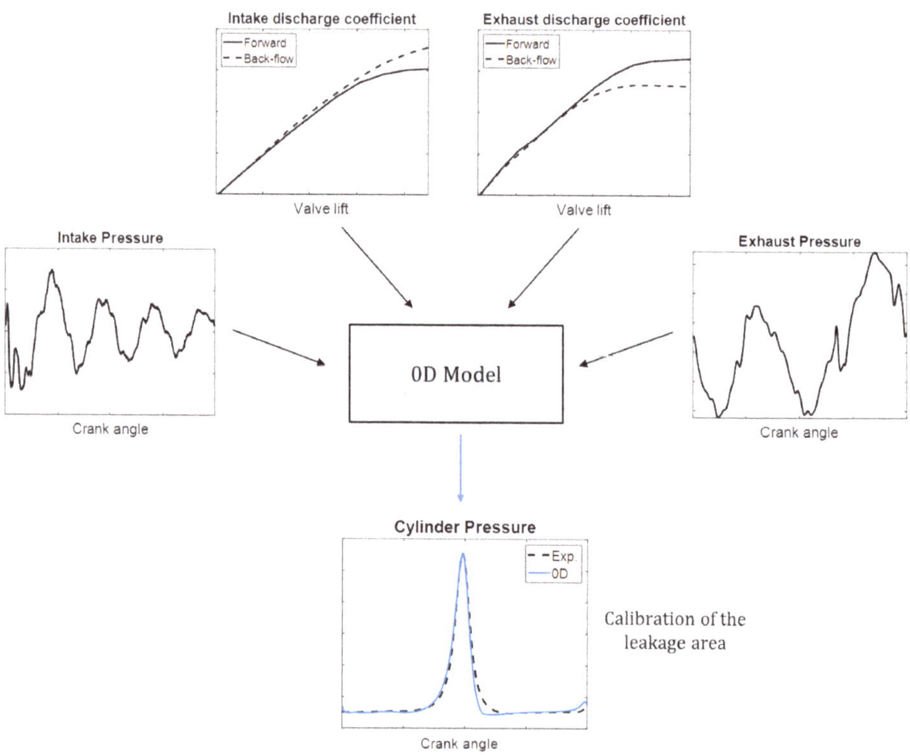

Figure 5. 0D modeling of the transparent engine—blow-by rate estimation.

By design, the transparent engine has a higher leakage rate between the piston and the cylinder liner than a conventional engine does. This is because the rings, which are partly made of graphite and vespel (polyimide-based plastics), do not scrape the cylinder walls as hard as conventional metal rings. Therefore, they allowed the engine to be sealed without damaging the transparent cylinder.

To be able to simulate the real phenomena occurring in the transparent engine, the blow-by flow mass rate must be estimated so that the numerical 3D model and the transparent engine can have a similar behavior (i.e., the same air mass trapped in the cylinder). In other words, the blow-by flow mass rate must be imposed as a boundary condition in the 3D CFD model.

As shown in Figure 5, the estimation of the blow-by flow mass rate is accomplished via 0D modeling, and, finally, the result is validated by comparing the pressure curve given by the 0D model [43] to the one experimentally read on the transparent motor.

This model takes the intake and exhaust pressure signals as boundary conditions, and it also considers the pressure drops at the valves of the transparent engine. The superposition of the two pressure curves is done by calibrating the leakage section, A_{bb}, and the theoretical blow-by flow rate, $\dot{m}_{th}$, is modeled similarly to the flow through valves in the usual way [44]. The value of the leakage section is taken differently depending on the value of the internal pressure in the cylinder. Indeed, the rings do not have the same behavior when the flow is incoming or outgoing:

$$A_{bb} = \begin{cases} A_{bb,\,sup}\ if\ P_{cyl} > P_{atm} \\ A_{bb,\,inf}\ if\ P_{cyl} < P_{atm} \end{cases}, \tag{2}$$

where $A_{bb,\,x} = \pi\left(D_{cyl}^2 - \left(D_{cyl} - 2gap_x\right)^2\right)/4$ with $gap_{sup} = 0.0657$ mm and $gap_{inf} = 0.66$ mm (for the case of the $VVT_{\text{intake/exhaust}} = 20/00°$ CA valve timing). The value gap_{sup} is used to calibrate the maximum pressure given by the model, and that of gap_{inf} is adjusted to have an admitted air flow rate estimated by the model that is equal to that observed experimentally. To have an idea of the order of magnitude of the leakage, the gap_{sup} creates an equivalent leakage surface of a circular section with a diameter of 4.35 mm, as shown in Figure 6.

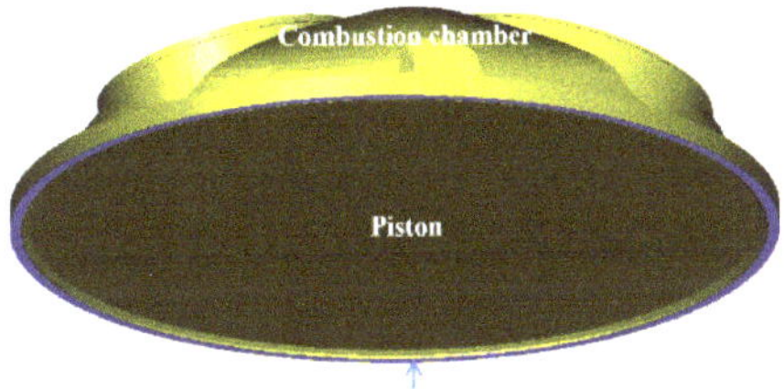

Figure 6. Imposing the blow-by flow rate as a boundary condition.

With respect to the geometry presented in Figure 4, the computational domain was slightly reduced because, as seen in Figure 7, the geometry beyond the areas corresponding to the inlet and outlet pressure sensors was not maintained (i.e., the computational domain starts and ends with the areas where the BCs are imposed).

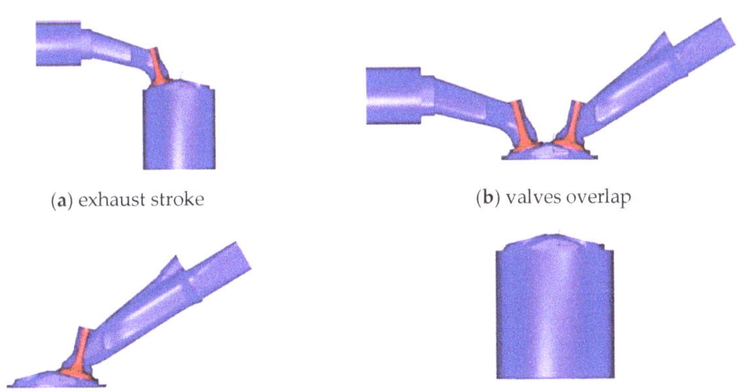

(**a**) exhaust stroke

(**b**) valves overlap

(**c**) intake stroke

(**d**) compression and expansion strokes

Figure 7. The computational domain.

To maintain the computational time within reasonable limits, on the one hand, the chosen timestep of the simulation was 1° CA, and, on the other hand, various grid zones were successively enabled and disabled according to the engine's physical strokes. In other words, during the calculation, the computational domain is divided into three main parts, and two of those parts, the intake and exhaust ones, are successively enabled/disabled depending on the simulation time (Figure 7):

- the exhaust duct, the exhaust ports, and the cylinder → exhaust stroke (Figure 7a);
- the exhaust, the cylinder, and the intake → valves overlap (Figure 7b);
- the intake duct, the intake ports, and the cylinder → intake stroke (Figure 7c);
- the cylinder → compression and power/expansion stroke (Figure 7d).

The discretization is created through the pre-processing software included in the AVL Fire, namely, the Fame Engine Plus module (FEP), which is an automatization used to set up a moving and deformable mesh engine project.

The turbulence model used for this application is of the high-Reynolds numbers type, which means it is not applicable in the near-wall region. In order to model the near-wall effects, the hybrid wall function was set, which was meant to ensure a gradual change between the viscous sublayer and the wall function. This hybrid wall function, which, as mentioned in [42], is recommended for use in conjunction with the k-ζ-f turbulence model, is actually a proprietary semi-empirical-based fitting function between the laminar and the turbulent part of the boundary layer obtained from experimental evidence. Therefore, the chosen wall treatment was an automatic one. Thus, the CFD code evaluated the y^+ (the usual dimensionless wall normal distance utilized in the boundary layer problem) and then selected the appropriate wall function on the fly without any user input. Hence, the computational domain was meshed automatically with a greater refinement in the areas susceptible to very high flow velocities such as the valve gaps, where the y^+ value is below 3 in the low lift periods, as can be seen from Figure 8.

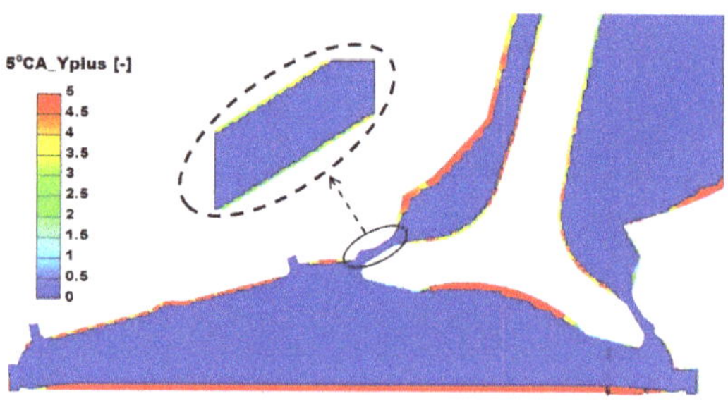

Figure 8. The y^+ dimensionless wall normal distance at the intake valve gap.

The details of the resulting mesh are given in Table 1, and an image of the meshed geometry is provided in Figure 9.

Table 1. Details of the geometry mesh.

Max. Cell Size	Max. Number of Cells	Repartition	(%)	Min. Number of Cells	Repartition	(%)
1 mm	7,527,028 (@180° CA)	hexahedral	82	2,866,060 (@360° CA)	hexahedral	81.5
		tetrahedral	0.67		tetrahedral	1.01
		prismatic	10.5		prismatic	13.89
		pyramidal	5.92		pyramidal	7.9

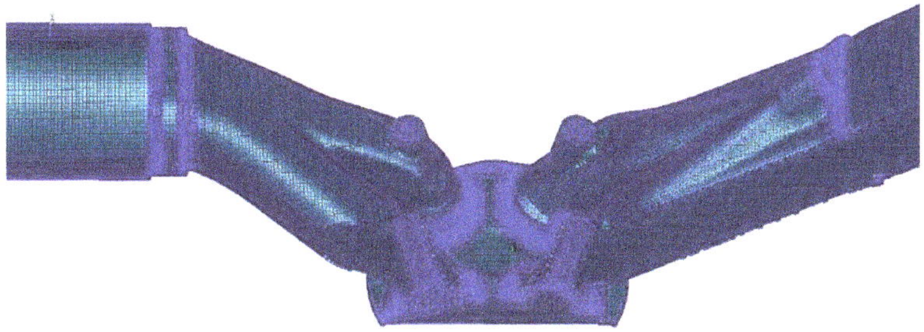

Figure 9. The meshed geometry.

As seen from Table 1, the predominant cell shape is hexahedral, but tetrahedral, prismatic, and pyramidal cells were also used for the areas with specific and complicated geometries. As the mesh is a moving and deformable one, the number of cells varies from 7,527,028 at the start of the intake stroke (Figure 7b) to 2,866,060 at the end of the compression stroke. To analyze the influence of the grid upon the numerical solution, different values for the maximum cell size were used (1.3, 1.0, and 0.7 mm). Obviously, this has a direct influence on the meshing and computational time of our Intel Xeon machine (2.3 GHz, 36-core processor, and 128 GB of RAM), as shown in Table 2.

Table 2. Computational time for one complete cold engine cycle.

Max. Cell Size [mm]	1.3	1.0	0.7
Meshing time [hours]	9	17	27
Simulation time [hours]	19	96	185

Since further decreasing the maximum cell size from 1.0 mm to 0.7 mm did not result in an obvious improvement (no apparent change in the large-scale structures), the maximum cell size used for the CFD study was 1.0 mm.

4. Results

This section focuses on comparing the experimental results with those obtained by the numerical CFD simulation. It starts by comparing the mean flows generated by the two Miller intake valve lift laws presented in Figure 2 and then ends by comparing the turbulence generated when using these two cases.

Obviously, this kind of methodology is typical for the flow studies. One could even say it has always been privileged because the two approaches (experimental and numerical) are complementary, as will be shown subsequently.

Thus, in order to obtain information about the internal aerodynamics of the engine at the end of the compression stroke, in the combustion chamber, which is not observable for the PIV investigation, our method relied on estimating these missing data with the help of the CFD results. More precisely, since both approaches (CFD and PIV) offered comparable trends for the mean kinetic energy and fluctuating velocity, the results of the 3D CFD numerical simulation were used to allow for the extrapolation of the data gathered through the PIV investigation and, finally, to complete the picture regarding the evolution of the aerodynamic phenomena occurring inside the cylinder for a Miller SI engine featuring two VVT offsets.

Therefore, this paper also aims to highlight that, despite the differences generated by the comparison between the experimental and the numerical results, useful conclusions may still be drawn because both approaches (CFD and PIV) offered comparable trends.

4.1. Mean Flow

Figures 10 and 11 present a side-by-side comparison of the average flow fields generated by the two Miller cases presented in Figure 2 ($VVT_{in/ex}$ = 20/00° CA and $VVT_{in/ex}$ = 30/30° CA). In these pictures, the intake valves are on the right. On the PIV results, the circled blue areas from the upper right corners and from the low left side are the effect of some reflections which could not be suppressed. Moreover, concerning the stair shape from the bottom of the PIV flow fields at 270° CA and 320° CA, this is because the piston blocks the laser sheet, and the closer it gets to TDC, the more accentuated this effect is.

When analyzing Figures 10 and 11, even though there are common features, one can first observe that the shapes of the structures are not identical between the numerical and experimental results.

Indeed, especially at 200° CA, there are noticeable differences in the large-scale structures between the two approaches (CFD and PIV). For this position, the CFD predicted a large-scale structure which is not characterized by a single center of rotation, as the PIV revealed. One may also notice a rotational flow around an elongated upper area (see the white curve from Figures 10 and 11). Despite these initial differences, at 270° CA, the large-scale flow fields may be considered close enough. Both approaches showed centers of rotation situated in the upper left regions (see the small white dots from Figures 10 and 11) with respect to the previous angular position. As for the 320° CA, ignoring the regions pointed out in Figures 10 and 11 with small vertical ellipses, which can be explained by a slight deposit of dust on the inner surface of the cylinder responsible for some reflections, once again, the large-scale flow fields featuring centers of rotations displaced even more to the upper left region with respect to the previous position may be considered comparable.

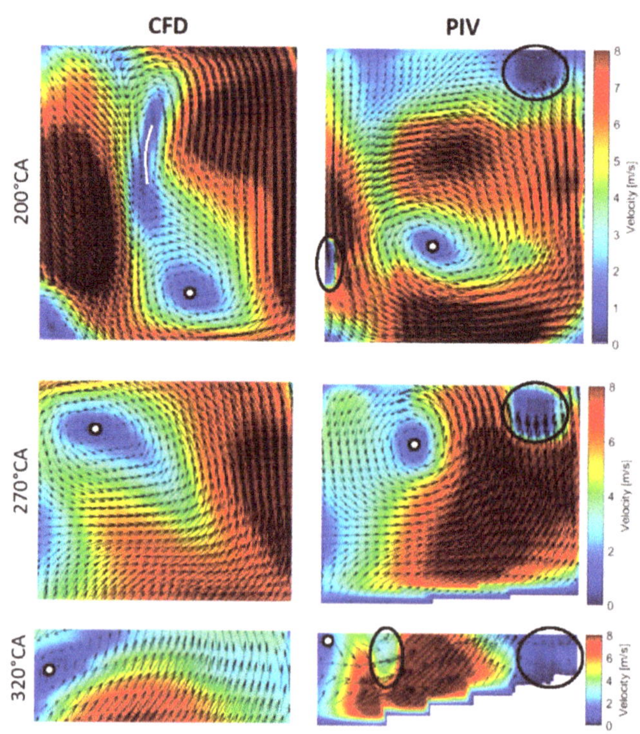

Figure 10. Mean flow velocity fields: simulation (**left**) and experiment (**right**)—$VVT_{in/ex}$ = 20/00.

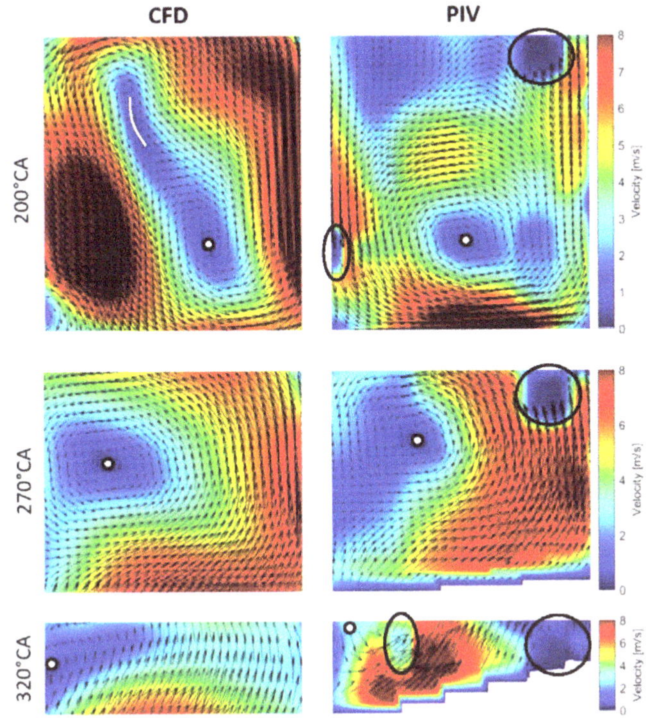

Figure 11. Mean flow velocity fields: simulation (**left**) and experiment (**right**) $VVT_{in/ex}$ = 30/30.

For all of the angular positions, the order of magnitude of the measured vs. simulated velocities is the same.

It is also interesting to notice from both data sets, CFD and PIV, that the effect of closing the intake valves earlier ($VVT_{in/ex}$ = 30/30) leads to a decrease in the intensity of the air movement.

Regarding the differences between the CFD prediction and the PIV reality, one reason could be related to a blow-by flow mass estimation not in full concordance with reality. This being set in the simulation as a boundary condition has a direct influence upon the numerical results. Another reason could be connected to the RANS model constants generally used for IC engine flow predictions, which were taken from the classical Otto cycle. Moreover, on the method of calculation of the average flow field, the RANS result is not the same average as the average of the PIV. An identical methodological approach would lead us to use the LES on as many cycles as possible so that the average was statistically converged. In the experiments, despite all the precision brought to the measurements, a discrepancy could be explained by a slight/inherent mechanical dissymmetry of the lift of the two intake valves, which would then generate a tumble plane not symmetrical to the direction of the flow. Furthermore, the particles seeded in the in-cylinder flow were then illuminated using a light sheet with a thickness of about 1 mm.

However, despite these differences, as will be shown subsequently, the trends illustrated by the two approaches (CFD and PIV) may be considered as close enough, which supports the idea of estimating the missing PIV data corresponding to the end of the compression stroke with the help of the CFD results, as already mentioned.

To quantify the difference between the velocity fields shown in Figures 10 and 11, the choice was made to calculate the mean kinetic energy (MKE) based on the velocities of the $2D$ grid's nodes going from 1 to N_m:

$$MKE_{2D} = \frac{1}{N_m} \sum_{k=1}^{N_m} \frac{U_k^2 + V_k^2}{2} \tag{3}$$

Figure 12 shows the result of the calculation for the two VVT offsets.

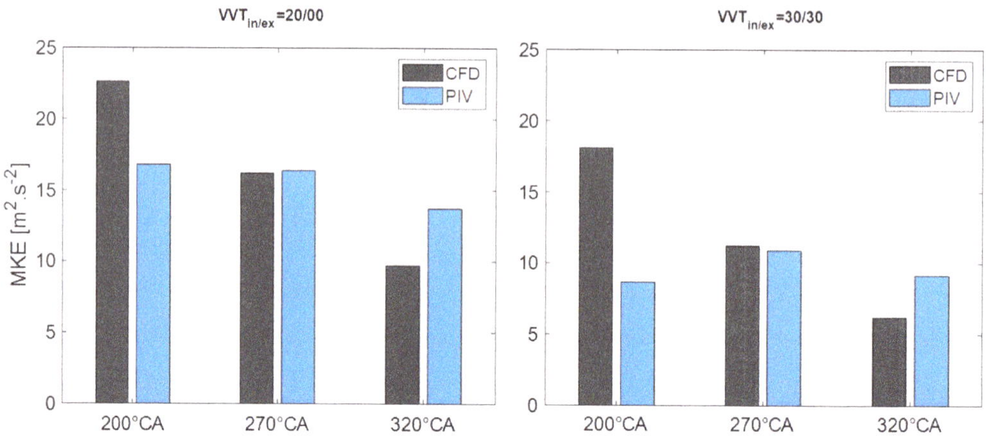

Figure 12. Comparison of the 2D mean kinetic energy: simulation vs. experiment.

As shown in Figure 12, the MKE differences observed at the beginning of the compression stroke (200° CA) are greatly reduced at mid-compression (270° CA). The values are then slightly different at the end of the compression (320° CA). Nonetheless, at least the differences are in the same way (at the PIV, lower at 200° CA and higher at 320° CA). An explanation for these differences could be the smaller vector number provided by the PIV measurements compared to the CFD measurements.

4.2. Turbulent Flow

The turbulence model of the CFD software can estimate the turbulent kinetic energy (TKE) during the engine cycle. This quantity, named k, corresponds to the temporal average of the kinetic energy per unit of mass associated with the turbulent flow. It is written as follows:

$$k_{CFD} = \frac{1}{2}\left(\overline{(u_1')^2} + \overline{(u_2')^2} + \overline{(u_3')^2}\right) \tag{4}$$

where $\overline{X}$ represents the temporal average of the quantity X and corresponds to the average during the engine cycles for the CFD data.

It is therefore necessary to calculate k from the PIV data. The turbulence model is based on the Reynolds decomposition. It will then be used to extract the fluctuations of the component i of the velocity fields:

$$u_i' = u_i - \overline{u}_i \tag{5}$$

A problem appears when comparing the CFD and PIV data. The PIV measurement is plane and thus gives access to only two components of the velocity. The turbulent kinetic energy will consequently be theoretically lower than the one calculated from the CFD model (apart from any other consideration):

$$k_{PIV\_2D} = \frac{1}{2}\left(\overline{(u_1')^2} + \overline{(u_2')^2}\right) \tag{6}$$

However, this difference can be minimized. Indeed, the turbulence model is based on the assumption of the isotropy of the turbulence, i.e., the fluctuations are similar in all directions $\left(\overline{(u_1')^2} = \overline{(u_2')^2} = \overline{(u_3')^2}\right)$. A second expression of the TKE can then be used for the PIV data by estimating the contribution of the third component as follows:

$$\overline{(u_3')^2} = \frac{\overline{(u_1')^2} + \overline{(u_2')^2}}{2} \tag{7}$$

The TKE from the PIV data, under the assumption of isotropic turbulence, is then written:

$$k_{PIV,\,iso} = \frac{1}{2}\left(\overline{(u_1')^2} + \overline{(u_2')^2} + \frac{\overline{(u_1')^2} + \overline{(u_2')^2}}{2}\right) \tag{8}$$

This assumption of isotropy also allows for the definition of a quantity u' to quantify the fluctuations, which is calculated as the average of the fluctuations in the three directions:

$$u' \equiv \sqrt{\frac{1}{3}\left(\overline{(u_1')^2} + \overline{(u_2')^2} + \overline{(u_3')^2}\right)} = \sqrt{\frac{2}{3}k} \tag{9}$$

Note that in the results presented later, k and u_0 are calculated at each point of the measurement plane in both cases (CFD/PIV). The spatial average of this set is then calculated to obtain values of TKE and fluctuating velocity at a specific time.

To estimate the order of magnitude of the TKE before combustion, the two CFD calculations for the two VVT offsets were performed in the transparent engine configuration. The mean value of the TKE and, therefore, u' was then extracted in the tumble plane at different crank angles.

As shown in Figure 13, which is about a trend predicted based on discrete crank angular positions, the results between CFD and PIV were compared at the same crank angles as before (200, 270, and 320° CA). Then, the TKE (i.e., u') values given by the CFD at the angles 340 and 355° CA were used to extrapolate those of the PIV to estimate the turbulence generated in the transparent engine configuration near TDC. The dashed curve, corresponding to the extrapolated PIV values, is calculated by shifting the CFD values by the offset k_{CFD} (320° CA) − k_{PIV} (320° CA). Note that the TKE of the PIV values is calculated under the assumption of turbulence isotropy. The left-hand side of Figure 13 shows the results for the $VVT_{in/ex}$ = 20/00° CA offset. The shape of the curves follows the same trend in both cases. The maximum TKE at 340° CA of the extrapolated PIV values is 9.44 $m^2 s^{-2}$, which gives a turbulent velocity of 2.35 $m s^{-1}$.The right-hand side of Figure 13 presents the results of the $VVT_{in/ex}$ = 30/30° CA offset, with a maximum TKE of 7.24 $m^2 s^{-2}$ and a velocity of 2.07 $m s^{-1}$. The curves appear less comparable. However, the expected effect of an advance of the closing time of the intake valves (VVT_{in} going from 20 to 30° CA) on the TKE is verified. This advance of closing decreases the order of magnitude of the TKE before combustion. These findings are also concordant with the results shown in [3,14].

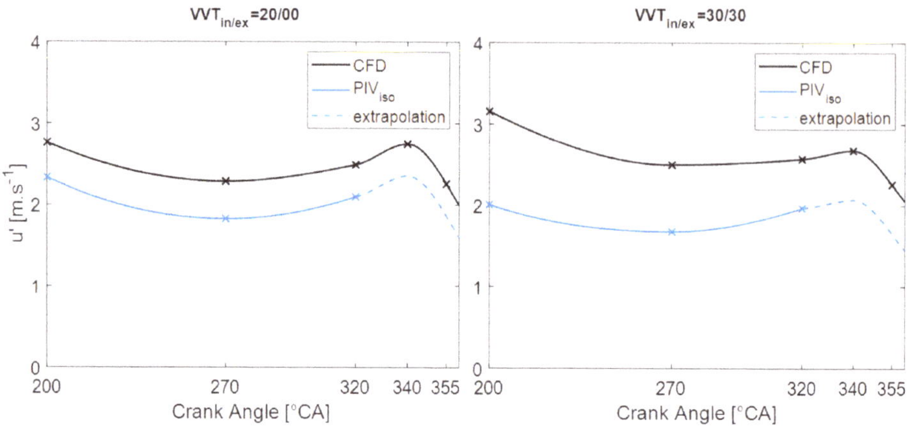

Figure 13. Comparison of the fluctuating velocities between the simulation and experiment.

These values of the fluctuating velocity are low. The values found by the PIV in the tumble plane, in a previous study with the same configuration but following an Otto cycle [17], were of the order of 3.5 m s $^{-1}$ at 330° CA. This further demonstrates the importance of intensifying the internal aerodynamics when applying the Miller cycle, as shown in paper [15].

Similar evolutions of the fluctuating velocities can also be found in paper [43], which provided an estimation of the in-cylinder turbulence during the engine cycle based on a $K - k - \varepsilon$ 0D model. The use of this model showed that advancing the intake valve closing comes to degrade the level of turbulence at the end of the compression.

Further work on the methods and means to raise the level of the internal aerodynamics must still be conducted to ensure a positive impact from the use of the Miller cycle. However, it should also be noted that the global paradigm of the thermal engine is changing and that a decrease in turbulence would be beneficial for the use of fuels such as hydrogen, which has a laminar combustion speed that is already very high.

5. Conclusions and Future Works

This paper dealt with turbulence estimation during the compression stroke in the case of an EIVC Miller SI engine.

This evaluation was performed by comparing the PIV results with the numerical ones obtained from a RANS-type model. More precisely, the results of the 3D CFD numerical simulation were used to allow for the extrapolation of the data gathered through the PIV investigation toward the end of compression, where the PIV investigation did not provide information.

This paper also aimed to highlight that, despite the differences generated by a desirably perfect comparison between the experimental and numerical results, useful conclusions were extracted, as both approaches (CFD and PIV) offered comparable trends.

Thanks to this approach, the following values were obtained for the fluctuating velocity at 340° CA: 2.35 m s $^{-1}$ for the first VVT offset and 2.07 m s $^{-1}$ for the second. These values are considered low since, for the Otto case in the same configuration, the fluctuating velocity was of the order of 3.5 m s $^{-1}$, as shown in [17].

These findings support the conclusion that, when using the EIVC-based Miller cycle, special attention needs to be paid to the intensification of the internal aerodynamics of the engine in order to finally benefit from clean and efficient combustion. On this subject, beyond the solutions discussed in paper [15], one may imagine an intensification of the aerodynamic movement by using an asymmetric lift of the intake valves. The addition of a swirl component to the tumble movement stabilizes the general movement during the compression phase. Another element that can be introduced as part of the optimization of the Miller cycle is the use of the engine at a high rpm to compensate for the loss of turbulent speed at the end of the compression.

However: (1) these findings are the consequences of working with the turbulent isotropic hypothesis, which was undertaken because only two velocity components can be measured with our PIV installation; the third component in the measurement plane could be obtained with a stereo PIV system [18], as presented in Section 2.1, meaning that a supplementary camera is needed; (2) the constants of the RANS model generally used for IC engine flow predictions were taken from the classical Otto cycle; this could lead to some inaccuracies when trying to predict the internal aerodynamics of a Miller cycle engine; (3) the blow-by flow mass estimated by the 0D approach (see Section 3.2) with its associated shortcomings, which was used as a boundary condition in the CFD simulation, could also lead to some imprecisions.

In light of these considerations, future works on this topic could be performed with a stereo PIV system and with a measurement of the blow-by flow mass. For the RANS-based CFD simulation, to study the influence of the RANS model constants on the predicted flow, parametric studies may be conducted. Another CFD approach would be to switch to LES with consideration of the boundary conditions well upstream of the intake ducts, which seems to deliver instantaneous velocity fields very close to the PIV results, as shown in papers [33,35,36]. Finally, another argument in favor of LES is that, since the RANS result is not the same average as the average of the PIV, an identical methodological approach would then impose the use of LES on as many cycles as possible so that the average is statistically converged.

Author Contributions: Conceptualization, P.G., A.C. and S.G.; methodology, M.P., P.G., A.C. and S.G.; PIV investigation, M.P.; CFD simulation, V.I.-S., M.N. and A.C.; writing—original draft preparation, M.P., A.C., M.N. and V.I.-S.; supervision, P.G., A.C. and S.G. All authors have read and agreed to the published version of the manuscript.

Funding: This research received no external funding.

Informed Consent Statement: Not applicable.

Data Availability Statement: Not applicable.

Acknowledgments: The authors want to express their gratitude to AVL List GmbH for providing the FIRE software for the CFD simulation.

Conflicts of Interest: The authors declare no conflict of interest.

Acronyms and Notations

0D	zero dimensional
3D	three dimensional
BC	boundary condition
BDC	bottom dead center
CA	crank angle
CFD	computational fluid dynamics
CPU	central processing unit
CO_2	carbon dioxide
EIVC	early intake valve closing
GB	gigabyte
GHz	gigahertz
IC	internal combustion
in/ex	intake/exhaust
LES	large eddy simulation
LIVC	late intake valve closing

MKE, K [$m^2 \cdot s^{-2}$]	mean kinetic energy
NO_x	nitrogen oxides
PIV	particle image velocimetry
RAM	random access memory
RANS	Reynolds-averaged Navier–Stokes
RDE	real driving emissions
rpm	revolutions per minute
RSM	Reynolds stress model
SI	spark ignition
TDC	top dead center
TKE, K [$m^2 \cdot s^{-2}$]	turbulent kinetic energy
URANS	unsteady Reynolds-averaged Navier–Stokes
VGT	variable geometry turbine
VVT	variable valve timing
A_{bb} [m^2]	leakage section/area
$\dot{m}_{th}$[kg/s]	blow-by mass flow rate
P_{cyl} [Pa]	in-cylinder pressure
P_{atm} [Pa]	atmospheric pressure
D_{cyl} [m]	bore
gap [m]	leakage gap
N_m	max. number of the grid's nodes
$U_k\left[m \cdot s^{-1}\right]$	X flow velocity at the k node
$V_k\left[m \cdot s^{-1}\right]$	Y flow velocity at the k node
$u_i'\left[m \cdot s^{-1}\right]$	velocity fluctuation of the i component
$u_i\left[m \cdot s^{-1}\right]$	velocity of the i component
$\overline{u}_i\left[m \cdot s^{-1}\right]$	mean velocity of the i component
$k_{CFD/PIV}\left[m^2 \cdot s^{-2}\right]$	mean TKE per mass unity for the CFD/PIV case
ζ	eddy viscosity
ε	rate of dissipation
ω	specific rate of dissipation
f	relaxation function
$v2$	velocity scale
g	the square root of a characteristic turbulent time scale

References

1. Miller, R. Miller cycle. Engine. U.S. Patent 2817322, 24 December 1957.
2. Mo, H.; Huang, Y.; Mao, X.; Zhuo, B. Investigations on the Potential of Miller Cycle for Performance Improvement of Gas Engine. *Glob. J. Res. Eng. B Automot. Eng.* **2016**, *16*, 36–46.
3. Braun, M.; Klaas, M.; Schröder, W. Influence of Miller Cycles on Engine Air Flow. *SAE Int. J. Engines* **2018**, *11*, 161–178. [CrossRef]
4. Huang, Z.-M.; Shen, K.; Wang, L.; Chen, W.-G.; Pan, J.-Y. Experimental study on the effects of the Miller cycle on the performance and emissions of a downsized turbocharged gasoline direct injection engine. *Adv. Mech. Eng.* **2020**, *12*, 1–9. [CrossRef]
5. Wang, Y.; Lin, L.; Zeng, S.; Huang, J.; Roskilly, A.P.; He, Y.; Huang, X.; Li, S. Application of the Miller cycle to reduce NOx emissions from petrol engines. *Appl. Energy* **2008**, *85*, 463–474. [CrossRef]
6. Transport & Environment. Road to Zero: The Last EU Emission Standard for Cars, Vans, Buses and Trucks. 2020. Available online: https://www.transportenvironment.org/wp-content/uploads/2021/06/2020_04_Road_to_Zero_last_EU_emission_standard_cars_vans_buses_trucks.pdf (accessed on 16 March 2022).
7. Budack, R.; Wurms, R.; Mendl, G.; Heiduk, T. The New Audi 2.0-l I4 TFSI Engine. *MTZ Worldw.* **2016**, *77*, 16–23. [CrossRef]
8. Demmelbauer-Ebner, W.; Persigehl, K.; Görke, M.; Werstat, E. The New 1.5-l Four-cylinder TSI Engine from Volkswagen. *MTZ Worldw.* **2017**, *78*, 16–23. [CrossRef]
9. Demmelbauer-Ebner, W.; Theobald, J.; Worm, J.; Scheller, P. The new 1.5-l EA211 TGI evo. *MTZ Worldw.* **2018**, *79*, 16–21. [CrossRef]
10. European Automobile Manufacturers Association. What Is the Real Driving Emissions (RDE) Test? Car Emissions Testing Facts. 2018. Available online: http://www.caremissionstestingfacts.eu/rde-real-driving-emissions-test/ (accessed on 17 March 2022).
11. Nose, H.; Inoue, T.; Katagiri, S.; Sakai, A.; Kawasaki, T.; Okamura, M. Fuel Enrichment Control System by Catalyst Temperature Estimation to Enable Frequent Stoichiometric Operation at High Engine Speed/Load Condition. In Proceedings of the SAE 2013 World Congress & Exhibition, Detroit, MI, USA, 16–18 April 2013; pp. 16–18. [CrossRef]
12. Morand, N.; Agnew, G.; Bontemps, N.; Jeckel, D. Variable Nozzle Turbine Turbocharger for Gasoline 'Miller' Engine. *MTZ Worldw.* **2016**, *78*, 40–45. [CrossRef]

13. Penzel, M.; Bevilacqua, V.; Böger, M. Meeting Future Emission Standards with Turbocharged High-Performance Gasoline Engines. *MTZ Worldw.* **2020**, *81*, 16–25. [CrossRef]
14. Scheidt, I.M.; Brands, I.C.; Kratzsch, M.; Günther, M. Combined Miller/Atkinson Strategy for Future Downsizing Concepts. *MTZ Worldw.* **2014**, *75*, 4–11. [CrossRef]
15. Niculae, M.; Clenci, A.; Iorga-Simăn, V.; Niculescu, R. Évaluation de l'impact du masquage des soupapes d'admission et de l'augmentation de la course du piston sur l'aérodynamique interne du moteur Miller. *Entropie Thermodyn. Énergie Environ. Économie* **2021**, *1*. [CrossRef]
16. Xu, J.; Guo, T.; Feng, Y.; Sun, M. Numerical investigation of Miller cycle with EIVC and LIVC on a high compression ratio gasoline engine. *Sci. Prog.* **2021**, *104*, 368504211023640. [CrossRef] [PubMed]
17. Cao, Y. Sensibilité d'un Ecoulement de Rouleau Compressé et des Variations Cycle à Cycle Associées à des Paramètres de Remplissage Moteur. Ph.D. Thesis, ENSMA, Chasseneuil-du-Poitou, France, 2014.
18. El-Adawy, M.; Heikal, M.; Aziz, A.R.A.; Opatola, R.A. Stereoscopic particle image velocimetry for engine flow measurements: Principles and applications. *Alex. Eng. J.* **2021**, *60*, 3327–3344. [CrossRef]
19. Matsuda, M.; Yokomori, T.; Shimura, M.; Minamoto, Y.; Tanahashi, M.; Iida, N. Development of cycle-to-cycle variation of the tumble flow motion in a cylinder of a spark ignition internal combustion engine with Miller cycle. *Int. J. Engine Res.* **2020**, *22*, 1512–1524. [CrossRef]
20. EBaum, E.; Peterson, B.; Surmann, C.; Michaelis, D.; Böhm, B.; Dreizler, A. Tomographic PIV measurements in an IC engine. In Proceedings of the 16th International Symposium on Applications of Laser Techniques to Fluid Mechanics Conference, Lisbon, Portugal, 16–19 July 2012; pp. 1–12.
21. Lee, H. Thermographic PIV for Turbulent Flux Measurement. Technische Universität Darmstadt. 2016. Available online: https://tuprints.ulb.tu-darmstadt.de/5532/ (accessed on 18 March 2022).
22. Jahanmiri, M. Particle Image Velocimetry: Fundamentals and Its Applications. Goteborg. 2011. Available online: https://www.academia.edu/2263312/Particle_Image_Velocimetry_Fundamentals_and_Its_Applications (accessed on 18 March 2022).
23. Lawson, J.M.; Dawson, J.R. A scanning PIV method for fine-scale turbulence measurements. *Exp. Fluids* **2014**, *55*, 1857. [CrossRef]
24. Adrian, R.J. Measurement Potential of Laser Speckle Velocimetry. NASA Conference Publication. October 1982. Available online: https://www.researchgate.net/publication/4675662_Measurement_potential_of_laser_speckle_velocimetry (accessed on 19 March 2022).
25. Koochesfahani, M.M.; Nocera, D.G. Molecular Tagging Velocimetry. In *Springer Handbook of Experimental Fluid Mechanics*; Tropea, H.C., Yarin, A., Foss, J.F., Eds.; Springer: Berlin/Heidelberg, Germany, 2007. [CrossRef]
26. Meier, A.H.; Roesgen, T. Imaging laser Doppler velocimetry. *Exp. Fluids* **2011**, *52*, 1017–1026. [CrossRef]
27. Hanjalić, K.; Popovac, M.; Hadžiabdić, M. A robust near-wall elliptic-relaxation eddy-viscosity turbulence model for CFD. *Int. J. Heat Fluid Flow* **2004**, *25*, 1047–1051. [CrossRef]
28. Basara, B. Analysis of the performance of linear and non-linear K-ε-ζ-F turbulence model on various applications. In Proceedings of the Fluids Engineering Division Summer Meeting, Miami, FL, USA, 17–20 July 2006; pp. 239–247. [CrossRef]
29. Alfonsi, G. Reynolds-Averaged Navier–Stokes Equations for Turbulence Modeling. *Appl. Mech. Rev.* **2009**, *62*, 040802. [CrossRef]
30. Yang, X.; Gupta, S.; Kuo, T.-W.; Gopalakrishnan, V. RANS and large Eddy simulation of internal combustion engine flows—A comparative study. *J. Eng. Gas Turbines Power* **2014**, *136*, 051507. [CrossRef]
31. Hasse, C.; Sohm, V.; Durst, B. Numerical investigation of cyclic variations in gasoline engines using a hybrid URANS/LES modeling approach. *Comput. Fluids* **2009**, *39*, 25–48. [CrossRef]
32. Hasse, C.; Sohm, V.; Durst, B. Detached eddy simulation of cyclic large scale fluctuations in a simplified engine setup. *Int. J. Heat Fluid Flow* **2009**, *30*, 32–43. [CrossRef]
33. SBuhl, S.; Dietzsch, F.; Buhl, C.; Hasse, C. Comparative study of turbulence models for scale-resolving simulations of internal combustion engine flows. *Comput. Fluids* **2017**, *156*, 66–80. [CrossRef]
34. Hemmingsen, C.S.; Ingvorsen, K.M.; Mayer, S.; Walther, J.H. LES And URANS simulations of the swirling flow in a dynamic model of a uniflow-scavenged cylinder. *Int. J. Heat Fluid Flow* **2016**, *62*, 213–223. [CrossRef]
35. Krastev, V.; Di Ilio, G.; Falcucci, G.; Bella, G. Notes on the hybrid URANS/LES turbulence modeling for Internal Combustion Engines simulation. *Energy Procedia* **2018**, *148*, 1098–1104. [CrossRef]
36. Enaux, B.; Granet, V.; Vermorel, O.; Lacour, C.; Thobois, L.; Dugué, V.; Poinsot, T. Large Eddy Simulation of a Motored Single-Cylinder Piston Engine: Numerical Strategies and Validation. *Flow Turbul. Combust.* **2010**, *86*, 153–177. [CrossRef]
37. Bakker, A. Lectures on Applied Computational Fluid Dynamics. Dartmouth College. 2008. Available online: https://www.bakker.org/Lectures-Applied-CFD.pdf (accessed on 15 January 2020).
38. Krastev, V.K.; Silvestri, L.; Falcucci, G.; Bella, G. A Zonal-LES Study of Steady and Reciprocating Engine-Like Flows Using a Modified Two-Equation des Turbulence Model. *SAE Technical Paper*, September 2017. [CrossRef]
39. Krastev, V.; Silvestri, L.; Bella, G. Effects of Turbulence Modeling and Grid Quality on the Zonal URANS/LES Simulation of Static and Reciprocating Engine-Like Geometries. *SAE Int. J. Engines* **2018**, *11*, 669–686. [CrossRef]
40. Karbon, M.; Sleiti, A.K. Turbulence Modeling Using Z-F Model and RSM for Flow Analysis in Z-SHAPE Ducts. *J. Eng.* **2020**, *2020*, 4854837. [CrossRef]
41. Durbin, P.A. Near-Wall Turbulence Closure Modeling without "Damping Functions". *Theor. Comput. Fluid Dyn.* **1991**, *3*, 1–13. [CrossRef]

42. Advanced Simulation Technologies. In *AVL FIRE—Turbulence Modeling (Brochure)*; AVL List GmbH: Graz, Austria, 2009.
43. Perceau, M.; Guibert, P.; Guilain, S. Zero-dimensional turbulence modeling of a spark ignition engine in a Miller cycle «Dethrottling» approach using a variable valve timing system. *Appl. Therm. Eng.* **2021**, *199*, 117535. [CrossRef]
44. Kirkpatrick, A.T.; Ferguson, C.R. *Internal Combustion Engines*, 3rd ed.; Wiley: Hoboken, NJ, USA, 2016. [CrossRef]

machines

Article

Stress Characteristics Analysis of Vertical Bi-Directional Flow Channel Axial Pump Blades Based on Fluid–Structure Coupling

Xinyi Liu [1], Fengyang Xu [2], Li Cheng [1,*], Weifeng Pan [3] and Weixuan Jiao [1,*]

[1] College of Hydraulic Science and Engineering, Yangzhou University, Yangzhou 225009, China; xinyiliu3@163.com
[2] Jiangsu Zhenjiang Jianbi Pumping Station Management Office, Zhenjiang 212006, China; xyliu98@126.com
[3] Luoyun Water Conservancy Project Management Divison in Jiangsu Province, Suqian 223800, China; jssqpwf@163.com
* Correspondence: chengli@yzu.edu.cn (L.C.); jiaowx@yzu.edu.cn (W.J.)

Abstract: The RANS equation and the RNG k-ε turbulence model were used in the three-dimensional non-constant numerical simulations of the full flow path of a vertical axial-flow pump which was carried out by applying CFX software. The velocity characteristics of the flow field and the pressure distribution of the impeller under different operating conditions were analyzed and verified by external characteristic tests. The fluid–structure interaction research was conducted for the stress distribution and deformation features of different surfaces of the blade under different working conditions. The result shows that where stress is most concentrated is at the point of the root of the blade near the hub. The low-pressure zone on the suction surface is mainly distributed near the rim, and the low-pressure area on the pressure side is mainly distributed near the hub.

Keywords: vertical axial-flow pumps; fluid–solid coupling; blade; equivalent force; deformation

Citation: Liu, X.; Xu, F.; Cheng, L.; Pan, W.; Jiao, W. Stress Characteristics Analysis of Vertical Bi-Directional Flow Channel Axial Pump Blades Based on Fluid–Structure Coupling. *Machines* **2022**, *10*, 368. https://doi.org/10.3390/machines10050368

Academic Editor: Antonio J. Marques Cardoso

Received: 28 March 2022
Accepted: 5 May 2022
Published: 12 May 2022

1. Introduction

In recent years, with the construction of many large water conservancy projects, unidirectional flow pumping stations have found difficulty in meeting the needs of the project in practice, such as in large water transfer projects, both in irrigation water delivery and also to take into account the drainage in years of major flooding. Vertical axial-flow pumps have the characteristics of high flow and low head, and are widely used in large pumping stations. Vertical bi-directional flow axial pump units are gradually being used in the field of drainage and irrigation due to their compact structure and space-saving advantages. The combination application of a bi-directional flow channel and vertical pump has achieved great social and economic benefits. In recent years, vertical bi-directional flow axial pump station operation is often found cracked at the pump blade, sometimes even blade fracture, damaging equipment. Therefore, the structural strength of the vertical bi-directional flow axial pump blade and the stress distribution calculation and analysis are particularly significant.

In the study of pump unit performance, fluid–solid coupling is a quite effective tool for analyzing the flow field properties of fluids. Qin et al. [1] used the sequential coupling method to study the strength of the impeller and guide vane of a large vertical shaft cross-flow pumping station in China under three different operating conditions. Zhang et al. [2] performed the same calculation method on a submersible Axial-flow pump under multiple operating conditions and analyzed the stress and strain distribution patterns of rotor components under fluid forces, centrifugal forces, and gravity. Shi et al. [3] analyzed the effect of fluid–solid coupling on pump head and efficiency through the joint solution of the internal flow field and the response of the structure of impeller of axial-flow pumps by bidirectional sequential fluid–solid coupling. Gao et al. [4] analyzed the impeller stress characteristics when cavitation did not occur, when cavitation was incipient, and when cavitation was severe by the sequential fluid–solid coupling method.

Impeller is a rather significant component inside the pump unit, and the parameters of the impeller have a certain regulatory effect on the hydraulic performance of the pump unit. As a result, scholars have conducted a great deal of research on the flow field characteristics of impeller blades. Persch et al. [5] studied and researched the highly unstable flow field characteristics in single-vane and two-vane pumps by supplementing the measured and simulated results to compare the pressure and flow undulation characteristics of the two pump types. Ju et al. [6] conducted a comparative analysis of different shapes of vanes from the internal vortex structure and found that 2D and 3D forward-angle vanes have a better effect on pump hydraulic performance than conventional radial straight vanes, producing more intense fore-and-aft swirls and smaller axial and radial eddy currents in the vane channel and higher pump efficiency. Numerous studies have shown that the structural strength of the impeller has a direct impact on the safe and stable operation of the pump. The numerical simulation of the impellers studied is grouped by Chen et al. [7], and different vane entrance angles are set, and it is found that the change in the vane inlet angle distribution has different effects on the pump performance, which is beneficial to the design of the vane centrifugal pump and its performance improvement. Zhang et al. [8], He et al. [9], and Pan et al. [10] applied this method in axial-flow pump to compute the intensity of the impeller and found that the maximum value of stress appeared at the junction of the blade and the hub, the most evident deformation appeared at the entrance side of the blade near the rim, and the point where the stress is prominently concentrated was at the combination of the root of the blade and the hub. Liang et al. [11] studied the blade stress distribution under different degrees of cavitation and found that with the occurrence of cavitation, the maximum equivalent force suffered by the impeller blade slowly reduces and the blade deformation increasingly decreases.

At present, relatively little research has been conducted on the strength of the impeller of vertical bi-directional flow channel axial pumps. This paper adopts model tests and numerical simulations, etc., takes a 3D solid modeling of a vertical axial-flow pump device as the research target, comprehensively analyzes the interior flow rate and stress field distribution of the axial-flow pump device, and a unidirectional fluid–solid coupling analysis is conducted for the distribution of stress field and deformation of the axial-flow pump blade under three operating conditions, which provides a basis for the design of the vertical axial-flow pump blade and makes reasonable recommendations for the stable operation of the axial-flow pumps.

2. Calculation Method

Fluid–solid coupling research explores various behaviors of solids that have been deformed under the effect of the flow and the interaction between the two [12]. The effect of solid deformation on the flow domain is also a significant topic in this subject [13]. The fluid–solid coupling fully reflects an interaction among two fields [14]. The solid is changed under the load of the flow field; meanwhile, the distribution of the flow field also changes under the deformation and movement of the former. Therefore, fluid–solid coupling is an undoubtedly important means to effectively research the impeller flow field.

The internal flow of an axial pump is 3D viscous turbulence, which cannot be compressed and follows the equation of continuity and momentum equation of the fluid domains; the continuity equation is also known as the mass equation [15].

$$\frac{\partial \rho}{\partial t} + \nabla(\rho u) = 0, \tag{1}$$

Momentum equation:

$$\frac{\partial(\rho u)}{\partial t} + \nabla(\rho u u - \tau) = F, \tag{2}$$

This paper cites the RNG k-ε [16] turbulence model for the derivation of the flow domain inside the pump during operation. RNG k-ε model can be well applied in a series of fluid calculations in pumping stations which are equipped with axial-flow pump [17–20]. The equation for k and the equation for ε are as follows.

The equation for k:

$$\frac{\partial(\rho k)}{\partial t}+\frac{\partial(\rho k u_i)}{\partial x_j}=\frac{\partial}{\partial x_j}\left(\alpha_k \mu_e \frac{\partial k}{\partial x_j}\right)+\rho(P_k-\varepsilon), \tag{3}$$

The equation for ε:

$$\frac{\partial(\rho\varepsilon)}{\partial t}+\frac{\partial(\rho\varepsilon u_j)}{\partial x_j}=\frac{\partial}{\partial x_j}\left(\alpha_\varepsilon \mu_\varepsilon \frac{\partial \varepsilon}{\partial x_j}\right)+\rho\frac{\varepsilon}{k}(C^*_{1\varepsilon}P_k-C_{2\varepsilon}\varepsilon)\eta=\left(2E_{ij}E_{ij}\right)^{\frac{1}{2}}\frac{k}{\varepsilon}, \tag{4}$$

In the formula, $C^*_{1\varepsilon}=C_{1\varepsilon}-\frac{\eta(1-\frac{\eta}{\eta_0})}{1+\beta\eta^3}$; $E_{ij}=\frac{1}{2}\left(\frac{\partial u_i}{\partial x_j}+\frac{\partial u_j}{\partial x_i}\right)$; the constants take the values: $\alpha_k=\alpha_\varepsilon=1.39$, $C_{1\varepsilon}=1.42$, $C_{2\varepsilon}=1.68$, $\mu_0=4.377$, $\beta=0.012$.

2.1. Solids Control Equations

The conservation equations for the solid domain can be derived from Newton's second law as follows.

$$\rho_s \ddot{d}_s=\nabla\cdot\sigma_s+f_s, \tag{5}$$

$$\frac{\partial(\rho h)}{\partial t}-\frac{\partial\rho}{\partial t}+\nabla\cdot(\rho_f v h)=\nabla\cdot(\lambda\nabla T)+\nabla\cdot(v\tau)+v\cdot\rho f_s+S_E, \tag{6}$$

2.2. Structural Dynamic Equations

When the axial-flow pump is in motion, the fluid which exists inside the pump applies load on the impeller, and it causes the impeller to produce response accordingly by Hamilton's (Hamilton) principle; the structural dynamic equation [21–23] is defined as follows:

$$M\ddot{u}+C\dot{u}+Ku=F, \tag{7}$$

The stress equation for the impeller structure calculation is as follows.

$$K\{u\}=\{F_s\}+\{F_t\}, \tag{8}$$

$$\{\sigma\}=DB\{u\}, \tag{9}$$

Calculation of the equivalent force based on the fourth strength theory combined with the σ is obtained from the above equation.

$$\sigma=\sqrt{\frac{1}{2}\left[(\sigma_1-\sigma_2)^2+(\sigma_2-\sigma_3)^2(\sigma_3-\sigma_1)^2\right]} \tag{10}$$

2.3. Computational Model and Simulation Method

In this study, 2500ZLQ-20-3.0 vertical axial-flow pump is the researched target, the fluid domain includes the inlet and outlet section, impeller, guide vane, the basic parameters of the pump are flow rate = 20 m^3/s head = 2.61 m, speed n = 150 r/min, number of vanes Z = 3, number of guide vane vanes = 5.

First, the 3D model of the axial pump impeller and the bi-directional flow channel was built in NX 9.0 as Figures 1 and 2 show. The exact dimensions of the model are shown in Figures 3 and 4. According to the running conditions of the pump, its hydraulic and force characteristics are studied. The RNG k-ε turbulence model is applied in the software that called CFX to make numerical simulation of the 3D non-constant turbulence in the flow channel, and make validation of the external features to receive the eigenvalues of velocity, pressure, and other distributions of the flow field.

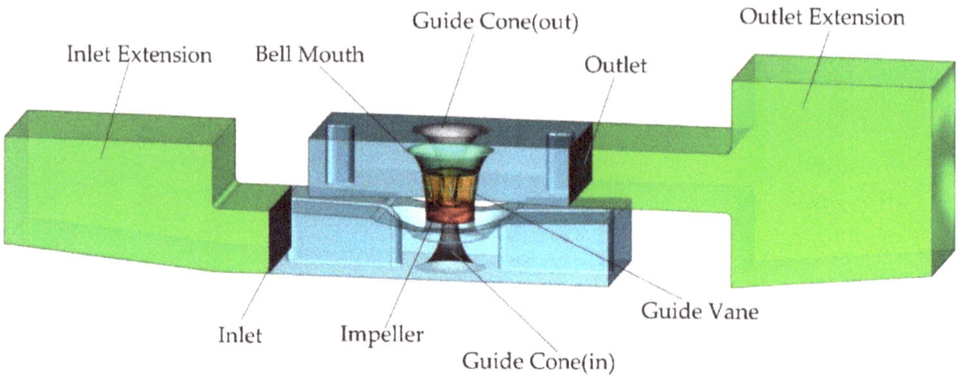

Figure 1. Three-dimensional model of the full flow channel of the pump.

Figure 2. Three-dimensional model of the Impeller.

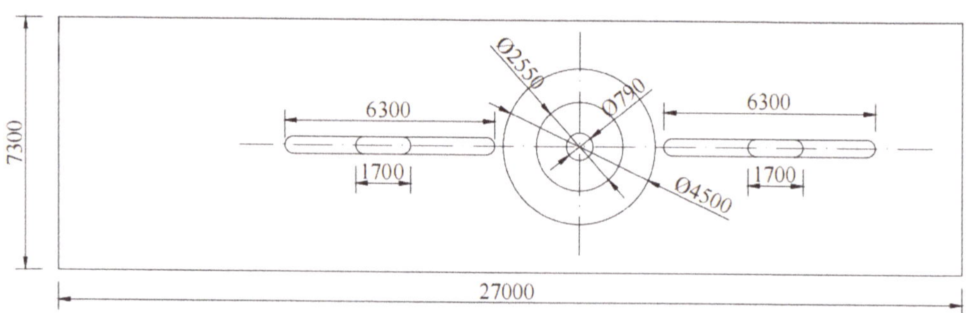

Figure 3. Top view of the full flow path of the pumping station.

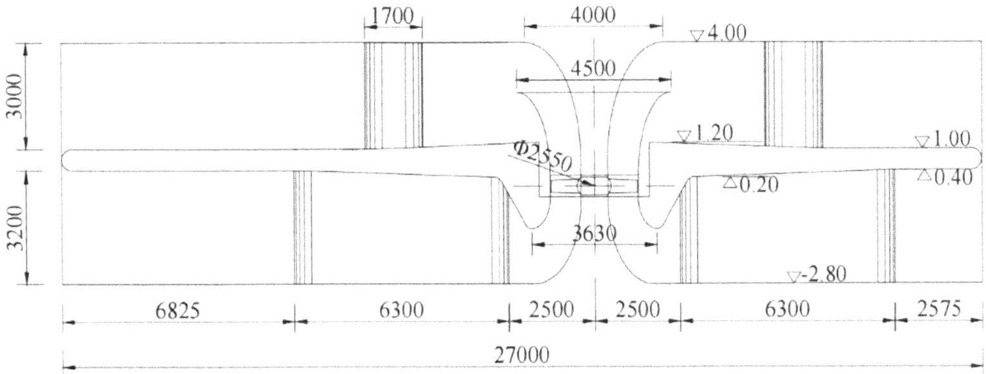

Figure 4. Cross-sectional view of the full flow path of the pumping station.

2.4. Grid Division

In this paper, the structured meshing of the full flow channel as well as the water column of the impeller and guide vane is used for the calculation of the fluid, using the meshing software ICEM. The design flow condition was selected to perform a grid independence analysis of the fluid domain of the pump unit, and the results are shown in Figure 5. When the total number of grids is greater than 9.56 million, the relative error of the pump unit head change is less than 1%, taking into account the efficiency and accuracy of the calculation. In this paper, the number of grids in the fluid domain of the pump unit is set at 9.56 million.

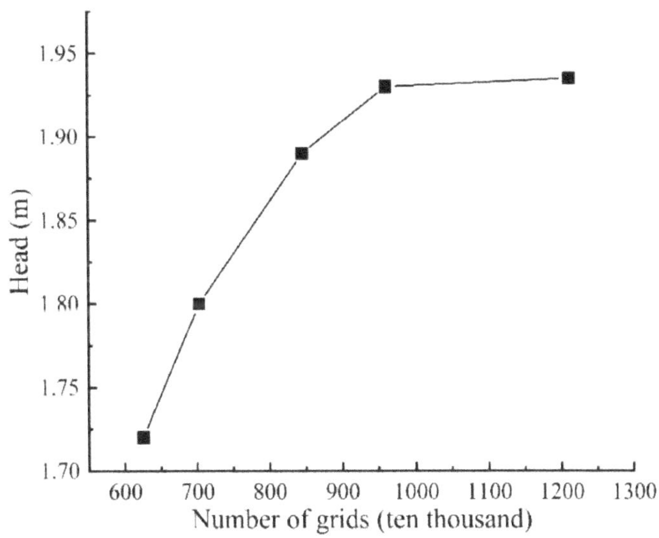

Figure 5. Grid independence analysis.

The meshing of the whole calculation area is shown in Figure 6. For the impeller solid domain, ANSYS Workbench was used to unstructured mesh the impeller solids and encrypt the wall mesh for the solids.

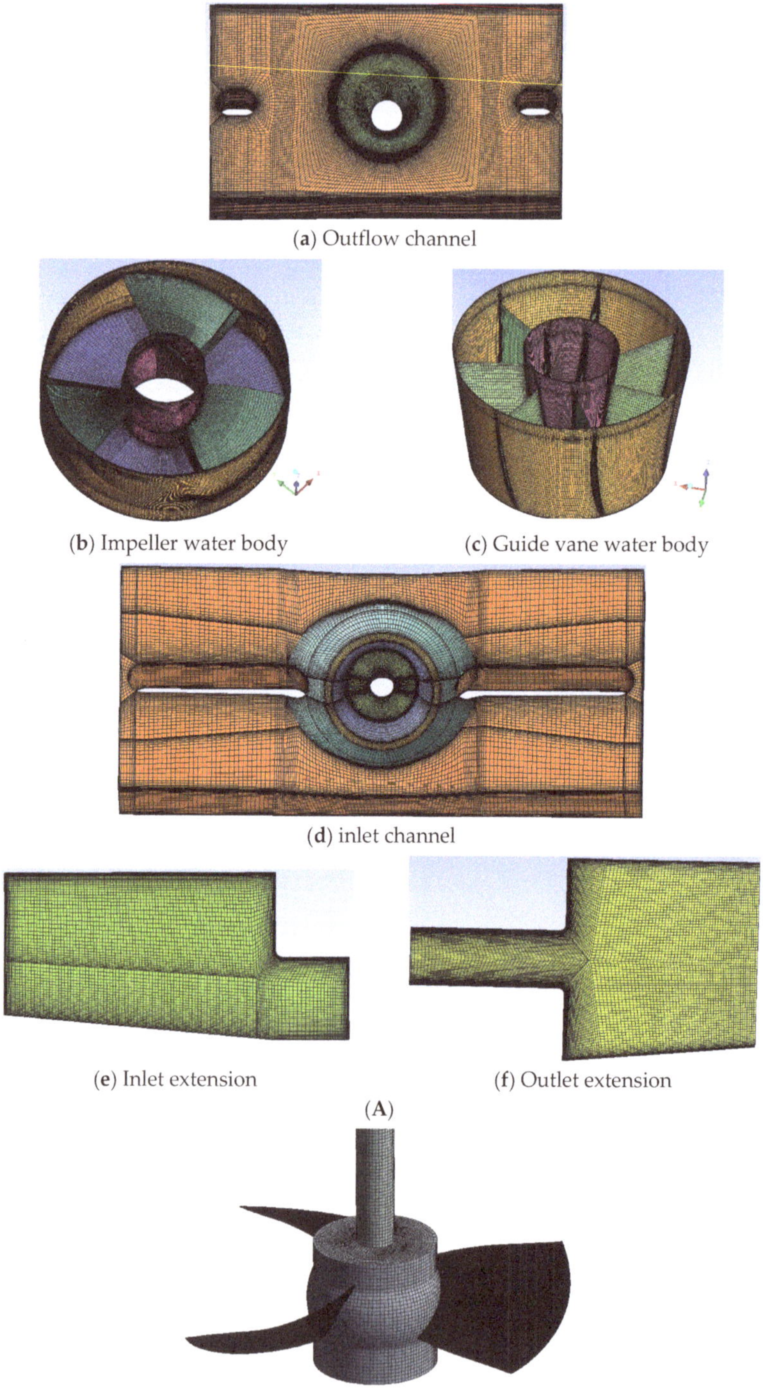

Figure 6. (**A**) Full flow channel structured grid, (**B**) Impeller grid.

The area of the fluid calculation includes the inlet channel and its extensions, the impeller, the guide vane, the outlet channel, and its extensions. The inlet of the inlet channel extension is the inlet boundary [24] of the entire pump unit and the boundary is mass flow rate; the inlet flow rates are 10 m^3/s (small flow condition), 20 m^3/s (design condition), and 28 m^3/s (high flow condition), respectively, with a turbulence intensity of 5%. The outlet boundary is located at the outlet of the outflow channel extension, where the turbulent motion is in relative equilibrium and the flow state does not affect the flow field in the upstream direction. The outlet boundary is taken as average static pressure with a pressure setting of 1 atm. No-slip condition for viscous fluid is used on solid wall, the impeller surface and the hub surface as moving wall, the inlet channel and its extensions; the outlet channel and its extensions are set up as static wall (wall). The intersections set up in this paper are of two types; one is the dynamic and static intersection, set up between the impeller and the guide vane, the impeller and the inlet guide cone with frozen rotor. The other is the static intersection, set at the junction of the rest of the chunking grid with general connection. The accuracy of calculation is 1.0×10^{-5}.

2.5. External Characteristics Test

The model test bench of the pump device as shown in Figure 7 was set up in the laboratory of the Engineering Hall of the College of Water Resources Science and Engineering of Yangzhou University. The whole test bench consists of inlet and outlet cistern, tested pump device, turbine flow meter, electromagnetic flow meter, booster pump, differential pressure meter, gate valve, etc. The pump unit test stand system is indicated in Figure 7. The breadth of the model impeller for the test is 120 mm, the diameter of the prototype impeller is 2.50 m, and the similar ratio λ_D is 20.83. Model pump device test speed is converted in light of the date received from prototype pump and model pump n_D value equal (i.e., equal head) for conversion, where n is the impeller speed and D is the impeller diameter, after a conversion model pump test speed of 3125 r/min.

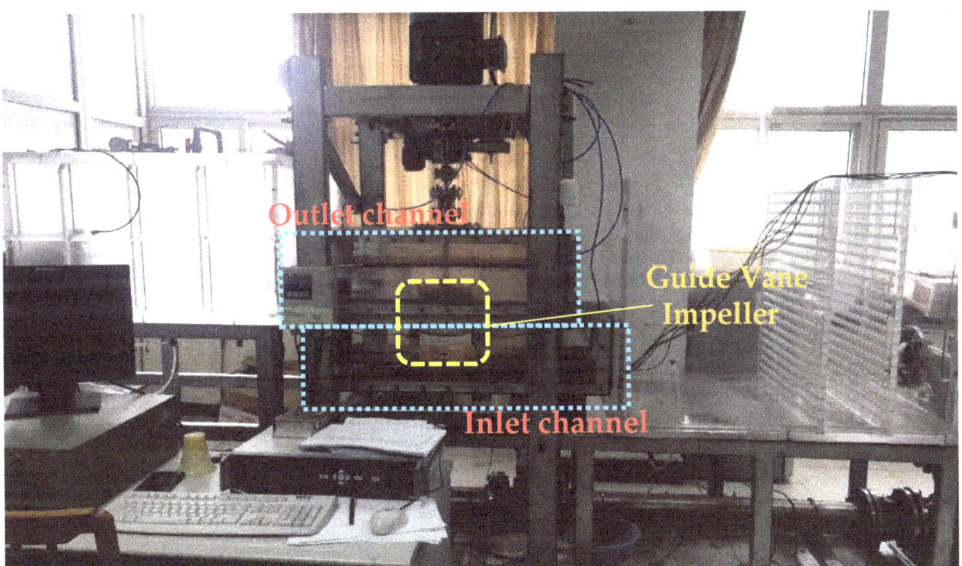

Figure 7. Diagram of the test setup.

The pump unit performance data measured by the test bench are converted in accordance with the mentioned method used to convert prototype pump performance in the

SL140-2006 'Pump Model and Device Model Acceptance Test Regulations'. The conversion formula is.

$$\frac{Q_p}{Q_m} = \frac{n_p D_p^3}{n_m D_m^3} \quad (11)$$

$$\frac{H_p}{H_m} = \frac{n_p^2 D_p^2}{n_m^2 D_m^2} \quad (12)$$

As shown in Figure 8, by comparing the performance data of the prototype pump deduced by conversions and the performance curve of the device received by numerical simulation, the overall tendency is similar. The efficiency of the pumping unit gradually decreases with the increase in the flow rate; the head shows a trend of increasing and then decreasing with the increase in the flow rate, and the average error of the numerical simulation is within 5%. However, the distinction is more apparent when the flow coefficient $K_Q < 1$, and the values are alike with less error when $K_Q > 1$. The flow coefficient K_Q in the figure is the ratio of calculated flow rate to design flow rate. Often, the prototype pumping station is operated under high-flow conditions, and the head of the device is 0.01 m when the calculated flow reaches 1.4 times the design flow, so it is often adopted in operating conditions with high flow for its fluid calculation and analysis, whose data are highly trustworthy.

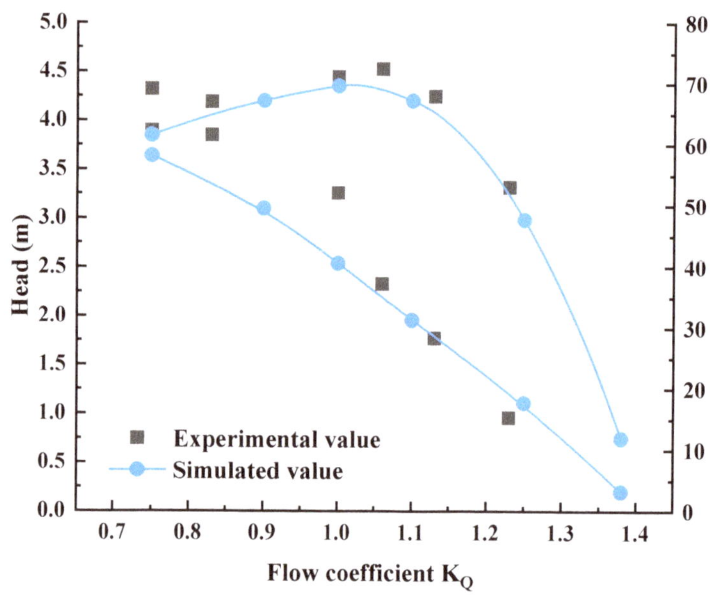

Figure 8. Numerical simulation and test hydraulic performance comparison.

3. Calculation Results and Analysis

3.1. Vertical Axial-Flow Pump Flow Field Flow Velocity Distribution

Figure 9 shows the flow diagrams of the full flow path of the axial-flow pump at 0.5, 1.0, and 1.4 times the design flow operating conditions.

From the flow diagram of the flow field, the flow velocity in the outflow channel gradually increases as the flow rate increases. The water flow in the inlet channel at different flow rates can be roughly divided into two parts. One is the stage where the water rushes into the intake channel and converges to the flapper pipe and the other is the adjustment stage. Most of the water flowing into the inlet channel enters the flapper pipe directly along the channel, with little water flow bypassing the sides and rear area into the flapper pipe. The water flow is constricted and adjusted in the inlet channel flapper pipe.

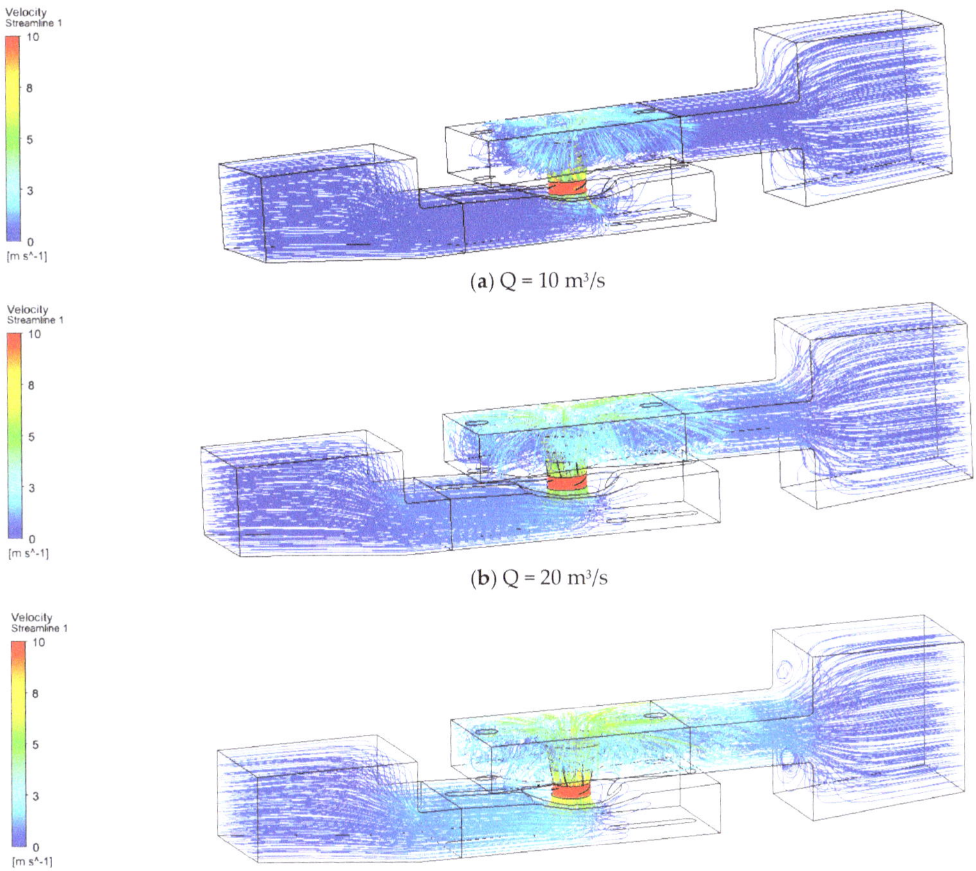

(**a**) Q = 10 m^3/s

(**b**) Q = 20 m^3/s

(**c**) Q = 28 m^3/s

Figure 9. Flow diagrams of the full flow channel under different working conditions.

The water flow at different flow rates obtains energy in the impeller and then rotates into the guide vane body, which still has a certain amount of velocity ring after recovering part of the energy by the guide vane and spirals into the outflow channel. The water flow in the space formed by the horn pipe and the cone is rotating and spreading around, and the water flowing out of the space formed by the horn pipe and the cone is radially and sharply turning and spreading around, and the flow velocity is further reduced. Part of the water flows to the space behind the outflow channel and forms a large area of stagnant water here. The stagnant area in the outflow channel often produces a large range of backflow, and the area of backflow expands more and more as the flow increases. A portion of the water flow is blocked by the left- and the right-side walls as well as the stagnant area at the rear and is diverted to flow out; another part of the water flows directly in the direction of the outlet of the outflow channel in a spiral. Figure 10 shows the flow diagrams of a cross-section of the full flow path of an axial pump at 0.5, 1.0, and 1.4 times the design flow rate.

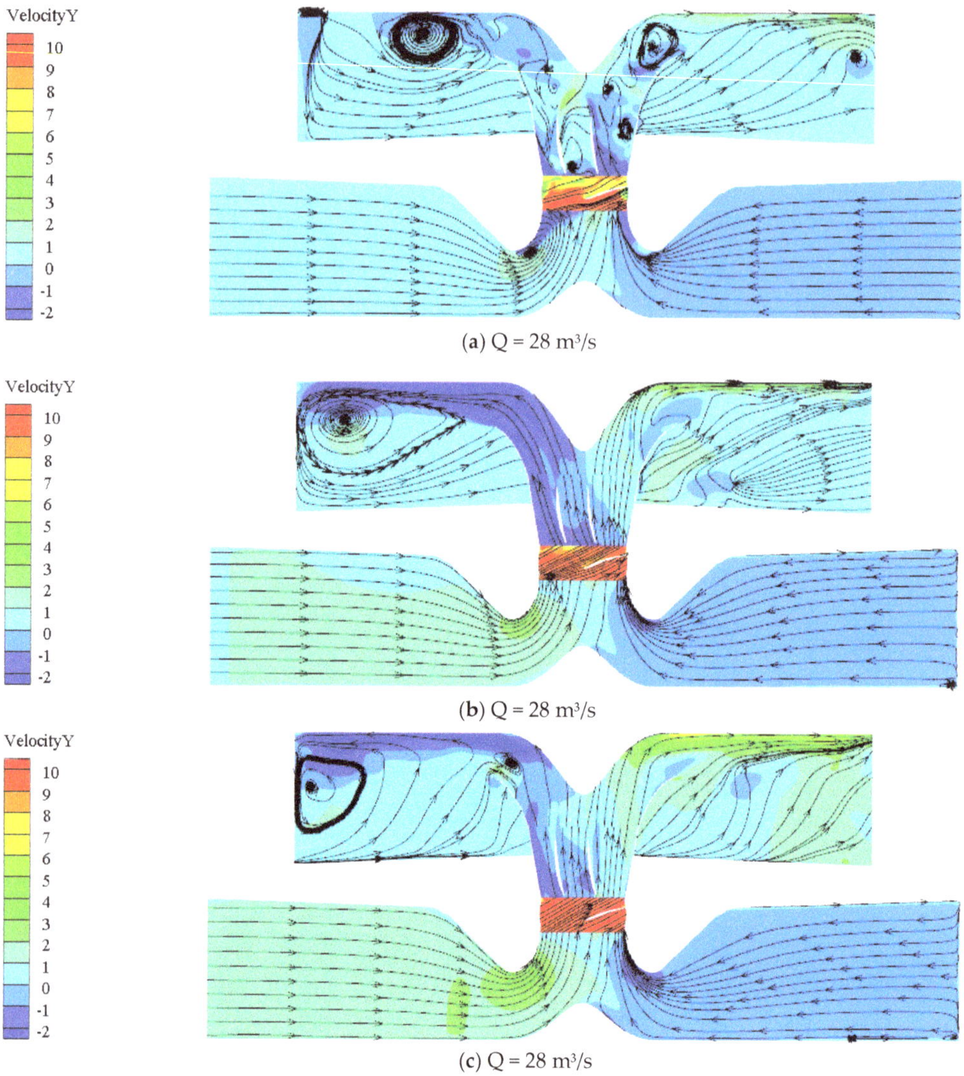

(**a**) Q = 28 m³/s

(**b**) Q = 28 m³/s

(**c**) Q = 28 m³/s

Figure 10. Cross-sectional flow diagrams of flow channels under different working conditions.

As can be seen from the flow diagrams, under the low-flow conditions, the velocity of water in the flow runner is relatively low and a backflow zone is formed on the side of the stagnant water zone of the flow channel. Under the design flow condition, the range of the backflow within the stagnant water zone continues to expand, but the flow line is relatively flat, and a small vortex appears; then, the flow line readjusts to a uniform state. Under high-flow conditions, the backflow zone is further expanded, causing a large area of cyclonic roll in the flow channel.

3.2. Static Pressure Distribution of the Blade

The impeller is the core of the axial-flow pump; it is a significant working part of the axial-flow pump, and the blade is the main unit of the impeller. In many instances, cracks are raised by excessive stress concentrations at the structural transition between the blade and the hub, which affects the safe operation of the axial-flow pump. By calculating the three-dimensional flow field inside the vertical axial-flow pump device, the pressure situation suffered by the blade is obtained. The pressure distribution diagrams of blade surfaces are shown in Figures 11–13.

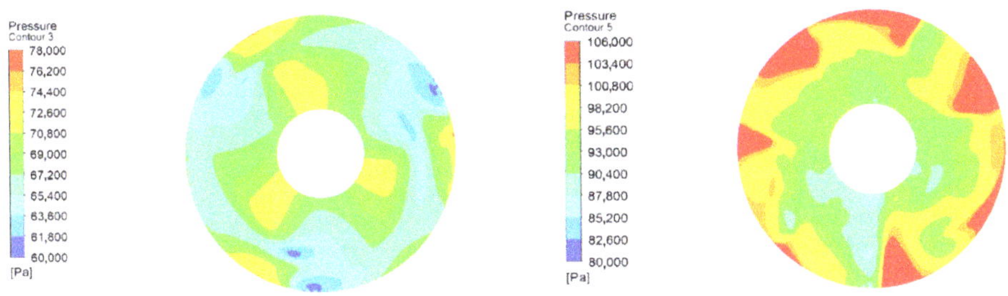

(**a**) Suction surface (**b**) Pressure surface

Figure 11. Q = 10 m^3/s Static pressure distribution of the blade.

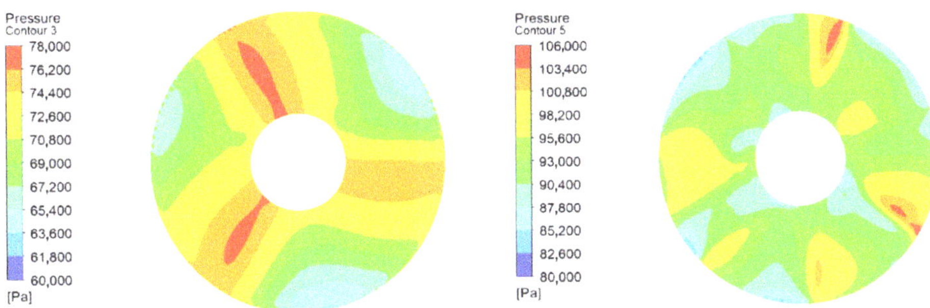

(**a**) Suction surface (**b**) Pressure surface

Figure 12. Q = 20 m^3/s Static pressure distribution of the blade.

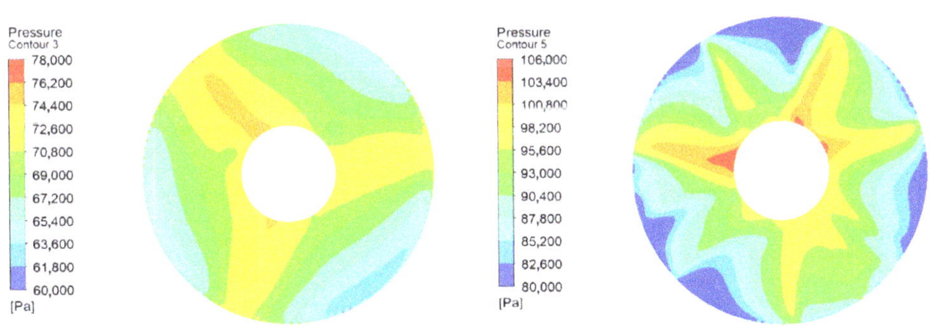

(**a**) Suction surface (**b**) Pressure surface

Figure 13. Q = 28 m^3/s Static pressure distribution of the blade.

It can be discovered from the above figures that the pressure on the surface constantly varies periodically with the rotation of the impeller, whether on the suction or pressure surface, or under different flow operating conditions, and the static pressure distribution takes the shape of three petals, which coincides with the number of blades. In the nearby area of the rim at the inlet of the impeller, there appears a distinct low-pressure zone; the pressure in the vicinity of the blade is relatively high; with the increased flow, the range of the high-pressure area shows a trend of first increasing and then decreasing; and the pressure gradient distribution under the design conditions is more favorable to the water intake action.

The high-pressure area of the water body in the impeller is mainly in the vicinity of the blades. In the impeller outlet, the role of the guide vane begins to appear; as the water rises, the pressure at the hub gradually increases with the increased flow rate, while the pressure at the edge of the blade gradually decreases. There is a large pressure differential between the suction and pressure surfaces of the blades: the water flow under low-flow conditions is more easily lifted up to the guide vane, thus creating low pressure at the hub and high pressure at the rim, whereas under high-flow conditions the pressure distribution is exactly the opposite.

3.3. Stress Distribution of the Blade

When the water pump is in operation, the blade will rotate accordingly. During the rotation, the centrifugal force exits which can cause tension. The contact between the water and the blade creates friction force, and at the same time, the water pressure will then generate bending and torsional stresses. So, it can be seen that the blade force is quite complex. In order to study the distribution of the impeller equivalent force, this paper analyzes the blade equivalent force of a vertical axial-flow pump for three flow conditions based on the flow–solid coupling principle. Figure 14 shows the equivalent force distribution of axial pump blade surfaces (suction and pressure surface) under 0.5 times the design flow condition. Under the small flow condition, the equivalent stress distribution spreads outward from the central point of concentrated stress at the entrance of the water flow, and the stress at the center position of the blade root to the middle part of the whole blade is relatively high, and the stress peak value is 280.4 MPa. The value of the equivalent force on the suction surface is slightly smaller compared to the pressure surface, while the stress value of the whole blade at the root and at the edge of the blade varies greatly.

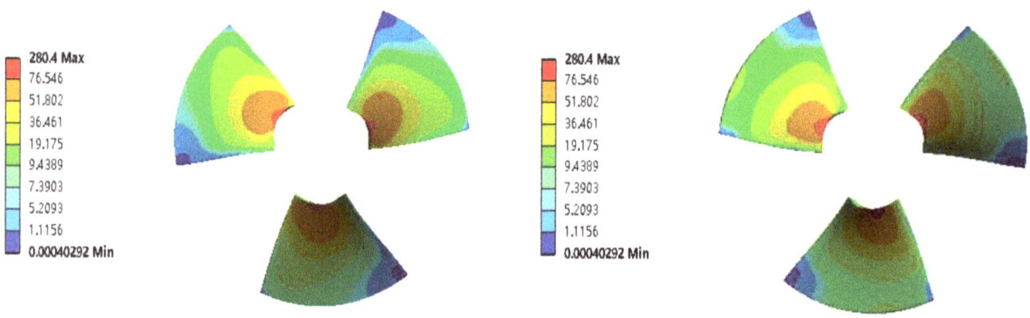

(**a**) Pressure surface (**b**) Suction surface

Figure 14. Q = 10 m^3/s Equivalent stress distribution of impeller, Unit: MPa.

Figure 15 shows the equivalent force distribution of the suction surface and pressure surface of the axial-flow pump blade under the design flow condition. Under this operating condition, the blade equivalent stress concentration point is located in the middle of the blade root; its maximum stress value is 195.03 MPa. The stress distribution tends to decrease in all directions, and the stress value transitions more gently in the diffusion process suction surface, and pressure surface stress distribution law is basically the same; the difference between the stress value at the root and edge of the entire blade is small, and the blade by the stress is more uniform.

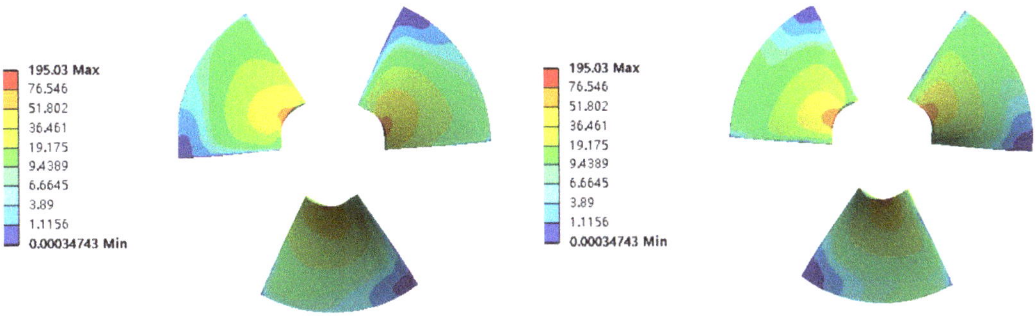

(**a**) Pressure surface (**b**) Suction surface

Figure 15. Q = 20 m^3/s Equivalent stress distribution of impeller, Unit: MPa.

Figure 16 shows the equivalent force distribution of the suction surface and pressure surface of the axial-flow pump blade under 1.4 times the design flow condition. The water flow energy is stronger under the large flow conditions; the force exerted on the blade is more obvious. The thinner the thickness of the blade, the greater the pressure surface equivalent force value with the stress peak value at 62.6 MPa, while the suction surface stress value by the thickness of the influence is not large. Overall, the stress distribution of the blade suction surface and pressure surface shows decreasing tendency from the root to the outer edge along the radial. The stress value of the suction surface relative to the pressure one is small, but on the suction surface, the difference between the peak and trough values of stress is larger.

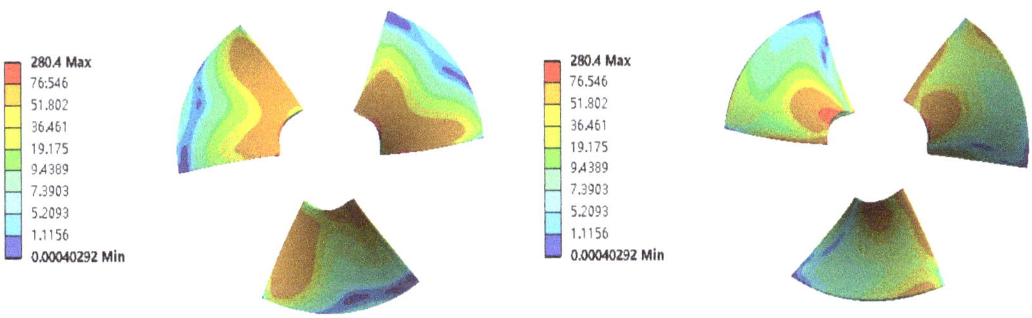

(**a**) Pressure surface (**b**) Suction surface

Figure 16. Q = 28 m^3/s Equivalent force distribution of impeller, Unit: MPa.

As can be seen from Figures 14–16, under the three working conditions, the peak values of the equivalent force are all occurring in the vicinity of the blade hub on the pressure surface of the blade, that is, the root of the blade on the water inlet side. The interface of blade and hub is a free surface, so the stress of the interface is not considered. There are more obvious stress concentration phenomena occurring on both surfaces of the blade near the hub. The stress distribution of the whole blade has a tendency to disperse outwards from the stress central point, which is at the location at which occurs the stress peak value near the hub on the water inlet, which is gradually decreasing stress outward; the rim of the blade of the water outlet side is where the stress is at its smallest. Under the same working condition, the difference between the stress value of the suction and pressure surface is not large and the law of the equivalent stress distribution is basically the same.

3.4. Deformation Analysis of the Blade

The results of the fluid–solid coupling revealed a slight deformation of the impeller blade, Figure 17 shows the deformation of the axial-flow pump impeller at 0.5 times the design flow condition. Under small-flow conditions, the deformation law of the suction surface and pressure surface of the blade are almost the same; the deformation peak is 4 mm, which occurs where the water inlet rim and outer edge intersect. The degree of the blade deformation decreases along the rim towards the hub, that is, the deformation condition becomes progressively better. There is no deformation of the blade at the hub.

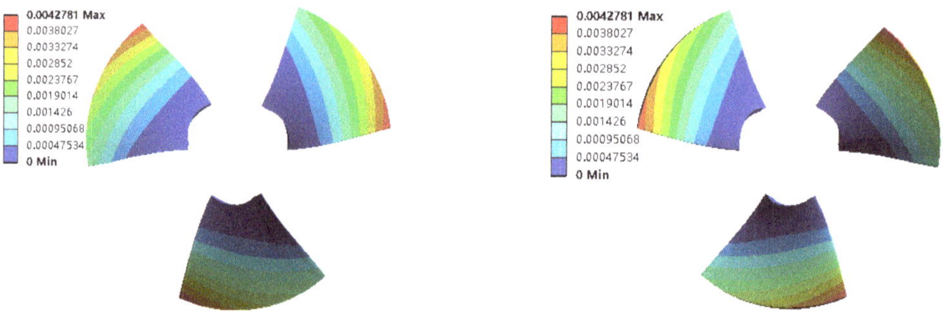

(a) Pressure surface (b) Suction surface

Figure 17. Q = 10 m^3/s Deformation of impeller, Unit: m.

Figure 18 shows the deformation of the axial-flow pump blade under the design flow condition. The suction surface and pressure surface are basically similar in deformation position, the blade in the stress value is smaller at the outer edge of the deformation which is more serious, the maximum deformation is 2 mm, the deformation is smaller in the area closer to the hub, and the deformation value at the hub is 0.

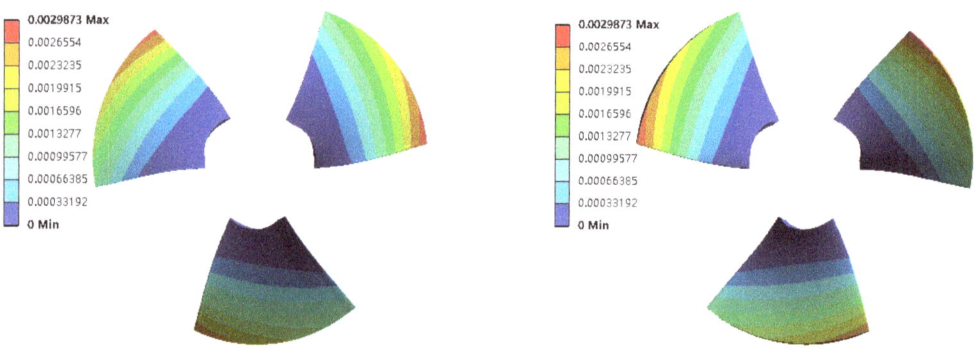

(a) Pressure surface (b) Suction surface

Figure 18. Q = 20 m^3/s Deformation of impeller, Unit: m.

Figure 19 shows the deformation condition of the impeller of the axial-flow pump at 1.4 times the design flow rate. Under high-flow conditions, the area of the blade deformation region is relatively large, and the maximum deformation is 0.7 mm. The deformation law of the suction surface is similar to the pressure surface. The blade is deformed near the center of the outer edge and the deformation gradually decreases from the rim to the hub, with no deformation near the hub.

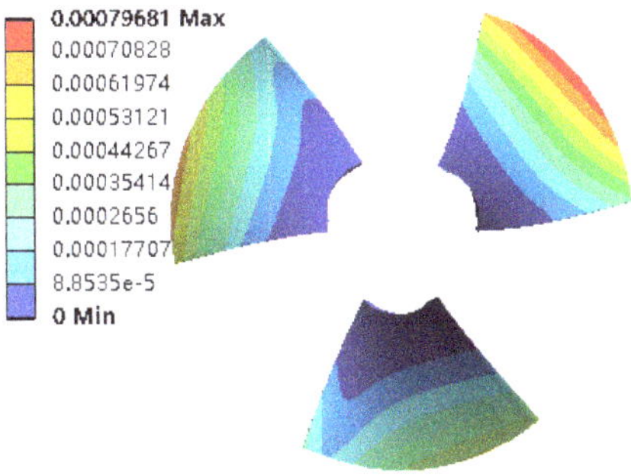

(a) Pressure surface

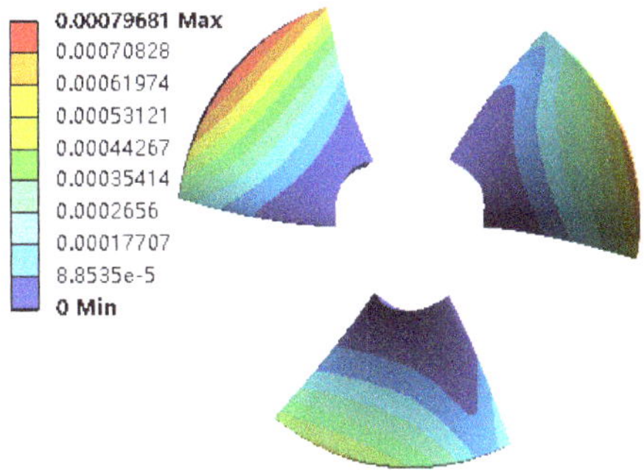

(b) Suction surface

Figure 19. Q = 28 m^3/s Deformation of impeller, Unit: m.

As can be seen from Figures 17–19, the most distinctive deformation of the blade mostly occurs at the outer edge of the inlet side of impeller. There is a gradual decrease in deformation from the edge to the hub, with no deformation occurring near the hub. The position of the greatest degree of deformation of the pressure surface and the suction surface is basically the same, and the size of the deformation is basically the same. Under the design working condition, the value of deformation will not be large, and the area where the deformation occurs will not be too large, which is conducive to making the blade maintain the original geometric parameters and ensuring the operation efficiency of the pump.

3.5. Comparative Analysis of Stress Characteristics

The maximum equivalent stress and maximum deformation of the blade under different working conditions are listed below.

The bar chart drawn according to Table 1 is shown in Figure 20.

Table 1. Comparison of maximum equivalent stress and maximum deformation (Q = 20 m^3/s).

Flow	$Q_1 = 0.5Q$	$Q_2 = Q$	$Q_3 = 1.4Q$
Maximum equivalent stress (MPa)	280.4	195.03	62.6
Maximum deformation (mm)	4	2	0.7

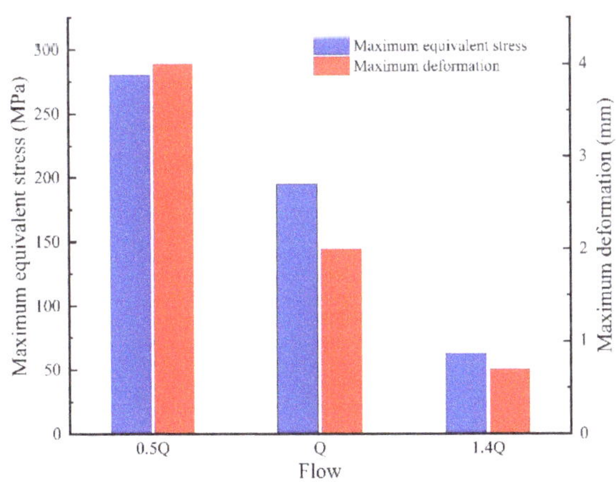

Figure 20. Comparison of maximum equivalent stress and maximum deformation (Q = 20 m^3/s).

As shown in Table 1 and Figure 20, the stress on the impeller blades varies significantly with the flow rate in the vertical bi-directional flow channel axial pump installations. With the increase in flow, the maximum equivalent stress and deformation progressively reduced. The shape characteristics of the impeller lead to a gradual increase in the equivalent force on the blade from the rim to the hub direction, with the maximum stress area appearing near the hub. As the flow rate increases, the maximum deformation of the blade shows a decreasing trend, and the range of obvious deformation in the blade increases with the flow rate.

The thickness of blades in the vertical bi-directional flow channel axial pump plays a dominant role in the amount of deformation of the blades, with the greatest thickness having the least amount of deformation near the root of the blades. Under the design working condition, the value of deformation will not be large, and the area where the deformation occurs will not be too large, which is conducive to making the blade maintain the original geometric parameters and ensuring the operation efficiency of the pump. The centrifugal force of the impeller is one of causes of the phenomenon of stress concentration at the root of the blade, and the complicated hydrodynamic operating environment is another significant reason for it. The combined effect of the two causes the blade to break. It has high probability to cause fatigue fraction. In engineering practice, the impeller generates cracks in the location of the blade root in most cases, which verifies the findings of this paper. There is only a fairly small gap between the blades and the pump casing; if the deformation of blade is too large, it will cause friction between the blade and the pump casing, affecting the normal work of the pump. In the design of an axial-flow pump impeller, the stress concentration near the root of the blades and the deformation at the rim of the blades should be fully considered; only by ensuring that both of these are within the permitted limits can the safe and stable operation of the impeller be guaranteed.

4. Conclusions

This paper presents a numerical simulation study of a vertical bi-directional flow channel axial pumping unit based on the continuity equation, the N-S equation and the RNG k-ε turbulence model, using a unidirectional fluid–structure coupling approach.

The blade stress characteristics of an axial-flow pump are analyzed in relation to the equivalent force distribution and deformation of the impeller blades in this study. This study can provide a theoretical reference for the optimization of the design of axial-flow pump vanes in order to prevent the fatigue fracture of the vanes that occurs more often during the operation of axial-flow pumps. The conclusions of this study are as follows.

(1) In the pressure distribution of the flow field of the axial-flow pump, a clear low-pressure area appears in the area near the edge of the impeller inlet, the pressure near the blades is relatively high, and as the flow rate increases, the high pressure area shows a trend of first increasing and then decreasing.
(2) Stress concentrations occur at the centre of the blade root–hub connection, with the maximum stress value at the stress concentration decreasing as the flow rate increases.
(3) The blade will be deformed in millimetres under stress, the maximum deformation is mainly distributed in the outer edge of the blade, and the closer to the hub, the smaller the deformation. As the flow rate increases, the blade deformation decreases.

For vertical bi-directional axial pumps, the conditions of the design flow conditions are more favorable for the safe and stable operation of the impeller blades.

Author Contributions: Data curation, L.C. and W.J.; Formal analysis, X.L.; Methodology, F.X. and W.P.; Writing—original draft, X.L.; Writing—review and editing, L.C., X.L. and W.J.; Supervision, L.C. All authors have read and agreed to the published version of the manuscript.

Funding: This research was funded by the National Natural Science Foundation of China (grant no. 51779214), A Project Funded by the Priority Academic Program Development of Jiangsu Higher Education Institutions (PAPD), the Key Project of Water Conservancy in Jiangsu Province (grant no. 2020030 and 2020027), and the Jiangsu Province South–North Water Transfer Technology Research and Development Project (SSY-JS-2020-F-45).

Institutional Review Board Statement: Not applicable.

Informed Consent Statement: Not applicable.

Data Availability Statement: Not applicable.

Conflicts of Interest: The authors declare no conflict of interest.

Nomenclature and Abbreviations

ρ	Fluid density
u	Speed vector
τ	Stress Tensor
ρ_s	Solid density
σ_s	Corsi stress tensor
f_s	Volumetric force tensor
$\ddot{d}_s$	Local acceleration vector in the solid domain
λ	Thermal conductivity
S_E	Energy source item
F	Volume force vector of fluid
M	Quality Matrix
C	Damping Matrix
K	Stiffness matrix
u	Displacement vector
$\dot{u}$	Speed vector
$\ddot{u}$	Acceleration vector
F	Combined force on the node diogravity, centrifugal force and the pressure of the flow field
d	Displacement
Q	Heat flow
T	Temperature
Q_p	Original pump flow
Q_m	Model pump flow
H_p	Original pump head
H_m	Model pump head
N_p	Original pump speed
n_m	Model pump speed
D_p	Original, model pump impeller diameter
D_m	Model pump impeller diameter
RANS	Reynolds-averaged systems analysis
CFD	Computational fluid dynamics
SST	Shaft frequency

References

1. Qin, J.; Liu, S.F. Strength Analysis of Vertical Shaft Through-Flow Pump Based on Fluid-Structure Interaction. *Northeast Water Resour. Hydropower* **2016**, *34*, 54–57.
2. Zhang, L.; Shi, W.D.; Zhang, D.S.; Zhang, J.J.; Guo, Y.L. Structural dynamic stress analysis of submersible axial flow pump. *J. Drain. Irrig. Mach. Eng.* **2015**, *33*, 1026–1032.
3. Shi, W.D.; Wang, G.T.; Zhang, D.S.; Jiang, X.P.; Xu, Y. Stress characteristics of axial pump blades based on fluid-solid coupling. *J. Drain. Irrig. Mach. Eng.* **2013**, *31*, 737–740.
4. Gao, J.Y.; Hou, Y.S.; Yang, Z.J.; Zhang, T.L. Static stress characteristics of axial flow pump impeller under cavitation conditions. *Agric. Eng.* **2018**, *8*, 84–90.
5. Pesch, A.; Melzer, S.; Schepeler, S.; Kalkkuhl, T.; Skoda, R. Pressure and Flow Rate Fluctuations in Single- and Two-Blade Pumps. *J. Fluids Eng.* **2021**, *143*, 011203. [CrossRef]
6. Ju, Y.P.; Liu, S.; Zhang, C.H. Effect of blade shape on hydraulic performance and vortex structure of vortex pumps. *Hydrodynamics* **2018**, *30*, 499–506. [CrossRef]
7. Chen, J.F.; Shi, W.D.; Zhang, D.S. Influence of blade inlet edge position on performance of single-blade centrifugal pump. *J. Drain. Irrig. Mach. Eng.* **2021**, *39*, 1093–1099.

8. Zhang, X.; Zheng, Y.; Mao, X.L.; Wu, Z.Q.; Kan, K.; Mou, T. Strength analysis of axial flow pump impeller based on fluid-solid coupling. *Hydropower Energy Sci.* **2014**, *32*, 137–139.
9. He, Y.Z.; Guo, Y.L. Study on the correlation between pressure pulsation and dynamic stress in axial flow pumps based on fluid-solid coupling. *China Rural. Water Conserv. Hydropower* **2020**, *12*, 158–163.
10. Pan, X.; Li, C.; Tie, E.; Zhang, W.Z. Analysis of the strength of fluid-solid coupling in axial pump blades. *J. Hydropower* **2012**, *31*, 221–226, 237.
11. Liang, W.K.; Ai, C.C.; Dong, W.; Wu, Z.J. Effect of cavitation on the fluid-solid coupling characteristics of axial pump impellers. *Large Electr. Mach. Technol.* **2021**, *5*, 92–98.
12. Zhang, L.; Yang, W.B.; Liang, S.F.; Wang, S.L. Research progress of impeller mechanical fluid-solid coupling problem. *Therm. Power Gener.* **2014**, *12*, 1–8.
13. Guo, J.Y.; Chen, J.H. Advances in fluid-solid coupling applications. *J. Jinan Univ. Nat. Sci. Ed.* **2004**, *2*, 123–126.
14. Ouyang, X.P.; Fang, X.; Yang, H.Y. An Investigation into the Swash Plate Vibration and Pressure Pulsation of Piston Pumps Based on Full Fluid—Structure Interactions. *J. Zhejiang Univ. Ser. A Appl. Phys. Eng.* **2016**, *17*, 202–214. [CrossRef]
15. Jiang, F.; Huang, P. *Fluent Advanced Applications and Case Studies*; Tsinghua University Press: Beijing, China, 2008; pp. 263–270.
16. Klaus-Jürgen, B.; Zhang, H. Finite element developments for general fluid flows with structural interactions. *Int. J. Numer. Methods Eng.* **2010**, *60*, 213–232.
17. Yang, C.X.; Guan, T.B.; Qian, C.C. Comparison of the applicability of large eddy simulation and RNG κ-ε model for numerical calculation of centrifugal pumps. *J. Gansu Sci.* **2015**, *27*, 111–114.
18. Shi, W.D.; Shao, P.P.; Zhang, D.S.; Zhang, H. Numerical simulation and PIV test of internal flow of axial flow pump. *J. Drain. Irrig. Mach. Eng.* **2015**, *33*, 277–282, 289.
19. Cao, Y.L.; He, G.; Ming, T.F.; Ding, J.M.; Su, Y.S. Non-constant analysis of cavitation in mixed-flow pumps with modified turbulent viscosity. *J. Harbin Eng. Univ.* **2016**, *37*, 678–683.
20. Guo, G.Q.; Zhang, R.H.; Hao, Y. Evaluation of different turbulence models on simulation of gas-liquid transient flow in a liquid-ring vacuum pump. *Vacuum* **2020**, *180*, 109586. [CrossRef]
21. Pei, J.; Yuan, S.Q.; Yuan, J.P. Numerical calculation of the effect of fluid-solid coupling on the internal flow field of centrifugal pumps. *J. Agric. Mach. J. Agric. Mach.* **2009**, *40*, 107–112.
22. Dang, S.J.; Liang, W.K.; Liao, W.L. Mathematical model for fluid-solid coupling of hydraulic units. *Mech. Strength* **2005**, *27*, 864–866.
23. Rubén, A.; Estrella, V.; Javier, C.; Cristina, A. Evolutionary optimization of compliant mechanisms subjected to non-uniform thermal effects. *Finite Elem. Anal. Des.* **2012**, *57*, 1–14.
24. Zhu, H.J.; Lin, Y.H.; Hsieh, L.H. *Fluent Fluid Analysis and Simulation Practical Tutorial*; Posts&Telecom Press Co., Ltd.: Beijing, China, 2010; pp. 253–259.

machines

MDPI

Article

A New Framework for the Harmonic Balance Method in OpenFOAM

Stefano Oliani [1,2,*], Nicola Casari [1], Mauro Carnevale [2]

[1] Department of Engineering, University of Ferrara, Via Saragat 1, 44122 Ferrara, Italy; nicola.casari@unife.it
[2] Department of Mechanical Engineering, University of Bath, Claverton Down, Bath BA2 7AY, UK; mc2497@bath.ac.uk
* Correspondence: stefano.oliani@unife.it

Abstract: The Harmonic Balance Method is one of the most commonly employed Reduced Order Models for turbomachinery calculations, since it leverages the signal sparsity in the frequency domain to cast the transient equations into a coupled set of steady-state ones. The present work aims at detailing the development and validation of a new framework for the application of the Harmonic Balance Method in the open-source software OpenFOAM. The paper is conceptually divided into building blocks for the implementation of the code. For each of these, theoretical notions and coding strategies are given, and an ad hoc validation test case is presented. This structure has been chosen with the aim of easing the reader in the understanding and implementation of such a method in a generic fluid dynamics solver. In a fully open source philosophy, the library files are freely accessible in the authors' repository (link provided below in the text).

Keywords: OpenFOAM; harmonic balance; spectral methods; turbomachinery; compressible flows; fourier series

Citation: Oliani, S.; Casari, N.; Carnevale, M. A New Framework for the Harmonic Balance Method in OpenFOAM. *Machines* **2022**, *10*, 279. https://doi.org/10.3390/machines10040279

Academic Editors: Chuan Wang, Li Cheng, Qiaorui Si and Bo Hu

Received: 16 March 2022
Accepted: 13 April 2022
Published: 14 April 2022

1. Introduction

Computational Fluid Dynamics (CFD) has become a landmark for academics and industries to analyze the flow field behaviour in complex domains. Part of its success is inherently due to CFD capacity of keeping pace with the increasingly demanding requirements of the scientific community in terms of performance and accuracy. Several challenges still remain about the simulation of large multi-scale domains, though. One striking example of such occurrence is the efficient simulation of unsteady turbulent flows in multi-stage turbomachinery. Time-domain solvers are still the most commonly employed, although they tend to be very computationally demanding. One of the methods CFD engineers and computer scientists are tackling these problems with, is the development of efficient algorithms to embed complex flow fields into a low-dimensional space (i.e., Reduced Order Models (ROMs)). Among ROMs, the Harmonic Balance Method (HBM) has established as a state-of-the-art tool for the calculation of periodic and quasi-periodic flows in time. While conventional Unsteady Reynolds Averaged Simulations (URANS) and Large Eddy Simulations (LES) can capture unsteady phenomena occurring at non-deterministic frequencies (e.g., vortex shedding), turbulent aeroacoustics interactions and transient non-periodic flows, many turbomachinery applications of interest exhibit periodic oscillations at known frequencies. In these cases, the HBM can be applied, resulting in a drop in computational time of typically one/two orders of magnitude compared to URANS analysis. It is evident that this could have a profound impact, especially in the design phase, when one may want to conduct a large number of calculations at a reduced cost to explore the design space, but unsteady phenomena still play an important role and must be accounted for. A thorough review of the numerical methods commonly employed for turbomachinery calculations with different levels of fidelity, including LES and hybrid methods, can be found in Tucker [1,2]. Compared to other ROMs, the HBM has the advantage of exploiting a universal basis (i.e., the Fourier basis) for the projection of the governing equation and therefore does not require any previous experimental or numerical data produced by the system. On the contrary, data-driven methods, such as the Proper Orthogonal Decomposition (POD) [3], employ a problem-dependent basis tailored on the specific dataset analyzed which will be optimal in representing the data in a desired sense. It is worth noticing that, as for every other ROM, the accuracy of the HBM is highly dependent on the number of non-zero entries retained in the chosen basis.

The HBM method originates from the ideas of He [4] and Hall [5] of leveraging the known spectral content of the flow to cast the equations into the frequency domain. In this way, the signal sparsity in the Fourier basis is exploited to approximate the time derivative in the governing equations. Applications of this method are manifold and range from limit cycle oscillation of aerodynamic components [6] to forced response problem in multi-row turbomachinery [7]. In the literature, many authors report different implementations of the method and its application to diverse problems. Hall et al. [5] and Thomas et al. [8] investigated the aerodynamics effects on flutter of a transonic compressor rotor and of airfoils, respectively. Jameson et al. [9] used a similar non-linear frequency domain method to study vortex shedding behind a cylinder and a pitching airfoil. They also modified their method that iteratively solves for the time period during the solution of the Navier–Stokes equations. van der Weide et al. [10] used their Time Spectral Method implemented in the TFLO2000 solver to compute the forced response in an axial compressor stage. One drawback of their method was that they had to use the same number of blades for the stator and the rotor. To overcome this issue, Ekici et al. [7] extended the harmonic balance technique to multi-stage turbomachinery employing a multi-frequencial HBM, where excitations at frequencies non-harmonically related to each other were considered. Later on, Gopinath et al. [11] and Sicot et al. [12] used a time-domain approach for inter-row coupling in axial compressors. Furthermore, Su and Yuan [13], Woodgate and Badcock [14], Sicot et al. [15] and Thomas et al. [16] proposed different implicit formulations of the HBM. More recently, Frey et al. proposed a frequency-domain formulation for multi-stage turbomachinery and implemented it in the TRACE solver. Nimmagaddda et al. [17] implemented an explicit version of the HBM in the open-source software SU2 and tested it on a pitching airfoil with multiple oscillation frequencies. Finally, Cvijetic et al. [18] and Oliani et al. [19] proposed a pressure-based HBM formulation for incompressible flows using *foam-extend* and OpenFOAM, respectively. In these works, an explicit formulation was used for the discretization of the source terms arising from the application of the HBM. On the contrary, in this paper we develop a novel density-based solver suitable for compressible, high-speed turbomachinery calculations and employ a fully-implicit formulation which significantly enhances the stability and convergence properties of the calculations. In addition, phase-lag boundary conditions allowing for single passage reduction were implemented, further speeding up the calculation. For a thorough review on the HBM and its variants, the reader is referred to the works of Hall et al. [20] and He [21].

The outline of the paper is as follows. In the next sections, the governing equations and the mathematical formulation of the HBM are illustrated. The implementation of a fully implicit HBM in the OpenFOAM (OF) framework, starting from the baseline formulation for a single fundamental frequency and its harmonics, is described. Attention is focused on the numerical solution of the resultant equation system and the related CFD code. We then move to more complex issues such as the multi-frequential formulation and the use of phase-lag boundary condition to reduce the computational domain to a single blade passage per row. For each step, the critical algorithmic issues are discussed, and the pseudo-code for the implementation is detailed. One of the main aims is to provide the reader with a thorough understanding of the building blocks necessary for the coding of a harmonic balance technique.

Furthermore, this work introduces several sources of novelty with respect to the state of the art. To begin with, this is the first implementation of a fully implicit HBM in a density-based solver in the context of open-source CFD softwares. We also illustrate how geometries with a rotational periodicity must be carefully treated when solving the equations in cartesian coordinates instead of cylindrical ones. Furthermore, we show how coupled interfaces typical of turbomachinery simulations can be integrated in the OF framework in an simple way. Additionally, as will be detailed in the next section, the very use of implicit density-based solvers in OF has been extremely limited so far, especially with respect to multi-row turbomachinery applications. All this being considered, the present work aims to represent a step forward in the growing application of open-source codes to industrial problems. Given also the wide diffusion of the C++ library OpenFOAM, the research community could benefit from this work as a starting point to enrich the current capabilites of the code with frequency domain methods for turbomachinery.

The code is available in the authors' repository: https://github.com/stefanoOliani/ICSFoam (accessed on 3 March 2022), and can be freely shared, edited and distributed by the users.

2. Implicit Density-Based Solver Implementation

2.1. Governing Equations

In the present work, we solve the unsteady three-dimensional compressible Reynolds-averaged Navier–Stokes (RANS) equations in conservation form:

$$\int_V \frac{\partial \boldsymbol{Q}}{\partial t} dV + \int_{\partial V} (\boldsymbol{F}_c - \boldsymbol{F}_v) dS = 0 \tag{1}$$

where V and ∂V denote the control volume and the related closed surface, respectively. For what concerns 2D simulations, in OF they are carried out on 3D grids with only one cell in the third (depth) direction. Therefore, there is no conceptual difference and we will maintain the 3D formulation throughout the paper for consistency. The vector of conservative variables $\boldsymbol{Q}$, the convective flux vector $\boldsymbol{F}_c$ and the diffusive flux vector $\boldsymbol{F}_v$ are given by

$$\boldsymbol{Q} = \begin{bmatrix} \rho \\ \rho \boldsymbol{u} \\ \rho E \end{bmatrix}, \boldsymbol{F}_c = \begin{bmatrix} \rho \boldsymbol{u} \cdot \boldsymbol{n} \\ (\rho \boldsymbol{u} \otimes \boldsymbol{u}) \cdot \boldsymbol{n} + p\boldsymbol{n} \\ \rho \boldsymbol{u} H \cdot \boldsymbol{n} \end{bmatrix}, \boldsymbol{F}_v = \begin{bmatrix} 0 \\ \boldsymbol{\tau} \cdot \boldsymbol{n} \\ (\boldsymbol{\tau} \cdot \boldsymbol{u} + \boldsymbol{q}) \cdot \boldsymbol{n} \end{bmatrix} \tag{2}$$

where $\boldsymbol{n}$ is the surface outward-pointing normal vector, ρ is the density, $\boldsymbol{u}$ is the velocity, E is the total internal energy, H is the total enthalpy, p is the static pressure, $\boldsymbol{\tau}$ is the viscous stress tensor and $\boldsymbol{q}$ is the heat flux vector. These last two terms, for a Newtonian fluid, are respectively given by

$$\boldsymbol{\tau} = (\mu + \mu_t)[(\nabla \boldsymbol{u} + \nabla \boldsymbol{u}^T) - \frac{2}{3}(\nabla \cdot \boldsymbol{u}) \cdot \mathbf{I}] \tag{3}$$

$$\boldsymbol{q} = -\left(\frac{\mu c_p}{Pr} + \frac{\mu_t c_p}{Pr_t}\right) \nabla T \tag{4}$$

where μ and μ_t are the molecular and turbulent viscosity, c_p is the specific heat at constant pressure, Pr and Pr_t are the laminar and turbulent Prandtl numbers, and T is the temperature. The resultant equations represent the conservation of mass, momentum and internal energy for an arbitrary control volume V. To obtain the turbulent viscosity, the turbulence equations are solved in a segregated manner in order to exploit the built-in OF structure. In this way, the turbulent quantities can be obtained without modifying the related part of the code, and the user can freely choose the desired turbulence model among the many present in OF. The set of equations is finally completed by the ideal gas hypothesis to relate the pressure and enthalpy to conservation variables.

2.2. Numerical Discretization

Using the finite volume method to carry out the spatial discretization of Equation (1), one obtains a set of semi-discretized equations:

$$VD_t\mathbf{Q} = \boldsymbol{R}(Q) \tag{5}$$

where $\boldsymbol{D_t}$ is the physical-time derivative operator and $\boldsymbol{R}(Q)$ is the numerical flux residual term. In density-based solvers, a dual time-stepping (DTS) technique is typically used to perform the time integration, adding a pseudo-time term τ in the equations:

$$V\frac{\partial Q}{\partial \tau} + VD_t\mathbf{Q} = \boldsymbol{R}(Q) \tag{6}$$

If an implicit method is employed to march the equations in pseudo-time to the iteration $n+1$, the residual is linearized about iteration n as

$$\boldsymbol{R}(Q^{n+1}) = \boldsymbol{R}(Q^n) + \left.\frac{\partial \boldsymbol{R}(Q)}{\partial Q}\right|_{Q=Q^n} \Delta Q^n + \mathcal{O}(\Delta Q^2) \tag{7}$$

$$\Delta Q^n = Q^{n+1} - Q^n \tag{8}$$

Since at each physical step, the system of equations is solved as a steady state problem in pseudo-time, a first-order backward scheme is used for the pseudo-time term. For the discretization in physical-time, we will see in the next section how the time derivative operator can be approximated with the HBM. For time-accurate simulations, on the other hand, a second-order backward scheme is employed in the present work. All this being considered, at each iteration one needs to solve the linear system of equations for the solution increment $\Delta \boldsymbol{Q^n}$:

$$\left[V\left(\frac{1}{\Delta\tau} - \frac{3}{2\Delta t}\right)\mathbf{I} - \frac{\partial \boldsymbol{R}(Q^n)}{\partial Q^n}\right]\Delta Q^n = \boldsymbol{R}(Q^n) + \left(V\frac{3Q^n - 4Q^k + Q^{k-1}}{2\Delta t}\right) \tag{9}$$

where k is the current physical time level and $\Delta\tau$ and Δt represent the pseudo and physical time-steps, respectively. If the numerical flux Jacobian is derived from an exact linearization of the numerical flux $\boldsymbol{R}(Q)$, Equation (9) represents a standard Newton iteration for the nonlinear system (6). Nevertheless, only approximate Jacobians are usually employed since an exact linearization of second-order inviscid fluxes requires large storage and can be excessively expensive to compute [22]. Here, for a generic interface between cell i and cell j we choose the approximate Jacobian as follows:

$$\frac{\partial \boldsymbol{R}(Q_i)}{\partial Q_i} = \frac{1}{2}(\boldsymbol{J}(Q_i) + |\lambda_{ij}|\mathbf{I}) \tag{10}$$

$$\frac{\partial \boldsymbol{R}(Q_i)}{\partial Q_j} = \frac{1}{2}(\boldsymbol{J}(Q_j) - |\lambda_{ij}|\mathbf{I}) \tag{11}$$

where $\boldsymbol{J}$ is the convective flux Jacobian and λ_{ij} is the sum of the spectral radii of the Roe and viscous flux matrices [23]:

$$|\lambda_{ij}| = |\boldsymbol{u}_{ij} \cdot \boldsymbol{n}_{ij}| + c_{ij} + \frac{1}{\|\boldsymbol{x}_i - \boldsymbol{x}_j\|}\max\left(\frac{4}{3\rho_{ij}}, \frac{\gamma}{\rho_{ij}}\right)\left(\frac{\mu}{Pr} + \frac{\mu_t}{Pr_t}\right) \tag{12}$$

where c_{ij} is the sound velocity and the subscript ij denotes quantities interpolated at the face between cell i and cell j. The GMRES linear solver [24] together with the LU-SGS preconditioner [25] is then used to find the solution of the system of equations. Since in OF all the implicit baseline solvers and the underlying code structure are based on a segregated pressure-based formulation, a new library has been implemented to accommodate the solution of the block-coupled system of Equation (9). The *HiSA* density-based library [26] has been uptaken as a starting point, to which the necessary features have been added. First of all, the structure for block-coupled matrix solution has been generalized, to accommodate the assembly of an arbitrary number of blocks and equations. In this way, a block-coupled solution of pressure-based or whatever type of solvers is possible as well. In addition, as detailed in the next section, this is necessary for the fully implicit version of the HBM, because the solution of Equation (6) must be carried out for different time instances simultaneously. Furthermore, the Roe approximate Riemann solver has been implemented for the computation of the inviscid fluxes, and has been employed in the numerical applications shown later. Eventually, Multiple Reference Frame (MRF) support for the solution in relative frames for rotating as well as translating domain motion has been integrated into the solvers. To achieve a high-order accuracy in space, a MUSCL reconstruction with the Van Leer limiter is applied on primitive variables in order to reconstruct the solution at cell faces. Viscous flux terms are computed using second-order central difference formulas.

2.3. Code Validation

The NASA rotor 37 test case has been used to validate the code. This is a widespread transonic compressor case used to validate turbomachinery CFD codes and it was also used by AGARD to test the performance of several numerical codes [27]. The detailed 3D geometry and design parameters can be found in the work of Reid and Moore [28]. The computational domain is composed of a structured hexahedral mesh of 1.1 million elements generated using Ansys Turbogrid. The tip gap between the blade and the shroud is also accounted for by meshing the tip clearance. At the inlet, an absolute total pressure of 101,325 Pa and an absolute total temperature of 288.15 K are imposed. A turbulent kinetic energy intensity of 3% is specified at the inlet, and the $k-\omega$ SST turbulence model is used as a closure for the RANS equations A steady-state calculation is employed, by solving the equation of motion in a relative frame of reference for the compressor row. The near peak efficiency condition, corresponding to a normalized mass flow rate ($\dot{m}/\dot{m}_{choking}$) of 0.98 has been used for the validation. To achieve this condition, the choking mass flow rate was initially computed by imposing a static pressure outlet boundary condition. The static pressure was then increased until the mass flow rate achieved the desired value. The obtained solution at 70% span compared to the experimental results from is reported in Figure 1a. It can be clearly seen that the bow shock position and strength are well captured. Furthermore, as reported in the experiments, shock/boundary layer interaction can be observed on the blade suction side. In Figure 1b, the relative Mach number profile at 70% span and 20% chord is reported for the numerical simulation and the experimental tests. The results obtained with OF compare fairly well with the experimental ones, so the code can be considered validated.

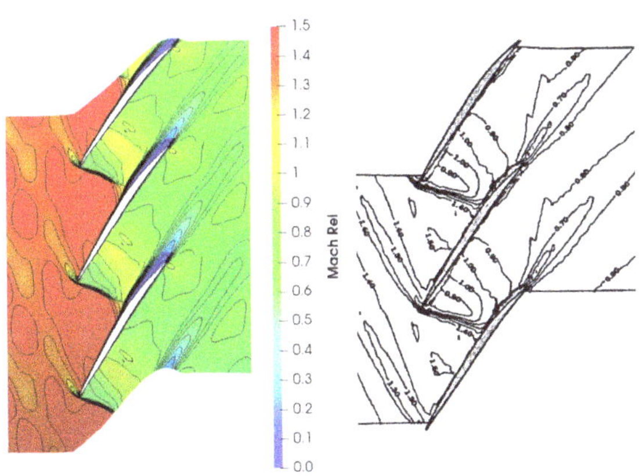

(a)

Figure 1. *Cont.*

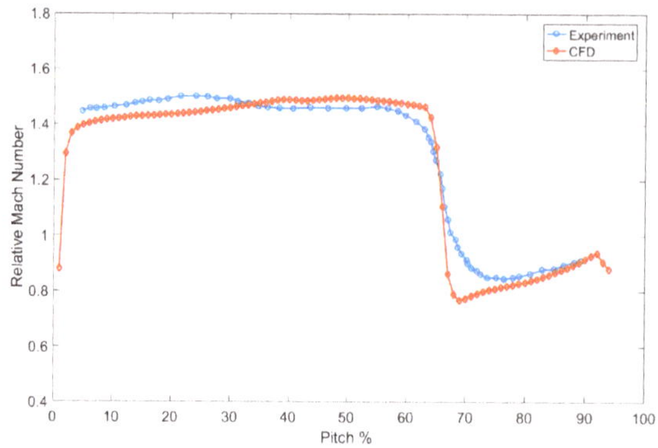

(b)

Figure 1. Numerical code validation against experimental results from [29]. (**a**) Relative Mach number contours at 70% span obtained with the CFD code (left) and experimentally (right). (**b**) Relative Mach number along the pitch at 70% span and 20% chord for the near peak efficiency condition.

3. Baseline Harmonic Balance Method

As already described in Section 1, several different approaches have been formulated in the last two decades that fall into the category of the so-called Fourier methods in CFD. Despite the seemingly fundamental differences, the common denominator of all these techniques is the frequency-domain approach to periodic unsteadiness. This idea basically harnesses the fact that a nonlinear harmonic solution with N_f harmonics is equivalent to $2N_f + 1$ coupled steady flow solutions. Therefore, the aim of an harmonic balance solution is to leverage the signal sparsity in the Fourier basis in order to approximate an unsteady solution with a coupled set of steady-state simulations. In the next paragraph, the mathematical formulation of the method is presented following the time-spectral approach by Hall et al. [5]. This approach is based on a time-domain formulation of the governing equations. Equivalent methods which are based on a frequency-domain formulation can be found in McMullen et al. [30] and Frey et al. [31].

3.1. Mathematical Formulation

Consider the vector of conservative variables $\boldsymbol{Q}$ in the control volume l that evolves periodically in time with a known single base frequency ω and its harmonics. Now suppose to expand the vector $\boldsymbol{Q}_l$ and the corresponding residual term $\boldsymbol{R}(\boldsymbol{Q}_l)$ in a Fourier series truncated to N_H harmonics

$$\boldsymbol{Q}_l \approx \hat{\boldsymbol{Q}}_{l,0} + \sum_{m=1}^{N_H} \hat{\boldsymbol{Q}}_{l,m} e^{-im\omega t} + \sum_{m=1}^{N_H} \hat{\boldsymbol{Q}}_{l,-m} e^{im\omega t} \tag{13}$$

$$\boldsymbol{R}(\boldsymbol{Q}_l) \approx \hat{\boldsymbol{R}}_{l,0} + \sum_{m=1}^{N_H} \hat{\boldsymbol{R}}_{l,m} e^{-im\omega t} + \sum_{m=1}^{N_H} \hat{\boldsymbol{R}}_{l,-m} e^{im\omega t} \tag{14}$$

where the terms $\hat{\boldsymbol{Q}}_{l,0}, \hat{\boldsymbol{Q}}_{l,m}, \hat{\boldsymbol{Q}}_{l,-m}$ constitute the $N_T = 2N_H + 1$ Fourier coefficients of the series for the time-average term and the positive and negative frequencies, respectively. This coefficients are unknown and represent an output of the calculation. Substituting Equations (13) and (14) into Equation (5), we obtain, for the control volume l

$$iV\mathbf{A}\widehat{\mathcal{Q}_l} = \widehat{\mathcal{R}_l} \tag{15}$$

where

$$\widehat{\mathcal{Q}_l} = \begin{Bmatrix} \hat{Q}_{l,0} \\ \hat{Q}_{l,1} \\ \vdots \\ \hat{Q}_{l,N_H} \\ \hat{Q}_{l,-N_H} \\ \vdots \\ \hat{Q}_{l,-1} \end{Bmatrix}, \widehat{\mathcal{R}_l} = \begin{Bmatrix} \hat{R}_{l,0} \\ \hat{R}_{l,1} \\ \vdots \\ \hat{R}_{l,N_H} \\ \hat{R}_{l,-N_H} \\ \vdots \\ \hat{R}_{l,-1} \end{Bmatrix}, \text{and } \mathbf{A} = \begin{bmatrix} 0 & \dots & \dots & \dots & \dots & \dots & 0 \\ \vdots & \omega & & & & & \\ \vdots & & \ddots & & & & \\ \vdots & & & N_H\omega & & & \\ \vdots & & & & -N_H\omega & & \\ \vdots & & & & & \ddots & \\ 0 & & & & & & -\omega \end{bmatrix} \tag{16}$$

It must be noticed that Equation (15) is a system of equations coupled through the residual flux term $\widehat{\mathcal{R}_l}$. Since this term is nonlinear, the k-th harmonic of the residual will, generally speaking, depend on all harmonics of the conservative variables vector $\boldsymbol{Q}_l$. In their *nonlinear harmonic approach*, He and Ning [4] propose to include this coupling only in the time-averaged residual equation (through a deterministic stress term), while neglecting the cross coupling of higher order harmonics. As shown by Hall et al. [5], a more convenient way to solve the equations is to model these terms implicitly, transforming them back into the time domain. This allows operating with time-domain solution stored at $2N+1$ time levels as working variables, and the nonlinear flux term can be computed in the usual way for each time level. Once again, this method leads to a set of coupled steady-state equations which are, hopefully, much cheaper to solve than a fully time-resolved simulation. Following this approach, we now proceed by expressing the Fourier coefficients vectors $\widehat{\mathcal{Q}}$ and $\widehat{\mathcal{R}}$ as

$$\widehat{\mathcal{Q}_l} = \mathbf{E}\mathcal{Q}_l, \quad \text{and} \quad \widehat{\mathcal{R}_l} = \mathbf{E}\mathcal{R}_l \tag{17}$$

where $\mathcal{Q}$ is the vector containing the variables $\boldsymbol{Q}$ stored at N_T discrete and equally spaced subtime levels over the fundamental period $T = 2\pi/\omega$ while $\mathbf{E}$ is the $N_T \times N_T$ discrete Fourier transform matrix. With the aid of Equation (17), and multiplying on the left Equation (15) by the inverse Fourier matrix $\mathbf{E}^{-1}$, one has finally

$$V\mathbf{D}\mathcal{Q}_l = \mathcal{R}_l \tag{18}$$

where $\mathbf{D} = \mathbf{E}^{-1}i\mathbf{A}\mathbf{E}$. Due to its definition, $\mathbf{D}$ is a skew-symmetric circulant matrix [20]. This system is coupled over N_T subtime levels (or snapshots of the flow field) because of the $V\mathbf{D}\mathcal{Q}_l$ term. It is worthwhile to notice that Equation (18) is a system of steady-state equations which is marched to convergence by adding a pseudo-time derivative term in the same way as in Equation (6):

$$\frac{\partial \mathcal{Q}_l}{\partial \tau} + V\mathbf{D}\mathcal{Q}_l = \mathcal{R}_l \tag{19}$$

A comparison with Equation (6) reveals that the aforementioned procedure has allowed finding a proxy for the time derivative operator $D_t\mathbf{Q}_{\mathrm{l,j}} \approx (\mathbf{D}\mathcal{Q}_l)_j$ for the flow field snapshot j and in the control volume l. The HBM is thus a way to replace the transient equations with a system of coupled steady-state equations. All the acceleration techniques for steady-state formulations can thus be exploited (e.g., multigrid and local time-stepping), significantly reducing the turnaround times of the simulation.

3.2. Numerical Solution

As explained in the previous section, we need now to solve a system of equations which couples all the subtime levels (i.e., snapshots) through the harmonic balance source term $V\mathbf{D}\mathcal{Q}$. There are various ways in which this term can be discretized numerically. If this term is treated implicitly, it can be easily included into the matrix of the linear system since the operator $\mathbf{D}$ is linear

$$V\mathbf{D}\mathcal{Q}^{n+1} = V\mathbf{D}\mathcal{Q}^{n} + V\mathbf{D}\Delta\mathcal{Q}^{n} \tag{20}$$

Including this linearization into Equation (19) and carrying out the discretization in pseudo-time one obtains:

$$\boldsymbol{M}\Delta\mathcal{Q}^{n} = \mathcal{R}^{n} - V\mathbf{D}\mathcal{Q}^{n} \tag{21}$$

where

$$\boldsymbol{M} = \begin{bmatrix} \left(\frac{V}{\Delta\tau_0}\mathbf{I} - \frac{\partial R_0^n}{\partial Q_0^n}\right) & VD_{0,1}\mathbf{I} & \dots & VD_{0,2N_H}\mathbf{I} \\ -VD_{0,1}\mathbf{I} & \left(\frac{V}{\Delta\tau_1}\mathbf{I} - \frac{\partial R_1^n}{\partial Q_1^n}\right) & & \\ \vdots & & \ddots & \\ -VD_{2N_H,0}\mathbf{I} & -VD_{2N_H,1}\mathbf{I} & & \left(\frac{V}{\Delta\tau_{2N_H}}\mathbf{I} - \frac{\partial R_{2N_H}^n}{\partial Q_{2N_H}^n}\right) \end{bmatrix} \tag{22}$$

In Equation (22) the fact that the matrix $\mathbf{D}$ is skew symmetric has been exploited to set $D_{k,k} = 0$ and $D_{i,j} = -D_{j,i}$. There are a few observations that are worth pointing out when solving Equation (21). First, if the off-diagonal terms in Equation (22) are neglected, this is equivalent to an explicit treatment of the HBM source term $\Omega\mathbf{D}\mathcal{Q}$, since now all the increments in the solution variables $\Delta\mathcal{Q}^n$ are not coupled anymore. Second, as pointed out by Su and Yuan [13], for large values of the angular frequency ω and a high number of wave modes N_H, the diagonal-dominance property of the matrix $\boldsymbol{M}$ may be lost. In these cases, algorithms that require diagonal-dominance to ensure the convergence (e.g., Jacobi and Gauss-Seidel method) can lead to stability problems during the iterative solution. Therefore, in the present work, LU-SGS is used only as a preconditioner, while the GMRES method is employed for the solution of the preconditioned system. With this method, once the linear solution achieves iterative convergence, the increment in the flow variables are found for all the time snapshots simultaneously. Similar methods, but with different preconditioners, have been used by Su and Yuan [13] and Woodgate and Badcock [14]. In our solver, the time derivatives of the turbulent quantities are not included in the HBM, but are treated in the usual segregated manner as steady-state quantities. However, this has been found to have only a minor impact on the final solution. Finally, in turbomachinery applications and, more generally, in all cases in which a dynamic mesh is involved, it is necessary to take into account the domain motion. Therefore, the relative fluxes with respect to the mesh motion are calculated and used in Equation (1).

The new block-coupled solver structure allows for an easy assembly of the matrix $\boldsymbol{M}$ for the linear system. Algorithm 1 below describes in detail all the steps of the solution for the baseline HBM.

3.3. Numerical Application: NASA Rotor 37

To showcase the implementation of the baseline HBM, the first numerical application presented here regards a radial slice of the NASA rotor geometry used in Section 2.3 to validate the solver. The midspan slice of the domain was projected onto a plane to perform a 2D simulation. Being a single row domain, steady-state simulations in a relative frame of reference are usually employed for this type of configuration. Nevertheless, the study of the downstream propagation of the wake of the rotor represents an excellent spot to validate the HBM on a simple test case with a single dominant frequency (i.e., the blade passing frequency). To this purpose, the domain has been divided into three zones: the intake, the rotor and the discharge, where the first and the last are solved in the absolute frame, while the rotor is solved in a relative frame that translates with it. A linear motion with a velocity of 200 m/s in the azimuthal direction has been set for the rotor. In addition, a wake-like perturbation has been imposed at the inlet, basing on the self-similarity condition proposed by Lakshminarayana and Davino [32]. Specifically, a Gaussian perturbation with the same azimuthal wavelength as the blade pitch has been imposed in terms of total pressure and temperature:

$$p_0(y) = p_{0,ref}[1 - \Delta p \cdot e^{-0.693(2\frac{y}{L})^2}]$$
$$T_0(y) = T_{0,ref}[1 - \Delta T \cdot e^{-0.693(2\frac{y}{L})^2}]$$

where $p_{0,ref} = 101\,325\,\mathrm{Pa}$ and $T_{0,ref} = 288.15\,\mathrm{K}$. Δp and ΔT are the total pressure and total temperature deficits, which have been selected according to the realistic values of 0.025 and -0.007, respectively, proposed by Gomar et al. [33]. L is the wake width and has been selected as 25% of the blade pitch. This non-uniform perturbation in the absolute frame is seen by the rotor as a travelling disturbance with a frequency equal to the BPF. At the same time, the rotor wake represents a travelling disturbance with the same frequency in the discharge region. We are interested in reproducing the correct behaviour of this entropy disturbances in the whole domain, as well as their interaction in the discharge region. This can be accomplished by the use of either time-resolved or HBM simulations.

Algorithm 1 Baseline HBM solution algorithm.

1: Select the base frequency ω and the number of harmonics N_H
2: Set the number of snapshots $N_T \leftarrow 2N_H + 1$
3: Uniformly sample $T \leftarrow \frac{\omega}{2\pi}$ at N_T instants
4: Create N_T instances of the mesh of the domain
5: **if** Mesh is dynamic **then**
6:     **for** $i \leftarrow 0, \ldots, N_T - 1$ **do**
7:         Update $Mesh_i$ position for time t_i
8:         Compute mesh flux due to $Mesh_i$ motion at position defined at step 7
9:     **end for**
10: **end if**
11: **while** Simulation not converged **do**
12:     Compute RHS part of the HBM source term $V\mathbf{D}\mathcal{Q}^n$
13:     Initialize implicit linear system $N_T \times N_T$ block matrix $\boldsymbol{M}$
14:     **for** $j \leftarrow 0, \ldots, N_T - 1$ **do**
15:         Compute the flux residual $\boldsymbol{R}_j$ and Jacobian $\frac{\partial \boldsymbol{R}_j^n}{\partial \boldsymbol{Q}_j^n}$ for $Mesh_j$
16:         **for** $k \leftarrow 0, \ldots, N_T - 1$ **do**
17:             **if** $j == k$ **then**
18:                 $\boldsymbol{M}[j][j] = \left(\frac{V}{\Delta\tau_j}\mathbf{I} - \frac{\partial \boldsymbol{R}_j^n}{\partial \boldsymbol{Q}_j^n}\right)$
19:             **else**
20:                 $\boldsymbol{M}[j][k] = VD_{j,k}\mathbf{I}$
21:             **end if**
22:         **end for**
23:     **end for**
24:     Solve block coupled system $\boldsymbol{M}\Delta\mathcal{Q}^n = \mathcal{R}^n - V\mathbf{D}\mathcal{Q}^n$
25:     $\mathcal{Q}^{n+1} = \mathcal{Q}^n + \Delta\mathcal{Q}^n$
26: **end while**

The aim is to compare the behaviour of the unsteady perturbations for three different modelling solutions: a steady-state, a time-accurate and an HBM simulation. With respect to the calculation, a second order MUSCL reconstruction with the van Leer limiter has been applied on the primitive variables for interpolation from cell centroids to face centers. The Roe scheme has been used to compute the inviscid fluxes, while a second-order centered scheme has been employed for the viscous fluxes. The $k - \omega$ SST model has been employed for turbulence closure. A pseudo Courant number of 50 has been set for the HBM, the steady-state and the inner loop of the DTS simulation. For DTS, 80 timesteps per blade passing period have been used, considering the inner loop as converged when a drop of the residuals of at least two orders of magnitude was achieved.

To compare the results for the different methods, the entropy field is calculated at the same reference relative position between the inlet wake disturbance and the blade (Figure 2). The results for the steady-state case are depicted in Figure 2a, where three blade passages are reported for the sake of clarity. In this case the interface between the different zones is treated as a "frozen rotor" Arbitrary Mesh Interface (AMI), meaning that the information is simply transferred between the two non conformal sides through a polygon-clipping conservative interpolation. It can be noticed that the wake incidence angle is discontinuous at the interfaces, both for the inlet disturbance and for the wake generated by the blade.

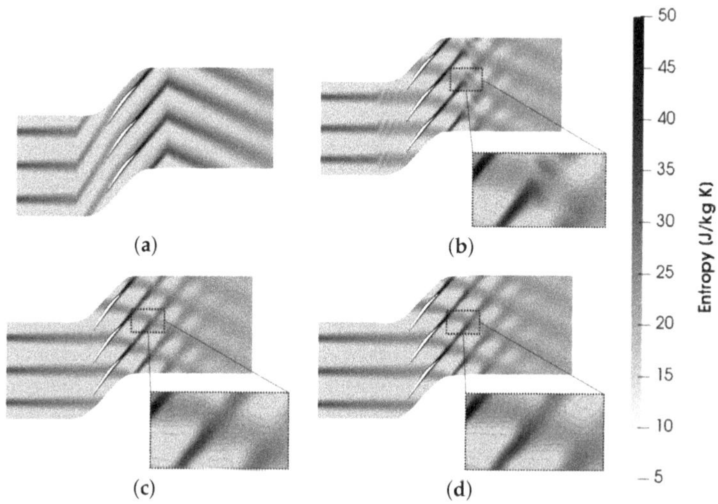

Figure 2. Entropy field obtained with the three different methods and different number of harmonics included in the flow spectrum. (**a**) Steady-state frozen rotor; (**b**) 1 Harmonic; (**c**) 3 Harmonics; (**d**) DTS.

This non-physical effect is inherently linked to the steady-state nature of the calculation and is due to the change in the relative frame of reference at the interfaces. On the other hand, for the transient case (Figure 2d), thanks to the mesh motion, the incidence angles are correctly preserved across the interface. The HBM aims to mimic this behaviour without resorting to a transient simulation, by simply coupling a set of steady-state simulations. When a single harmonic is retained in the flow spectrum, the results are reported in Figure 2b. It can be noticed that the mere addition of a single harmonic to the mean flow is not sufficient to reconstruct the wakes shape, although they are now correctly aligned across the interfaces. In general, the thinner are the wake, the greater is the number if harmonics necessary to reconstruct their shape [33]. When three harmonics are included in the frequency set, the situation improves dramatically, as can be seen by the comparison of Figure 2c with the reference solution in Figure 2d. The inlet entropy wave and the rotor wake are correctly transferred across interfaces and their interaction in the downstream region is well captured. The entropy "bubbles" due to the poor resolution of the flow spectrum disappear, as highlighted by the close-ups in the figures. When a small number of harmonics are retained, under and overshoots can be produced, leading to non-physical values of flow quantities. This can be seen in Figure 3, where the entropy at location A (Figure 4) is plotted as a function of the pitch. The HBM solution with one harmonic is not capable of reproducing the entropy distribution across the pitch, resulting in some values falling below zero at the wake edge. On the contrary, excellent agreement is found between the three harmonics and the DTS solutions, although the former has been found to be about seven times cheaper than the latter in terms of CPU times.

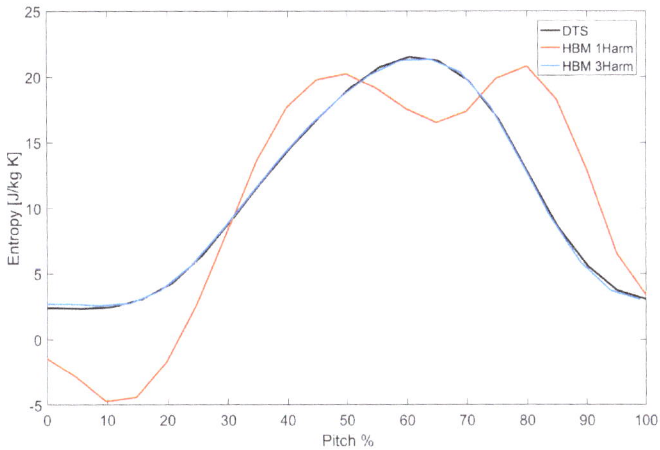

Figure 3. Entropy profile in the wake at plane A for HBM and DTS.

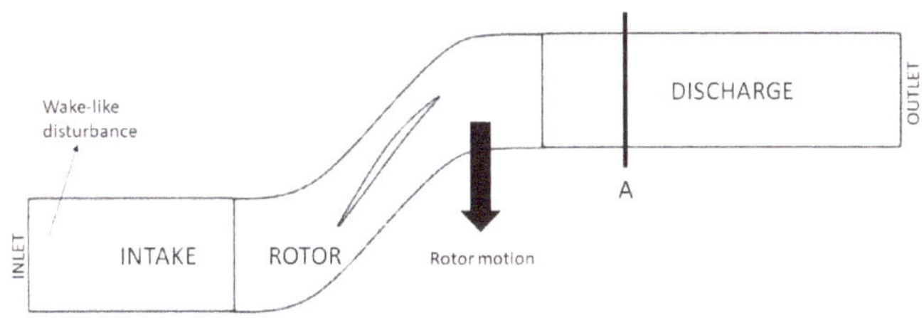

Figure 4. Domain of the NASA rotor 37 test case for the baseline HBM.

4. Multi-Frequencial Harmonic Balance Method

4.1. Mathematical Formulation and Numerical Implementation

In the baseline method presented in the previous section—sometimes referred to as the time-spectral method [10]—a single fundamental frequency ω and its harmonics are present. Nevertheless, in several applications of industrial interest, multiple discrete dominant frequencies exist, which need not be integral multiples of each other. This kind of signal is defined as almost-periodic. Examples of such occurrences are multi-rows turbomachinery, in which the number of blades and vanes are typically not multiples. In such a scenario, the flow variables can still be projected into a set of arbitrary, non-harmonically related frequencies. It is important to underline that in this case the chosen frequency set does not form, in general, an orthogonal basis. Furthermore, we do not have an analytical expression for the operator $\mathbf{D}$, nor for the discrete Fourier matrix $\mathbf{E}$. Instead, the common approach is to define the matrix $\mathbf{E}^{-1}$ analytically, and then compute numerically its inverse $\mathbf{E}$ [11,34]. In this way, the harmonic balance operator can still be defined similarly to the baseline case. Denoting with $\mathbb{F} = [\omega_0, \omega_1, \ldots, \omega_K, \omega_{-K}, \ldots, \omega_{-1}]$ the vector of selected input frequencies, where $\omega_0 = 0$ and $\omega_{-k} = -\omega_k$, we can define the inverse transform matrix by its components as $E^{-1}_{n,k} = e^{i\omega_k t_n}$. Finally, the harmonic balance operator becomes simply

$$\mathbf{D} = \mathbf{E}^{-1} i\mathbf{A}\mathbf{E}, \quad \text{where} \quad A_{k,k} = \omega_k \tag{23}$$

where $\mathbf{E}$ has been computed by numerically inverting $\mathbf{E}^{-1}$. With this procedure, the same formulation as the baseline case can be used. Unfortunately, the presence of non-harmonically related frequencies entails that we can not always work out a fundamental period for uniform sampling of the subtime levels. Furthermore, even if a common fundamental period is present (e.g., for frequencies which are integers and coprime, see Section 4.2), we will be hardly interested in uniformly sampling that period. Indeed, this would be equivalent to a baseline HBM in which the base frequency corresponds to the common fundamental period and the discrete frequencies we are interested in are represented as harmonics of this base frequency. If the original discrete frequencies are far from each other, a huge number of harmonics needs to be considered into the spectrum (i.e., a huge number of snapshots). Therefore, we seek a way more general approach, even if the selection of sampling points typically becomes a thorny issue. Indeed, the time levels of the snapshots used during the calculation have a pivotal role in the convergence of the method. This is because the condition number of the matrix **E** is responsible for amplifying errors coming from the HB source term during the iterative solution of the equations [34].

In particular, the following conceptual differences with the baseline approach apply:

1. The frequency set is composed of N_O frequencies that, in general, are no longer harmonics of a common base frequency. Please notice that, in spite of this, the following relation between the minimum necessary number of samples/snapshots ($N_{T,min}$) and the number of frequencies still holds: $N_{T,min} = 2N_O + 1$
2. Now the matrix **D** is no longer skew-symmetric. Therefore, the components on its diagonal are not zero and step 18 of Algorithm 1 must be modified according to:

$$M[j][j] = \left(\frac{V}{\Delta\tau_j}\mathbf{I} - \frac{\partial \boldsymbol{R}_j^n}{\partial \boldsymbol{Q}_j^n} \right) + VD_{j,j}\mathbf{I} \tag{24}$$

It is worth noticing that for periodic flows, the Fourier matrix is always well-conditioned since the uniform sampling leads to a condition number equal to 1. This is mainly due to the orthogonality of the complex exponential basis. Conversely, when the frequency set is arbitrary, finding a set of time instants over which the matrix **E** is well conditioned is much more difficult. A variety of approaches have been proposed by several authors to overcome this issue. For example, Ekici and Hall [7] proposed an oversampling of the solution to achieve better conditioning of the matrix, while Guedeney et al. [34] used a gradient-based optimization algorithm to find a suitable set of subtime levels. In this work, we pursue the OptTP approach described by Nimmagadda et al. [17]. This approach has proved to be robust and easy to implement, and with a negligible performance drop compared to gradient-based optimization methods. The proposed method is de facto equal to the first step of the optimization method proposed by Guedeney et al. [34]. It uses a brute-force research to find the time period T^* which minimizes the condition number of matrix **E** inside a user-defined range of possible periods, and then uniformly samples it. This operation is performed as a pre-processing step of the simulation and, due to its simplicity, it represents a negligible portion of the total CPU time. The method is detailed in Algorithm 2.

Unfortunately, this algorithm can lead to a sampling period which is much bigger than the periods corresponding to the frequencies in the set. This, in turn, could traduce in an amplification of aliasing errors in the flow spectrum. Moreover, there is no guarantee to achieve a sufficiently low value for the condition number. For these reasons, the support for oversampling was also added to our solver. The formulation in this case is exactly the same, except that now the number of time snapshots N_T can be greater than $2N_O + 1$. As a consequence, one must work with rectangular Fourier matrices to obtain the HB operator:

$$\underbrace{\mathbf{D}}_{N_T \times N_T} = \underbrace{\mathbf{E}^{-1}}_{N_T \times N_O} \underbrace{\mathbf{A}}_{N_O \times N_O} \underbrace{\mathbf{E}^{+}}_{N_O \times N_T} \tag{25}$$

where $\mathbf{E}^{-1}$ is defined as usual and $\mathbf{E}^{+}$ is the Moore–Penrose pseudo inverse of $\mathbf{E}^{-1}$. Once the HB operator has been defined in this manner, no other differences exist in the solver due to oversampling, except that the code structure must allow for a different length of the snapshots vector $\mathcal{Q}$ and the frequency vector $\mathbb{F}$.

Algorithm 2 OptTP algorithm for the minimisation of the Fourier matrix condition number. As per Nimmagadda et al. [17].

```
Compute ω_min = min{ω_1, ω_2, ..., ω_K} and T_0 = ω_min / 2π
Form a set of test time periods 𝕋 starting from T_0 and ending at 5T_0, with a step size of 0.01T_0
Initialize condNumber = 10^10
Initialize snapshots = [0, 0, ..., 0]   (N_T entries)
Initialize selectedSnapshots = [0, 0, ..., 0]   (N_T entries)
for i ← 0, ..., length(𝕋) − 1 do
    Compute T_f = 𝕋[i]
    Compute ΔT = T_f / N_T
    for k ← 0, ..., N_T − 1 do
        snapshots[k] = kΔT
    end for
    Calculate the inverse Fourier matrix corresponding to the time instances in snapshots and the vector of frequencies 𝔽
    Calculate thisCondNumber as the condition number of the inverse Fourier matrix just computed
    if thisCondNumber < condNumber then
        condNumber = thisCondNumber
        selectedSnapshots = snapshots
    end if
end for
```

4.2. Numerical Application: Channel Flow

A test case very similar to the one used by Guedeney et al. [34] is used to validate the multi-frequential HBM implementation. It consists of a 1D channel of length 1000 m and height 1 m discretized with 2500 elements in the streamwise direction. At the inlet, a constant total pressure of 101,325 Pa and a total temperature of 288.15 K are imposed. An oscillating pressure is set at the outlet

$$p_s(t) = \bar{p} \cdot [1 + A(\sin 2\pi f_1 t + \sin 2\pi f_2 t)] \tag{26}$$

where $\bar{p}$ is the 60% of the inlet total pressure, $A = 0.001$ $f_1 = 3$ Hz and $f_2 = 17$ Hz. Please notice that these two frequencies are not harmonically related but they are integer so that the flow is periodic with $T = 1$ s. In any case, the method is completely general and can be used for any set of harmonically and non-harmonically related frequencies. The flow is considered as inviscid and the convective fluxes are calculated by the second order Roe scheme with van Leer limiter. The time resolved simulation employs the DTS second order backward scheme with a time step of 0.001 s. A drop in the residuals of at least 10^{-3} was used for the iterative convergence in pseudo time. Twenty seconds of physical time were simulated to obtain a periodic flow.

Since the flow is subsonic throughout the channel, the objective is to observe the propagation of pressure waves from the outlet as depicted in the sketch of the domain in Figure 5. The nonlinearity of the flow triggers other frequencies in the flow spectrum as the waves propagate towards the inlet. These frequencies need to be included in the input frequency set for the HBM simulation. Strictly speaking, due to the quadratic nonlinearity of the NS equations, infinitely many frequencies are produced in the flow spectrum. These frequencies are all the linear combination of f_1 and f_2 and all their harmonics. For practical purposes, the array of solved frequencies is truncated to $\mathbb{F}^* = [f_1, 2f_1, 3f_1, f_2 - 2f_1, f_2 - f_1, f_2, f_2 + f_1, f_2 + 2f_1, f_2 + 3f_1]$. In this case, the algorithm described in Section 4.1 produces a condition number of the Fourier matrix equal to 1.69. In the authors' experience, values around 2.5 or below can be considered satisfying for the simulation. The pressure signal is probed at three locations at 25%, 50% and 75% of the channel length, respectively, as show in Figure 5. Then the outcomes from the DTS and the HBM simulation are compared. The flow field at the desired time instants and locations for the HBM calculation can be easily reconstructed from the resultant flow spectrum through an inverse Fourier transform once the simulation has converged. The comparison is reported for each probe in Figure 6 over one period 1 s. It can be seen that the overall agreement between the two methods is good, especially for the third probe. This is to be expected, since it is the closest to the origin of pressure disturbance and therefore its spectrum is the sparser than the other two probes [34]. On the other hand, the spectrum at probe 1 location is richer and more complex, resulting in pressure peaks slightly overpredicted with respect to the DTS method. Still, the agreement is considered satisfactory throughout the channel and the method is considered validated.

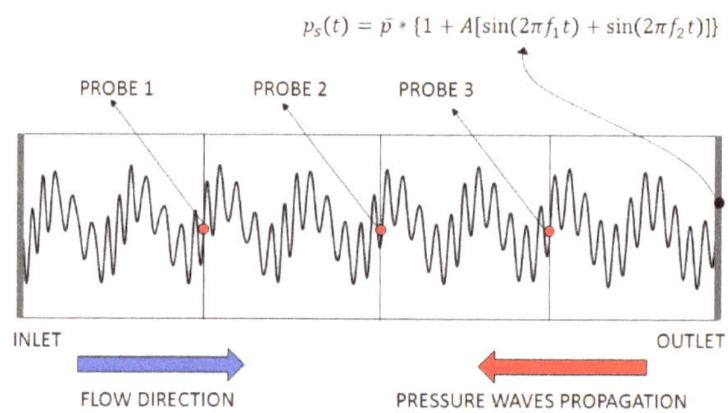

Figure 5. Domain and boundary conditions for the channel flow test case.

4.3. Coupling between Different Zones

It is well known that in multi-stage turbomachinery, the frequencies of the unsteady flow field are linear combinations of the blade passing frequencies of the neighbouring rows [35]. For this reason, the code structure must accommodate the possibility to impose a different set of frequencies in each zone. For example, in a single stage configuration, the fundamental frequency of the HBM for either row is the passing frequency of the opposite row $\omega = N_B\Omega$. The common approach to deal with single and multi-stage turbomachinery is to set different snapshots for each blade row, so that the time instants solved in each row do not necessarily match to each other. The information between adjacent zones is generated for the time levels of the receiving side through spectral interpolation of the flow field on the donor side [12], or through spatial Fourier coefficients matching [7]. The advantage of the former methods is that the coupling is performed purely in the time domain, allowing the use of non-matching radial grid lines at the interfaces. Both these techniques, however, if not properly handled through non-reflecting boundary conditions [7] or oversampling at the interface [11,12], can generate aliasing and spurious wave reflections. Furthermore, the application of non-reflecting boundary conditions to filter spurious frequencies at blade row interfaces often leads to a significant complexity in the source code. Therefore, in this work we follow another strategy, which has recently been devised by Crespo and Contreras [36] and is called the Synchronized HBM. This method bolsters the flow continuity between stator/rotor interfaces by using the same physical subtime levels all the way through the different blade rows. Since now spectral interpolation is no longer necessary at the interface, fluxes continuity is easily ensured through the use of fully conservative non-conformal interfaces. In addition, Fourier interpolation of different time instants can lead to oscillations in the solution due to the Runge phenomenon. Since in the Synchronized

HBM, time instances represent a snapshot of the flow field throughout all the rows at the same time, this technique automatically refrains Runge phenomenon without the necessity to use non-reflecting boundary conditions at the interface [36]. The main drawback is that now an optimized set of instants must be retrieved to accommodate the necessity of a low condition number of the matrix **E** for all the blade rows at the same time, even for single-stage configurations. To tackle this problem, the strategy highlighted in Section 4.1 was used, and the subtime levels were selected in such a way to minimize the maximum condition number among all blade rows:

$$\mathbb{T}^* = \min(\max(\kappa(\mathbb{T}))) \tag{27}$$

Following Algorithm 2, the implementation of the coupling is straightforward, since no complex non-reflecting boundary conditions or interpolation steps are necessary at blade row interfaces. It is instead sufficient to include a loop to find the maximum condition number among all rows for each test time period $\mathbb{T}$. This is the variable to minimize using Algorithm 2. Another minor drawback is that, in theory, with the non-synchronized approach, one could use a different number of snapshots in each zone (i.e., for each row) of the domain. This could speed up the calculation when it is known that the flow spectrum in some region is more sparse. Of course, this is no longer possible with the synchronized approach, where the selected number of snapshots is used throughout the whole domain.

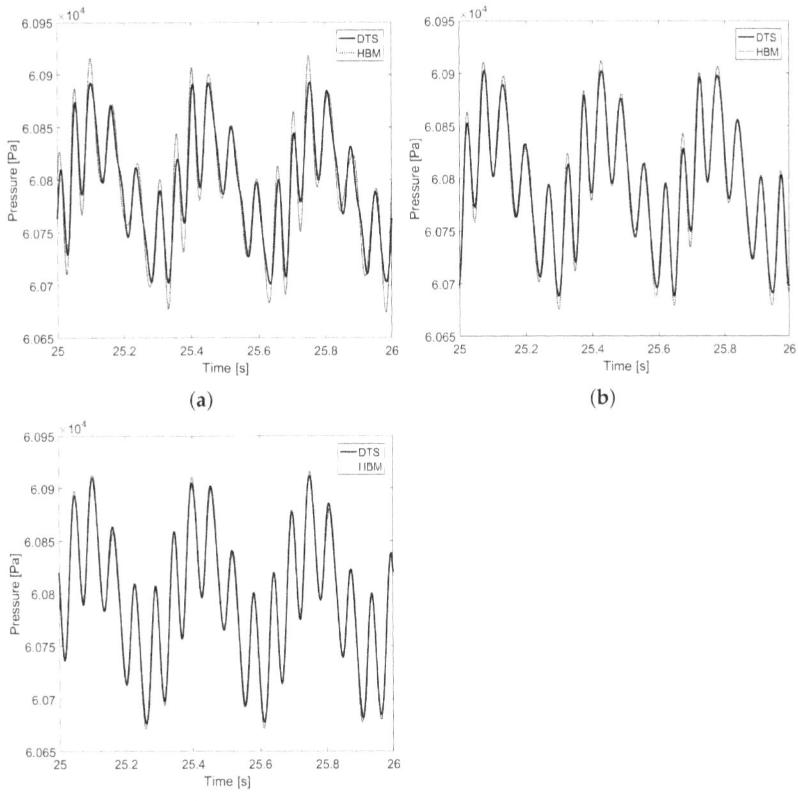

Figure 6. Comparison between HBM and DTS for the pressure signal registered by the three probes for the channel flow test case. (**a**) Probe 1. (**b**) Probe 2. (**c**) Probe 3.

4.4. Numerical Application: E^3 Axial Turbine Stage

It is now time to test the HBM on a case study of greater engineering relevance. Specifically, the midspan section of the first stage of E^3 engine axial turbine is studied.

The case is 2D but contains the combined complexity of coupling between different zones and the presence of different base frequencies in the flow spectrum. It must be emphasized that, as described in the previous section, if each row was sampled at its own set of time levels, the multi-frequencial approach would not be necessary. The use of the synchronized approach, on the other hand, implies that the solved time instants must be the same for the entire domain. Therefore, the columns of the Fourier matrix are not orthogonal anymore, and the multi-frequencial approach must be applied. The peculiarity is that the non-orthogonality of the basis is now caused by the selection of time instants not corresponding to the uniform sampling of any particular frequency in the spectrum, rather than the presence of multiple frequencies itself in single mesh regions.

A two vane/three blades configuration was employed in order to match the pitchwise extension of the two rows. The structured mesh was generated with ICEM CFD and is composed of 141,074 elements, selected after a grid convergence study. A turbulent viscous grid is used, ensuring a maximum y^+ value of ≈ 5 on the airfoil walls. The boundary conditions imposed for the simulation are reported in Table 1. Roe upwind scheme and van Leer limiter were used to achieve a second-order spacial accuracy for the convective fluxes. The central difference scheme was employed for the viscous terms. Eighty physical timesteps per blade passing were ensured for the DTS simulation, where a second-order backward scheme is employed for the temporal derivative term. A pseudo Courant number of 50 was set for the DTS as well as for the HBM simulations.

Table 1. Boundary conditions for the E^3 axial stage simulation.

	Quantity	Value
Inlet	Total pressure	350 kPa
	Total temperature	683 K
	Turbulence intensity	3%
Outlet	Pressure	144 kPa

Up to five harmonics of the blade passing frequency and the vane passing frequency were retained in the spectrum of the stator and rotor regions, respectively. Interestingly, due to the particular choice of the stator/rotor blade count, the condition number obtained with the brute-force algorithm detailed in Section 4.1 was sufficiently low for all but the four harmonics case. To overcome this issue, oversampling with $3N_H + 1 = 13$ snapshots was employed for the four harmonics simulation in order to have a small condition number of the Fourier matrix. In this way, a condition number of approximately 1 was obtained for all cases. The results for the four harmonics case are reported for the sake of completeness, although the computations is more computationally demanding than the five harmonics solution, due to the oversampling. A comparison of the entropy contours in Figure 7 reveals a good agreement between the HBM and the DTS solutions. The wakes are correctly aligned across the interface, preserving their position and their width. Remarkably, the effect of wake stretching and tilting while it is convected across the blade passages is well resolved by the HBM solution. A small difference is present in the wake of the rotor causing a slightly larger wake pattern downstream of the row. This is probably due to some self-induced vortex shedding frequencies which are not properly captured by the HBM solution. Spectral convergence is shown by means of a plot of the density and the absolute velocity magnitude as a function of the domain pitch in front of the rotor (Figure 8). Despite the complex trend across the domain pitch, the global error between the DTS and the HBM solutions gradually decreases when the spectral content of the solution is augmented. Figure 9 represents the actual amount of computational resources needed for the simulations. The plot is non-dimensionalized by the DTS simulation values, so that the values represent the CPU saving and the memory consumption factors. As expected, the latter one grows almost linearly with the number of snapshots, while the former shows a decreasing trend. It can be observed that even when 13 snapshots are used for the solution, the HBM solution is still about three times faster than the time-resolved simulation, but the memory consumption is about 19 times as much.

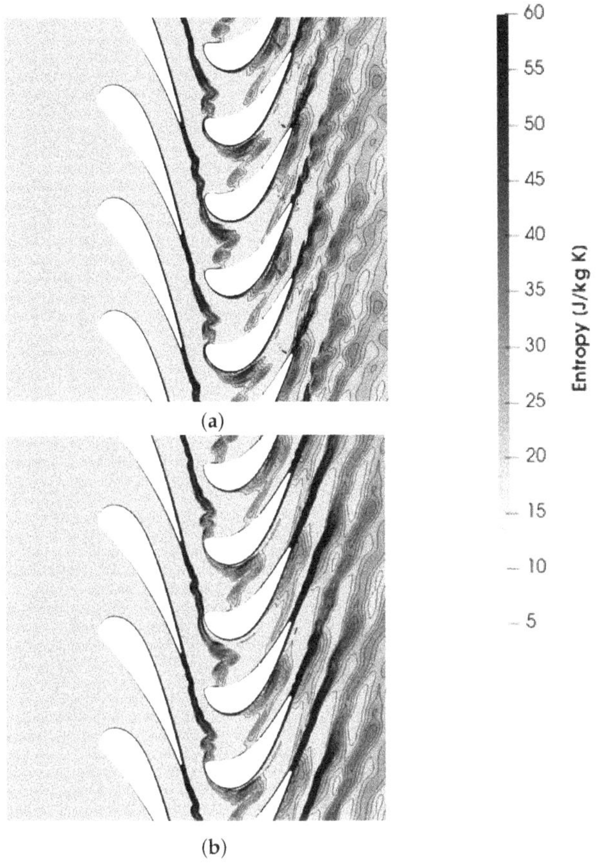

Figure 7. Comparison of entropy contours for the E^3 axial turbine stage test case. (**a**) Five harmonics. (**b**) DTS.

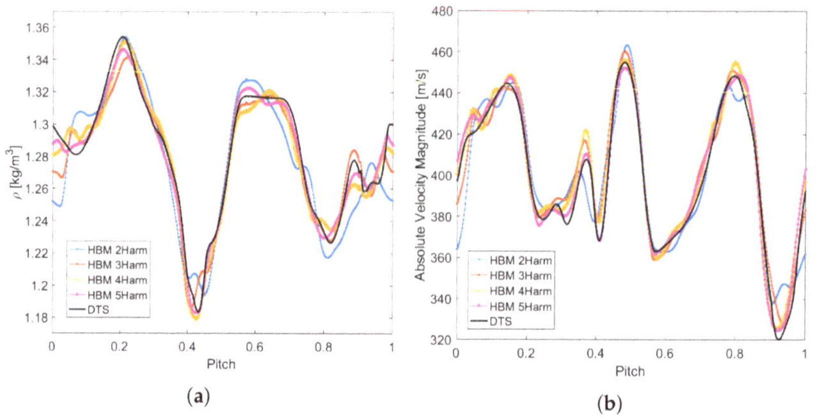

Figure 8. Instantaneous flow quantities in front of the rotor blades as a function of the blade pitch for the DTS and HBM simulations. (**a**) Density. (**b**) Absolute velocity magnitude.

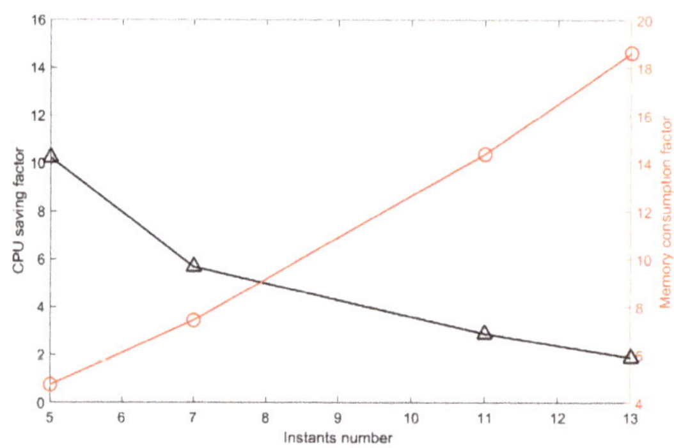

Figure 9. Comparison of the CPU factors required for the HBM and the DTS calculations. The plot is non-dimensionalised by the DTS simulation values.

5. Other Relevant Issues for Turbomachinery Applications

5.1. Three-Dimensional Simulations in Cartesian Coordinates

Being a general purpose C++ library, OpenFOAM stores and computes the solution of the flow equations in cartesian coordinates. For this reason, caution should be applied when using the HBM for turbomachinery applications, in which the relevant coordinate system is the cylindrical one. This traduces in some differences in the code when applying the HB operator to a vector that rotates about the machine axis with an angular velocity equal to the revolution speed of the machine. Think for example of the time evolution of the velocity vector inside a cell of a rotating mesh during a transient turbomachinery simulation. In this case, the zeroth harmonic of the vector corresponds to a steady term in a cylindrical frame of reference rotating with the machine. On the other hand, if expressed in a cartesian reference in the absolute frame, this term represents a periodic oscillation of the velocity components with a frequency corresponding to the machine rotation frequency. In other words, the term is no longer steady if seen in the absolute frame of reference. If an HB simulation is set up with the snapshots corresponding to one or more base frequencies and their harmonics, the presence of this spurious unsteadiness will produce a non-zero component for the time derivative when it is projected on the selected frequencies through the HB operator. This is true in general even if there is only one base frequency that is a higher harmonic of the rotation frequency (e.g., the blade passing frequency and the rotation frequency), as can be easily verified numerically.

A workaround this issue is transforming the vector into a cylindrical coordinate system, then applying the HB operator and transforming the result back to the original cartesian coordinate system. In this way, the real unsteadiness is captured by the HB operator, while the spurious one related to the mesh rotation is filtered out. Finally, it is necessary to include the effect of the mesh rotation in the time derivative operator. From simple kinematic considerations, the time derivative in the absolute frame of reference of a generic vector $\mathbf{v}$ rotating with angular velocity $\mathbf{\Omega}$ can be expressed as [37]

$$\frac{\partial \mathbf{v}}{\partial t} = \mathbf{\Omega} \times \mathbf{v} \tag{28}$$

Please notice that this term is not required in unsteady simulations where the mesh *actually* undergoes a rotational displacement. On the contrary, since in our version of the HBM each dynamic mesh instance (one per snapshots) is moved in the position of the corresponding time sample and then kept fixed during the simulation, this term must be included in the momentum equation. This should not be surprising since exactly the same term ($\mathbf{\Omega} \times \rho \mathbf{u_k}$) is used for MRF steady-state simulations to include the effect of the frame rotation. This term will therefore be labelled as the "steady" term. For a generic cell of the k-th snapshot's mesh, the time derivative of the momentum can finally be expressed as

$$\frac{\partial \rho \mathbf{u_k}}{\partial t} = \underbrace{\mathbf{\Omega} \times \rho \mathbf{u_k}}_{\text{"Steady"}} + \underbrace{(D_{kj}\rho_j u_j^r)\mathbf{e_k^r} + (D_{kj}\rho_j u_j^\theta)\mathbf{e_k^{\theta}} + (D_{kj}\rho_j u_j^z)\mathbf{e_k^z}}_{\text{Unsteady}} \tag{29}$$

where $\mathbf{u_k}$ is the k-th snapshot velocity vector, u^r, u^θ and u^z are the radial, tangential and axial components of the velocity vector, respectively, while $\mathbf{e_k^r}$, $\mathbf{e_k^{\theta}}$ and $\mathbf{e_k^z}$ represent the radial, tangential and axial versors, respectively. The terms inside the round brackets denote the scalar product between the k-th row of the $\mathbf{D}$ matrix and the vector containing the momentum of all the snapshots in the selected cell, following the Einstein notation on repeated indices. All this being considered, the calculation of the HB source term in the code is modified as follows:

1. Transform the momentum vector in cylindrical coordinates.
2. Apply the HB operator $\mathbf{D}$ to the momentum vector to find the time derivative approximation in cylindrical coordinates (second term on the rhs of Equation (29)).
3. Transform the just derived source term back to cartesian coordinates.
4. Add the "steady" source term as in Equation (29). This can be easily included exploiting the MRF support of the solver.
5. Compute the source term for the continuity and energy equation as usual, since for scalar quantities no changes are needed.

5.2. Single-Passage Reduction

Spatial periodicity in the azimuthal direction is often employed in turbomachinery simulations in order to reduce the computational domain to only a portion of the whole annulus. Nevertheless, in transient calculations, it is required to have the same pitch for the different rows, so that a large fraction of the annulus must often be simulated. This typically results in a highly increased computational effort. One technique to avoid such numerical issue is the usage of phase-lagged boundary conditions [38] and the exploitation of the chorochronic (spatio-temporal) periodicity of turbomachinery flows [39]. In this way, the computational domain can be reduced to only a single passage per blade row, independently of the blade count ratio. The method hinges upon the fact that the flow field inside a blade passage at a certain time t must be equal to the flow field in an adjacent passage at another time $t + \mathcal{T}$

$$Q(r, \theta + \Delta\theta, z, t) = Q(r, \theta, z, t + \mathcal{T}) \tag{30}$$

where $\mathcal{T}$ is the time lag between the two passages, θ is the azimuthal coordinate and $\Delta\theta$ is the row pitch. Let us consider for simplicity a single-stage machine. In this case the time lag is calculated basing on the phase of a wave that travels at a rotational speed $\omega_k = 2\pi k f_{BP}$ in the azimuthal direction: $\mathcal{T} = k\sigma/\omega_k$, where σ is the so called interblade phase angle (IBPA). For a single stage, the IBPA in a specific row is computed basing on the relative blade count and the rotation sense [39]:

$$\sigma = -2\pi \mathrm{sgn}(\Omega)\left(1 - \frac{N_{B_2}}{N_{B_1}}\right) \tag{31}$$

where N_{B_2} and N_{B_1} are the number of blades in the opposite and the current blade rows, respectively. To apply phase-lagged conditions on periodic boundaries, a common approach is to update the Fourier coefficients on the boundaries at each time-step. For this reason, phase-lagged boundary conditions are very well-suited and easier to implement in frequency-domain solvers (e.g., HBM) since this kind of solver is intrinsically related to the Fourier decomposition of the solution. Indeed, an equivalent formulation of Equation (30) in the frequency domain is

$$\sum_{k=-K}^{K} \hat{\mathbf{Q}}_{\boldsymbol{k}}(r, \theta + \Delta\theta, z) e^{i\omega_k t} = \sum_{k=-K}^{K} \hat{\mathbf{Q}}_{\boldsymbol{k}}(r, \theta, z) e^{i\omega_k t} e^{i\omega_k \mathcal{T}} \tag{32}$$

Which entails that the flow spectrum in one passage is equal to the one of the next passage modulated by the IBPA:

$$\hat{\mathbf{Q}}_{\boldsymbol{k}}(r, \theta + \Delta\theta, z) = \hat{\mathbf{Q}}_{\boldsymbol{k}}(r, \theta, z) e^{i\omega_k \mathcal{T}} = \hat{\mathbf{Q}}_{\boldsymbol{k}}(r, \theta, z) e^{ik\sigma} \tag{33}$$

All this being considered, and following very similar arguments as in Section 3, one can express the relation between the snapshots of the flow field on the two periodic boundaries through a linear operator as [11]

$$\mathcal{Q}(r, \theta + \Delta\theta, z) = \mathbf{E}^{-1}\mathbf{S}\mathbf{E}\mathcal{Q}(r, \theta, z) \tag{34}$$

where $\mathbf{S}$ is a diagonal matrix whose components are $S_{kk} = e^{ik\sigma}$. The implementation of phase-lagged boundary conditions is relatively straightforward once the structure for the HBM has been built. It is sufficient to implement a new type of boundary condition which is very similar to a cyclic periodicity, here called *phase-lag cyclic*. The only difference is that now, when the variables are reconstructed on the faces of the periodic boundaries, Equation (34) must be used to maintain the correct phase-shift between the two sides of the coupled patches.

Now that the pitches of the various rows need not be equal to each other, the logic underpinning the exchange of information at the interface between two rows must be modified. Specifically, one needs to replicate each of the two sides of the interface an integer number of times to provide information where they do not overlap. Due to the phase-lag between the different passages, the replication needs to be carried out accordingly. Therefore, we construct a new type of interface called phase-lag Arbitrary Mesh Interface (AMI). As an example, the set up of the boundary condition is sketched in Figure 10 for a radial slice of an axial compressor stage. As can be seen from the picture, one aim of the combination of phase-lag cyclics and AMI interfaces is to obtain information about the wake from the cyclic boundaries, even if the wake is not received from the AMI interface.

The implementation follows the steps described in Algorithm 3 below and sketched in Figure 11.

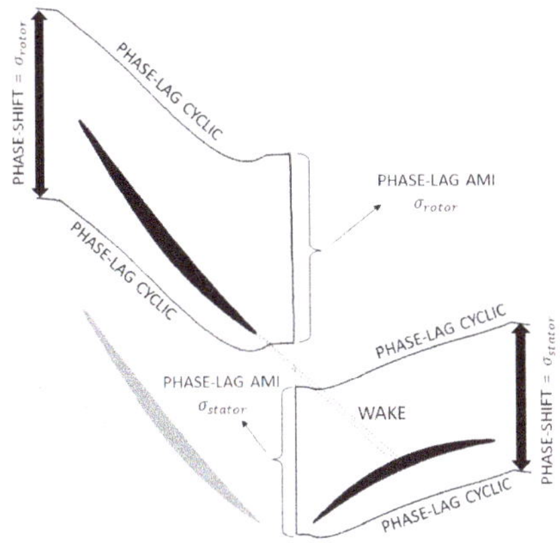

Figure 10. Illustration of the phase-lagged boundary condition behaviour.

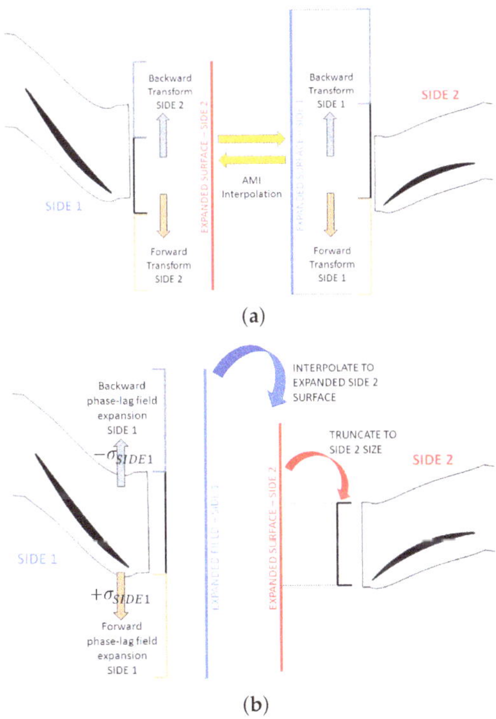

Figure 11. Illustration of Algorithm 3 for the implementation of phase-lag AMI interfaces. (**a**) Steps 1–16. (**b**) Steps 17–25.

Steps 1–15 are needed to calculate the weights for the interpolation on the two sides of the interface. To calculate the weights is necessary that both sides are fully paved. This is obtained by replicating the other side according to a periodic transformation (e.g., rotation by the pitch angle). These steps are performed only once as a preprocessing part of the simulation since the mesh does not change its relative position thereafter. Since the geometry replication can be performed either in the forward or the backward direction, step 8 is a way to understand if we are replicating the geometry in such a way to increase the paved area of the other side of the interface. Once we obtain as many forward and backward transformations as needed for each side, the fields can be interpolated through the interface in either direction. Therefore, during steps 16–24, the required information is exchanged from one side to the other through fields interpolation. These steps are performed once per solver iteration.

Here we presented the algorithm for a rotational geometry, but please notice that the aforementioned algorithm can be equally applied to a translational periodicity such as the E^3 stage presented in Section 4.4. Naturally, in this case, it is no longer necessary to transform vectors in a cylindrical frame before applying the HB and the phase-lag operators.

Algorithm 3 Phase-lag AMI algorithm.

1: Save original geometries for Side 1 and Side 2 of the interface
2: Compute weights for AMI interpolation between the two original sides
3: Initialize $NTransfForward_{Side1} = 0$ and $NTransfBackward_{Side1} = 0$
4: **while** Side 2 of the interface is not fully paved **do**
5:     Replicate the *original* geometry of Side 1 of the interface according to the periodic transformation (e.g., rotate by pitch periodicity angle, Figure 11a)
6:     Append the transformed geometry to Side 1 of the interface
7:     Recalculate weights for AMI interpolation between the two sides
8:     **if** the sum of the new weights is lower than the previous one **then** reverse the direction of the transformation in step 4
9:     **end if**
10:     **if** We are transforming in the forward direction **then** $NTransfForward_{Side1} + +$
11:     **else** $NTransfBackward_{Side1} + +$
12:     **end if**
13: **end while**
14: Repeat steps 3–13 for the other side of the interface
15: Compute the AMI weights for the two expanded sides of the interface (Figure 11a)

16: ...
17: **for** $i \leftarrow 0, \ldots, NTransfForward_{Side1} - 1$ **do**
18:     Expand the field on Side 1 of the interface according to the phase-lag boundary conditions (34) with a positive IBPA (Figure 11b)
19: **end for**
20: **for** $i \leftarrow 0, \ldots, NTransfbackward_{Side1} - 1$ **do**
21:     Expand the field on Side 1 of the interface according to the phase-lag boundary conditions (34) with a negative IBPA (Figure 11b)
22: **end for**
23: Interpolate the expanded field on Side 1 of the interface onto the expanded surface of Side 2 thanks to the AMI weights previously calculated (Figure 11b).
24: Truncate the interpolated field to the size of the Side 2 of the interface to obtain the required information (Figure 11b).
25: Repeat steps 17–23 for the other side of the interface.

5.3. Numerical Application: NASA Stage 37

The last application presented aims to include all the building blocks of the HBM that were illustrated in the previous sections. The 3D geometry of the NASA stage 37 [28] is selected for such purpose. The geometry consists of 36 blades rotating at 18,000 rpm and 46 vanes. Only one blade passage per row is modelled according to the single passage reduction method illustrated in Section 5.2. Up to four harmonics of the passing frequencies of each row have been considered in the spectrum. Once again, the Roe upwind scheme was employed for the convective fluxes, in combination to a second-order MUSCL reconstruction with van Leer limiter. A pseudo Courant number of 25 was used for the local time stepping integration. Since the mesh does not change topology from one snapshot to another, the code parallelization has been carried out in a simple and efficient way. The domain decomposition is performed only on the original mesh, and maintained for all the snapshots. In this way, corresponding cells from different snapshots are kept on the same processor. Since the calculation of the HBM source term does not require the values of the conservative variables in adjacent cells, no communication between processors is necessary during the computation of this term.

For the calculation with four harmonics of the blade passing frequency, the HBM required approximately 600 core-hours. Please notice that a comparable time-resolved simulation would require modelling half the annulus (18 blades and 23 vanes), which is approximately twenty times the size of the current simulation. Assuming to use 80 physical time steps per blade passing and 20 subiterations for the inner loops, at least 2/3 complete revolutions of the rotor are usually necessary to reach a periodic behaviour. This results in an estimated simulation time about two orders of magnitude larger than the HBM, depending on the number of frequencies included in the spectrum. No oversampling was necessary in this case, since the maximum condition number was obtained for the simulation with four harmonics and was equal to 1.77.

Figure 12 illustrates the results for the simulations with two and four harmonics of the passing frequencies in terms of instantaneous pressure and entropy contours in a midspan section of the annulus. The flow field has been reconstructed at the desired time and four adjacent passages are reported taking into account the phase-lag between them. It can be noticed that the wakes are resolved with a better resolution across the interface and in the downstream blade row for the four harmonics simulation, as well as the pressure disturbances due to the bow waves interaction with the upstream row. Overall, an improved resolution of the periodic disturbance is obtained across the compressor span, as depicted in Figure 13, where the entropy contours on an axial plane cutting the vane at 20% of the chord are shown. As observed in the previous sections, the use of a higher number of harmonics leads to a better localization in time and space of the disturbances, thanks to a lower degree of under/overshoots in the flow quantities. As a final comparison, the time evolution of the total forces over one passing period has been reported for the rotor and the stator in Figure 14. It can be observed that, by increasing the number of harmonics, the solution tends to converge to a fixed distribution over the time period. Please notice that conventional steady-state methods (e.g., mixing plane) would predict a constant value over time which is not necessary close to the time-averaged value resulting from a time-resolved simulation.

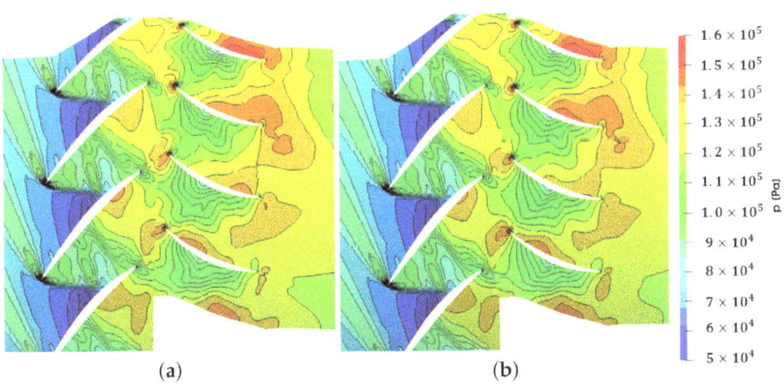

Figure 12. *Cont.*

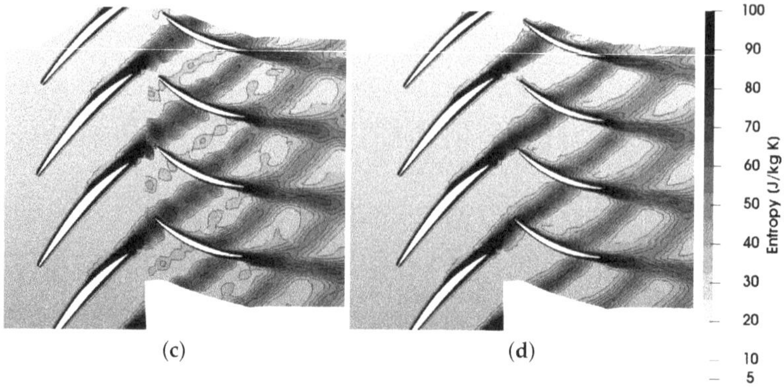

Figure 12. Contours on a radial slice at midspan for the NASA stage 37 test case. (**a**) Pressure contours for 2 harmonics. (**b**) Pressure contours for 4 harmonics. (**c**) Entropy contours for 2 harmonics. (**d**) Entropy contours for 4 harmonics.

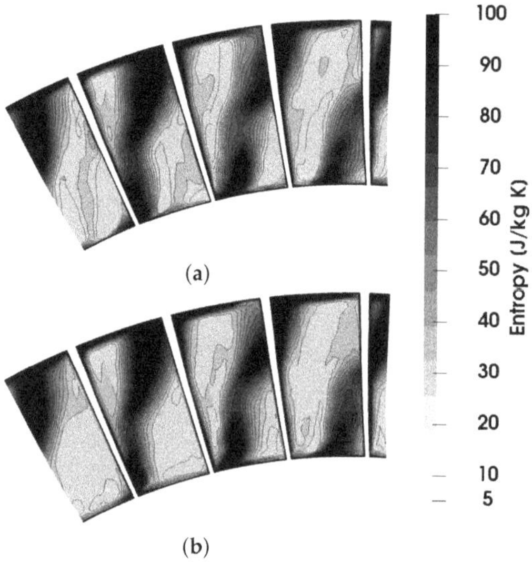

Figure 13. Entropy contours on an axial plane at 20% of the vane chord. (**a**) Two harmonics. (**b**) Four harmonics.

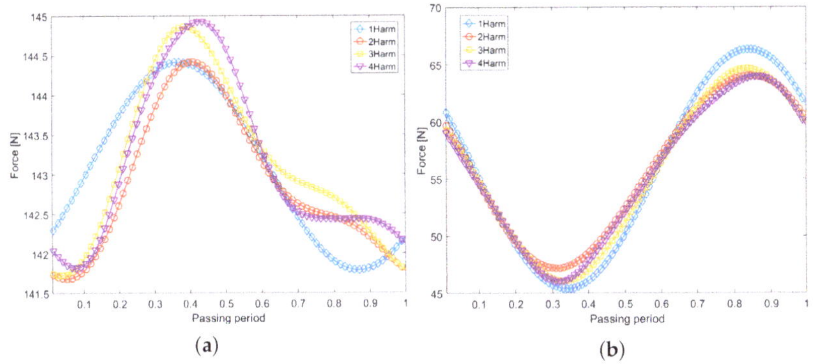

Figure 14. Evolution of forces on stator and rotor blades over one passing period. (**a**) Rotor. (**b**) Stator.

6. Conclusions

This work has illustrated how a fully implicit HBM can be implemented step-by-step in a CFD code. The choice has fallen upon the open-source software OpenFOAM, due to its wide diffusion and high-standard C++ compliance, trying to propose an implementation as far-reaching as possible. First of all, a new density-based solver is presented and validated in the OF framework, forming the basis for the HB support. The paper then proceeds on a combined theoretical and practical approach, starting from the fully-implicit baseline HBM and progressively adding other important features such as multi-frequencial support and phase-lagged boundary conditions. A research algorithm combined with a synchronized approach are used to improve flow continuity at the interfaces. This also allows to minimize the propagation of errors in the solution that possibly arises from a bad conditioning of the Fourier matrix. Four test cases have been proposed to verify the correct implementation of the different features. For the baseline HBM, disturbances propagation in a multi-region domain has been studied, showing the differences between the steady-state and the HB methods. It is shown that retaining three harmonics of the blade passing frequency in the spectrum is sufficient to reproduce the transient solution with a negligible error. The multi-frequencial approach has been tested at first on a simplified channel flow and then has been applied to a real axial turbine stage. Both test cases have shown a satisfactory agreement with the reference unsteady solutions. For the axial turbine case, it is reported that five harmonics of the passing frequency are adequate to capture transient effects. Finally, the addition of phase-lag boundary conditions on periodic boundaries and blade-row interfaces has been illustrated for an axial compressor stage. It is also shown that the correct formulation of the HB operator in cartesian coordinates allows to produce the correct solution for rotationally periodic geometries. Conclusions are also drawn with respect to the performance of the HB solver compared to time-resolved simulations. Specifically, the CPU saving factor strongly depends on the specific application and the number of frequencies considered in the spectrum. Broadly speaking, in case of identical domains for time-resolved and HBM solutions, a reduction in the CPU time of an order of magnitude can be expected, as well as an increase of one order of magnitude in the memory consumption. For turbomachinery cases where the domain is reduced to a single passage per row through phase-lagged boundary conditions, a reduction in CPU time of nearly two order of magnitude can be easily achieved.

Author Contributions: Software development and CFD simulations, S.O.; resources, review and supervision, N.C.; resources and review, M.C. All authors have read and agreed to the published version of the manuscript.

Funding: This research received no external funding.

Institutional Review Board Statement: Not applicable.

Informed Consent Statement: Not applicable.

Data Availability Statement: The source code developed and used in this work can be found in the authors' repository: https://github.com/stefanoOliani/ICSFoam (accessed on 3 March 2022). Data presented in the paper can be obtained upon reasonable request by contacting the authors.

Conflicts of Interest: The authors declare no conflict of interest.

Abbreviations and Nomenclature

CFD	Computational Fluid Dynamics
CPU	Central Processing Unit
HBM	Harmonic Balance Method
OF	OpenFOAM
Q	Vector of conservative variables
F	Vector of fluxes
V	Control volume
$\mathbf{D}$	Harmonic balance operator
$\mathbf{D_t}$	Time derivative operator
$\mathbf{E}$	Discrete Fourier transform matrix
$R(Q)$	Discretized fluxes residual term
$J(Q)$	Convective flux Jacobian
$\mathbf{\Omega}$	Domain angular velocity vector
ω	Angular frequency corresponding to a known flow periodicity
σ	Interblade phase angle
θ	Azimuthal coordinate
N_B	Number of blades in a row
N_T	Number of snapshots for the HBM simulation
$\hat{Q}_m$	m-th Fourier coefficient of Q
ρ	Flow density
u	Flow velocity
λ_{ij}	Spectral radius of Roe matrix between cell i and j
$\Delta\tau$	Pseudo time-step
Δt	Physical time-step
p	Flow pressure
c	Sound speed
H	Flow total enthalpy
E	Flow total energy
T	Flow temperature
μ	Dynamic viscosity
Pr	Prandtl number
C_p	Specific heat at constant pressure
τ	Viscous stress tensor
q	Heat flux vector
n	Surface normal vector
M	Matrix of the equations system

References

1. Tucker, P. Computation of unsteady turbomachinery flows: Part 1—Progress and challenges. *Prog. Aerosp. Sci.* **2011**, *47*, 522–545. [CrossRef]
2. Tucker, P. Computation of unsteady turbomachinery flows: Part 2—LES and hybrids. *Prog. Aerosp. Sci.* **2011**, *47*, 546–569. [CrossRef]
3. Berkooz, G.; Holmes, P.; Lumley, J. The Proper Orthogonal Decomposition in the Analysis of Turbulent Flows. *Annu. Rev. Fluid Mech.* **2003**, *25*, 539–575. [CrossRef]
4. He, L.; Ning, W. Efficient Approach for Analysis of Unsteady Viscous Flows in Turbomachines. *AIAA J.* **1998**, *36*, 2005–2012. [CrossRef]
5. Hall, K.C.; Thomas, J.P.; Clark, W.S. Computation of Unsteady Nonlinear Flows in Cascades Using a Harmonic Balance Technique. *AIAA J.* **2002**, *40*, 879–886. [CrossRef]
6. Ekici, K.; Hall, K.C. Harmonic Balance Analysis of Limit Cycle Oscillations in Turbomachinery. *AIAA J.* **2011**, *49*, 1478–1487. [CrossRef]

7. Ekici, K.; Hall, K.C. Nonlinear Analysis of Unsteady Flows in Multistage Turbomachines Using Harmonic Balance. *AIAA J.* **2007**, *45*, 1047–1057. [CrossRef]
8. Thomas, J.P.; Dowell, E.H.; Hall, K.C. Nonlinear Inviscid Aerodynamic Effects on Transonic Divergence, Flutter, and Limit-Cycle Oscillations. *AIAA J.* **2002**, *40*, 638–646. [CrossRef]
9. Jameson, A.; Alonso, J.; McMullen, M. Application of a non-linear frequency domain solver to the Euler and Navier-Stokes equations. In Proceedings of the 40th AIAA Aerospace Sciences Meeting & Exhibit, Reno, NV, USA, 14–17 January 2002.
10. Van der Weide, E.; Gopinath, A.; Jameson, A. Turbomachinery Applications with the Time Spectral Method. In Proceedings of the 35th AIAA Fluid Dynamics Conference and Exhibit, Toronto, ON, Canada, 6–9 June 2005; p. 4905.
11. Gopinath, A.; van der Weide, E.; Alonso, J.; Jameson, A.; Ekici, K.; Hall, K. Three-Dimensional Unsteady Multi-stage Turbomachinery Simulations Using the Harmonic Balance Technique. In Proceedings of the 45th AIAA Aerospace Sciences Meeting and Exhibit, Reno, NV, USA, 14–17 January 2007.
12. Sicot, F.; Dufour, G.; Gourdain, N. A time-domain harmonic balance method for rotor/stator interactions. *J. Turbomach.* **2012**, *134*. [CrossRef]
13. Su, X.; Yuan, X. Implicit solution of time spectral method for periodic unsteady flows. *Int. J. Numer. Methods Fluids* **2010**, *63*, 860–876. [CrossRef]
14. Woodgate, M.A.; Badcock, K.J. Implicit Harmonic Balance Solver for Transonic Flow with Forced Motions. *AIAA J.* **2009**, *47*, 893–901. [CrossRef]
15. Sicot, F.; Puigt, G.; Montagnac, M. Block-Jacobi Implicit Algorithms for the Time Spectral Method. *AIAA J.* **2008**, *46*, 3080–3089. [CrossRef]
16. Thomas, J.P.; Custer, C.H.; Dowell, E.H.; Hall, K.C.; Corre, C. Compact Implementation Strategy for a Harmonic Balance Method Within Implicit Flow Solvers. *AIAA J.* **2013**, *51*, 1374–1381. [CrossRef]
17. Nimmagadda, S.; Economon, T.D.; Alonso, J.J.; da Silva, C.R.I. Robust uniform time sampling approach for the harmonic balance method. In Proceedings of the 46th AIAA Fluid Dynamics Conference, Washington, DC, USA, 13–17 June 2016.
18. Cvijetić, G.; Gatin, I.; Vukčević, V.; Jasak, H. Harmonic Balance developments in OpenFOAM. *Comput. Fluids* **2018**, *172*, 632–643. [CrossRef]
19. Oliani, S.; Casari, N.; Pinelli, M.; Suman, A.; Carnevale, M. Numerical study of a centrifugal pump using Harmonic Balance Method in OpenFOAM. In Proceedings of the 34th International Conference on Efficiency, Cost, Optimization, Simulation and Environmental Impact of Energy Systems, Taormina, Italy, 28 June–2 July 2021.
20. Hall, K.C.; Ekici, K.; Thomas, J.P.; Dowell, E.H. Harmonic balance methods applied to computational fluid dynamics problems. *Int. J. Comput. Fluid Dyn.* **2013**, *27*, 52–67. [CrossRef]
21. He, L. Fourier methods for turbomachinery applications. *Prog. Aerosp. Sci.* **2010**, *46*, 329–341. [CrossRef]
22. Luo, H.; Baum, J.D.; Löhner, R. A Fast, Matrix-free Implicit Method for Compressible Flows on Unstructured Grids. *J. Comput. Phys.* **1998**, *146*, 664–690. [CrossRef]
23. Blazek, J. *Computational Fluid Dynamics: Principles and Applications*, 3rd ed.; Butterworth-Heinemann: Woburn, MA, USA, 2015.
24. Saad, Y.; Schultz, M.H. GMRES: A generalized minimal residual algorithm for solving nonsymmetric linear systems. *Siam J. Sci. Stat. Comput.* **1986**, *7*, 856–869. [CrossRef]
25. Yoon, S.; Jameson, A. Lower-upper Symmetric-Gauss-Seidel method for the Euler and Navier-Stokes equations. *AIAA J.* **1988**, *26*, 1025–1026. [CrossRef]
26. Heyns, J.; Oxtoby, O. Modelling high-speed viscous flow in OpenFOAM®. In Proceedings of the 9th South African Conference on Computational and Applied Mechanics, Somerset West, South Africa, 14–16 January 2014.
27. Dunham, J. *CFD Validation for Propulsion System Components*; Technical Report AR 355; AGARD: Neuilly-sur-Seine, France, 1998.
28. Reid, L.; Moore, R.D. *Design and Overall Performance of Four Highly Loaded, High Speed Inlet Stages for an Advanced High-Pressure-Ratio Core Compressor*; Technical Report 1337; NASA: Washington, DC, USA, 1978.
29. Suder, K.L. *Experimental Investigation of the Flow Field in a Transonic, Axial Flow Compressor with Respect to the Development of Blockage and Loss*; Technical Report 107310; NASA: Washington, DC, USA, 1996.
30. McMullen, M.S.; Jameson, A.; Alonso, J.J. Demonstration of Nonlinear Frequency Domain Methods. *AIAA J.* **2006**, *44*, 1428–1435. [CrossRef]
31. Frey, C.; Ashcroft, G.; Kersken, H. Simulations of Unsteady Blade Row Interactions Using Linear and Non-Linear Frequency Domain Methods. In Proceedings of the ASME Turbo Expo 2015: Turbomachinery Technical Conference and Exposition, Montreal, QC, Canada, 15–19 January 2015.
32. Lakshminarayana, B.; Davino, R. Mean Velocity and Decay Characteristics of the Guidevane and Stator Blade Wake of an Axial Flow Compressor. *J. Eng. Power* **1980**, *102*, 50–60. [CrossRef]
33. Gomar, A.; Bouvy, Q.; Sicot, F.; Dufour, G.; Cinnella, P.; François, B. Convergence of Fourier-based time methods for turbomachinery wake passing problems. *J. Comput. Phys.* **2014**, *278*, 229–256. [CrossRef]
34. Guédeney, T.; Gomar, A.; Gallard, F.; Sicot, F.; Dufour, G.; Puigt, G. Non-uniform time sampling for multiple-frequency harmonic balance computations. *J. Comput. Phys.* **2013**, *236*, 317–345. [CrossRef]
35. Tyler, J.M.; Sofrin, T.G. *Axial Flow Compressor Noise Studies*; SAE Technical Paper; SAE International: Warrendale, PA, USA, 1962.

36. Crespo, J.; Contreras, J. On the Development of a Synchronized Harmonic Balance Method for Multiple Frequencies and its Application to LPT Flows. In Proceedings of the ASME Turbo Expo 2020: Turbomachinery Technical Conference and Exposition, London, UK, 22–26 June 2020.
37. Greitzer, E.M.; Tan, C.S.; Graf, M.B. *Internal Flow: Concepts and Applications*; Cambridge Engine Technology Series; Cambridge University Press: New York, NY, USA, 2004.
38. Erdos, J.I.; Alzner, E.; McNally, W. Numerical Solution of Periodic Transonic Flow through a Fan Stage. *AIAA J.* **1977**, *15*, 1559–1568. [CrossRef]
39. Gerolymos, G.A.; Michon, G.J.; Neubauer, J. Analysis and Application of Chorochronic Periodicity in Turbomachinery Rotor/Stator Interaction Computations. *J. Propuls. Power* **2002**, *18*, 1139–1152. [CrossRef]

machines

MDPI

Article

Flow Loss Analysis and Optimal Design of a Diving Tubular Pump

Xiao Yang [1], Ding Tian [2], Qiaorui Si [2,*], Minquan Liao [2], Jiawei He [2], Xiaoke He [1] and Zhonghai Liu [3]

1 School of Civil Engineering and Communication, North China University of Water Resources and Electric Power, Zhengzhou 450045, China; sendeyouxiang@163.com (X.Y.); hexiaoke@ncwu.edu.cn (X.H.)
2 Research Center of Fluid Machinery Engineering and Technology, Jiangsu University, Zhenjiang 212013, China; 18848968152@163.com (D.T.); liaominquanujs@163.com (M.L.); hejiaweiujs@163.com (J.H.)
3 Hunan Neptune Pump Co., Ltd., Changsha 410100, China; lzh@hnnep.com
* Correspondence: siqiaorui@ujs.edu.cn

Abstract: As important parts of underground water conveyance equipment, diving tubular pumps are widely used in various fields related to the national economy. Research and development of submersible pumps with better performance have become green goals that need to be achieved urgently in low-carbon development. This paper provides an effective approach for the enhancement of the performance of a diving tubular pump by adopting computational fluid dynamics, one-dimensional theory, and response surface methodology. First, the flow loss characteristics of the pump under several flow rate conditions are analyzed by entropy production theory, and then the impeller and guide vanes are redesigned using the traditional one-dimensional theory. Then, the surface response experimental method is used to improve pump hydraulic efficiency. The streamline angle (A) of the front cover of the impeller blade, the placement angle (B) of the middle streamline inlet, and the placement angle (C) of the rear cover flowline inlet are the response variables to optimize the design parameters of the diving tubular pump. Results show that wall entropy production and turbulent kinetic energy entropy production play the leading role in the internal flow loss of the diving tubular pump, while viscous entropy production can be ignored. The flow loss inside the impeller is mainly concentrated at the inlet and the outlet of the impeller blade, and the flow loss inside the guide vane is mainly concentrated in the area near the guide vane and the entrance of the guide vane. A, B, and C are all significant factors that affect efficiency. The order of the influencing factors from strong to weak is as follows: A^2 (p = 0.000) > C (p = 0.007) = A × B (p = 0.007) > B (p = 0.023) > B^2 (p = 0.066) > A × C (p = 0.094) > A (p = 0.162) > C^2 (p = 0.386) > A × B (p = 0.421). The best combination of response variables after surface response test design is A = 9°, B = 31°, and C = 36°. After optimization, the pump efficiency and the head of the model pump are increased by 32.99% and 18.71%, respectively, under the design flow rate. The optimized model pump is subjected to tests, and the test data and the simulation data are in good agreement, which proves the feasibility of using the surface response method to optimize the design of the model pump.

Keywords: diving tubular pump; energy loss; entropy generation; surface response method; pump hydraulic performance

Citation: Yang, X.; Tian, D.; Si, Q.; Liao, M.; He, J.; He, X.; Liu, Z. Flow Loss Analysis and Optimal Design of a Diving Tubular Pump. *Machines* **2022**, *10*, 175. https://doi.org/10.3390/machines10030175

Academic Editor: Davide Astolfi

Received: 2 January 2022
Accepted: 23 February 2022
Published: 25 February 2022

1. Introduction

A tubular pump is a pump with a built-in motor. Because the inlet and outlet channels of the pump are in a straight line, the tubular pump is widely used in the fields of emergency drainage of mines, water lifting in large-scale water conservancy projects, water supply to high-rise buildings, and so on [1]. Due to the complex internal flow structure, the pump efficiency is usually low. With the rapid economic growth and the continuous increase in population, the problem of energy is becoming increasingly prominent. At present, the international community is trying to solve the problem of energy shortage; the excessive consumption of energy will inevitably lead to climate problems. Energy conservation and emission reduction are an important strategy to achieve the global carbon peak goal [2,3]. As general-purpose machinery, pumps are widely used in various fields related to the national economy. According to statistics, the electrical energy consumed by pumps accounts for about 20% of the world's total power consumption. Therefore, improving the efficiency of the pump is of great significance in achieving energy saving, consumption reduction, and alleviation of energy shortages [4]. The research objects in this study are diving tubular pumps, which are widely used in municipal engineering, flood drainage, and emergency rescue in mine flooding accidents because of their high flow pumping characteristic. Due to power shortage on some emergency rescue occasions, identifying a method for its performance optimization is of great value.

It is necessary to analyze the flow loss of a pump before optimizing the design of the pump. The traditional pressure drop method has limitations regarding the calculation of the energy loss of a centrifugal pump, and it is impossible to accurately visualize the location and distribution of energy loss inside the centrifugal pumps. According to Denton [5], entropy generation is an effective method to explain the energy loss of fluid machinery. Therefore, this method can be used to analyze the internal energy loss of centrifugal pumps effectively and provide a reference for the optimal design of centrifugal pump flow components. Kock and Herwig [6] proposed an enhanced model to calculate the direct dissipation entropy generation based on the entropy generation theory. In recent years, the theory of entropy generation has been applied to the visualization of the location and distribution of energy loss in high-power mechanical pumps [7], the study of internal energy loss in multistage pumps with or without rings [8], the study of the energy loss area of the side channel pump [9], and the energy loss analysis of the large-flow fast-start self-priming centrifugal pump [10]. In addition, the theory of entropy generation has been applied to the study of energy loss in hydraulic turbines [11] and cavitation flow in cryogenic liquid submersible pumps [12]. It is an intuitive and feasible method to analyze the internal flow loss of a centrifugal pump based on the theory of entropy generation.

At present, methods such as intelligent optimization algorithms and experimental design are widely used in the optimal design of pumps. Si et al. [13] optimized the design of electronic water pumps based on orthogonal experiments and multi-island genetic algorithms (MIGA) with efficiency as the optimization goal, which increased the efficiency of the electronic water pump design operating point by 4.2%. Zhang et al. [14] combined a genetic algorithm with a three-dimensional hydraulic design (Q3DHD) and boundary vorticity diagnosis (BVF) to optimize the design of multiphase pump impellers and used experiments to verify the feasibility of the optimization method. The total pressure ratio and the isentropic efficiency of the compressor were improved. Kim et al. [15] optimized the four parameters of the shape of the front and rear cover plates of a centrifugal compressor impeller with the optimization goals of total pressure ratio and isentropic efficiency, combined with numerical simulation and hybrid multi-objective optimization algorithm, and the goal of significantly improving the isentropic efficiency and total pressure in the full range of working conditions was achieved. Although the optimal solution can be found theoretically by using intelligent optimization algorithms, the optimization cycle is long and computational resources are consumed. The use of experimental design methods can avoid the drawbacks of intelligent optimization algorithms. Hyun-Su et al. [16] optimized the impeller of the centrifugal compressor by the fluid–structure coupling approach and the response surface method, thereby improving the aerodynamic performance and structural stability of the centrifugal compressor. Shi et al. [17] adopted a multi-disciplinary optimization design method based on an approximation model to improve the comprehensive performance of an axial-flow pump, and the results show that the mass of a single blade was reduced from 0.947 to 0.848 kg (a decrease of 10.47%) and the efficiency of the design condition increased from 93.91% to 94.49% (an increase rate of 0.61%). Liu et al. [18] optimized the geometric parameters of the flow components of the multiphase pump based on the orthogonal optimization design method of five factors and four levels. The optimized multiphase pump has a more uniform internal gas volume fraction and pressure distribution, and its performance has been significantly improved. Bonaiuti et al. [19] optimized the impeller design of a centrifugal compressor based on a combination of experimental design, surface response, and multi-objective optimization algorithms. Lee et al. [20] optimized the design of the axial fan impeller based on the three-dimensional inverse problem design method and used the response surface method. After the optimization, the fan pressure increased by 28.2%. Thakkar et al. [21] presented an effective approach for sanitary centrifugal pump performance enhancement by adopting computational fluid dynamics and a response surface methodology with a multi-objective optimization algorithm; the efficiency improved by 10.15% at the design point compared to the initial pump. Nataraj et al. [22] implemented a response surface methodology complemented with CFD simulations to improve the performance of a centrifugal pump by modifying the impeller design. The total head was increased, and the power consumption was minimized with the optimized impeller parameters. Wang et al. [23] optimized a centrifugal pump impeller based on a numerical simulation, Latin hypercube sampling (LHS), a surrogate model, and a genetic algorithm. Three different surrogate models were compared, namely the RSM, kriging, and the radial basis neural network. At the design point, the RSM model predicted the highest efficiency while the optimization technique increased the efficiency by 8.34%.

It can be observed from the literature survey that most research only focuses on rotors in fluid machinery. Based on the energy loss analysis of the submersible tubular pump, this study combined computational fluid dynamics, one-dimensional theory, and response surface methods to modify the parameters of the guide vane and the impeller to improve the performance of the submersible tubular pump. Firstly, the CFD method is used to numerically simulate the hydraulic characteristic of the initial pump and the entropy analysis based on the calculation results are processed to identify the key factors that lead to hydraulic efficiency reduction. Secondly, the impeller and the guide vanes of the prototype pump are initially optimized in terms of the excellent hydraulic model. Next, the surface response method is used to carry out a multi-objective optimization design on the impeller and the space guide vane of the diving tubular pump to achieve the best performance. After that, the optimized model is manufactured to test pump performance. Finally, the reliability of the optimized design method is verified by experimental results, which provide a reference for the optimized design of diving tubular pumps.

2. Numerical Method

2.1. Flow Control Equations

The internal flow of the diving tubular pump studied in this paper is a three-dimensional incompressible turbulent flow. The flow control equation in the Cartesian coordinate system without considering the heat transfer of the fluid is [24]:

$$\frac{\partial \overline{u}_i}{\partial x_i} = 0 \tag{1}$$

$$\frac{\partial(\rho\overline{u}_i)}{\partial t} + \frac{\partial(\rho\overline{u}_i\overline{u}_j)}{\partial x_j} = -\frac{\partial \overline{p}}{\partial x_i} + \frac{\partial}{\partial x_j}(\mu\frac{\partial \overline{u}_i}{\partial u_j} - \tau_{ij})(i,j = 1,2,3) \tag{2}$$

Since the number of unknowns in the governing equation is more than the number of equations, it is necessary to supplement the turbulence model to close the equations for a solution. The numerical simulation in this paper adopts the SST k-ω turbulence model with higher accuracy and reliability, and its equations are as follows [25–27]:

$$\frac{\partial}{\partial t}(\rho k) + \frac{\partial}{\partial x_i}(\rho w u_i) = \frac{\partial}{\partial x_j}(\Gamma_k\frac{\partial_k}{\partial x_j}) + G_k - Y_k + S_k \tag{3}$$

$$\frac{\partial}{\partial t}(\rho w) + \frac{\partial}{\partial x_i}(\rho w u_i) = \frac{\partial}{\partial x_j}(\Gamma_w\frac{\partial_w}{\partial x_j}) + G_w - Y_w + D_w + S_w \tag{4}$$

where G_k and G_w are the production terms for k and w, respectively; Γ_k and Γ_w are the diffusion terms of k and w, respectively; Y_k and Y_w are the destruction terms of k and w, respectively; and D_w represents the orthogonal divergent term.

2.2. Theory of Entropy Generation

The entropy generation analysis method is based on the second law of thermodynamics. It combines heat transfer and fluid mechanics to analyze energy loss and describes the irreversible conversion of high-quality energy, such as mechanical energy, to low-quality energy, such as internal energy. Diving tubular pumps will inevitably produce energy dissipation during operation; so this study uses the entropy generation analysis method to qualitatively and quantitatively analyze the energy loss of diving tubular pumps. Since entropy is a state quantity, the entropy generation and transport equation of a single-phase incompressible ideal fluid is [28]:

$$\rho(\frac{\partial s}{\partial t}+u\frac{\partial s}{\partial x}+v\frac{\partial s}{\partial y}+w\frac{\partial s}{\partial z})=div(\frac{q}{T})+\frac{\phi}{T}+\frac{\phi_\theta}{T^2} \quad (5)$$

where $div(\frac{q}{T})$ represents reversible heat transfer; t is for time; ρ is for density; T is the system temperature; s is the entropy generation rate; x, y, and z are coordinate components; and u, v, and w are velocity components in the Cartesian coordinate system. ϕ is the viscous dissipation term of mechanical energy; ϕ_θ is the dissipation term generated by heat transfer due to temperature difference, and the two items on the right are the source terms, so both are positive values. The first term is the entropy generation generated by viscous dissipation, and the second term is the entropy generation generated by the heat transfer process. Reynolds averaging the Navier–Stokes equations decomposes unsteady variables into time-averaged and fluctuating terms, so the calculation of entropy generation is also divided into average term and pulsating term. Herwig [29] gave the formula for calculating entropy generation per unit volume after Reynolds time-average processing:

$$S_{PRO,\overline{D}}=\frac{\mu}{T}\left\{2\left[(\frac{\partial u}{\partial x})^2+(\frac{\partial \overline{v}}{\partial y})^2+(\frac{\partial \overline{w}}{\partial z})^2\right]+(\frac{\partial \overline{u}}{\partial y}+\frac{\partial \overline{v}}{\partial x})^2+(\frac{\partial \overline{u}}{\partial z}+\frac{\partial \overline{w}}{\partial x})^2+(\frac{\partial \overline{v}}{\partial z}+\frac{\partial \overline{w}}{\partial y})^2\right\} \quad (6)$$

where "-" represents the time-average term and $S_{PRO,\overline{D}}$ represents the entropy generation due to viscous dissipation.

$$S_{PRO,D'}=\frac{\mu}{T}\left\{2\left[\overline{(\frac{\partial u'}{\partial x})^2}+\overline{(\frac{\partial v'}{\partial y})^2}+\overline{(\frac{\partial w'}{\partial z})^2}\right]+\overline{(\frac{\partial u'}{\partial y}+\frac{\partial v'}{\partial x})^2}+\overline{(\frac{\partial u'}{\partial z}+\frac{\partial w'}{\partial x})^2}+\overline{(\frac{\partial \overline{v}}{\partial z}+\frac{\partial w'}{\partial y})^2}\right\} \quad (7)$$

where "'" represents the pulsation term and $S_{PRO,D'}$ represents the entropy generation caused by turbulent energy dissipation.

Since it is difficult to directly solve the pulsating velocity in the numerical calculation process, Herwig [30] proposed a method to directly solve the entropy generation of turbulent kinetic energy based on the turbulent energy dissipation rate. The calculation formula is as follows:

$$S_{PRO,D'}=\frac{\rho\varepsilon}{T} \quad (8)$$

where ρ is the density, ε is the turbulent energy dissipation rate, and T is the system temperature.

Volume analysis of the above formula is performed in the calculation domain to obtain the viscous dissipation entropy generation and turbulent kinetic energy entropy generation of the calculation domain. The calculation formulas are [30]:

$$\Delta S_{PRO,\overline{D}}=\int_V S_{PRO,\overline{D}}dV \quad (9)$$

$$\Delta S_{PRO,D'}=\int_V S_{PRO,D'}dV \quad (10)$$

In addition, wall entropy generation cannot be ignored but because the medium is viscous, there is a large velocity gradient inside the boundary layer and the entropy generation near the wall is calculated by the viscous entropy generation formula with large errors. Therefore, Zhang [31] gave a method to calculate the entropy generation near the wall. The total wall entropy generation was calculated by dividing the area of the calculation domain wall by the entropy generation in the wall area. The calculation formula is as follows:

$$S_{PRO,W} = \int_A \frac{\tau_w v_p}{T} dA \tag{11}$$

where $S_{PRO,W}$ is the wall surface entropy generation, τ_w is the wall shear stress, and v_p is the average velocity of the fluid at the center of the first layer of the grid away from the wall.

3. Numerical Simulation and Optimization Potential Analysis of Internal Flow

3.1. Computational Domain Model

This paper takes a diving tubular pump as the optimization object, and its main performance design parameters are shown in Table 1.

Table 1. Main performance design parameters of diving tubular pumps.

Parameters	Symbols	Value
Rate flow	Q_d	240 m^3/h
Head	H_d	120 m
Rotational speed	n	3600 r/min
Impeller suction diameter	D_j	276 mm
Impeller outlet diameter	D_2	110 mm
Impeller outlet width	b_2	18 mm
Blade number	z	6

To conduct a preliminary analysis of the hydraulic performance of the diving tubular pump, three-dimensional modeling of the water body is built as the computational domain of the diving tubular pump. As shown in Figure 1, the computational domain is composed of five parts: the inlet pipe, the impeller, the spatial guide vane, the ducted, and the outlet pipe.

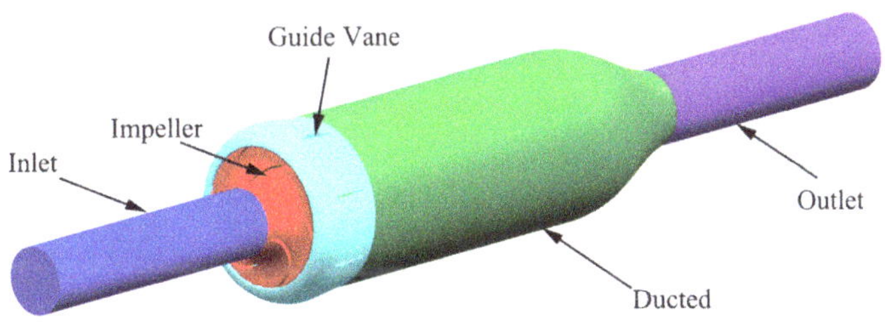

Figure 1. Three-dimensional model of a diving tubular pump computational domain.

3.2. Grid Division of a Computational Domain

To improve the calculation accuracy and speed up the convergence speed, ANSYS ICEM was used to divide the water body in the calculation domain of the diving tubular pump with a structured grid. As the structured meshes only contained quadrilaterals or hexahedrons in this case, their topological structure was equivalent to a uniformly orthogonal mesh within a rectangular domain. Accordingly, the nodes on each layer of the mesh lines can be effectively adjusted to ensure a high quality [32]. For inlet and outlet pipes, ICEM with an O-Block strategy was adopted to discretize the domains. In addition, to be able to capture more flow details inside the diving tubular pump, the meshes closed to the wall part were locally encrypted. The quality of meshes within all computational domains was above 0.4. The mesh division result and the partial mesh enlargement are shown in Figure 2. To treat the high-velocity gradient, all near-wall surfaces were refined with prism layers. The expansion ratio of near-wall grids is 1.2. Because wall loss is directly affected by the quality of the wall mesh, the boundary layer was applied to improve the quality of the wall mesh. Different models have different requirements for y^+ and do not have a fixed value [33,34]. The average values of y^+ of each flow passage component are listed in Figure 2b, and it is less than 50 [9]. Because the calculation results were generally satisfactory, the y^+ value can be considered as appropriate in this paper.

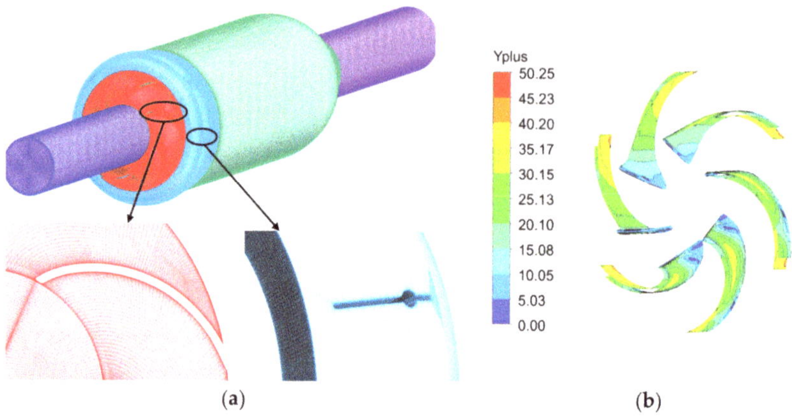

Figure 2. Results of grid division. (**a**) Mesh detail. (**b**) $y+$ distribution.

A grid-independent check (GIC) was conducted to make sure that the simulations in the optimization process are free from errors caused by the element count. It can be seen from Table 2 that when the element count increases from 7.68 to 8.56 million, the relative error of head and efficiency are 0.04% and 0.16%. Finally, the element count of 7.68 million was used for subsequent optimizations and simulations. Pump head and efficiency are defined as below.

$$H = \frac{p_2 - p_1}{\rho g} + \frac{v_2^2 \quad v_1^2}{\rho g} + (z_2 - z_1) \tag{12}$$

$$\eta = \frac{\rho g H Q}{P} \tag{13}$$

Table 2. Calculation and analysis of grid independence.

Element Count (Million)	H (m)	Relative Error (%)	Efficiency (%)	Relative Error (%)
5.79	113.97	-	61.35	-
6.36	114.64	0.58%	61.04	0.51%
7.68	115.52	0.76%	61.91	1.42%
8.56	115.57	0.04%	62.01	0.16%
9.72	115.56	0.01%	61.98	0.04%

3.3. Boundary Condition Setting

To analyze the reasons for the low efficiency of the diving tubular pump and the internal flow loss, before the optimization design, numerical calculation was completed based on the commercial CFD software ANSYS CFX. The shear stress transport (SST) k-ω model of the Reynolds time-averaged equation can usually provide satisfactory results in the analysis of the internal flow field and the separation flow in the centrifugal pump [8,35]. In the paper, the SST k–ω turbulence model is applied in the numerical simulation of the model pump. The medium is clean water under normal temperature and pressure. The inlet boundary condition is set to the total pressure inlet (1 atm), the outlet boundary condition is the mass flow outlet, and the wall boundary condition is a non-slip wall. The wall function is set to an automatic wall function, the dynamic and static interface is set to a frozen rotor model, and the convection term uses a high-resolution difference format. The timestep for the transient case was 3.33×10^{-4}, which corresponds to 3° of the impeller rotation. The root mean square (RMS) residual is set to 10^{-5}.

3.4. Performance Analysis of the Initial Model

To analyze the reasons for the low efficiency of diving tubular pumps, this paper determines the specific location of the internal flow loss and analyzes the cause of the flow loss according to the numerical calculation results under the design conditions. Figure 3 shows the cross-sectional pressure distribution cloud diagram, the velocity streamline diagram, and the guide vane velocity streamline diagram in the impeller under design conditions. It can be seen from Figure 3a that the pressure distribution in the inlet area of the impeller blade is uneven and the pressure at the inlet of the impeller blade fluctuates greatly. This shows that due to the unreasonable selection of the blade inlet angle, the incoming flow has a greater impact on the blade inlet edge. It can be seen from Figure 3b that there is a serious stall phenomenon on the back of the blade in the runner channel and vortices and recirculations develop on the back of the blade. It can be seen from Figure 3c that the direction of the liquid flow will change from radial to axial after the liquid flow enters the guide vane from the rotating impeller. However, since the guide vane blades of the original plan are straight, the angle of attack between the incoming flow and the inlet placement angle of the guide vane is too large. After the liquid flow enters the guide vane, there will be a large stall area and the flow state in the guide vane flow passage will show considerable disarrangement.

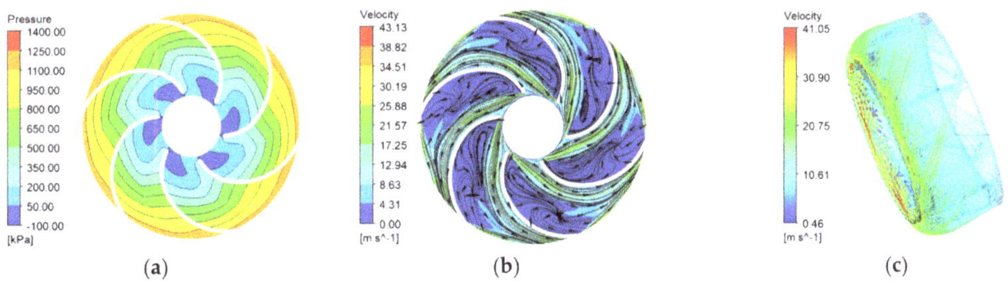

Figure 3. Analysis of the internal flow field of the impeller and the guide vanes. (**a**) Pressure cloud diagram of the midsection of the impeller. (**b**) Velocity streamline diagram of the midsection of the impeller. (**c**) Velocity streamline diagram of guide vanes.

3.5. Analysis of Entropy Generation

The entropy production analysis in this paper is used to analyze the flow loss of the initial pump so as to identify the parts of the original pump with large energy loss, which provides a reference for the subsequent redesign. Figure 4 shows the entropy generation distribution of the impeller and the guide vane of the diving tubular pump. It can be seen from Figure 4 that the viscous entropy generation value and the turbulent kinetic energy entropy generation value of the impeller and the guide vane first decrease and then increase with the increase in the flow, while the wall entropy generation value increases with the increase in the flow. At the same flow rate, in terms of viscous entropy generation value, turbulent kinetic energy entropy generation value, and wall entropy generation value, the values corresponding to the guide vanes are all greater than the values corresponding to the impeller, which shows that the turbulent energy dissipation in the flow field inside the guide vane is more serious than the turbulent energy dissipation in the impeller. Under different flow conditions, the wall entropy generation value of both impeller and guide vane is greater than the entropy generation value of turbulent kinetic energy and the magnitude of wall entropy generation and turbulent kinetic energy entropy generation is much larger than the viscous entropy generation value. Therefore, the effect of viscous dissipation and entropy generation on the flow loss of diving tubular pumps can be ignored. Wall entropy generation and turbulent kinetic energy entropy generation play a leading role in the flow loss of diving tubular pumps. The subsequent analysis also mainly focuses on wall entropy generation and turbulent kinetic energy entropy generation.

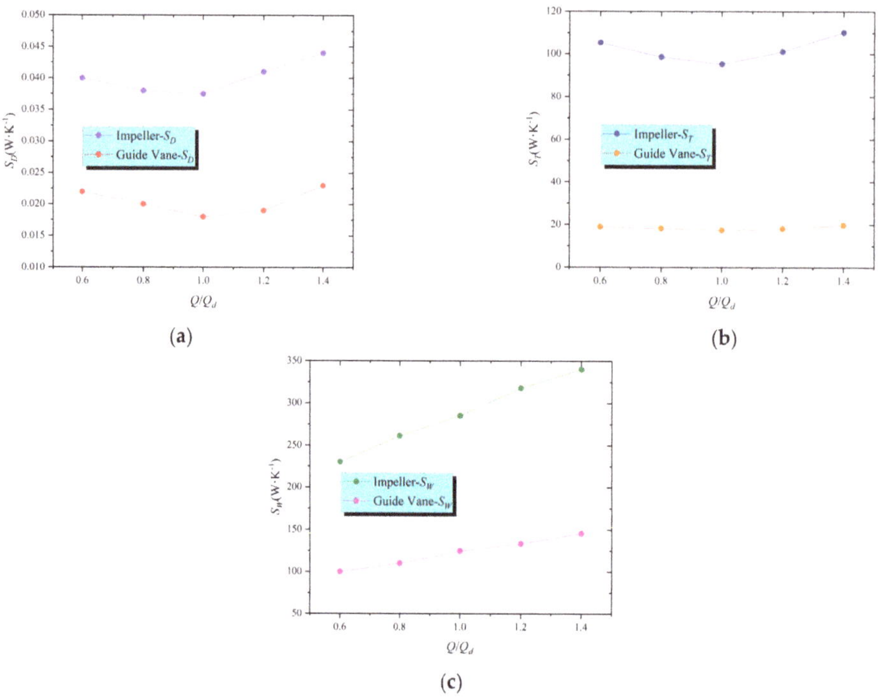

Figure 4. Entropy generation of the impeller and the guide vane. (**a**) Viscous entropy generation. (**b**) Turbulent kinetic energy entropy generation. (**c**) Wall entropy generation.

Figure 5 shows the entropy generation distribution of turbulent kinetic energy and wall entropy generation on the blade surface and the guide vane surface. It can be seen that there is turbulent kinetic energy entropy generation and wall entropy generation in the blade inlet area and the blade outlet area under the three flow conditions and the entropy generation is mainly concentrated on the blade working surface. The entropy production at the blade inlet indicates that the incoming flow has an impact on the blade inlet, which results in flow loss, and the entropy generation at the blade exit is caused by the dynamic and static interference between the impeller and the guide vane. Therefore, the flow loss inside the impeller is mainly concentrated in the inlet and outlet areas of the impeller. It can also be seen from Figure 6 that there is turbulent kinetic energy entropy generation and wall entropy generation in the guide vane inlet area under the three flow conditions. The strength of the guide vane entropy generation and the area size of the generated area increase with the increased inflow. The entropy production of turbulent kinetic energy is mainly concentrated in the inlet area of the guide vane, while the wall entropy production is mainly concentrated in the inlet area of the guide vane, and the intensity of the two entropies along the axial direction shows a decreasing trend. So the internal flow loss of the guide vane is mainly concentrated in the guide vane inlet area and the guide vane blade inlet area.

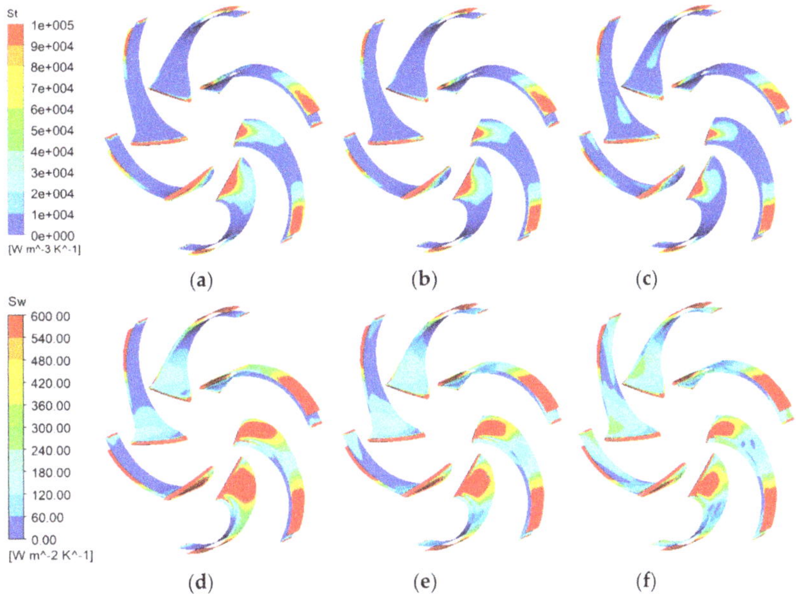

Figure 5. Entropy generation distribution of the impeller. Turbulent kinetic energy entropy: (**a**) $0.8Q_d$, (**b**) $1.0Q_d$, and (**c**) $1.2Q_d$. Wall entropy generation: (**d**) $0.8Q_d$, (**e**) $1.0Q_d$, and (**f**) $1.2Q_d$.

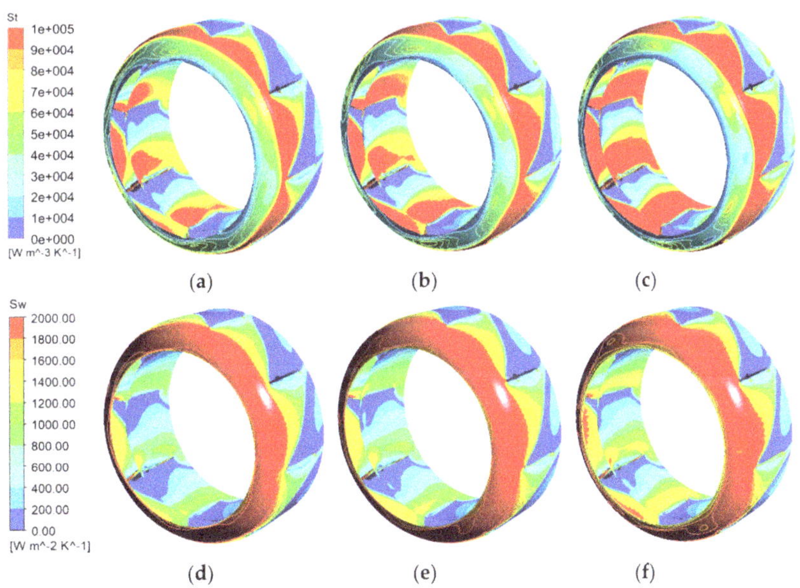

Figure 6. Entropy generation distribution of the guide vane. Turbulent kinetic energy entropy: (**a**) $0.8Q_d$, (**b**) $1.0Q_d$, and (**c**) $1.2Q_d$. Wall entropy generation: (**d**) $0.8Q_d$, (**e**) $1.0Q_d$, and (**f**) $1.2Q_d$.

From the numerical calculation results above, it can be seen that due to the deviation in the selection of the impeller blade inlet angle and the unreasonable design of the guide vane structure, the matching of the impeller and the space guide vane is problematic, which leads to large flow losses and low operating efficiency of the diving tubular pump. Therefore, this paper intends to first redesign the impeller and guide vane shape according to the excellent hydraulic model without changing the size of the impeller and guide vane axial surface. Then, based on the full-factor design of the experiment and the response surface method, the impeller blade inlet angle is used as the design variable and the efficiency and head are used as the response to perform the initial optimization of the impeller. Finally, the geometric parameters of the impeller and guide vanes that meet the expected goals are determined to achieve the improvement of the overall performance of the diving tubular pump.

4. Optimal Design

4.1. Optimal Design Based on the Empirical Method

4.1.1. Hydraulic Design of the Impeller and Guide Vanes

The impeller is designed by the empirical method [36]. Based on the traditional one-dimensional theory, the impeller is redesigned hydraulically by referring to the excellent hydraulic model of the same specific speed based on the prototype pump impeller. Same as the hydraulic design of the impeller, based on the prototype pump guide vane, the guide vane is redesigned according to the excellent hydraulic model with the same specific speed. First, the straight guide vane of the prototype pump is designed as a twisted blade and the outlet angle of the guide vane is 90°. Then, the area of the inlet of the guide vane is reduced to shift the maximum efficiency point of the diving tubular pump to the design working condition point. The research results of the literature [37,38] show that the extension of the outlet edge of the guide vane can inhibit the evolution and growth of the vortex core due to the stall, thereby improving the liquid flow in the guide vane channel. In addition, the extension of the outlet edge of the guide vane increases the static pressure of each section at the outlet of the guide vane, which can improve the rectification ability of the guide vane. The geometric parameters of the redesigned impeller and guide vanes are shown in Table 3.

Table 3. The geometric parameters of the redesigned pump.

Parameters	Symbols	Value
Hub diameter	d_h	0 mm
Impeller inlet diameter	D_j	110 mm
Impeller outlet diameter	D_2	276 mm
Impeller outlet width	b_2	18 mm
Blade wrap angle	Φ	140°
Blade outlet angle	φ_2	29°
Number of blades	Z	6
Guide vane inlet width	b_3	38 mm
Maximum diameter of an inner streamline of the guide vane	D_3	144 mm
Maximum diameter of an outer streamline of the guide vane	D_4	176 mm
Inside diameter of the guide vane outlet	D_5	132.9 mm
Outside diameter of the guide vane outlet	D_6	167 mm
Axial length of the guide vane	L	147 mm
Number of guide vanes	z	8
Guide vane inlet angle	α_3	12°
Guide vane outlet angle	α_4	89.71°
Guide vane wrap angle	φ_0	65.2°

4.1.2. The Results of Numerical Calculation

To compare and analyze the performance of the prototype pump, the optimized model adopts the same meshing method and the same boundary conditions as the prototype pump. Numerical calculations are completed also in ANSYS CFX. The calculation results of the pump performance are shown in Table 4. It can be seen that the optimal working condition point is at its design point, and the head of the design point is 134.59 m, which meets the design requirements. The subsequent optimization does not need to consider the head.

Table 4. Calculation results of pump performance.

Q/Q_d	Head (m)	Efficiency (%)
0.75	140.80	71.09
1	134.95	75.38
1.2	130.81	73.59

4.2. Optimal Design Based on the Full-Factorial Experiment

4.2.1. Optimization Object

Based on the hydraulic design of the impeller and the guide vanes in Section 4.1, the design is optimized only for the hydraulic performance of the impeller. With the streamline placement angle (A) of the front cover of the impeller blade, the placement angle (B) of the middle streamline inlet, and the placement angle (C) of the rear cover flowline inlet as the response variables and efficiency as the target variable, multi-objective optimization of the impeller of the mining pump is carried out to optimize the efficiency under the condition of meeting the head. Minitab software is used for a linearity test (full-factorial experiment) and a secondary test (response surface experiments) in this paper.

4.2.2. Full-Factor Design of the Experiment

A full-factorial experimental design is an experimental design in which all combinations of all factors and all levels must be tested at least once. Since all combinations are included, the full-factor experiment requires more time, but its advantage is that all main effects and interaction effects of various orders can be estimated. So the factorial experimental design is suitable for situations where the number of factors is small and the interaction effects of each order need to be investigated. In the full-factorial experiment design, the streamline placement angle (A) of the front cover of the impeller blade, the placement angle (B) of the middle flowline inlet, and the placement angle (C) of the back cover flowline inlet are selected as the three factors of the full-factorial experiment design factor. The level of each factor is shown in Table 5. In the table, −1, 0, and 1, respectively, represent the low level, the medium level, and the high level of the experimental factors.

Table 5. Table of experimental factors.

Levels \ Factors	A	B	C
	β_1	β_2	β_3
−1	9	17	28
0	15	24	32
1	21	31	36

Since the full-factorial experiment with three factors and two levels requires at least 8 experiments (under the condition of no repeatability test), three center points are added during optimization to reduce the error of fitting the regression equation, so a total of 11 full-factorial experiments are set. The experiment design scheme and the corresponding calculated CFD data are shown in Table 6. Analysis based on CFD calculation results and the variance of the full-factor experimental efficiency are shown in Table 7. It can be seen from the table that the p-values of A, B, and C are 0.015, 0.002, and 0.007, respectively, which means all of them are significant factors because they are less than 0.05. The order of impact factors from strong to weak is B (p = 0.002) > C (p = 0.007) > A (p = 0.015). The p-value of the two-factor interaction A * B and A * C is also less than 0.05, which also belongs to significantly affected items. The p-value of bending is 0.003, which is less than 0.005, so there is bending, indicating that the results of the linear regression equation are distorted, there is the influence of the quadratic factor, and further design experiments are needed.

Table 6. Full-factorial experimental design scheme and results of calculation.

Standard Order	Operation Order	Center Point	Zone	A	B	C	Efficiency (%)	Head (m)
9	1	0	1	15	24	32	81.91	136.92
10	2	0	1	15	24	32	81.91	136.92
7	3	1	1	9	31	36	82.38	136.81
6	4	1	1	21	17	36	81.65	133.50
2	5	1	1	21	17	28	81.32	134.04
1	6	1	1	9	17	28	79.88	130.93
5	7	1	1	9	17	36	81.66	137.19
3	8	1	1	9	31	28	78.86	129.05
11	9	0	1	15	24	32	81.91	136.92
8	10	1	1	21	31	36	77.65	126.66
4	11	1	1	21	31	28	77.06	126.71

Table 7. Analysis of variance for the efficiency of the full-factorial experimental design.

Source	Degree of Freedom	Adj SS	Adj MS	F	*p*
Model	6	35.066	5.84437	30.14	0.003
Linearity	3	17.2534	75.75113	29.93	0.003
A	1	3.2704	3.27040	17.02	0.015
B	1	9.1485	9.14850	47.61	0.002
C	1	4.8345	4.83450	25.16	0.007
Two-factor interaction	2	10.3243	5.16217	26.86	0.005
A × B	1	7.9142	7.91423	41.18	0.003
A × C	1	2.4101	2.41011	12.54	0.024
Bending	1	7.4885	7.48848	38.97	0.003
Error	4	0.7687	0.19217		
Lack of fit	2	0.7687	0.38434		
Pure error	2	0.0000	0.00000		
Total	10	35.8349			

4.3. *Optimal Design Based on Surface Response Experimental Method*

According to the results of the full-factor experimental method, the fitted bending judgment coefficient $p < 0.05$ ($p > 0.05$ proves that the model is available and no bending occurs), the equation has bending, and the main effect (single factor) coefficient $p < 0.05$, which proves that A, B, and C are all significant influencing factors. When the number of factors does not exceed three and they are all significant influencing factors, a more detailed surface response design analysis method should be adopted. The response surface experimental method adopts a central composite design, adding nine groups based on the full-factor experimental method. The experimental design and the corresponding experimental data are shown in Table 8.

The variance in the efficiency of the surface response experimental is shown in Table 9. It can be seen from the table that the *p*-values of blades A, B, and C are 0.0162, 0.023, and 0.007, respectively. The *p*-values of B and C are both less than 0.05, which are significant influencing factors, and the *p*-value of A is greater than 005, which is an insignificant influencing factor. In the square term, the *p*-value of A^2 is 0.000, which is a strong significant influence factor, and the *p*-values of B^2 and C^2 are both greater than 0.05, which are weakly significant factors. In the two-factor interaction, the *p*-value of A * B is 0.007 < 0.05, and A * B is a significant influence factor, while the *p*-value of A * C is 0.094 and the *p*-value of B * C is 0.421, both greater than 0.05, which are weakly significant impact factors. It can be seen that A, B, and C are all significant factors (single term, square term, and interaction term) that affect the regression equation. The order of impact factors from strong to weak is:

$$\begin{aligned} &A^2(P=0.000) > C(P=0.007) = A\times B(P=0.007) > B(P=0.023) > B^2(P=0.066) \\ &> A\times C(P=0.094) > A(P=0.162) > C^2(P=0.386) > A\times B(P=0.421) \end{aligned} \quad (14)$$

Table 8. Design plan and calculation results of the surface response experimental method.

Standard Order	Operation Order	Center Point	Zone	A	B	C	Efficiency (%)	Head (m)
18	1	0	1	15	24	32	81.91	136.92
19	2	0	1	15	24	32	81.91	136.92
8	3	1	1	21	31	36	77.64	126.66
12	4	−1	1	15	35.8	32	80.70	131.09
9	5	−1	1	4.9	24	32	78.41	127.51
14	6	−1	1	15	24	38.7	81.76	135.13
1	7	1	1	9	17	28	79.88	130.93
3	8	1	1	9	31	28	78.86	129.05
7	9	1	1	9	31	36	82.38	136.81
6	10	1	1	21	17	36	81.65	133.50
11	11	−1	1	15	12.2	32	80.55	131.55
15	12	0	1	15	24	32	81.91	136.92
10	13	−1	1	25	24	32	78.67	127.96
13	14	−1	1	15	24	25.2	80.94	133.47
2	15	1	1	21	17	28	81.32	134.04
16	16	0	1	15	24	32	81.91	136.92
17	17	0	1	15	24	32	81.91	136.92
5	18	1	1	9	17	36	81.66	137.19
20	19	0	1	15	24	32	81.91	136.92
4	20	1	1	21	31	28	77.06	126.71

Table 9. Analysis of variance of the design efficiency of the surface response experiment.

Source	Degree of Freedom	Adj SS	Adj MS	F	*p*
Model	9	43.9376	4.8820	6.94	0.003
Linearity	3	10.8785	3.6262	5.16	0.021
A	1	1.6022	1.6022	2.28	0.162
B	1	5.0415	5.0415	7.17	0.023
C	1	4.8345	4.83450	25.16	0.007
Square	3	22.2402	7.4134	10.54	0.002
A^2	1	20.5424	20.5424	29.2	0.000
B^2	1	2.9996	2.9996	4.26	0.066
C^2	1	0.5769	0.5769	0.82	0.386
Two-factor interaction	3	10.8189	3.6063	5.13	0.021
A × B	1	7.9142	7.9142	11.25	0.007
A × C	1	2.4101	2.4101	3.43	0.094
B × C	1	0.4945	0.4945	0.70	0.421
Error	10	7.0342	0.7034		
Lack of fit	5	7.0342	1.4068		
Pure error	5	0.0000	0.0000		
Total	19	50.9718			

From the above analysis, it can be seen that the model is effective and there is no lack of fit. The multiple linear regression equation of efficiency is as follows:

$$\begin{aligned}\eta = 42.1 + 2.238A + 0.431B + 1.069C - 0.03316A2 - 0.00931B2 \\ -0.0125C2 - 0.02368AB - 0.0229AC + 0.0089BC\end{aligned} \quad (15)$$

Figure 7 shows the two-factor interaction diagram of efficiency of A, B, and C. It can be seen from Figure 7a that the slopes of the curved surfaces of A and C are steep and parabolic and the efficiency value changes caused by these two factors are relatively large. When C remains unchanged, the efficiency first increases and then decreases with A. When A remains unchanged, the efficiency increases with an increase in C and the contour is presented as an ellipse with large curvature, indicating that the interaction between A and C is significant. It can be seen from Figure 7b that the influence of A and B on efficiency is more complicated. When B is at a low level, the efficiency increases with an increase in A. When A is at a low level, the efficiency increases with an increase in B. The contour lines are presented as ellipses with greater curvature. It shows that the interaction between A and C is more significant; It can be seen from Figure 7c that when C is at a low level, the efficiency decreases with an increase in B. When C is at a high level, the efficiency first increases and then decreases with an increase in B. It can also be seen from the contour lines that C and B have significant interactive effects.

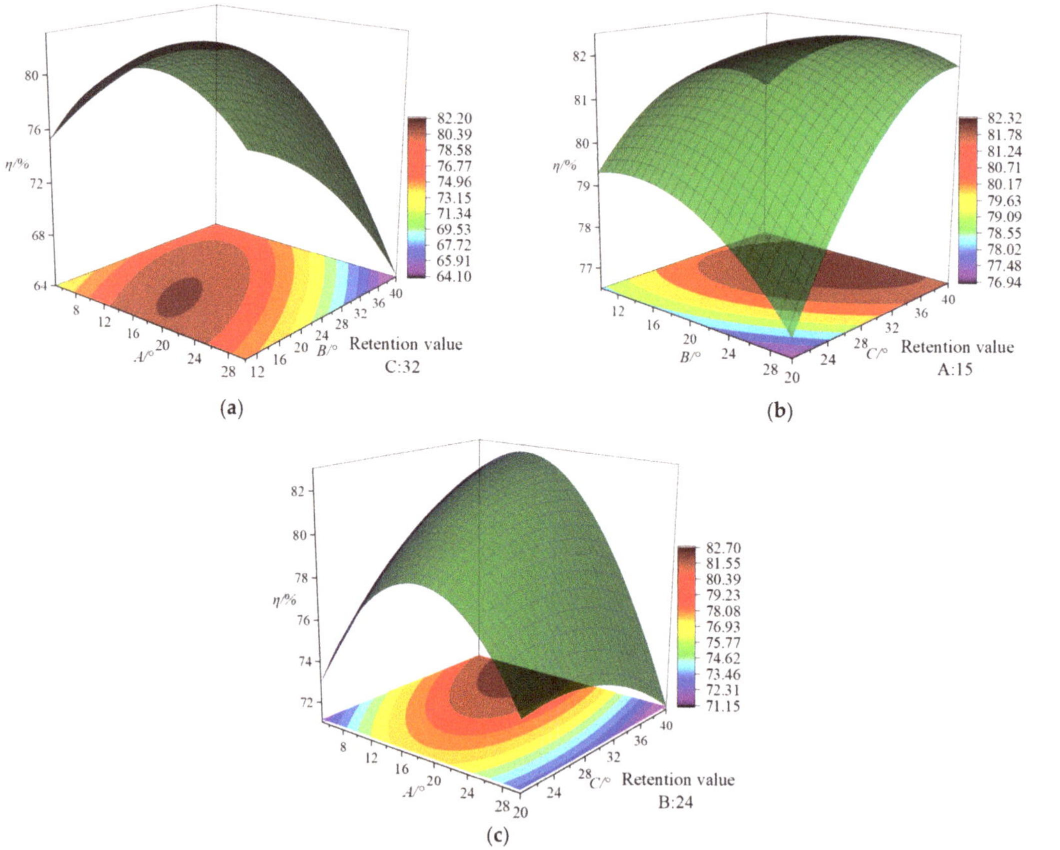

Figure 7. Contour plots and response surface plots of the interaction terms of efficiency and variables. (**a**) Contour plots and response surface plots of efficiency and A, B. (**b**) Contour plots and response surface plots of efficiency and B, C. (**c**) Contour plots and response surface plots of efficiency and A, C.

4.4. Optimization Results

4.4.1. Comparison of Optimization Effects on Pump Performance

The optimal solutions of the response optimization variables obtained by the response optimizer are A = 10°, B = 28.87°, and C = 38.72°, and the efficiency value is 82.7%. To compare and analyze the performance of the prototype pump again, the same meshing method and the same boundary conditions are set as those of the prototype pump. The efficiency of the design operating point obtained after the numerical calculation is 81.9%, which is lower than the 82.7% obtained by the response optimizer, which is caused by the error of the fitting equation. The calculation efficiency of the 9th group in the surface response experiment design scheme is 82.34%, and the error of the efficiency value obtained by the response optimizer is only 0.435%. Therefore, the variables A = 9°, B = 31°, and

C = 36° of the 9th group of experiments are selected as the best combination for solving the surface response. According to the optimal solution of the response variable, the head and efficiency values of the diving tubular pump are calculated as shown in Table 10.

Table 10. The results of surface response optimization.

	Initial		Optimized	
Q/Q_d	Head (m)	Efficiency (%)	Head (m)	Efficiency (%)
0.4	120.46	46.34	153.09	56.99
0.6	117.43	55.19	149.09	71.08
0.8	115.80	59.23	143.63	79.31
1.0	114.64	61.91	136.09	82.34
1.2	113.28	63.22	127.99	80.52
1.4	111.12	64.66	118.36	78.32
1.6	109.27	65.69	107.08	74.54

To verify the accuracy of the numerical calculation results, a performance test was carried out on the model pump. Figure 8 is the schematic diagram of the pump test and the test site diagram, and it includes the tested pump, motor, pipes, multifunctional data collector, and so on. The pump was a vertical installation, and an elbow pipe was used to connect the pump and the outlet pipe. The pump outlet pressure was collected by a pressure sensor whose accuracy was better than 0.1%. The flow was measured by the electromagnetic flowmeter with an uncertainty of 0.2%, and the motor speed and other parameters were collected by a multifunctional data collector. The systematic uncertainty of the test rig was 0.36%, which meets the level 1 accuracy requirements specified in ISO9906-2012.

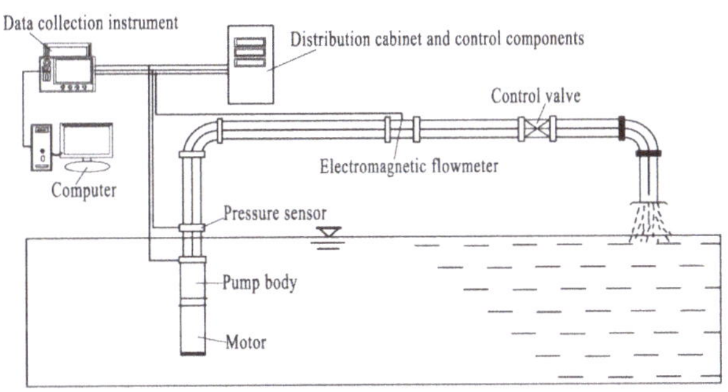

(**a**)

(**b**)

Figure 8. Experimental measurement. (**a**) The schematic diagram of the pump test. (**b**) The test site diagram.

Figure 9 presents the optimization effect of the numerical simulation on the pump performance, and Figure 10 shows the comparison between the test and the numerical simulation after the optimization. It can be seen from Figure 9 that the head and the efficiency of the pump after optimization have been greatly improved compared with those before optimization. The efficiency improvement is most obvious near the design operating point. The performance curve has no camelback phenomenon; the efficiency curve declines slowly under large flow rate conditions, and the range of the high-efficiency zone has been broadened. In addition, the optimal working condition point of the optimized diving tubular pump efficiency is located at the design working condition ($1.0Q_d$), with an efficiency value of 82.34% and head of 136.09 m. The efficiency of the optimized model is increased by 32.99%, and that of the head is increased by 18.71% compared with the initial scheme under design conditions, which fully meets the design requirements. The pump adopts a variable frequency motor, so the frequency converter is used to drive the pump to 3600 r/min during the engineering application. However, the frequency converter will interfere with the test sensor during the experiment. Therefore, the pump test does not use the frequency converter to drive the motor instead of the 50 Hz AC, which could drive the pump to 2900 r/min. To compare the numerical simulation and the test data, the pump flow rate and the pump head are made dimensionless as per Equation (16) and Equation (17), and the results are shown in Figure 10.

$$\varphi = \frac{Q_l}{2\pi R_2 b_2 u_2} \tag{16}$$

$$\psi = \frac{gH}{u_2^2} \tag{17}$$

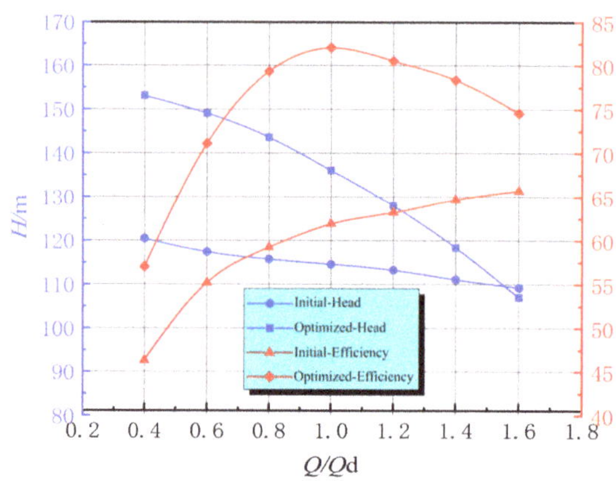

Figure 9. Optimization effect of the numerical simulation on the pump performance.

As the mechanical and volumetric efficiency was not considered in simulations, the results obtained by CFD are generally higher than the experimental values. It can be seen from Figure 10 that the trends of the head coefficient curve and the efficiency curve of the test and the numerical simulation are consistent with the flow rate increase. The maximum relative errors of the head coefficient and hydraulic efficiency are lower than 5%. The absolute predicted deviations for the head coefficient and hydraulic efficiency at the design condition are 2.47% and 2.92%, which proves that the optimization method and the numerical simulation method adopted in this research are correct and feasible.

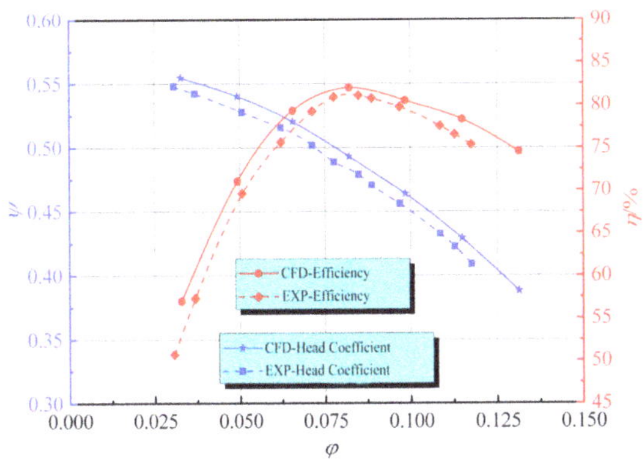

Figure 10. Comparison of the pump performance between the test and the numerical simulation.

4.4.2. Analysis of Internal Flow Characteristics

Figure 11 shows the velocity streamline distribution diagram of the middle section of the impeller outlet. Before optimization, there is a large-scale separation vortex on the back of the blade and backflow occurs. The vortex area and the backflow area occupy more than half of the flow channel, and the flow pattern is turbulent. The flow velocity in the vortex area and the backflow area is low, and the velocity distribution in the entire flow channel is uneven, which leads to a reduction in the blade's functional power, which seriously affects the hydraulic performance of the diving tubular pump. The large-scale separation vortices in the optimized impeller flow channel are significantly reduced. Only small-scale vortices are generated on the blade working surface, but there is no backflow phenomenon in the flow channel. Meanwhile, the flow velocity distribution in the flow channel after optimization is more uniform, and the vortex located on the blade working surface gradually disappears with the flow increase. Overall, the flow loss after optimization is significantly reduced compared to that before optimization, so the hydraulic performance of the diving tubular pump has been improved.

Figure 12 shows the static pressure distribution cloud diagram of the middle section of the impeller. Before optimization, there is a large low-pressure area at the blade inlet and the static pressure distribution is extremely uneven. The pressure distribution at the inlet presents a phenomenon of alternating high pressure and low pressure under different flow rates, and the low-pressure zone gradually expands with the flow rate increase. It shows that the improper setting of the blade inlet angle before optimization resulted in a serious impact between the incoming flow and the blade inlet, resulting in greater flow loss. In addition, due to the existence of a large low-pressure region, the possibility of cavitation at the blade inlet side was increased. The low-pressure region at the inlet of the optimized impeller is reduced, the static pressure at the inlet is increased compared to before optimization, and the static pressure distribution is more even. The pressure at the outlet of the impeller is reduced compared to that before optimization, so the energy loss caused by the liquid flow from the impeller into the guide vane is reduced.

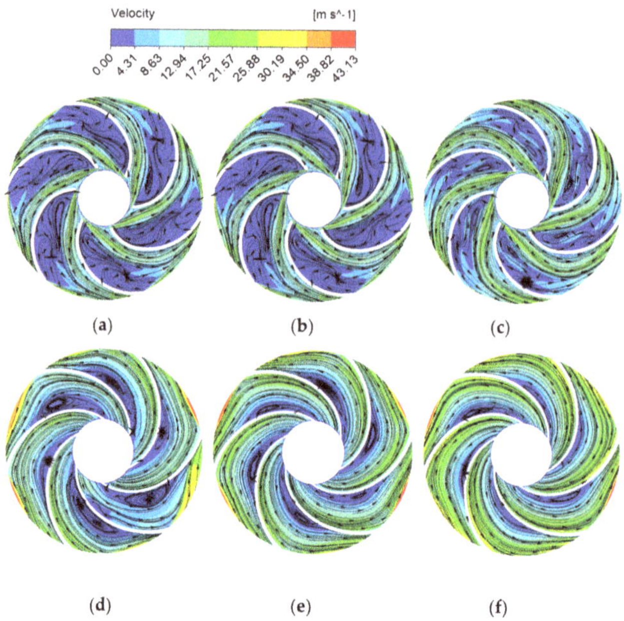

Figure 11. Comparison of the velocity and the streamline of the midsection of the impeller. (**a**) $0.8Q_d$ of the initial pump, (**b**) $1.0Q_d$ of the initial pump, (**c**) $1.2Q_d$ of the initial pump, (**d**) $0.8Q_d$ of the optimized pump, (**e**) $1.0Q_d$ of the optimized pump, and (**f**) $1.2Q_d$ of the optimized pump.

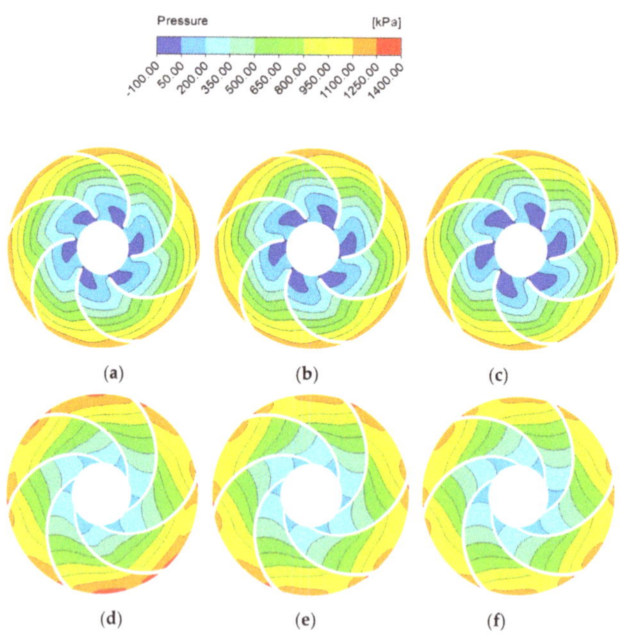

Figure 12. Comparison of static pressure distribution in the middle section of the impeller. (**a**) $0.8Q_d$ of the initial pump, (**b**) $1.0Q_d$ of the initial pump, (**c**) $1.2Q_d$ of the initial pump, (**d**) $0.8Q_d$ of the optimized pump, (**e**) $1.0Q_d$ of the optimized pump, and (**f**) $1.2Q_d$ of the optimized pump.

Figure 13 shows the internal velocity streamline distribution diagram of the guide vane. Before optimization, the guide vane blades of the original plan were straight, resulting in an excessively large attack angle between the incoming flow and the entrance angle of the guide vane and a large area of stall occurred after the liquid flow entered the guide vane. In addition, the flow pattern in the guide vane shows disarrangement and the velocity at the inlet of the guide vane is unevenly distributed, which is more obvious under small flow conditions. After optimization, the stall area of the guide vane is significantly reduced, the streamline distribution is more uniform, and the flow pattern is improved. In addition, the flow velocity in the guide vane inlet area is stable, indicating that the selection of a reasonable guide vane inlet angle and design of the guide vane blade profile improves the internal flow of the guide vane and its internal flow loss is reduced.

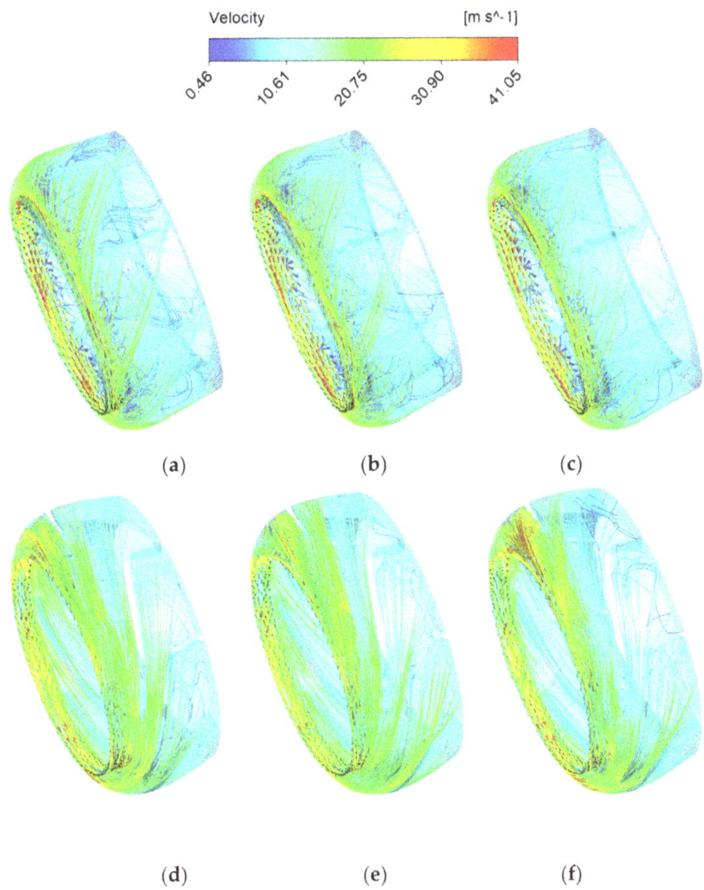

Figure 13. Comparison of guide vane velocity streamlines. (**a**) $0.8Q_d$ of the initial pump, (**b**) $1.0Q_d$ of the initial pump, (**c**) $1.2Q_d$ of the initial pump, (**d**) $0.8Q_d$ of the optimized pump, (**e**) $1.0Q_d$ of the optimized pump, and (**f**) $1.2Q_d$ of the optimized pump.

5. Conclusions

Based on the theory of entropy production, the energy loss analysis of a submersible tube pump was carried out and the reasons for its low efficiency were analyzed. The investigation on the diving tubular pump with low hydraulic efficiency performance improvement was carried out with the parameters of the impeller blade and the guide vane as design variables by coupling the numerical simulation, one-dimensional theory, and response surface methodology. Three design variables, the streamline placement angle (A) of the front cover of the impeller blade, the placement angle (B) of the middle streamline inlet, and the placement angle (C) of the streamline inlet of the rear cover, with two objectives, pump head and efficiency, were considered to construct the RSM model. The reliability of the numerical simulation was validated by the experimental test. The results show that:

(1) Entropy generation can effectively visualize the flow loss distribution caused by turbulent dissipation and flow separation. The internal flow loss of the diving tubular pump is mainly concentrated in the inlet and outlet area of the impeller and the inlet area of the guide vane. The main cause of flow loss is that the angle of attack between the relative liquid flow angle and the blade placement angle at the inlet of the impeller blade is too large; the matching between the guide vane and the impeller is poor, and the guide vane design is unreasonable.

(2) The streamline placement angle (A) of the front cover of the impeller blade, the placement angle (B) of the middle streamline inlet, and the placement angle (C) of the streamline inlet of the rear cover are all significant factors that affect the efficiency. The order of the influencing factors from strong to weak is as follows: A^2 (p = 0.000) > C (p = 0.007) = A * B (p = 0.007) > B (p = 0.023) > B^2 (p = 0.066) > A * C (p = 0.094) > A (p = 0.162) > C^2 (p = 0.386) > A * B (p = 0.421). The best combination of response variables after surface response test design is A = 9°, B = 31°, and C = 36°.

(3) The optimization process successfully improves the head and efficiency by 32.99% and 18.71%, respectively, compared to those of the initial pump. The optimized simulation data of the diving tubular pump are in good agreement with the test data. After optimization, the large-scale separation vortex inside the impeller is significantly reduced and no backflow occurs. The internal outflow area of the guide vane is significantly reduced after optimization, and the internal flow is greatly improved because the flow is more uniform and smoother.

(4) The optimization method in this study is universal and can be applied to conduct optimization of other fluid machinery. This study focuses on only some parameters of the impeller, and more parameters can be studied as variables in the future.

Author Contributions: Data curation, D.T.; Formal analysis, X.H.; Funding acquisition, Q.S. and Z.L.; Investigation, M.L.; Supervision, Q.S.; Validation, J.H. and Z.L.; Writing – original draft, X.Y. All authors have read and agreed to the published version of the manuscript.

Funding: This research was funded by the National Key Research and Development Program of China (No. 2020YFC1512403), the National Natural Science Foundation of China (No. 51976079), Research Project of State Key Laboratory of Mechanical System and Vibration MSV202201, and Industry University Research Cooperation Project of Jiangsu Province (No. BY2019059).

Institutional Review Board Statement: Not applicable.

Informed Consent Statement: Not applicable.

Data Availability Statement: Not applicable.

Conflicts of Interest: The authors declare no conflict of interest.

Nomenclature

$u\ v\ w$	Cartesian velocity components
$x\ y\ z$	Coordinate components
t	Time
p	Pressure
μ	Dynamic viscosity
v	Kinematic viscosity
ε	Turbulent eddy dissipation
T	Temperature
s	Entropy
p_1	Pressure of the pump inlet
p_2	Pressure of the pump outlet
v_1	Velocity of the pump inlet
v_2	Velocity of the pump outlet
z_1	Installation height of the pump inlet
z_2	Installation height of the pump outlet
P	Shaft power
ϕ	The viscous dissipation term of mechanical energy
ϕ_θ	The dissipation term generated by heat transfer due to temperature difference
$S_{PRO,\overline{D}}$	Viscous entropy generation
$S_{PRO,D'}$	Turbulent kinetic energy entropy generation
$S_{PRO,W}$	Wall entropy generation
τ_w	Wall shear stress
v_p	The average velocity
Q	Rate flow
H	Head
n	Rotational speed
η	Efficiency
D_{h}	Hub diameter
D_{j}	Impeller inlet diameter
D_2	Impeller outlet diameter
b_2	Impeller outlet width
ϕ	Blade wrap angle
φ_2	Blade outlet angle
Z	Blade number
b_3	Guide vane inlet width
D_3	Maximum diameter of an inner streamline of the guide vane
D_4	Maximum diameter of an outer streamline of the guide vane
D_5	Inside diameter of the guide vane outlet
D_6	Outside diameter of the guide vane outlet
L	Axial length of the guide vane
Z	Number of guide vanes
α_3	Guide vane inlet angle
α_4	Guide vane outlet angle
φ_0	Guide vane wrap angle
φ	Flow coefficient
ψ	Head coefficient
A	The streamline placement angle of the front cover of the impeller blade
B	The placement angle of the middle streamline inlet
C	The placement angle of the rear cover flowline inlet
Superscripts	
-	Time-averaged value
′	Fluctuating component

References

1. Shi, L.J.; Zhang, W.P.; Jiao, H.F.; Tang, F.P.; Wang, L.; Sun, D.D.; Shi, W. Numerical simulation and experimental study on the comparison of the hydraulic characteristics of an axial-flow pump and a full tubular pump. *Renew. Energy* **2020**, *153*, 1455–1464. [CrossRef]
2. Abdelaziz, E.A.; Saidur, R.; Mekhilef, S. A review on energy saving strategies in industrial sector. *Renew. Sustain. Energy Rev.* **2011**, *15*, 150–168. [CrossRef]
3. Edwards, G.; Spence, D. The European Commission. *Dev. Eur.* **1997**, *8*, 14–26.
4. Shankar, A.; Kalaiselvan, V.; Subramaniam, U.; Shanmugam, P.; Hanigovszki, N. A comprehensive review on energy efficiency enhancement initiatives in centrifugal pumping system. *Appl. Energy* **2016**, *181*, 495–513. [CrossRef]
5. Denton, J.D. Loss Mechanisms in Turbomachines. *Trans ASME J. Turbomach.* **1993**, *115*, 10–23. [CrossRef]
6. Herwig, H.; Kock, F. Direct and indirect methods of calculating entropy generation rates in turbulent convective heat transfer problems. *Heat Mass Transf.* **2007**, *43*, 207–215. [CrossRef]
7. Gu, Y.; Pei, J.; Yuan, S.; Wang, W.; Zhang, F.; Wang, P. Clocking Effect of Vaned Diffuser on Hydraulic Performance of High-Power Pump by Using the Numerical Flow Loss Visualization Method. *Energy* **2019**, *170*, 986–997. [CrossRef]
8. Ren, Y.; Zhu, Z.C.; Wu, D.H.; Li, X.J. Influence of guide ring on energy loss in a multistage centrifugal pump. *J. Fluids Eng.* **2019**, *141*, 0613021–06130213.
9. Zhang, F.; Appiah, D.; Hong, F.; Zhang, J.; Wei, X. Energy loss evaluation in a side channel pump under different wrapping angles using entropy production method. *Int. Commun. Heat Mass Transf.* **2020**, *113*, 104526. [CrossRef]
10. Hao, C.; Shi, W.D.; Li, W.; Liu, J.R. Energy loss analysis of novel self-priming pump based on the entropy production theory. *J. Therm. Sci.* **2019**, *28*, 150–162.
11. Li, D.; Wang, H.; Qin, Y.; Han, L.; Wei, X.; Qin, D. Entropy production analysis of hysteresis characteristic of a pump-turbine model. *Energy Convers. Manag.* **2017**, *149*, 175–191. [CrossRef]
12. Wang, C.; Zhang, Y.; Hou, H.; Zhang, J.; Xu, C. Entropy production diagnostic analysis of energy consumption for cavitation flow in a two-stage lng cryogenic submerged pump. *Int. J. Heat Mass Transf.* **2019**, *129*, 342–356. [CrossRef]
13. Si, Q.; Lu, R.; Shen, C.; Xia, S.; Sheng, G.; Yuan, J. An intelligent cfd-based optimization system for fluid machinery: Automotive electronic pump case application. *Appl. Sci.* **2020**, *10*, 366. [CrossRef]
14. Zhang, J.Y.; Cai, S.J.; Li, Y.J.; Zhou, X.; Zhang, Y.X. Optimization design of multiphase pump impeller based on combined genetic algorithm and boundary vortex flux diagnosis. *J. Hydrodyn.* **2017**, *29*, 1023–1034. [CrossRef]
15. Kim, J.H.; Choi, J.H.; Husain, A.; Kim, K.Y. Multi-objective optimization of a centrifugal compressor impeller through evolutionary algorithms. *Proc. Inst. Mech. Eng. J. Power Energy* **2010**, *224*, 711–721. [CrossRef]
16. Kang, H.S.; Song, Y.J.; Kim, Y.J. Optimal Design of Impeller for Centrifugal Compressor Under the Influence of Fluid-Structure Interaction. *J. Mech. Sci. Technol.* **2016**, *30*, 3953–3959. [CrossRef]
17. Shi, L.; Zhu, J.; Tang, F.; Wang, C. Multi-Disciplinary optimization design of axial-flow pump impellers based on the approximation model. *Energies* **2020**, *13*, 779. [CrossRef]
18. Liu, M.; Tan, L.; Cao, S. Design method of controllable blade angle and orthogonal optimization of pressure rise for a multiphase pump. *Energies* **2018**, *11*, 1048. [CrossRef]
19. Bonaiuti, D.; Arnone, A.; Ermini, M.; Baldassarre, L. Analysis and optimization of transonic centrifugal compressor impellers using the design of experiments technique. *J. Turbomach.* **2006**, *128*, 647–652. [CrossRef]
20. Lee, K.Y.; Choi, Y.S.; Kim, Y.L.; Yun, J.H. Design of axial fan using inverse design method. *J. Mech. Sci. Technol.* **2008**, *22*, 1883–1888. [CrossRef]
21. Thakkar, S.; Vala, H.; Patel, V.K.; Patel, R. Performance improvement of the sanitary centrifugal pump through an integrated approach based on response surface methodology, multi-objective optimization and cfd. *J. Braz. Soc. Mech. Sci. Eng.* **2021**, *43*, 1–15. [CrossRef]
22. Nataraj, M.; Ragoth, S.R. Analyzing pump impeller for performance evaluation using RSM and CFD. *Desalination Water Treat.* **2014**, *52*, 6822–6831. [CrossRef]
23. Wang, W.; Pei, J.; Yuan, S.; Zhang, J.; Yuan, J.; Xu, C. Application of different surrogate models on the optimization of centrifugal pump. *J. Mech. Sci. Technol.* **2016**, *30*, 567–574. [CrossRef]
24. Yang, F.; Jin, Y.; Liu, C.; Tang, F.; Cheng, L.; Yang, H. Numerical analysis and performance test on diving tubular pumping system with symmetric aerofoil blade. *Trans. Chin. Soc. Agric. Eng.* **2012**, *28*, 60–67.
25. Lu, R.; Yuan, J.; Wei, G.; Zhang, Y.; Lei, X.; Si, Q. Optimization Design of Energy-Saving Mixed Flow Pump Based on MIGA-RBF Algorithm. *Machines* **2021**, *9*, 365. [CrossRef]
26. Shi, Y.; Zhu, H.; Zhang, J.; Zhang, J.; Zhao, J. Experiment and numerical study of a new generation three-stage multiphase pump. *J. Pet. Sci. Eng.* **2018**, *169*, 471–484. [CrossRef]
27. Zhou, L.; Bai, L.; Li, W.; Shi, W.; Wang, C. PIV validation of different turbulence models used for numerical simulation of a centrifugal pump diffuser. *Eng. Computations* **2018**, *35*, 2–17. [CrossRef]
28. Al-Obaidi, A.R. Effects of different turbulence models on three-dimensional unsteady cavitating flows in the centrifugal pump and performance prediction. *Int. J. Nonlinear Sci. Numer. Simul.* **2019**, *20*, 487–509. [CrossRef]
29. Lu, R.; Yuan, J.P.; Wang, L.Y.; Fu, Y.X.; Hong, F.; Wang, W.J. Effect of volute tongue angle on the performance and flow unsteadiness of an automotive electronic cooling pump. *Proc. Inst. Mech. Eng. Part A J. Power Energy* **2020**, *235*, 227–241. [CrossRef]

30. Herwig, H.; Gloss, D.; Wenterodt, T. A new approach to understanding and modelling the influence of wall roughness on friction factors for pipe and channel flows. *J. Fluid Mech.* **2008**, *613*, 35–53. [CrossRef]
31. Xiang, Z. Energy conversion characteristic within impeller of low specific speed centrifugal pump. *Trans. Chin. Soc. Agric. Mach.* **2011**, *42*, 75–81.
32. Wang, H.; Long, B.; Wang, C.; Han, C.; Li, L. Effects of the impeller blade with a slot structure on the centrifugal pump performance. *Energies* **2020**, *13*, 1628. [CrossRef]
33. Tao, R.; Zhao, X.; Wang, Z. Evaluating the transient energy dissipation in a centrifugal impeller under rotorstator interaction. *Entropy* **2019**, *21*, 271. [CrossRef]
34. Liu, Y.; Tan, L. Spatial-temporal evolution of tip leakage vortex in a mixed flow pump with tip clearance. *J. Fluids Eng.* **2019**, *141*, 081302. [CrossRef]
35. Ran, H.; Luo, X.; Zhu, L.; Zhang, Y.; Wang, X.; Xu, H. Experimental study of the pressure fluctuations in a pump turbine at large partial flow conditions. *Chin. J. Mech. Eng.* **2012**, *25*, 1205–1209. [CrossRef]
36. Lu, L.G.; Chen, J.; Liang, J.D.; Leng, Y. Optimal hydraulic design of bulb tubular pump system. *J. Hydraul. Eng.* **2008**, *39*, 355–360.
37. Zhou, L.; Yang, Y.; Shi, W.; Lu, W.; Ye, D. Influence of outlet edge position of diffuser vane on performance of deep-well centrifugal pump. *J. Drain. Irrig. Mach. Eng.* **2016**, *34*, 1028–1034.
38. Cheng, X.; Zhang, X.; Wei, Y.; Zhang, S.; Wang, P. Influence of outlet edge position of guide vane on performance of well submersible pump. *Trans. Chin. Soc. Agric. Eng.* **2018**, *34*, 68–75.

machines

Article

Theoretical and Numerical Investigations on Static Characteristics of Aerostatic Porous Journal Bearings

Yandong Gu [1,*], Jinwu Cheng [1], Chaojie Xie [2], Longyu Li [1] and Changgeng Zheng [1]

1 College of Hydraulic Science and Engineering, Yangzhou University, Yangzhou 225009, China; c736351469@163.com (J.C.); lly2044828313@163.com (L.L.); zcg2034450942@163.com (C.Z.)
2 CAE Department, Shanghai Branch, AVIC Xinxiang Aviation Industry (Group) Co., Ltd., Shanghai 201299, China; vsxcjxcj@163.com
* Correspondence: guyandong@yzu.edu.cn

Abstract: To investigate the static characteristics of aerostatic journal bearings with porous bushing, the flow model—in which the compressibility of lubricating gas is considered—is established based on the Reynolds lubrication equation, Darcy equation for porous material, and continuity equation. With the finite difference method, difference schemes for non-uniform grids, relaxation method, and virtual node method, the numerical method for the governing equations of compressible flow in porous journal bearings is proposed. The effects of nominal clearance of bearings and compressibility of gas on the static characteristics are analyzed. Under the same minimum film thickness and the same gas compressibility, as the nominal clearance widens, the load capacity, mass flow rate, and power consumption increase. Under the same minimum film thickness and the same nominal clearance, with the increase in gas polytropic index, the load capacity strengthens, while the mass flow rate and power consumption decline. This study could provide a reference for the design of porous journal bearings.

Keywords: aerostatic porous journal bearing; compressible flow; theoretical modeling; numerical solution; nominal clearance; gas polytropic index

Citation: Gu, Y.; Cheng, J.; Xie, C.; Li, L.; Zheng, C. Theoretical and Numerical Investigations on Static Characteristics of Aerostatic Porous Journal Bearings. *Machines* **2022**, *10*, 171. https://doi.org/10.3390/machines10030171

Academic Editor: Davide Astolfi

Received: 14 January 2022
Accepted: 21 February 2022
Published: 24 February 2022

1. Introduction

Aerostatic journal bearings are normally lubricated by externally pressurized gas, holding advantages of high cleanliness, environmental protection, and low viscosity. They are frequently used in high-precision, high-rotating-speed, and extreme conditions, such as high-precision machine tools and high-speed machines [1]. Several restrictor types for aerostatic journal bearings have been developed, such as porous bushing, orifices, and slots. In contrast, porous bushing forms an approximately uniform gas film on the supporting surface, which offers significant load capacity and stable support [2,3].

Due to a micron-scale lubrication gap, shafting alignment, and shafting rotation, it is difficult to conduct experimental research on the static characteristics of the aerostatic porous journal bearing [4,5]. Presently, theoretical models and numerical solutions are the main methods to obtain static characteristics [6,7]. Sneck et al. [8] assumed that the flow in porous bushing was one-dimensional and compressible. The radial integral of the Darcy equation for porous material was substituted into the Reynolds lubrication equation, and a two-dimensional nonlinear partial differential equation was obtained. Stiffening occurs with an increasing eccentricity ratio, especially in short bearings. Majumdar et al. [9] established a three-dimensional flow equation for the porous bushing and a two-dimensional flow equation for the gas film. The finite difference method was used to solve the theoretical model. The steady-state performance characteristics of a stationary and a rotating journal bearing at various design conditions were studied. Singh et al. [10] considered the slip velocity on the interface between porous bushing and gas film for an aerostatic porous journal bearing. Prakash et al. [11] reported that the roughness and slip significantly affect the performance of hydrodynamic bearings. Naduvinamani et al. [12] investigated the effect of constant and cyclic load on the performance of porous short journal bearings. Under a cyclic load, the couple stress fluids provide a reduction in the journal velocity and an increase in the minimum permissible height of squeeze films. Saha et al. [13] found that the restrictive layer significantly influences the performance of porous journal bearing. Elsharkawy et al. [14] investigated the effects of journal misalignment on the performance of porous journal bearings. The effect of journal misalignment is negligible in porous journal bearings when the permeability is high. Lu et al. [15] studied the compressible flow in an aerostatic spherical bearing. Ruan et al. [16] used the perturbation method to solve the gas lubrication equation, and analyzed the operation stability of aerostatic porous journal bearings. It is found that the stability decreases with the increase in the supply pressure and permeability. Miyatake et al. [17] applied a surface-restricted layer into an aerostatic porous journal bearing, which improves the stiffness and stability at high rotating speed. Nicoletti et al. [18,19] adopted the Newton–Raphson method to solve the flow model of aerostatic porous journal bearings. Lee et al. [20] introduced the under-relaxation method to ensure the solution stability of compressible flow and analyzed the geometrical parameters of gas-lubricated porous bearings. The maximum load capacity occurs in the range of feed parameters between 0.5 and 1.0. Nishitani et al. [21] used the perturbation method to solve the dynamic and static characteristics of hydrostatic porous thrust bearings under small eccentricity. The hydrostatic porous thrust bearing has a higher maximum load capacity and slightly lower stiffness in comparison with bearings with capillary restrictors. Chien et al. [22] established a compressible thermohydrodynamic model of journal bearings lubricated with supercritical carbon dioxide. Feng et al. [23,24] considered the effect of temperature on gas properties and material deformation and established a temperature model of porous bearings with a restricted layer. The maximum gas pressure increases and the minimum film thickness decreases when the thermal deformation of the rotor is considered. Böhle et al. [25] established two-dimensional and three-dimensional flow models of hydrostatic porous journal bearings. Bhattacharjee et al. [26] investigated double-layer porous bearings lubricated with micro-polar fluids. Compared with Newtonian lubricants, the micropolar fluid significantly improves lubrication quality. Substantial studies have been contributed to the understanding of porous journal bearings. However, the compressible flow model and numerical solution of porous journal bearings are still open questions. The effects of nominal clearance of bearings and lubricant compressibility on the static characteristics are not yet completely understood.

Commercial computational fluid dynamics software based on three-dimensional Navier–Stokes equations is another method to predict the static characteristics of porous journal bearings [27]. However, considerable grids are required to fill the micron-scale lubrication gap to ensure a good quality of grids. Moreover, it would consume enormous computational resources and time [28,29]. In contrast, the solver based on the Reynolds lubrication equation can readily solve lubrication problems, showing advantages of low computing resource consumption and fast solution.

This study aims to model and solve the compressible flow in porous journal bearings and investigate the effects of nominal clearance and lubricant compressibility on static characteristics. First, a theoretical flow model of the porous bushing and aerostatic film is established. Then, a numerical solution method of the flow model is proposed, and a bearing solver is programmed. Finally, analysis of lubricant nominal clearance and compressibility is performed with the solver.

2. Porous Journal Bearings

Figure 1 shows the nomenclature of porous journal bearings. Aerostatic lubrication means that the journal is supported by externally pressurized gas fed at the bearing inlet, and then the lubricating gas flow exits from the bearing outlet. The nominal clearance is one of the dominant parameters determining static characteristics such as load capacity and power consumption. Five porous journal bearings are designed, and the parameters are shown in Table 1. The only difference in the bearings is the nominal clearance.

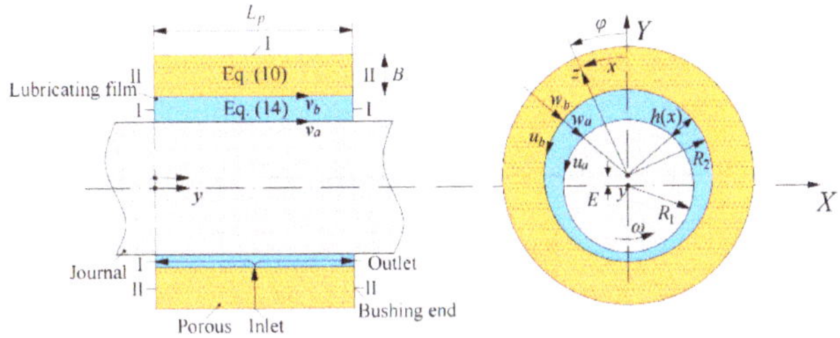

Figure 1. Nomenclature of porous journal bearings.

Table 1. Parameters of porous journal bearings.

Parameter	Case A	Case B	Case C	Case D	Case E
Journal radius R_1 (mm)	15	15	15	15	15
Nominal clearance h_0 (mm)	0.01	0.02	0.03	0.04	0.05
Bearing inside radius R_2 (mm)	15.01	15.02	15.03	15.04	15.05
Porous bushing thickness B (mm)			5		
Porous bushing length L_p (mm)			38		
Dynamic viscosity of lubricant μ (Pa·s)			1.7894×10^{-5}		
Permeability α (m^2)			1×10^{-13}		
Rotating speed n (r·min^{-1})			5000		
Feeding pressure p_s (bar)			4		
Background pressure p_b (bar)			1		
Feeding pressure difference $\Delta p = p_s - p_b$ (bar)			3		

3. Theoretical Modeling

3.1. Reynolds Lubrication Equation

The flow in porous journal bearings is assumed to be isothermal, compressible, and laminar. The Reynolds lubrication equation is used to describe the lubricating film flow of porous journal bearings, and the general complete form under steady condition is

$$\frac{\partial}{\partial x}\left(\frac{\rho \cdot h^3}{12 \cdot \mu} \cdot \frac{\partial p}{\partial x}\right) + \frac{\partial}{\partial y}\left(\frac{\rho \cdot h^3}{12 \cdot \mu} \cdot \frac{\partial p}{\partial y}\right) = \frac{\partial}{\partial x}\left(\rho \cdot \frac{u_a + u_b}{2} \cdot h\right) + \frac{\partial}{\partial y}\left(\rho \cdot \frac{v_a + v_b}{2} \cdot h\right) + \rho \cdot (w_a - w_b) - \rho \cdot u_a \cdot \frac{\partial h}{\partial x} - \rho \cdot v_a \cdot \frac{\partial h}{\partial y} \quad (1)$$

where x, y, and z stand for the circumferential, axial, and radial coordinates, respectively. h is the radial clearance function [30]. μ is the dynamic viscosity. u, v, and w represent the circumferential, axial, and radial velocities, respectively. The subscripts a and b represent the components on the journal bearing and bushing, respectively. The following process is to simplify Equation (1) according to the physical condition of a porous journal bearing.

The non-slip boundary condition is applied to the journal, and the squeezing effect is considered [30,31]. Consequently, the velocity boundary conditions on the journal bearing are given as

$$u_a = R_1 \cdot \omega \quad (2)$$

$$v_a = 0 \quad (3)$$

$$w_a = u_a \cdot \frac{\partial h}{\partial x} \quad (4)$$

By substituting Equations (2)–(4) into Equation (1), a simple form for the flow in the lubrication clearance is written as

$$\begin{aligned} &\frac{\partial}{\partial x}\left(\frac{\rho \cdot h^3}{12 \cdot \mu} \cdot \frac{\partial p}{\partial x}\right) + \frac{\partial}{\partial y}\left(\frac{\rho \cdot h^3}{12 \cdot \mu} \cdot \frac{\partial p}{\partial y}\right) = \\ &\frac{\rho \cdot u_a}{2} \cdot \frac{\partial h}{\partial x} + \frac{\partial}{\partial x}\left(\frac{\rho \cdot u_b}{2} \cdot h\right) + \frac{\partial}{\partial y}\left(\frac{\rho \cdot v_b}{2} \cdot h\right) - w_b \end{aligned} \tag{5}$$

3.2. Flow Model of Porous Bushing

The flow model of the porous bushing is established based on the Darcy equation for porous material and continuity equation. The assumptions are given as:

(a) The porous bushing is a three-dimensional laminar flow, and the flow inertia effect is neglected.

(b) The permeability of porous material is isotropic.

(c) The lubricant viscosity is constant, and the lubricant density is only related to pressure.

The Darcy equation is given as

$$\frac{\partial p}{\partial x_i} = -\frac{\mu}{\alpha} \cdot u_i \tag{6}$$

$$\frac{\partial(\rho \cdot u_j)}{\partial x_j} = 0 \tag{7}$$

where α is the permeability, i is the free index, and j is the dummy index. Equation (6) is substituted into the compressible continuity Equation (7), and the governing equation of porous bushings is obtained, as shown in Equation (8).

$$\frac{\partial}{\partial x_j}\left(\rho \cdot \frac{\alpha}{\mu} \cdot \frac{\partial p}{\partial x_j}\right) = 0 \tag{8}$$

According to the assumption (b), the permeability term is cancelled out, and then Equation (8) is rewritten as

$$\frac{\partial}{\partial x_j}\left(\rho \cdot \frac{\partial p}{\partial x_j}\right) = 0 \tag{9}$$

In the compressible calculation of porous journal bearings, the non-conservative form of Equation (9) is adopted and converted to the coordinate described in Figure 1, namely the pressure Laplace equation in the cylindrical coordinate is given as

$$\frac{\partial \rho}{\partial x} \cdot \frac{\partial p}{\partial x} + \frac{\partial \rho}{\partial y} \cdot \frac{\partial p}{\partial y} + \frac{\partial \rho}{\partial z} \cdot \frac{\partial p}{\partial z} + \rho \cdot \frac{\partial^2 p}{\partial x^2} + \rho \cdot \frac{\partial^2 p}{\partial y^2} + \rho \cdot \frac{\partial^2 p}{\partial z^2} + \rho \cdot \frac{1}{z} \cdot \frac{\partial p}{\partial z} = 0 \tag{10}$$

3.3. Flow Model of Lubricating film

The flow model of the aerostatic lubricating film is established based on the Reynolds lubrication equation and Darcy equation. The interface between the porous and film domains is assumed to be a slip boundary condition [32]. Consequently, the circumferential and axial slip velocities are controlled by the Darcy equation, as shown in Equations (11) and (12).

$$u_{b,s} = -\frac{\alpha}{\mu} \cdot \frac{\partial p}{\partial x} \tag{11}$$

$$v_{b,s} = -\frac{\alpha}{\mu} \cdot \frac{\partial p}{\partial y} \tag{12}$$

The injection flow from the porous bushing to the lubrication clearance is modeled by the term of radial velocity on the bushing wall, and it is also described by the Darcy equation, as shown in Equation (13).

$$w_b = \frac{\alpha}{\mu} \cdot \frac{\partial p}{\partial z} \tag{13}$$

Equations (11)–(13) are substituted into the Equation (5), the governing equation of aerostatic lubricating film of porous journal bearings is expressed as

$$\begin{gathered}\frac{\partial}{\partial x}\left(\frac{\rho \cdot h^3}{12 \cdot \mu} \cdot \frac{\partial p}{\partial x}\right) + \frac{\partial}{\partial y}\left(\frac{\rho \cdot h^3}{12 \cdot \mu} \cdot \frac{\partial p}{\partial y}\right) = \frac{\partial}{\partial x}\left(\rho \cdot \frac{u_a}{2} \cdot h\right) - \frac{\partial}{\partial x}\left(\frac{\rho \cdot h}{2} \cdot \left(\frac{\alpha}{\mu} \cdot \frac{\partial p}{\partial x}\right)\right) - \\ \frac{\partial}{\partial y}\left(\frac{\rho \cdot h}{2} \cdot \left(\frac{\alpha}{\mu} \cdot \frac{\partial p}{\partial y}\right)\right) - \rho \cdot \left(\frac{\alpha}{\mu} \cdot \frac{\partial p}{\partial z}\right)\end{gathered} \tag{14}$$

The governing equations for the entire flow in porous journal bearings are composed of Equations (10) and (14), and the application scope is shown in Figure 1.

3.4. Boundary Condition

The unknown variables in the above elliptic partial differential equation are pressure and density, which can be solved with the given boundary conditions of pressure and the correlation between pressure and density. Lubricant with certain pressure is supplied at the inlet of the porous bushing and then discharged to the working environment from the bearing outlet. Thus, the Dirichlet boundary condition of pressure (denoted by I in Figure 1) is specified at the bearing inlet and outlet. There is no lubricant passing through the porous bushing end, on which the normal velocity (axial velocity) is 0 m·s^{-1}, as shown in Equation (15).

$$v = -\frac{\alpha}{\mu} \cdot \frac{\partial p}{\partial y} = 0 \text{ m} \cdot \text{s}^{-1} \tag{15}$$

Subsequently, the Neumann boundary condition of pressure (denoted by II in Figure 1) on the porous bushing end is deduced as

$$\frac{\partial p}{\partial y} = 0 \text{ Pa} \cdot \text{m}^{-1} \tag{16}$$

It should be noted that the pressure on the porous bushing end is controlled by both Equations (10) and (16).

3.5. Ideal Gas State Equation

The effect of gas compressibility on the static characteristics of porous journal bearings is considered. The temperature increase is neglected in the aerostatic lubrication [1,2,31]. The density of ideal gas is positively correlated with pressure under isothermal conditions. By considering the variable process of ideal gas, the gas state equation is expressed as

$$\rho^n = \frac{p}{R_g \cdot T} \tag{17}$$

where R_g is the gas constant, and n is the gas polytropic index. The gas polytropic index varies from 1 to 1.4. If the pressure stays constant, the larger the index is, the lower the density is.

4. Numerical Solution

4.1. Grid Generation

The governing equations of porous journal bearings are solved synchronously by the finite difference method. The computational domain is discretized into structured grids, as shown in Figure 2. Uniform node distribution is adopted in the circumferential and axial directions. Non-uniform node distribution with geometric growth is employed in the radial direction, which is written as

$$\Delta z_{k+1} = q \cdot \Delta z_k \tag{18}$$

where i, j, and k are the indexes for circumferential, axial, and radial difference schemes, respectively.

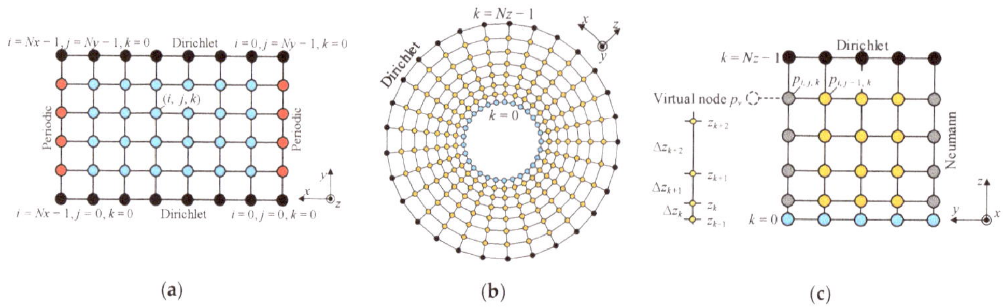

Figure 2. Structured grids; (**a**) Grids for lubricating film; (**b**) Grids on xz-section; (**c**) Grids on yz-section.

4.2. Difference Scheme

The central difference schemes are used for the first- and second-order derivatives in the circumferential and axial directions in the governing equations, as shown in Equations (19)–(20).

$$\left(\frac{\partial p}{\partial x}\right)_{i,j,k} \approx \frac{p_{i+1,j,k} - p_{i-1,j,k}}{2 \cdot \Delta x} \tag{19}$$

$$\left(\frac{\partial p}{\partial y}\right)_{i,j,k} \approx \frac{p_{i,j+1,k} - p_{i,j-1,k}}{2 \cdot \Delta y} \tag{20}$$

$$\left(\frac{\partial^2 p}{\partial x^2}\right)_{i,j,k} \approx \frac{p_{i+1,j,k} - 2 \cdot p_{i,j,k} + p_{i-1,j,k}}{\Delta x^2} \tag{21}$$

$$\left(\frac{\partial^2 p}{\partial y^2}\right)_{i,j,k} \approx \frac{p_{i,j+1,k} - 2 \cdot p_{i,j,k} + p_{i,j-1,k}}{\Delta y^2} \tag{22}$$

The three-node forward difference scheme is used for the first-order radial derivative in Equation (14) [33], as shown in Equation (23).

$$\begin{gathered}\left(\frac{\partial p}{\partial z}\right)_{i,j,k} \approx -\frac{2 \cdot \Delta z_{k+1} + \Delta z_{k+2}}{\Delta z_{k+1} \cdot (\Delta z_{k+1} + \Delta z_{k+2})} \cdot p_{i,j,k} + \\ \frac{\Delta z_{k+1} + \Delta z_{k+2}}{\Delta z_{k+1} \cdot \Delta z_{k+2}} \cdot p_{i,j,k+1} - \frac{\Delta z_{k+1}}{\Delta z_{k+2} \cdot (\Delta z_{k+1} + \Delta z_{k+2})} \cdot p_{i,j,k+2}\end{gathered} \tag{23}$$

The central difference schemes are used for the first- and second-order radial derivatives in Equation (10) [33], as shown in Equations (24) and (25).

$$\begin{gathered}\left(\frac{\partial p}{\partial z}\right)_{i,j,k} \approx -\frac{\Delta z_{k+1}}{\Delta z_k \cdot (\Delta z_k + \Delta z_{k+1})} \cdot p_{i,j,k-1} + \\ \frac{\Delta z_{k+1} - \Delta z_k}{\Delta z_k \cdot \Delta z_{k+1}} \cdot p_{i,j,k} + \frac{\Delta z_k}{\Delta z_{k+1} \cdot (\Delta z_k + \Delta z_{k+1})} \cdot p_{i,j,k+1}\end{gathered} \tag{24}$$

$$\left(\frac{\partial^2 p}{\partial z^2}\right)_{i,j,k} \approx \frac{\Delta z_{k+1} \cdot p_{i,j,k-1} - (\Delta z_{k+1} + \Delta z_k) \cdot p_{i,j,k}}{\frac{1}{2} \cdot ({\Delta z_{k+1}}^2 \cdot \Delta z_k + \Delta z_{k+1} \cdot {\Delta z_k}^2)} + \frac{\Delta z_k \cdot p_{i,j,k+1}}{\frac{1}{2} \cdot ({\Delta z_{k+1}}^2 \cdot \Delta z_k + \Delta z_{k+1} \cdot {\Delta z_k}^2)} \tag{25}$$

Equations (19)–(22) are of second-order accuracy, while the accuracy of Equation (25) is between first-order and second-order, which depends on the grid growth ratio q. The grid growth ratio is set as 1.1, showing a quasi-second-order accuracy.

The pressure on the bushing end can be calculated by the one-side difference scheme of Equation (16). However, Equation (16) with difference schemes of either first or second-order accuracy converges slowly. As described in Section 3.4, the pressure on the bushing end is simultaneously controlled by the Equations (10) and (16). A Laplace–Neumann virtual node method is proposed as follows:

(a) The virtual node method is used to deal with the Neumann boundary condition [34]. A virtual node is created outside the flow domain, as shown in Figure 2c. The virtual node combined with the central difference scheme is used to discretize Equation (16), which holds second-order accuracy, and then the pressure at the virtual node is yielded, as shown in Equations (26) and (27).

$$\frac{\partial p}{\partial y} = \frac{p_v - p_{i,j-1,k}}{2 \cdot \Delta y} + O\left(\Delta x^2\right) = 0 \tag{26}$$

$$p_v = p_{i,j-1,k} \tag{27}$$

(b) The pressure on the bushing end is also governed by Equation (16). Equations (21), (22), (24), and (25) combined with the virtual node are used to discretize Equation (10), and then the pressure at the virtual node is replaced by Equation (27). Therefore, the second-order axial partial derivative in Equation (10) is discretized by Equation (28).

$$\left(\frac{\partial^2 p}{\partial y^2}\right)_{i,j,k} \approx \frac{-2 \cdot p_{i,j,k} + 2 \cdot p_{i,j-1,k}}{\Delta y^2} \tag{28}$$

A linear algebraic equation for the flow model of porous journal bearings is created with the above difference schemes. Then, the relaxation method and root mean square of residuals are introduced to solve the linear algebraic equation iteratively. The Laplace–Neumann virtual node method shows fast convergence rates and small residuals through massive code tests.

4.3. Pressure–Density Coupling Calculation Method

Figure 3 shows the pressure–density coupling calculation strategy. First, the pressure field is calculated with the numerical method described in Sections 4.1 and 4.2, and this iterative process is called internal iteration. Then, the density field is updated based on the pressure field and Equation (17), and the updated density field is entered to the following pressure calculation, which is an external iteration. The internal iteration is converged when the root mean square of pressure residuals is less than 1×10^{-6}. When the current pressure and density fields are both less by 1×10^{-4} than the previous fields, the external iterative process is terminated.

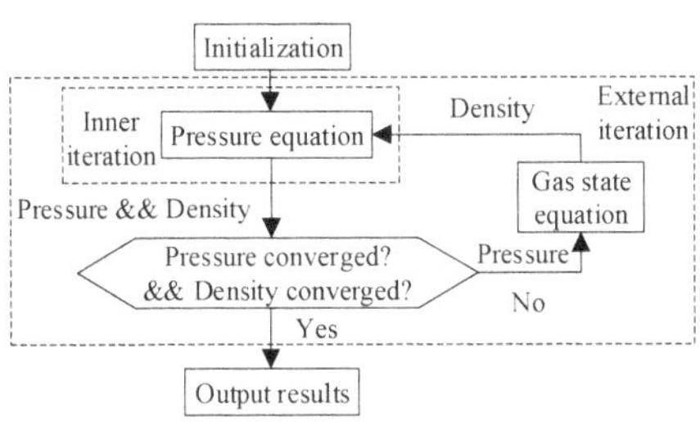

Figure 3. Coupling solutions for pressure and density.

4.4. Static Characteristics

The load capacity, mass flow rate and power consumption are investigated. After the flow field converged, the static characteristics are calculated by the following equations.

Load capacity:

$$F_X = \int_0^L \int_0^{2\cdot\pi} p \cdot \sin(\varphi) \cdot R_2 \cdot d\varphi \cdot dy \tag{29}$$

$$F_Y = \int_0^L \int_0^{2\cdot\pi} p \cdot \cos(\varphi) \cdot R_2 \cdot d\varphi \cdot dy \tag{30}$$

$$F = \sqrt{F_X^2 + F_Y^2} \tag{31}$$

Mass flow rate:

$$\dot{m} = \int_{L/2-L_p/2}^{L/2+L_p/2} \int_0^{2\cdot\pi\cdot(R_2+B)} \rho \cdot \frac{\alpha}{\mu} \cdot \left.\frac{\partial p}{\partial z}\right|_{z=R_2+B} \cdot dx \cdot dy \tag{32}$$

Volumetric flow rate:

$$\dot{Q} = \int_{L/2-L_p/2}^{L/2+L_p/2} \int_0^{2\cdot\pi\cdot(R_2+B)} \frac{\alpha}{\mu} \cdot \left.\frac{\partial p}{\partial z}\right|_{z=R_2+B} \cdot dx \cdot dy \tag{33}$$

Power consumption:

$$P = \dot{Q} \cdot \Delta p \tag{34}$$

Case B under the minimum film thickness of 0.001 mm is selected to conduct grid sensitivity analysis. The circumferential, axial, and radial nodes are adjusted with the same growth ratio. Figure 4 presents the load capacity versus the number of nodes. The load capacity is changeless when the nodes reach 240,000. To balance the calculation accuracy and time efficiency, the number of nodes in the circumferential, axial, and radial directions are determined as 120, 80, and 25, respectively.

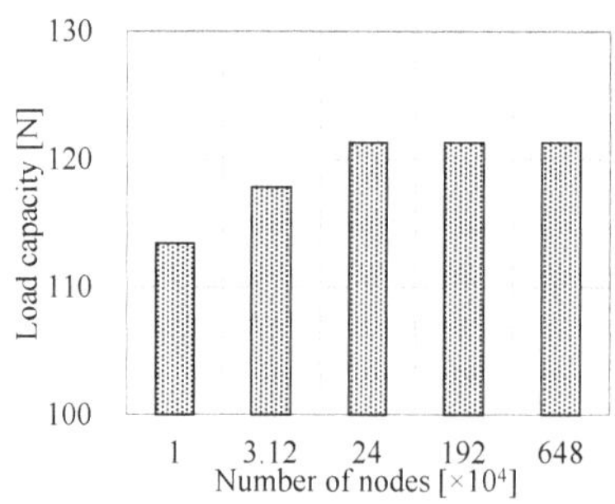

Figure 4. Grid sensitivity analysis.

5. Results and Discussion

5.1. Effects of Nominal Clearance on Static Characteristics

Figure 5 shows the effect of nominal clearance on the load capacity. The gas polytropic index is set as 1, which means that density linearly relates to pressure. Under the same minimum film thickness, the load capacity increases with the increase in nominal clearance. The eccentricity ratio is defined as the ratio of eccentricity to nominal clearance. For example, when the minimum film thickness is 0.001 mm, the load capacity of nominal clearance of 0.02 mm (the eccentricity ratio is 0.95) is around three times that of nominal clearance of 0.01 mm (the eccentricity ratio is 0.9). However, with the uniform increase in nominal clearance, the growth amplitude of load capacity gradually decreases. For instance, when the minimum film thickness keeps 0.001 mm, the load capacity increases by nearly 80 N as the nominal clearance is increased from 0.01 mm to 0.02 mm; in contrast, with the rise in nominal clearance from 0.04 mm (the eccentricity ratio is 0.975) to 0.05 mm (the eccentricity ratio is 0.98), the load capacity increases only by around 6 N. Under the same nominal clearance, the load capacity decreases with the increase in the minimum film thickness. When no eccentricity occurs, the load capacity is 0 N.

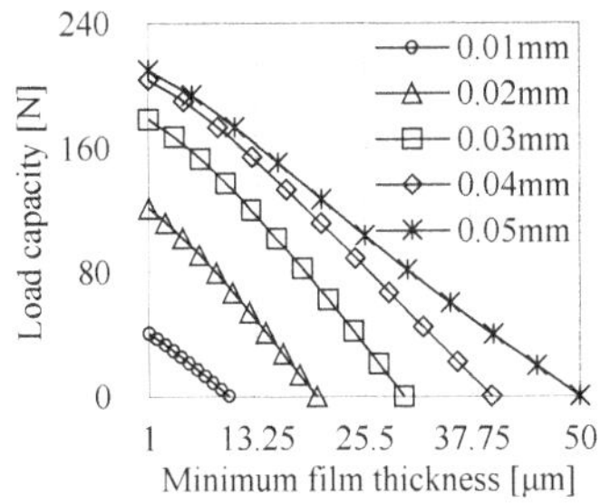

Figure 5. Load capacity under different nominal clearances.

To understand why the load capacity amplifies with nominal clearance for compressible flow in porous journal bearings, the minimum film thickness of 0.001 mm is taken as an example, and the pressure fields of lubricating film with nominal clearances of 0.01 mm, 0.03 mm (the eccentricity ratio is 0.967), and 0.05 mm are compared, as shown in Figure 6. With the increase in nominal clearance, the maximum pressure shows little change and occurs around the circumferential position of 180°, where the minimum film thickness is located. The maximum pressure reaches approximately 4.01 bar under each nominal clearance, which is slightly greater than the feeding pressure because of the aerodynamic effect. However, the pressure at the circumferential position of 0° or 360° decreases remarkably as the nominal clearance increases, which means the circumferential gradient of pressure varies significantly against the nominal clearance. Thus, a greater load capacity is generated.

Figure 7 shows the effect of nominal clearance on mass flow rate. Under the same minimum film thickness, the mass flow rate increases with the increase in nominal clearance. For example, when the minimum film thickness is 0.001 mm, the mass flow rate of nominal clearance of 0.02 mm is approximately 2.3-times higher than that of nominal clearance of 0.01 mm. However, as the nominal clearance rises uniformly, the increased amplitude of mass flow rate gradually decreases. For instance, when the minimum film thickness stands at 0.001 mm, as the nominal clearance is adjusted from 0.01 mm to 0.02 mm, the mass flow rate increases by around 0.05 $kg \cdot min^{-1}$; by comparison, with the nominal clearance expanded from 0.04 mm to 0.05 mm, the mass flow rate rises by nearly 0.02 $kg \cdot min^{-1}$. The mass flow rate of nominal clearances of 0.01, 0.02, and 0.03 mm is negatively correlated with the minimum film thickness. Conversely, the mass flow rate of nominal clearances of 0.04 mm and 0.05 mm presents a positive correlation with the minimum film thickness.

Figure 8 shows the effect of nominal clearance on power consumption. Standing at the same minimum film thickness, the power consumption rises with the increase in nominal clearance. The power consumption of nominal clearances of 0.01, 0.02, and 0.03 mm declines as the minimum film thickness increases. However, for nominal clearances of 0.04 mm and 0.05 mm, the power consumption positively correlates with the minimum film thickness.

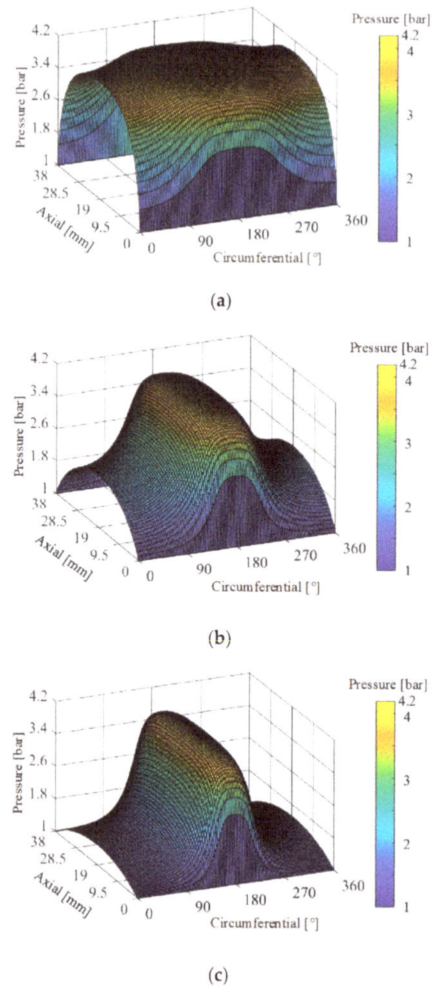

Figure 6. Comparison of lubricating film pressure under different nominal clearances with the minimum film thickness of 0.001 mm; (**a**) h_0 = 0.01 mm; (**b**) h_0 = 0.03 mm; (**c**) h_0 = 0.05 mm.

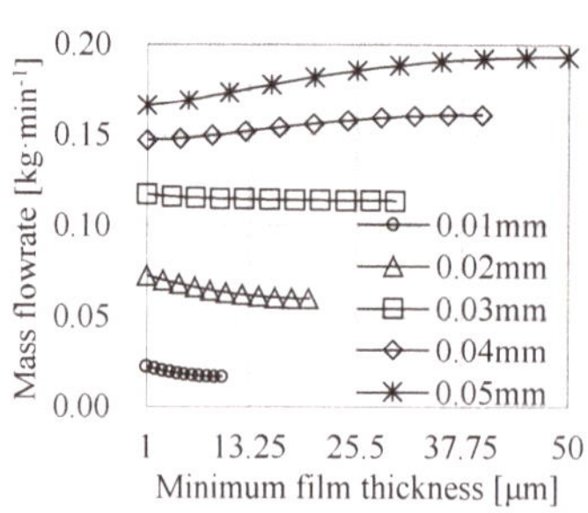

Figure 7. Mass flow rate under different nominal clearances.

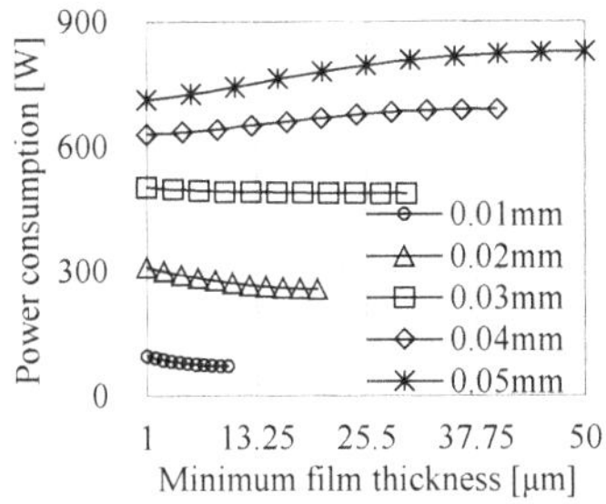

Figure 8. Power consumption under different nominal clearances.

5.2. Effects of Gas Polytropic Index on Static Characteristics

The nominal clearances of 0.01, 0.03, and 0.05 mm are presented to investigate the effect of compressibility on static characteristics. The nominal clearances of 0.02 mm and 0.04 mm show a consistent conclusion. The gas polytropic indices are set as 1, 1.2, and 1.4. The load capacity is shown in Figure 9. Under the same minimum film thickness, the load capacity strengthens with the increase in gas polytropic index. However, as the nominal clearance increases, the increased amplitude of load capacity with regard to the gas polytropic index diminishes. For nominal clearance of 0.05 mm, curves of load capacity versus minimum film thickness for different gas polytropic indexes almost coincide.

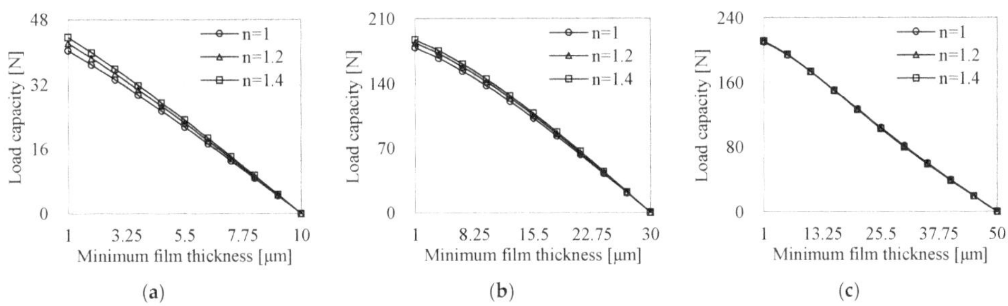

Figure 9. Load capacity under different gas polytropic indexes; (**a**) h_0 = 0.01 mm; (**b**) h_0 = 0.03 mm; (**c**) h_0 = 0.05 mm.

Figure 10 shows the effect of gas polytropic index on mass flow rate (denoted as M) and power consumption (denoted as P) under nominal clearances of 0.01, 0.03, and 0.05 mm. Under the same minimum film thickness, the mass flow rate and power consumption decrease with the increase in gas polytropic index. It is opposed to the effect of gas polytropic index on load capacity. That is to say, within the scope of this study, the static performance can be improved by using lubricant with less compressibility. For nominal clearances of 0.01 mm and 0.03 mm, the mass flow rate and power consumption under different gas polytropic indexes are negatively correlated with the minimum film thickness. However, for nominal clearance of 0.05 mm, the mass flow rate and power consumption expand as the minimum film thickness widens.

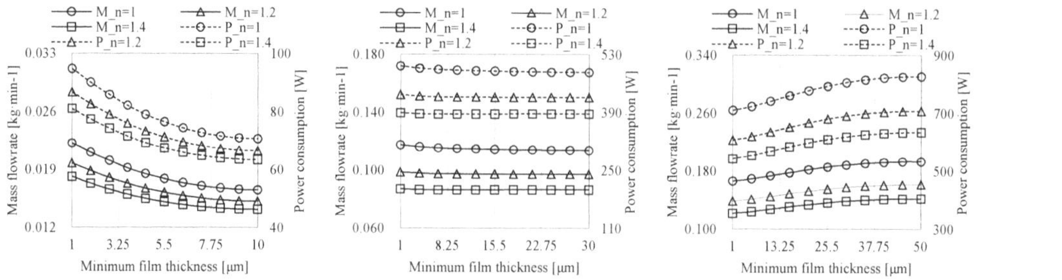

(**a**) (**b**) (**c**)

Figure 10. Mass flow rate and power consumption under different gas polytropic indexes; (**a**) h_0 = 0.01 mm; (**b**) h_0 = 0.03 mm; (**c**) h_0 = 0.05 mm.

6. Conclusions

The theoretical model of aerostatic porous journal bearings is established based on the Reynolds lubrication equation, Darcy equation, and continuity equation. The numerical method for the bearing model is proposed with the finite difference method, difference schemes, relaxation method, and virtual node method. Under the same minimum film thickness and the same gas polytropic index, with the increase in nominal clearance, the load capacity strengthens due to the rise in pressure circumferential gradient. Meanwhile, the mass flow rate and power consumption expand. As the nominal clearance widens uniformly, the increased amplitude of load capacity gradually decreases. Under the same minimum film thickness and the same nominal clearance, as the gas polytropic index rises, the load capacity amplifies because the pressure circumferential gradient increases, while the mass flow rate and power consumption decline. The effect of gas polytropic index on load capacity diminishes with the increase in nominal clearance.

Author Contributions: Conceptualization, Y.G. and J.C.; Data curation, Y.G. and J.C.; Formal analysis, Y.G. and J.C.; Investigation, J.C.; Project administration, Y.G.; Resources, Y.G.; Software, Y.G.; Supervision, Y.G.; Validation, J.C., L.L. and C.Z.; Visualization, C.X., L.L. and C.Z.; Writing—original draft, Y.G. and J.C.; Writing—review & editing, Y.G. and J.C. All authors have read and agreed to the published version of the manuscript.

Funding: This research was funded by the Postdoctoral Research Fund of Jiangsu Province, China (grant no. 2021K569C), and National Natural Science Foundation of China (grant no. 51779214).

Institutional Review Board Statement: Not applicable.

Informed Consent Statement: Not applicable.

Data Availability Statement: Data on the analysis and reporting results during the study can be obtained by contacting the authors.

Acknowledgments: The authors thank the College of Hydraulic Science and Engineering, Yangzhou University. The authors acknowledge the funding support from the Postdoctoral Research Fund of Jiangsu Province and National Natural Science Foundation of China. The authors are very grateful for the discussion with Martin Böhle and Artur Schimpf from the Technical University of Kaiserslautern, Germany. A huge thanks is due to the editor and reviewers for their valuable comments to improve the quality of this paper.

Conflicts of Interest: The authors declare no conflict of interest.

Nomenclature

B	porous bushing thickness
E	eccentricity
F_X	X-direction load capacity

F_Y	Y-direction load capacity
F	load capacity
h	radial gap function
h_0	nominal clearance
i	index for circumferential difference scheme
j	index for axial difference scheme
k	index for radial difference scheme
Lp	porous bushing length
$\dot{m}$	mass flow rate
n	rotating speed
N_x	circumferential grid node number
N_y	axial grid node number
N_z	radial grid node number
P	power consumption
p	pressure
p_s	feeding pressure
p_b	backing pressure
p_v	virtual node
$p_{i,j,k}$	internal node
Δp	feeding pressure difference ($\Delta p = p_s - p_b$)
q	grid growth ratio
R_1	journal radius
R_2	bearing inside radius
R_g	gas constant
T	temperature
u	circumferential velocity
u_a	journal circumferential velocity component
u_b	bushing circumferential velocity component
$u_{b,s}$	bushing circumferential slip velocity component
v	axial velocity
v_a	journal axial velocity component
v_b	bushing axial velocity component
$v_{b,s}$	bushing axial slip velocity component
w	radial velocity
w_a	journal radial velocity component
w_b	bushing radial velocity component
$w_{b,s}$	bushing radial slip velocity component
x	circumferential coordinate
y	axial coordinate
z	radial coordinate
Δz_k	grid thickness of layer k
Greek symbols	
ρ	density
μ	lubricant dynamic viscosity
v	lubricant kinematic viscosity
α	permeability
φ	circumferential angle
ω	journal angular velocity
Superscripts	
n	gas polytropic index
Subscripts	
a	journal component
b	bushing component
i	free index
j	dummy index
s	slip velocity
x	circumferential
y	axial
z	radial

References

1. Gu, Y.; Böhle, M.; Schimpf, A.; Yuan, S. Aerostatic bearing with porous restrictor: Research status and future perspectives. *J. Drain. Irrig. Mach. Eng.* **2021**, *39*, 818–825. [CrossRef]
2. Gu, Y.; Böhle, M.; Schimpf, A.; Yuan, S. Theoretical modeling and numerical solution of hydrostatic radial bearing with porous restrictor. *J. Vib. Shock.* **2021**, *40*, 16–24. [CrossRef]
3. Lee, C.C.; You, H.I. Characteristics of externally pressurized porous gas bearings considering structure permeability. *Tribol. Trans.* **2009**, *52*, 768–776. [CrossRef]
4. San Andres, L.; Cable, T.A.; Zheng, Y.; De Santiago, O.; Devitt, D. Assessment of porous type gas bearings: Measurements of bearing performance and rotor vibrations. In Proceedings of the ASME Turbo Expo 2016: Turbomachinery Technical Conference and Exposition, Seoul, Korea, 13–17 June 2016. [CrossRef]
5. Liu, W.; Feng, K.; Huo, Y.; Guo, Z. Measurements of the rotordynamic response of a rotor supported on porous type gas bearing. *J. Eng. Gas Turbines Power* **2018**, *140*, 102501. [CrossRef]
6. Wang, B.; Wang, Y.; Liu, Y.; Zhu, C.; Yuan, X. Static and dynamic characteristics of tilting-pad thrust bearing on inclined shaft. *J. Xi'an Jiaotong Univ.* **2020**, *54*, 129–138. [CrossRef]
7. Hu, X.; Lü, P.; Feng, K.; Zhao, X. Static and dynamic performance of laminated gas foil thrust bearing. *J. Aerosp. Power* **2018**, *37*, 236–241. [CrossRef]
8. Sneck, H.J.; Yen, K.T. The externally pressurized, porous wall, gas-lubricated journal bearing. *ASLE Trans.* **1964**, *7*, 288–298. [CrossRef]
9. Majumdar, B.C. Analysis of externally pressurized porous gas journal bearings—I. *Wear* **1975**, *33*, 25–35. [CrossRef]
10. Singh, K.C.; Rao, N.S.; Majumdar, B.C. Effect of slip flow on the steady-state performance of aerostatic porous journal bearings. *J. Tribol.* **1984**, *106*, 156–162. [CrossRef]
11. Prakash, J.; Gururajan, K. Effect of velocity slip in an infinitely long rough porous journal bearing. *Tribol. Trans.* **1999**, *42*, 661–667. [CrossRef]
12. Naduvinamani, N.B.; Hiremath, P.S.; Gurubasavaraj, G. Squeeze film lubrication of a short porous journal bearing with couple stress fluids. *Tribol. Int.* **2001**, *34*, 739–747. [CrossRef]
13. Saha, N.; Majumdar, B.C. Study of externally-pressurized gas-lubricated two-layered porous journal bearings: A steady state analysis. *Proc. Inst. Mech. Eng. Part J J. Eng. Tribol.* **2002**, *216*, 151–158. [CrossRef]
14. Elsharkawy, A.A. Effects of misalignment on the performance of flexible porous journal bearings. *Tribol. Trans.* **2003**, *46*, 119–127. [CrossRef]
15. Lu, Z.; Du, J.; Sun, Y. Analysis on aerostatic porous spherical bearings static performance. *Chin. J. Mech. Eng.* **2004**, *2004*, 115–119. [CrossRef]
16. Ruan, H.; Zhang, D.; Jing, X. Analysis on operation stability of porous aerostatic journal bearing. *Bearing* **2006**, *2006*, 1–3. [CrossRef]
17. Miyatake, M.; Yoshimoto, S.; Sato, J. Whirling instability of a rotor supported by aerostatic porous journal bearings with a surface-restricted layer. *Proc. Inst. Mech. Eng. Part J J. Eng. Tribol.* **2006**, *220*, 95–103. [CrossRef]
18. Nicoletti, R.; Silveira, Z.C.; Purquerio, B.M. Modified Reynolds equation for aerostatic porous radial bearings with quadratic Forchheimer pressure-flow assumption. *J. Tribol.* **2008**, *130*, 031701. [CrossRef]
19. Nicoletti, R.; Purquerio, B.D.M.; Silveira, Z.D.C. The effect of permeability distribution on the numerical analysis of aerostatic ceramic porous bearings. *Lubr. Sci.* **2013**, *25*, 185–194. [CrossRef]
20. Lee, C.C.; You, H.I. Geometrical design considerations on externally pressurized porous gas bearings. *Tribol. Trans.* **2010**, *53*, 386–391. [CrossRef]
21. Nishitani, Y.; Yoshimoto, S.; Somaya, K. Numerical investigation of static and dynamic characteristics of water hydrostatic porous thrust bearings. *Int. J. Autom. Technol.* **2011**, *5*, 773–779. [CrossRef]
22. Chien, S.Y.; Cramer, M.; Untaroiu, A. A compressible thermohydrodynamic analysis of journal bearings lubricated with supercritical CO_2. In Proceedings of the ASME 2017 Fluids Engineering Division Summer Meeting, Waikoloa, HI, USA, 30 July–3 August 2017. [CrossRef]
23. Feng, K.; Wu, Y.; Liu, W.; Zhao, X.; Li, W. Theoretical investigation on porous tilting pad bearings considering tilting pad motion and porous material restriction. *Precis. Eng.* **2018**, *53*, 26–37. [CrossRef]
24. Feng, K.; Li, W.; Huo, Y.; Huo, C. Thermal characteristic analysis of aerostatic porous journal bearings with surface-restricted layer. *J. Mech. Eng.* **2018**, *54*, 216–224. [CrossRef]
25. Böhle, M.; Gu, Y.; Schimpf, A. Two flow models for designing hydrostatic bearings with porous material. In Proceedings of the ASME-JSME-KSME 2019 8th Joint Fluids Engineering Conference, San Francisco, CA, USA, 28 July–1 August 2019. [CrossRef]
26. Bhattacharjee, B.; Chakraborti, P.; Choudhuri, K. Evaluation of the performance characteristics of double-layered porous micropolar fluid lubricated journal bearing. *Tribol. Int.* **2019**, *138*, 415–423. [CrossRef]
27. Wu, D.; Tao, J. Analysis of static performance of porous graphite aerostatic thrust bearing. *China Mech. Eng.* **2010**, *21*, 2296–2301.
28. Cui, H.; Wang, Y.; Yue, X.; Huang, M.; Wang, W. Effects of manufacturing errors on the static characteristics of aerostatic journal bearings with porous restrictor. *Tribol. Int.* **2017**, *115*, 246–260. [CrossRef]
29. Pei, H.; Long, W.; Yang, S.; Gong, L. Formation mechanism of micro-vibration in aerostatic bearings. *J. Vib. Shock.* **2018**, *37*, 71–78. [CrossRef]

30. Böhle, M. Numerical investigation of the flow in hydrostatic journal bearings with porous material. In Proceedings of the ASME 2018 5th Joint US-European Fluids Engineering Division Summer Meeting, Montreal, Qc, Canada, 15–20 July 2018. [CrossRef]
31. Hamrock, B.J.; Schmid, S.R.; Jacobson, B.O. *Fundamentals of Fluid Film Lubrication*, 2nd ed.; Marcel Dekker, Inc.: New York, NY, USA, 2004. [CrossRef]
32. Sparrow, E.M.; Beavers, G.S.; Hwang, I.T. Effect of velocity slip on porous-walled squeeze films. *J. Lubr. Technol.* **1972**, *94*, 260–264. [CrossRef]
33. Ferziger, J.H.; Perić, M.; Street, R.L. *Computational Methods for Fluid Dynamics*, 3rd ed.; Springer: Berlin/Heidelberg, Germany, 2002. [CrossRef]
34. Tao, W. *Numerical Heat Transfer*, 2nd ed.; Xi'an Jiaotong University Press: Xi'an, China, 2001. (In Chinese)

machines

Article

Experimental Study on the Internal Pressure Pulsation Characteristics of a Bidirectional Axial Flow Pump Operating in Forward and Reverse Directions

Xiaowen Zhang [1], Fangping Tang [1,*], Yueting Chen [1], Congbing Huang [2], Yujun Chen [3], Lin Wang [1] and Lijian Shi [1]

[1] College of Hydraulic Science and Engineering, Yangzhou University, Yangzhou 225009, China; zhangxw_yzu@163.com (X.Z.); chenyt_yzu@163.com (Y.C.); wanglin.yzu@gmail.com (L.W.); shilijian@yzu.edu.cn (L.S.)
[2] Jiangsu Aerospace Hydraulic Equipment Co., Ltd., Yangzhou 225009, China; huangcb_jsht@163.com
[3] Jiangsu Qinhuai River Water Conservancy Project Management Office, Nanjing 210001, China; chenyj_qhhglc@163.com
* Correspondence: tangfp@yzu.edu.cn; Tel.: +86-139-5105-9552

Abstract: A bidirectional axial flow pump can realize bidirectional pumping, which has a wide application prospect in coastal low-head pumping stations and water jet propulsion systems. In this paper, a typical bidirectional axial flow pump is taken as the research object, and the hydraulic model of the bidirectional axial flow pump is manufactured. The hydrodynamic characteristics of the bidirectional axial flow pump are tested on the high-precision hydraulic mechanical test bench, including the positive and negative directions. In the experiment, multiple pressure pulsation monitoring points were arranged in the impeller chamber, and the pressure fluctuations in the pump under a total of 42 flow conditions were measured by a micro pressure pulsation sensor, involving 21 working conditions of forward operation and 21 working conditions of reverse operation. According to the experimental results, the hydrodynamic characteristics, especially the pressure pulsation characteristics in the pump, of the two-way axial flow pump under positive and negative operation are comprehensively compared and analyzed, and the energy characteristics and the propagation law of pressure pulsation of the two-way axial flow pump under positive and negative operation are revealed. The research results provide an important reference for the safe and stable operation of coastal bidirectional axial flow pump stations.

Keywords: bidirectional axial flow pump; bidirectional operation; experiment; hydrodynamic characteristics, pressure pulsation; spectrum

Citation: Zhang, X.; Tang, F.; Chen, Y.; Huang, C.; Chen, Y.; Wang, L.; Shi, L. Experimental Study on the Internal Pressure Pulsation Characteristics of a Bidirectional Axial Flow Pump Operating in Forward and Reverse Directions. *Machines* **2022**, *10*, 167. https://doi.org/10.3390/machines10030167

Academic Editor: Antonio J. Marques Cardoso

Received: 19 January 2022
Accepted: 17 February 2022
Published: 23 February 2022

1. Introduction

In recent years, axial flow pumps have gradually played an increasingly important role in water jet propulsion systems and coastal low-head pumping stations [1–4]. At present, most coastal low-head pumping stations adopt axial flow pumps.

In the application of axial flow pump in coastal pumping stations, researchers and managers found that the pumping stations in coastal areas have low head and large water level variations in upstream and downstream. Many coastal low-head pumping stations need bidirectional operation to meet the needs of both drainage and irrigation. There are three main ways to achieve bidirectional operation of coastal pumping stations. The first is equipped with a two-way pumping impeller for direct reverse pumping. The second is a two-way channel. The bidirectional pumping is realized by the special layout of the bidirectional flow channel [5]. The third one is to disassemble the impeller and install it after turning it 180 degrees. Since it is very difficult to disassemble and reinstall impellers frequently in the actual operation of pumping stations, the third method has not been applied. Compared with the first method, the second method uses the special arrangement of two-way channel to realize two-way pumping, which is expensive and inconvenient to maintain. So the most reliable and convenient way to realize the bidirectional operation of a coastal pumping station is to equip the axial flow pump with bidirectional pumping for direct reverse pumping.

However, due to the one-way impeller basically equipped in the previous axial flow pump station [6–8], research on the hydrodynamic characteristics of the bidirectional axial flow pump is still in a very preliminary stage. The research is mainly aimed at the energy characteristics of the bidirectional axial flow pump. Ma et al. [9] designed a new type of bidirectional axial flow pump by reducing the camber of the curved airfoil and thickening the airfoil skeleton, which significantly improved the hydraulic efficiency and cavitation performance of the bidirectional axial flow pump. Pei et al. [10] studied the influence of the distance between the impeller and the guide vane on the forward and reverse performance of the bidirectional axial flow pump by using the entropy generation method based on the numerical calculation results. Meng et al. [11] used the two-way fluid-solid coupling method to quantitatively analyze the deformation stress of the impeller of the two-way axial flow pump. It was found that the total deformation continued to rise from the inlet to the outlet of the impeller, whether the impeller was in positive or negative rotation. Ma et al. [12] studied the internal flow field of a bidirectional axial flow pump during reverse operation. It was found that the flow separation at the trailing edge of the blade was the main reason for the significant decline of the performance of the bidirectional axial flow pump during reverse operation. Meng et al. [13] optimized the bidirectional axial flow pump based on the method of double-layer artificial neural network. The forward operation efficiency of the bidirectional axial flow pump obtained by the optimized design was reduced, but the reverse operation efficiency was greatly improved.

Although the energy characteristics of the two-way axial flow pump have been improved to some extent, the hydrodynamic performance of the axial flow pump with two-way airfoil is poor, and the inlet flow pattern of the pump is disturbed by the guide vane during the reverse operation. This means the two-way pump in operation will produce more serious vibration and noise [14], and if equipped with a two-way axial flow pump the pumping station system security and stability is hugely at risk. In recent years, more and more researchers have pointed out that the pressure pulsation induced by unstable flow in the pump is the most critical factor affecting the safety and stability of the pumping station system [15–18]. Therefore, it is urgent to study the hydrodynamic characteristics of the bidirectional pump, especially the internal pressure pulsation characteristics.

Considering the high cost and long test cycle of the pressure pulsation experiment [19–21], it is convenient to study the pressure pulsation characteristics of the pump by using numerical simulation [22–25]. However, some existing studies on the two-way pump [26] show that the numerical simulation method is not a particularly reliable means to study the hydrodynamic characteristics of the two-way pump because the vortex, reflux, and other unstable flow phenomena in the two-way pump are obvious under reverse operation. In this paper, the hydrodynamic characteristics of the bidirectional axial flow pump are studied by the experimental method. Firstly, the hydraulic model of a typical two-way axial flow pump was manufactured, and its hydrodynamic characteristics were tested on a high-precision hydraulic mechanical test bench, including positive and negative directions. In the experiment, multiple pressure pulsation monitoring points were arranged in the impeller chamber, and the pressure fluctuations in the pump under a total of 42 flow conditions were measured by using a micro pressure pulsation sensor, involving 21 working conditions of forward operation and 21 working conditions of reverse operation. Based on the experimental results, the hydrodynamic characteristics of the bidirectional axial flow pump under positive and negative operation, especially the pressure pulsation characteristics in the pump, are comprehensively analyzed and compared, and the energy characteristics and the propagation law of pressure pulsation of the bidirectional axial flow pump under positive and negative operation are revealed. The research results provide an important reference for the safe and stable operation of coastal bidirectional axial flow pump stations.

2. Experimental Model and System

2.1. Research Object

This experiment takes a typical bidirectional axial flow pump as the research object. The impeller diameter of the bidirectional axial flow pump hydraulic model is 300 mm. The speed is 1450 r/min. The blade placement angle was 0°. The design lift of forward operation is 3.32 m, and that of reverse operation is 4.03 m. The tip clearance is 0.15 mm. The number of impeller blades is four. The impeller blade is formed by numerical control machining with thick copper plate, which avoids the defects such as sand holes and pores caused by casting. The hub body was formed by numerical control machining with copper bars. The number of guide leaves is five. The vanes of guide vanes are processed by mould and welded. Figure 1 shows the experimental figure of the hydraulic model of the bidirectional axial flow pump. Figure 2 shows the rotation direction of the impeller of the bidirectional axial flow pump.

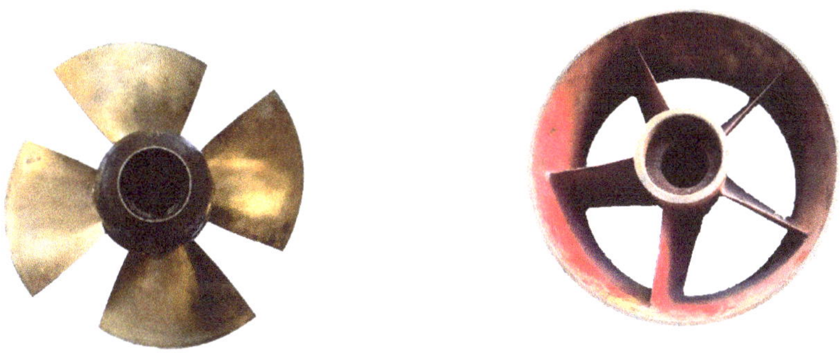

(**a**) Impeller (**b**) Guide vane

Figure 1. Hydraulic model of bidirectional axial flow pump.

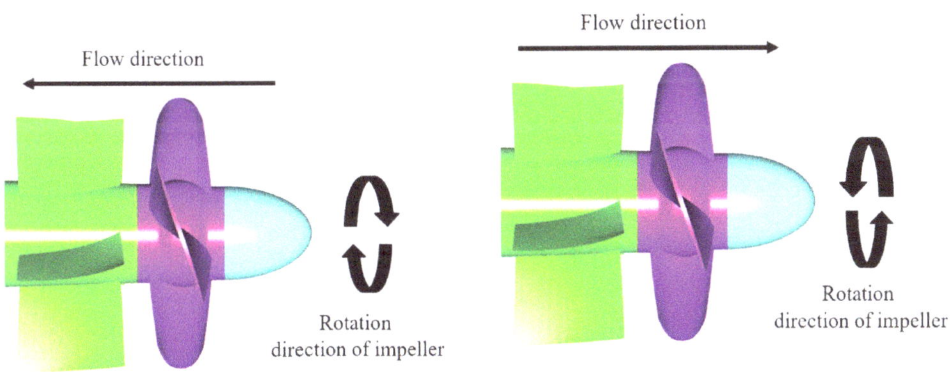

(**a**) Forward operation (**b**) Reverse operation

Figure 2. Rotation direction of impeller of bidirectional axial flow pump.

2.2. Experimental Model

The experimental test of the standard pump section was carried out on the hydraulic model of the bidirectional pump completed by processing. The standard pump section included straight pipe section, impeller, guide vane, and 60° elbow. Figure 3 shows the schematic diagram of the standard pump section in the experiment. Figure 4 shows the schematic diagram of the experimental pump section after installation and splicing.

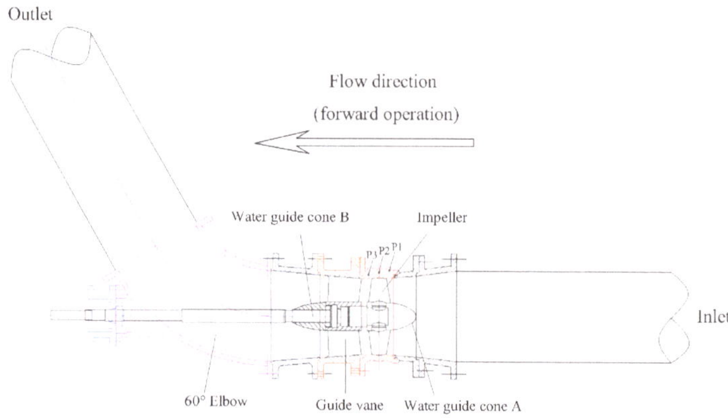

(**a**) Forward operation

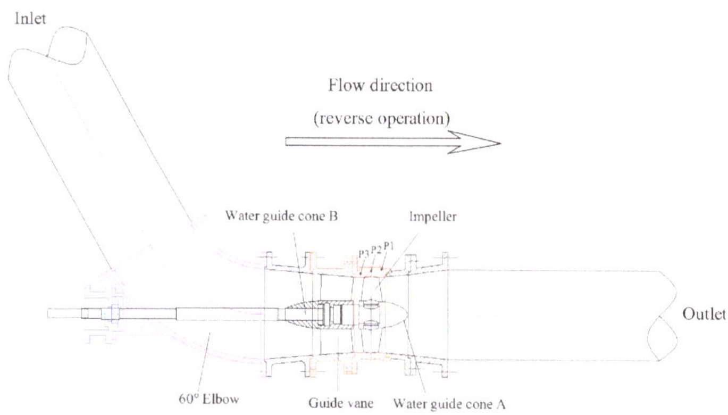

(**b**) Reverse operation

Figure 3. Schematic diagram of standard pump section in experiment.

Figure 4. Schematic diagram of the experimental pump section after installation and splicing.

2.3. Experimental Test System

In this paper, a high-precision hydraulic mechanical test bench is built. The standard pump section equipped with the bidirectional pump hydraulic model was tested on the test bench, including energy characteristics experiment and pressure pulsation characteristics experiment. During the test, strict reference was made to 'water pump model and device model acceptance test procedures' (SL 140-2006) requirements [27–29]. The high-precision hydraulic mechanical test bench is a large vertical sealing system. The experimental system includes forward and reverse operation, control gate valve, auxiliary pump unit, working condition adjustment gate valve, intake tank, pressure outlet tank, stable pressure rectifier cylinder, etc. The schematic diagram of the experimental system is shown in Figure 5. The test accuracy of the test system is 0.39%. The system uncertainty of pump performance test is the square and root of each single system uncertainty. The calculation formula is as follows [30]:

$$(E_\eta)_s = \pm\sqrt{E_Q^2 + E_H^2 + E_M^2 + E_n^2} = \pm 0.274\% \tag{1}$$

where E_Q is the system uncertainty of flow measurement and the calibration result is ±0.2%. E_H is the uncertainty of the static head measurement system, the calibration results of the full range of ±0.10%. E_M is the system uncertainty of torque measurement, the uncertainty of torque speed sensor is ±0.15%. E_n is the system uncertainty of speed measurement. When the sampling period is 2 s and the speed is not less than 1000 r/min, the uncertainty is ±0.05%.

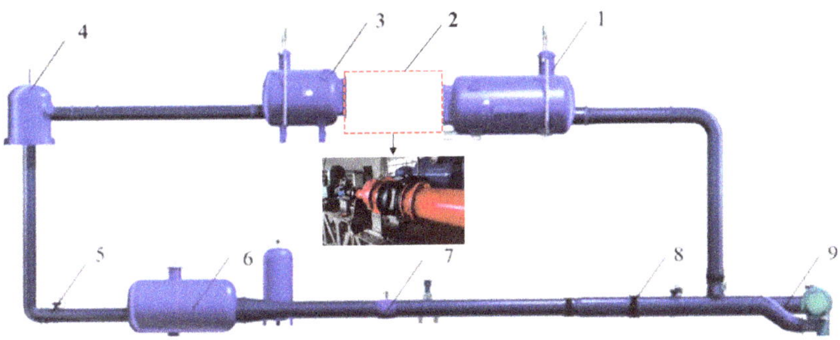

1. Intake tank. 2. Tested pump unit and drive motor. 3. Pressure outlet tank. 4. Bifurcation tank. 5. Condition regulating gate valve. 6. Voltage regulating rectifier. 7. Electromagnetic flowmeter. 8. System forward and reverse operation control gate valve. 9. Auxiliary pump unit.

Figure 5. Physical schematics of the test bench.

The main measuring instruments of the test system include differential pressure transmitter, electromagnetic flowmeter, torque meter, speed torque sensor, absolute pressure transmitter. and so on. Table 1 shows the specific parameters of the main measuring instruments for the energy characteristic test. Table 2 shows the specific parameters of pressure sensor in pressure pulsation test. The specific parameters of the pressure sensor in the pressure pulsation test are shown in Table 2. The sampling frequency of the pressure pulsation sensor is 124 times that of the impeller rotation frequency (RF), which can meet the requirements of the pressure pulsation acquisition in the axial flow pump system [31]. The pressure pulsation experiment uses synchronous channel record, monitoring points p1, p2, and p3 pressure pulsation signal acquisition is synchronous. Figure 6 shows the schematic layout of pressure pulsation measuring points. Three pressure fluctuation monitoring points are located near the impeller inlet, the impeller middle, and near the impeller outlet.

Table 1. Main measuring instruments for energy characteristic experiment.

Measuring Items	Instrument Name	Instrument Types	Instrument Range	Calibration Accuracy
Head	Difference pressure transmitter	EJA 110A	0~200 kPa	±0.1%
Flow	Electromagnetic flowmeter	E-mag type	DN400 mm	±0.20%
Torque	Torsiograph	JW-3	200 Nm	±0.15%
Rotation speed	The speed and torque sensor	JC	0~10,000 r/min	±0.15%

Table 2. The specific parameters of the pressure sensor used for the pressure field measurement in the pump.

Item	Parameter
Model	HM90A
Precision	0.1%
Range	0~200 kPa
Output signal	0~5 V
Sampling frequency	3 kHz

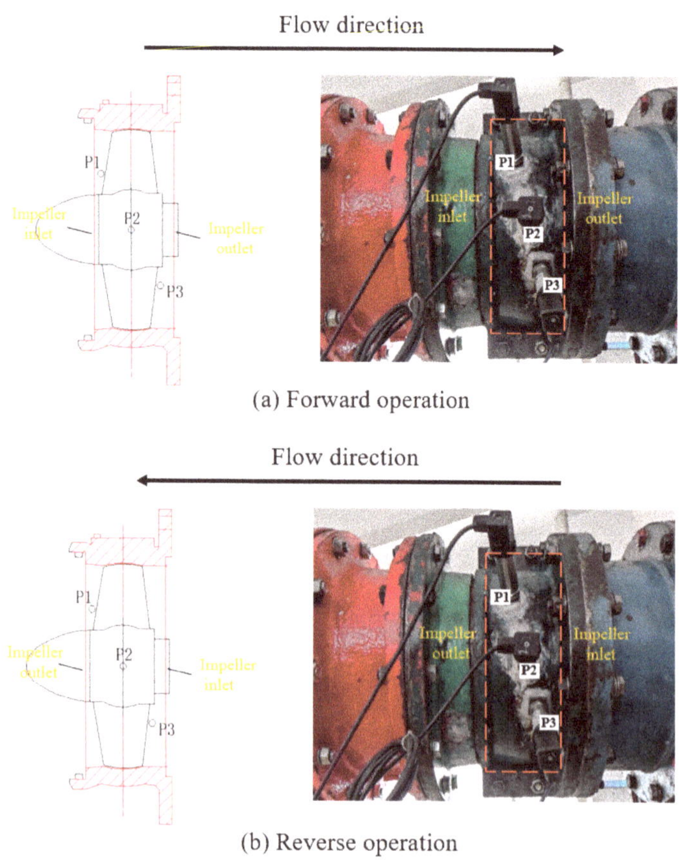

Figure 6. Schematic layout of pressure pulsation measuring points.

3. Experimental Results of Energy Characteristics

The bidirectional axial flow pump was experimentally tested on a high-precision hydraulic mechanical test bench, including forward operation and reverse operation. The test speeds of forward and reverse operation condition were both 1450 r/min. The flow rate of the pump defining the highest efficiency point of the forward operation condition is Q_{bep1}, and the flow rate of the pump defining the highest efficiency point of the reverse operation condition is Q_{bep2}. The experimental energy characteristic curve of bidirectional axial flow pump is shown in Figure 7. The characteristic parameters of the highest efficiency point under forward operation condition and reverse operation condition are shown in Table 3.

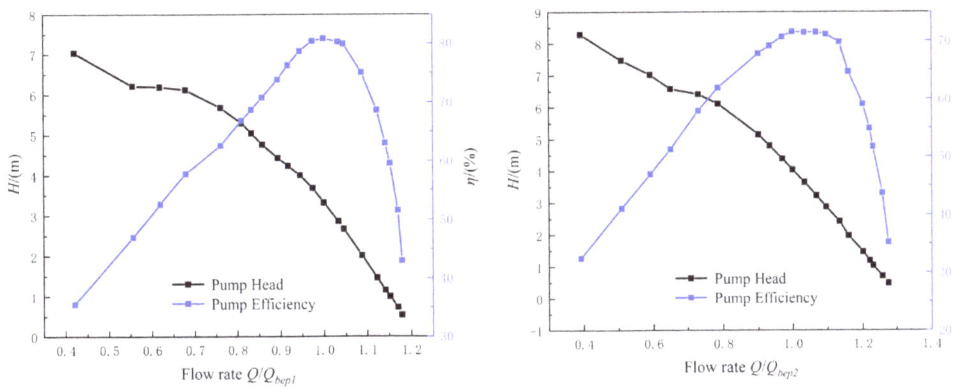

(**a**) Forward operation (**b**) Reverse operation

Figure 7. Experimental energy characteristic curve of bidirectional axial flow pump.

Table 3. Characteristic parameters of highest efficiency points for forward and reverse operation.

Parameters at the Highest Efficiency Point Q_{bep}	Forward Operation Condition	Reverse Operation Condition
Flow Q_d	376.53	308.05
Head H (m)	3.32	4.03
Shaft power N (kW)	15.13	16.98
Efficiency η (%)	80.85	71.55

The variation trend of efficiency and head with flow rate of bidirectional axial flow pump in reverse operation is similar to that in forward operation. The efficiency increases first and then decreases with the increase of flow rate, and the head decreases gradually with the increase of flow rate. Under the forward operation condition, the flow rate at the highest efficiency point of the bidirectional axial flow pump is 376.53 L/s, the head is 3.32 m, the axial power is 15.13 kW, and the efficiency is 80.85%. Under the reverse operation condition, the flow rate at the highest efficiency point of the bidirectional axial flow pump is 308.05 L/s, the head is 4.03 m, the axial power is 16.98 kW, and the efficiency is 71.55%.

In order to more intuitively compare the energy characteristics of forward and reverse operating conditions, the conversion coefficient K is introduced. By comparing the flow, head, efficiency at the highest efficiency point, and the flow range in the high efficiency zone (limited by the drop of 5% at the highest efficiency point) of the forward and reverse operating conditions, the ratio of the flow, head, efficiency, and the flow range in the high-efficiency zone of the reverse operating condition is obtained, namely, the conversion coefficient K. The comparison of energy characteristics between forward and reverse operating conditions based on conversion coefficient is shown in Figure 8. It can be seen from Figure 8 that under the reverse operation condition, the flow and efficiency corresponding to the optimal point of the bidirectional axial flow pump are smaller than those under the forward operation condition. The flow rate is about 81.81% of the forward operation, which is reduced by 18.19%. The efficiency is about 88.50% of the forward operation, which is reduced by 11.50%. Under the reverse operation condition, the range of the high efficiency zone and the corresponding head of the optimum of the bidirectional axial flow pump are greater than those of the forward operation condition. The head is about 121.39% of the forward operation condition, increased by 21.39%. The range of high efficiency zone is about 131.25% of the forward operation condition, increased by 31.25%. From the perspective of energy characteristics, the efficiency of bidirectional axial flow pump in reverse operation is lower than that in forward operation, but the range of high-efficiency area is large. In general, the bidirectional axial flow pump can ensure good hydraulic performance in both forward and reverse operation, and can meet the needs of bidirectional pumping.

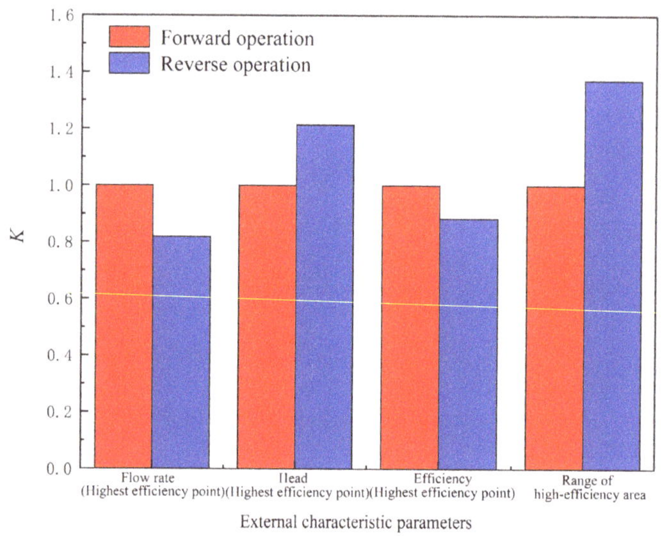

Figure 8. Comparison of energy characteristics between forward and reverse operation conditions based on conversion coefficient.

4. The Experimental Results and Analysis of Pressure Pulsation Characteristics

4.1. Time Domain Analysis of Pressure Fluctuation

In the experiment, the micro pressure pulsation sensor was used to measure the internal pressure fluctuations of the bidirectional axial flow pump under a total of 42 flow conditions, involving 21 forward operating conditions and 21 reverse operating conditions. In order to compare and analyze the pressure fluctuation characteristics of bidirectional axial flow pump under forward and reverse operation conditions, the time domain analysis of pressure fluctuation in pump under different flow conditions is carried out. In the time domain analysis, the pressure fluctuation coefficient is introduced, and the pressure fluctuation in four impeller rotation cycles (one impeller rotation cycle is about 0.041 s) is dimensionlessly processed to eliminate the static pressure interference. The pressure pulsation coefficient formula is as follows [31]:

$$C_p = \frac{p - \overline{p}}{0.5\rho u_2^2} \quad (2)$$

where p is the transient pressure value. $\overline{p}$ is the average pressure value. u_2 is the impeller outlet circumferential velocity.

The experimental data of pressure pulsation at 16 working conditions (including eight working conditions of forward operation and eight working conditions of reverse operation) from $0.4Q_{bep}$ to $1.2Q_{bep}$ are selected for processing. Figure 9 shows the peak-to-peak value of the pressure pulsation coefficient at the monitoring points in the bidirectional axial flow pump under different flow conditions. It can be seen from Figure 9 that no matter whether the bidirectional axial flow pump is in the forward or reverse operation, the maximum value of the peak-to-peak value of the pressure pulsation coefficient in the pump always appears at the inlet of the impeller, and the minimum value of the peak-to-peak value of the pressure pulsation coefficient in the pump appears at the outlet of the impeller in most cases. By comparing Figure 9a,b, it can be found that the variation trend of the peak value of the pressure fluctuation at the inlet of the impeller with the flow rate is, basically, the same under the forward and reverse operation conditions. The variation trend of the peak-to-peak value of the pressure fluctuation at the middle and outlet of the impeller with the flow rate is similar under the small flow condition, and there are some differences near the large flow condition. Compared with the unidirectional axial flow pump, the peak value of the maximum pressure fluctuation at the key position of the bidirectional axial flow pump is slightly offset. According to the experimental results of the pressure pulsation of the unidirectional axial flow pump in the reference [31], the maximum peak-to-peak value of pressure pulsation coefficient of conventional unidirectional axial flow pump appears in the vicinity of $0.6Q_{bep}$–$0.7Q_{bep}$. In this experiment, the maximum peak-to-peak value of pressure pulsation coefficient of bidirectional axial flow pump appears in the vicinity of $0.8Q_{bep}$–$0.9Q_{bep}$ at the key positions under the forward and reverse operation. The maximum peak-to-peak value of the pressure pulsation coefficient at the inlet of the impeller appears near $0.90Q_{bep}$ under the forward and reverse operation conditions, and the maximum peak-to-peak value of the pressure pulsation coefficient at the middle and outlet of the impeller appears near $0.80Q_{bep}$.

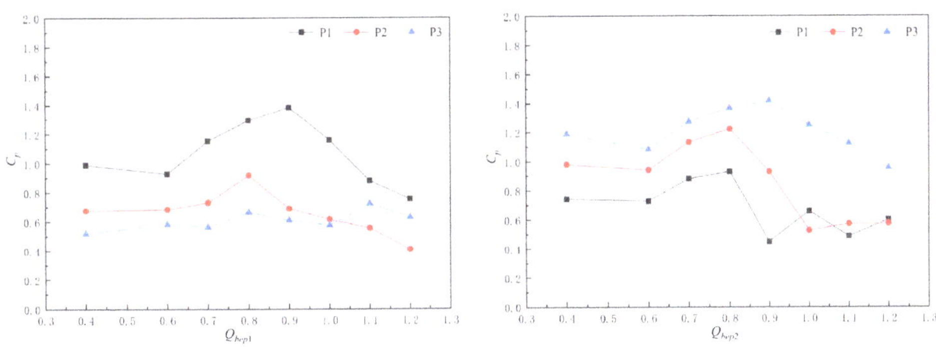

(**a**) Forward operation (**b**) Reverse operation

Figure 9. Peak-to-peak value of the pressure pulsation coefficient in the monitoring points of bidirectional axial flow pump under different flow conditions.

Taking $0.70Q_{bep}$, $1.00Q_{bep}$, and $1.20Q_{bep}$ as typical operating points, the time domain analysis of pressure fluctuation at six operating points under forward and reverse operation conditions was carried out. The time domain diagram of pressure fluctuation coefficient at each monitoring point under forward operation condition is given in Figure 10. Figure 11 shows the pressure fluctuation coefficient time domain diagram of each monitoring point under reverse operation condition. It can be seen from Figures 10 and 11 that under the two-way operation condition, the regularity of the waveform of the pressure pulsation in the two-way axial flow pump is good under the optimal working condition of $1.00Q_{bep}$ and the large flow condition of $1.20Q_{bep}$. There are obviously four main wave peaks and four main wave troughs at each monitoring point in the pump during one impeller rotation cycle. Under 0.70 Q_{bep} small flow conditions, the two-way axial flow pump in the forward and reverse operation of the impeller inlet reflux and tip leakage vortex and other unstable flow, leading to the phenomenon of vortex induced harmonics in the pump. The regularity of pressure fluctuation waveform of each monitoring point in the pump is obviously poor, and more sub-peak amplitude appears in an impeller rotation cycle.

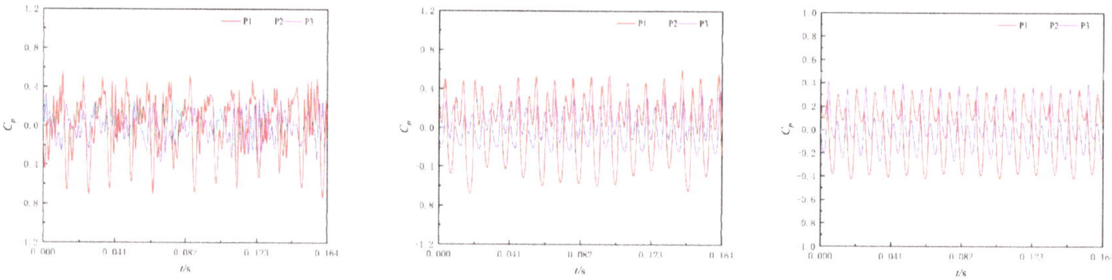

(a) 0.70Q_{bep1} (b) 1.00Q_{bep1} (c) 1.20Q_{bep1}

Figure 10. Time domain diagram of pressure fluctuation coefficient of each monitoring point under forward operation condition.

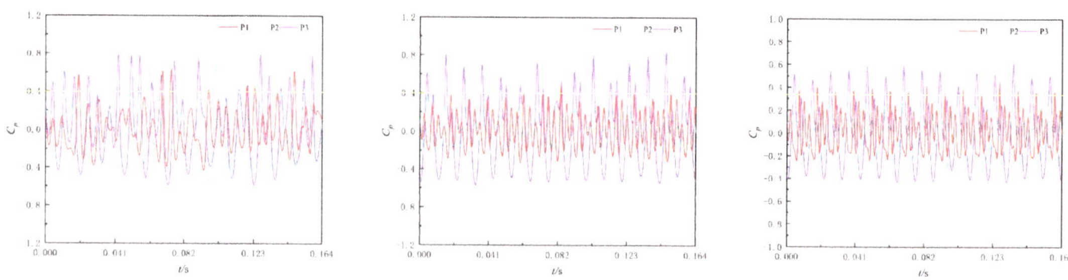

(a) 0.70Q_{bep2} (b) 1.00Q_{bep2} (c) 1.20Q_{bep2}

Figure 11. Time domain diagram of pressure fluctuation coefficient of each monitoring point under reverse operation condition.

4.2. Frequency Domain Analysis of Pressure Fluctuation at the Highest Efficiency Point

In order to analyze the local characteristics of the pressure pulsation signal and accurately identify the frequency component of the pressure pulsation, the pressure pulsation data are processed by fast Fourier transform (FFT). In order to ensure that the Fourier transform has enough high resolution, 3000 sampling points are selected for Fourier transform. The frequency resolution of this experiment is 1 Hz, and the definition of frequency resolution is as follows [31]:

$$\Delta f = \frac{f_s}{M} \tag{3}$$

where f_s is the sampling frequency of the pressure sensor, 3000 Hz. M is the number of sample points selected number, 3000 in total.

The data in the four rotation cycles of the impeller are taken for fast Fourier transform analysis. The X axis of the transverse axis of the spectrum diagram is the rotation frequency multiple, the Y axis is the different monitoring points, and the Z axis of the longitudinal axis is the pressure pulsation amplitude. The frequency conversion multiple formula is as follows [31]:

$$N_F = \frac{f}{f_n} = \frac{60F}{n} \tag{4}$$

where F is the frequency after Fourier transform. n is impeller speed.

Figure 12 shows the pressure pulsation distribution of the bidirectional axial flow pump at the optimal point of $1.00Q_{bep}$. The following conclusions can be drawn from Figure 12. Firstly, the pressure pulsation of the bidirectional axial flow pump at the optimum point of $1.00Q_{bep}$ is dominated by the blade frequency (BPF = 96.67 Hz) or the high-order harmonics of the blade frequency. In forward operation, the main frequency of pressure pulsation at impeller inlet monitoring point P1 is blade frequency, the main frequency of pressure pulsation at impeller middle monitoring point P2 and impeller outlet monitoring point P3 is twice blade frequency (2BPF = 193.33 Hz). In reverse operation, the main frequency of pressure pulsation at the impeller inlet monitoring point P3 and the impeller central monitoring point P2 is blade frequency, and the main frequency of pressure pulsation at the impeller outlet monitoring point P1 is three times blade frequency (3BPF = 290.00 Hz). It can be found that the results of the main frequency components of the pressure pulsation of the bidirectional axial flow pump obtained in this experiment are different from the experimental conclusions obtained in the reference that the pressure pulsation of the key monitoring points of the unidirectional axial flow pump is controlled by the blade frequency [31]. The research shows that when the bidirectional axial flow pump operates under the internal design conditions, in addition to the frequency components caused by the impeller rotation cycle, there are other high-frequency components, such as noise caused by fluid excitation and static and dynamic interference in the pump [32–34].

Secondly, under the optimal condition of $1.00Q_{bep}$, the composition of the pressure pulsation signal in the pump is relatively simple, and the pulsation of each monitoring point in the low-frequency region and high-frequency region is small. The pulsating components in the low-frequency region are mainly concentrated in the axial frequency (SF = 24.17 Hz). The pulsation in high-frequency region is mainly concentrated in the high order harmonics of the blade frequency, but it is almost not observed in the high-frequency region above four times the blade frequency. Thirdly, the pressure pulsation frequency components of the impeller inlet are very close in the forward and reverse operation conditions, and the pressure pulsation frequency components in the middle of the impeller change greatly. The main frequency of the monitoring points in the middle of the impeller changes from two times the blade frequency in the forward operation to the blade frequency in the reverse operation, and the secondary main frequency changes from the blade frequency to two times the blade frequency in the reverse operation.

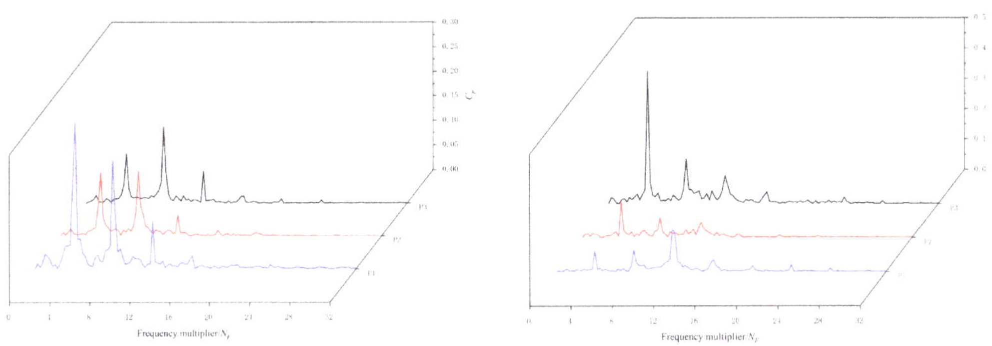

(**a**) Forward operation (**b**) Reverse operation

Figure 12. Pressure pulsation distribution of bidirectional axial flow pump in optimal condition of $1.00Q_{bep}$.

Figure 13 shows the C_p amplitude of the main frequency of the pressure pulsation of the bidirectional axial flow pump under the optimal condition of $1.00Q_{bep}$. The following conclusions can be drawn from Figure 13. Firstly, no matter whether the bidirectional axial flow pump is in the forward or reverse operation, the C_p amplitude of the main frequency of the pressure pulsation in the pump decreases first and then increases from the impeller inlet to the impeller outlet. The C_p amplitudes from the impeller inlet to the outlet are 0.30, 0.13, and 0.16, respectively, under the forward operation condition. Under the reverse operation condition, the C_p amplitudes from the impeller inlet to outlet are 0.44, 0.12 and 0.14, respectively. The maximum value of C_p amplitude in both forward and reverse operation occurs at the inlet of the impeller, which is due to the suction of the impeller in the large range of low-pressure area formed by the suction surface of the blade. There is also a local impact on the flow and the inlet edge of the impeller. The velocity gradient and pressure gradient of the fluid particles are also large, and the final rotating pressure gradient shows a large pressure pulsation. Secondly, no matter whether the bidirectional axial flow pump is in the forward or reverse operation, the maximum C_p amplitude of the main frequency of the pressure pulsation in the pump appears at the inlet of the impeller, and the minimum C_p amplitude appears at the middle of the impeller. The C_p amplitude of the main frequency of the pressure pulsation at the inlet of the impeller is 2.31 times that at the outlet of the impeller under the positive rotation operation condition, and it reaches 3.67 times under the reverse operation condition. Thirdly, compared with the forward operation condition, the C_p amplitude of the main frequency of the pressure pulsation at the inlet of the impeller under the reverse operation condition increases by 46.67%, the central part of the impeller decreases by 7.69%, and the outlet of the impeller decreases by 12.50%.

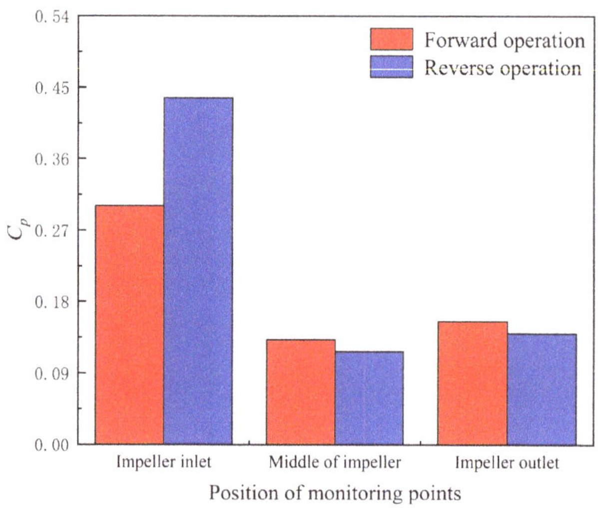

Figure 13. C_p amplitude of main frequency of pressure pulsation of bidirectional axial flow pump under optimal condition of $1.00Q_{bep}$.

4.3. Frequency Domain Analysis of Pressure Pulsation under Small Flow Conditions

Figure 14 shows the pressure pulsation distribution of the bidirectional axial flow pump at a small flow rate of $0.70Q_{bep}$. The following conclusions can be drawn from Figure 14. Firstly, the pressure fluctuation of the bidirectional axial flow pump at a small flow rate of $0.70Q_{bep}$ is dominated by the blade frequency (BPF = 96.67 Hz) or the higher order harmonics of the blade frequency. In the positive operation, the main frequency of pressure pulsation at each monitoring point in the pump is the blade frequency. In reverse operation, the main frequency of pressure fluctuation at impeller inlet monitoring point P3 and impeller middle monitoring point P2 is blade frequency, and the main frequency of pressure fluctuation at impeller outlet monitoring point P1 is twice blade frequency (2BPF = 193.33 Hz). Secondly, the low-frequency signals induced by impact and reflux are very rich in the bidirectional axial flow pump at small flow rate of $0.70Q_{bep}$. The pulsations of different frequency components can be observed in the low-frequency region, whether in the forward or reverse operation. This phenomenon is related to the rotational stall in the pump under small flow conditions, indicating that low-frequency local vortices appear at the inlet and outlet of the impeller.Thirdly, compared with the reverse operation, the frequency band of the pressure fluctuation spectrum is significantly wider in the forward operation, and the pulsation amount in the high-frequency region is significantly increased. A certain amount of pulsation can still be observed at 4BPF~7BPF. The reason may be that under the condition of small flow in forward operation, there is a relatively unstable vortex area in the pump, which causes the water flow in the pump to cause different degrees of impact on the walls of the blades and the flow channel, forming pressure waves of different frequencies.

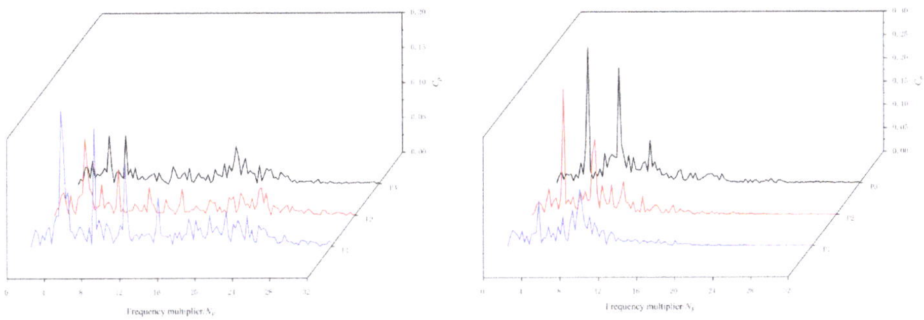

(**a**) Forward operation. (**b**) Reverse operation.

Figure 14. Pressure pulsation distribution of bidirectional axial flow pump in small flow rate of $0.70Q_{bep}$.

Figure 15 shows the C_p amplitude of the main frequency of the pressure pulsation of the bidirectional axial flow pump at a small flow rate of $0.70Q_{bep}$. The following conclusions can be drawn from Figure 15. Firstly, no matter whether the bidirectional axial flow pump is in the forward or reverse operation, the C_p amplitude of the main frequency of the pressure pulsation in the pump decreases gradually from the impeller inlet to the impeller outlet. The C_p amplitudes from the impeller inlet to outlet are 0.20, 0.11, and 0.07, respectively, under the forward operation condition. Under the reverse operation condition, the C_p amplitudes from the impeller inlet to outlet are 0.29, 0.27 and 0.12, respectively. Secondly, no matter whether the bidirectional axial flow pump is in the forward or reverse operation, the maximum value of C_p amplitude of the main frequency of pressure pulsation in the pump appears at the impeller inlet, and the minimum value appears at the impeller outlet. Under the forward operation condition, the C_p amplitude of the main frequency of the pressure pulsation at the impeller inlet is 2.86 times that at the impeller outlet, and the reverse operation condition reached 2.42 times. Third, compared with the forward running condition, the C_p amplitude of the main frequency of pressure pulsation at each monitoring point is larger under the reverse running condition. The C_p amplitude of the main frequency of the pressure fluctuation at the inlet of the impeller increased by 45.00%, the central part of the impeller increased by 145.45%, and the outlet of the impeller increased by 71.43%. It indicates that the energy conversion of water flow in the pump is relatively intense under the condition of reverse operation with small flow rate.

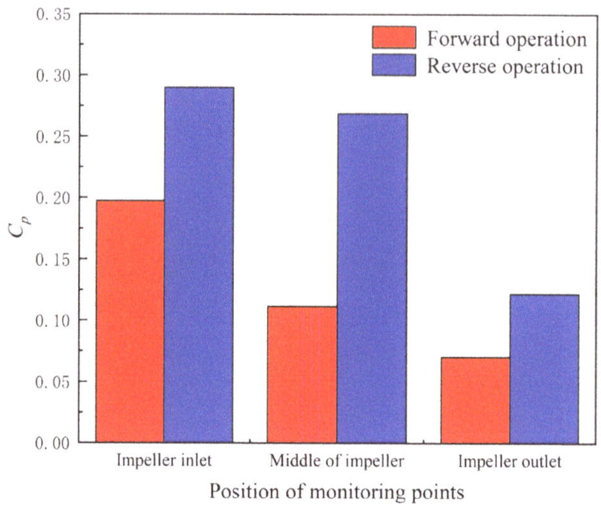

Figure 15. C_p amplitude of main frequency of pressure pulsation of bidirectional axial flow pump at small flow rate of $0.70Q_{bep}$.

4.4. Frequency Domain Analysis of Pressure Pulsation under Large Flow Conditions

Figure 16 shows the pressure pulsation distribution of the bidirectional axial flow pump at a large flow rate of $1.20Q_{bep}$. The following conclusions can be drawn from Figure 16. Firstly, the pressure pulsation of the bidirectional axial flow pump at large flow rate of $1.20Q_{bep}$ is dominated by the high-order harmonic of the blade frequency (BPF = 96.67 Hz) or the blade frequency. In forward operation, the main frequency of pressure fluctuation at impeller inlet monitoring point P1 and impeller middle monitoring point P2 is blade frequency, and the main frequency of pressure fluctuation at impeller outlet monitoring point P3 is twice blade frequency (2BPF = 193.33 Hz). In reverse operation, the main frequency of pressure fluctuation at impeller inlet monitoring point P3 and impeller middle monitoring point P2 is blade frequency, and the main frequency of pressure fluctuation at impeller outlet monitoring point P1 is four times blade frequency (4BPF = 386.68 Hz). Secondly, the pressure fluctuation spectrum of the bidirectional axial flow pump under the condition of large flow rate of $1.20Q_{bep}$ has few frequency components except the frequency doubling of the blade frequency, and the pressure fluctuation signal component tends to be simple. This shows that the flow pattern of the bidirectional axial flow pump is relatively good under the condition of large flow rate, and the unstable flow phenomenon is relatively weak. Thirdly, compared with the forward operation, a certain amount of pulsation can be observed in the low-frequency and high-frequency regions of the pressure pulsation spectrum at the middle and outlet of the impeller under the reverse operation. The occurrence of pulsation in low-frequency region is related to the fact that there is no guide vane recycling ring at the outlet of the impeller under reverse operation. The flow of high velocity near the middle and outlet of the impeller is unstable, resulting in a certain amount of pulsation in the low-frequency region of the pressure pulsation spectrum at the middle and outlet of the impeller. The occurrence of the pulsation in the high-frequency region may be due to the fact that under the condition of large flow rate, the axial flow pump is more likely to induce the early generation of cavitation in the reverse direction, and the occurrence of cavitation pulsation leads to large pulsation in the reverse pressure pulsation spectrum [32–34].

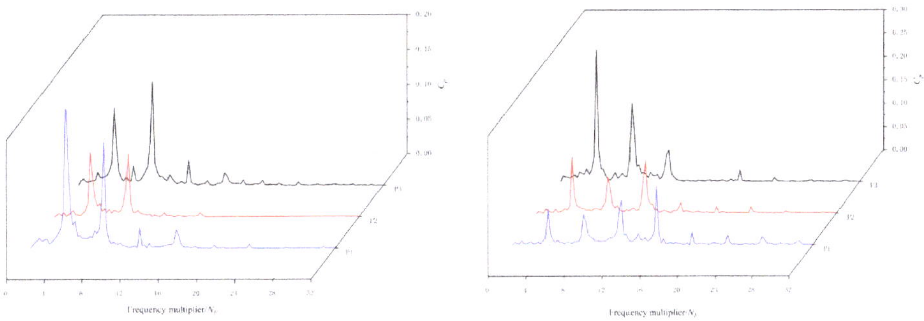

(**a**) Forward operation (**b**) Reverse operation

Figure 16. Pressure pulsation distribution of bidirectional axial flow pump in large flow rate of $1.20Q_{bep}$.

Figure 17 shows the C_p amplitude of the main frequency of the pressure pulsation of the bidirectional axial flow pump at a large flow rate of $1.20Q_{bep}$. The following conclusions can be drawn from Figure 17. Firstly, no matter whether the two-way axial flow pump is running in the forward or reverse direction, the change rule of the Cp amplitude of the main frequency of the pressure pulsation in the pump is the same, which first decreases and then increases from the impeller inlet to the impeller outlet. The C_p amplitudes from the impeller inlet to the outlet are 0.22, 0.09, and 0.15, respectively, under the forward operation condition. Under the reverse operation condition, the C_p amplitudes from the impeller inlet to outlet are 0.28, 0.12, and 0.13, respectively. Secondly, no matter whether the bidirectional axial flow pump is in the forward or reverse operation, the maximum value of C_p amplitude of the main frequency of pressure pulsation in the pump appears at the inlet of the impeller, and the minimum value appears at the middle of the impeller. The C_p amplitude of the main frequency of the pressure pulsation at the inlet of the impeller under the forward operation condition is 1.47 times that at the outlet of the impeller, and the reverse operation condition reaches 2.15 times. Thirdly, compared with the forward operation condition, the C_p amplitude of the main frequency of the pressure pulsation at the inlet of the impeller under the reverse operation condition increases by 27.27%, the central part of the impeller increases by 33.33%, and the outlet of the impeller decreases by 13.33%.

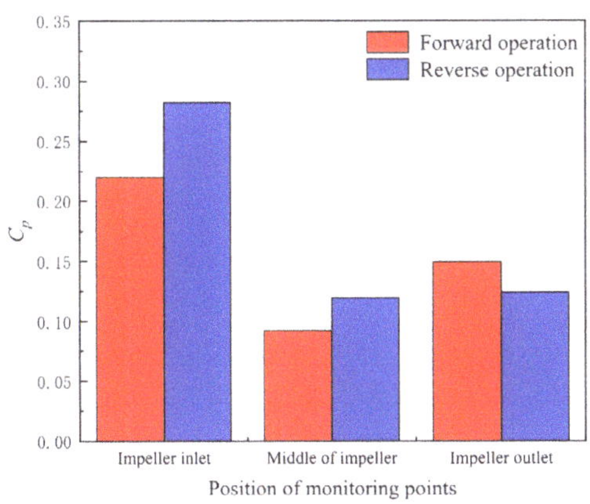

Figure 17. C_p amplitude of main frequency of pressure pulsation of bidirectional axial flow pump at large flow rate of $1.20Q_{bep}$.

4.5. Frequency Domain Analysis of Pressure Pulsation under Different Flow Conditions

Figure 18 shows the pressure pulsation distribution of monitoring points P1 and P3 in the pump under different operating conditions. The following conclusions can be drawn from Figure 18. Firstly, no matter whether the two-way axial flow pump is running in the forward or reverse direction, the main frequency of pressure pulsation at impeller inlet is always dominated by the blade frequency (BPF = 96.67 Hz). It shows that the rotation of the impeller is the main cause of the pressure pulsation at the inlet of the bidirectional pump impeller. Secondly, when running in the forward direction, the main frequency of the pressure pulsation at the impeller outlet under different flow conditions is always dominated by twice the impeller frequency (2BPF = 193.33 Hz). During reverse operation, with the increase of flow rate, the main frequency of pressure pulsation at the impeller outlet has a tendency to gradually move to the high-frequency region, from two times the impeller frequency under the $0.70Q_{bep2}$ flow condition, to the $1.00Q_{bep2}$ flow condition three times the blade frequency (3BPF = 290.00 Hz) at the time, and then transfer to four times the blade frequency (4BPF = 386.68 Hz) when the flow rate is $1.20Q_{bep2}$. Thirdly, when running in forward rotation, as the flow rate increases, the significant pulsation in the high-frequency region of the impeller inlet and outlet gradually disappears. In reverse operation, the pressure pulsation at the impeller outlet is sensitive to the change of flow rate. With the increase of flow rate, the pulsation in the high-frequency region increases. It shows that with the increase of flow rate, the unstable flow phenomenon in the pump gradually disappears in the forward operation, and the unstable flow such as cavitation in the pump tends to be severe under the condition of large flow rate in the reverse operation [35–37].

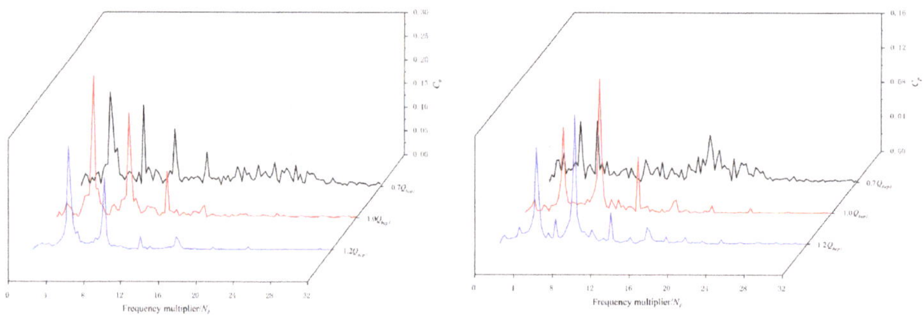

P1 (Impeller inlet) P3 (impeller outlet)

(**a**) Forward operation

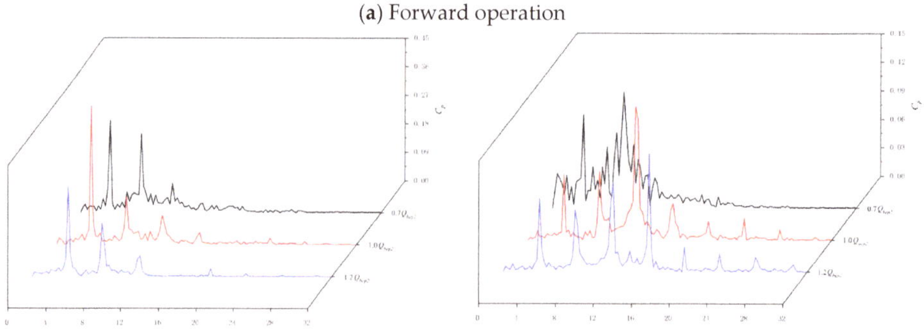

P3 (Impeller inlet) P1 (impeller outlet)

(**b**) Reverse operation

Figure 18. Pressure pulsation distribution of monitoring points P1 and P3 in pump under different operating conditions.

Figure 19 shows the main frequency C_p amplitude of the impeller inlet and outlet of the bidirectional axial flow pump under different operating conditions. The following conclusions can be drawn from Figure 19. Firstly, the C_p amplitude of the main frequency of the impeller inlet in the reverse operation condition is greater than that in the forward operation condition under different flow conditions, and the percentage of the difference between the two changes with the flow rate is relatively small, which is always maintained at about 30–45%. Secondly, the C_p amplitude of the main frequency at the outlet of the impeller is very sensitive to the change of the flow rate at the forward operation. As the flow rate increases from $0.70Q_{bep1}$ to $1.20Q_{bep1}$, the C_p amplitude of the main frequency at the outlet of the impeller increases from 0.29 times of 1.15 times of the reverse operation condition. Thirdly, the main frequency C_p amplitude of the impeller inlet and outlet in the forward and reverse operation conditions of the bidirectional axial flow pump has the same trend with the flow rate, which increases first and then decreases with the increase of the flow rate. In forward operation, compared with the optimal $1.00Q_{bep1}$ condition, the C_p amplitude of the impeller inlet main frequency under the condition of small flow $0.70Q_{bep1}$ decreases by 34.14% and the outlet decreases by 54.68%. Under the condition of large flow $1.20Q_{bep1}$, the C_p amplitude of the impeller inlet main frequency decreased by 26.70% and the outlet decreased by 3.78%. In reverse operation, compared with the optimal $1.00Q_{bep2}$ condition, the C_p amplitude of the impeller inlet main frequency under the condition of small flow $0.70Q_{bep2}$ decreases by 33.63% and the outlet decreases by 13.09%. Under the condition of large flow $1.20Q_{bep2}$, the C_p amplitude of the impeller inlet main frequency decreased by 35.36%, and the outlet decreased by 11.32%.

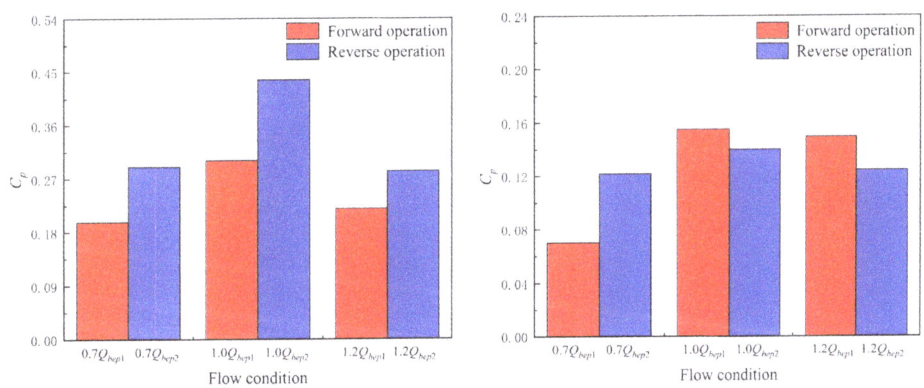

(**a**) Impeller inlet (**b**) Impeller outlet

Figure 19. C_p amplitude of main frequency at impeller inlet and impeller outlet of bidirectional axial flow pump under different operating conditions.

5. Conclusions

In order to explore the hydrodynamic characteristics of bidirectional axial flow pump, this paper carried out experiments on a bidirectional axial flow pump on a high-precision hydraulic mechanical test bench, including positive and negative directions. In the experiment, a micro pressure pulsation sensor was used to measure the pressure fluctuation in the pump under a total of 42 flow conditions, involving 21 forward operation conditions and 21 reverse operation conditions. Based on the experimental results, the hydrodynamic characteristics of the two-way axial flow pump under forward and reverse operation, especially the pressure pulsation characteristics in the pump, are comprehensively analyzed and compared. The main conclusions are as follows:

(1) Compared with the forward operation condition, the flow rate and efficiency corresponding to the optimal point under the reverse operation condition of bidirectional axial

flow pump are smaller than those under the forward operation condition, and the flow rate is reduced by 18.19%. The efficiency decreased by 11.50%. However, the range of head and high efficiency zone is larger than that of forward operation. The head increased by 21.39% and the scope of the efficient zone increased by 31.25%. From the point of view of energy characteristics, the hydraulic performance of bidirectional axial flow pump is relatively balanced in forward and reverse operation, which can meet the needs of bidirectional pumping.

(2) Under the optimal condition of $1.00Q_{bep}$, the composition of the pressure pulsation signal in the pump is simple, and the pressure pulsation signal in the pump is mainly controlled by the blade frequency (BPF = 96.67 Hz), two times the blade frequency (2BPF = 193.33 Hz) or three times the blade frequency (3BPF = 290.00 Hz). This is different from the previous conclusion that the pressure pulsation of the key monitoring points of the unidirectional axial flow pump is controlled by the blade frequency. Compared with the optimal $1.00Q_{bep1}$ working condition of forward operation, the Cp amplitude of the main frequency of the pressure pulsation at the inlet of the impeller under the optimal $1.00Q_{bep2}$ working condition of reverse operation increases by 46.67%, the central part of the impeller decreases by 7.69%, and the outlet of the impeller decreases by 12.50%.

(3) In forward operation, with the increase of flow rate, the obvious pulsation in the high-frequency region of the impeller inlet and outlet gradually disappears. In reverse operation, the pressure pulsation at the impeller outlet is sensitive to the change of flow rate. With the increase of flow rate, a certain degree of pulsation gradually appears in the high-frequency region. It shows that with the increase of flow rate, the unstable flow phenomenon in the pump gradually disappears in the forward operation, and the energy conversion and flow pattern change at the impeller outlet tend to be intense in the reverse operation.

(4) The main frequency C_p amplitude of the impeller inlet and outlet in the forward and reverse operation conditions of the bidirectional axial flow pump has the same trend with the flow rate, which increases first and then decreases with the increase of the flow rate. In forward operation, compared with the optimal $1.00Q_{bep1}$ condition, the C_p amplitude of the impeller inlet main frequency under the condition of small flow $0.70Q_{bep1}$ decreases by 34.14% and the outlet decreases by 54.68%. Under the condition of large flow $1.20Q_{bep1}$, the C_p amplitude of the impeller inlet main frequency decreased by 26.70% and the outlet decreased by 3.78%. In reverse operation, compared with the optimal $1.00Q_{bep2}$ condition, the C_p amplitude of the impeller inlet main frequency under the condition of small flow $0.70Q_{bep2}$ decreases by 33.63% and the outlet decreases by 13.09%. Under the condition of large flow $1.20Q_{bep2}$, the C_p amplitude of the impeller inlet main frequency decreased by 35.36%, and the outlet decreased by 11.32%.

The current work is mainly to reveal the hydrodynamic characteristics of the bidirectional axial flow pump through experimental methods, especially the pressure pulsation characteristics in the pump. By comparing the forward operation condition with the reverse operation condition, the safety and stability of the bidirectional axial flow pump system in the bidirectional utilization are evaluated. The research results can provide important reference for safe and stable operation of bidirectional axial flow pump station system in bidirectional operation. However, how to eliminate or improve the pressure pulsation in the bidirectional axial flow pump has not been well solved. In the further study, more physical analysis of bidirectional axial flow pump should be carried out based on CFD method to reveal the damage mechanism of pressure pulsation on bidirectional axial flow pump station under bidirectional operation conditions.

Author Contributions: Formal analysis and writing original draft preparation, X.Z.; validation and writing review and editing the paper, F.T. and C.H.; data curation, Y.C. (Yujun Chen) and L.S.; paper translation, Y.C. (Yueting Chen) and L.W. All authors have read and agreed to the published version of the manuscript.

Funding: This research work was supported by the National Natural Science Foundation of China (funder: National Natural Science Foundation of China. funding number: 51376155), the Natural Science Foundation of Jiangsu Province (funder: Jiangsu Provincial Department of Science and

Technology. funding number: BK20190914), the China Postdoctoral Science Foundation (funder: China Postdoctoral Science Foundation. funding number: 2019M661946), and the University Science Research Project of Jiangsu Province (funder: People's Government of Jiangsu Province. funding number: 19KJB570002), the Water Conservancy Science and Technology Projects of Jiangsu Province (funder: Department of Water Resources of Jiangsu Province. funding number: 2021012).

Data Availability Statement: Not applicable.

Acknowledgments: A project funded by the Priority Academic Program Development (PAPD) of Jiangsu Higher Education Institutions Support for construction and assembly of the facility was also provided by the Hydrodynamic Engineering Laboratory of Jiangsu Province.

Conflicts of Interest: The authors declare no conflict of interest.

Nomenclature

C_p	Pressure pulsation coefficient
f	Frequency (Hz)
F	Frequency after Fourier transform (Hz)
g	Local acceleration of gravity (m/s^2)
H_{exp}	Experimental head (m)
L	Hydraulic loss coefficient
K	Conversion coefficient
M	The impeller torque (N·m)
n	Rated speed (r/min)
N_F	Multiple of rotational frequency
P	Transient pressure (Pa)
$\overline{p}$	Average pressure (Pa)
Q	Flow rate of the model pump system (m^3/s)
Q_{bep}	Optimal condition
t	Time (s)
u_2	Circumferential velocity of impeller outlet (m/s)
ρ	The density of flow (kg/m^3)
ω	The angular velocity of the impeller (rad/s)
η	Efficiency (%)
η_{exp}	Experimental Efficiency (%)
Abbreviations	
CFD	Computational fluid dynamics
SF	Shaft frequency
SST	Shear–stress transport
BPF	Impeller rotation frequency

References

1. Kang, C.; Yu, X.; Gong, W.; Li, C.; Huang, Q. Influence of stator vane number on performance of the axial-flow pump. *J. Mech. Sci. Technol.* **2015**, *29*, 2025–2034. [CrossRef]
2. Feng, J.; Luo, X.; Guo, P.; Wu, G. Influence of tip clearance on pressure fluctuations in an axial flow pump. *J. Mech. Sci. Technol.* **2016**, *30*, 1603–1610. [CrossRef]
3. Kan, K.; Zheng, Y.; Chen, Y.; Xie, Z.; Yang, G.; Yang, C. Numerical study on the internal flow characteristics of an axial-flow pump under stall conditions. *J. Mech. Sci. Technol.* **2018**, *32*, 4683–4695. [CrossRef]
4. Shi, L.; Zhang, W.; Jiao, H.; Tang, F.; Wang, L.; Sun, D.; Shi, W. Numerical simulation and experimental study on the comparison of the hydraulic characteristics of an axial-flow pump and a full tubular pump. *Renew. Energy* **2020**, *153*, 1455–1464. [CrossRef]
5. Shi, L.; Tang, F.; Xie, R.; Zhang, W. Numerical and experimental investigation of tank-type axial-flow pump device. *Adv. Mech. Eng.* **2017**, *9*, 2071938656. [CrossRef]
6. Zhu, H.; Bo, G.; Zhou, Y.; Zhang, R.; Cheng, J. Pump Selection and Performance Prediction for the Technical Innovation of an Axial-Flow Pump Station. *Math. Probl. Eng.* **2018**, *2018*, 1–9. [CrossRef]
7. Zhang, X.; Tang, F.; Liu, C.; Shi, L.; Liu, H.; Sun, Z.; Hu, W. Numerical Simulation of Transient Characteristics of Start-Up Transition Process of Large Vertical Siphon Axial Flow Pump Station. *Front. Energy Res.* **2021**, *9*, 382. [CrossRef]
8. Shi, L.; Zhu, J.; Yuan, Y.; Tang, F.; Huang, P.; Zhang, W.; Liu, H.; Zhang, X. Numerical Simulation and Experiment of the Effects of Blade Angle Deviation on the Hydraulic Characteristics and Pressure Pulsation of an Axial-Flow Pump. *Shock Vib.* **2021**, *2021*, 1–14. [CrossRef]

9. Ma, P.; Wang, J.; Wang, H. Investigation of performances and flow characteristics of two bi-directional pumps with different airfoil blades. *Sci. China Technol. Sci.* **2018**, *61*, 1588–1599. [CrossRef]
10. Pei, J.; Meng, F.; Li, Y.; Yuan, S.; Chen, J. Effects of distance between impeller and guide vane on losses in a low head pump by entropy production analysis. *Adv. Mech. Eng.* **2016**, *8*, 2071837044. [CrossRef]
11. Meng, F.; Yuan, S.; Li, Y. Fluid—Structure coupling analysis of impeller in unstable region for a reversible axial-flow pump device. *Adv. Mech. Eng.* **2018**, *10*, 2071944264. [CrossRef]
12. Ma, P.; Wang, J. An analysis on the flow characteristics of bi-directional axial-flow pump under reverse operation. *Proc. Inst. Mech. Eng. Part A J. Power Energy* **2017**, *231*, 239–249. [CrossRef]
13. Meng, F.; Li, Y.; Yuan, S.; Wang, W.; Zheng, Y.; Osman, M.K. Multiobjective Combination Optimization of an Impeller and Diffuser in a Reversible Axial-Flow Pump Based on a Two-Layer Artificial Neural Network. *Processes* **2020**, *8*, 309. [CrossRef]
14. Kang, C.; Mao, N.; Pan, C.; Zhou, Y. Turbulent Flow Characteristics in an Axial-flow Pump at Direct and Reverse Modes. *J. Appl. Sci. Eng.* **2016**, *19*, 447–458. [CrossRef]
15. Widmer, C.; Staubli, T.; Ledergerber, N. Unstable Characteristics and Rotating Stall in Turbine Brake Operation of Pump-Turbines. *J. Fluids Eng.* **2011**, 133. [CrossRef]
16. Bai, L.; Zhou, L.; Han, C.; Zhu, Y.; Shi, W. Numerical Study of Pressure Fluctuation and Unsteady Flow in a Centrifugal Pump. *Processes* **2019**, *7*, 354. [CrossRef]
17. Liu, H.; Xia, R.; Wang, K.; Jing, Y.; He, X. Experimental Analysis on Pressure Fluctuation Characteristics of a Centrifugal Pump with Vaned-Diffuser. *Water* **2020**, *12*, 126. [CrossRef]
18. Wang, W.; Chen, Q.; Yan, D. Hydraulic Stability Analysis of a Large Prototype Francis Turbine Based on Field Test. *J. Fluids Eng.* **2018**, *140*, 114501. [CrossRef]
19. Ran, H.; Luo, X.; Zhu, L.; Zhang, Y.; Wang, X.; Xu, H. Experimental study of the pressure fluctuations in a pump turbine at large partial flow conditions. *Chin. J. Mech. Eng.* **2012**, *25*, 1205–1209. [CrossRef]
20. Arndt, N.; Acosta, A.J.; Brennen, C.E.; Caughey, T.K. Experimental Investigation of Rotor-Stator Interaction in a Centrifugal Pump with Several Vaned Diffusers. *J. Turbomach.* **1990**, *112*, 98–108. [CrossRef]
21. GUO, S.; MARUTA, Y. Experimental Investigations on Pressure Fluctuations and Vibration of the Impeller in a Centrifugal Pump with Vaned Diffusers. *JSME Int. J. Ser. B* **2005**, *48*, 136–143. [CrossRef]
22. Ma, P.; Wang, J.; Li, H. Numerical Analysis of Pressure Pulsation for a Bidirectional Pump under Positive and Reverse Operation. *Adv. Mech. Eng.* **2014**, *6*, 730280. [CrossRef]
23. SHI, W. Numerical and Experimental Investigation of High-efficiency Axial-flow Pump. *Chin. J. Mech. Eng.* **2010**, *23*, 38. [CrossRef]
24. Zhang, L.; Zhang, J.; Yu, X.; Lv, J.; Zhang, X. Transient Simulation for a Pumped Storage Power Plant Considering Pressure Pulsation Based on Field Test. *Energies* **2019**, *12*, 2498. [CrossRef]
25. Xiao, Y.; Wang, Z.; Zhang, J.; Luo, Y. Numerical predictions of pressure pulses in a Francis pump turbine with misaligned guide vanes. *J. Hydrodyn.* **2014**, *26*, 250–256. [CrossRef]
26. Ma, P.F. *Study on Optimal Design and Internal Flow Characteristics of Bidirectional Axial Flow Pump*; Huazhong University of Science and Technology: Wuhan, China, 2016.
27. Xu, Y.; Tan, L.; Liu, Y.; Hao, Y.; Zhu, B.; Cao, S. Pressure fluctuation and flow pattern of a mixed-flow pump under design and off-design conditions. *Proc. Inst. Mech. Eng. Part C J. Mech. Eng. Sci.* **2018**, *232*, 2430–2440. [CrossRef]
28. Miyabe, M.; Maeda, H.; Umeki, I.; Jittani, Y. Unstable head-flow characteristic generation mechanism of a low specific speed mixed flow pump. *J. Therm. Sci.* **2006**, *15*, 115–120. [CrossRef]
29. Zhang, D.; Shi, W.; Chen, B.; Guan, X. Unsteady Flow Analysis and Experimental Investigation of Axial-Flow Pump. *J. Hydrodyn.* **2010**, *22*, 35–43. [CrossRef]
30. Wang, G.; Wang, Z.; Song, L. Uncertainty analysis for different virtual pump water flow meters. *Sci. Technol. Built Environ.* **2019**, *25*, 297–308. [CrossRef]
31. Zhang, D.S.; Wang, H.Y.; Shi, W.D.; Pan, D.Z.; Shao, P.P. Experimental investigation of pressure fluctuation with multiple flow rates in scaled axial flow pump. *Trans. Chin. Soc. Agric. Mach.* **2014**, *45*, 139–145.
32. Zhu, G.J.; Li, K.; Feng, J.J.; Luo, X.Q. Effects of cavitation on pressure fluctuation of draft tube and runner vibration in a Kaplan turbine. *Trans. Chin. Soc. Agric. Eng.* **2021**, *37*, 40–49.
33. Li, W.; Ji, L.L.; Shi, W.D.; Zhou, L.; Ping, Y.F. Experiment on pressure fluctuation in mixed-flow pump under different flow rate conditions. *Trans. Chin. Soc. Agric. Mach.* **2016**, *47*, 70–76.
34. Zhang, D.S.; Wang, C.C.; Dong, Y.G.; Shi, L.; Jin, Y.X. Tests for inner pressure fluctuation features in an oblique flow pump with high ratio rotating speed. *J. Vib. Shock* **2019**, *38*, 27–34.
35. Barrio, R.; Blanco, E.; Parrondo, J.; González, J.; Fernández, J. The effect of impeller cutback on the fluid-dynamic pulsations and load at the blade-passing frequency in a centrifugal pump. *J. Fluids Eng.* **2008**, *130*, 111102. [CrossRef]
36. Parrondo-Gayo, J.L.; Gonzalez-Perez, J.; Fernandez-Francos, J. The effect of the operating point on the pressure fluctuations at the blade passage frequency in the volute of a centrifugal pump. *Trans. ASME* **2002**, *124*, 784–790. [CrossRef]
37. Shi, L.; Yuan, Y.; Jiao, H.; Tang, F.; Cheng, L.; Yang, F.; Jin, Y.; Zhu, J. Numerical investigation and experiment on pressure pulsation characteristics in a full tubular pump. *Renew. Energ.* **2021**, *163*, 987–1000. [CrossRef]

machines

Article

Design and Optimization of High-Pressure Water Jet for Coal Breaking and Punching Nozzle Considering Structural Parameter Interaction

Lihuan Chen [1,2], Muzheng Cheng [2], Yi Cai [2], Liwen Guo [3] and Dianrong Gao [1,*]

1 School of Mechanical Engineering, Yanshan University, Qinhuangdao 066004, China; chenlihuan@nciae.edu.cn
2 School of Mechanical and Electrical Engineering, North China Institute of Aerospace Engineering, Langfang 065000, China; cmuzheng@nciae.edu.cn (M.C.); cayi@nciae.edu.cn (Y.C.)
3 College of Mining Engineering, North China University of Science and Technology, Tangshan 063210, China; guoliwen@ncst.edu.cn
* Correspondence: gaodr@ysu.edu.cn

Abstract: The technology of increasing coal seam permeability by high-pressure water jet has significant advantages in preventing and controlling gas disasters in low-permeability coal seam. The structural parameters of a nozzle are the key to its jet performance. The majority of the current studies take strike velocity as the evaluation index, and the influence of the interaction between the nozzle's structural parameters on its jet performance is not fully considered. In practice, strike velocity and strike area will affect gas release in the process of coal breaking and punching. To further optimize the structural parameters of coal breaking and punching nozzle, and improve water jet performance, some crucial parameters such as the contraction angle, outlet divergence angle, and length-to-diameter ratio are selected. Meanwhile, the maximum X-axis velocity and effective Y-axis extension distance are used as evaluation indexes. The effect of each key factor on the water jet performance is analyzed by numerical simulation using the single factor method. The significance and importance effect of each factor and their interaction on the water jet performance are quantitatively analyzed using the orthogonal experiment method. Moreover, three optimal combinations are selected for experimental verification. Results show that with an increase in contraction angle, outlet divergence angle, and length-to-diameter ratio, the maximum X-axis velocity increases initially and decreases thereafter. The Y-direction expansion distance of the jet will be improved significantly with an increase in the outlet divergence angle. Through field experiments, the jet performance of the improved nozzle 3 is the best. After optimization, the coal breaking and punching diameter of the nozzle is increased by 118%, and the punching depth is increased by 17.46%.

Keywords: nozzle; high-pressure water jet; structural parameter; numerical simulation; interaction; structure optimization

Citation: Chen, L.; Cheng, M.; Cai, Y.; Guo, L.; Gao, D. Design and Optimization of High-Pressure Water Jet for Coal Breaking and Punching Nozzle Considering Structural Parameter Interaction. *Machines* **2022**, *10*, 60. https://doi.org/10.3390/machines10010060

Academic Editor: Davide Astolfi

Received: 15 December 2021
Accepted: 12 January 2022
Published: 14 January 2022

1. Introduction

An increase in coal mining depth results in an increase in in-situ stress and gas pressure, thereby resulting in gas accumulation in local areas. Consequently, the threat of gas disasters has increased substantially, and gas prevention and control have become increasingly difficult. Gas extraction is one of the main technical measures to prevent and control gas disasters. At present, the common methods of increasing coal seam permeability include deep-hole blasting, rotary hydraulic hole drilling, and dense long-hole drilling, which play a positive role in preventing gas disasters in low-permeability coal seam [1–3]. However, these methods are limited by a variety of conditions, including such problems as narrow scope of application and high cost. Accordingly, high-pressure water jet technology to punch coal has emerged, which has the advantages of high safety factor, low energy consumption, and low cost compared with other pressure relief anti-reflective technology [4–6]. High-pressure water jet technology has been successfully applied in Pingdingshan, Kailuan, Fengfeng, and other mining areas of low-permeability coal seam, thereby accumulating considerable field application.

As the executing element of a high-pressure water jet system, the nozzle converts the pressure energy of the water into the kinetic energy of the water jet, and shoots out in the form of a high-speed water jet to break and punch the coal; therefore, the nozzle is one of the core components of the entire system. In recent years, local and international scholars have conducted substantial research on the structural parameters and working conditions of coal breaking and punching nozzle that affect water jet performance [7–10]. Wen et al. [11] conducted a theoretical analysis and numerical simulation of computational fluid dynamics (CFD) to study the water jet performance of nozzles, particularly cone-straight nozzles, with different flow channel shapes. They performed an orthogonal experiment to determine the nozzle structure parameters when the water jet performance was optimal. Li et al. [12] used the finite element method (FEM) and smooth particle–fluid dynamics (SPH) to investigate the rock-breaking performance of a self-excited oscillating pulsed waterjet (SOPW). They likewise determined the mechanism of crack formation, and propagation and formation of the fracture zone, and analyzed the effects of pulse amplitude, pulse frequency, and pulse circumference on the rock-breaking ability. Chen et al. [13] used the CFD numerical simulation method to calculate a fully developed submerged impingement water jet, and studied the flow field structure and velocity distribution of the jet under different impact angles. Mohammad, Huang et al. [14,15] conducted a numerical simulation and experiment to study the influence of the nozzle exit shape on jet velocity, cluster property, and central impact force. Liu et al. [16] conducted an experiment on SC-CO_2 jet coal breaking based on conical convergent and Laval nozzles, and concluded that the latter had a higher energy conversion rate and a stronger coal breaking effect than the former. They also determined that expansion ratio was the key factor affecting the jet performance of the Laval nozzle. Zhang et al. [17] studied the characteristics of complex unsteady flow when pulsed jet impinges on the rotating wall using the W-A turbulence model. They concluded that when the water hammer effect occurs, the impact pressure of a pulsed jet on the impacting wall is greater than that of a continuous jet in a certain period. Yang et al. [18] numerically analyzed a square centroid nozzle and found an optimal shrinkage degree at the nozzle outlet, thereby leading to a strong cavitation phenomenon in the jet and improving the performance of the cavitation jet. Hong et al. [19] proposed a method of using nitrogen gas as an abrasive jet medium for coalbed methane mining. Their experiments indicated that the nozzle diameter and target distance are the key factors in improving the performance of nitrogen jet coal breaking. Qiang et al. [20] used the DPM model to study particle movement trajectory in an abrasive water jet, and concluded that a high particle inlet position and a large convergence angle of the focus tube can improve jet performance and prolong the service life of the nozzle. Wang et al. [21] utilized the W-A turbulent model as a basis in analyzing the effects of water jets with different impinging heights on the flow field characteristics, and impinging pressure of submerged impinging. They concluded that the maximum velocity of the axis decreases rapidly with the increase in impact height, but the impact height has a minimal effect on the velocity of the wall jet zone. Peng et al. [22] analyzed the internal flow characteristics of the slurry pump under the conditions of clear water and solid–liquid two-phase by the Euler–Euler multiphase flow model. Ekiciler et al. [23] studied the influence of surface shape on impinging jet performance at different Reynolds numbers, and concluded that for small Reynolds numbers, a sinusoidal corrugated surface has a higher performance evaluation criterion (PEC). Liu et al. [24] investigated the effects of cutting depth and water pressure on rock-breaking performance using a conical pick assisted by an abrasive water jet (AWJ). The results show that a conical pick can efficiently break hard rock when assisted by the strong impact of an AWJ, which can provide a reference for practical application. Chen et al. [25] measured the shock parameters of a coal–water medium with different mass concentrations and obtained the effect of coal particle diameter on shock pressure and shock distance.

At present, a straight-taper or conical nozzle is mostly used as a punching nozzle for breaking coal. Moreover, the related research has mainly focused on nozzle inner flow- channel shape, contraction angle, outlet diameter, and other factors [26–29]. The experimental method is likewise relatively simple, the effect of the interaction of various factors on the experiment index is not considered, and the effective strike area is excluded in the evaluation index.

The reminder of this study is organized as follows: In Section 2, the construction features and principle of the nozzle are introduced, and two evaluation indexes of the X-axis velocity and effective Y-axis extension distance are proposed. In Section 3, the effect of key structural parameters on water jet performance is studied by single factor method and CFD numerical simulation, and the corresponding results are analyzed and discussed. In Section 4, the significance of key structural parameters and their interaction on water jet performance are further analyzed by the orthogonal experiment, and three structure combinations of the nozzle with better jet performance are selected. In Section 5, field experiments are carried out on the selected nozzles to verify the correctness of the theoretical analysis. Finally, the main conclusions are drawn, and the future work is prospected in Section 6.

2. Construction Features and Principle

The conical or cone-straight nozzle has good hydraulic performance, such as good jet bunching and energy concentration. Furthermore, they are machined conveniently, so are widely used in water jet cutting and impact technology. The structure of coal breaking and punching nozzle in this study is shown in Figure 1. In general, the design parameters of the nozzle are based on empirical values [30], the contraction angle is set to 12°–14°, and the length-to-diameter ratio is set to 2 to 4. However, the total length of the nozzle in actual production is limited owing to limitations of construction conditions and of water jet equipment. Hence, the water jet effect of the nozzle designed according to the empirical value is not ideal. According to the field experiment in the early stage, the inlet diameter of all nozzles studied in this paper is $D = 10$ mm, the outlet diameter is $d = 2$ mm and the length of outlet divergence angle $l' = 1$ mm. Of particular note is $l' = 0$ mm when the outlet divergence angle $\gamma = 0°$, other specific parameters are shown in Table 1, and the formula for L is as follows:

$$L = \frac{D}{2}\cot\frac{\theta}{2} - \frac{d}{2}\cot\frac{\theta}{2} \tag{1}$$

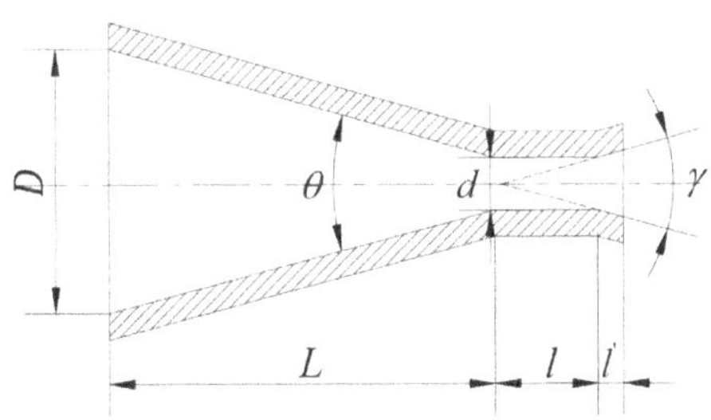

Figure 1. Diagram of the coal breaking and punching nozzle structure.

Table 1. Structure parameter of the nozzle.

Structure Parameter	Inlet Contraction Angle θ (°)	Outlet Divergence Angle γ (°)	Length-to-Diameter Ratio l/d	Inlet Diameter D (mm)	Outlet Diameter d (mm)
Initial nozzle	30	0	2	10	2
Improved nozzle	30, 50, 70, 90	0	2	10	2
	30	0, 10, 20, 30	2	10	2
	30	0	0, 2, 2.5, 3	10	2

The research shows that when a continuous water jet impinges vertically on the surface of an object, it actually converts the kinetic energy of the water jet into dynamic pressure, that is, the stagnation pressure P_s. Furthermore, P_s is the direct force of the crushing development of the object [31,32]:

$$P_s = \frac{1}{2}\rho v^2 \quad (2)$$

where, ρ is the density of the water (kg/m^3), and v is the propagation speed of the water (m/s).

The following conclusion can be drawn from formula (1): fluid velocity is one of the important indexes for breaking objects, and also a significant evaluation index of water jet performance. This study selects the maximum X-axis velocity of a water jet 100 mm away from the end face of the nozzle outlet as the first quantitative evaluation index. Moreover, this research refers to the attenuation of the axis velocity of water jet in the flow field. Given that the axis velocity attenuation is extremely complicated, quantitative analysis is difficult to conduct and such an attenuation should not be used as a sole specific evaluation index.

The stagnation pressure describes the force per unit area. In addition, the effective expansion characteristic of the water jet in the horizontal water jet field is also an important indicator of the water jet performance, which represents the effective impact area of the water jet. Given that the numerical model in this study is axisymmetric, the extension distance of the Y-axis was calculated according to the method of Azad et al. [33], specifically by selecting the data with a velocity attenuation within 15% of the target distance of 100 mm, which can be used as the second quantitative evaluation index of water jet performance.

3. Effect of Key Structural Parameters on Water Jet Performance

3.1. Computational Model

Water jet is a multiphase flow problem. A relatively large flow field calculation area outside the nozzle should be established to accurately reflect the actual water jet performance. In this paper, CFD is used to simulate the process of water jet which is produced by the coal breaking and punching nozzle impact on the target plate surface.

The effect of key structural parameters on the water jet performance is explored through the impact effect of different nozzles' water jets, and then the structural parameters can be optimized. In this CFD numerical simulation, the injection distances are all set to 200 mm. In addition, the ICEM software is used for meshing, and the FLUENT software is used for setting parameters and computing.

Given that the computational grid has a significant effect on the time and accuracy of the CFD numerical simulation, the computational model is divided into parts, namely nozzle and flow field, [34] and divided by the hybrid grid technology. Inside the nozzle, the structure size is small, local pressure is large, and turbulence is intense. Hence, an unstructured grid is used to divide the nozzle. The external flow field is large and has a regular shape, and is divided by structured grids, thereby ensuring calculation accuracy and saving on calculation time.

According to the characteristics of the model, the nozzle's inlet is set to pressure-inlet, and inlet pressure is 20 MPa. The outer flow field boundary is set to pressure-out, except for the rightmost boundary, and pressure is set to 1 atmosphere. The rightmost boundary and other boundaries of the flow field are set to wall. Given that the process of high-pressure water jet is ejected from the nozzle's outlet into the air, turbulent diffusion and momentum exchange will occur between the water and ambient air. Air at the boundary is sucked in by the water jet, and droplets are "torn apart" by aerodynamic forces. As they have different velocities, the mixture model of the multiphase flow model is selected for numerical simulation. The main and second phases are set to air and water, respectively, and the transient and implicit pressure solvers are used for calculation. The two-equation model of RNG k-ε is the best choice for this study, which can clearly simulate the separation, secondary, swirl, and other complex flows in the water jet; a simple algorithm is used in this research to solve the coupling of pressure and velocity [35–37]; the computational model is shown in Figure 2.

Figure 2. Computational Model.

3.2. Numerical Simulation

In the CFD simulation analysis, grid size has a substantial effect on the accuracy of results. To ensure the accuracy of results, the initial nozzle model was taken as the experiment object to obtain the best grid size. Grid sizes are set to 0.2, 0.3, 0.4, and 0.5 mm. Figure 3 shows the velocity distribution of the flow field under different grid sizes. Note that velocity distribution at the nozzle outlet and flow field change with an increase in grid size. Velocity distribution within the 0–100 mm target distance of the outflow field is shown in Table 2. Analysis indicates that when grid size is 0.2–0.3 mm, velocity changes of the flow field are the same, which has minimal effect on the calculation results. Although the calculation result is considerably accurate when the grid size is 0.2 mm, the calculation cost will increase several times compared with when the grid size is 0.3 mm. Therefore, the water jet performance of coal breaking and punching nozzle is eventually simulated and analyzed with a grid size of 0.3 mm. The structural parameters selected in this section are shown in Table 1.

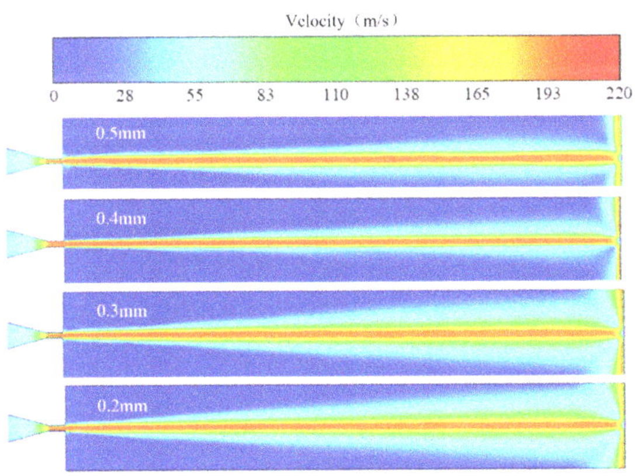

Figure 3. Velocity cloud of the flow field of different grid sizes.

Table 2. Water jet velocity distribution of different grid sizes.

Grid Size (mm)	Grid Quantity	Velocity Distribution (m/s)
0.5	26,528	198.0234–200.9876
0.4	41,419	198.4325–201.0324
0.3	74,391	199.9132–201.4522
0.2	176,040	199.9246–201.4703

3.2.1. Effect of Contraction Angle on Water Jet Performance

To explore the effects of the previously mentioned three key structural parameters on the water jet performance of coal breaking and punching nozzle, the single factor method is used for simulation analysis. Under the condition that other parameters are the same, the value of contraction angle is changed, and the specific value is shown in Table 2.

When contraction angle increases, the maximum velocity at the nozzle exit increases significantly. When contraction angle is 50°, outlet velocity reaches the maximum value of 203.95 m/s. Outlet velocity decreases as contraction angle continues to increase, but the velocity is still higher than the contraction angle $\theta = 30°$, as shown in Figure 4. When the target distance is below 100 mm, the X-axis velocity of water jet attenuation is slowest with the contraction angle $\theta = 30°$, and its X-axis velocity is evidently higher than that of other contraction angles. Within a 40-mm target distance, the X-axis velocity of the water jet of the nozzle with contraction angle $\theta = 70°$ is higher than that of the nozzle with contraction angle $\theta = 30°$. However, within the target distance of 40–100 mm, the X-axis velocity of the water jet of the two contraction angles is the same. Although the exit velocity of the nozzle is larger when the contraction angle $\theta = 90°$, the X-axis velocity of the water jet decays rapidly with an increase in the target distance, particularly in the first 100-mm target distance, which decays nearly linearly. In the range of 100–200-mm target distance, the X-axis velocity of the nozzles with four different contraction angles continues to decay. The X-axis velocity of the water jet of the nozzles with contraction angles $\theta = 30°$ and $\theta = 50°$ decreases gradually, and the X-axis velocity of the two nozzles remains the same. The X-axis velocity of the water jet of the two nozzles with contraction angles $\theta = 70°$ and $\theta = 90°$ attenuates sharply, and are evidently lower than those of the other two nozzles. When the target distance X = 100 mm, the maximum X-axis velocity increases initially and decreases thereafter with an increase in the contraction angle. When the contraction angles $\theta = 50°$ and $\theta = 90°$, X-axis velocities are the maximum and minimum, respectively. When the contraction angles $\theta = 30°$ and $\theta = 70°$, the X-axis velocity is the same.

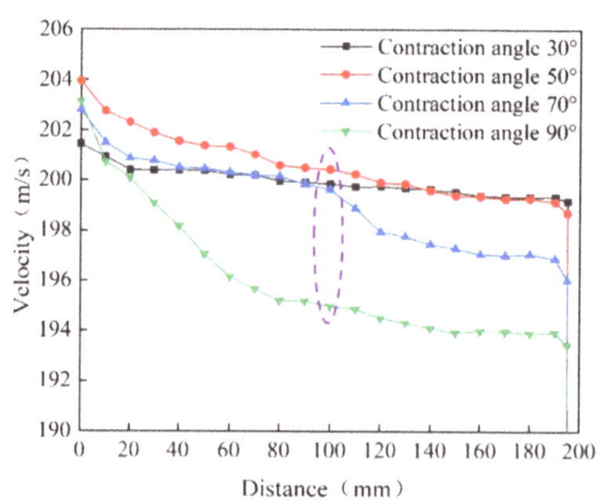

Figure 4. Effect of contraction angle on the X-axis velocity attenuation of water jet.

When the contraction angle increases, the effective Y-axis extension distance of the water jet increases initially and decreases thereafter, as shown in Figure 5. When the contraction angle increases from 30° to 50°, the effective Y-axis extension distance of the water jet increases by 2.48%. However, when contraction angle increases from 50° to 70° and 90°, the Y-axis extension distance of the water jet decreases by 5.08% and 9.62%, respectively. An increase in the effective Y-axis extension distance can increase the effective punching area of the water jet.

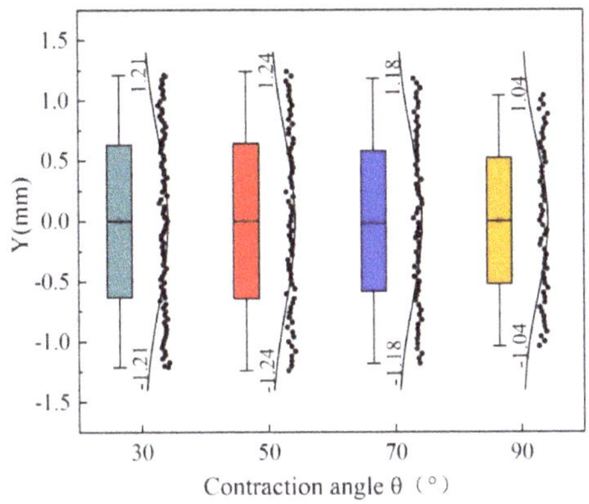

Figure 5. Effect of contraction angle on the effective Y-axis extension distance of water jet at X = 100 mm.

3.2.2. Effect of Outlet Divergence Angle on Water Jet Performance

To explore the effect of the divergence angle on the water jet performance of the nozzle, under the condition that the other parameters are the same, the value of the divergence angle is changed. The specific value is shown in Table 2.

The existence of the divergence angle makes the nozzle become a converging–diverging nozzle, and the outlet velocity increases significantly, as shown in Figure 6a. When the divergence angle $\gamma = 0°$, the outlet velocity of the water jet is 201.45 m/s. With an increase in the divergence angle, the outlet velocity of the water jet increases gradually. When the divergence angle $\gamma = 30°$, the outlet velocity of the water jet reaches 229.45 m/s, which is increased by 13.90% compared with that without the divergence angle. To better observe the change in the X-axis velocity of the water jet, the initial target distance starts from 20 mm, as shown in Figure 6b. With an increase in the divergence angle, the axis water jet velocity increases initially and decreases thereafter. In the whole flow field, the X-axis velocity of the water jet of the nozzle with divergence angle $\gamma = 30°$ decays faster than that of other nozzles. Among the other three nozzles, the X-axis velocity of the nozzle with divergence angle $\gamma = 10°$ is slightly higher. When the target distance is X = 100 mm, the sequence of velocity is as follows: $\gamma = 10° > \gamma = 20° > \gamma = 0° > \gamma = 30°$.

When there is an outlet divergence angle of the nozzle, the outlet diameter of the water jet increases, and the effective Y-axis extension distance of the water jet increases significantly. When the divergence angle $\gamma = 30°$, the effective Y-axis extension distance of the water jet reaches 3.24 mm, which is 33.88% more than that of the divergence angle $\gamma = 0°$. Moreover, the number of data points reaching the effective water jet velocity increases significantly, as shown in Figure 7.

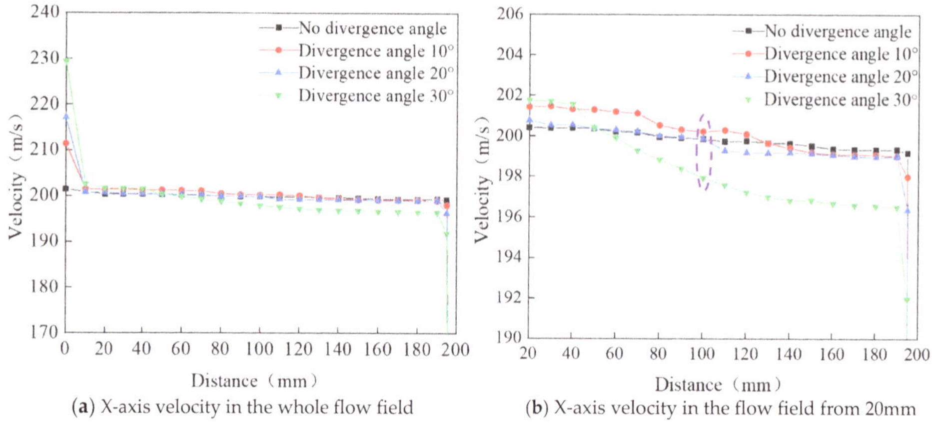

Figure 6. Effect of divergence angle on the X-axis velocity attenuation of the water jet.

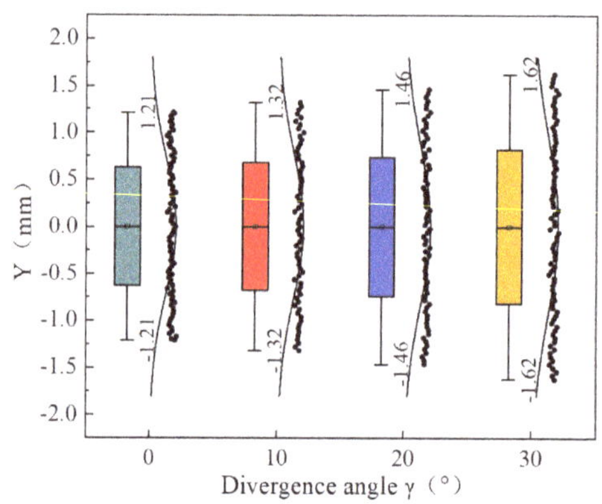

Figure 7. Effect of divergence angle on the effective Y-axis extension distance of the water jet at X = 100 mm.

3.2.3. Effect of the Length-to-Diameter Ratio on Water Jet Performance

Under the condition that the other parameters remain the same, the value of the length-to-diameter ratio is changed, and the specific value is shown in Table 2.

When the nozzle has no outlet cylindrical section, Figure 8 shows that the outlet and X-axis velocities of the water jet are significantly lower than those of the nozzle with an outlet cylinder. When the target distance exceeds 100 mm, the X-axis velocity of the water jet of the nozzle without an outlet cylinder decreases rapidly. When there is an outlet cylinder of the nozzle, outlet velocity of water jet increases initially and decreases thereafter with an increase in the length-to-diameter ratio. When $l/d = 2.5$, the outlet velocity of the water jet is the largest, X-axis velocity attenuation is the slowest, and water jet performance is the best. When $l/d = 2$ and $l/d = 3$, the X-axis velocity and attenuation of the water jet are the same. When the target distance is X = 100 mm, the X-axis velocity of the nozzle with outlet cylinder is considerably large, and maximum velocity is achieved when $l/d = 2.5$. This result indicates that when some parameters are constant, there is an optimal length-to-diameter ratio to optimize the water jet performance.

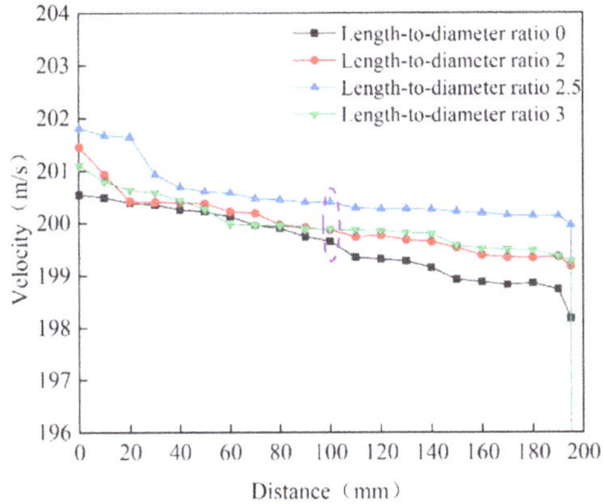

Figure 8. Effect of length-to-diameter ratio on X-axis velocity attenuation of water jet.

The non-zero value of the length-to-diameter ratio will increase the effective Y-axis extension distance of the water jet. When the length-to-diameter ratio is $l/d = 2.5$, the effective Y-axis extension distance is the maximum, which is increased by 8.3% compared with the minimum value, as shown in Figure 9. Note that the non-zero value of length-to-diameter increases the effective area of the water jet and improves the effects of coal breaking and punching.

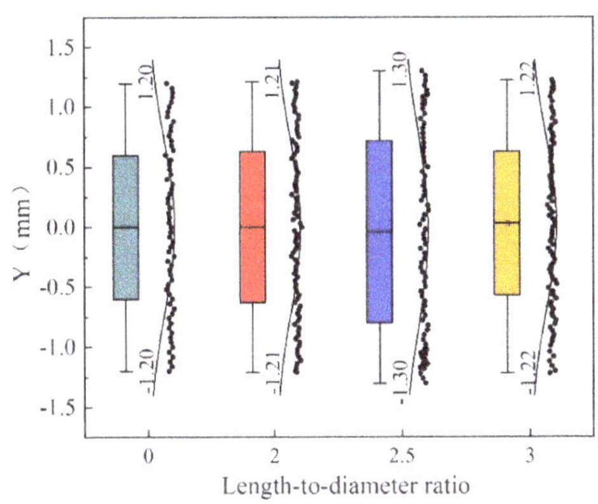

Figure 9. Effect of length-to-diameter on the effective Y-axis extension distance of water jet at X = 100 mm.

3.3. Analysis and Discussion

In the process of jetting, the water jet will have a strong momentum exchange with the surrounding air, and increasing air will move forward with the water jet under the suction of the longitudinal vortex, making its axial speed continuously attenuated. With the involvement of air, jet flow will gradually increase and, in a certain range, will realize the increase in the effective Y-axis extension distance, thereby increasing the effective strike area. The change in the nozzle's key structural parameters will affect the air entrainment rate of the water jet, and affect the jet effect thereafter.

When the contraction angle increases, flow resistance decreases, nozzle exit velocity increases, and high-speed water jet will have a strong entrainment effect on the nearby air. When $t = 0.02$ ms, the water jet has just squirted out, and with an increase in contraction angle, the maximum vortex of the jet moves backward and the front of the jet fans out and spreads around, as shown in Figure 10a–c. When the contraction angle is 90°, the nozzle exit velocity is considerably large, air entrainment rate is the largest, and the air volume fraction of the jet is the highest, thereby attaining the axial velocity attenuation fastest. Figure 10d shows that the air around the jet indicates a scattered strong volume of suction state, the consistency of the fluid moving forward is poor, and the front end of the jet is strongly disturbed. When $t = 0.8$ ms, the jet has been injected into the middle of the flow field, as shown in Figure 11. At this point, the velocity vector distribution of the flow field outside the nozzle of θ = 30° and θ = 50° is similar. When θ = 70° and θ = 90°, the air entrainment rate at the front end of the jet is substantially large. When $t = 2$ ms, the jet has been injected to the end of the flow field, as shown in Figure 12. That is, the larger the injection angle, the more evident the effect on air coiling. The air coiling rate of the entire flow field can be expressed by the volume fraction of air in the jet, as shown in Figure 13.

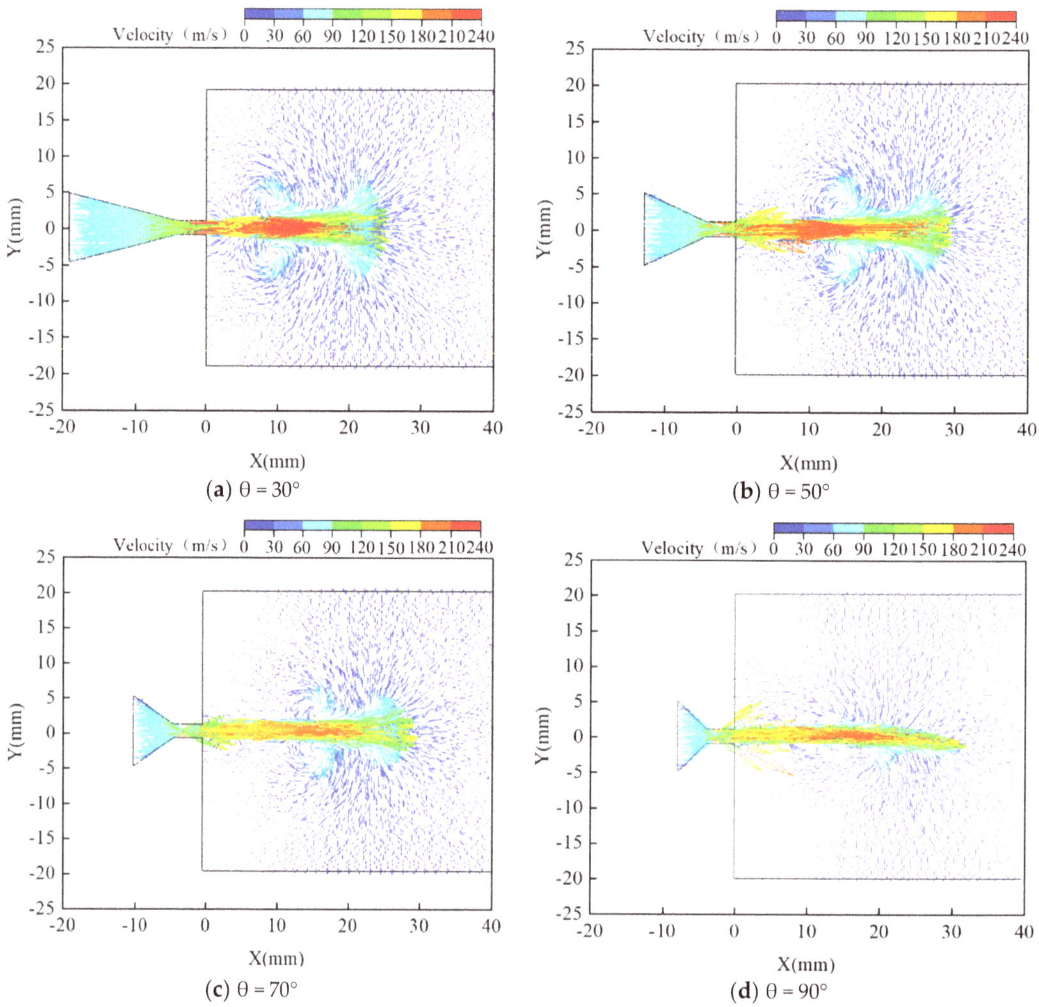

Figure 10. Velocity vector of the flow field when $t = 0.02$ ms.

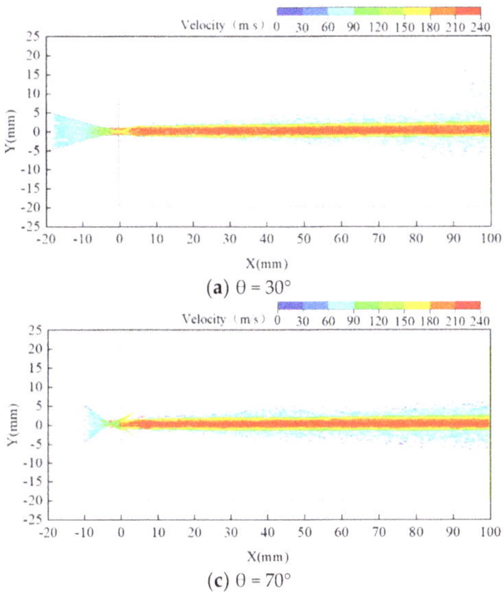

Figure 11. Velocity vector of the flow field when $t = 0.8$ ms.

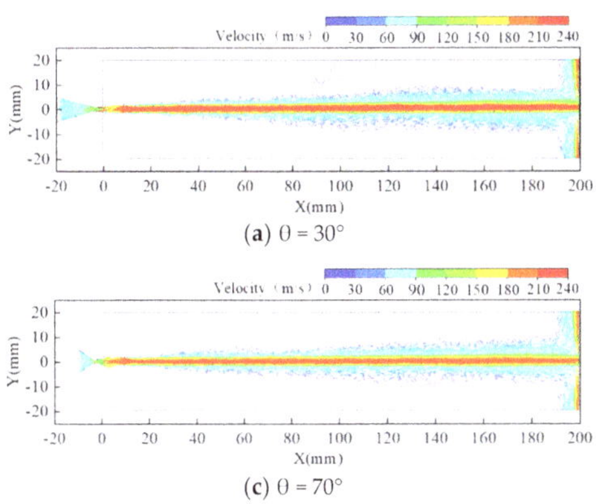

Figure 12. Velocity vector of the flow field when $t = 2$ ms.

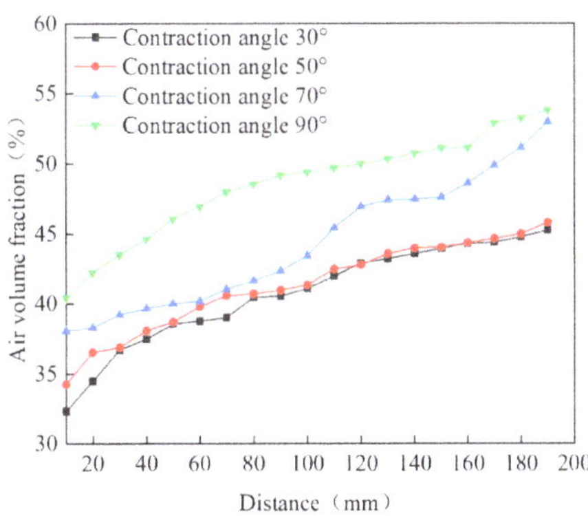

Figure 13. Effect of contraction angle on air volume fraction.

A comparison of Figures 4 and 13 shows that the axial velocity decay of the water jet is negatively related to the volume fraction of air in the water jet. The larger the volume fraction, the more momentum exchange between the water jet and the air, and the more rapid the velocity decay. When the volume fraction of air is large, the Y-axis extension distance of the water jet will increase accordingly. For coal breaking and punching, only when the velocity reaches a certain level can an effective striking force be formed. This study calculates the effective Y-axis extension distance within a 15% decay of the Y-axis velocity. Figure 5 shows that an effective Y-axis extension distance does not increase constantly, although the larger the contraction angle, the higher the air roll absorption rate.

When there is an outlet divergence angle of the nozzle, a converging–diverging-type nozzle will form, resulting in a certain cavitation effect and significantly increasing the exit velocity. Moreover, an increase in the divergence angle will also reduce the wall attachment effect of the water jet, thereby further enhancing the exit velocity and producing strong entrainment to the surrounding air, and changing the jet effect. When the divergence angle is changed, the volume fraction of air in the water jet is shown in Figure 14. A comparison of Figures 6, 7 and 14 indicates that the larger the divergence angle, the higher the air enrolling rate. Moreover, axial velocity attenuation increases. However, the presence of the divergence angle increases the nozzle exit velocity. Thus, the effective Y-axis extension distance remains considerably large when the target distance X = 100 mm.

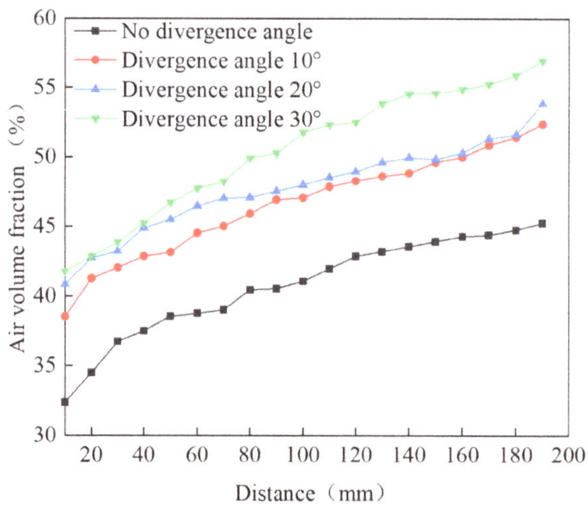

Figure 14. Effect of divergence angle on air volume fraction.

The presence of an outlet cylindrical section of the nozzle will play a certain stabilizing effect on the water jet, thereby reducing the turbulent flow pattern and improving the water jet cluster effect. Consequently, the axial velocity attenuation is reduced. However, when the cylindrical length is markedly long, flow resistance will increase, thereby reducing the exit velocity. When the length-to-diameter ratio is changed, the volume fraction of air in the water jet is shown in Figure 15. Figures 8 and 9 show an optimal length-to-diameter ratio, which makes the water jet to air volume absorption rate considerably moderate, thereby obtaining superior axial velocity and an effective Y-axis extension distance.

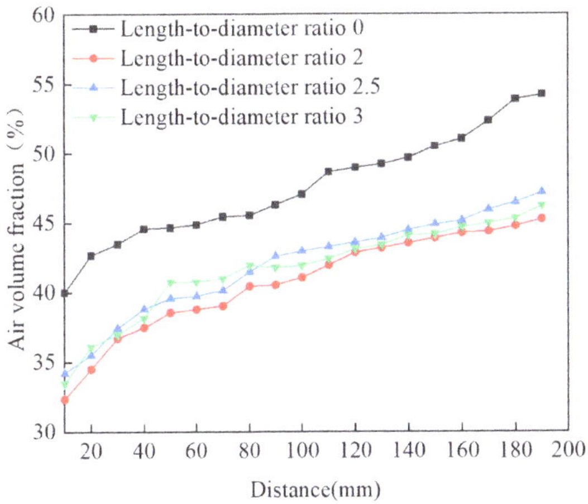

Figure 15. Effect of length-to-diameter on air volume fraction.

4. Orthogonal Experiment

To obtain the optimal combination of nozzle structural parameters when the X-axis velocity and effective Y-axis extension distance of the water jet are the largest, a numerical simulation and an orthogonal experiment are performed to further study the relationship among key parameters and their effects on water jet performance. According to the preceding analysis results, three good levels are selected for each factor. Factors A, B, and C are the contraction angle, divergence angle, and length-to-diameter ratio, respectively. The specific values are shown in Table 3.

Table 3. Level table of each factor.

Level	A	B	C
	θ (°)	γ (°)	*l/d*
1	30	0	2
2	50	20	2.5
3	70	30	3

4.1. Experiment Design

In the study of the effect of nozzle key structural parameters on water jet performance, the effect of the interaction between various factors on the experiment index is fully considered. Combined with research rules and practical principles, the first-order interaction between factors is mainly investigated in this study. According to the number of selected factors, the orthogonal experiment table of L27 (3^{13}) was selected to make a reasonable arrangement for all factors in the experiment. The experimental data are shown in Table 4. In particular, single factors A, B, and C are placed in columns 1, 2, and 5, respectively. Moreover, A × B is in columns 3 and 4, A × C is in columns 6 and 7, and B × C is in columns 8 and 11. The interaction of the three factors A × B × C is placed in columns 9, 10, 12, and 13, and is taken as the error. When the target distance is X = 100 mm, the maximum X-axis velocity and effective Y-axis extension distance of the water jet are selected as inspection indexes. Given that the model is axisymmetric, the effective Y-axis extension distance is replaced by a half-value.

Table 4. Orthogonal experiment table L_{27} (3^{13}) and experimental data.

Factors	A	B	AB	A^2B	C	AC	A^2C	BC	ABC	A^2BC	B^2C	AB^2C	A^2B^2C	Maximum X-Axis Velocity (m/s)	Half of Effective Y-Axis Extension Distance (mm)
Experimental Number	1	2	3	4	5	6	7	8	9	10	11	12	13		
1	30	0	1	1	2	1	1	1	1	1	1	1	1	199.865	1.212
2	30	0	1	1	2.5	2	2	2	2	2	2	2	2	200.393	1.303
3	30	0	1	1	3	3	3	3	3	3	3	3	3	199.879	1.221
4	30	20	2	2	2	1	1	2	2	2	3	3	3	199.868	1.323
5	30	20	2	2	2.5	2	2	3	3	3	1	1	1	199.424	1.503
6	30	20	2	2	3	3	3	1	1	1	2	2	2	200.024	1.402
7	30	30	3	3	2	1	1	3	3	3	2	2	2	198.397	1.463
8	30	30	3	3	2.5	2	2	1	1	1	3	3	3	198.118	1.563
9	30	30	3	3	3	3	3	2	2	2	1	1	1	198.350	1.532
10	50	0	2	3	2	2	3	1	2	3	1	2	3	200.484	1.242
11	50	0	2	3	2.5	3	1	2	3	1	2	3	1	199.382	1.283
12	50	0	2	3	3	1	2	3	1	2	3	1	2	199.426	1.281
13	50	20	3	1	2	2	3	2	3	1	3	1	2	199.181	1.341
14	50	20	3	1	2.5	3	1	3	1	2	1	2	3	198.277	1.522
15	50	20	3	1	3	1	2	1	2	3	2	3	1	198.390	1.401
16	50	30	1	2	2	2	3	3	1	2	2	3	1	197.160	1.502
17	50	30	1	2	2.5	3	1	1	2	3	3	1	2	198.476	1.582
18	50	30	1	2	3	1	2	2	3	1	1	2	3	195.748	1.441
19	70	0	3	2	2	3	2	1	3	2	1	3	2	198.331	1.183
20	70	0	3	2	2.5	1	3	2	1	3	2	1	3	198.936	1.191
21	70	0	3	2	3	2	1	3	2	1	3	2	1	199.225	1.101
22	70	20	1	3	2	3	2	2	1	3	3	2	1	195.775	1.361
23	70	20	1	3	2.5	1	3	3	2	1	1	3	2	195.603	1.381
24	70	20	1	3	3	2	1	1	3	2	2	1	3	195.284	1.321
25	70	30	2	1	2	3	2	3	2	1	2	1	3	188.234	1.421
26	70	30	2	1	2.5	1	3	1	3	2	3	2	1	193.189	1.602
27	70	30	2	1	3	2	1	2	1	3	1	3	2	191.606	1.522

4.2. *Results and Discussion*

4.2.1. Range Analysis of the Orthogonal Experiment

In the orthogonal experiment, the degree of effect of the different factors on the results is often compared through the intuitive analysis method–range method, and the range value R can be calculated using Formula (3) [38,39]. The greater the R value, the greater the effect of this factor on the results.

$$R = \max(T_1, T_2, T_3) - \min(T_1, T_2, T_3) \tag{3}$$

where, R is the range value and T_i (i = 1, 2, 3) represents the sum of the 9 experimental data corresponding to the ith level in each column.

According to the preceding formula, range analysis was conducted on the factors corresponding to the maximum X-axis velocity and effective Y-axis extension distance at the target distance of 100 mm and their first-order interaction. The results are shown in Table 5.

Table 5. Range analysis of the water jet performance.

Factors	A	B	C	A × B	A × C	B × C
Maximum X-axis velocity (m/s)	38.135	36.643	14.603	18.178	8.967	6.934
Half of effective Y-axis extension distance (mm)	0.512	2.611	0.882	0.317	0.212	0.252

Range analysis indicates that the primary and secondary sequence of each factor and interaction is as follows for the maximum X-axis velocity:

$$A > B > A \times B > C > A \times C > B \times C$$

For the effective Y-axis extension distance, primary and secondary sequences of each factor and interaction are as follows:

$$B > C > A > A \times B > B \times C > A \times C$$

4.2.2. Variance Analysis of the Orthogonal Experiment

In the orthogonal experiment analysis, although the range analysis is simple, the effect of the experimental error on the results cannot be excluded, and the accuracy of the analysis cannot be verified. Therefore, the *F* function should be used to conduct variance analysis for each factor and the interaction between factors, further excluding the experimental error and exploring the effect of each factor on the index [40,41]. The *F* function is as follows:

$$F_a(n_1, n_2) = F \quad (4)$$

where a is the significant level, n_1 is the degree of freedom corresponding to each factor, and n_2 is the sum of the degree of freedom of error.

When the significant level and degree of freedom are determined, the specific value of the *F* function can be obtained using the *F* distribution table [42]. In this experiment, $F_{0.01}$ and $F_{0.05}$ are taken as critical values. If $F > F_{0.05}$ of a factor, then there is 95% confidence that this factor has a significant impact on the index value, which is statistically significant. If $F > F_{0.01}$, then there is 99% confidence that this factor has a significant impact on the index value, which is highly statistically significant. In the variance analysis table of this paper, * * stands for highly significant and * stands for significant.

Factors A × C and B × C are close to the sum of the squares of the error column A × B × C. Thus, they are combined into the experimental error column, and a new sum of squares and degree of freedom are recalculated thereafter. The variance analysis of the maximum X-axis velocity is shown in Table 6.

Table 6. Variance analysis of the maximum X-axis velocity.

Source of Variance	Deviation Sum of Squares	Degree of Freedom	Variance	Variance Ratio	Significant Level
A	90.207	2	45.1035	44.22725	* *
B	75.918	2	37.959	37.22155	* *
C	10.32	2	5.16	5.059754	*
A × B	32.482	4	8.1205	7.962738	* *
A × C	5.765	4	1.44125		
B × C	5.346	4	1.3365		
A × B × C	5.206	8	0.65075		
Error	16.317	16	1.019813		
Sum	255.244	26			
F_a	$F_{0.05}(2, 16) = 3.63$, $F_{0.05}(4, 16) = 3.01$, $F_{0.05}(8, 16) = 2.59$ $F_{0.01}(2, 16) = 6.23$, $F_{0.01}(4, 16) = 4.77$, $F_{0.01}(8, 16) = 3.89$				

Table 7 shows that the sum of the square of the factor A × C and error column A × B × C is close. Adding it to the error and the final results are as follows:

Table 7. Variance analysis of effective Y-axis extension distance.

Source of Variance	Deviation Sum of Squares	Degree of Freedom	Variance	Variance Ratio	Significant Level
A	0.017	2	0.0085	20.4	* *
B	0.383	2	0.1915	459.6	* *
C	0.048	2	0.024	57.6	* *
A × B	0.011	4	0.00275	6.6	* *
A × C	0.003	4	0.00075		
B × C	0.006	4	0.0015	3.6	*
A × B × C	0.002	8	0.00025		
Error	0.005	12	0.000417		
Sum	0.473	26			
F_a		$F_{0.05}(2, 12) = 3.88$, $F_{0.05}(4, 12) = 3.26$, $F_{0.05}(8, 12) = 2.85$ $F_{0.01}(2, 12) = 6.93$, $F_{0.01}(4, 12) = 5.41$, $F_{0.01}(8, 12) = 4.5$			

4.3. Comprehensive Analysis of Experimental Data Based on Orthogonal Experiment

According to the results of the orthogonal experiment, when the X-axis velocity is maximum, optimal levels of each factor are A_2, B_1, $A_2 \times B_1$, C_1, $A_2 \times C_1$, and $B_1 \times C_1$, comprehensively considering that $A_2 \times B_1 \times C_1$ is the optimal combination. A nozzle of this structure is shown in the orthogonal experiment table, and the maximum X-axis velocity is 200.484 m/s at 100 mm. When the effective Y-axis extension distance is the largest, the optimal levels of each factor are B_3, C_2, A_3, $A_3 \times B_3$, $B_3 \times C_2$, and $A_3 \times C_2$. Hence, $A_3 \times B_3 \times C_2$ is the optimal combination. A nozzle of this structure is shown in the orthogonal table, and its effective Y-axis extension distance is 3.204 mm.

Further analysis of the data in Table 4 shows that the maximum X-axis velocity and maximum effective Y-axis extension distance do not appear simultaneously, but the two evaluation indexes should be considered comprehensively in the process of high-pressure water jet breaking coal and punching. The effective Y-axis extension distance of the $A_2 \times B_1 \times C_1$ combination is 2.484 mm, which is below that of other combinations. The maximum X-axis velocity of the $A_3 \times B_3 \times C_2$ combination is 193.189 m/s, which is the lowest among the different combinations. Therefore, a comprehensive analysis of the results of the single factor method in Part 2 and orthogonal experiment method in Part 3 indicates that the combination of $A_2 \times B_2 \times C_2$ is the optimal parameter combination of the water jet performance. At this point, the maximum X-axis velocity is 198.277 m/s, and effective Y-axis extension is 3.044 mm, which are relatively superior in their respective index values. The improved nozzles processed with the three parameter combinations will be verified by field experiments and compared with the original nozzle.

5. Field Experiments

To further verify the water jet performance after nozzle optimization, a water jet experiment was conducted in the Qianjiaying mining area of the Kailuan Group using the self-developed equipment of a high-pressure water jet. The schematic of the equipment and water jet performance are shown in Figures 16 and 17, respectively.

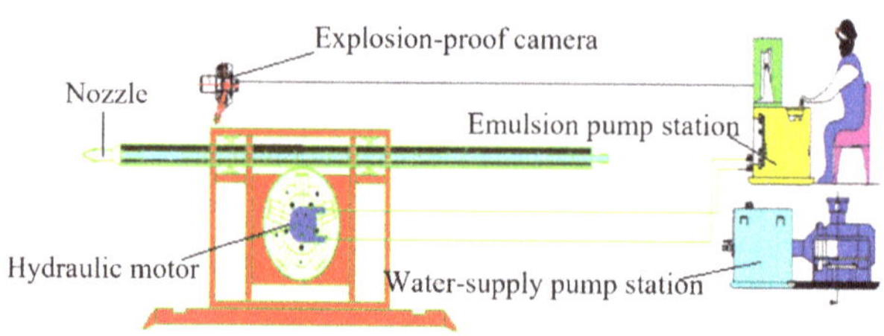

Figure 16. Schematic of the self-developed equipment of high-pressure water jet.

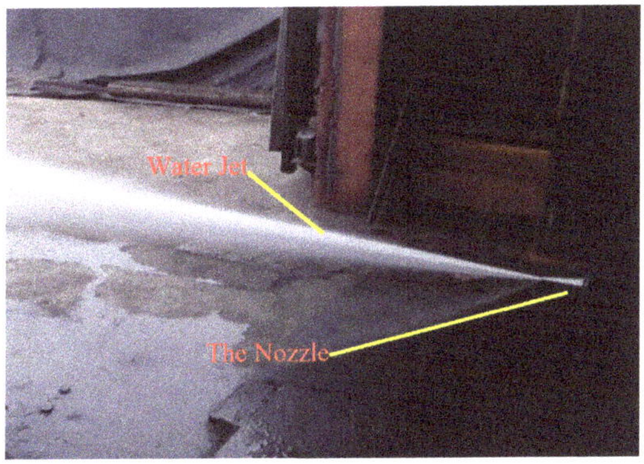

Figure 17. Performance of the water jet.

5.1. Determination of Coal Firmness Coefficient

To eliminate the interference of other factors on the effect of coal breaking and punching and ensure the accuracy of the results, the firmness coefficient of the four wall surfaces selected in the punching experiment was measured. Taking wall 1 as an example, after removing the floating coal with a thickness of approximately 0.3 m on the surface, coal samples were collected in three locations: the upper left area of the coal wall, the area of the coal wall directly opposite of the high-pressure water jet equipment, and the upper right area of the coal wall. The coal samples are labeled A_1, B_1, and C_1. According to the standard of the determination method of the coal firmness coefficient [43], the coal with label A_1 is sampled and measured.

The three groups were measured in parallel at each position, and the arithmetic mean value of the 3 positions was taken as the final firmness coefficient of wall 1. Coal samples in other areas were also measured strictly according to the preceding method. The obtained data are shown in Table 8. The difference in the firmness coefficient within 0.86% indicates that the firmness coefficient of the selected three wall surfaces is consistent. According to the orthogonal analysis experiment results, the initial nozzle, the improved 1, improved 2, and improved 3 nozzles were arranged to perform coal breaking and punching experiments at walls 1, 2, 3, and 4, respectively, to further reduce experimental error.

Table 8. Firmness coefficient of the coal wall.

Coal Wall	Sample Position	Firmness Coefficient of Group 1	Firmness Coefficient of Group 2	Firmness Coefficient of Group 3	Average of Firmness Coefficient
1	A_1	0.51	0.53	0.52	0.5222
	B_1	0.52	0.51	0.53	
	C_1	0.52	0.52	0.54	
2	A_2	0.53	0.52	0.52	0.5256
	B_2	0.52	0.53	0.54	
	C_2	0.54	0.51	0.52	
3	A_3	0.52	0.53	0.53	0.5244
	B_3	0.51	0.53	0.53	
	C_3	0.52	0.51	0.54	
4	A_4	0.52	0.52	0.53	0.5267
	B_4	0.53	0.53	0.54	
	C_4	0.53	0.52	0.52	

5.2. Experimental Analysis

Before the coal breaking and punching experiment, the pressure of the pump was adjusted, and the nozzle inlet pressure was stabilized at 20 MPa according to the pressure gauge. The erosion time of each nozzle was 300 s. According to the analysis result in Part 3, three groups of parameters were selected to improve the nozzle. The field coal breaking and punching experiment is shown in Figure 18. The parameters of the nozzle and measured data are shown in Table 9. The punching depth is the maximum hole depth measured perpendicular to the punching end face. Given that all the holes are quasi-circular, the punching diameter is the arithmetic average of the maximum and minimum diameters.

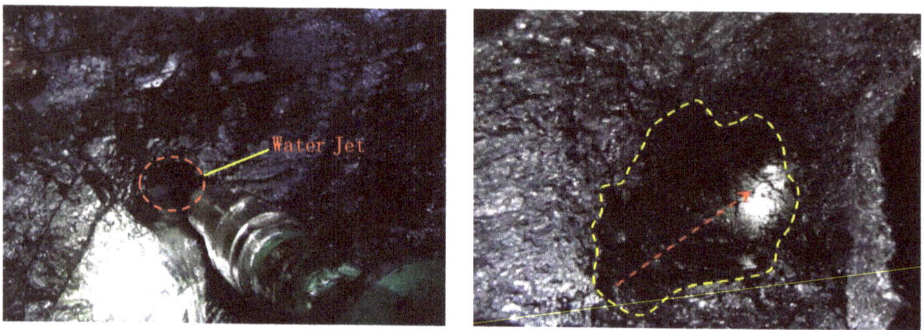

(**a**) Field experiments (**b**) The hole

Figure 18. Coal breaking and punching experiment.

Table 9. Experimental results of coal breaking and punching before and after nozzle optimization.

Experimental Group	Structural Parameters			Punching Depth (m)	Punching Diameter (m)
	A (°)	B (°)	C		
Initial nozzle	30	0	2	0.63	0.11
Improved 1	50	0	2	0.71	0.15
Improved 2	70	30	2.5	0.58	0.21
Improved 3	50	20	2.5	0.74	0.24

According to the experimental data, compared with the initial nozzle, the punching depth and diameter of improved nozzles 1 and 2 has been significantly improved. However, the diameter and depth of punching cannot reach the maximum value simultaneously. The reason is that improved nozzle 1 is selected based on the maximum X-axis velocity, while improved nozzle 2 is selected based on the maximum effective Y-axis extension distance. This result also confirms that the two indexes represent the depth and area, respectively, of punching, which is consistent with the analysis result of the orthogonal experiment. Moreover, this result indicates the reliability of the orthogonal experiment. When the maximum X-axis velocity is optimal, punching depth is increased by 12.70% compared with the initial value. When the effective Y-axis extension distance is optimal, punching diameter is increased by 90.91% compared with the initial value. Note that improved nozzle 3 is worth focusing on, the maximum X-axis velocity and effective Y-axis extension distance of which are less than those of improved nozzles 1 and 2. However, the punching depth and punching diameter are optimal, which are 4.22% and 14.29% higher than the two nozzles on the punching depth and diameter, respectively. Both are improved by 17.46% and 118% compared with the initial nozzle. This result shows that punching depth and punching area are mutually affected. That is, when the punching depth deepens, it is easier to obtain a larger punching area. Similarly, when the punching area increases, it is easier to obtain a deeper punching depth. Therefore, to achieve the overall improvement of the water jet performance, the maximum X-axis velocity and effective Y-axis extension distance must be considered simultaneously. In future research, multi-parameter and multi-objective collaborative optimization of the nozzle can be conducted to improve the water jet performance comprehensively.

6. Conclusions

In this study, the effect of nozzle structure parameters on water jet performance is explored by CFD simulation and verified by field experiments. The research results provide a certain reference value for nozzle structure optimization. The major findings include:

(1) The key structural parameters of the nozzle have a significant impact on the performance of the water jet. Among the key parameters, the value of the contraction angle has the greatest impact on the maximum X-axis velocity, and the value of the divergence angle has the greatest impact on the effective Y-axis extension distance.

(2) Based on the comprehensive results of the orthogonal experiment, range analysis, and variance analysis, three optimal combinations of nozzle structural parameters are selected. Through the field experiment, the optimized nozzles can improve the coal breaking and punching ability of the water jet.

(3) The orthogonal experiment shows that interaction among multiple parameters will affect the water jet performance, and the X-axis velocity and effective Y-axis extension distance of nozzle do not reach the optimal value at the same time. According to the field experiment result in Part 4, the punching depth and area also affect each other. Therefore, in future research, multi-parameter and multi-objective collaborative optimization of the nozzle can be carried out to improve the water jet performance more comprehensively.

Author Contributions: Conceptualization, L.C.; Methodology, L.C.; Software, M.C.; Validation, L.G.; Formal analysis, Y.C.; Writing—original draft preparation, L.C. and M.C.; Writing—review and editing, L.C. and D.G.; Supervision, L.C. and D.G.; Funding acquisition, L.G. All authors have read and agreed to the published version of the manuscript.

Funding: This work was supported by the National Natural Science Foundation of China (Grant No.51874012).

Institutional Review Board Statement: Not applicable.

Informed Consent Statement: Not applicable.

Data Availability Statement: Not applicable.

Conflicts of Interest: The authors declare no conflict of interest.

References

1. Shen, Z.H. *Water Jet Theory and Technology;* China University of Petroleum Press: Dongying, China, 1998.
2. Chen, X.C.; Deng, S.S.; Guan, J.F.; Hua, W.X. Experiment and simulation research on abrasive water jet nozzle wear behavior and anti-wear structural improvement. *J. Braz. Soc. Mech. Sci. Eng.* **2017**, *39*, 2023–2033. [CrossRef]
3. Njock, P.G.A.; Chen, J.; Modoni, G.; Arulrajah, A.; Kim, Y.H. A review of jet grouting practice and development. *Arab. J. Geosci.* **2018**, *11*, 459. [CrossRef]
4. Singh, D.; Premachandran, B.; Kohli, S. Effect of nozzle shape on jet impingement heat transfer from a circular cylinder. *Int. J. Therm. Sci.* **2015**, *96*, 45–69. [CrossRef]
5. Vinze, R.; Chandel, S.; Limaye, M.D.; Prabhu, S.V. Influence of jet temperature and nozzle shape on the heat transfer distribution between a smooth plate and impinging air jets. *Int. J. Therm. Sci.* **2016**, *99*, 136–151. [CrossRef]
6. Zhang, X.Z.; Wiśniewski, P.; Dykas, S.; Zhang, G. Permeability enhancement properties of high-pressure abrasive water jet flushing and its application in a soft coal seam. *Front. Energy Res.* **2021**, *9*, 679623. [CrossRef]
7. Chen, C.; Nie, S.L.; Wu, Z.J.; Li, Z.Y. A study of high pressure water jet characteristics by CFD simulation. *Mach. Tool Hydraul.* **2006**, *2*, 103–105.
8. Wen, J.; Chen, C. Multifunctional experimental device of teaching model and its application of water jet testing and breaking rock. *Sci. Technol. Eng.* **2017**, *17*, 168–172. [CrossRef]
9. Kong, L.; Wang, Y.; Lei, X.; Feng, C.; Wang, Z. Integral modeling of abrasive waterjet micro-machining process. *Wear* **2021**, *482–483*, 203987. [CrossRef]
10. Nedelcu, D.; Cojocaru, V.; Avasiloaie, R.C. Numerical investigation of nozzle jet flow in a pelton microturbine. *Machines* **2021**, *9*, 158. [CrossRef]

11. Wen, J.W.; Chen, C. Optimizing the structure of the straight cone nozzle and the parameters of borehole hydraulic mining for huadian oil shale based on experimental research. *Energies* **2017**, *10*, 2021. [CrossRef]
12. Li, H.S.; Liu, S.Y.; Jia, J.G.; Wang, F.C.; Guo, C.W. Numerical simulation of rock-breaking under the impact load of self-excited oscillating pulsed water jet. *Tunn. Undergr. Space Technol.* **2020**, *96*, 179–192. [CrossRef]
13. Chen, X.X.; Wang, C.; Shi, W.D.; Zhang, Y.C. Numerical simulation of submerged impinging water jet at different impact angles. *J. Drain. Irrig. Mach. Eng.* **2020**, *38*, 658–662.
14. Rahman, M.S.; Tay, G.F.K.; Mark, F.T. Effects of nozzle geometry on turbulent characteristics and structure of surface attaching jets. *Flow Turbul. Combust.* **2019**, *103*, 797–825. [CrossRef]
15. Huang, F.; Hu, B.; Zuo, W.Q.; Li, S.Q. Experiments on the impact pressure of high-pressure water jet under different nozzle shapes. *J. Chongqing Univ.* **2019**, *42*, 124–133.
16. Liu, Y.; Cui, J.W.; Jian, P.; Liu, X.T. Effect of nozzle structure on coal breakage of SC-CO_2 used for well drilling. *Geomech. Geophys. Geo-Energy Geo-Resour.* **2020**, *6*, 67. [CrossRef]
17. Zhang, L.; Wang, C.; Zhang, Y.; Xiang, W.; He, Z.; Shi, W. Numerical study of coupled flow in blocking pulsed jet impinging on a rotating wall. *J. Braz. Soc. Mech. Sci. Eng.* **2021**, *43*, 496–508. [CrossRef]
18. Yang, M.G.; Xiao, S.N.; Kang, C.; Wang, Y.L. Effect of geometrical parameters on submerged cavitation jet discharged from profiled central-body nozzle. *Chin. J. Mech. Eng.* **2013**, *26*, 476–482. [CrossRef]
19. Hong, C.Y.; Yang, R.Y.; Huang, Z.W.; Liu, W.; Chen, J.X.; Cong, R.C. Experimental investigation on coal-breakage performances by abrasive nitrogen-gas jet with a conical nozzle. *Int. J. Rock Mech. Min. Sci.* **2021**, *142*, 104781. [CrossRef]
20. Qiang, Z.R.; Wu, M.P.; Miao, X.J.; Rupy, S. CFD research on particle movement and nozzle wear in the abrasive water jet cutting head. *Int. J. Adv. Manuf. Technol.* **2018**, *95*, 4091–4100. [CrossRef]
21. Wang, H.; Qian, Z.; Zhang, D.; Wang, T.; Wang, C. Numerical study of the normal impinging water jet at different impinging height, based on Wray-Agarwal turbulence model. *Energies* **2020**, *13*, 1744. [CrossRef]
22. Peng, G.J.; Tian, L.; Hao, C.; Hong, S.M.; Ye, D.X.; You, B.J. Numerical and experimental study of hydraulic performance and wear characteristics of a slurry pump. *Machines* **2021**, *9*, 373. [CrossRef]
23. Ekiciler, R.; Cetinkaya, M.S.; Arslan, K. Convective heat transfer investigation of a confined air slot-jet impingement cooling on corrugated surfaces with different wave shapes. *J. Heat Transf.* **2019**, *141*, 22202. [CrossRef]
24. Liu, S.Y.; Zhou, F.Y.; Li, H.S.; Chen, Y.Q.; Wang, F.C.; Guo, C.W. Experimental investigation of hard rock breaking using a conical pick assisted by abrasive water jet. *Rock Mech. Rock Eng.* **2020**, *53*, 4221–4230. [CrossRef]
25. Chen, J.; Guo, L.W.; Zhang, J.Y. Relationship between shock parameter and coal particle parameter of advanced premixed micro-soft abrasive coal-water jet. *Powder Technol.* **2021**, *379*, 393–406. [CrossRef]
26. Fang, M.H.; Yu, T.; Feng, F.; Jeff, X. An experimental investigation of abrasive suspension flow machining of injector nozzle based on orthogonal test design. *Int. J. Adv. Manuf. Technol.* **2020**, *110*, 1071–1082. [CrossRef]
27. Zhou, J.R.; Zhao, M.M.; Wang, C.; Gao, Z.J.; Zhu, Y. Optimal design of diversion piers of lateral intake pumping station based on orthogonal test. *Shock. Vib.* **2021**, *2021*, 6616456. [CrossRef]
28. Zhu, Y.; Li, G.; Wang, R.; Tang, S.; Su, H.; Cao, K. Intelligent fault diagnosis of hydraulic piston pump combining improved lenet-5 and PSO hyper parameter optimization. *Appl. Acoust.* **2021**, *183*, 108336. [CrossRef]
29. Wang, H.; Long, B.; Wang, C.; Han, C.; Li, L. Effects of the impeller blade with a slot structure on the centrifugal pump performance. *Energies* **2020**, *13*, 1628. [CrossRef]
30. Shen, H.J. Structural optimization and test of nozzle for coal seam slotting. *Saf. Coal Mines* **2020**, *51*, 10–13. [CrossRef]
31. Lu, Y.Y.; Huang, F.; Liu, X.; Ao, X. On the failure pattern of sandstone impacted by high-velocity water jet. *Int. J. Impact Eng.* **2015**, *76*, 67–74. [CrossRef]
32. Yu, Y.F.; Li, C.X.; Meng, H.B.; Wang, Y.F.; Wu, J.H. Flow and entrainment characteristics of jet from different shape nozzles. *Chin. J. Process Eng.* **2014**, *14*, 549–555.
33. Azad, M.; Quinn, W.R.; Groulx, D. Mixing in turbulent free jets issuing from isosceles triangular orifices with different apex angles. *Exp. Therm. Fluid Sci.* **2012**, *39*, 237–251. [CrossRef]
34. Shi, L.; Zhu, J.; Tang, F.; Wang, C. Multi-Disciplinary optimization design of axial-flow pump impellers based on the approximation model. *Energies* **2020**, *13*, 779. [CrossRef]
35. Wen, J.W.; Qi, Z.W.; Seyed, S.B.; Pei, X.J.; Tom, I. Research on the structures and hydraulic performances of the typical direct jet nozzles for water jet technology. *J. Braz. Soc. Mech. Sci. Eng.* **2019**, *41*, 558–570. [CrossRef]
36. Chen, B.; Gao, D.R.; Li, Y.B.; Chen, C.Q.; Yuan, X.M.; Wang, Z.S.; Sun, P. Investigation of the droplet characteristics and size distribution during the collaborative atomization process of a twin-fluid nozzle. *Int. J. Adv. Manuf. Technol.* **2020**, *107*, 1625–1639. [CrossRef]
37. Humaira, Y.; Naveed, I.; Anum, T. Engineering applications of peristaltic fluid flow with hall current, thermal deposition and convective conditions. *Mathematics* **2020**, *8*, 1710. [CrossRef]
38. Wang, H.; Hu, Q.; Yang, Y.; Wang, C. Performance differences of electrical submersible pump under variable speed schemes. *Int. J. Simul. Model.* **2021**, *20*, 76–86. [CrossRef]
39. Tang, S.; Zhu, Y.; Yuan, S. An improved convolutional neural network with an adaptable learning rate towards multi-signal fault diagnosis of hydraulic piston pump. *Adv. Eng. Inform.* **2021**, *50*, 101406. [CrossRef]

40. Liang, B.J.; Gao, D.R. Optimization of structural parameters of fan—Shaped high-pressure nozzle. *J. Drain. Irrig. Mach. Eng.* **2020**, *38*, 69–75.
41. Zhu, Y.; Li, G.; Wang, R.; Tang, S.; Su, H.; Cao, K. Intelligent fault diagnosis of hydraulic piston pump based on wavelet analysis and improved Alex Net. *Sensors* **2021**, *21*, 549. [CrossRef]
42. He, W.; Xue, W.D.; Tang, B. *Optimization Method of Experimental Design and Data Analysis*; Chemical Industry Press: Beijing, China, 2012.
43. *GB/T 23561.12-2010*; Methods for Determining the Physical and Mechanical Properties of Coal and Rock—Part 12: Methods for Determining Coal Hardiness Coefficient. Standards Press of China: Beijing, China, 2011.

machines

MDPI

Article

Research on the Influence of Tip Clearance of Axial-Flow Pump on Energy Characteristics under Pump and Turbine Conditions

Yanjun Li *, Qixu Lin, Fan Meng, Yunhao Zheng and Xiaotian Xu

Research Center of Fluid Machinery Engineering and Technology, Jiangsu University, Zhenjiang 212013, China; 2211911013@stmail.ujs.edu.cn (Q.L.); mf@ujs.edu.cn (F.M.); 2112111018@stmail.ujs.edu.cn (Y.Z.); 2221911027@stmail.ujs.edu.cn (X.X.)
* Correspondence: lyj782900@ujs.edu.cn

Abstract: In order to study the influence of tip clearance on the performance and energy dissipation of the axial-flow pump and the axial-flow pump as a turbine, and find the location of high dissipation rate, this study took an axial-flow pump model as its research object and designed four tip radial clearance schemes (0, 0.2, 1 and 2 mm). The unsteady calculation simulation of each tip clearance scheme was carried out based on CFD technology. The calculated results were compared with the experimental results, and the simulation results were analyzed using entropy production analysis theory. The results showed that, under both an axial-flow pump and axial-flow pump as turbine operating conditions, increasing the blade tip clearance led to a decrease in hydraulic performance. Compared with the 0 mm clearance, the maximum decreases in pump efficiency, head and shaft power under 2 mm tip clearance were 15.3%, 25.7% and 12.3% under the pump condition, and 12.7%, 18.5% and 28.8% under the turbine condition, respectively. Under the axial-flow pump operating condition, the change in blade tip clearance had a great influence on the total dissipation of the impeller, guide vane and outlet passage, and the maximum variation under the flow rate of $1.0Q_{des}$ was 53.9%, 32.1% and 54.2%, respectively. Under the axial-flow pump as a turbine operating condition, the change in blade tip clearance had a great influence on the total dissipation of the impeller and outlet passage, the maximum variation under the flow rate of $1.0Q_{des}$ was 22.7% and 17.4%, respectively. Under the design flow rate condition, with the increase in tip clearance, the dissipation rate of the blade surface showed an increasing trend under both the axial-flow pump and axial-flow pump as turbine operating conditions, and areas of high dissipation rate were generated at the rim and clearance.

Keywords: axial-flow pump as turbine; tip clearance; entropy production; loss distribution; numerical simulation

Citation: Li, Y.; Lin, Q.; Meng, F.; Zheng, Y.; Xu, X. Research on the Influence of Tip Clearance of Axial-Flow Pump on Energy Characteristics under Pump and Turbine Conditions. *Machines* **2022**, *10*, 56. https://doi.org/10.3390/machines10010056

Academic Editor: Davide Astolfi

Received: 17 December 2021
Accepted: 10 January 2022
Published: 12 January 2022

1. Introduction

The axial-flow pump is a kind of highly specific speed pump with a large flow rate and low head, which is widely used in irrigation, flood control, large-scale water diversion and industrial circulating water systems and other fields of national economic importance [1]. Some axial-flow pumps with good performance can be used as hydraulic turbines, which can convert the pressure energy of a high-pressure medium into the kinetic energy of a rotating machine to generate electricity for energy recovery [2].

At present, research on the pump and the pump as a turbine is mostly based on numerical simulation [3–5] and experimental methods [6–8]. With the development of modern computer technology and the improvement of simulation analysis theory, numerical computer simulation has been increasingly used in the research of pumps and turbines. Qian et al. [9] used numerical simulation to study the performance of the axial-flow pump both under pump and turbine conditions. In the study of hydraulic dissipation within the axial-flow pump and turbine, the spatial distribution of hydraulic dissipation in the axial-flow pump and turbine can be described by using entropy production theory for analysis; therefore, entropy production theory has an obvious advantage in hydraulic dissipation assessment, compared with the usual calculation method of total pressure loss [10]. Zhang et al. [11] used numerical simulation to obtain the value and spatial distribution of hydraulic dissipation inside a centrifugal pump by introducing the entropy production analysis theory, which proves the applicability of entropy production analysis theory in the study of centrifugal pump energy loss. Pei et al. [12] studied the influence of the distance between the impeller and guide vane on the performance of the bidirectional shaft tubular pump by using entropy production analysis theory; turbulent dissipation was found to be the main energy dissipation of the total dissipation in the bidirectional shaft tubular pump. Li et al. [13] introduced entropy production analysis theory for the numerical simulation of pump turbines in pump mode, analyzed the hydraulic dissipation and entropy production of pump turbines under different guide vanes, and obtained the hump characteristic of pump turbines in pump mode; the results show that the peak value of entropy production corresponds to the minimum value of efficiency in the external characteristic curve, and the entropy production of the pump turbine mainly occurs on the impeller and guide vane under the pump condition. Gong et al. [14] studied the distribution of turbine hydraulic dissipation based on entropy production analysis theory. It was found that the energy dissipation in the guide vane area accounted for nearly 25% of the total dissipation, which mainly occurred at the head and tail areas of the vanes, and the dissipation in the impeller accounted for more than half of the whole flow passage, which mainly occurred at the leading edge of the blade and the trailing edge of blade, and nearly 20% of the energy dissipation occurred at the elbow. The results show that entropy production theory is suitable for evaluating the performance of the turbine, and has the advantages of determining the amount and the location of energy dissipation.

There is a clearance between the impeller edge and the shell of the axial-flow pump, which can cause clearance leakage flow and result in cavitation, which will influence the operation stability and reduce the operation efficiency of the axial-flow pump under both pump and turbine conditions, and increase vibration and noise during operation [15,16]. Shi [17] and Shen et al. [18] studied the influence of tip clearance size on the internal flow and pressure pulsation performance of the axial-flow pump by using numerical calculation; the results show that the increase in tip clearance leads to the decrease in axial-flow pump efficiency and the increase in pressure pulsation. Chen et al. [19] established a simplified model of clearance flow and used it to evaluate the trajectory of tip leakage vortices. Li et al. [20] studied the influence of five different passive turbine tip clearance flow control methods on axial turbines based on a numerical calculation; the results show that by controlling the passive turbine tip clearance flow, the secondary flow loss including the passage vortex losses and the tip leakage losses near the outer casing can be reduced. Han et al. [21] studied the influence of tip clearance on the hydraulic and thrust performance of the contra-rotating axial-flow water-jet pumps by numerical calculation; the results show that with the increase in tip clearance, the tip clearance leakage of contra-rotating axial-flow water-jet pumps increases and the hydraulic performance decreases obviously.

Tip clearance has a great influence on the operation performance, energy characteristics and manufacturing difficulty of the axial-flow pump and axial-flow pump as a turbine; however, there are few works in the literature about the effect of tip clearance on the internal energy dissipation characteristics of axial flow pump as turbine based on entropy production theory, and a lack of research on the spatial distribution law of its internal hydraulic dissipation. This paper takes the axial-flow pump model as the research object; on the basis of numerical simulation, and the influence of tip clearance on hydraulic loss of the axial-flow pump under both pump and turbine conditions is analyzed using the method of entropy production analysis theory. It can obtain the specific location of high dissipation rate and solve the limitation of optimization design, providing a reference for coupling the optimization design of the axial-flow pump under pump and turbine conditions.

2. Calculation Model and Numerical Simulation Calculation

2.1. Three-Dimensional (3D) Modeling and Mesh Division

In this paper, an axial-flow pump model with a specific speed of 1000 was selected as the prototype of the research object. Its components include the inlet passage (straight pipe in pump operation and elbow in turbine operation), impeller, guide vane and outlet passage (elbow in pump operation and straight pipe in turbine operation), as shown in Figure 1. The main hydraulic parameters of the model are as follows: pump design flow rate is $Q_{des} = 0.39\ m^3/s$; turbine design flow rate is $Q'_{des} = 0.47\ m^3/s$; rotational speed is $n = 1450r/\text{min}$. The main structural parameters are as follows: the number of impeller blades is 3; the diameter of the impeller is 300 mm; the number of guide vane blades is 5; and the bending degree of the elbow is 60°. Four tip radial clearance schemes R were designed, namely, 0, 0.2, 1 and 2 mm, respectively.

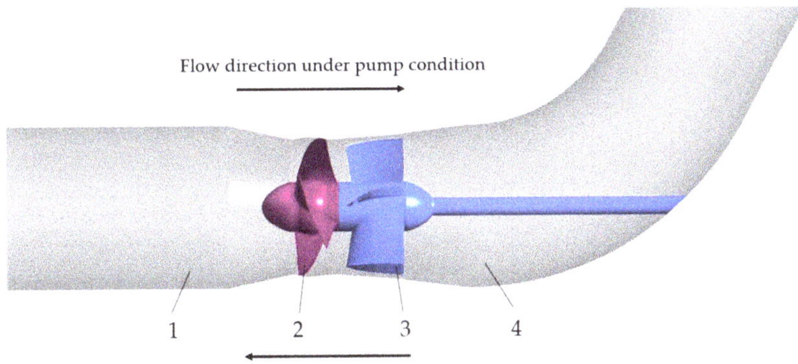

Figure 1. Three-dimensional model of axial-flow pump. 1—inlet passage; 2—impeller; 3—guide vane; 4—outlet passage.

In order to ensure the computational accuracy and convergence of the grid model, the computational domain was divided into structural grids [22–24]. ICEM was used to generate the grid of the inlet passage, impeller and outlet passage, and TuboGrid was used to generate the grid of the guide vane, as shown in Figure 2. Local mesh encryption was carried out on the wall surface around the blade and the rim clearance area to ensure that the wall Y+ value met the solution requirements of the turbulence model [25]. In order to ensure the accuracy of the calculation, a total of 15 layers of grids were divided at the blade tip clearance. The number of grid nodes of the inlet passage, guide vane and outlet passage was 1 million, 1.2 million and 1.4 million, respectively, and the number of grid nodes of the impeller under the four tip clearance schemes was controlled at approximately 1.8 million.

2.2. Setting of Boundary Conditions

Based on Newton's continuity equation and momentum conservation equation for incompressible fluid, the SST k−ω [26,27] turbulence model was used to calculate the internal flow field and hydraulic performance of the axial-flow pump with different tip clearance schemes under axial-flow pump and turbine operating conditions. In both axial-flow pump and turbine operating conditions, the inlet passage adopts a mass flow rate condition, and the outlet passage adopts a static pressure condition [28]. The boundary details adopt a non-slip and smooth wall. The mixing model with stationary components is composed of "None", and the mixing model with rotating and stationary interfaces is composed of a "Transient Rotor Stator". The residual type of convergence criterion was RMS, which was set as 10^{-5}, and the time step was 0.000344828 s, that is, the time required for the impeller to rotate 3°. The calculation cycle was set as 10 laps, and the total time was 0.413793 s.

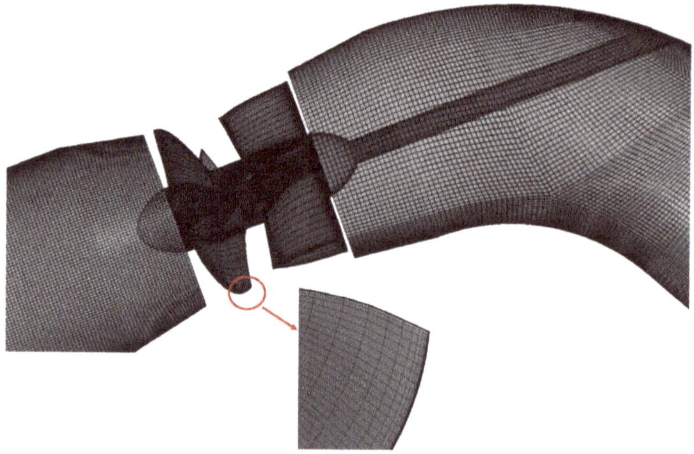

Figure 2. Axial-flow pump and impeller grid.

2.3. Comparative Analysis of Calculation Results and Experiments

Figure 3 shows the test bench of axial-flow pump. In order to verify the feasibility of the model simulation, the external characteristics of the model pump simulation calculation were compared with the test results of the test bench of axial-flow pump. The flow rate was measured by a cologne intelligent electromagnetic flowmeter, the measurement uncertainty of which was ±0.2%; the head was measured by the YOKOGAWA EJA intelligent differential pressure transmitter, the measurement uncertainty of which was ±0.1%; torque speed was measured by the NJL2/500 Nm intelligent torque speed sensor, the measurement uncertainty of which was ±0.1%. Figure 4 shows the comparison between the experimental results and numerical calculation results of the model pump under different flow rates. The trend of CFD numerical calculation results is the same as that of the test results. The simulated operating efficiency of the axial-flow pump was slightly lower than the experimental value, and the error under the condition of a small flow rate was greater than that under the condition of a large flow rate, and the maximum relative error was 3.1%. The simulated head of the axial-flow pump was slightly lower than the test value under the condition of a small flow, and slightly higher than the test value under the condition of the design flow and large flow, and the maximum relative error was 2.1%. In conclusion, the error between the simulation results and the test results is small, and the reliability of the calculation results is good.

(a)

Figure 3. *Cont.*

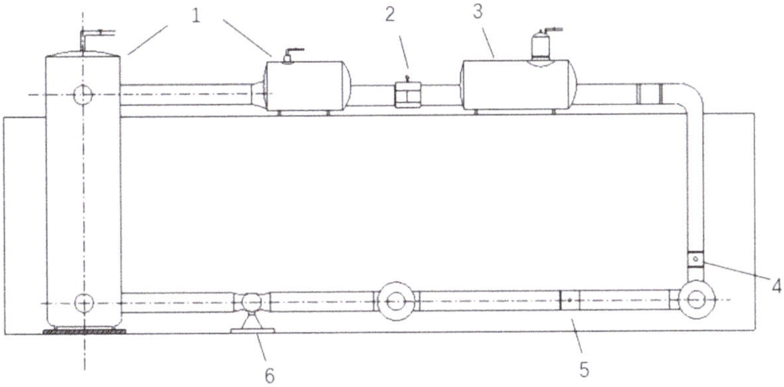

1. High pressure tank 2. Model pump 3. Low pressure tank 4. Regulating valve
5. Electromagnetic flowmeter 6. Circulating pump

(b)

Figure 3. Test bench of axial-flow pump. (**a**) Test instruments. (**b**) Layout of test bench.

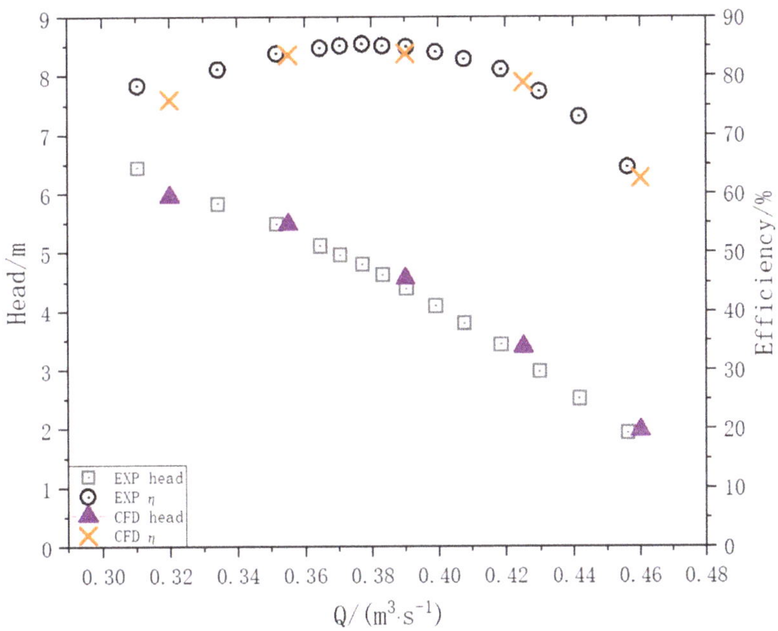

Figure 4. Comparison of model pump external characteristics test and numerical calculation results.

3. Entropy Production Analysis of Axial-Flow Pump with Different Tip Clearances under Pump and Turbine Conditions

3.1. Entropy Production Theory

According to the second law of thermodynamics, the fluid in the system is always accompanied by energy dissipation in the process of flow. Reynolds stress and the viscosity of the fluid itself will lead to irreversible energy dissipation in the flow process of the fluid inside the pump, while the eddy phenomenon generated in the flow will increase the energy dissipation, which will lead to an increase in entropy generation and hydraulic dissipation. The distribution of hydraulic loss in the axial-flow pump can be analyzed by entropy production theory.

Axial-flow pump internal entropy production $\overline{\left(\frac{\Phi}{T}\right)}$ is caused by direct entropy production $S_{PRO,\overline{D}}$ and the turbulent entropy production $S_{PRO,D'}$; the calculation formula is as follows:

$$\overline{\left(\frac{\Phi}{T}\right)} = S_{PRO,\overline{D}} + S_{PRO,D'} \tag{1}$$

$$S_{PRO,\overline{D}} = \frac{\mu}{\overline{T}} \cdot \left\{ 2\left[\left(\frac{\partial \overline{v_1}}{\partial x}\right)^2 + \left(\frac{\partial \overline{v_2}}{\partial y}\right)^2 + \left(\frac{\partial \overline{v_3}}{\partial z}\right)^2 \right] + \left(\frac{\partial \overline{v_1}}{\partial y} + \frac{\partial \overline{v_2}}{\partial x}\right)^2 + \left(\frac{\partial \overline{v_1}}{\partial z} + \frac{\partial \overline{v_3}}{\partial x}\right)^2 + \left(\frac{\partial \overline{v_2}}{\partial z} + \frac{\partial \overline{v_3}}{\partial y}\right)^2 \right\} \tag{2}$$

$$S_{PRO,\,D'} = \frac{\mu}{\overline{T}} \cdot \left\{ 2\left[\left(\frac{\partial v_1'}{\partial x}\right)^2 + \left(\frac{\partial v_2'}{\partial y}\right)^2 + \left(\frac{\partial v_3'}{\partial z}\right)^2 \right] + \left(\frac{\partial v_1'}{\partial y} + \frac{\partial v_2'}{\partial x}\right)^2 + \left(\frac{\partial v_2'}{\partial z} + \frac{\partial v_3'}{\partial x}\right)^2 + \left(\frac{\partial v_2'}{\partial z} + \frac{\partial v_3'}{\partial y}\right)^2 \right\} \tag{3}$$

Among them, $\overline{v_1}$, $\overline{v_2}$, $\overline{v_3}$ are the components of the average time velocity on the coordinate x, y, z axes; v_1', v_2', v_3' are velocity pulsations on the x, y, z axes; μ is the dynamic viscosity; $\overline{T}$ is the time average temperature.

In order to solve the problem that $S_{PRO,D'}$ in the formula cannot be calculated by solving the Reynolds equation, Kock [29,30] and Herwig [31,32] proposed Equation (4),

$$S_{PRO,D'} = \frac{\rho \varepsilon_1}{\overline{T}} \tag{4}$$

where ρ is the fluid density and ε_1 is the dissipation rate of turbulent kinetic energy.

The direct dissipation rate $\Phi_{\overline{D}}$ and turbulent dissipation rate $\Phi_{D'}$ can be calculated from Equations (5) and (6):

$$\Phi_{\overline{D}} = \overline{T} \cdot S_{PRO,\overline{D}} \tag{5}$$

$$\Phi_{D'} = \overline{T} \cdot S_{PRO,D'} \tag{6}$$

Thus, the direct dissipation and turbulent dissipation of each hydraulic component can be obtained by volume integration of Equations (5) and (6), as shown in Equations (7) and (8).

$$P_{\overline{D}} = \int_V \Phi_{\overline{D}} dV \tag{7}$$

$$P_{D'} = \int_V \Phi_{D'} dV \tag{8}$$

Therefore, the total dissipation rate Φ_T and total dissipation P_T can be calculated from Equations (9) and (10):

$$\Phi_T = \Phi_{\overline{D}} + \Phi_{D'} \tag{9}$$

$$P_T = P_{\overline{D}} + P_{D'} \tag{10}$$

3.2. *Analysis of Calculation Results*

3.2.1. Influence of Tip Clearance on External Characteristics

Figure 5 shows the comparison of the influences of different tip clearances on the external characteristic parameters of the axial-flow pump operating under pump conditions. Under the pump condition, both the head and shaft power of the axial-flow pump decreased with the increase in the flow rate. Under the same flow condition, both the head and shaft power decreased with the increase in tip clearance. Under the conditions of $0.8Q_{des}$, $1.0Q_{des}$ and $1.2Q_{des}$, the head decreased by 10.0%, 15.5% and 25.7%. At $0.8Q_{des}$, $1.0Q_{des}$ and $1.2Q_{des}$, the shaft power decreased by 3.2%, 6.3% and 12.3%. When the flow rate increased from $0.8Q_{des}$ to $1.0Q_{des}$, the efficiency increased; when the flow rate increased from $1.0Q_{des}$ to $1.2Q_{des}$, the efficiency decreased.

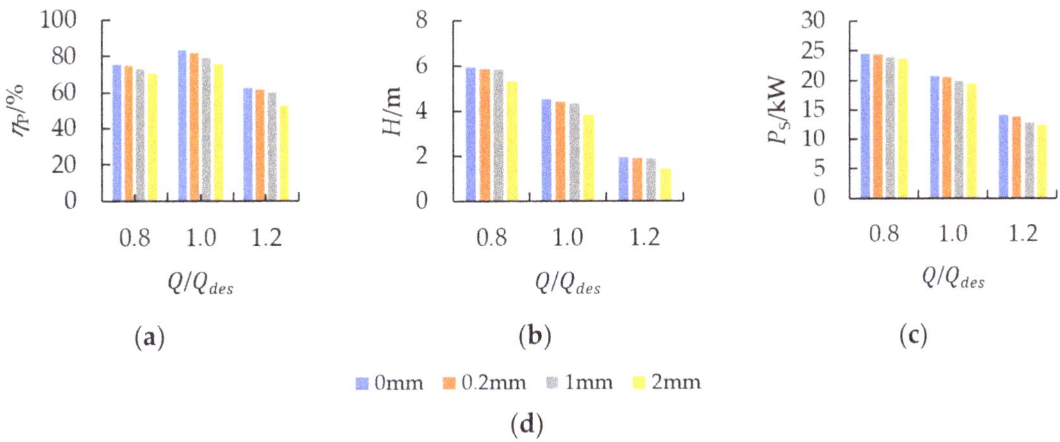

Figure 5. Comparison of (**a**) efficiency; (**b**) head; (**c**) shaft power under different tip clearances in pump condition; (**d**) Legend.

Figure 6 shows the comparison of the influences of different tip clearances on the external characteristic parameters of the axial-flow pump under turbine conditions. Under the turbine condition, both the head and shaft power showed a rising trend with the increase in flow rate. Under the same flow condition, both the head and shaft power decreased with the increase in tip clearance. The decreases in the head were 18.5%, 12.0% and 7.9% under $0.8Q_{des}$, $1.0Q_{des}$ and $1.2Q_{des}$ conditions. At $0.8Q_{des}$, $1.0Q_{des}$ and $1.2Q_{des}$, the shaft power decreased by 28.8%, 15.9% and 11.5%. When the flow rate increased from $0.8Q_{des}$ to $1.0Q_{des}$, the efficiency increased; when the flow rate increased from $1.0Q_{des}$ to $1.2Q_{des}$, the efficiency decreased. According to the formula:

$$H = \frac{P_{out} - P_{in}}{\rho g} \tag{11}$$

where H—head, m; P_{out}—the total pressure of outlet, Pa; P_{in}—the total pressure of inlet, Pa; ρ—the fluid density, kg/m^3; g—the acceleration of gravity, m/s^2; P_S—the shaft power, kW; η_P—the efficiency of axial-flow pump; η_T—the efficiency of axial-flow pump as a turbine; Q—the flow rate, kg/s; M—the torque of impeller, N·m; ω—the rotating speed, r/min.

It can be seen that the efficiency of the axial-flow pump was directly proportional to the head and inversely proportional to the shaft power, and the efficiency of axial-flow pump as a turbine was directly proportional to the shaft power and inversely proportional to the head. The efficiency of the axial-flow pump decreased greatly under the condition of a large flow rate, while the efficiency of the axial-flow pump as a turbine decreased slightly under the condition of a large flow rate. With the increase in tip clearance, the efficiency of the axial-flow pump showed a decreasing trend under both pump and turbine condition. In the pump condition, the maximum decrease was 7.1%, 9.8% and 15.3% under $0.8Q_{des}$, $1.0Q_{des}$ and $1.2Q_{des}$ conditions, respectively. Under turbine conditions, the maximum decline was 12.7%, 4.5% and 3.9% under $0.8Q_{des}$, $1.0Q_{des}$ and $1.2Q_{des}$ conditions, respectively.

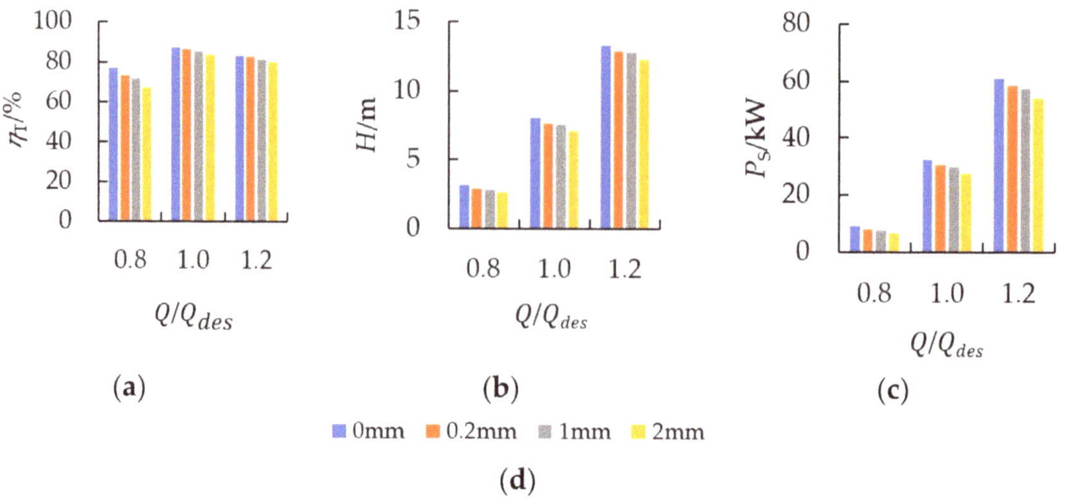

Figure 6. Comparison of (**a**) efficiency; (**b**) head; (**c**) shaft power under different tip clearances in turbine condition; (**d**) Legend.

3.2.2. Dissipation Distribution Based on Entropy Production Theory

Figure 7 shows the influence of tip clearance on total dissipation under the axial-flow pump condition. In all flow rate conditions, the total dissipation increased with the increase in tip clearance. With the increase in flow rate, the influence of tip clearance change on total dissipation decreased. Under the $0.8Q_{des}$ condition, the total dissipation increased by 47.7%; under the $1.0Q_{des}$ condition, the total dissipation increased by 45.0%; under the $1.2Q_{des}$ condition, the total dissipation increased by 11.2%.

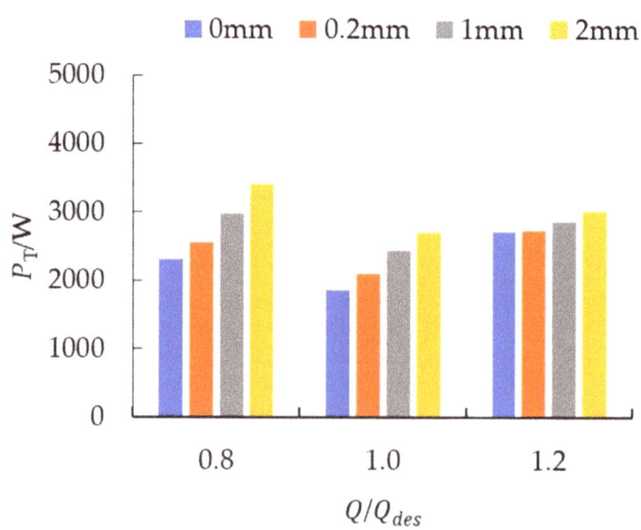

Figure 7. Influence of tip clearance on total dissipation under pump condition.

Figure 8 shows the influence of tip clearance on total dissipation under the turbine condition of the axial-flow pump. Under all flow rate conditions, the total dissipation increased with the increase in tip clearance, and with the increase in flow rate, the influence caused by tip clearance changes also increased. Under the $0.8Q_{des}$ condition, the total dissipation increased by 7.9%; under the $1.0Q_{des}$ condition, the total dissipation increased by 10.5%; under the $1.2Q_{des}$ condition, the total dissipation increased by 14.9%.

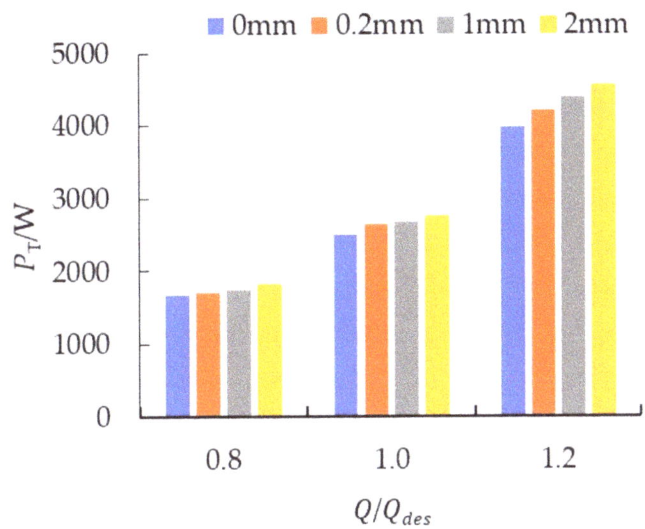

Figure 8. Influence of tip clearance on total dissipation under turbine condition.

Figure 9 shows the influence of tip clearance on the distribution of total dissipation inside the pump under different flow rate conditions. Tip clearance had an obvious influence on the impeller, guide vane and outlet passage, but had little influence on the inlet passage. Under all three flow conditions, the total dissipation of the impeller and outlet passage increased with the increase in tip clearance, and the influence of tip clearance was more obvious with the increase in flow rate. Under $0.8Q_{des}$,$1.0Q_{des}$ and $1.2Q_{des}$, the total dissipation of the impeller increased with the increase in tip clearance of 114.8%, 53.9% and 7.6%, respectively. The total dissipation of the outlet passage increased by 18.5%, 54.2% and 23.5% with the increase in tip clearance. For the guide vane, the total dissipation decreased with the increase in tip clearance at $0.8Q_{des}$, with a maximum decrease of 8.4%. Under the condition of $1.0Q_{des}$, the total dissipation increased with the increase in tip clearance, with a maximum increase of 32.1%. Under the condition of $1.2Q_{des}$, the total dissipation of the guide vane decreased first and then increased with the increase in tip clearance.

Figure 10 shows the influence of tip clearance on the distribution of the internal dissipation of the pump turbine under different flow rate conditions. Tip clearance had an obvious influence on the impeller and outlet passage, but had little influence on the guide vane and inlet passage. Under the three flow rate conditions, the total dissipation in the impeller increased with the increase in tip clearance, and the smaller the flow rate was, the more obvious the influence of tip clearance was. The total dissipation increased by 4.9%, 22.7% and 42.2% with tip clearance at $0.8Q_{des}$, $1.0Q_{des}$ and $1.2Q_{des}$, respectively. For the outlet passage, the total dissipation increased with the increase in tip clearance at $0.8Q_{des}$, and the maximum increase was 31.1%. However, under the conditions of $1.0Q_{des}$ and $1.2Q_{des}$, the total dissipation decreased first and then increased with the increase in tip clearance.

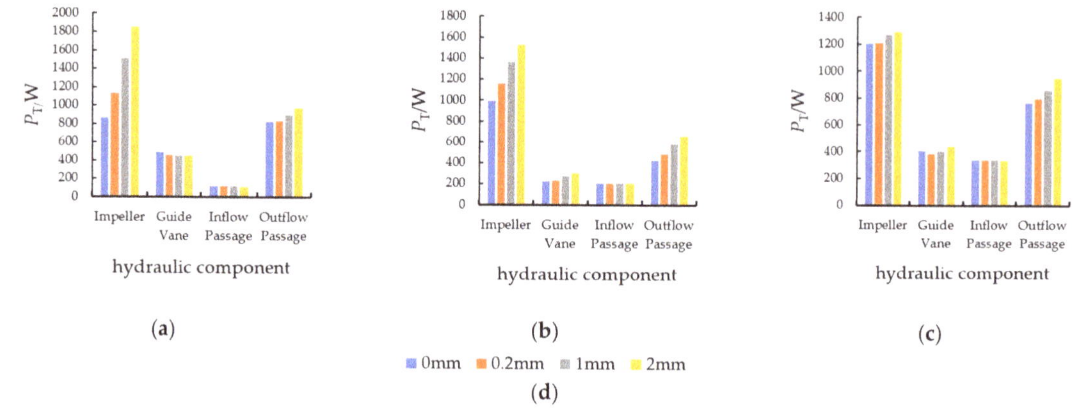

Figure 9. Influence of tip clearance on total dissipation under pump condition; (**a**) $0.8Q_{des}$; (**b**) $1.0Q_{des}$; (**c**)$1.2Q_{des}$; (**d**) Legend.

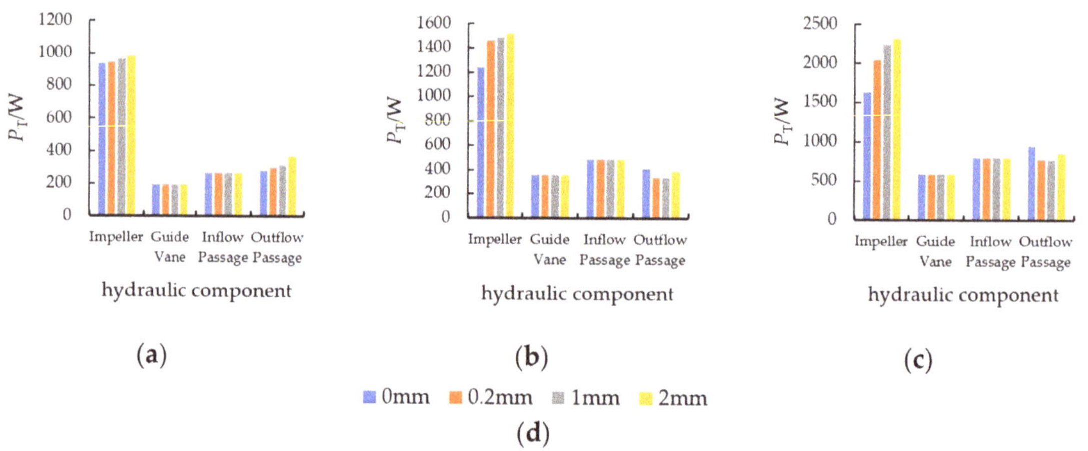

Figure 10. Influence of tip clearance on total dissipation under turbine condition; (**a**) $0.8Q_{des}$; (**b**) $1.0Q_{des}$; (**c**) $1.2Q_{des}$; (**d**) Legend.

Figure 11 shows the comparison of turbulence dissipation and direct dissipation caused by different tip clearance schemes in the pump and turbine conditions of the axial-flow pump under the design flow rate condition. As can be seen from the figure, the total dissipation generated by the direct loss of the axial-flow pump under pump and turbine conditions was very small compared with the turbulence loss. The maximum direct dissipation of the pump and turbine conditions was only 3.1% of the total dissipation, and the maximum direct dissipation of the turbine condition was only 2.8% of the total dissipation. Therefore, the dissipation inside the pump was mainly affected by turbulence dissipation.

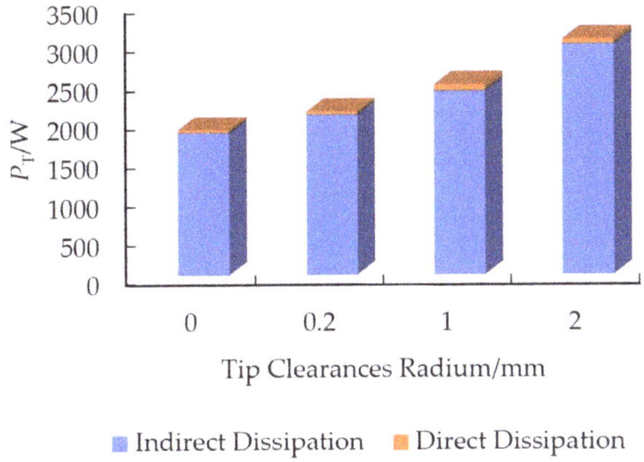

(**a**)

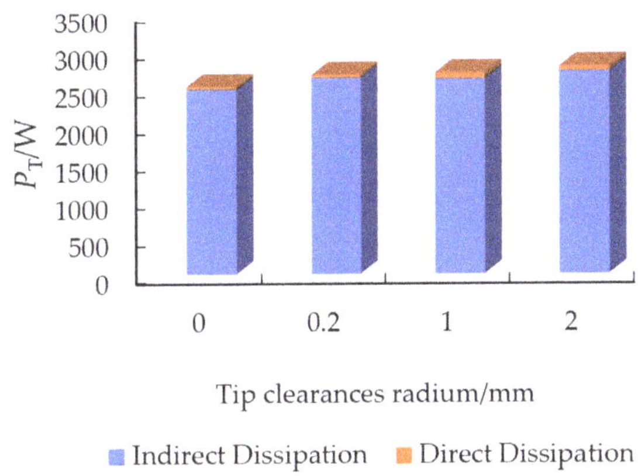

(**b**)

Figure 11. Comparison of direct dissipation and turbulence dissipation; (**a**) Pump condition; (**b**) Turbine condition.

Figure 12 shows the distribution of the blade surface dissipation rates with different tip clearances under the design flow rate of the pump condition. The dissipation rate from the hub to the rim on the blade surface increased gradually. A high dissipation area was generated at the rim on the pressure side. At the angle between the leading edge and the rim, a high dissipation area was formed on the suction side. On the suction side, the high dissipation area was significantly greater than the pressure side. This is due to the fact that the pressure side of the blade was mainly affected by the fluid circumferential velocity, and the suction side was mainly affected by fluid axial velocity; the fluid in the impeller had a great impact on the suction side of the blade. In addition, due to the increase in blade tip clearance, leakage and backflow phenomena at the clearance were intensified, leading to a gradual increase in the speed loss at the rim, and then the dissipation rate of the rim.

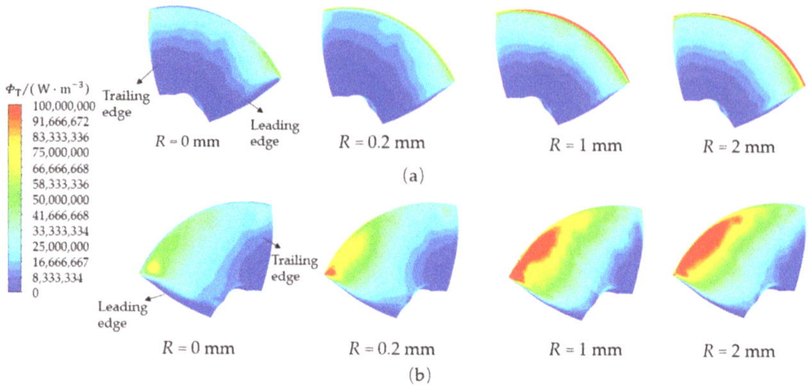

Figure 12. Dissipation rate distribution on blade surface under pump condition. (**a**) Pressure side. (**b**) Suction side.

In order to more intuitively explain the influence of tip clearance on the axial-flow pump, the impeller and guide vanes were expanded radially on a cylindrical surface, as shown in Figure 13. The radial coefficient of the expanded surface r^* can be calculated by Equation (12):

$$r^* = \frac{(r - r_h)}{(r_t - r_h)} \tag{12}$$

where r is the radius of the cylinder, mm; r_h is the radius of the hub, mm; r_t is the radius of the rim, mm.

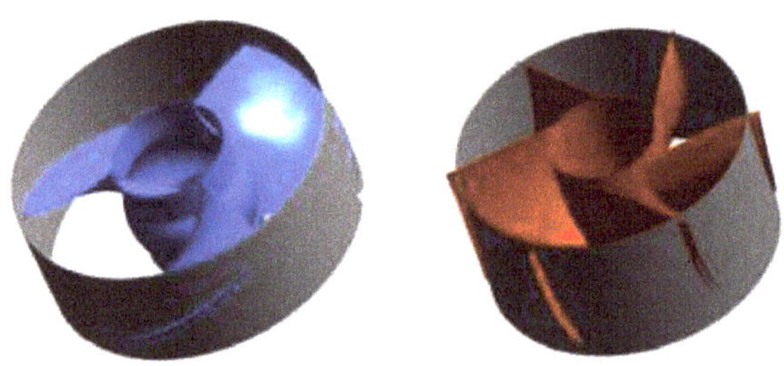

Figure 13. Cylinder section of impeller and guide vane.

Figure 14 shows the cloud diagram of velocity distribution on the cylindrical section of the impeller under different tip clearance schemes and the pump design flow rate condition ($r^* = 0.97$). The flow velocity inside the impeller decreased from the impeller inlet to the impeller outlet, and as a result of the existence of the blade tip clearance, part of the fluid through the blade tip clearance produced return loss and leakage caused by impeller import flow velocity. The bigger the tip clearance, the greater the reflux and leakage phenomena; therefore, the increase in tip clearance speed reduced the impeller inlet from the front.

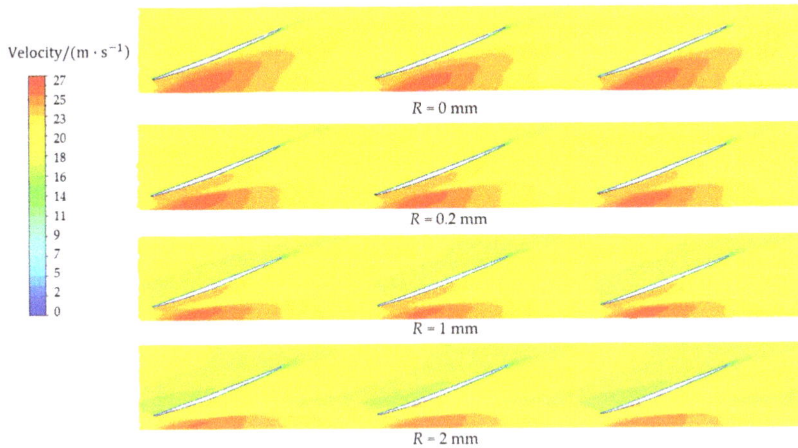

Figure 14. Cylindrical distribution of impeller internal velocity under pump condition ($r^* = 0.97$).

Figure 15 shows the distribution of the total dissipation rate on the cylindrical section of the impeller under different tip clearance schemes and the pump design flow rate condition ($r^* = 0.97$). The velocity loss caused by the backflow and leakage of some fluid in the impeller passing through the tip clearance led to an increase in the rate of dissipation; therefore, a high dissipation area was generated on the suction side of the impeller. With the increase in tip clearance, the dissipation rate on the suction side of the impeller increased gradually.

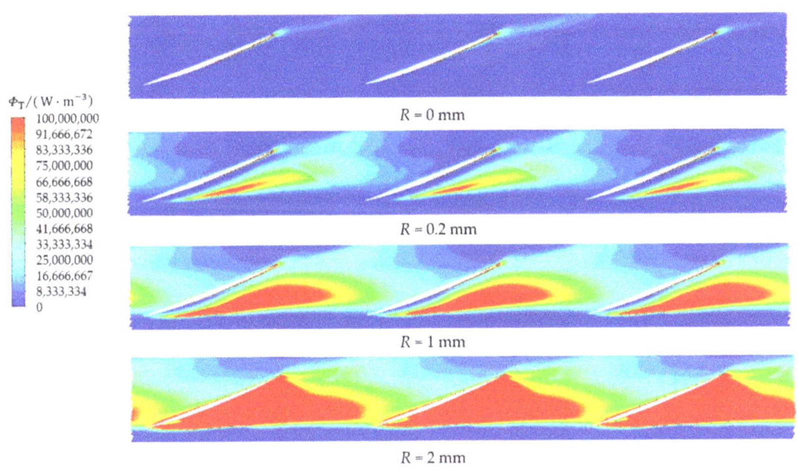

Figure 15. Cylinder distribution of internal dissipation rate of impeller under pump condition ($r^* = 0.97$).

Figure 16 shows the cloud diagram of velocity distribution on the cylindrical section of the guide vane under different tip clearance schemes and the pump design flow rate condition ($r^* = 0.99$). The kinetic energy of the fluid was transformed into pressure energy after passing the guide vane; therefore, the flow velocity inside the guide vane decreased from the inlet to the outlet; there was a large amount of backflow at the outlet of the guide vane, which led to a low-speed zone here, and a point with the lowest speed was formed at the guide vane inlet due to the mixture of backflow and mainstream. The impeller clearance also caused speed loss, resulting in a gradual decrease in the internal flow velocity of the guide vane with the increase in tip clearance.

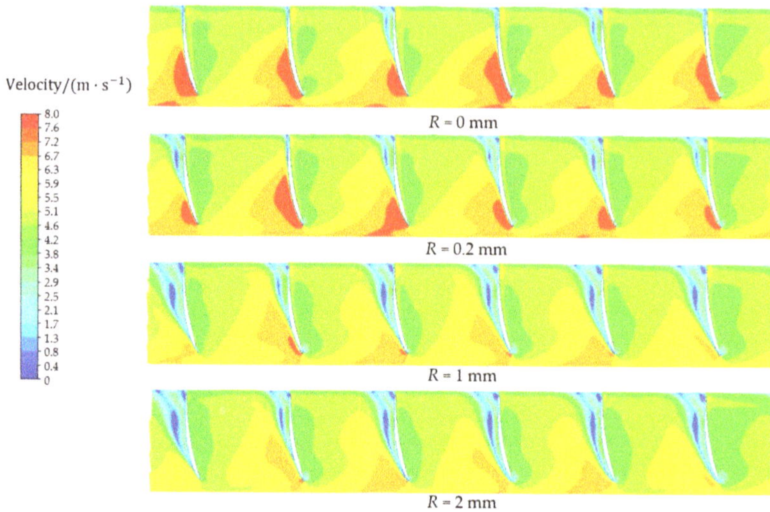

Figure 16. Cylindrical distribution of guide vane internal velocity under pump conditions ($r^* = 0.99$).

Figure 17 shows the distribution of the total dissipation rate on the cylindrical section of the guide vane under different tip clearance schemes and the pump design flow rate condition ($r^* = 0.99$). The high dissipation area inside the guide vane was the suction side and the inlet of the guide vane, where the velocity gradient was large, leading to the generation of high dissipation. The high dissipation area at the suction side was caused by the impact of backflow; the velocity gradient near the inlet of guide vane was large, which also resulted in high entropy production.

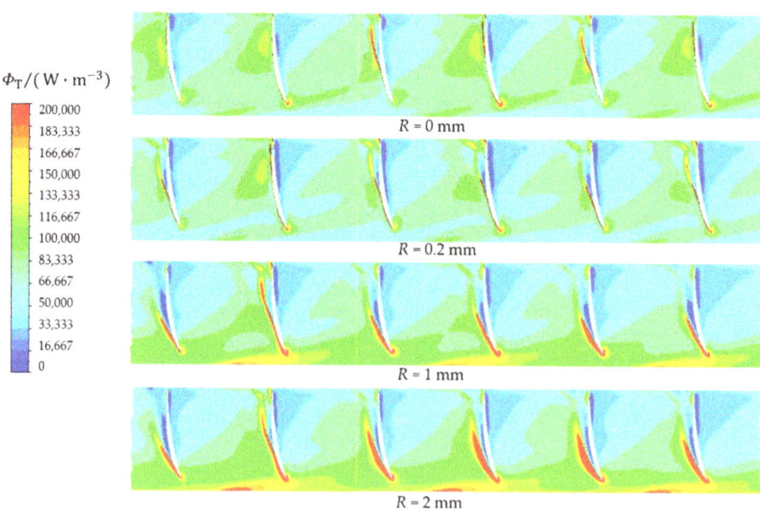

Figure 17. Cylinder distribution of internal dissipation rate of guide vane under pump condition ($r^* = 0.99$).

Figure 18 shows the internal dissipation rate distribution of the inlet passage with different tip clearances under the pump design flow rate condition. Energy dissipation in the inlet passage mainly occurred at the inlet, the bend of the inner wall of the elbow and the junction of the rotating shaft and the outer wall. As the fluid in the inlet passed through the inlet transition section, the bend of the passage wall and the rotation around the rotating shaft, a certain degree of flow separation occurred, which resulted in energy dissipation.

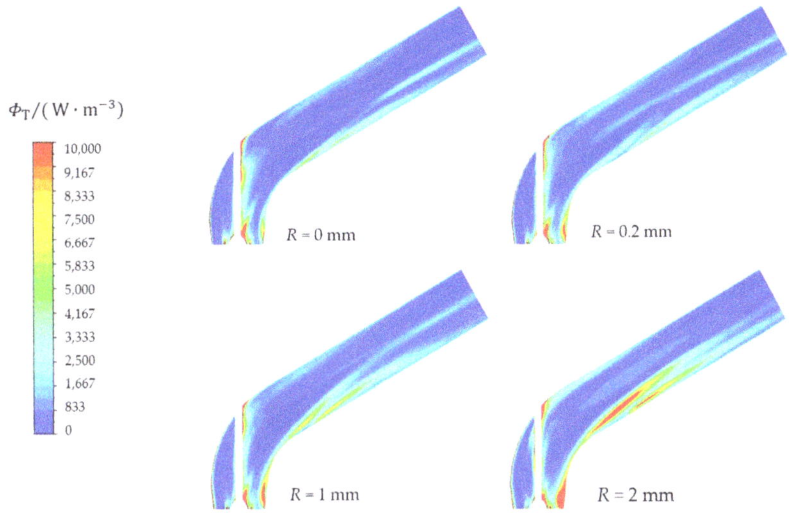

Figure 18. Distribution of internal dissipation rate in outflow passage under pump condition.

Figure 19 shows the distribution of blade surface dissipation rates with different tip clearances under the design flow rate of the turbine condition. The surface dissipation rate of the blade pressure side increased gradually from the hub to the rim and produced a higher dissipation at the rim. The suction side produced a high dissipation area at the angle between the rim and the leading edge; the area of the high dissipation rate on the suction side was obviously larger than that on the pressure side. This shows that when the impeller runs under the turbine condition, the fluid cannot pass through the suction side properly, which has a great impact on it. In addition, with the increase in blade tip clearance, the leakage and backflow in the clearance were intensified, and as a result, the speed loss at the rim increased gradually, and then the dissipation rate of the rim increased gradually.

Figure 20 shows the cloud diagram of velocity distribution on the cylindrical section of the impeller under different tip clearance schemes and the turbine design flow rate condition ($r^* = 0.97$). The flow velocity on the suction side of the impeller was obviously greater than that on the pressure side; under the turbine condition, the fluid could not flow well over the suction side surface, and obvious flow separation occurred at the leading and trailing edges of the impeller; therefore, a low-speed zone was generated at the leading and trailing edges of the impeller.

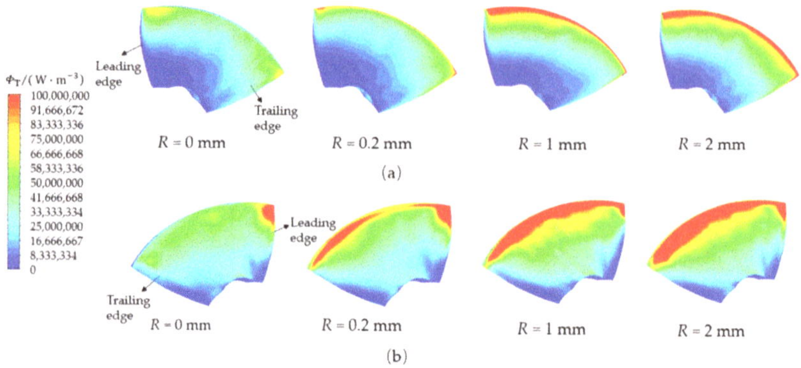

Figure 19. Dissipation rate distribution on blade surface under turbine condition. (**a**) Pressure side. (**b**) Suction side.

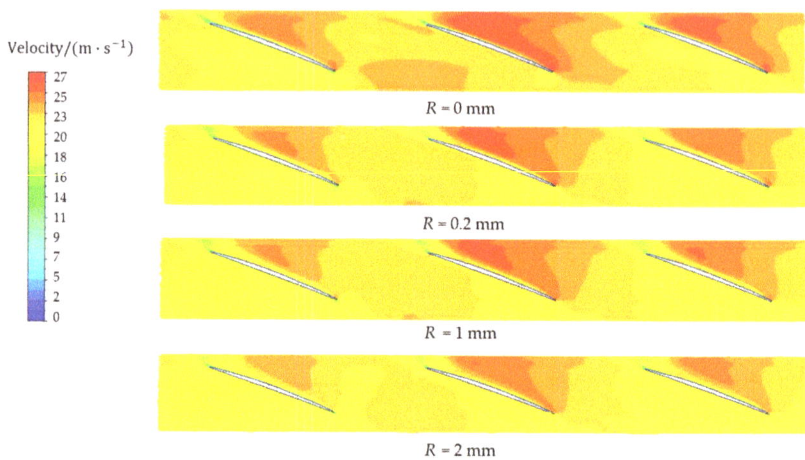

Figure 20. Cylindrical distribution of impeller internal velocity under turbine condition ($r^* = 0.97$).

Figure 21 shows the distribution of the total dissipation rate on the cylindrical section of the impeller under different tip clearance schemes and the turbine design flow rate condition ($r^* = 0.97$). As the fluid near the blade suction side could not fit the blade surface effectively. This had a great impact and flow separation phenomenon and caused a large velocity gradient on the blade suction side. There was a high dissipation area on the suction side of the impeller. High energy dissipation also occurred at the blade outlet due to flow separation. With the increase in tip clearance, the dissipation rate of impeller working surface and outlet increased gradually.

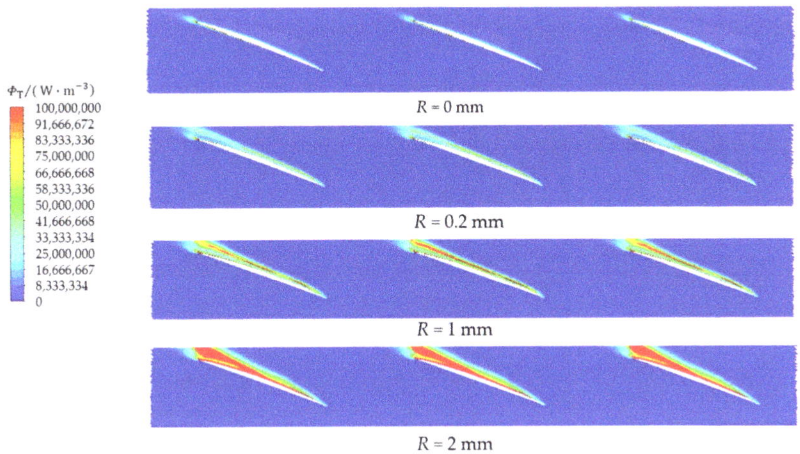

Figure 21. Cylinder distribution of internal dissipation rate of impeller under turbine condition ($r^* = 0.97$).

Figure 22 shows the distribution of the dissipation rate of the outlet passage with different tip clearances under the turbine design flow rate condition. There was no rear guide vane when the axial-flow pump operated as a turbine. The fluid at the outlet of the impeller lacked the rectifier action of the guide vane. There was a high dissipation area at the inlet of the outlet passage. With the increase in tip clearance, the dissipation rate at the inlet of outlet passage increased. Moreover, due to the effect of the passage wall, the kinetic energy flowing out of the impeller was converted into pressure energy; therefore, the dissipation rate of the outlet passage decreased from the inlet to the outlet.

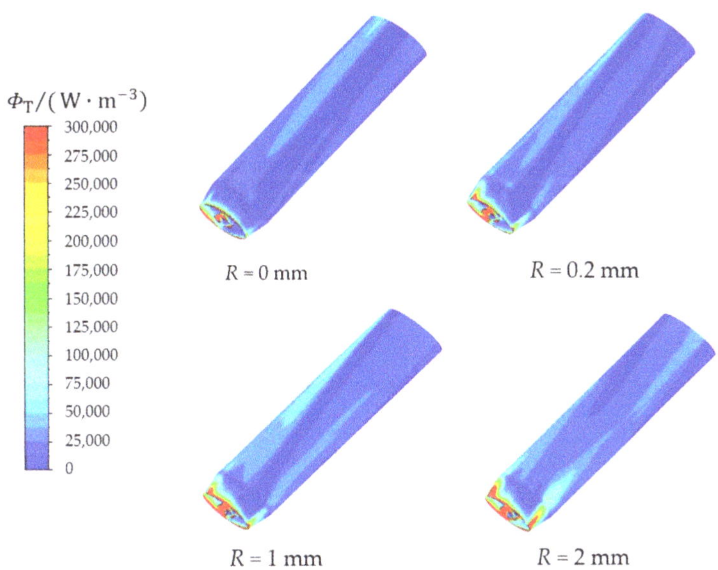

Figure 22. Distribution of internal dissipation rate in outflow passage under turbine condition.

4. Conclusions

In this paper, a numerical simulation method was adopted to study the energy loss of the axial-flow pump under both pump and turbine conditions based on entropy production theory, and four different tip clearance schemes were analyzed. The following conclusions can be drawn:

1. The efficiency, head and shaft power of the axial-flow pump under pump and turbine conditions all showed a downward trend with the increase in tip clearance. The maximum decline occurred at the $1.2Q_{des}$ flow rate under pump condition and the $0.8Q_{des}$ flow rate under turbine condition. Compared with the 0 mm clearance, the maximum decreases in pump efficiency, head and shaft power under 2 mm tip clearance were 15.3%, 25.7% and 12.3%, respectively, under the pump condition, and 12.7%, 18.5% and 28.8% under the turbine condition.
2. Under the pump condition of the axial-flow pump, the change in tip clearance had an obvious effect on the total dissipation of the impeller, guide vane and outlet passage. Under the flow rate of $1.0Q_{des}$, the maximum variation in the total dissipation of the impeller, guide vane and outlet passage was 53.9%, 32.1% and 54.2%, respectively. Under the turbine condition of the axial-flow pump, the change in tip clearance had an obvious effect on the total dissipation of the impeller and outlet passage. Under the flow rate of $1.0Q_{des}$, the maximum variation in the total dissipation of the impeller and outlet passage was 22.7% and 17.4%, respectively. The total dissipation of the axial-flow pump increased with the increase in tip clearance under both pump and turbine conditions, in addition, under the flow rate of $1.0Q_{des}$, the change in the total dissipation under the pump condition with the increase in tip clearance was significantly greater than that under the turbine condition, and the total dissipation was 45.0% and 10.5%, respectively.
3. At the design flow rate, when the axial-flow pump ran under both pump and turbine conditions, the dissipation rate of the pressure side was lower than that of the suction side. When the axial-flow pump operated under the pump condition, there was a high dissipation area at the rim of the impeller, the impeller inlet. When the axial-flow pump operated under the turbine condition, at the rim impeller, the impeller outlet produced a high dissipation area. With the increase in tip clearance, total dissipation in the impeller increased gradually.

Therefore, a smaller tip clearance should be selected when the pump condition is the main condition; however, a larger tip clearance can be selected when the turbine condition is the main condition. The study result shows the relationship between blade tip clearance and dissipation loss in the pump and its spatial distribution under both pump and turbine conditions. This provides suggestions for the optimization of the axial-flow pump as a turbine, to find suitable tip clearance in the axial-flow pump as a turbine, and provides a reference for solving the problems of large vibration and noise, low operating efficiency caused by clearance that is too large, and manufacturing difficulty caused by clearance that is too small.

Author Contributions: Conceptualization, Y.L. and Q.L.; methodology, Y.L. and F.M.; software, Q.L.; validation, Y.L. and Q.L.; formal analysis, Q.L.; investigation, Q.L.; resources, Q.L.; data curation, Y.L. and Q.L.; writing—original draft preparation, Q.L.; writing—review and editing, Q.L. and F.M.; visualization, Q.L.; supervision, Y.Z. and X.X.; project administration, Y.L.; funding acquisition, Y.L. All authors have read and agreed to the published version of the manuscript.

Funding: This research was funded by National Natural Science Foundation of China, grant number 51809120 and Industry-University-Research Collaboration of Jiangsu Province, grant number BY2020373.

Institutional Review Board Statement: Not applicable.

Informed Consent Statement: Not applicable.

Data Availability Statement: Not applicable.

Conflicts of Interest: The authors declare no conflict of interest.

Nomenclature

Q_{des} (m^3/s)	design flow rate under pump condition
Q'_{des} (m^3/s)	design flow rate under turbine condition
n (r/min)	rotational speed
R (mm)	tip clearance radius
$S_{PRO,\overline{D}}(W \cdot K/m^3)$	direct dissipation
$S_{PRO,\,D'}(W \cdot K/m^3)$	turbulent dissipation
$\overline{v_1}, \overline{v_2}, \overline{v_3}$ (m/s)	time average velocity on the coordinate x, y, z axes
v'_1, v'_2, v'_3(m/s)	velocity pulsations on the x, y, z axes
μ(Pa·s)	dynamic viscosity
$\overline{T}$(K)	time average temperature
ρ (kg/m^3)	fluid density
ε_1 (W/kg)	dissipation rate of turbulent kinetic energy
$\Phi_{\overline{D}}$ (W/m^3)	direct dissipation rate
$\Phi_{D'}$$(W/m^3)$	turbulent dissipation rate
$P_{\overline{D}}$ (W)	direct dissipation
$P_{D'}$ (W)	turbulent dissipation
Φ_T (W/m^3)	total dissipation rate
P_T (W)	total dissipation
H(m)	head
P_{out}(Pa)	total pressure of outlet
P_{in}(Pa)	total pressure of inlet
$\rho(kg/m^3)$	fluid density
$g(m/s^2)$	acceleration of gravity
P_S(kW)	shaft power
η_P(%)	efficiency of axial-flow pump
η_T(%)	efficiency of axial-flow pump as a turbine
Q(kg/s)	flow rate
M(N·m)	torque of impeller
ω(r/min)	rotating speed
r(mm)	radius of the cylinder
r_h(mm)	radius of the hub
r_t(mm)	radius of the rim

References

1. Wang, W.; Wang, W.; Zhang, L.; Zhao, L.; Lu, J.; Feng, J.; Luo, X. Mechanism for end-wall slots to improve hump in an axial flow pump. *Trans. Chin. Soc. Agric. Eng.* **2020**, *36*, 12–20. (In Chinese)
2. Wang, X.H.; Yang, J.H.; Shi, F.X. Status and prospect of study on energy recovery hydraulic turbines. *J. Drain. Irrig. Mach. Eng.* **2014**, *32*, 742–747.
3. Yang, S.S.; Derakhshan, S.; Kong, F.Y. Theoretical, numerical and experimental prediction of pump as turbine performance. *Renew. Energy* **2012**, *48*, 507–513. [CrossRef]
4. Yang, S.S.; Kong, F.Y.; Jiang, W.M.; Qu, X.Y. Effects of impeller trimming influencing pump as turbine. *Comput. Fluids* **2012**, *67*, 72–78. [CrossRef]
5. Yang, S.S.; Liu, H.L.; Kong, F.Y.; Xia, B.; Tan, L.W. Effects of the radial gap between impeller tips and volute tongue influencing the performance and pressure pulsations of pump as turbine. *J. Fluids Eng.* **2014**, *136*, 054501. [CrossRef]
6. Singh, P.; Nestmann, F. An optimization routine on a prediction and selection model for the turbine operation of centrifugal pumps. *Exp. Therm. Fluid Sci.* **2010**, *34*, 152–164. [CrossRef]
7. Singh, P.; Nestmann, F. Internal hydraulic analysis of impeller rounding in centrifugal pumps as turbines. *Exp. Therm. Fluid Sci.* **2011**, *35*, 121–134. [CrossRef]
8. Joshi, S.; Gordon, A.; Holloway, L.; Chang, L. Selecting a high specific speed pump for low head hydro-electric power generation. In Proceedings of the Canadian Conference on Electrical and Computer Engineering, Saskatoon, SK, Canada, 1–4 May 2005; pp. 603–606.

9. Qian, Z.; Wang, F.; Guo, Z.; Lu, J. Performance evaluation of an axial-flow pump with adjustable guide vanes in turbine mode. *Renew. Energy* **2016**, *99*, 1146–1152. [CrossRef]
10. Li, D.; Wang, H.; Qin, Y.; Han, L.; Wei, X.; Qin, D. Entropy production analysis of hysteresis characteristic of a pump-turbine model. *Energy Convers. Manag.* **2017**, *149*, 175–191. [CrossRef]
11. Zhang, Y.X.; Hou, H.C.; Xu, C.; He, W.; Li, Z. Application of entropy production method to centrifugal pump energy loss evaluation. *J. Drain. Irrig. Mach. Eng.* **2017**, *288*, 277–282.
12. Pei, J.; Meng, F.; Li, Y.; Yuan, S.; Chen, J. Effects of distance between impeller and guide vane on losses in a low head pump by entropy production analysis. *Adv. Mech. Eng.* **2016**, *8*, 1687814016679568. [CrossRef]
13. Li, D.; Gong, R.; Wang, H.; Xiang, G.; Wei, X.; Qin, D. Entropy Production Analysis for Hump Characteristics of a Pump Turbine Model. *Chin. J. Mech. Eng.* **2016**, *29*, 803–812. [CrossRef]
14. Gong, R.; Wang, H.; Chen, L.; Li, D.; Zhang, H.; Wei, X. Application of entropy production theory to hydro-turbine hydraulic analysis. *Sci. China Technol. Sci.* **2013**, *56*, 1636–1643. [CrossRef]
15. Shi, W.; Li, T.; Zhang, D.; Wang, G.; Zhou, L. Numerical simulation on cavitating characteristic in impeller of axial-flow pump. *Trans. CSAE* **2012**, *28*, 88–93. (In Chinese)
16. Li, W.; Qiao, W.; Xu, K.; Luo, H. Numerical simulation of active control on tip leakage flow in axial turbine. *Chin. J. Aeronaut.* **2009**, *22*, 129–137. [CrossRef]
17. Shi, W.D.; Li, T.T.; Zhang, D.S.; Tian, F.; Zhang, G.J. *Numerical Simulation of Tip Clearance Leakage Vortex Flow Characteristic in Axial Flow Pump*; Institute of Physics Publishing: Bristol, UK, 2012.
18. Shen, S.; Qian, Z.; Ji, B.; Agarwal, R.K. Numerical investigation of tip flow dynamics and main flow characteristics with varying tip clearance widths for an axial-flow pump. *Proc. Inst. Mech. Eng. Part. A J. Power Energy* **2018**, *233*, 476–488. [CrossRef]
19. Chen, G.T.; Greitzer, E.M.; Tan, C.S.; Marble, F.E. Similarity Analysis of Compressor Tip Clearance Flow Structure. *J. Turbomach.* **1991**, *113*, 260–271. [CrossRef]
20. Li, W.; Qiao, W.Y.; Xu, K.F.; Luo, H.L. Numerical simulation of tip clearance flow passive control in axial turbine. *J. Therm. Sci.* **2008**, *17*, 147–155. [CrossRef]
21. Han, W.; Liu, Y.; Gong, C.; Su, Y.; Guo, P.; Su, M.; Wei, Z. Effect of tip clearance on performance of contra-rotating axial flow water-jet propulsion pump. *Mod. Phys. Lett. B* **2020**, *34*, 2050094. [CrossRef]
22. Shi, L.; Zhu, J.; Tang, F.; Wang, C. Multi-Disciplinary optimization design of axial-flow pump impellers based on the approximation model. *Energies* **2020**, *13*, 779. [CrossRef]
23. Wang, H.; Long, B.; Wang, C.; Han, C.; Li, L. Effects of the impeller blade with a slot structure on the centrifugal pump performance. *Energies* **2020**, *13*, 1628. [CrossRef]
24. Zhang, L.; Wang, C.; Zhang, Y.; Xiang, W.; He, Z.; Shi, W. Numerical study of coupled flow in blocking pulsed jet impinging on a rotating wall. *J. Braz. Soc. Mech. Sci. Eng.* **2021**, *43*, 508. [CrossRef]
25. Fujun, W. *Computational Fluid Dynamics Analysis—CFD Principles and Application*; Tsinghua University Press: Beijing, China, 2004.
26. Yang, Y.; Zhou, L.; Zhou, H.; Lv, W.; Wang, J.; Shi, W.; He, Z. Orthogonal optimum design method for high specific-speed axial-flow pumps. *J. Vib. Shock.* **2018**, *37*, 115–121. (In Chinese)
27. Feng, J.; Luo, X.; Guo, P.; Wu, G. Influence of tip clearance on pressure fluctuations in an axial flow pump. *J. Mech. Sci. Technol.* **2016**, *30*, 1603–1610. [CrossRef]
28. Yang, S.; Kong, F.; Chen, B. Research on Pump Volute Design Method Using CFD. *Int. J. Rotating Mach.* **2011**, *2011*, 137860. [CrossRef]
29. Kock, F.; Herwig, H. Entropy production calculation for turbulent shear flows and their implementation in cfd codes. *Int. J. Heat Fluid Flow* **2005**, *26*, 672–680. [CrossRef]
30. Kock, F.; Herwig, H. Local entropy production in turbulent shear flows: A high-Reynolds number model with wall functions. *Int. J. Heat Mass Transf.* **2004**, *47*, 2205–2215. [CrossRef]
31. Herwig, H.; Kock, F. Direct and indirect methods of calculating entropy generation rates in turbulent convective heat transfer problems. *Heat Mass Transf.* **2007**, *43*, 207–215. [CrossRef]
32. Herwig, H.; Kock, F. Local entropy production in turbulent shear flows: A tool for evaluating heat transfer performance. *J. Therm. Sci.* **2006**, *15*, 159–167. [CrossRef]

machines

MDPI

Article

Correlation between the Internal Flow Pattern and the Blade Load Distribution of the Centrifugal Impeller

Bo Chen *, Xiaowu Chen, Zuchao Zhu and Xiaojun Li

Key Laboratory of Fluid Transmission Technology of Zhejiang Province, Zhejiang Sci-Tech University, Hangzhou 310018, China; cxpkevinsyt@gmail.com (X.C.); zhuzuchao@zstu.edu.cn (Z.Z.); lixj@zstu.edu.cn (X.L.)
* Correspondence: chenbofluid@zstu.edu.cn

Abstract: The blade load distributions reflect the working characteristics of centrifugal impellers, and the vortexes in the impeller channel affect the blade load distribution, but the mechanism of this phenomenon is still unclear. In this study, particle image velocimetry (PIV) was adopted to clarify the correlation between the internal flow pattern and the blade load distribution. The internal flow pattern and the blade load distribution were presented under different working conditions to study the influence of the internal flow pattern on the blade load. Results showed that the vortexes in the flow channel redistributed the blade load. The clockwise vortex made the position of the maximum blade load closer to the outlet, while the counterclockwise vortex had the opposite effect. Meanwhile, the vortexes caused the blade load distribution to be steeper, which reduced energy conversion efficiency. Moreover, the mean absolute flow angle was introduced to explain the mechanism of the effects of vortexes on blade load. The results can be used as a theoretical basis for the design of high-performance impellers.

Keywords: centrifugal impeller; blade load distribution; particle image velocimetry (PIV); internal flow patterns; the mean absolute flow angle

Citation: Chen, B.; Chen, X.; Zhu, Z.; Li, X. Correlation between the Internal Flow Pattern and the Blade Load Distribution of the Centrifugal Impeller. *Machines* **2022**, *10*, 40. https://doi.org/10.3390/machines10010040

Academic Editor: Davide Astolfi

Received: 22 November 2021
Accepted: 28 December 2021
Published: 5 January 2022

1. Introduction

The centrifugal pump is an important energy conversion device. Many researchers adopted various methods to analyze the flow characteristics of the pump and its influence on the pump performance [1–3]. The blade load distribution is a main parameter that reflects the working characteristics of centrifugal impellers. The blade load has gradually gained people's attention [4–6]. Studying the correlation between internal flow pattern and blade load will deepen the understanding of impeller flow pattern and stable operation of the pump.

Many studies have been carried out to explain the flow pattern in pumps. Wang et al. [7] adopted the PIV experiment to measure the flow characteristics in the five-blade centrifugal impeller. The results showed that when the flow rate was reduced to the flow rate of 0.8 Q_{BEP}, flow separation began to form on the suction side of the flow channel. Atkins et al. [8] applied the PIV technology to measure the velocity fields of the upstream, gap, and downstream areas around five side-by-side cylinders. Zhang et al. [9] used the PIV technology to measure the flow structure in a low-specific-speed centrifugal pump in detail. The results showed that when the flow velocity was lower than the nominal conditions, a typical structure of jet-wake was observed at the blade outlet. Zhang et al. [10] studied the internal flow characteristics and hydraulic performance of the twin-screw pump by CFD. Results showed that the screw pressure gradually increased from the inlet face to the outlet face along the screw axis under different heads. Bennacer et al. [11] applied the PIV technology to conduct an in-depth analysis of the internal flow. They found that the dominant mode accounted for more than half of the total kinetic energy. Guo et al. [12] used an experiment and numerical simulation to characterize the performance of pumps with variable pitch inducers at different speeds. Si et al. [13] adopted the PIV technology to study the influence of internal flow characteristics on the overall pump performance. Results showed that the main loss was occurred inside the vaneless part of the diffuser under the low flow rates. Wu et al. [14] used the PIV technology to measure the internal flow field under design flow working conditions, which revealed the distribution of the principal Reynolds normal stress and the principal Reynolds shear stress. Li et al. [15] studied the influence of internal flow patterns on hydraulic performance and energy conversion characteristics through visual experiments. Detailed studies showed that a counterclockwise vortex would increase energy losses, while a clockwise vortex would have a positive effect on the pump head. Westra et al. [16] used the PIV technology to study the internal velocity field distribution characteristics of centrifugal pumps, especially the secondary flow structure in the impeller flow channel. Yu et al. [17] used a laser doppler anemometer to obtain velocity measurements from a model of bio-centrifugal pump. Almost all the aforementioned studies have focused on the effects of flow rates on the flow pattern. In the present study, the effect of the rotational speeds on internal flow pattern were analyzed. Several researchers have aimed to investigate the internal flow pattern by analyzing the absolute flow angle. Liu et al. [18] applied PIV technology to study the unstable internal flow of a double-blade centrifugal pump impeller at a design flow rate. It focused on the analysis of the change trend of the mean absolute flow angle and the average dynamic pressure. Gaetani et al. [19] studied the pump performance and focused on computing spanwise mean flow angles and velocity distribution. Wang et al. [20] studied the formation mechanism of the reverse flow. Their results showed that the absolute flow angle at the blade inlet nearing the shroud was close to 180° because of the joint action of the leakage flow and blade inlet impact under low flow rates. Poujol et al. [21] tested the centrifugal compressor under the staggered angle of three inlet guide vanes. It showed that the absolute flow angle at the diffuser inlet decreases with the increase of IGV stagger angle near the surge point.

At the same time, many studies have been conducted to explain the effect of blade load on pump performance. Chalghoum et al. [22] analyzed the unsteady value of the strong interaction between the impeller blade and the volute tongue, then studied the correlation between the dynamic load and the unstable radial force. Zhang et al. [23] used a new method to design a new impeller based on the blade load. Li et al. [24] studied the effect of blade load distribution on the head, radial force, and pressure pulsation of a low-specific-speed centrifugal pump with cylindrical impeller blades. It showed that the front-loaded impeller produced a more uniform flow than the rear-loaded impeller. Li et al. [25] discussed the impact of monitoring points and blade load variations on the unstable behavior of the impeller through sensor experiment. Zhang et al. [26] studied the pressure fluctuation characteristics of a reversible pump turbine under various blade load conditions by using pressure signals. They found that pressure fluctuations were very obvious under the low load conditions. Sun et al. [27] revealed the optimal load distribution mechanism of the hub by different methods. Durante et al. [28] introduced a comparison for the analysis of the unsteady loads among different hydrodynamic models. It highlighted the ability to describe high frequency hydrodynamic loads. Li et al. [29] proposed the blade load distribution based on PIV technology to analyze the energy conversion mechanism in the centrifugal pump impeller. Few works have investigated the correlation between the blade load distribution and the flow pattern in the impeller. The current study aimed to clarify the effect of vortexes in the impeller channel on the blade load distribution. In this paper, the correlation between the flow pattern in the impeller and the blade load distribution was further discussed with the help of PIV.The current study aimed to clarify the mechanism of how vortexes in the impeller channel affect the blade load distribution. Results include the theory of blade load, the process of blade load data extraction, the size of vortexes in the flow channel, and the movement laws of the vortex core. Finally, the reasons for the special trend of some blade load distributions were analyzed.

2. Implementation Process Based on PIV

2.1. Theoretical Basis of Blade Load

The blade load distribution is an important factor to measure the hydraulic performance of the impeller in blade design. The blade load is defined as the pressure difference between the pressure side and the suction side in the length direction of the blade. By changing the distribution law of the blade load, the pressure difference distribution between the pressure side and the suction side of the blade can be effectively changed. Therefore, the reasonable change of the blade load distribution can improve the performance of the impeller.

As for the inverse design method, the distribution represents the change of the velocity loop rV_θ along the streamline. V_θ is the circumferential component of the absolute velocity. The distribution of the velocity circulation along the streamline reflects the distribution change of the blade load.

Based on this feature of impeller machinery, the distribution of $\partial r\overline{V_\theta}/\partial m$ along the streamline is used as the flow filed condition in the inverse design method. Therefore, the F (the value of blade load) can be rewritten as Equation (1):

$$F = p^+ - p^- = \frac{2\pi}{B}\rho W \frac{\partial(r\overline{V_\theta})}{\partial m} \tag{1}$$

p^+ and p^- are the pressure at the pressure side and suction side of the blade respectively, B is the number of blades, W is the relative velocity at the blade side, ρ is the density of water, rV_θ is the velocity loop quantity, m is the relative axial side flow length.

The blades are represented by vorticity sheets, which depend on the specific distribution of circumferential mean angular momentum $r\overline{V_\theta}$ defined as follows:

$$r\overline{V_\theta} = \frac{Z}{2\pi}\int_0^{\frac{2\pi}{Z}} rV_\theta d\theta \tag{2}$$

According to the velocity triangle, we have the following conclusions:

$$V_\theta = \frac{Q}{A\ tan\ \alpha} \tag{3}$$

$$A = 2\pi rb \tag{4}$$

where, Q is the flow rate through the impeller, A is the cross-sectional area, b is the width of the channel in the cross-section, and α (the absolute flow angle) is the angle between the circumferential direction and the absolute velocity.

In this paper, the blade load distribution was obtained by an experimental method. The data obtained from the PIV test were processed by MATLAB to obtain the relative velocity and absolute velocity in flow field, and then the blade load was obtained based on the blade load Equation (1). The following is an introduction to the experimental device.

2.2. Methodology

In this paper, PIV experimental method was used to extract blade load value after post-processing. Experiments of the test pump were conducted on a closed test rig as shown in Figure 1. The main parameters of the impeller are listed in Table 1. The valves were used to adjust the flow rate of the pump. The flow rate was measured by using an electromagnetic flowmeter. The water head was measured by two pressure sensors. The rotational speed was regulated by a frequency inverter.

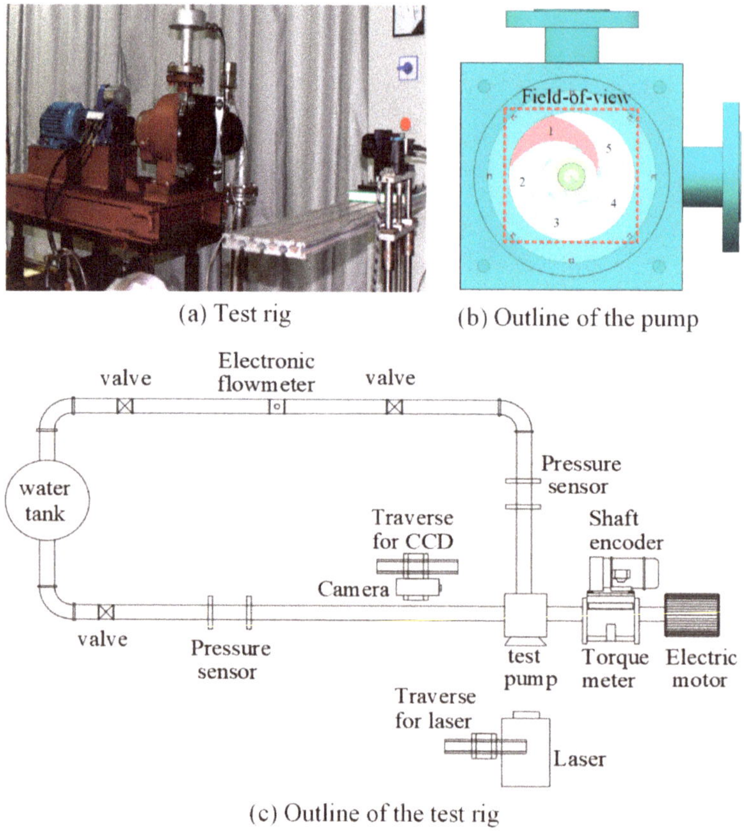

Figure 1. Outline of the test rig.

Table 1. Main design parameters of the impeller.

Parameter Name	Value
Inlet diameter D_1 (mm)	56
Outlet diameter D_2 (mm)	142
Inlet angle of blade β_1 (°)	28
Outlet angle of blade β_2 (°)	34
Blade width (mm)	7
Blade thickness (mm)	4
Wrap angle of blade φ (°)	96
Specific speed n_s	23.8
Blade number Z	5

The PIV measurement system was described in Li et al. [29]. The PIV system (TSI) consisted of a double-pulsed Nd: YAG laser (λ = 532 nm), a CCD camera (spatial resolution of 2048 px × 2048 px, time delay of 130 μs), a synchronizer, and a data-processing system (Insight 3 G). The image pair's acquisition was synchronized with the impeller rotation using an optical trigger which was located at the pump shaft. The spatial resolution of the acquired images was 102.38 μm/px. The seeding particles' (soild alumina particles) diameter was 20–60 μm, and their density was 1050 kg/m^3. A multi-pass method with 50% overlap was chosen with a final interrogation region of 32 × 32 px^2. According to Li et al. [29], the uncertainty of the phase-averaged velocities was lower than 1–3% of the measured velocity.

Before the PIV experiment, the performance characteristics of the test pump were firstly tested. The optimum flow rates (Q_{BEP}) and the corresponding heads at different rotational speeds are listed in Table 2.

Table 2. Optimum flow rates and the corresponding heads for different rotational speeds.

Rotational Speeds (r/min)	Optimum Flow Points (m^3/h)	Corresponding Heads (m)
600 r/min	1.0	1.0
800 r/min	1.7	1.5
1000 r/min	1.8	2.67
1200 r/min	2.3	3.7
1400 r/min	2.5	5.05

Based on these optimum flow rates, the PIV experiments were conducted at flow rates of 1.2 Q_{BEP}, 1.0 Q_{BEP}, 0.8 Q_{BEP}, 0.6 Q_{BEP}, 0.4 Q_{BEP}, and 0.2 Q_{BEP} at different rotational speeds. During the measuring process, the images were acquired at the same section by using a triggering system. The velocity fields were obtained from 300 pairs of images which were obtained at the same blade position. The experimental data were then processed as follows: The velocities are a function of position x, y and time t, including phase average and pulsation, shown in Equation (5):

$$V(x,y,t) = \overline{V}(x,y) + V'(x,y,t) \tag{5}$$

Then the velocity field corresponding to each working condition captured in the test was averaged to obtain the mean absolute speed field shown in the following Equation (6):

$$\overline{v}(x,y) = \frac{1}{n}\sum_{i=1}^{n} V_i(x,y,t_0 + i\Delta t)(n = 300) \tag{6}$$

The phase-averaged relative velocity field was obtained by subtracting the local circumferential velocity of the impeller from the absolute velocity at each point:

$$W = V - \Omega \times r_a \tag{7}$$

W is the relative velocity of the particles. Ω is the -rotational speed of the impeller and r_a is the local radius.

After the relative velocity field was obtained, the blade load value could be obtained by combining Equation (1).

2.3. Results of Blade Load

Figure 2a shows a schematic of the flow channel meridian. The V_θ distribution on the same radius in the flow channel was averaged to obtain the mean V_θ distribution along the meridian. The channel meridian was consistent with the blade profile. The blade load distribution and the mean absolute flow angle distribution were analyzed along this meridian. Figure 2b shows the side view of the impeller.

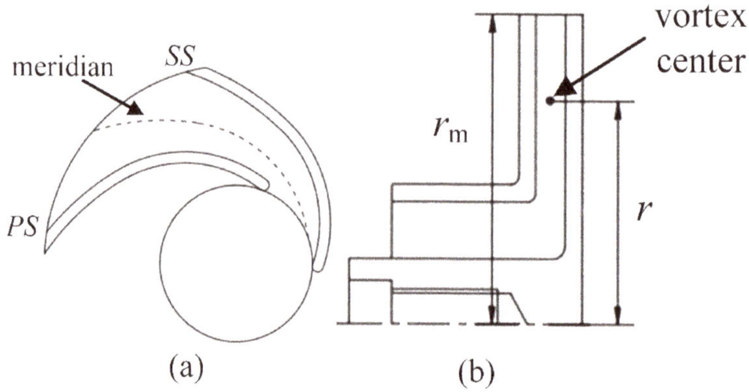

Figure 2. Schematic of flow channel meridian (**a**) and side view of the impeller (**b**); r is the distance from the vortex center to the origin of the impeller, r_m is the distance from the impeller outlet to the origin of the impeller.

Figure 3 shows the distributions of the blade load for different rotational speeds. The blade load distribution curves behaved as an aft-loaded type at each flow rate. At different rotational speeds, the blade load distributions behaved inconsistently. The corresponding blade load value of each flow rate increased accordingly with the increase of rotational speed. Comparing the blade load distributions of the five rotational speeds, it can be found that the load peak area was distributed along r/r_m = 0.75–0.85, and then the blade load started to decrease until the impeller outlet.

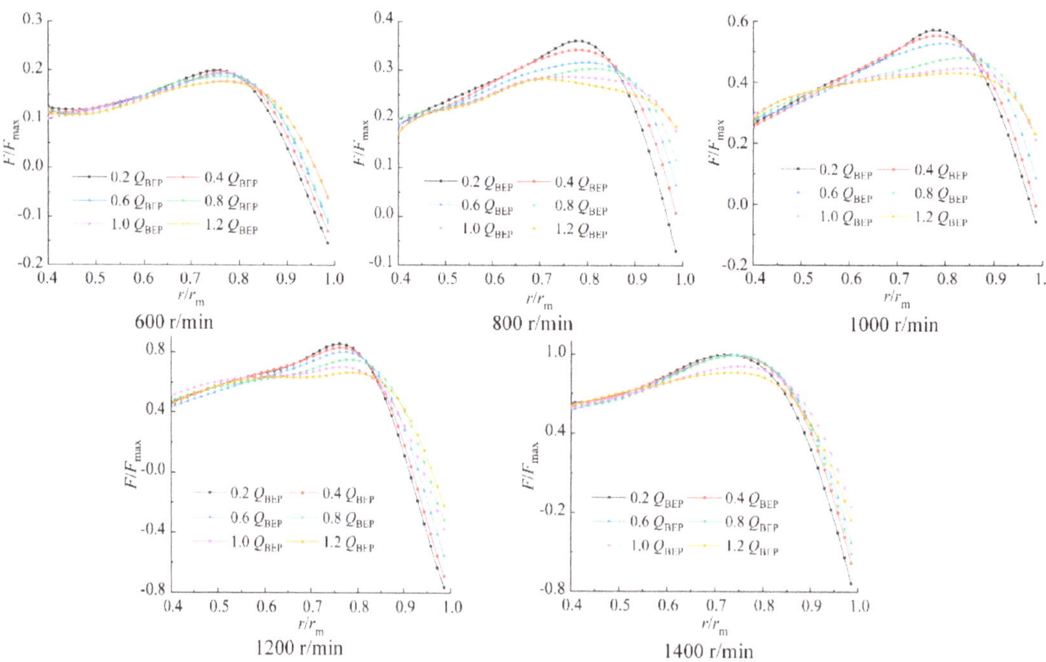

Figure 3. Blade load distribution curve under the five rotational speeds.

3. Analysis of Internal Flow Patterns

3.1. Discussion on the Relative Velocity Field

Li et al. [15,29] studied the flow field in the impeller channel at different rotational speeds. At different rotational speeds, their results showed that the developing trend of internal flow pattern with flow rate is consistent. The internal flow pattern behaved steadily near the optimum flow rates, while it vortexes appeared at low flow rates. However, at the same flow rates, the flow field behaved more stably with the increase in the rotational speed. Figure 4 shows the flow streamlines at the flow rate of 0.2 Q_{BEP} for five rotational speeds. At the flow rate of 0.2 Q_{BEP}, there were two vortexes in the channel for all the five rotational speeds. The clockwise vortex appeared on the suction side, whereas the counterclockwise vortex appeared on the pressure side. The two vortexes occupied almost the entire flow channel. When the rotational speed increased, the zone and intensity of the low-velocity area decreased with the increase of rotational speed. At the rotational speed of 600 r/min, the flow channel was almost completely covered by the low-velocity area. While at the rotational speed of 1200 r/min and 1400 r/min, a secondary vortex appeared at the inlet of suction side. Figure 4 shows that the primary vortex was accompanied by a secondary vortex, which also led to a variation in the radial scale and the position of the vortex center of the primary vortex. On the other hand, for the flow rate of 1.0 Q_{BEP}, no separation vortexes were observed on the suction side or pressure side for all the five rotational speeds, and the flow pattern behaved in good condition.

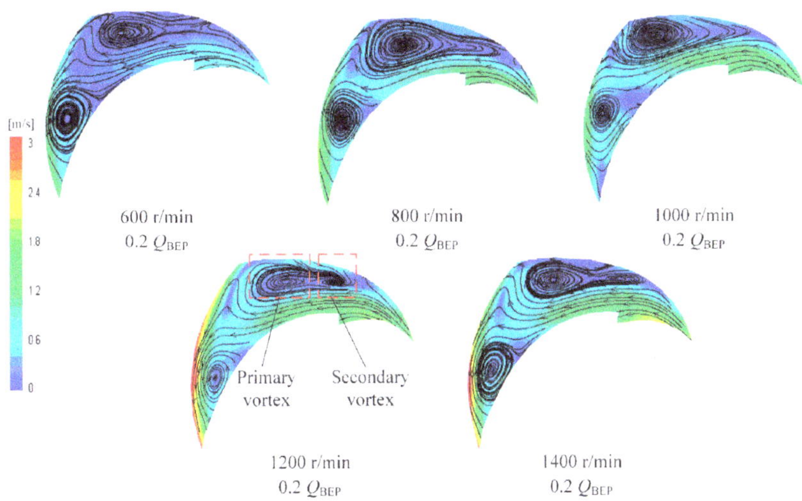

Figure 4. Flow streamlines for the five rotational speeds at the flow rate of 0.2 Q_{BEP}.

3.2. Discussion on the Vortex in the Flow Channel

Figure 5 shows the position of the arcs over the vortex center and radial area of the vortex: arc 1 represents the position of the arc that through the center of the clockwise vortex, and arc 2 represents the position of the arc that through the center of the counterclockwise vortex. The radial scale is the length of the meridian between edge 1 and edge 2 [30]. Edge 1 and edge 2 represent the initial and end radial position of the vortexes, respectively.

Figure 6 shows the radial scales of the clockwise vortex and the counterclockwise vortex under the five rotational speeds. It can be seen that the clockwise vortex was larger in scale than the counterclockwise vortex in the radial direction. The clockwise vortex was caused by the increase of the incidence impulse angle, while the counterclockwise vortex was generated by the blocking effect of the clockwise vortex [7]. The clockwise vortex and counterclockwise vortex tended to increase in scale with the decrease of the flow rates. The scale of counterclockwise vortex behaved more consistently and increased slowly with the decrease of flow rates.

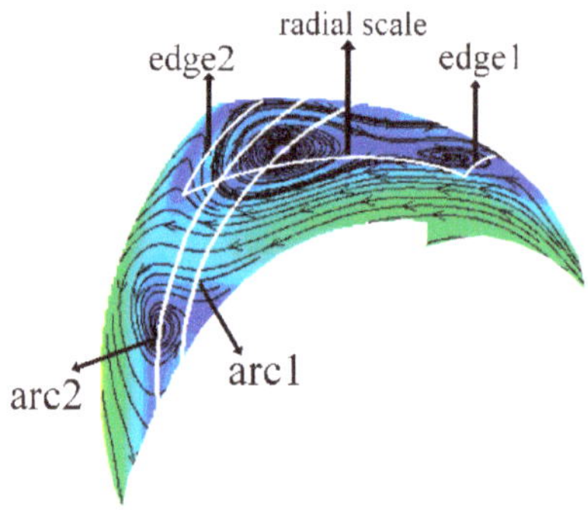

Figure 5. Radial scale of the vortex and the arc that through the vortex center.

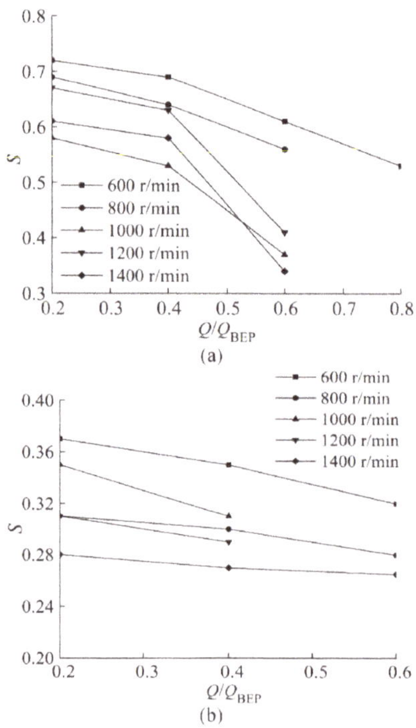

Figure 6. Radial scales of the clockwise vortex (**a**) and the counterclockwise vortex (**b**) at the five rotational speeds.

The relationship between the distance from the vortex center to the impeller rotation axis (r) under different flow rates and rotational speeds is shown in Figure 7. The diagram of r and r_m is shown in Figure 2b. For the clockwise vortex, r decreased with the decrease in the flow rate, and vice versa for the counterclockwise vortexes. At the same time, it can be seen that the position of clockwise vortex center showed a tendency to move toward the impeller inlet with the decrease of the flow rate at different rotational speeds. For the rotational speed of 600 r/min, 800 r/min, and 1000 r/min, it appears that the position of the clockwise vortex center was closer to the impeller outlet with the decrease of rotational speed. Compared with other rotational speeds, the position of the clockwise vortex center was closer to the impeller outlet under the rotational speed of 1200 r/min and 1400 r/min. This is due to the appearance of the secondary vortex at the rotational speed of 1200 r/min and 1400 r/min. The position of the counterclockwise vortex center gradually moved toward the impeller inlet with the increase of rotational speed. At the rotational speed of 600 r/min, the counterclockwise vortex center appeared much closer to the impeller outlet than that at other rotational speeds.

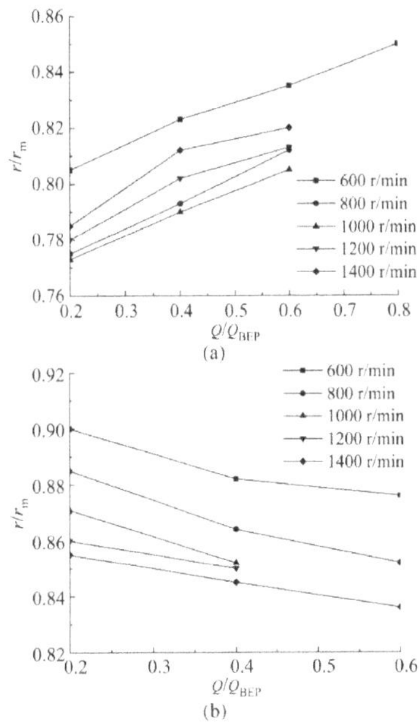

Figure 7. Movement of the clockwise vortex center (**a**) and the counterclockwise vortex (**b**) at different rotational speeds.

In this paper, we focused on the effect of rotational speeds and flow rates on the internal flow pattern inside the impeller. Additionally, we concentrated on the analysis of the scale size of the vortex and position of the vortex center in the flow channel, and specifically analyzed how vortexes affect the blade load distribution.

4. Effect of Flow Pattern in the Flow Channel on the Blade Load Distribution

4.1. Correlation of Internal Flow Pattern and Blade Load

From the above analysis, it can be seen that the internal flow patterns were basically the same with the variation of flow rates for different rotational speeds. Therefore, the flow patterns at the rotational speed of 800 were selected as examples for analysis. The correlation between the internal flow pattern and the blade load distribution is shown in Figure 8. For the flow rates of 0.2 Q_{BEP}, 0.4 Q_{BEP}, and 0.6 Q_{BEP}, the distribution curves can be divided into two sections: the first section of the blade load distribution curves was in the rising stage, which manifested as a slow increase, and then it started to reach the highest point linearly and rapidly. The second section decreased until the outlet. While, for the flow rates of 0.8 Q_{BEP} to 1.2 Q_{BEP}, the first section of the blade load curves increased slowly with a smaller slope, and the second section of the blade load curves decreased slightly until the outlet. This is also consistent with the conclusion mentioned in the paper by Li et al. [24]: the higher slope of the load curve, the lower pump efficiency. This indicates that the flow blockage and hydraulic losses in the impeller channel will increase when the slope of the blade load curve increases.

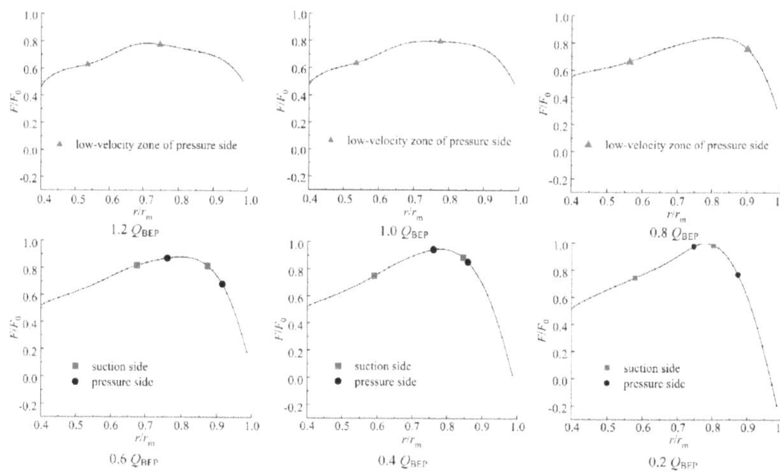

Figure 8. Blade load distribution and the position of the vortexes under different flow rates. For 0.2–0.6 Q_{BEP}: the front point is the position of the leading edge of vortex, the back point is the position of the vortex center. For 0.8–1.2 Q_{BEP}: the two points are the position of the leading edge and back edge of low-velocity-zone in the pressure side, respectively.

To analyze the correlation between the flow pattern inside the impeller and the blade load, the location of low-velocity-zone of pressure side, the vortex leading edge (the front point) and the vortex center (the latter point) are also shown in Figure 8. It can be seen that for the flow rates of 0.8 Q_{BEP} to 1.2 Q_{BEP}, the position where the blade load distribution curve began to increase rapidly coincided with the position where the low-velocity-zone started to appear on the pressure side. The peak position coincided with the end of the low-velocity zone. For the flow rates from 0.2 Q_{BEP} to 0.6 Q_{BEP}, the position of the blade load distribution curve began to increase rapidly, coincided with the initial position of the clockwise vortex on the suction side, and the peak position coincided with the position of core area of the clockwise vortex. The position where the blade load starts to decrease coincided with the initial position of the counterclockwise vortex on the pressure side. The blade load kept increasing when the clockwise vortex appeared. When the counterclockwise vortex appeared, the blade load increased a little, and at this time the strength of the clockwise vortex was greater than that of the counterclockwise vortex. When the strength of counterclockwise vortex was greater than that of the clockwise vortex, the blade load curve started to decrease rapidly at core area of the counterclockwise vortex. It is inferred that the vortex on the suction side increases the blade load, while the vortex on the pressure side decreases the blade load.

Combining Equations (1)–(4), the blade load is related with the absolute flow angle. Therefore, the absolute flow angle was used as an intermediate medium to correlate the flow pattern with the blade load. Figure 9 shows the mean absolute flow angle distribution along the meridian. It can be seen that the overall trends of the mean absolute flow angle distribution behaved similarly under different rotational speeds. When the rotational speed increased, the curve became steeper under the same flow rates. For the flow rates of 0.2 Q_{BEP} to 0.6 Q_{BEP}, the mean absolute flow angle curve dropped slowly from the inlet to the outlet. After the position of $r/r_m = 0.8$, the mean absolute flow angle's behavior was almost unchanged. For the flow rates from 0.8 Q_{BEP} to 1.2 Q_{BEP}, the mean absolute flow angle curve decreased more rapidly than that at the flow rates from 0.2 Q_{BEP} to 0.6 Q_{BEP}. In addition, combined with the analysis of the blade load curve, it can be seen that the blade load curve had an upward trend in the first half of the curve, which was opposite to the distribution trend of the mean absolute flow angle. This is because that as the decrease of absolute flow angle α, the $tan\alpha$ decreases, resulting in the increase of V_θ, which coincides with Equation (1). The mean absolute flow angle decreased slowly near the trailing edge of the blade. As a result, the mean absolute flow angle had little effect on the blade load near the impeller outlet. Therefore, the blade load was mainly affected by the variation of the blade profile near the impeller outlet.

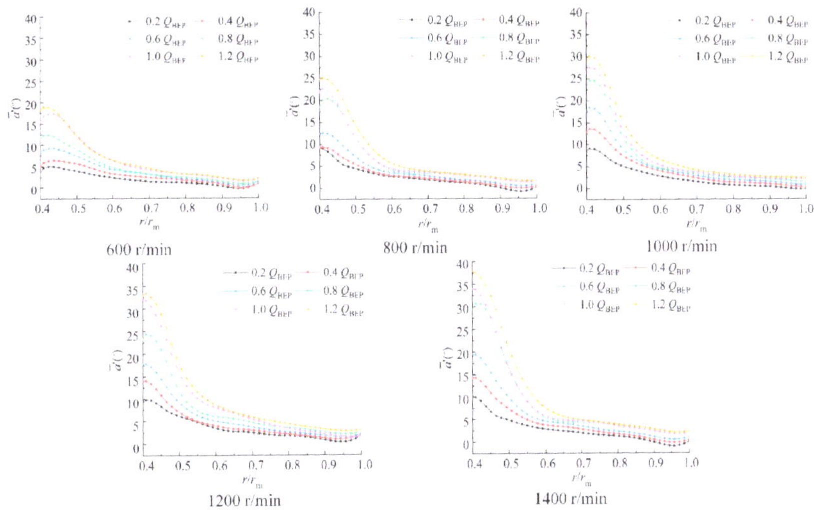

Figure 9. The mean absolute flow angle distribution curve under the five rotational speeds.

In addition, it was found that the blade load distribution trends for some flow rates and some rotational speeds were different from the overall distribution trend. At the rotational speeds of 600 r/min and 1400 r/min, the blade load distribution curves for the flow rate of 0.8 Q_{BEP} were closer to the flow rate of 0.6 Q_{BEP} compared with the situation at other rotational speeds. For the rotational speeds of 1000 r/min and 1200 r/min, at flow rate of 0.6 Q_{BEP}, the development trend of the first half blade load distribution curves were consistent with that at the flow rate of 0.2 Q_{BEP} and 0.4 Q_{BEP}. However, the latter half of the curves began to move away from that at the flow rates of 0.2 Q_{BEP} and 0.4 Q_{BEP}, and the positions of the maximum values were significantly closer to the impeller outlet. The specific values are shown in Table 3.

Table 3. The position of the maximum blade load and the vortex centers.

Rotational Speed	Flow	Position (Suction Side)	Position (Pressure Side)	Position (Maximum Load)
600 r/min	0.2	0.805	0.9	0.76
	0.4	0.823	0.882	0.775
	0.6	0.835	0.876	0.785
800 r/min	0.2	0.775	0.885	0.78
	0.4	0.793	0.864	0.785
	0.6	0.812	0.852	0.79
1000 r/min	0.2	0.773	0.871	0.785
	0.4	0.79	0.852	0.8
	0.6	0.805		0.81
1200 r/min	0.2	0.782	0.86	0.785
	0.4	0.8	0.85	0.79
	0.6	0.813		0.805
1400 r/min	0.2	0.785	0.855	0.74
	0.4	0.812	0.845	0.745
	0.6	0.82	0.836	0.75

4.2. Reasons for the Difference in Blade Load

The blade load is related with the absolute flow angle, which related to the flow pattern in turn. Therefore, it is necessary to describe how the vortexes affect the distribution of the absolute flow angle. Figure 10a shows the distributions of the absolute flow angle along arc 1 (across the position of the clockwise vortex center) at the flow rates of 0.2 Q_{BEP}, 0.4 Q_{BEP}, 0.6 Q_{BEP}, and 1.0 Q_{BEP} under the rotational speed of 1000 r/min, respectively. Arc 1 and arc 2 are shown in Figure 5. When the vortex appears on the suction side (at flow rates of 0.2 Q_{BEP}, 0.4 Q_{BEP}, 0.6 Q_{BEP}), the absolute flow angle curves decreased slightly at first, then increased rapidly, and then decreased until the outlet. An interesting phenomenon was also found: the first half of the curve basically coincided with the spanwise area of the clockwise vortex. At the flow rate of 1.0 Q_{BEP}, it can be seen that the absolute flow angle distribution curve was much flatter. In addition, when the vortex appeared, the sharp increase and decrease process occurred for the absolute flow angle. Although the maximum value of the absolute flow angle was larger than that without vortex, the growth gradient was so large that the average value was smaller than that without vortex. Therefore, the overall blade load was increased. Figure 10b shows the distribution of the absolute flow angle along arc 2 (across the position of the counterclockwise vortex center). When the counterclockwise vortex appeared (0.2 Q_{BEP}, 0.4 Q_{BEP}), the absolute flow angle fluctuated along the curve at first. Then, in the region corresponding to the counterclockwise vortex, it became negative, decreased sharply, and rose again after reaching the minimum value. At the flow rate of 0.6 Q_{BEP}, only the clockwise vortex appeared, and the position of arc 2 could not be determined. For comparison, the positions of arc 2 at the flow rates of 0.2 Q_{BEP} and 0.4 Q_{BEP} were selected for the flow rate of 0.6 Q_{BEP}. It can be seen that their distribution trend was consistent with that along arc 1 when the clockwise vortex appeared. When the flow rate was 1.0 Q_{BEP}, the absolute flow angle distribution changed a little, and the value remained around zero. From the above analysis, it can be seen that the appearance of the counterclockwise vortex causes the absolute flow angle to become negative and make the overall blade load become negative, resulting in the unloading of the blade load.

Figure 11 shows the flow patterns of the flow rates of 0.4 Q_{BEP}, 0.6 Q_{BEP}, and 0.8 Q_{BEP} at the rotational speeds of 600 r/min, 1000 r/min, 1200 r/min, and 1400 r/min. At the rotational speeds of 600 r/min and 1400 r/min, for the flow rate of 0.8 Q_{BEP}, it can be seen that the clockwise vortex appeared, which caused the blade load distribution curve to be closer to the curve at flow rate of 0.6 Q_{BEP}. At the rotational speeds of 1000 r/min and 1400 r/min, only the clockwise vortex appeared, while at the flow rate of 0.4 Q_{BEP}, both vortexes appeared. For the flow rate of 0.6 Q_{BEP}, the unloading effect of the counterclockwise vortex disappeared, which made the blade load curve smoother and the length of its growth section longer than that at the flow rate of 0.4 Q_{BEP}.

According to the previous analysis, the vortex in the channel affected the absolute flow angle distribution and redistributed the blade load. The clockwise vortex made the position of maximum blade load move toward the blade outlet. It also made the value of the maximum blade load become larger. The counterclockwise vortex made the position of maximum blade load move toward the blade inlet, and caused the unloading of the blade load.

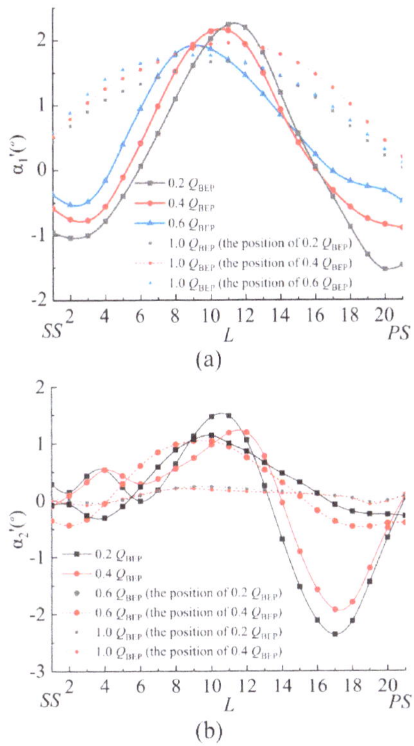

Figure 10. Absolute flow angle distributions along arc 1 and arc 2 under the rotational speed of 1000 r/min. Suction side (**a**) and pressure side (**b**).

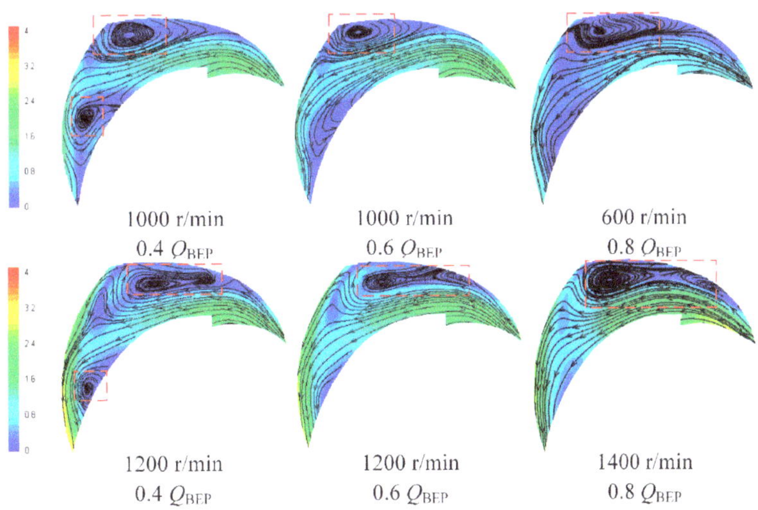

Figure 11. Flow streamlines.

5. Conclusions

In this paper, the blade load distribution in the pump was obtained through a PIV experiment. The characteristics of blade load were analyzed, and the specific correlation between blade load distribution and internal flow pattern were discussed.

(i) Based on the PIV method, the images taken in the test were converted into usable relative velocity field data through a series of post-processing. On this basis, the concept of the circumferential mean angular momentum was innovatively used to obtain the blade load value.
(ii) The scale of counterclockwise vortex was smaller than that of clockwise vortex. In general, the scale of both type vortexes increased with the decrease of rotational speed. The positions of the vortex centers were distributed in the range of r/r_m = 0.77–0.9, and the peak blade loads were located at the position of r/r_m = 0.75–0.85.
(iii) The blade load distribution along the impeller channel was interrelated with the flow pattern. The vortexes redistributed blade load by changing the absolute flow angle of the flow. Additionally, the clockwise vortex made the position of the maximum blade load closer to the outlet, while the counterclockwise vortex had the opposite effect. Additionally, the number of vortexes in the flow channel influenced the blade load distribution.

Author Contributions: Conceptualization, B.C. and Z.Z.; methodology, X.L.; validation, B.C. and Z.Z.; investigation, X.C.; data curation, X.C.; writing—original draft preparation, X.C.; writing—review and editing, B.C.; visualization, X.C.; funding acquisition, B.C., Z.Z. and X.L. All authors have read and agreed to the published version of the manuscript.

Funding: Project supported by the National Natural Science Foundation of China (Grant Nos. 51806197, 51976202), the Natural Science Foundation of Zhejiang Province (Grant No. LY20E060006) and the Key Research and Development Program of Zhejiang Province (Grant No. 2021C05006), Top-notch Talent Support Program of Zhejiang Province (2019R51002), the Fundamental Research Funds of Zhejiang Sci-Tech University (2021Q017).

Informed Consent Statement: Informed consent was obtained from all individual participants included in the study.

Data Availability Statement: The data used to support the findings of this study are included within the article.

Conflicts of Interest: The authors declare that they have no conflict of interest.

Ethical Approval: This article does not contain any studies with human participants or animals performed by any of the authors.

References

1. Shigemitsu, T.; Ogawa, Y.Y.; Nakaishi, E. PIV Measurement of Internal Flow in Mini Centrifugal Pump. *Int. J. Fluid Mach. Syst.* **2019**, *12*, 251–260. [CrossRef]
2. Zhang, L.; Wang, C.; Zhang, Y.C.; Xiang, W.; He, Z.M.; Shi, W.D. Numerical study of coupled flow in blocking pulsed jet impinging on a rotating wall. *J. Braz. Soc. Mech. Sci. Eng.* **2021**, *43*, 508. [CrossRef]
3. Zhou, J.R.; Zhao, M.M.; Wang, C.; Gao, Z.J. Optimal design of diversion piers of lateral intake pumping station based on orthogonal test. *Shock Vib.* **2021**, *5*, 6616456. [CrossRef]
4. Wang, H.L.; Hu, Q.X.; Yang, Y.; Wang, C. Performance differences of electrical submersible pump under variable speed schemes. *Int. J. Simul. Model.* **2021**, *20*, 76–86. [CrossRef]
5. Shi, L.J.; Zhu, J.; Tang, F.P.; Wang, C. Multi-Disciplinary optimization design of axial-flow pump impellers based on the approximation model. *Energies* **2020**, *13*, 779. [CrossRef]
6. Zhu, Y.; Li, G.P.; Wang, R.; Tang, S.N.; Su, H.; Cao, K. Intelligent fault diagnosis of hydraulic piston pump combining improved LeNet-5 and PSO hyperparameter optimization. *Appl. Acoust.* **2021**, *183*, 108336. [CrossRef]
7. Wang, Y.P.; Yang, H.; Chen, B.; Gao, P.L.; Zhu, Z.C. Analysis of vortexes formed in flow channel of a five-bladed centrifugal water pump by means of PIV method. *AIP Adv.* **2019**, *9*, 075011. [CrossRef]
8. Atkins, M.D.; Kim, T. Isotropic-planar illumination for PIV experiments. *Exp. Fluids* **2015**, *56*, 63. [CrossRef]
9. Zhang, N.; Gao, B.; Li, Z.; Daisy, N. Unsteady flow structure and its evolution in a low specific speed centrifugal pump measured by PIV. *Exp. Therm. Fluid Sci.* **2018**, *97*, 133–144. [CrossRef]

10. Zhang, D.; Cheng, L.; Li, Y.Y.; Jiao, W.X. The hydraulic performance of twin-screw pump. *J. Hydrodyn.* **2020**, *32*, 605–615. [CrossRef]
11. Bennacer, R.; Sefiane, K. Proper orthogonal decomposition (POD) analysis of flow structure in volatile binary droplets. *Int. Commun. Heat Mass Transf.* **2016**, *71*, 172–175. [CrossRef]
12. Guo, X.M.; Zhu, Z.C.; Shi, G.P.; Huang, Y. Effects of rotational speeds on the performance of a centrifugal pump with a variable-pitch inducer. *J. Hydrodyn.* **2017**, *29*, 854–862. [CrossRef]
13. Si, Q.R.; Dupont, P.; Dazin, A.; Roussette, O.; Yuan, S.Q.; Bayeul, A. An Experimental Study of the Flow Field Inside the Diffuser Channel of a Laboratory Centrifugal Pump. *J. Fluids Eng.* **2015**, *137*, 061105. [CrossRef]
14. Wu, Y.L.; Liu, S.H.; Yuan, H.J.; Shao, J. PIV measurement on internal instantaneous flows of a centrifugal pump. *Sci. China Technol. Sci.* **2011**, *54*, 270–276. [CrossRef]
15. Li, X.J.; Chen, H.; Chen, B.; Luo, X.W.; Yang, B.F.; Zhu, Z.C. Investigation of flow pattern and hydraulic performance of a centrifugal pump impeller through the PIV method. *Renew. Energ.* **2020**, *162*, 561–574. [CrossRef]
16. Westra, R.W.; Broersma, L.; Kruyt, N.P.; Van, K. PIV Measurements and CFD Computations of Secondary Flow in a Centrifugal Pump Impeller. *J. Fluids Eng.* **2010**, *132*, 061104. [CrossRef]
17. Yu, S.C.M.; Chua, L.P.; Leo, H.L. The angle-resolved velocity measurements in the impeller passages of a model bio-centrifugal pump. *Proc. Inst. Mech. Eng. Part C J. Mech. Eng. Sci.* **2001**, *215*, 547–568. [CrossRef]
18. Liu, H.L.; Wang, K.; Kim, H.B.; Tan, M.G. Experimental investigation of the unsteady flow in a double-blade centrifugal pump impeller. *Sci. China Technol. Sci.* **2013**, *56*, 812–817. [CrossRef]
19. Gaetani, P.; Boccazzi, A.; Sala, R. Low field in the vaned diffuser of a centrifugal pump at different vane setting angles. *J. Fluids Eng.* **2012**, *134*, 031101. [CrossRef]
20. Wang, L.K.; Lu, J.L.; Liao, W.L.; Wang, W. Numerical analysis of the formation mechanism and suppression method of the reverse flow in a semi-open centrifugal pump. *J. Mech. Sci. Technol.* **2020**, *34*, 3667–3678. [CrossRef]
21. Poujol, N.; Trébinjac, I.; Duquesne, P. Effects of inlet guide vanes on the performance and stability of an aeronautical centrifugal compressor. *J. Turbomach.* **2020**, *143*, 101010. [CrossRef]
22. Chalghoum, I.; Elaoud, S.; Kanfoudi, H.; Akrout, M. The effects of the rotor-stator interaction on unsteady pressure pulsation and radial force in a centrifugal pump. *J. Hydrodyn.* **2018**, *30*, 672–681. [CrossRef]
23. Zhang, R.H.; Guo, R.; Yang, J.H.; Luo, J.Q. Inverse Method of Centrifugal Pump Impeller Based on Proper Orthogonal Decomposition (POD) Method. *Chin. J. Mech. Eng.* **2017**, *30*, 1025–1031. [CrossRef]
24. Li, X.J.; Gao, P.L.; Zhu, Z.C.; Li, Y. Effect of the blade loading distribution on hydrodynamic performance of a centrifugal pump with cylindrical blades. *J. Mech. Sci. Technol.* **2018**, *32*, 1161–1170. [CrossRef]
25. Li, J.W.; Zhang, Y.N.; Yu, J.X. Experimental investigations of a prototype reversible pump turbine in generating mode with water head variations. *Sci. China Technol. Sci.* **2018**, *61*, 604–611. [CrossRef]
26. Zhang, Y.N.; Chen, T.; Li, J.W.; Yu, J.X. Experimental Study of Load Variations on Pressure Fluctuations in a Prototype Reversible Pump Turbine in Generating Mode. *J. Fluids Eng.* **2017**, *139*, 074501. [CrossRef]
27. Sun, W.Y.; Huang, G.F. Integrated lifting line/surface panel method for optimal propeller design with consideration of hub effect. *J. Hydrodyn.* **2019**, *31*, 1218–1230. [CrossRef]
28. Durante, D.; Dubbioso, G.; Testa, C. Simplified hydrodynamic models for the analysis of marine propellers in a wake-field. *J. Hydrodyn.* **2013**, *25*, 954–965. [CrossRef]
29. Li, X.J.; Chen, B.; Luo, X.W.; Zhu, Z.C. Effects of flow pattern on hydraulic performance and energy conversion characterisation in a centrifugal pump. *Renew. Energy* **2020**, *151*, 475–487. [CrossRef]
30. Fujiwara, K.; Sriram, R.; Kontis, K. Experimental investigations on the sharp leading-edge separation over a flat plate at zero incidence using particle image velocimetry. *Exp. Fluids* **2020**, *61*, 205. [CrossRef]

machines

MDPI

Article

Analysis of the Formation Mechanism of Secondary Tip Leakage Vortex (S-TLV) in an Axial Flow Pump

Hu Zhang [1,2], Jianbo Zang [1], Desheng Zhang [2,*], Weidong Shi [3,*] and Jiean Shen [1]

1 School of Mechanical Technology, Wuxi Institute of Technology, Wuxi 214121, China; zhanghutianxia@126.com (H.Z.); zangjb@126.com (J.Z.); sjiean@126.com (J.S.)
2 Research Center of Fluid Machinery Engineering and Technology, Jiangsu University, Zhenjiang 212013, China
3 School of Mechanical Engineering, Nantong University, Nantong 226019, China
* Correspondence: zds@ujs.edu.cn (D.Z.); wdshi@ujs.edu.cn (W.S.)

Abstract: Studies on the tip leakage vortex (TLV) are extensive, while studies on the secondary tip leakage vortex (S-TLV) are rare. To advance the understanding of the formation mechanism of the S-TLV, turbulent cavitating flows were numerically investigated using the shear stress transport (SST) turbulence model and the Zwart–Gerber–Belamri cavitation model. The morphology and physical quantity distribution of the S-TLV under two cavitation conditions were compared, and its formation mechanism was analyzed. The results reveal that in the lower cavitation number case, there is a low-velocity zone of circumferential flow near the tip in the back half of the blade. The shear vortices formed by the leakage jet gradually accumulate and concentrate in the low-velocity area, which is one of the main sources of the S-TLV. Meanwhile, the radial jet pushes the vortices on the suction surface to the tip, which mixes with the S-TLV. The flow path formed by the radial jet and the leakage jet is in accordance with the rotation direction of the S-TLV, which promotes the S-TLV's further development. Under the conditions of a small cavitation number and low flow rate, the circumferential velocity and radial velocity of the fluid near the gap have altered significantly, which is conducive to the formation of the S-TLV.

Keywords: tip leakage vortex (TLV); secondary tip leakage vortex (S-TLV); cavitation; axial flow pump; vortex; leakage jet

Citation: Zhang, H.; Zang, J.; Zhang, D.; Shi, W.; Shen, J. Analysis of the Formation Mechanism of Secondary Tip Leakage Vortex (S-TLV) in an Axial Flow Pump. *Machines* **2022**, *10*, 41. https://doi.org/10.3390/machines10010041

Academic Editor: Davide Astolfi

Received: 22 November 2021
Accepted: 30 December 2021
Published: 5 January 2022

1. Introduction

The axial flow pump is widely used in water diversion, nuclear power, irrigation, marine water jet propulsion and other fields. When the axial flow pump is working, a leakage flow will be formed in the gap between the tip and the end wall, accompanied by the generation of vortices [1,2]. When cavitation conditions are reached, cavity bubbles are formed in the tip clearance and above the suction surface, which will affect the hydraulic performance of the axial flow pump, resulting in noise and vibration [3–8]. There are various vortices in the leakage flow field near the tip clearance. According to the position and formation mechanism, they can be divided into the tip leakage vortex (TLV), the tip separation vortex, the induced vortex, the secondary tip leakage vortex (S-TLV) and the perpendicular cavitation vortex (PCV), etc. [9–12].

Numerous attempts have been devoted to analyzing the formation and evolution mechanism of the TLV. The study of the gap flow in the in-line cascade found that the velocity gradient is an important reason for the formation of vorticity, turbulence kinetic energy (TKE) and Reynolds' stress [13,14]. A comparative study of the numerical simulations on right-angle and rounded tip geometries found that the TLV originates from the continuous shear action between the leakage jet and the low-velocity flow on the suction surface to form and transport vorticity [15]. In the three-dimensional particle image velocimetry (PIV) experiment, it was observed that the tip leakage flow extends to the TLV in the form of a jet, and a series of vortex filaments are generated. When the TLV is delivered to the second half of the blade, the vortices gradually collapse, resulting in an increase in number but a decrease in size. Turbulence within the TLV is highly anisotropic and spatially non-uniform [16–18].

The tip geometry has a great influence on the TLV. The experimental study on NACA0009 hydrofoil found that the gap size had a strong correlation with the TLV, which affected the vortex core trajectory and its strength, and the existence of a specific gap width makes the vortex strength the largest [19–21]. Experimental studies on square and round blade tips have found that the square blade tips are responsible for the inception of unstable secondary vortices, which affect the main vortex structure of the TLV [22]. The research found that when the pressure edge of the blade is rounded in an axial flow pump, the clearance cavitation is capable of being eliminated, but the change of the gap geometry has no significant effect on the leakage vortex cavitation [23].

The development and evolution of the TLV are also affected by cavitation [24,25]. It was found that the amount of the TLV circulation declines with the decrease in the cavitation number, especially near the trailing edge, which is affected by the reduction in the lift coefficient [26]. Experimental and numerical studies of the axial flow propulsion pump under different cavitation conditions have found that cavitation reduces the leakage flow rate and increases the amplitude of pressure pulsation [27].

The flow rate has a great impact on the TLV. Under the condition of a high flow rate, the TLV intensity and the migration velocity magnitude of the low-pressure axial flow fan are fairly smaller. When the flow rate decreases, the TLV wanders around, and the vortex is unstable [28]. It is found that the initiation of the TLV trajectory under high flow rates is delayed backward, and the angle between the trajectory and the chord length diminishes by analyzing the axial flow pumps [29].

In general, the S-TLV is regarded as a series of small vortex structures attached to the blade suction surface, which is induced by the interaction of the tip clearance jet flow and mainstream in the blade-to-blade channel. The S-TLV interacts with the TLV and generates a helical entangle effect [6]. As the cavitation number decreases, the S-TLV becomes significant gradually. The formation of the perpendicular cavitation vortex (PCV) is mainly influenced by the TLV and is believed to be formed by the evolution of the S-TLV [30]. When the cavitation number is low, there are large-scale cloud-like cavitation vortex structures near the blade tip of the axial flow pump, which will cause a blockage of the flow channel, inducing instability and a rapid drop in performance [31–34]. However, there are still a few related studies on the S-TLV, and its formation mechanism requires further investigation.

In this article, the formation mechanism of the S-TLV was studied by comparing the two cavitation conditions combined with cavitation experiments and numerical calculation methods. The Shear-Stress-Transfer (SST) k-ω turbulence model and the Zwart–Gerber–Belamri cavitation model were used for numerical simulation to study the cavitation turbulent flows in the axial flow pump.

2. Numerical Method and Experimental Setup

2.1. Pump Geometry and Mesh

The main design parameters of the axial flow pump are depicted in Table 1.

The whole computational domain was divided into the following five subdomains: the inlet passage subdomain, the impeller subdomain, the guide vane subdomain, the ribs subdomain and the outlet passage subdomain, as presented in Figure 1.

Figure 2 shows the relevant parameters of the impeller. The radial coefficient is defined as $r^* = r/R$. The circumferential direction is the tangential direction corresponding to the circle of revolution, and it is perpendicular to the radial direction. The axial direction is aligned with the y-axis and the axial coefficient is defined as $\gamma = y/R$, where R is the radius of the impeller chamber, with a size of 100 mm. The main flow direction is from the pump inlet to the outlet. The chordwise direction is from leading edge (LE) to trailing edge (TE).

The blade chord coefficient is also defined as λ = s/c, where s refers to the distance from leading edge to the chord section.

Table 1. Pump geometry and reference data.

Parameters	Value
Number of rotor blades (Z_i)	3
Number of stator blades (Z_d)	7
Optimum flow rate (Q_{BEP})	0.101 $m^3\ s^{-1}$
Chord length (c)	113.7 mm
Rotor diameter (d_3)	199 mm
Hub diameter (d_t)	90 mm
Inlet diameter (d_1)	200 mm
Outlet diameter (d_2)	250 mm
Tip clearance (τ)	0.5 mm
Tip velocity (U_{tip})	15.18 $m\ s^{-1}$
Rotor angular velocity (Ω)	151.84 rad s^{-1} (1450 rpm)

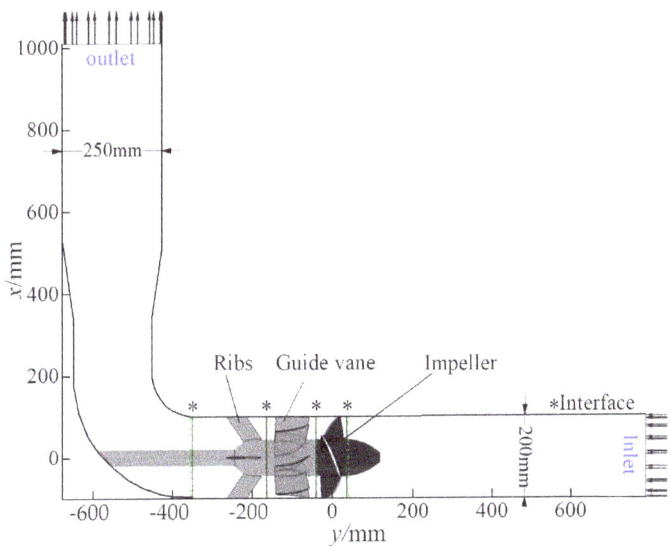

Figure 1. Axial flow pump computational domains.

The circumferential vorticity w_c is calculated by $w_c = w_z \sin\theta - w_x \cos\theta$, where w_x and w_z are the vorticity in the x and y directions, respectively. θ is the angle between the radial line and the z axis.

The pump head H is calculated by $H = (p_{out} - p_{in})/\rho_l\ g$, where p_{in} and p_{out} represent the total inlet and outlet pressure, respectively; and ρ_l represents the density of the liquid. The pump heads calculated by using three groups of structural grids with 5.24 million, 7.26 million and 9.46 million nodes were 3.07, 3.08 and 3.08, respectively, which verified the grid independence of the computing domain. In order to capture the details of flow and the small-scale vortex structures, the grid with a total number of 9.46 million was selected as the computing domain grid model. Thirty nodes were set in the tip gap with a size of 0.5 mm. The details of the computational grid are depicted in Figure 3.

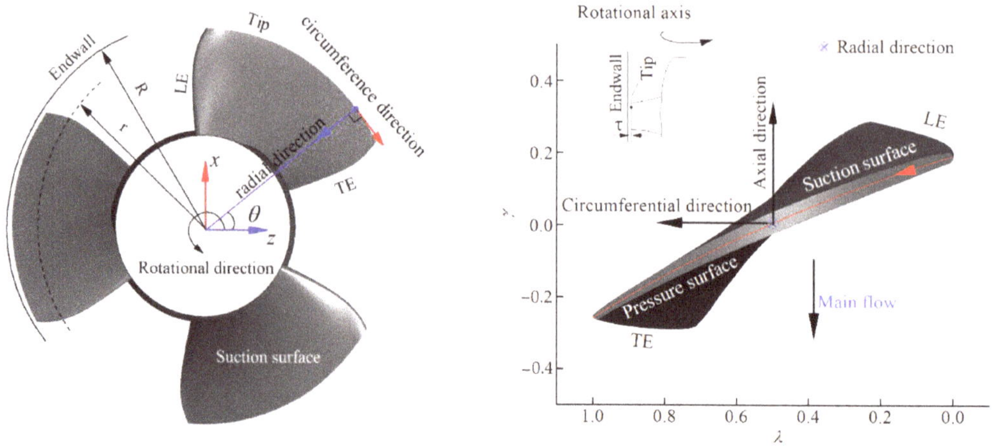

(a) Radial coefficient r* = r/R.

(b) Axial coefficient γ and chord length coefficient λ

Figure 2. (a,b) Geometric definition of the impeller.

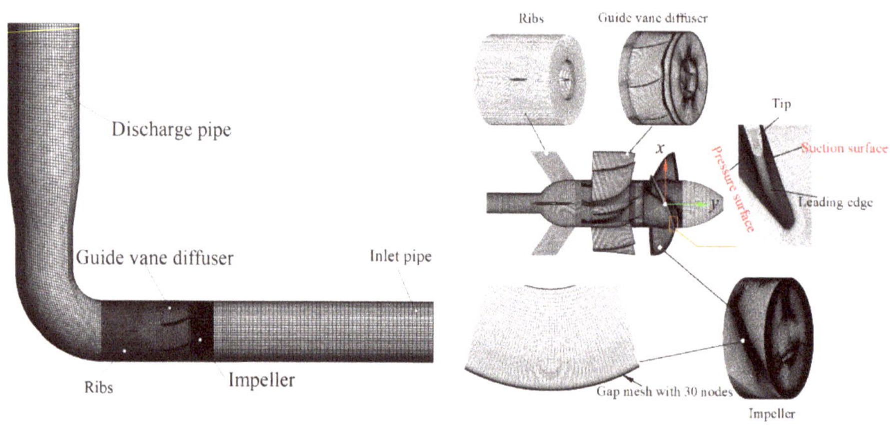

Figure 3. Computational grid of main components.

2.2. Experimental Setup

The external characteristic experiment and the cavitation experiment of the axial flow pump were carried out on the closed test loop, as shown in Figure 4a. The experimental device was mainly composed of a gate valve, an axial flow pump model, a butterfly valve, a boosting pump, a turbine flow meter and a water tank, etc. Before the experiment, connect all the equipment, adjust the inlet and outlet valves to the maximum, fill the pipe with water and discharge the air in the pipe. Then, adjust the flow rate through the outlet gate valve and vacuum the cavitation tank to obtain different inlet pressures. After the pump is running stably, record the flow rate, head and other data. Repeat the measurement three times and take the arithmetic average so as to minimize the random error in the measurement. The cavitation image is derived by a high-speed imaging system, as shown in Figure 4c.

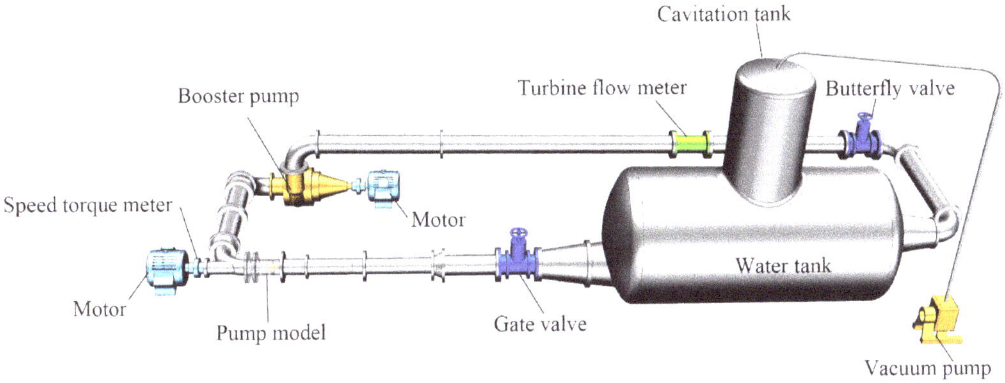

(a) Axial flow pump loop

(b) Test section

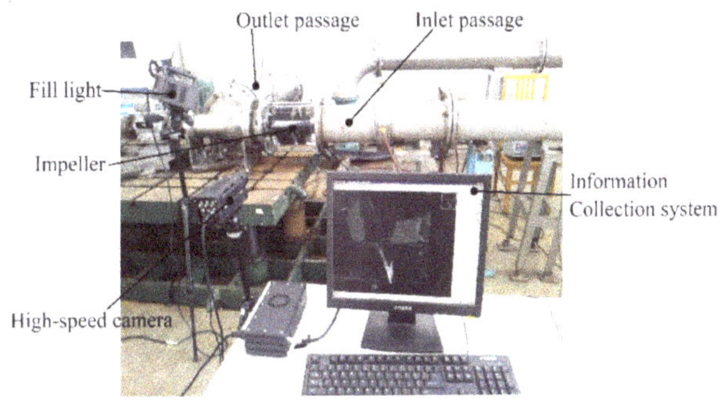

(c) High-speed imaging system

Figure 4. (**a–c**) Experimental equipment for external characteristic and cavitation of the axial flow pump.

2.3. *Governing Equations, Turbulence Model and Cavitation Model*

The continuity and momentum equations are given by the following expression [35]:

$$\frac{\partial \rho}{\partial t} + \frac{\partial}{\partial x_j}(\rho u_j) = 0 \tag{1}$$

$$\frac{\partial}{\partial t}(\rho u_i) + \frac{\partial}{\partial x_j}(\rho u_i u_j) = -\frac{\partial p}{\partial x_i} + \frac{\partial}{\partial x_j}\left[(\mu + \mu_t)\left(\frac{\partial u_i}{\partial x_j} + \frac{\partial u_j}{\partial x_i} - \frac{2}{3}\frac{\partial u_k}{\partial x_k}\delta_{ij}\right)\right] \tag{2}$$

where u_i is the velocity component in the ith direction, p is the pressure and μ_T is the turbulent viscosity. The dynamic viscosity μ and density ρ were defined as $\mu = \alpha_v \mu_v + (1 - \alpha_v)\mu_l$ and $\rho = \alpha_v \rho_v + (1 - \alpha_v)\rho_l$, respectively, where the subscripts l and v represent the liquid and vapor phases, respectively. α_v is the vapor volume fraction.

The SST k-ω model can accurately predict the flow separation, which has been proven to be highly adaptable and feasible for the tip leakage flow [36–38].

The cavitation model was proposed by Zwart, Gerber and Belamri [39], and can be expressed as follows:

$$\frac{\partial(\rho_v \alpha_v)}{\partial t} + \frac{\partial(\rho_v \alpha_v u_j)}{\partial x_j} = \dot{m}^+ + \dot{m}^- \tag{3}$$

The source terms for the specific mass transfer rate corresponding to the vaporization $\dot{m}^{+}$ and condensation $\dot{m}^{-}$ are defined by

$$\dot{m}^{+} = F_{\text{vap}} \frac{3\alpha_{\text{nuc}}(1-\alpha_{\text{v}})\rho_{\text{v}}}{R_{\text{b}}} \sqrt{\frac{2}{3}\frac{p_{\text{v}}-p}{\rho_{\text{l}}}} \quad p \le p_{\text{v}} \tag{4}$$

$$\dot{m}^{-} = F_{\text{cond}} \frac{3\alpha_{\text{v}}\rho_{\text{v}}}{R_{b}} \sqrt{\frac{2}{3}\frac{p-p_{\text{v}}}{\rho_{\text{l}}}} \quad p \ge p_{\text{v}} \tag{5}$$

where F_{vap} and F_{cond} are the empirical coefficients for the mass transfer process with the recommended values of 50 and 0.01, respectively; R_b is the typical bubble radius with a value of 1×10^{-6} m; p_v is the saturation vapor pressure, with a value of 3169 Pa; and α_{nuc} is the nucleation site volume fraction with a value of 5×10^{-4}. These parameters were validated for the simulations of cavitating flow in pumps [29–32].

The commercial code ANSYS CFX 17.1, which is widely used in engineering applications, was adopted in this study.

2.4. Numerical Calculation Result Verification

In the experiment, the outlet flow rate was adjusted to make the outlet flow rate equal to 1.0 Q_{BEP}. By reducing the inlet pressure, the axial flow pump could be manipulated under different cavitation conditions. The cavitation number was written as follows:

$$\sigma = 2(p_{\text{in}} - p_{\text{v}})/\rho_{\text{l}} U_{\text{tip}}{}^{2} \tag{6}$$

where p_{in} is the inlet pressure. The transient rotor–stator was used for the frame change of the impeller. All the physical surfaces of the pump were set as no-slip walls. The automatic near-wall function was used as the wall treatment method. The impeller rotational speed was 1450 r/min. The time required for the impeller to complete one revolution was denoted as T. The convergence accuracy was set to 1×10^{-5}. The time step was set to 1/360 T, about 1.1494×10^{-4} s.

The time-averaged value of the pump head H obtained via the transient calculation was in comparison with that from the experiment, as depicted in Figure 5. The applicability of the mesh and the turbulence model were verified by a comparison of the numerical simulation and the external characteristic experiment.

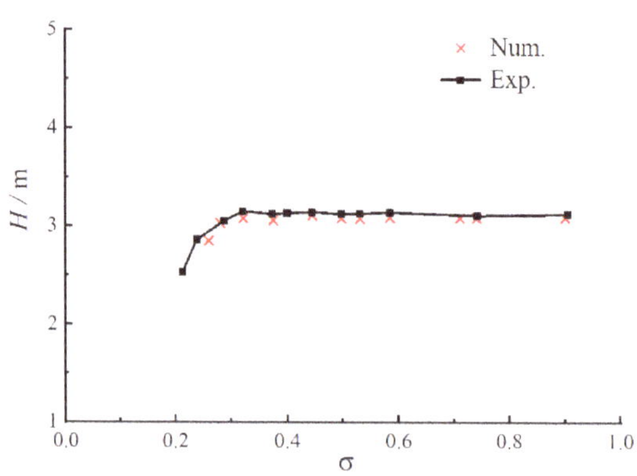

Figure 5. Comparison of the axial flow pump's σ–*H* curves from the numerical simulation and experiment.

3. Results and Discussion

3.1. TLV Formation and Evolution Mechanism

For a better presentation of the velocity field, Figure 6 illustrates the 3D streamlines along three measurement planes under non-cavitating conditions. The leakage velocity magnitude is larger at the leading edge, and subsequently, a vortex structure with concentrated vorticity is formed, which is transmitted to the trailing edge. The vortex scale gradually expands along the chord length, and the value of vorticity gradually decreases.

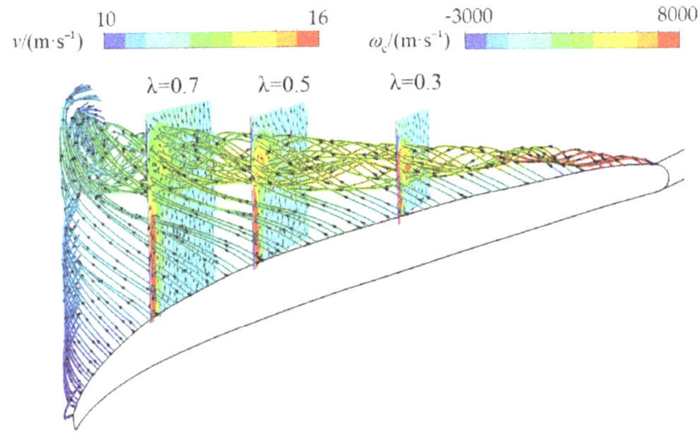

Figure 6. 3D streamlines of the tip leakage flow. The colors in the three measurement planes show the norm of the circumferential vorticity.

As observed from Figure 7, in the section of λ = 0.5, there are two opposite flows in the axial direction, the leakage jet flow and the axial main flow. A shear layer is formed between the two opposite flows, where the vorticity is relatively large. The vortex and vorticity formed by shearing are ultimately transported to the TLV core region.

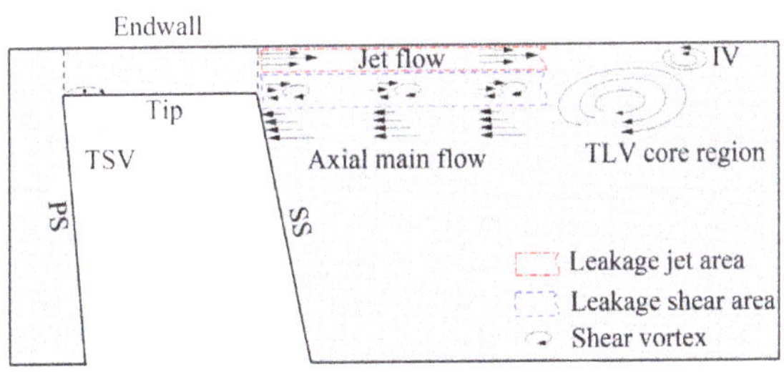

Figure 7. Schematic diagram of axial main flow, tip leakage flow and vortex distribution (λ = 0.5 section).

3.2. S-TLV Structure and Vorticity Distribution

Liu et al. proposed a new vector Liutex $\vec{R}$ to describe the local fluid rotational motion [40,41]. This method specifies the Z_1 direction as the vortex axis direction $\vec{r}$ and R represents the rigid rotation strength of local fluid in a new $X_1Y_1Z_1$-frame. $\vec{R}$ is defined as follows:

$$\vec{R} = R\vec{r} \tag{7}$$

$$R = \begin{cases} 2(\beta - \alpha), \text{ if } \alpha^2 - \beta^2 < 0, \beta > 0, \\ 2(\beta + \alpha), \text{ if } \alpha^2 - \beta^2 < 0, \beta < 0, \\ 0, \text{ if } \alpha^2 - \beta^2 \geq 0, \end{cases} \tag{8}$$

$$\alpha = \frac{1}{2}\sqrt{(\frac{\partial V}{\partial Y_1} - \frac{\partial U}{\partial X_1})^2 + (\frac{\partial V}{\partial X_1} + \frac{\partial U}{\partial Y_1})^2} \quad (9)$$

$$\beta = \frac{1}{2}(\frac{\partial V}{\partial X_1} - \frac{\partial U}{\partial Y_1}) \quad (10)$$

Under the standard flow rate condition ($Q = 1.0Q_{BEP}$), adjust the inlet pressure so that the cavitation number σ is equal to 0.53 and 0.37, case A and case B, respectively. A numerical calculation and experiment were carried out, and the leakage flow characteristics and cavities are depicted in Figure 8. The numerical calculation results choose the $R = 1000$ isosurface to characterize the vortex and the cavitation volume fraction $\alpha_v = 0.1$ isosurface to illustrate the cavity bubbles. As can be observed in the figure, both cases have a TLV structure. In case B, a significant vortex structure is formed near the blade tip at about $\lambda = 0.7$, which is defined as the secondary tip leakage vortex, S-TLV. In case A, this vortex structure is not found, and it is worth analyzing this phenomenon.

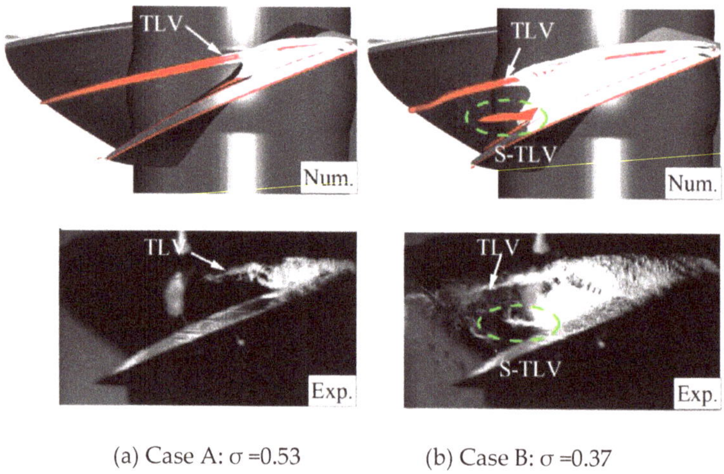

Figure 8. (**a**,**b**) Cavitation and vortex distributions under numerical simulation and experiment.

Take three sections at $\lambda = 0.5/0.7/0.9$ of the blade to analyze the vortex distribution to further understand the S-TLV's evolution, as demonstrated in Figure 9. Overall, case A and case B are similar in the vorticity w_c distribution, and significant shear vorticity is formed near the tip of the blade. The main vortex structure of the TLV gradually moves away from the suction surface along the chord length, and the vorticity of the vortex center gradually decreases. Simultaneously, there is an induced vortex with a negative vorticity at the end wall.

However, there are also obvious differences between the two cases. The double-vortex structure appears in the sections of $\lambda = 0.7$ and $\lambda = 0.9$ in case B, while this structure does not occur in case A.

In case A, there is a large vorticity in the leakage shear area of $\lambda = 0.7-0.9$, but it is difficult to form a significant S-TLV. The tip leakage channel does not have chance to transfer the vorticity generated by being sheared to the TLV core area.

In case B, the double-vortex structure changes the leakage jet and shear channels, and the vorticity distribution is quite different to case A. More noticeable differences are observed that the TLV vorticity in the two sections of $\lambda = 0.7$ and $\lambda = 0.9$ is larger, while the vorticity in the leakage shear area is smaller. This is triggered by the S-TLV transporting the vorticity formed in the leakage shear area to the TLV core region. Therefore, the TLV has a higher vorticity, and the S-TLV expands its vortex scale.

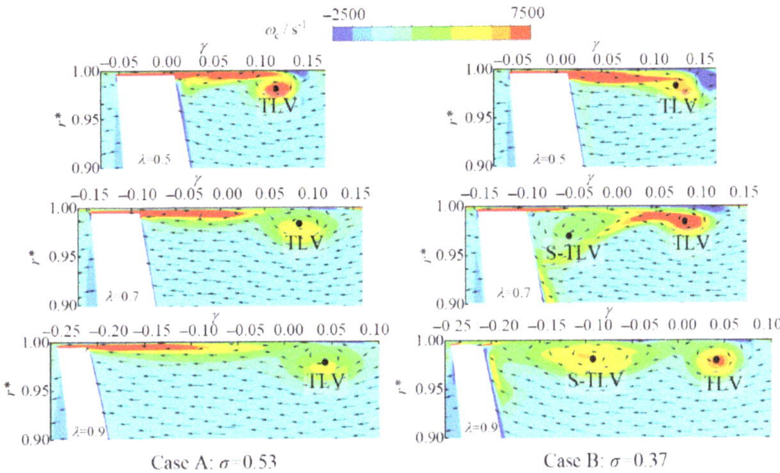

Figure 9. Distributions of circumferential vorticity on three planes.

3.3. *Mechanism of S-TLV Formation*

(1) In the back half of the blade, continuous tip leakage jet helps to form the S-TLV.

The tip leakage flow and the TLV arise from the pressure difference between the pressure surface and the suction surface of the blade. In order to obtain the distribution of the blade pressure difference, sampling lines r* = 0.95/0.9/0.85/0.8/0.75/0.70 are taken at equal intervals. After taking the average of the pressure difference between the pressure surface and the suction surface, the C_p and C_{pd} are derived by non-dimensional processing. The pressure coefficient C_p and the pressure difference coefficient C_{pd} are defined as follows:

$$C_p = p/0.5\rho U_{tip}^2 \tag{11}$$

$$C_{pd} = (p_{ps} - p_{ss})/0.5\rho U_{tip}^2 \tag{12}$$

where p is the pressure, and p_{ps} and p_{ss} are the pressure on the pressure side and suction side, respectively.

The white line frame in Figure 10a indicates the cavitation isoline $\alpha_v = 0.1$. A low-pressure area is generated at the leading edge above the suction surface. Affected by the cavities, the pressure difference at the leading edge is fairly large, and then gradually decreases. As the TLV vortex cavitation increases, the pressure difference gradually rises at λ = 0.05–0.15.

Due to the influence of cavitation, the pressure difference of case B was larger than that of case A at λ = 0.25–0.7. In the r* = 0.997 section, a line with a distance of 1 mm parallel to the suction side is used as a sampling line to obtain the axial velocity curve, as observed in Figure 10b. The axial velocity in case B is larger than in case A at λ = 0.35–0.7, which provides sufficient leakage flow and momentum for the formation of the S-TLV.

Figure 11 depicts the axial velocity distributions in the radial sections. As shown in the figure, the axial velocity near the tip of case B is higher than that of case A. The reason relies on the fact that the cavitation on the suction surface increases the pressure difference, which prompts the continuous generation of leakage jet flow. The axial velocity of case B above the double-vortex structure near the blade tip is higher than that of case A.

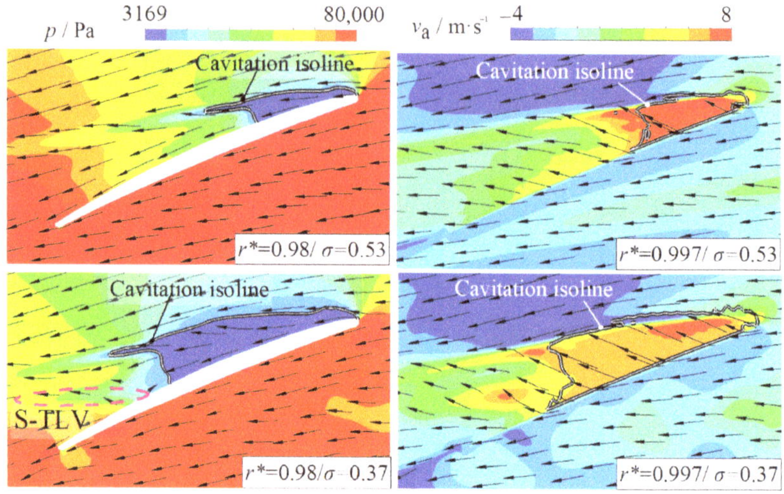

(**a**) Distributions of pressure and axial velocity.

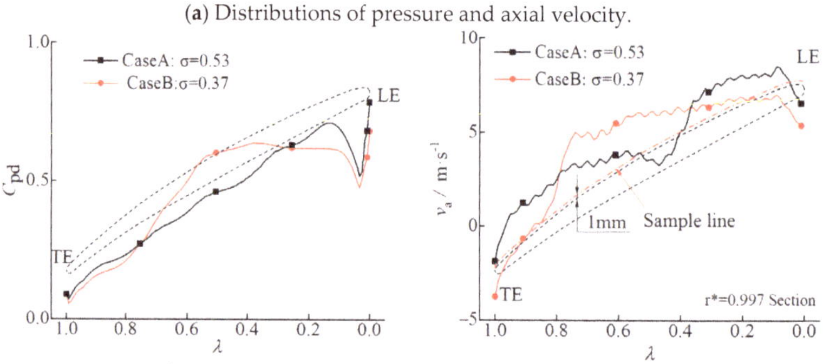

(**b**) Curves of pressure difference and axial velocity.

Figure 10. (**a**,**b**) Distributions of blade pressure difference and axial velocity.

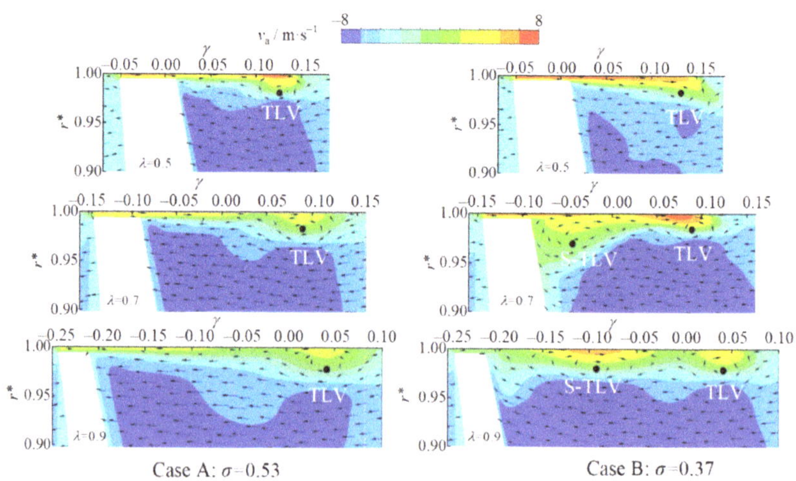

Figure 11. Distributions of axial velocity on three planes.

(2) The low-velocity zone M is where the vortices are concentrated.

In the r* = 0.98 section, take the curve with a distance of 1mm parallel to the suction side as the sampling line to obtain the circumferential velocity curve, as shown in Figure 12. In case B, the fluid on the suction surface is blocked by the cavity at the front of the blade, and the circumferential velocity is low in the range of λ = 0.5 to λ = 1.0. Particularly, a local low-velocity zone M emerges, which is located in the position of λ = 0.6–0.7. Combined with the pressure distributions in Figure 10a, the re-entrant jet flow is caused by the circumferential pressure difference. Since the cavity bubbles are blocked in front of the blade, the circumferential flow bypasses, which further promotes the formation of the zone M.

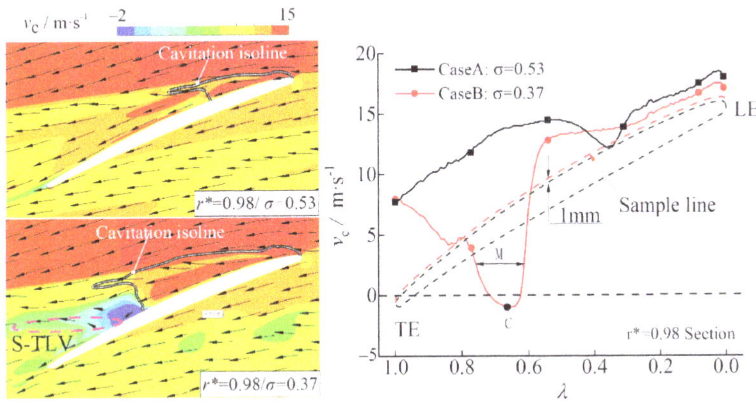

Figure 12. Circumferential velocity distributions and low velocity zone M.

The zone M is an essential area for the development of the S-TLV, in detail, where the axial leakage velocity is high, and the formed shear vortices are gradually accumulated, which appear as the prominent source of the S-TLV.

As observed from Figure 13, in the sections of λ = 0.7 and λ = 0.9, the circumferential velocity of the S-TLV area of case B is lower than that of the corresponding position of case A.

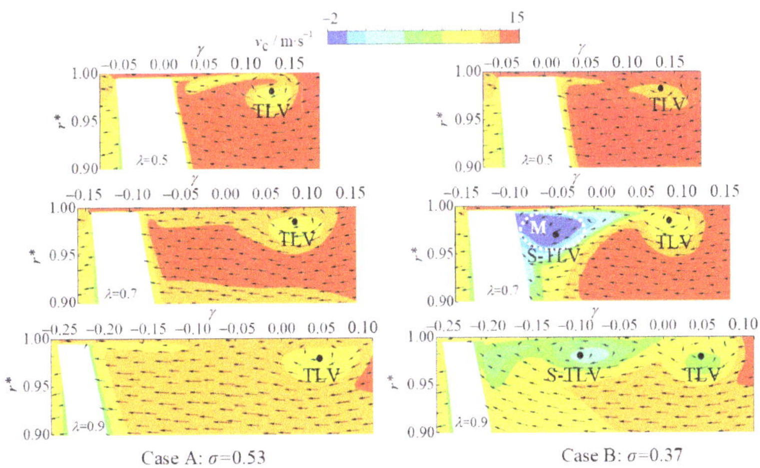

Figure 13. Distributions of circumferential velocity on three planes.

In case B, a large-scale low-velocity zone M is formed in the section of λ = 0.7, where the leakage jet is strong and the S-TLV is generated, accumulated and evolved. Subsequently, in the section of λ = 0.9, the circumferential velocity increases.

(3) The radial jet promotes the development of the S-TLV.

Figure 14 shows the distributions of velocity streamlines and cavity bubbles on the suction surface. In case B, due to the low cavitation number, a large number of cavity bubbles (TLV cavitation, shear cavitation, sheet cavitation, etc.) are formed in the vicinity of the tip above the suction surface. The pressure in this position is lower, yielding a radial flow under the action of the pressure difference between the hub and the tip, as demonstrated in Figure 14. Additionally, the vapor–liquid separation triggered by the rotation of the impeller also contributes to the formation of a radial jet under the action of centrifugation. In case A, the radial jet is not observed due to insufficient cavities.

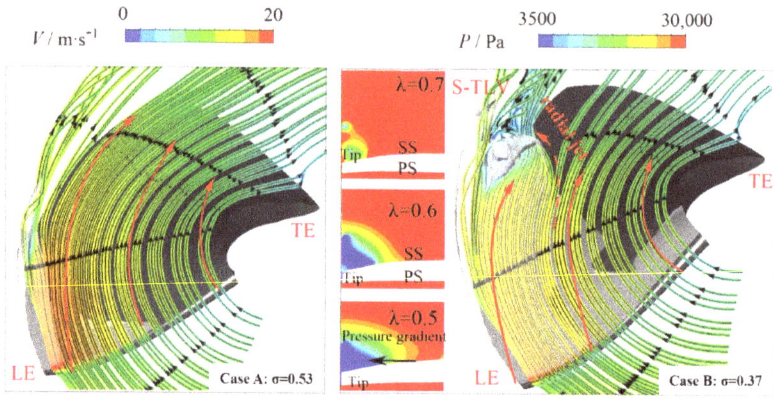

Figure 14. Velocity streamlines and cavitation distributions on the suction surface of the blade.

As shown in case B in Figure 15, the existence of radial jet is easy to capture at the two cross-sections, λ = 0.7 and λ = 0.9. Since the flow direction formed by the radial jet and the leakage flow is in accordance with the rotation direction of the S-TLV. Therefore, this flow path is beneficial for the development of the S-TLV.

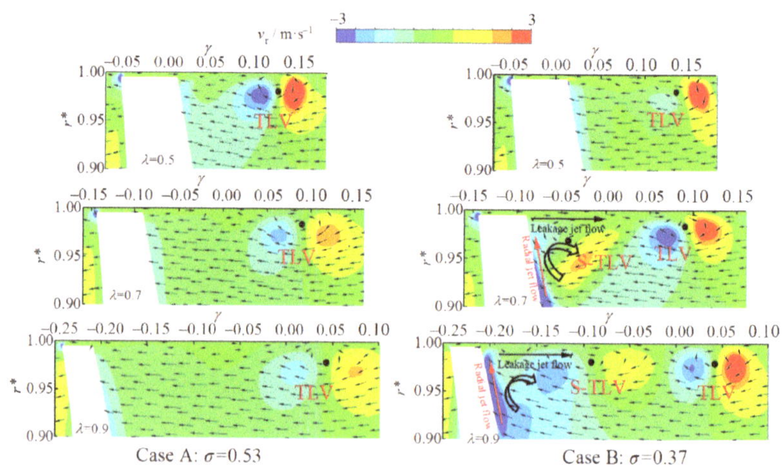

Figure 15. Distributions of radial velocity on three planes.

Figure 16 depicts the vortex intensity distributions acquired according to the velocity swirling strength criterion [42]. In the two cases, the distributions of vortex intensity α are basically similar in the cross section of λ = 0.5. There are early forms of the S-TLV near the suction surface, and a small number of vortices cover the suction surface of the blade.

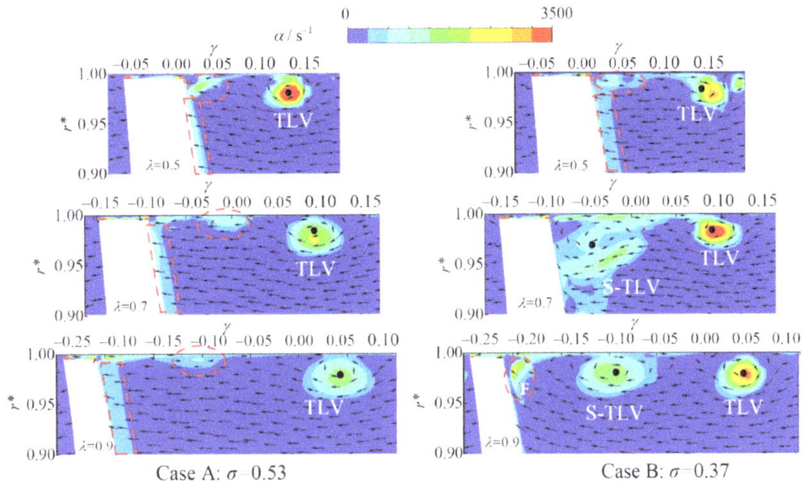

Figure 16. Distributions of vortex intensity on three planes.

In the λ = 0.7 section of case A, the leakage velocity is small, and the TLV center is pulled away from the suction surface. The leakage jet cannot wrap the TLV, thus accelerating the TLV dissipation. In addition, the vortex intensity transported by the leakage jet in the leakage channel is not sufficient to compensate for the TLV dissipation, and the TLV vortex intensity gradually diminishes. In case B, the vortex intensity of the TLV is enhanced. Since the leakage jet flow rate at this location is still high, and the leakage jet can wrap around the TLV area, which can reduce the dissipation of the TLV. Additionally, under the action of the radial jet, the vortices attached on the wall of the suction surface are successively pushed to the tip of the blade. Under the combined action of the leakage jet and radial jet, they are mixed and absorbed with each other, leading to the further development and growth of the TLV and the S-TLV. In this process, the S-TLV played a key role in transit and transmission.

In the λ = 0.9 section of case B, there are vortices marked F in Figure 16 originating from the suction surface near the tip, which migrate to the blade tip under the action of the radial jet. It can be inferred that in the follow-up process, under the successive promotion of the leakage jet, the vortices will gradually mix with the S-TLV and evolve into a part of the S-TLV. Compared with case A, the vorticity on the suction surface in case B is almost absorbed by the S-TLV. The TLV and the S-TLV are basically equivalent in vortex intensity and scale, maintaining a relative balance. The vortices attached on the wall of the suction surface are transported to the S-TLV, which is conducive to the S-TLV's development.

As shown in Figure 17, the formation mechanism of the S-TLV is illustrated. Compared with case A, a significant S-TLV can be formed in case B for the following reasons. Firstly, the leakage jet and shear vortices are formed by the larger pressure difference in the back half of the blade. Furthermore, a low-velocity zone is formed attributed to the blocking of the cavity bubbles and the effect of the re-entrant jet flow, which provides a favorable condition for the vortices to gather together. Simultaneously, the cavity bubbles also lift the core of the TLV, making it difficult for the S-TLV to be entrained and absorbed by the TLV. Finally, the flow path formed by the radial jet and the leakage jet promotes the development of the S-TLV, and the radial jet pushes the vortices attached on the suction surface to the blade tip, which strengthens the S-TLV.

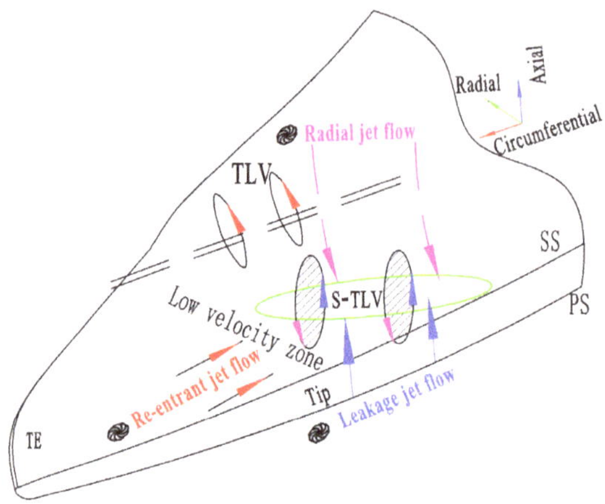

Figure 17. Schematic diagram of the S-TLV formation mechanism.

In case A, the external conditions of the low-velocity zone and strong radial jet failed to be generated. At the same time, with its small pressure difference, the leakage jet was not sufficient to give birth to the large-scale S-TLV. Meanwhile, the tip leakage flow is also difficult to wrap around the TLV area, and consequently, the TLV is more prone to dissipating.

3.4. TKE, Pressure Pulsation and Vortex Core Trajectory

Figure 18 depicts the distributions of the TKE. In the figure, due to the mutual shear between the main flow and the leakage jet in the axial direction, the TKE is relatively large. In contrast, it is relatively small in other positions with proximity to zero.

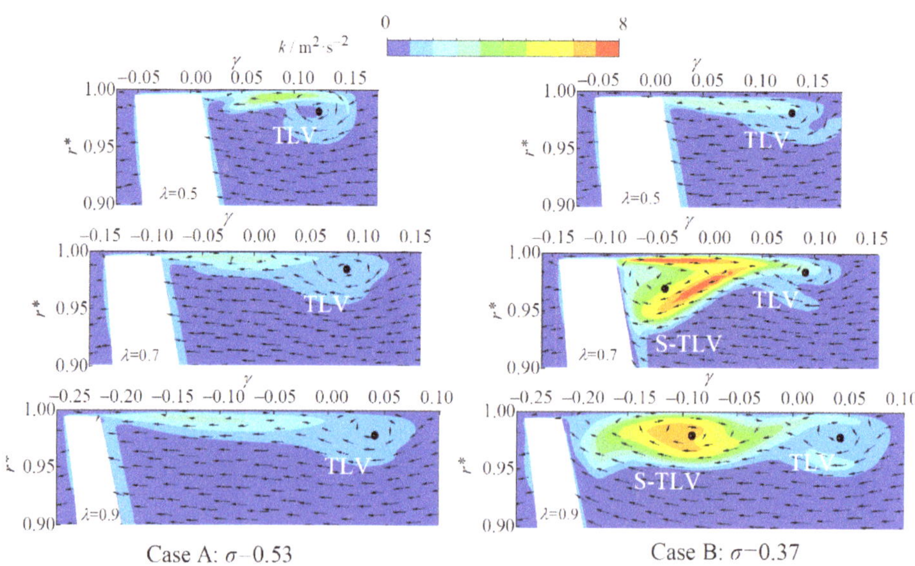

Figure 18. Distributions of TKE on three planes.

In the λ = 0.5 section, the TKE distribution of case B basically coincides with that of case A. However, in the λ = 0.7 and λ = 0.9 sections, the TKE distribution of case B is concentrated in the S-TLV area, which varies much with case A. Under the combined action of the re-entrant jet and the radial jet, the velocity pulsation is large, and the flow shear in multiple directions occurs at the same time. Consequently, the TKE is relatively large and messy.

In general, the generation of TKE is accompanied by the generation of vorticity. In the section of λ = 0.9, the re-entrant jet weakens, whereas the radial jet is still strong. There is still circumferential vorticity being generated, and the TKE is also large in case B.

Compared with case A, the axial vorticity and its distribution in case B are more significant in the λ = 0.7 section, as demonstrated in Figure 19a. There is a pressure gradient in circumferential direction, which gives birth to the re-entrant jet in case B. When the re-entrant jet flow and the circumferential main flow form convection, a pair of axial vortices in opposite directions are formed, as demonstrated in Figure 19b.

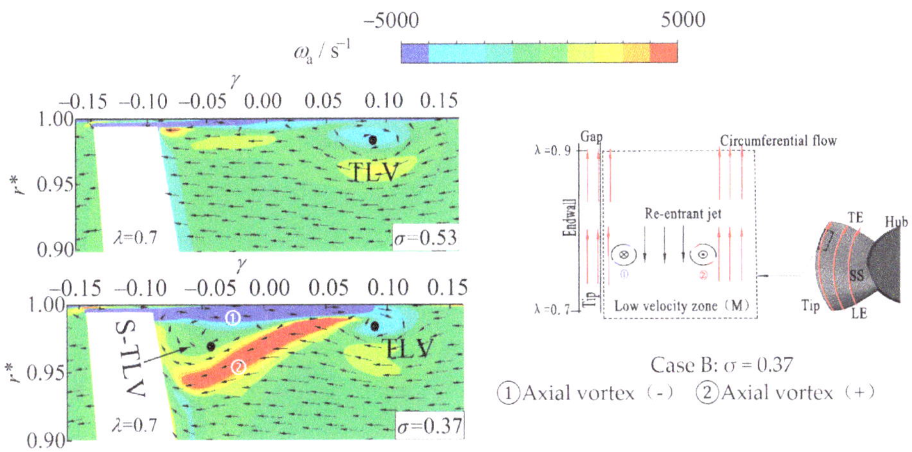

(**a**) Distributions of axial vorticity(λ = 0.7) (**b**) Formation mechanism of axial vorticity in case B

Figure 19. (**a**,**b**) Axial vorticity distributions in the section of λ = 0.7 and its formation mechanism.

The S-TLV is sandwiched between the two axial vortices and remains stable. The axial vortices may play a certain auxiliary role in the generation and development of the S-TLV.

Figure 20 shows three monitoring points and the time-pressure coefficient curves. As illustrated in Figure 20, case A has a higher pressure in the same position than case B.

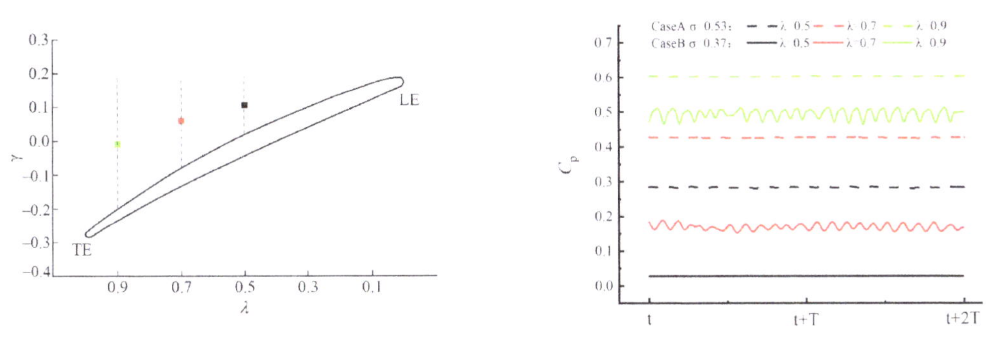

Figure 20. Monitoring points and curves of pressure pulsation.

The pressure at the $\lambda = 0.5$ point is higher than the saturated vapor pressure, indicating that case A is free from the influence of cavitation. There is no cavitation shedding and vortex influence, and as a result, the pressure pulsation remains small.

The pressure gradually rises along the chord length ($p_{\lambda=0.9} > p_{\lambda=0.7} > p_{\lambda=0.5}$) in the two cases, which verifies that there is a pressure gradient in the circumferential direction in the back half of the blade.

In case B, an obvious pressure pulsation appears at $\lambda = 0.7$ and $\lambda = 0.9$ points, which are triggered by cavitation shedding and vortex formation. Correspondingly, the TKE at these locations is also larger, as demonstrated in Figure 18.

The vortex can wrap and entrain the cavity bubbles, prevent the cavity bubbles from falling off and collapsing, thereby avoiding the increase in pressure pulsation. The vortices gradually dissipate along the chord length. Therefore, the pressure pulsation amplitude of the $\lambda = 0.9$ point appears slightly higher than that of the $\lambda = 0.7$ point. Simultaneously, the pressure increases, further advancing the process of the collapse of the cavities.

Figure 21 shows the distribution of the vortex core trajectory. In the axial direction, the TLV core of case B is slightly higher than that of case A at $\lambda = 0-0.7$, which is affected by the cavities. Near the trailing edge, under the influence of the S-TLV, the height of the vortex core in case B drops faster. The S-TLV primarily stays and accumulates in the low-velocity zone M on the suction surface, and subsequently driven by the TLV entrainment and jet, the vortex core trajectory develops upward.

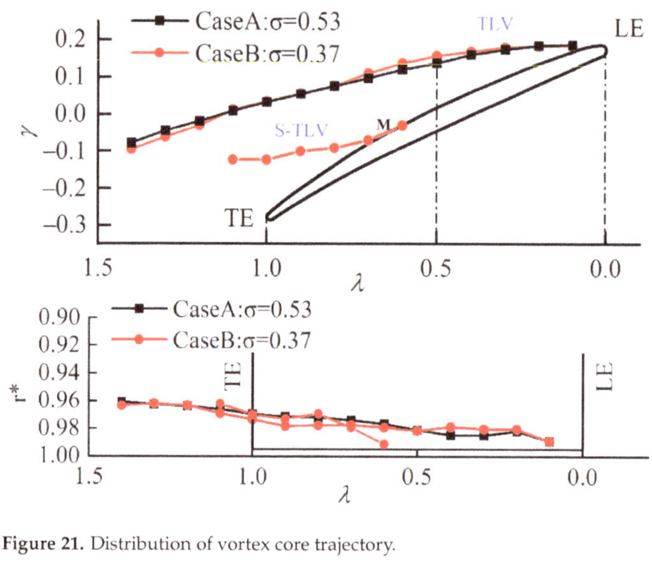

Figure 21. Distribution of vortex core trajectory.

In the radial direction, at $\lambda = 0-0.5$, the cavity bubbles make the TLV's vortex core trajectory in case B closer to the hub and away from the blade tip; and at $\lambda = 0.5-1$, driven by the radial jet, it migrates to the blade tip. The S-TLV is subject to the combined action of the TLV and leakage flow in case B, and the radial position of the S-TLV is basically in agreement with that of the TLV.

3.5. *Influence Factors on the S-TLV's Formation*

As observed from Figure 22, the cavitation number is used as the unique variable in the experiments and numerical simulations. The Liutex method is used for vortex identification, the isosurface $R = 1000$. In the figure, red represents the circumferential vortex, and blue represents the axial vortex.

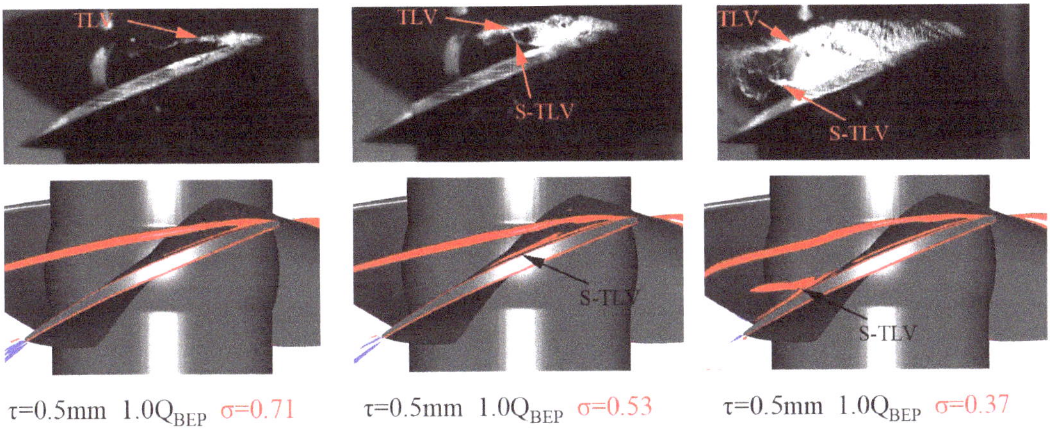

Figure 22. Cavitation images and vortex isosurface (Liutex $R = 1000$, Red—circumferential vortex, Blue—axial vortex) under different cavitation numbers.

When the cavitation number decreases, the S-TLV gradually develops and extends to the middle and trailing edge of the blade. When the cavitation is enhanced, the formed cavity bubbles reduce the pressure on the suction surface, increase the pressure difference and accelerate the leakage flow velocity. Cavitation heightens the TLV vortex core and promotes the formation of a low-velocity zone. Cavitation changes the radial pressure distribution and promotes the formation of a radial jet. The lower the cavitation number decreases, the more significant the S-TLV becomes.

As depicted in Figure 23, take the flow rate as a variable and analyze its impact on the S-TLV. When the flow rate decreases, the axial mainstream velocity decreases. The main stream's constraint on the leakage flow is weakened, and the leakage flow appears to be in a wandering state. The shear cavitation and vortex cavitation emerge and occupy the tip side of the blade, flow instability and cavitation gradually develop and finally, the S-TLV is gradually transformed to the PCV. The lower the flow rate drops, the higher the vortex intensity is generated.

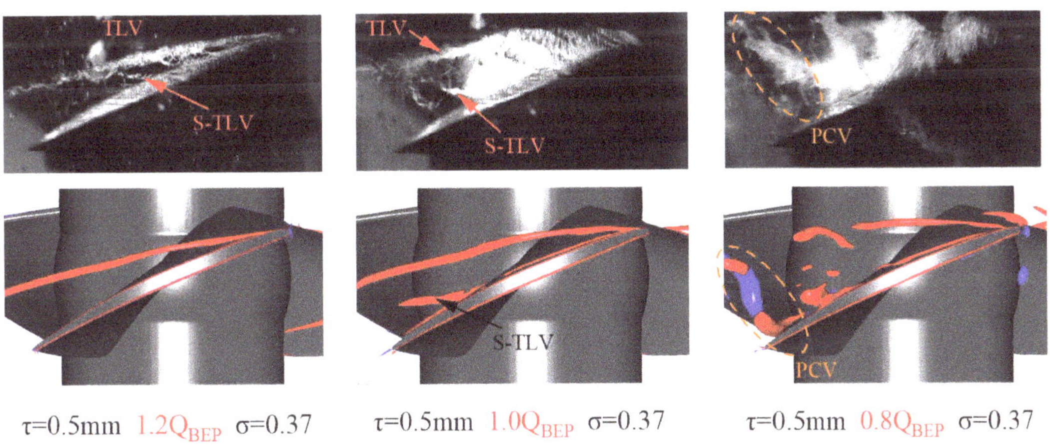

Figure 23. Cavitation images and vortex isosurface (Liutex $R = 1000$, Red—circumferential vortex, Blue—axial vortex) under different flow rates.

As shown in Figure 24, take the tip clearance size as a variable and analyze its impact on the S-TLV. When the tip clearance size is reduced, the leakage jet is enhanced in the back half of the blade. With a small clearance size, the S-TLV seems to be easier to generate.

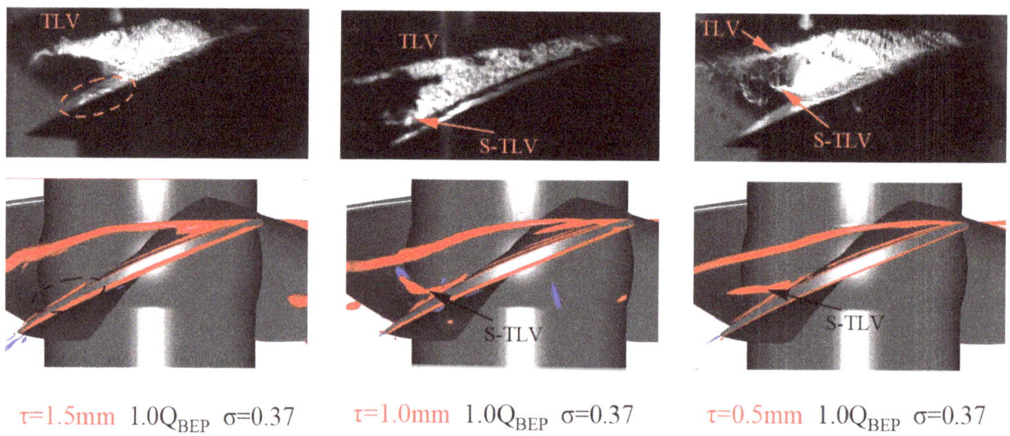

Figure 24. Cavitation images and vortex isosurface (Liutex R = 1000, Red—circumferential vortex, Blue—axial vortex) under different tip clearance sizes.

4. Conclusions

In this study, the cavitation number σ = 0.37 of the axial flow pump with a clearance of 0.5 mm is taken to study the formation mechanism of the S-TLV. In order to facilitate the study, the case of the cavitation number σ = 0.53 is used as the control group. The main conclusions can be drawn as follows:

(1) The formation of the S-TLV is attributed to the existence of a low-velocity zone M, which is located at the λ = 0.7 section near the tip above the suction surface of the blade. The formation of the low-velocity zone is conducive to the gradual accumulation of the vortices formed by the shearing between the leakage jet flow and the axial main flow at this location. Driven by the radial jet, the small vortices attached on the suction surface are pushed to the blade tip and mixed with the S-TLV by entrainment, which contribute to the development of the TLV and the S-TLV.

(2) The S-TLV contributes to the leakage flow transporting the shear vorticity to the TLV, which can delay the dissipation of the TLV. The S-TLV changes the TLV's vortex core trajectory in the axial direction.

(3) The S-TLV is wrapped by a pair of axial vortices located above the suction surface at λ = 0.7, which may be beneficial for its stability. The S-TLV's generation and development process is accompanied by drastic changes in the TKE and pressure pulsation.

(4) The condition of the low cavitation number and small flow rate is conducive to the formation of the S-TLV, which can be transformed into the PCV under a certain condition.

Author Contributions: Conceptualization, H.Z. and W.S.; investigation and writing, H.Z. and D.Z.; validation, H.Z. and D.Z.; visualization, H.Z., W.S. and J.S.; methodology, H.Z., W.S. and J.Z.; software, H.Z., J.S. and J.Z.; formal analysis, H.Z., W.S. and J.Z.; resources, D.Z. and H.Z.; funding acquisition, D.Z. and W.S.; supervision, W.S. All authors have read and agreed to the published version of the manuscript.

Funding: This research was funded by the National Natural Science Foundation of China (51776087), the Equipment Pre-Research Fund Project (8151440002) and the Intelligent Manufacturing Equipment Design and Engineering Application Program of Excellent Scientific and Technological Innovation Team of Colleges and Universities in Jiangsu Province (201907).

Data Availability Statement: The numerical and experimental data used to support the findings of this study are included within the article.

Conflicts of Interest: The authors declare no conflict of interest.

Nomenclature

Q_{BEP}	Design flow rate
H	Pump head
d_2	Outlet diameter
d_1	Inlet diameter
d_t	Hub diameter
c	Chord length
τ	Tip clearance size
r	Radius
R	Radius of the impeller chamber
$r*$	Radial coefficient
γ	Axial coefficient
λ	Chord length coefficient
U_{tip}	Tip velocity
V	Velocity
V_c	Circumferential velocity
V_a	Axial velocity
V_r	Radial velocity
k	Turbulence kinetic energy
p	Pressure
μ	Laminar viscosity
μ_t	Turbulent eddy viscosity
Ω	Rotor angular velocity
C_{pd}	Pressure differential coefficient
ω	Vorticity
C_p	Pressure coefficient
σ	Cavitation number
α_v	Vapor volume fraction
R	Rigid rotation strength
α	Vortex intensity
ω_c	Circumferential vorticity

References

1. Guan, X. *Axial Flow Pump and Mixed Flow Pump*; Aerospace Press: Beijing, China, 2009.
2. Rains, D.A. Tip Clearance Flows in Axial Compressors and Pumps. Ph.D. Thesis, California Institute of Technology, Pasadena, CA, USA, 1954.
3. Hsiao, C.T.; Chahine, G.L. Scaling of tip vortex cavitation inception noise with a bubble dynamics model accounting for nuclei size distribution. *J. Fluids Eng.* **2005**, *127*, 55–65. [CrossRef]
4. You, D.; Wang, M.; Moin, P.; Mittal, R. Vortex Dynamics and Low-Pressure Fluctuations in the Tip-Clearance Flow. *J. Fluids Eng.* **2007**, *129*, 1002–1014. [CrossRef]
5. Tan, D.; Li, Y.; Wilkes, I.; Vagnoni, E.; Miorini, R.L.; Katz, J. Experimental Investigation of the Role of Large Scale Cavitating Vortical Structures in Performance Breakdown of an Axial Waterjet Pump. *J. Fluids Eng.* **2015**, *137*, 111301. [CrossRef]
6. Liu, Y.; Tan, L. Tip clearance on pressure fluctuation intensity and vortex characteristic of a mixed flow pump as turbine at pump mode. *Renew. Energy* **2018**, *129*, 606–615. [CrossRef]
7. Miorini, R.L.; Wu, H.; Katz, J. The internal structure of the tip leakage vortex within the rotor of an axial Water jet pump. *J. Turbomach.* **2012**, *134*, 031018. [CrossRef]
8. Chen, H.; Doeller, N.; Li, Y.; Katz, J. Experimental Investigations of Cavitation Performance Breakdown in an Axial Waterjet Pump. *J. Fluids Eng.* **2020**, *142*, 091204. [CrossRef]
9. Zhang, D.; Shi, L.; Shi, W.; Zhao, R.; Wang, H.; Van Esch, B. Numerical analysis of unsteady tip leakage vortex cavitation cloud and unstable suction-side-perpendicular cavitating vortices in an axial flow pump. *Int. J. Multiph. Flow* **2015**, *77*, 244–259. [CrossRef]
10. Moghadam, S.; Meinke, M.; Schröder, W. Analysis of tip-leakage flow in an axial fan at varying tip-gap sizes and operating conditions. *Comput. Fluids* **2019**, *183*, 107–129. [CrossRef]

11. Shi, L.; Zhang, D.; Zhao, R.; Shi, W.; Van Esch, B. Visualized observations of trajectory and dynamics of unsteady tip cloud cavitating vortices in axial flow pump. *J. Fluid Sci. Technol* **2017**, *12*, JFST0007. [CrossRef]
12. Zhang, H.; Zang, J.; Shi, W.; Zhang, D. Analysis of the Formation Mechanism and Evolution of the Perpendicular Cavitation Vortex of Tip Leakage Flow in an Axial-Flow Pump for Off-Design Conditions. *J. Mar. Sci. Eng.* **2021**, *9*, 1045. [CrossRef]
13. You, D.; Wang, M.; Moin, P.; Mittal, R. Effects of tip-gap size on the tip-leakage flow in a turbo machinery cascade. *J. Phys. Fluids* **2006**, *18*, 105102. [CrossRef]
14. You, D.; Wang, M.; Moin, P.; Mittal, R. Large-eddy simulation analysis of mechanisms for viscous losses in a turbomachinery tip-clearance flow. *J. Fluid Mech.* **2007**, *586*, 177–204. [CrossRef]
15. Zhang, H.; Zuo, F.; Zhang, D.; Shi, W. Formation mechanism and geometric influence of tip clearance vortex structure around hydrofoil. *J. Zhejiang Uni. (Eng. Sci.)* **2020**, *54*, 2344–2355.
16. Wu, H.; Miorini, R.L.; Tan, D.; Katz, J. Turbulence within the tip-leakage vortex of an axial waterjet pump. *AIAA J* **2012**, *50*, 2574–2587. [CrossRef]
17. Wu, H.; Miorini, R.L.; Katz, J. Measurements of the tip leakage vortex structures and turbulence in the meridional plane of an axial water-jet pump. *Exp. Fluids* **2011**, *50*, 989–1003. [CrossRef]
18. Wu, H.; Tan, D.; Miorini, R.L.; Katz, J. Three-dimensional flow structures and associated turbulence in the tip region of a waterjet pump rotor blade. *Exp. Fluids* **2011**, *51*, 1721–1737. [CrossRef]
19. Dreyer, M. Mind the Gap: Tip Leakage Vortex Dynamics and Cavitation in the Axial Turbines. Ph.D. Thesis, Swiss Federal Institute of Technology in Lausanne (EPFL), Lausanne, Switzerland, 2015.
20. Decaix, J.; Dreyer, M.; Balarac, G.; Farhat, M.; Münch, C. RANS computations of a confined cavitating tip-leakage vortex. *Eur. J. Mech. B-Fluid* **2018**, *67*, 198–210. [CrossRef]
21. Dreyer, M.; Decaix, J.; Münch-Alligné, C.; Farhat, M. Mind the gap: A new insight into the tip leakage vortex using stereo-PIV. *Exp. Fluids* **2014**, *55*, 1849. [CrossRef]
22. Giuni, M.; Green, R.B. Vortex formation on squared and rounded tip. *Aerosp. Sci. Technol.* **2013**, *29*, 191–199. [CrossRef]
23. Laborde, R.; Chantrel, P.; Mory, M. Tip clearance and tip vortex cavitation in an axial flow pump. *J. Fluids Eng.* **1997**, *119*, 680–685. [CrossRef]
24. Zhang, H.; Wang, J.; Zhang, D.; Shi, W.; Zang, J. Numerical Analysis of the Effect of Cavitation on the Tip Leakage Vortex in an Axial-Flow Pump. *J. Mar. Sci. Eng.* **2021**, *9*, 775. [CrossRef]
25. Chen, H.Y.; Bai, X.R.; Long, X.P.; Ji, B.; Farhat, M. Large eddy simulation of the tip-leakage cavitating flow with an insight on how cavitation influences vorticity and turbulence. *Appl. Math. Model* **2019**, *77*, 788–809. [CrossRef]
26. Zhao, Y.; Wang, G.; Jiang, Y.; Huang, B. Numerical analysis of developed tip leakage cavitating flows using a new transport-based model. *Int. Commun. Heat Mass Transf.* **2016**, *78*, 39–47. [CrossRef]
27. Guo, Q.; Huang, X.; Qiu, B. Numerical investigation of the blade tip leakage vortex cavitation in a waterjet pump. *Ocean Eng.* **2019**, *187*, 106170. [CrossRef]
28. Lee, H.; Park, K.; Choi, H. Experimental investigation of tip-leakage flow in an axial flow fan at various flow rates. *J. Mec. Sci. Tec.* **2019**, *33*, 1271–1278. [CrossRef]
29. Zhang, D.; Shi, W.; van Esch, B.; Shi, L.; Dubuisson, M. Numerical and experimental investigation of tip leakage vortex trajectory and dynamics in an axial flow pump. *Comput. Fluids* **2015**, *112*, 61–71. [CrossRef]
30. Zhang, H.; Zuo, F.; Zhang, D.; Shi, W. Formation and Evolution Mechanism of Tip Leakage Vortex in Axial Flow Pump and Vortex Cavitation Analysis. *Trans. Chin. Soc. Agr. Mach.* **2020**, *52*, 157–167.
31. Shi, L.; Zhang, D.; Jin, Y.; Shi, W.; van Esch, B.P.M. A study on tip leakage vortex dynamics and cavitation in axial-flow pump. *Fluid Dyn. Res.* **2017**, *49*, 035504. [CrossRef]
32. Zhang, D.; Shi, W.; Pan, D.; Dubuisson, M. Numerical and Experimental Investigation of Tip Leakage Vortex Cavitation Patterns and Mechanisms in an Axial Flow Pump. *J. Fluids Eng. Trans. ASME* **2015**, *137*, 121103. [CrossRef]
33. Tan, D.Y.; Miorini, R.L.; Keller, J.; Katz, J. Flow Visualization Using Cavitation Within Blade Passage of an Axial Waterjet Pump Rotor. In Proceedings of the ASME 2012 Fluids Engineering Summer Meeting, Rio Grande, PR, USA, 8–12 July 2012.
34. Shen, X.; Zhang, D.; Xu, B.; Jin, Y.; Shi, W.; Esch, B. Experimental investigation of the transient patterns and pressure evolution of tip leakage vortex and induced-vortices cavitation in an axial flflow pump. *J. Fluid Eng.* **2020**, *142*, 101206.
35. ANSYS, Inc. *ANSYS CFX-Solver Theory Guide, Release 17.1*; SAS IP, Inc.: Canonsburg, PA, USA, 2016.
36. Menter, F.R. Review of the shear-stress transport turbulence model experience from an industrial perspective. *Int. J comput. Fluid. D.* **2009**, *23*, 305–316. [CrossRef]
37. Heydari, M.; Sadat-Hosseini, H. Analysis of propeller wake field and vortical structures using k-Omega SST Method. *Ocean Eng.* **2020**, *204*, 107247. [CrossRef]
38. Geng, L.; Escaler, X. Assessment of RANS turbulence models and Zwart cavitation model empirical coefficients for the simulation of unsteady cloud cavitation. *Eng. Appl. Comp. Fluid Mech.* **2020**, *14*, 151–167. [CrossRef]
39. Zwart, P.; Gerber, A.G.; Belamri, T. A two-phase model for predicting cavitation dynamics. In Proceedings of the ICMF 2014 International Conference on Multiphase Flow, Yokohama, Japan, 30 May–3 June 2004; p. 152.
40. Bai, X.; Cheng, H.; Ji, B.; Long, X.; Qian, Z.; Peng, X. Comparative Study of different vortex identification methods in a tip-leakage cavitating flow. *Ocean Eng.* **2020**, *207*, 107373. [CrossRef]

41. Liu, C.Q.; Gao, Y.S.; Tian, S.L.; Dong, X.R. Rortex-A new vortex vector definition and vorticity tensor and vector decompositions. *Phys. Fluids* **2018**, *30*, 12. [CrossRef]
42. Liu, Y.; Tan, L. Method of C groove on vortex suppression and energy performance improvement for a NACA0009 hydrofoil with tip clearance in tidal energy. *Energy* **2018**, *155*, 448–461. [CrossRef]

machines

MDPI

Article

A Study on the Cavitation and Pressure Pulsation Characteristics in the Impeller of an LNG Submerged Pump

Wei Li [1,2,*], Shuo Li [1], Leilei Ji [1], Xiaofan Zhao [1], Weidong Shi [3], Ramesh K. Agarwal [4], Muhammad Awais [1] and Yang Yang [5]

1 National Research Center of Pumps, Jiangsu University, Zhenjiang 212013, China; 2222011024@stmail.ujs.edu.cn (S.L.); leileiji@ujs.edu.cn (L.J.); 2221511006@ujs.edu.cn (X.Z.); agri.engg@outlook.com (M.A.)
2 Institute of Fluid Engineering Equipment Technology, Jiangsu University, Zhenjiang 212009, China
3 College of Mechanical Engineering, Nantong University, Nantong 226019, China; wdshi@ujs.edu.cn
4 Department of Mechanical Engineering & Materials Science, Washington University in St. Louis, St. Louis, MO 63130, USA; rka@wustl.edu
5 College of Hydraulic Science and Engineering, Yangzhou University, Yangzhou 225009, China; yangyang_ujs@outlook.com
* Correspondence: lwjiangda@ujs.edu.cn

Abstract: Based on CFD analysis technology, this paper studies the cavitation performance of an LNG submerged pump and the pressure pulsation characteristics under cavitation excitation. The variation laws of the waveform, amplitude and frequency of the pressure pulsation in the impeller of the LNG submerged pump under different flow rates and $NPSH_a$ are also analysed. By calculating the root mean square of the pressure coefficient of the low-frequency pulsation, the influence of the aggravation process of cavitation on the low-frequency pulsation in the LNG submerged pump is quantitatively analysed, and the characteristics of the pressure pulsation in the LNG submerged pump under the cavitation condition are revealed. The results show that with the increase in flow rate, the pressure pulsation in the impeller becomes stronger, periodically, and the amplitude decreases. The influence of cavitation on the pressure pulsation in the primary impeller is greater than that in the secondary impeller. When critical cavitation occurs, the low-frequency signal amplitude of pressure pulsation in the primary impeller increases and exceeds the blade frequency, becoming the main frequency.

Keywords: LNG submerged pump; numerical calculation; cavitation; pressure pulsation

Citation: Li, W.; Li, S.; Ji, L.; Zhao, X.; Shi, W.; Agarwal, R.K.; Awais, M.; Yang, Y. A Study on the Cavitation and Pressure Pulsation Characteristics in the Impeller of an LNG Submerged Pump. *Machines* **2022**, *10*, 14. https://doi.org/10.3390/machines10010014

Academic Editor: Davide Astolfi

Received: 11 November 2021
Accepted: 21 December 2021
Published: 24 December 2021

1. Introduction

As the core power transmission equipment of the LNG transmission industry chain, the LNG submerged pump's performance affects the pump's service life and affects the safety and economy of the LNG station. Due to the low temperature and flammable characteristics of the LNG, the submersible pump should be able to withstand the efficient operation performance at low temperatures and have higher requirements for the airtightness and reliability of the pump. Therefore, it is important to develop the key design technology of high-efficiency LNG submerged pumps, master the cavitation characteristics and pressure pulsation characteristics under cavitation conditions, and solve the key scientific problems in the process of developing LNG storage tank technology and auxiliary device design technology for the effective utilisation of natural gas resources and alleviating the energy crisis.

Cavitation occurs when the liquid pressure is less than the saturation pressure at the corresponding temperature. The liquid changes to gas and destroys the flow passage parts to varying degrees [1]. Since the low-temperature medium LNG transported by the LNG submerged pump is often saturated, the low-temperature medium LNG is very easy to vaporise [2]. Cavitation will cause cavitation damage to the surface of overflow parts and cause the vibration and noise of the pump. When the cavitation range becomes larger, the performance parameters of the pump will be seriously reduced. Rayleigh takes spherical symmetric cavitation as the research object and puts forward the famous Rayleigh equation [3]. Plesset et al. [4,5] considered that the effects of gas, fluid viscosity and surface tension contained in the cavitation improved the cavitation dynamics theory and formed the Rayleigh Plesset equation. The derived cavitation models of Singhal, Kunz and Merkle have been integrated into commercial software, such as fluent, CFX and Pumplinx, which provides a theoretical and technical basis for the research of cavitation numerical simulation [6]. Therefore, it is a trend to predict the cavitation region and evolution process using the hydrodynamic method based on CFD technology. In order to improve the accuracy of the calculation results and better predict the occurrence and evolution of cavitation, some scholars have tried to modify the cavitation model in numerical calculations [7–13].

Jun Li et al. [14] used the improved cavitation model coupled with RANS equation solving technology to simulate the internal flow of the centrifugal pump, and studied the effect on the shape of the cavitation bubbles attached to the surface centrifugal pump blades with the change of the cavitation coefficient. Yong Wang [15], Rakibuzhaman [16] and others analysed the centrifugal pump's three-dimensional turbulent cavitation flow field using the steady numerical simulation method and the two-phase flow mixing model, which improved the accuracy of the numerical simulation. Jian Wang et al. [17] studied the cavitation phenomenon of centrifugal pumps at room temperature, improved the cavitation model on this basis, and put forward the prediction method of centrifugal pump cavitation according to the research results. Huili Jiang [18], Naijian Liao [19], Dongping Shen [20] and Wei Li [21,22] analysed the cavitation performance of engine cooling water pumps, discussed the causes of the cavitation of engine cooling water pumps under different working conditions and put forward measures to improve the cavitation performance of the automobile engine and cooling system of the water pump.

Compared with high-temperature cavitation, there are few numerical studies on low-temperature cavitation, at this stage. Hucan Hou et al. [23] used Reynold's stress turbulence equation and energy equation model to simulate the LNG submersible pump and evaluated its energy consumption using local entropy production. Haitao Cao [24] simulated hydrofoil cavitation using liquid hydrogen as a medium and the two-dimensional axisymmetric steady numerical simulation method. Then, the influence of flow field parameters, such as temperature and pressure on the cavitation performance, was analysed. Xiangfu Ma [25] and Hord [26] studied the cavitation characteristics of a two-dimensional hydrofoil and symmetrical rotating bodies in different liquid nitrogen and liquid hydrogen media. Franc et al. [27,28] experimentally studied the cavitation phenomenon of inducers under thermodynamic effect with low-temperature R-114 and low-temperature water as a flow medium. Gustavsson et al. [29] studied the cavitation flow characteristics of NACA0015 airfoils at different speeds and angles of attack in fluoroketone.

Scholars at home and abroad have carried out extensive research on the pressure pulsation induced by dynamic and static interference in centrifugal pumps. Guo S.J. et al. [30] studied the centrifugal pump with a guide vane structure, collected the pressure pulsation signal by arranging pressure pulsation sensors in the impeller and guide vane channel, analysed the pressure pulsation spectrum characteristics at different positions under different working conditions and systematically and comprehensively revealed the dynamic and static interference excitation characteristics. Parrondo J.W. et al. [31] studied the pressure pulsation of a centrifugal pump under different working conditions by numerical calculation. Through the analysis of the calculation results, the frequency spectrum characteristics of the pressure pulsation were obtained, and the circumferential distribution characteristics of the pressure pulsation with an amplitude near the blade frequency under different working conditions were predicted. Jianping Yuan et al. [32] analysed the flow in the centrifugal pump, based on the unsteady numerical calculation method of large eddy simulation. The results show that the pressure pulsation changes periodically, due to the dynamic and static interference between the impeller and the guide vane. The amplitude at the blade frequency is the largest and becomes the main frequency of the pressure pulsation spectrum. Fujun Wang et al. [33] pointed out that the blade frequency of hydraulic machinery plays a dominant role in the pressure fluctuation spectrum, through the large eddy simulation method. Compared with the pressure pulsation induced by dynamic and static interference, few reports on the pressure pulsation were induced by cavitation. Yongyan Ni et al. [34] used pressure mean square deviation to describe the pulsation characteristics of outlet pressure and analysed the relationship between $NPSH_a$ and outlet pressure pulsation through experiments. Ning Zhang [35] studied the cavitation excitation characteristics of centrifugal pumps and considered that cavitation-induced vibration and noise can be an effective criterion for cavitation monitoring. Cheng et al. [36] summarise the research progress of cavitation, including numerical methods, cavitation characteristics, the influence of cavitation on flow field and cavitation control strategies, and some frontier topics are suggested, which is of great significance to promote the research of cavitation and deepen the understanding of cavitation.

In previous studies, there has been little research on the low-temperature cavitation performance of the submersible pump. In this paper, the cavitation performance and cavitation-induced pressure pulsation characteristics of an LNG submerged pump are analysed based on the numerical calculation method. By understanding the influence of the cavitation on low-frequency pulsation and radial force in LNG submersible pumps, building the internal relationship between the cavitation performance and pressure fluctuation characteristics, the present study provides a theoretical basis for the cavitation monitoring of an LNG submersible pump.

2. Simulation Model

2.1. Computational Model

The main components of the LNG submerged pump studied in this paper are inducer, impeller and guide vane, and their geometric parameters directly determine the pump's performance. Considering the performance of the pump and the rationality of the structure, the model of a low-temperature LNG submerged pump is designed, and three-dimensional modelling is carried out. The main design parameters of the LNG submerged pump studied in this paper are shown in Table 1. The impeller design parameters are shown in Table 2. The overall structure of the LNG submerged pump is shown in Figure 1.

Table 1. Overall design parameters of the LNG submerged pump.

Content	Parameters
Structural	Submerged centrifugal pump
Series	Two-stage
Design speed n	6000 r/min
Specific speed n_s	89.4
Flow at design operating point Q	60 m^3/h
Head at design operating point H	200 m
NPSH at design operating point $NPSH_a$	$\leq$5.5 m
Conveying medium	Liquified natural gas (LNG)

The impeller inlet and the low-pressure area in the channel are where cavitation easily occurs, and the near-wall flow is complex. Numerical simulation can better simulate the flow in the main flow area of the channel, and accurately capture the distribution characteristics of pressure pulsation. Therefore, in order to analyse the pressure pulsation characteristics in the impeller during the transient cavitation flow of the LNG submerged pump, six monitoring points are set in the middle of the primary impeller and the secondary impeller channel, respectively. Since cavitation mostly occurs near the centreline of the flow channel [37], the monitoring points are evenly arranged along the centreline, from the inlet to the outlet of the impeller flow channel, and the distribution is shown in Figure 2.

Table 2. Main parameters of the LNG submersible pump impeller.

Content	Parameters
Impeller inlet diameter D_s	68 mm
Hub diameter d_h	30 mm
Blade inlet diameter D_1	68 mm
Blade outlet diameter D_2	150 mm
Impeller outlet width b_2	10 mm
Blade inlet angle β_1	24°
Blade outlet angle β_2	30°
Number of blades Z	7
Blade wrap angle φ	120°

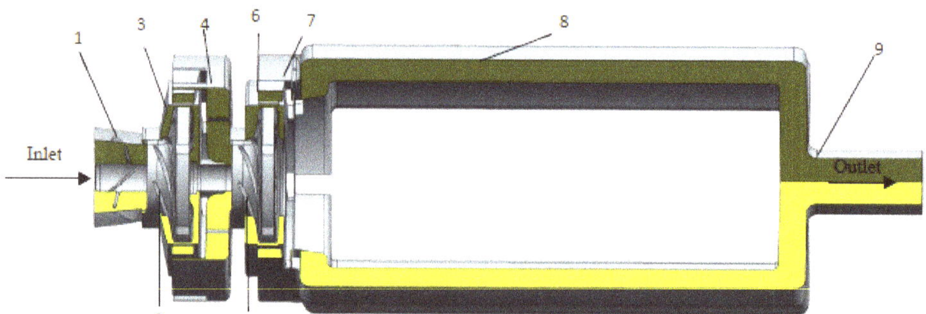

1—Inducer and inlet section; 2—Primary impeller; 3—Primary pump chamber; 4—Primary guide vane; 5—Secondary impeller; 6—Secondary pump chamber; 7—Secondary guide vane; 8—Outlet pipeline; 9—Outlet pipe

Figure 1. Water body assembly drawing of the LNG submersible pump.

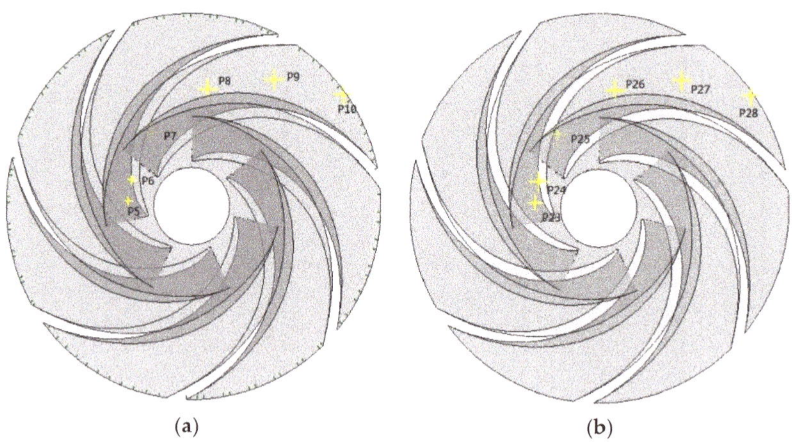

Figure 2. Distribution of the monitoring points in the impeller. (**a**) Primary impeller and (**b**) secondary impeller.

2.2. Meshing

The overflow area of the LNG submersible pump is taken as the calculation domain, generated in blocks according to different fluid domains by using ICEM software. When meshing, the blades can be separated from the impeller according to the blade streamline, and the meshes of each part can be processed respectively. The entire division process generates lines from points, and then the surface is composed of lines. Finally, the volume is divided by surface, and the divided impeller grid adopts a structural grid. The calculation domain of the impeller, inlet section and outlet section adopts a structural grid, while other calculation domains adopt an unstructured tetrahedral grid. The y+ on the blade wall is lower than 50, which meets the requirements of the turbulence model for y+ near the wall. The grid is shown in Figure 3 [38,39].

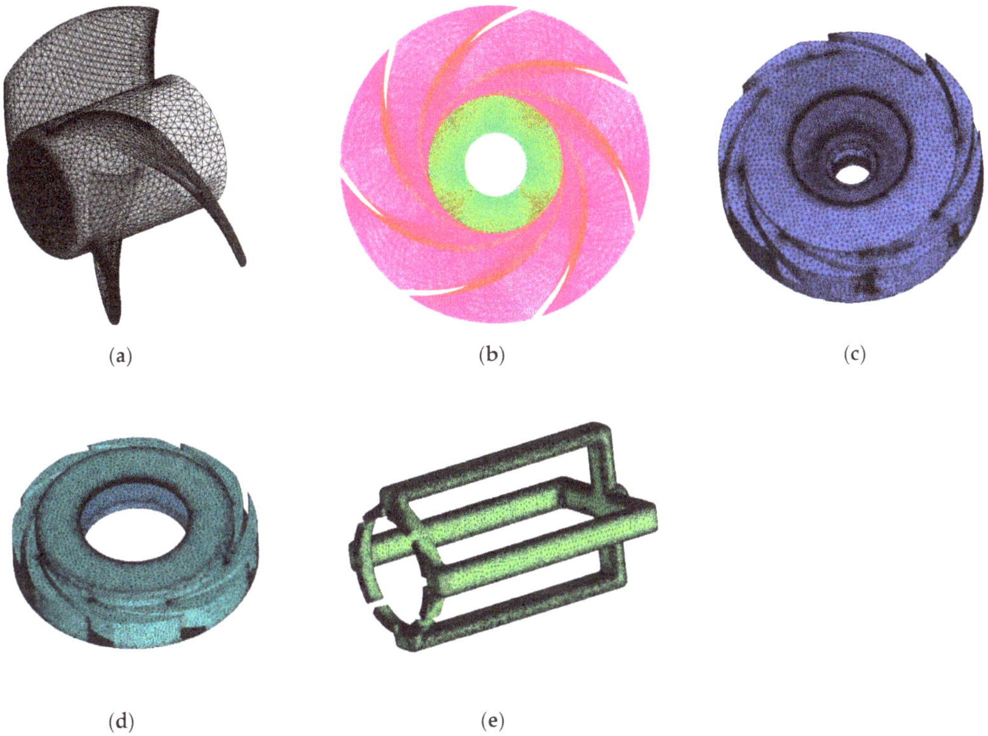

Figure 3. Grid diagram of the computing domain. (**a**) Inducer; (**b**) impeller; (**c**) primary pump chamber; (**d**) secondary guide vane and (**e**) outlet pipeline.

2.3. Grid Independence Verification

In numerical calculations, the number of grid nodes greatly impacts the calculation of the time and accuracy. The main reason is that the conversion of the differential equation into a discrete equation will produce discrete errors in numerical calculations. The denser the grid under the same discrete format, the smaller the discrete error will be. Usually, the more grids, the higher the calculation accuracy of simulation, but too many grids will increase the calculation cost. With the increase in the grid density, when the performance parameters of the pump fluctuate stably, it can be considered that the numerical calculation results become stable, which is called the grid independence in theory [40–42].

In order to determine the grid independence under the design flow condition ($Q = 60\ m^3/h$), the grid quality is changed by encrypting the grid, and the numerical calculation of four grid schemes is carried out for the entire pump. Table 3 shows the LNG submerged pump head calculated using different grid numbers, but using the same boundary conditions, fluid control equation and turbulence model.

Table 3. Comparison of the calculated head under different grid numbers.

Scheme	Total Number of Grids	Head (m)
1	2,439,179	227.8
2	3,252,239	228.2
3	4,065,299	236.7
4	4,878,358	236.2

It can be seen from Table 3 that when the number of global grids is around 4.06 million, the change of the calculation head obtained by increasing the number of grids by encrypting the grid is small, and the error is less than 3%. It shows that its hydraulic efficiency and head tend to be stable and meet the requirements of grid independence. For convergence analysis, four different grids are calculated; the ratio of the difference between them is greater than zero and less than one. Therefore, monotonic convergence can be achieved [43]. In order to ensure the calculation accuracy and save the calculation cost, scheme 3, with a grid number of about 4.06 million, is finally selected for calculation.

2.4. Boundary Conditions and Turbulence Model

In order to study the distribution position of cavitation, the unsteady numerical simulation of cavitation performance is carried out. The initial conditions are the results of steady cavitation calculation, and the boundary conditions of the total pressure inlet and mass flow outlet are adopted. The dynamic and static interface between the inducer and impeller, impeller and guide vane, is set as a "transient rotor–stator" interface. Set the calculation attribute to instantaneous calculation and set the total time and time step of its calculation. The simulation takes the impeller rotation of 360° as a calculation cycle T, which can be calculated as T = 0.01 s. Take the impeller rotation of 2° in each step and 180 steps in a cycle. When simulating the cavitation performance, liquid phase LNG and gas phase LNG at −162 °C are used for calculation and a homogeneous multiphase model is used. The pump inlet adopts the total pressure inlet and the initial inlet pressure is set to 1.38 times the atmospheric pressure. The $NPSH_a$ is controlled by continuously changing the inlet pressure, that is, the occurrence and development process of cavitation, the reference pressure is set to 0 and the outlet adopts a mass flow outlet. The inlet boundary condition with the volume fraction of the initial medium is set as 1 and the volume fraction of the bubble is set as 0; the smooth wall is selected, the standard wall function is adopted near the wall and the adiabatic non-slip wall boundary condition is adopted [44–49].

According to Li's research results, the standard k-ε turbulence model is used as the turbulence model [50]. The governing equation is:

$$\frac{\partial(\rho_m k)}{\partial t} + \frac{\partial(\rho_m u_j k)}{\partial x_j} = P_t - \rho_m \varepsilon + \frac{\partial}{\partial x_j}\left[\left(\mu + \frac{\mu_t}{\sigma_k}\right)\frac{\partial k}{\partial x_j}\right] \tag{1}$$

$$\frac{\partial(\rho_m \varepsilon)}{\partial t} + \frac{\partial(\rho_m u_j \varepsilon)}{\partial x_j} = C_{\varepsilon 1}\frac{\varepsilon}{k}P_t - C_{\varepsilon 2}\frac{\varepsilon^2}{k}\rho_m + \frac{\partial}{\partial x_j}\left[\left(\mu + \frac{\mu_t}{\sigma_\varepsilon}\right)\frac{\partial \varepsilon}{\partial x_j}\right]. \tag{2}$$

$$\mu_t = \frac{C_\mu \rho_m k^2}{\varepsilon}. \tag{3}$$

where P_t is the turbulent kinetic energy generation term; μ_t is the turbulent viscosity coefficient, which is a function of k and ε; K is the turbulent kinetic energy; ε is the turbulent dissipation rate and C_μ is the empirical coefficient. Constant coefficient: $Cs_1 = 1.44$, $Cs_2 = 1.92$, $\sigma_s = 1.3$, $\sigma_k = 1.0$, $C_\mu = 0.09$.

According to Li's research results, the cavitation model adopts the Zwart model [50], and its evaporation term and condensation term are:

$$m^{+} = C_{vap}\frac{3\alpha_{auc}(1-\alpha_{v})\rho_{v}}{R_{B}}\sqrt{\frac{2}{3}\frac{|p_{v}-p|}{\rho_{l}}}. \quad (4)$$

$$m^{-} = C_{cond}\frac{3\alpha_{v}\rho_{v}}{R_{B}}\sqrt{\frac{2}{3}\frac{|p_{v}-p|}{\rho_{l}}}. \quad (5)$$

where a_{nuc} is the initial vapor core volume fraction, usually taken as 5×10^{-4}; R_B is the bubble radius, taken as 1×10^{-6}; p_v is the vaporisation pressure; p is the pressure of the liquid around the bubble and the evaporation and condensation coefficient $C_{vap} = 50$; $C_{cond} = 0.01$.

3. External Characteristic Test

The external characteristic test of the LNG submerged pump is carried out on a closed test bench. The medium is constant temperature water. The external characteristic experimental system comprises of a water tank, inlet valve, inlet pressure transmitter, submerged pump, outlet pressure transmitter, electromagnetic flowmeter, outlet valve, vacuum pump, speed and torque measuring instrument and cavitation tank. The distance between the inlet and outlet pressure transmitter and the inlet and outlet of the submersible pump is twice the inlet and outlet pipe diameter of the submersible pump. In order to measure the stable flow, the electromagnetic flowmeter is more than 20 times the outlet pipe diameter from the adjacent flange. The model used in the experiment is shown in Figure 4a. In the figure, 1 is the inlet valve; 2 is the inlet pressure transmitter; 3 is the submerged pump; 4 is the outlet pressure transmitter; 5 is the electromagnetic flowmeter; 6 is the outlet valve; 7 is the cavitation tank; 8 is the extraction valve; 9 is the vacuum pump; 10 is the exhaust valve and 11 is the water tank. The pressure sensor is used to measure the pressure at the inlet and outlet of the pump, the measurement range of the pressure sensor at the inlet is −0.1 MPa to 0.1 MPa, and the measurement range of the pressure sensor at the outlet is 0 MPa to 1.6 MPa; the measurement accuracy of both sensors is grade 0.1. Change the pump's flow through the flow regulating valve, and use the electromagnetic flowmeter to measure the flow in the pump. The measurement accuracy of the electromagnetic flowmeter is grade 0.5.

(a)

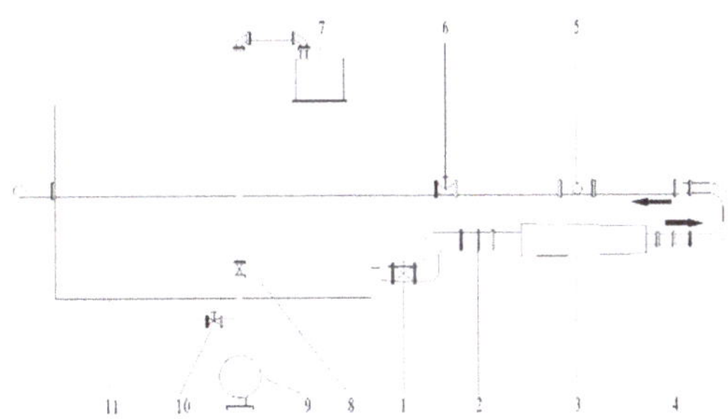

(b)

Figure 4. Experimental images of the submerged pump characteristics. (**a**) Physical model and (**b**) schematic diagram of the experimental system.

4. Analysis of Numerical Results

4.1. External Characteristic Test Verification

It can be seen from the external characteristic curve of the pump in Figure 5, that the test head and efficiency of the submerged pump under the design working condition are 205.56 m and 72.93%, respectively. The numerical calculation head and efficiency are 211.09 m and 71.12%, respectively, so the errors were 2.56 % and 2.48%. The highest efficiency point is near the working condition of 1.1 Q_{des}, and the head and efficiency are

197.2 m and 73.52%, respectively. Under the condition of small flow, the head and efficiency of numerical simulation are lower than those measured by the test, and the simulation results are between 0.4 Q_{des} and 0.8 Q_{des}. There is a "hump area" in the head due to the rotating stall. With the increase in flow, the simulated head becomes highter than the test head from the working condition of 0.9 Q_{des}; from the 1.2 Q_{des} working condition, the simulation efficiency is higher than the test efficiency, and the simulation value is same as the test value as a whole. The head and efficiency of the pump under 0.8 Q_{des}~1.4 Q_{des} are high, which is the high-efficiency area of the pump. In conclusion, this numerical calculation can accurately simulate the actual performance of the pump.

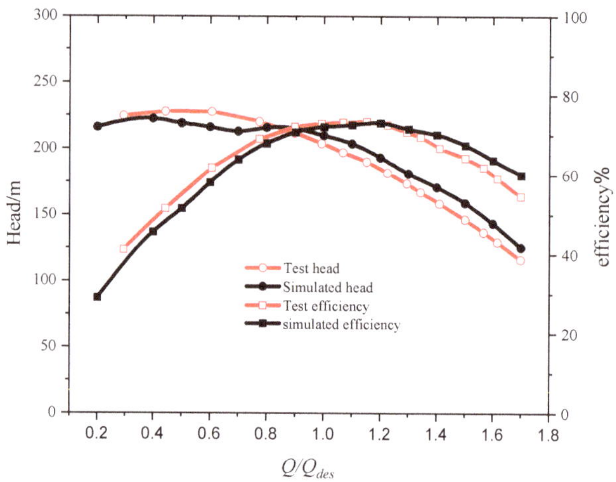

Figure 5. The characteristic curve of the LNG submerged pump.

4.2. Cavitation Characteristic Curve

The cavitation performance curve of the LNG submerged pump is obtained by adjusting $NPSH_a$ under different inlet total pressure. The $NPSH_a$ expression of the pump is:

$$NPSH_a = \frac{P_{in} - Pv}{\rho g} \tag{6}$$

The cavitation performance prediction curve of the LNG submerged pump under design conditions, is shown in Figure 6. As with ordinary centrifugal pumps, the critical *NPSH* is defined as $NPSH_a$ when the head decreases by 3%. It can be seen from the figure that the critical *NPSH* of the LNG submersible pump is $NPSH_c$ = 4.92 m.

4.3. Analysis of the Cavitation in the Impeller

Figure 7 shows the pressure distribution of the primary impeller under different $NPSH_a$. It can be seen from the figure that the pressure in the impeller presents a gradient distribution. With the decrease in $NPSH_a$, the low-pressure area first appears on the suction surface of the blade at the inlet end, and the circumferential distribution is asymmetric. With the decrease in the inlet pressure, the low-pressure area gradually expands. With the further decrease in the inlet pressure, the low-pressure inlet zone gradually develops into the flow channel. A high-pressure area is evenly distributed along the circumferential direction at the junction of the blade outlet and rim. With the decrease in the inlet pressure, the uniformity of the circumferential distribution gradually decreases, and the area of the high-pressure area gradually decreases. When $NPSH_a$ = 4 m, the high-pressure area disappears.

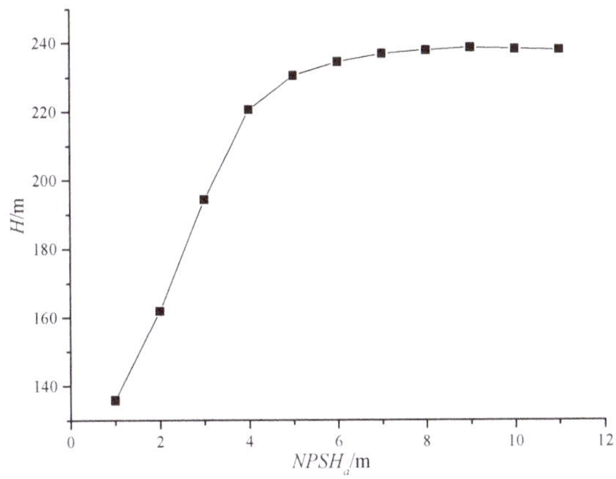

Figure 6. Cavitation performance curve of the LNG submersible pump.

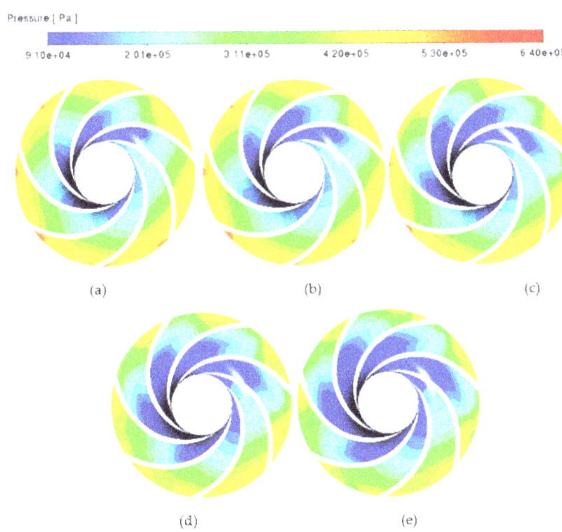

Figure 7. Pressure distribution of the primary impeller. (**a**) $NPSH_a$ = 9 m; (**b**) $NPSH_a$ = 7 m; (**c**) $NPSH_a$ = 6 m; (**d**) $NPSH_a$ = 5 m and (**e**) $NPSH_a$ = 4 m.

Table 4 shows the distribution of the cavitation volume in different flow surfaces of the impeller under different $NPSH_a$, where S is the dimensionless distance from the front cover plate to the rear cover plate. When $S = 0$, the position is the rear cover plate, and when $S = 1$, the position is the front cover plate. It can be seen from the figure, that the distribution position of cavitation in the impeller is consistent with the pressure distribution. With the decrease in $NPSH_a$, cavitation gradually occurs at the inlet of the impeller. Under the four $NPSH_a$, there is a high void volume fraction on the section with the dimensionless distance, $S = 0.75$. With the decrease in $NPSH_a$, the cavitation gradually develops towards the rear cover plate of the impeller. It shows that cavitation always occurs first at the inlet end of the blade suction surface and diffuses along with the blade suction towards the outlet, gradually blocking the entire flow channel. Under the same $NPSH_a$, the void volume fraction in the flow surface of the front cover plate is the greatest. With the decrease in the dimensionless distance, the volume fraction of the cavitation gradually decreases. At the same time, the cavitation in each channel of the impeller is asymmetrically distributed. The reason can be the dynamic and static interference between the number of the asymmetric impeller and guide vane, resulting in the asymmetric pressure distribution on the surface of the impeller blade.

Table 4. Bubble volume fraction in the primary impeller.

$NPSH_a$	S = 0.1	S = 0.25	S = 0.5	S = 0.75
9 m				
6 m				
5 m				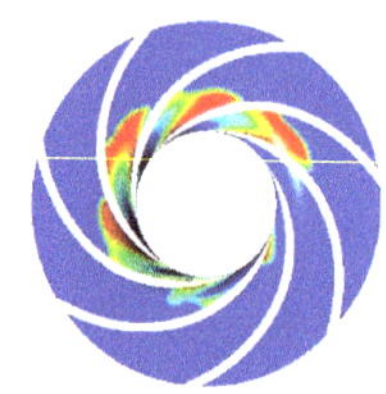
4 m				

Figure 8 shows the distribution and evolution law of transient cavitation in the impeller channel under the design flow rate and $NPSH_a$ = 7 m; the time interval between the adjacent pictures is 10. It can be seen from the numerical calculation results that, due to the existence of the inducer, there is not much bubble distribution in the impeller under the same $NPSH_a$, which shows that the internal cavitation performance of the impeller is good at this time. A small number of bubbles first appear at the inlet of the leading edge of the impeller. With the rotation of the impeller, the bubbles gradually diffuse along the working surface of the blade. The bubble distribution in each impeller channel is not uniform, which can be caused by the dynamic and static interference between the impeller and the guide vane.

4.4. Analysis of the Pressure Fluctuation Characteristics in the Impeller

In order to facilitate the comparative analysis of the pressure pulsation, the pressure pulsation data obtained by numerical calculation are processed to dimensionless, in this paper. The pressure fluctuation coefficient C_P is introduced, which is defined as:

$$C_P = \frac{p - \overline{p}}{\frac{1}{2}\rho u_2^2} \quad (7)$$

where p is the pressure at the monitoring point; $\overline{p}$ is the mean value of the pressure at the monitoring point during one cycle of impeller rotation; ρ is the density of the conveying medium and u_2 is the circumferential speed of the impeller outlet.

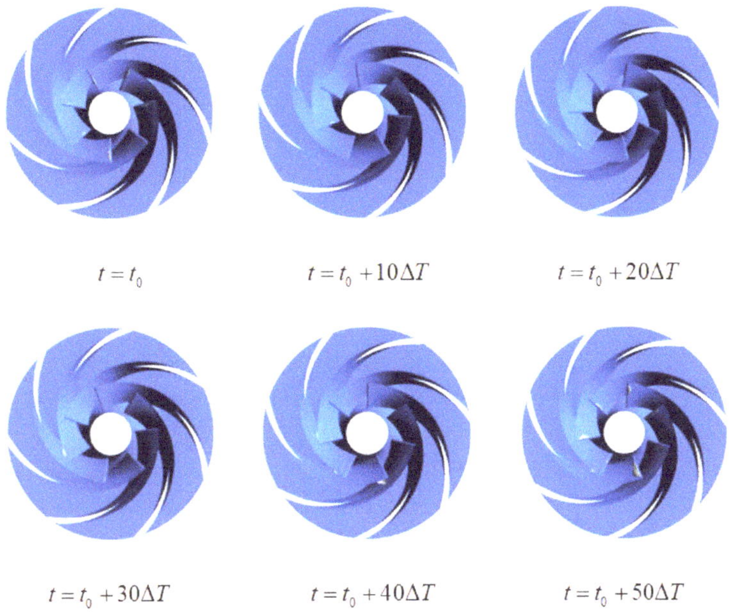

Figure 8. Bubble distribution at the impeller inlet ($NPSH_a$ = 7 m).

4.4.1. Pressure Pulsation under Non-Cavitation Conditions

(1) Time Domain analysis

Figure 9 shows the time domain distribution of pressure pulsations at each monitoring point in the primary and secondary impellers, under three different flow conditions. It can be seen from the figure that, under various working conditions, the amplitude of pressure pulsation in the primary and secondary impeller increases from inlet to outlet, and the pulsation at the outlet is biger than that in the impeller channel. Under the same flow rate, the amplitude of pressure fluctuation of the primary impeller and secondary impeller is close, but the waveform is quite different. When the impeller rotates for one circle, the pressure pulsation in the impeller presents eight similar waveforms, and The higher of the flow rate, the more similar of the waveform. In addition to the eight similar waveforms, the pressure pulsation in the secondary impeller also contains a long-period low-frequency waveform, which can be caused by the uneven incoming flow of the secondary impeller. The low-frequency waveform weakens with the flow increase and disappears under high flow rate conditions.

The above analysis shows that the number of waveforms of pressure pulsation in each stage of the impeller, is related to the number of blades of the positive guide vane. The closer to the inlet edge of the positive guide vane, the higher the pulsation amplitude, indicating that the pulsation source of pressure pulsation in each stage of the impeller mainly comes from the dynamic and static interference between the impeller and the positive guide vane. The absolute pressure difference between the primary impeller and the secondary impeller is great, but the pressure fluctuation amplitude is close under the same flow, indicating that there is no absolute correlation between the pressure fluctuation intensity in the impeller and the ambient pressure.

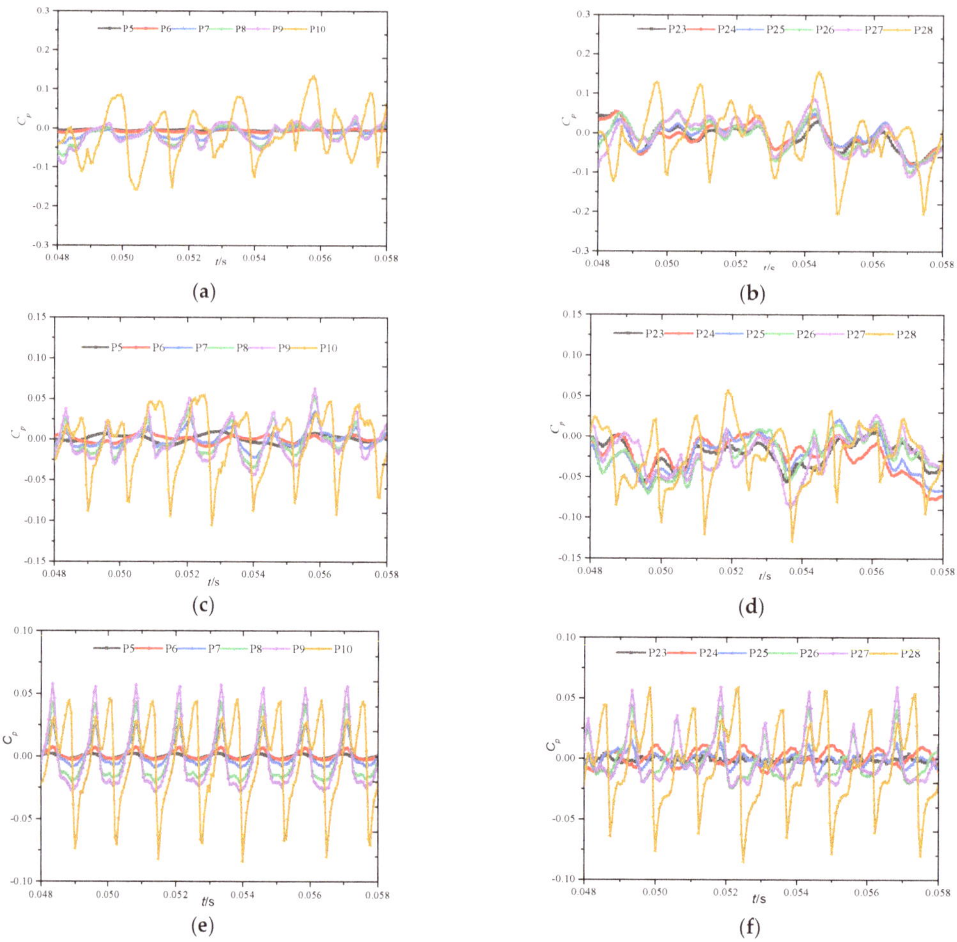

Figure 9. Time domain diagram of the pressure pulsation of impellers at all levels under non-cavitation conditions: (**a**) 0.6 Q_{des}−primary impeller; (**b**) 0.6 Q_{des}−secondary impeller; (**c**) 1.0 Q_{des}−primary impeller; (**d**) 1.0 Q_{des}−secondary impeller; (**e**) 1.4 Q_{des}−primary impeller and (**f**) 1.4 Q_{des}−secondary impeller.

(2) Frequency Domain Analysis

Figure 10 shows the frequency domain diagram of the pressure pulsation of the impellers at all levels under non-cavitation conditions. It can be seen from the figure that the main frequency of the pressure pulsation in the primary impeller and secondary impeller is always the blade frequency of the guide vane, the amplitude of the main frequency gradually attenuates from the impeller outlet to the impeller inlet, and the deceleration rate of the main frequency in the secondary impeller is much higher than that in the primary impeller. The low-frequency component of pressure pulsation in the secondary impeller is significantly greater than that in the primary impeller, and the low-frequency signal amplitude of the pressure pulsation in the two impellers decreases with the increase in flow. Due to the pre-rotation of the outlet flow of the primary anti-guide vane, the flow field in the secondary impeller is smoother than that in the primary impeller, indicating that the low-frequency signal in the secondary impeller does not come from its internal vortex, but is caused by the instability of the upstream flow.

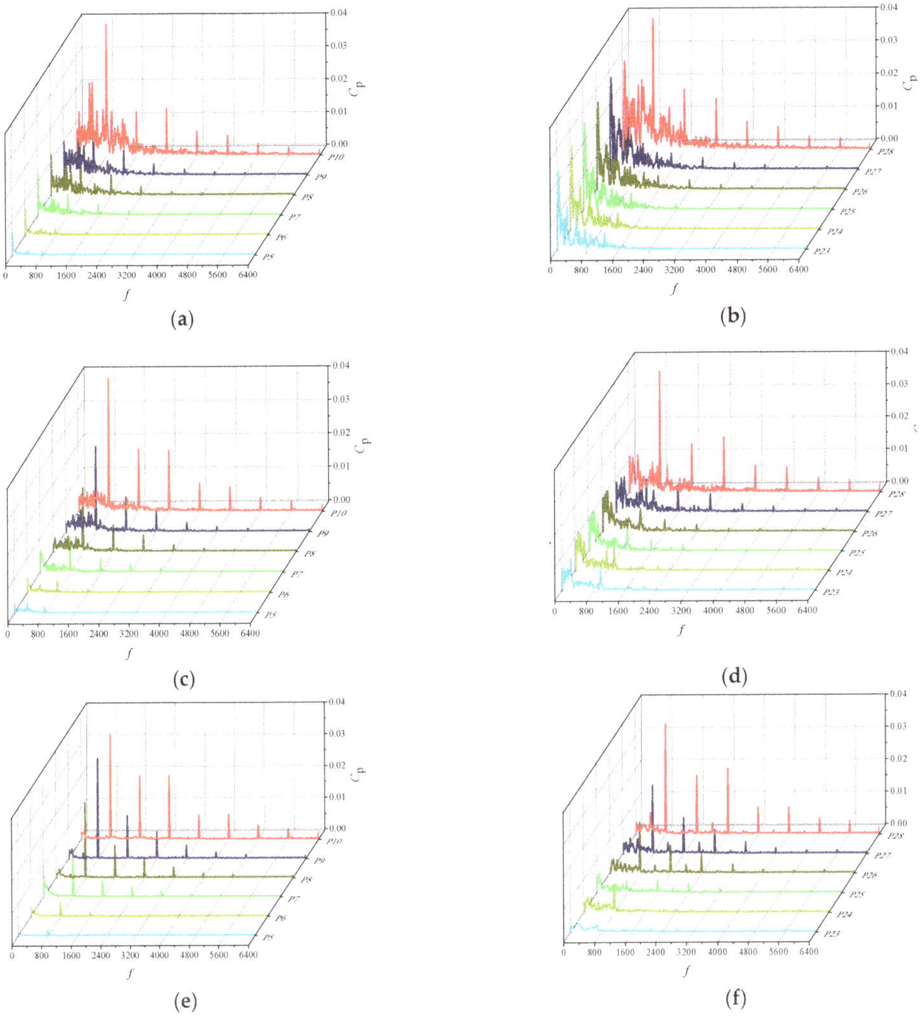

Figure 10. Frequency domain diagram of pressure pulsation of the impellers at all levels under non-cavitation conditions: (**a**) 0.6 Q_{des}−primary impeller; (**b**) 0.6 Q_{des}−secondary impeller; (**c**) 1.0 Q_{des}−primary impeller; (**d**) 1.0 Q_{des}−secondary impeller; (**e**) 1.4 Q_{des}−primary impeller and (**f**) 1.4 Q_{des}−secondary impeller.

4.4.2. Pressure Pulsation under Cavitation Condition

(1) Time Domain Analysis

Figure 11 shows the time domain distribution of the pressure pulsation in the primary and secondary impellers under different $NPSH_a$. It can be seen from the figure that, during one revolution of the impeller, the pressure pulsation at each monitoring point under different $NPSH_a$ still presents eight complete peaks and troughs, indicating that the dynamic and static interference under different cavitation conditions is still the main pulsation source of pressure fluctuation in the impeller. With the decrease in $NPSH_a$, the pressure fluctuation waveform at the monitoring points in the primary impeller and secondary impeller is almost unchanged. Only the amplitude of low-frequency fluctuation at the monitoring points P7, P8 and P9 in the primary impeller increases.

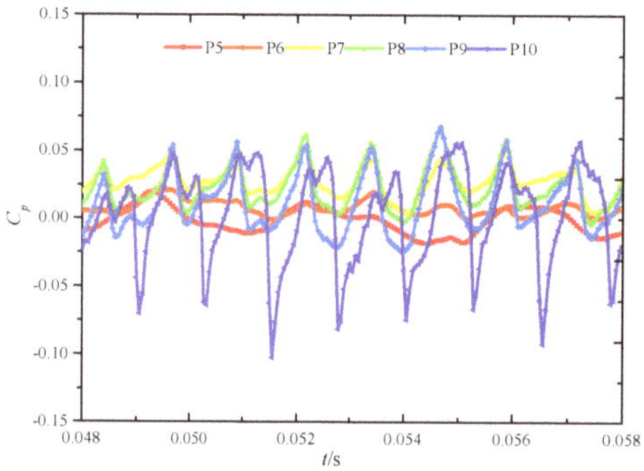

(a)

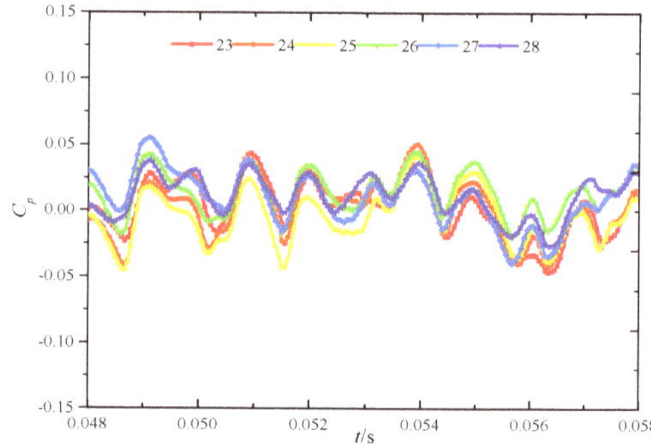

(b)

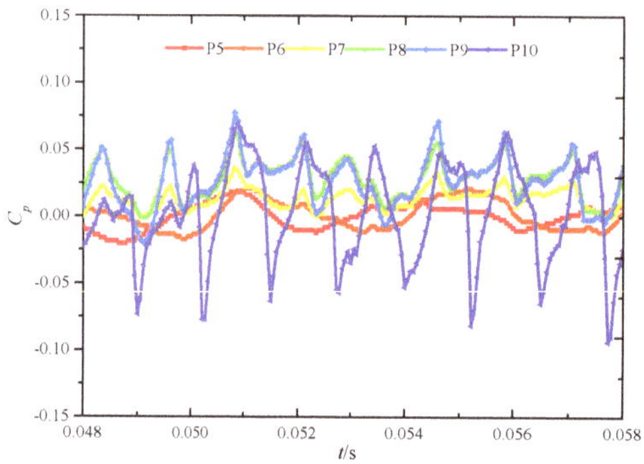

(c)

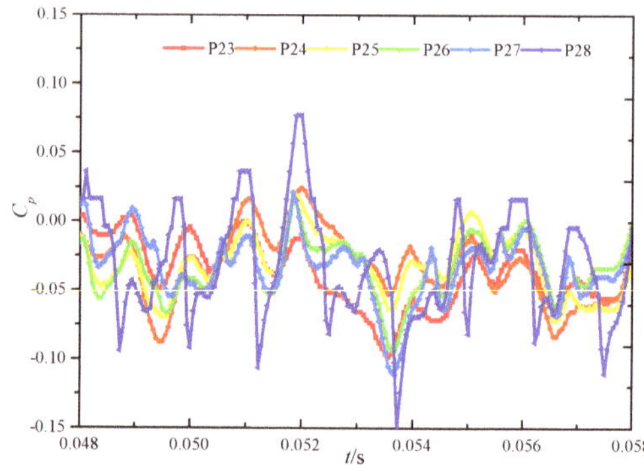

(d)

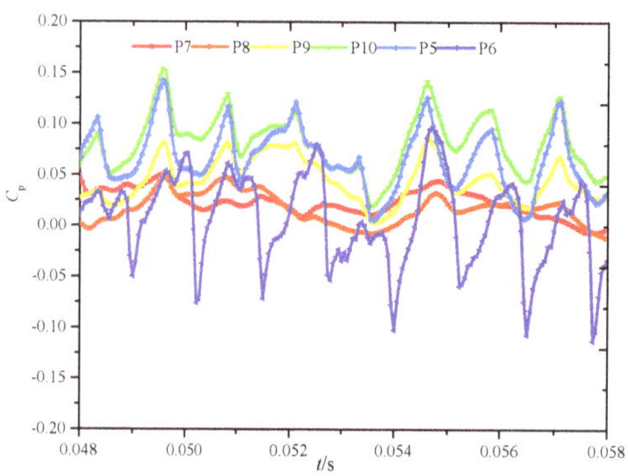

(e)

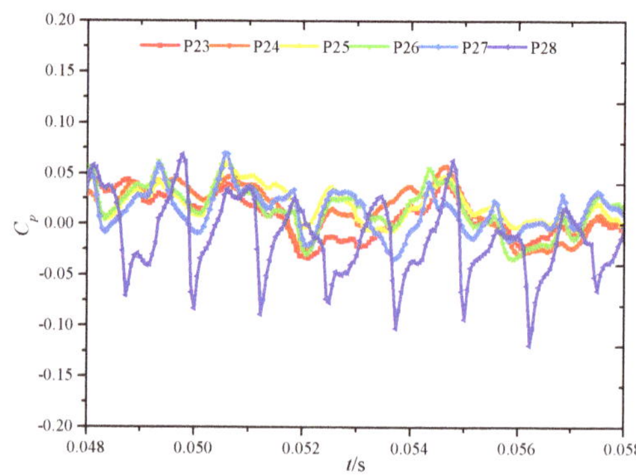

(f)

Figure 11. *Cont.*

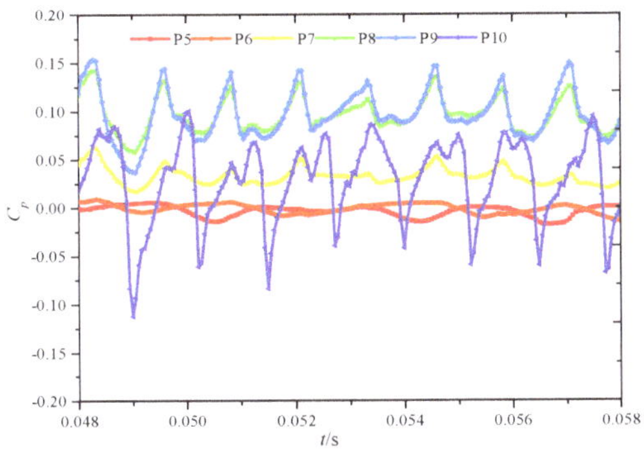

(g)

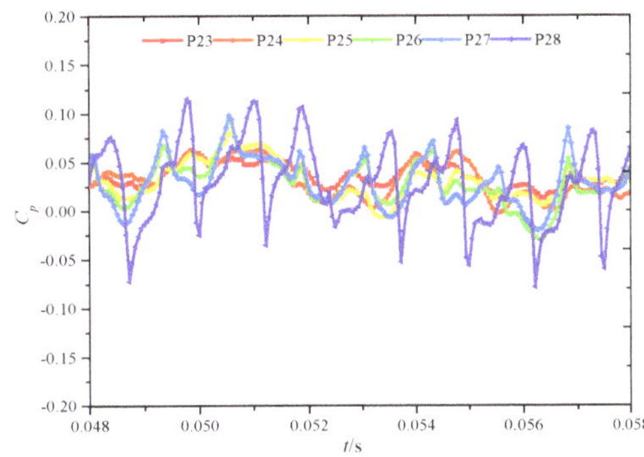

(h)

Figure 11. Time domain diagram of the pressure pulsation of the impellers at all levels under cavitation conditions. (**a**) Primary impeller ($NPSH_a$ = 11 m); (**b**) secondary impeller ($NPSH_a$ = 11 m); (**c**) primary impeller ($NPSH_a$ = 9 m); (**d**) secondary impeller ($NPSH_a$ = 9 m); (**e**) primary impeller ($NPSH_a$ = 7 m); (**f**) secondary impeller ($NPSH_a$ = 7 m); (**g**) primary impeller ($NPSH_a$ = 4 m) and (**h**) secondary impeller ($NPSH_a$ = 4 m).

(2) Frequency Domain Analysis

Figure 12 shows the frequency spectrum of pressure pulsation in each impeller stage under different $NPSH_a$. It can be seen from the figure that the amplitude of guide vane frequency (800 Hz) and its harmonic frequency hardly changes with the change of $NPSH_a$. The pulsating low-frequency signal at each monitoring point is affected by $NPSH_a$. This is because the falling off of cavitation on the blade surface will produce low-frequency disturbance in the flow channel. At the same time, the cavitation group near the blade will form a local blockage to the flow channel and change the flow state in the impeller flow channel. The frequency domain characteristics of pulsation in the primary impeller are affected by $NPSH_a$, and the signal lower than the impeller axial frequency (100 Hz) increases significantly with the decrease in $NPSH_a$, especially at the monitoring points P7, P8 and P9 in the middle of the flow channel. The amplitude of the low-frequency signal when $NPSH_a$ = 7 m is five times that when $NPSH_a$ = 11 m, due to the high ambient pressure; therefore, the pressure pulsation hardly changes under different $NPSH_a$. Nonetheless, the low-frequency signal becomes more concentrated with the decrease in $NPSH_a$.

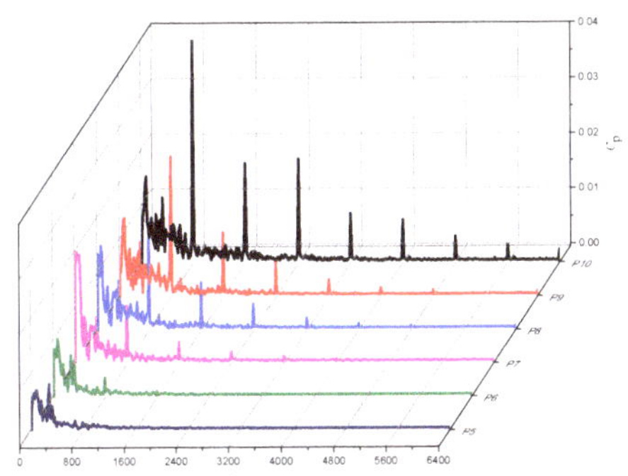

(a)

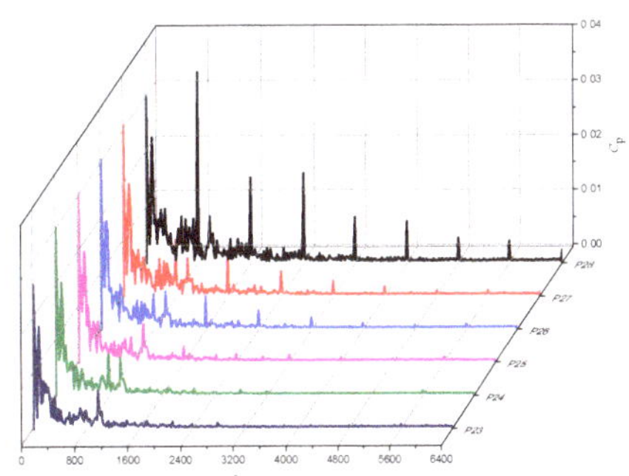

(b)

Figure 12. *Cont.*

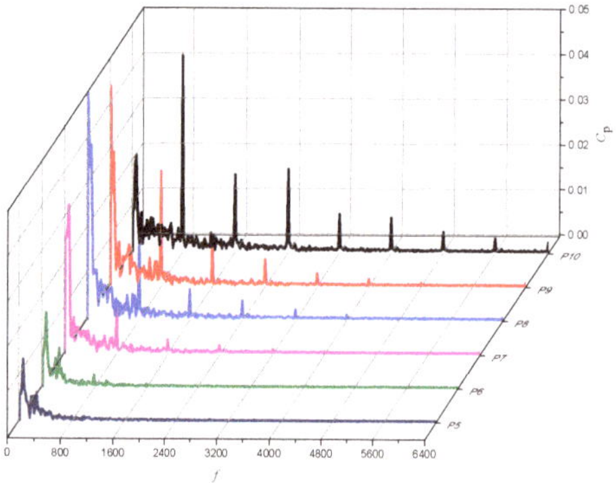

(c)

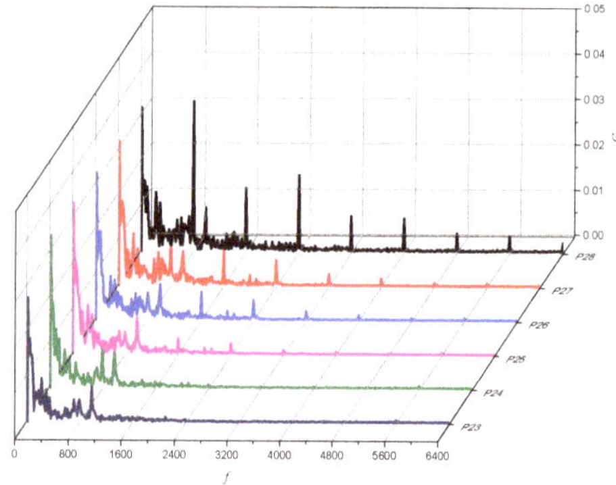

(d)

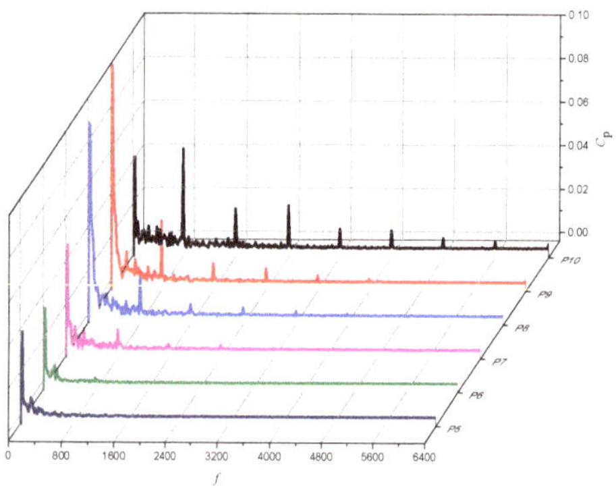

(e)

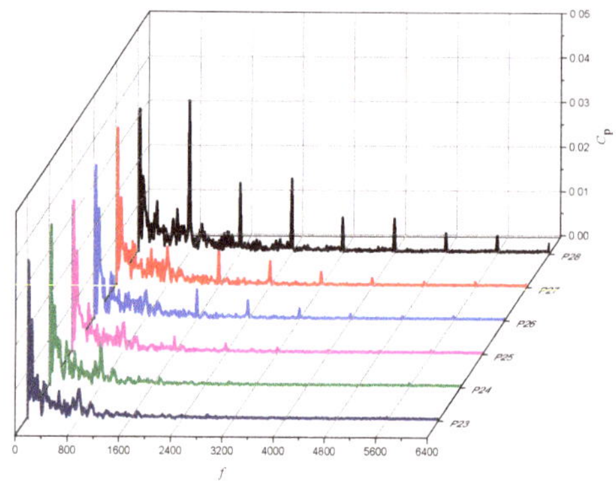

(f)

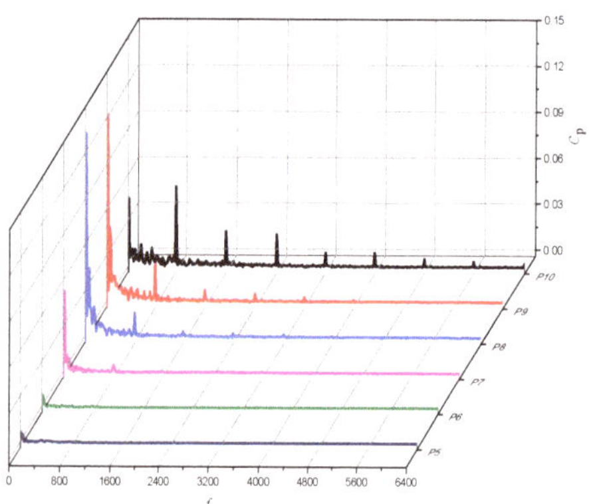

(g)

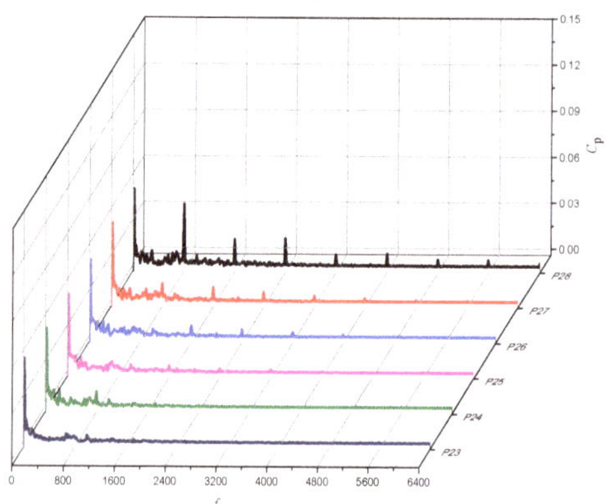

(h)

Figure 12. Frequency domain diagram of pressure pulsation of the impellers at all levels under cavitation conditions. (**a**) Primary impeller ($NPSH_a$ = 11 m); (**b**) secondary impeller ($NPSH_a$ = 11 m); (**c**) primary impeller ($NPSH_a$ = 9 m); (**d**) secondary impeller ($NPSH_a$ = 9 m); (**e**) primary impeller ($NPSH_a$ = 7 m); (**f**) secondary impeller ($NPSH_a$ = 7 m); (**g**) primary impeller ($NPSH_a$ = 4 m) and (**h**) secondary impeller ($NPSH_a$ = 4 m).

Figure 13 shows the variation process of the pulsating root mean square value, C_{RMS}, of a low-frequency signal (0~700 Hz), at the monitoring points in the primary impeller and secondary impeller with $NPSH_a$. It can be seen from the figure, that the influence of cavitation on the low-frequency pulsation intensity in the primary impeller is greater than that in the secondary impeller. At the middle and rear of the primary impeller (P8 and P9), the C_{RMS} value during critical cavitation increases from 0.004 to 0.02, under non-cavitation conditions, an increase of about five times. At the first half of the impeller channel (P5, P6 and P7) and the outlet (P10), the C_{RMS} value of the low-frequency pressure pulsation shows a slow growth trend. This is because cavitation mainly starts from the suction surface of the impeller blade close to the inlet edge. With the influence of the mainstream, it extends downstream along the suction surface of the blade and begins to fall off when it is close to the middle. Therefore, the pressure pulsation at the middle and rear of the impeller channel is affected by the falling off of the cavitation and generates a low-frequency signal. It can be seen from Figure 13b that, under the non-cavitation condition, the pressure pulsation intensity of the secondary impeller in the low-frequency band is higher than that of the primary impeller, and the flow in the secondary impeller is relatively disordered at this time. With the decrease in $NPSH_a$, the intensity of the low-frequency pulsation at each monitoring point slowly increases. Among them, the upper incoming pulsation affects the inlet monitoring point, P23, and the increase in C_{RMS} is the greatest.

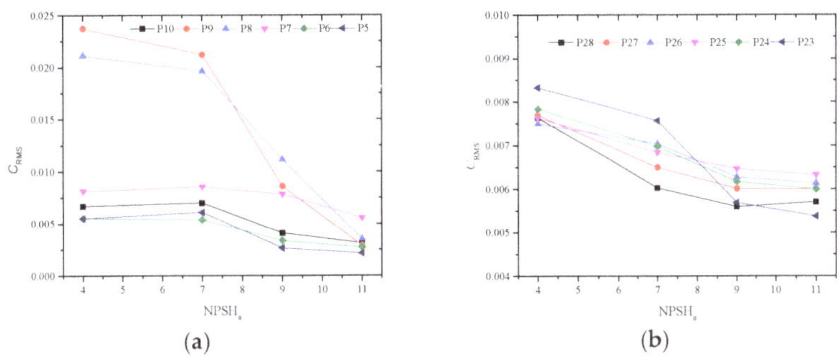

Figure 13. Low-frequency pulsation intensity of the impellers at all levels under different cavitation conditions. (**a**) Primary impeller and (**b**) secondary impeller.

5. Conclusions

The distribution and evolution of cavitation in the impeller of the LNG submersible pump are analysed through unsteady calculation. The variation laws of the pressure pulsation waveform, amplitude and frequency in the impeller of the LNG submersible pump, under different flow rates and $NPSH_a$, are studied by calculating the root mean square of the pressure coefficient of the low-frequency pulsation. The influence of the aggravation process of cavitation on the low-frequency pulsation in the LNG submersible pump is quantitatively analysed, and the internal pressure pulsation characteristics of the LNG submersible pump under cavitation state are revealed.

(1) The cavitation in the impeller is first generated near the inlet edge at the back of the blade. With the decrease in $NPSH_a$, the cavitation diffuses along the back of the blade to the outlet, and with the increase in impeller channel area, the bubble gradually expands to the blade working surface. Due to the dynamic and static interference between the impeller and guide vane, the distribution of cavitation in each impeller channel is asymmetric.

(2) The pressure pulsation in the impeller of the LNG submerged pump is greatly affected by the flow. With the increase in flow, the pressure pulsation in the impeller becomes periodically stronger, the amplitude decreases, and the low-frequency and broadband signals gradually disappear. The pulsation source of the internal pressure of the impeller is in the inlet edge of the positive guide vane. The amplitude of the internal pressure pulsation of the impeller is the largest at the dynamic and static interface,

and the pulsation amplitude decreases gradually with the distance away from the pulsation source.

(3) The time domain and frequency domain characteristics of pressure pulsation in the primary impeller and secondary impeller are the same. The channel pressure does not affect the amplitude of pressure pulsation. Due to the uneven inflow of the primary guide vane, the low-frequency component of the pressure fluctuation spectrum in the secondary impeller is greater than that in the primary impeller.

(4) The occurrence of cavitation greatly influences the frequency domain distribution of pressure pulsation in the impeller of the LNG submerged pump. With the decrease in $NPSH_a$, a low-frequency signal appears in the frequency domain of the pressure pulsation in the impeller, and the amplitude of the low-frequency signal continuously increases. The influence of cavitation on the pressure pulsation in the primary impeller is greater than that in the secondary impeller. When critical cavitation occurs in the pump, the amplitude of the low-frequency signal in the frequency domain of the pressure pulsation in the primary impeller increases and exceeds the blade frequency, becoming the main frequency.

Author Contributions: Conceptualization, W.L.; formal analysis, L.J.; data curation, X.Z.; writing—original draft preparation, S.L.; writing—review and editing, S.L., W.L., M.A., Y.Y., W.S. and R.K.A. All authors have read and agreed to the published version of the manuscript.

Funding: The work was sponsored by the Key International Cooperative research of the National Natural Science Foundation of China (No.52120105010), the National Natural Science Foundation of China (No.52179085), the National Key R&D Program Project (No.2020YFC1512405), the Fifth "333 High Level Talented Person Cultivating Project" of Jiangsu Province and the funded projects of "Blue Project" in Jiangsu colleges and universities.

Data Availability Statement: The data used to support the findings of this study are included within the article.

Conflicts of Interest: The authors declare no conflict of interest.

References

1. Guan, X.F. *Modern Pump Theory and Design*; China Aerospace Publishing House: Beijing, China, 2011.
2. Gu, A. *LNG Technical Manual*; Machinery Industry Press: Beijing, China, 2010.
3. Lindsay, R.B. On the pressure developed in a liquid during the collapse of a spherical cavity. In *Men of Physics Lord Rayleigh—The Man & His Work*; Elsevier: Amsterdam, The Netherlands, 1970; pp. 221–226.
4. Plesset, M.S.; Devine, R.E. Effect of exposure time on cavitation damage. *ASME J. Basic Eng. Trans.* **1965**, *88*, 31. [CrossRef]
5. Plesset, M.S.; Prosperetti, A. Bubble dynamics and cavitation. *Annu. Rev. Fluid Mech.* **1977**, *9*, 145–185. [CrossRef]
6. Huang, B.; Wang, G.; Zhang, B.; Yu, Z. Evaluation of cavitation model for numerical simulation of cloud cavitation flow. *J. Beijing Inst. Technol.* **2009**, *29*, 785–789.
7. Avdeev, A.A. Laws of vapor bubble growth in the superhearted liquid volume (Thermal growth scheme). *High Temp.* **2014**, *52*, 588–602. [CrossRef]
8. Zhang, Y.; Luo, X.; Xu, H.; Wu, Y. Improvement of thermodynamic cavitation model and its numerical application. *J. Eng. Thermophys.* **2010**, *31*, 1671–1674.
9. Ji, B.; Luo, X.; Wu, Y.; Zhang, Y.; Xu, H. Simulation of high temperature water cavitation considering thermodynamic effect. *J. Tsinghua Univ.* **2010**, *50*, 262–265.
10. Wang, W.; Lin, Y.; Wang, X.; Wang, Y. Influence of thermodynamic effect on cavitation at high temperature. *J. Drain. Irrig. Mech. Eng.* **2014**, *32*, 835–839.
11. Shi, S.; Wang, G.; Hu, C.; Gao, D. Experimental study on hydrodynamic pulsation characteristics of cavitation in water at different temperatures. *J. Mech. Eng.* **2014**, *8*, 174–181. [CrossRef]
12. Shi, S.; Wang, G.; Ma, R. Numerical study on cavitation characteristics of low temperature fluid. *Eng. Mech.* **2012**, *5*, 61–67.
13. Shi, S.; Wang, G.; Chen, G. Application evaluation of mass transfer model in cavitation flow calculation considering thermodynamic effects. *J. Appl. Mech.* **2011**, *6*, 589–594+672.
14. Li, J.; Liu, L.; Li, G.; Feng, Z. Numerical prediction of cavitation flow in centrifugal pump impeller. *J. Eng. Thermophys.* **2007**, *28*, 948–950.
15. Wang, Y.; Liu, H.; Liu, D.; Yuan, S.; Wang, J.; Jiang, L. Application of the two-phase three-component computational model to predict cavitating flow in a centrifugal pump and its validation. *Comput. Fluids* **2016**, *131*, 142–150. [CrossRef]

16. Rakibuzzaman, M.; Kim, K.; Suh, S.H. Numerical and experimental investigation of cavitation flows in a multistage centrifugal pump. *J. Mech. Sci. Technol.* **2018**, *32*, 1071–1078. [CrossRef]
17. Wang, J. Numerical Calculation and Experimental Study on Cavitation Erosion of Hydraulic Equipment. Master's Thesis, Jiangsu University, Zhenjiang, China, 2015.
18. Jiang, H. Research on assembly height detection method of engine cooling water pump impeller. *Intern. Combust. Engine Parts* **2018**, *8*, 40–42.
19. Liao, N.; Xie, D. Discussion on cavitation of water pump of lj465q series engine. *Equip. Manuf. Technol.* **2010**, *2*, 177–179.
20. Shen, D. CFD Numerical Simulation of Engine Cooling Water Pump. Master's Thesis, Jilin University, Changchun, China, 2014.
21. Li, W.; Shi, W.; Pei, B.; Zhang, H.; Lu, W. Numerical simulation and improvement of cavitation characteristics of engine cooling water pump. *Trans. Csice* **2013**, *2*, 165–170.
22. Li, W.; Shi, W.; Zhang, H.; Pei, B.; Lu, W. Prediction of cavitation performance of engine cooling water pump based on CFD. *J. Drain. Irrig. Mach. Eng.* **2012**, *30*, 176–180.
23. Hou, H.; Zhang, Y.; Li, Z.; Jiang, T.; Zhang, J.; Xu, C. Numerical analysis of entropy production on a LNG cryogenic submerged pump. *J. Nat. Gas Sci. Eng.* **2016**, *36*, 87–96. [CrossRef]
24. Cao, H. Numerical Study on Cavitation Flow Characteristics at Low Temperature. Master's Thesis, Harbin Institute of Technology, Harbin, China, 2010.
25. Ma, X.; Wei, Y.; Wang, C.; Huang, W. Numerical simulation of low temperature cavitation flow of liquid nitrogen around hydrofoil. *J. Ship Mech.* **2012**, *12*, 1345–1352.
26. Hord, J. *Cavitation in Liquid Cryogenics III—Ogive*; Technical Report No. NASA-CR-2242; NASA: Washington, DC, USA, 1973; pp. 10–60.
27. Franc, J.P.; Rebattet, C.; Coulon, A. An experimental investigation of thermal effects in a cavitating inducer. *J. Fluids Eng.* **2004**, *126*, 716–723. [CrossRef]
28. Franc, J.P.; Pellone, C. Analysis of thermal effects in a cavitating inducer using rayleigh equation. *J. Fluids Eng.* **2007**, *129*, 974–983. [CrossRef]
29. Gustavsson, J.P.R.; Denning, K.C.; Segal, C. Hydrofoil cavitation under strong thermodynamic effect. *J. Fluids Eng.* **2008**, *130*, 091303. [CrossRef]
30. Guo, S.; Maruta, Y. Experimental investigations on pressure fluctuations and vibration of the impeller in a centrifugal pump with vaned diffusers. *JSME Int. J.* **2005**, *48*, 136–143. [CrossRef]
31. Fernández, J.; Marcos, A.; Barrio, R.; Blanco, E.; Parrondo, J. Studies of the flow of air in a mixed-flow pump using numerical simulations. *Proc. Inst. Mech. Eng. Part A J. Power Energy* **2011**, *225*, 647–654. [CrossRef]
32. Yuan, J.; Fu, Y.; Liu, Y.; Zhang, J.; Pei, J. Analysis of pressure fluctuation characteristics in volute of centrifugal pump based on large eddy simulation. *J. Drain. Irrig. Mech. Eng.* **2010**, *28*, 310–314.
33. Wang, F.; Zhang, L.; Zhang, Z. Study on pressure fluctuation characteristics of unsteady flow field in axial flow pump. *J. Hydraul. Eng.* **2007**, *38*, 1003–1009.
34. Ni, Y.; Pan, Z.; Li, H.; Yuan, J. Application of outlet pressure fluctuation characteristics in cavitation monitoring of centrifugal pump. *J. Drain. Irrig. Mech. Eng.* **2006**, *24*, 40–43.
35. Zhang, N. Study on Excitation Characteristics of Unsteady Flow in Centrifugal Pump. Master's Thesis, Jiangsu University, Zhenjiang, China, 2016.
36. Cheng, H.Y.; Ji, B.; Long, X.P.; Huai, W.; Farhat, M. A review of cavitation in tip-leakage flow and its control. *J. Hydrodyn.* **2021**, *33*, 226–242. [CrossRef]
37. Deng, Z.; Jiang, J.; Liu, X. Analysis of unsteady cavitation in the internal flow field of a centrifugal pump. *Water Resour. Power* **2021**, *39*, 166–170.
38. Li, W.; Ji, L.; Li, E.; Shi, W.; Agarwal, R.; Zhou, L. Numerical investigation of energy loss mechanism of mixed-flow pump under stall condition. *Renew. Energy* **2020**, *167*, 740–760. [CrossRef]
39. Ji, L.; Li, W.; Shi, W.; Chang, H.; Yang, Z. Energy characteristics of mixed-flow pump under different tip clearances based on entropy production analysis. *Energy* **2020**, *199*, 1–15. [CrossRef]
40. Feng, J.; Tang, X.; Wang, W.; Ying, R.; Zhang, T. Verification method of numerical simulation reliability based on grid independence and time independence. *J. Shihezi Univ.* **2017**, *35*, 52–56.
41. Li, W.; Li, E.; Ji, L.; Zhou, L.; Shi, W.; Zhu, Y. Mechanism and propagation characteristics of rotating stall in a mixed-flow pump. *Renew. Energy* **2020**, *153*, 74–92. [CrossRef]
42. Ji, L.; Li, W.; Shi, W.; Tian, F.; Agarwal, R. Effect of blade thickness on rotating stall of mixed-flow pump using entropy generation analysis. *Energy* **2021**, *236*, 121–381. [CrossRef]
43. Kang, S.; Shi, L.; Dai, L.; Fan, F. Error analysis and grid convergence of CFD simulation. *J. Eng. Thermophys.* **2010**, *31*, 2009–2013.
44. Wang, H.; Hu, Q.; Yang, Y.; Wang, C. Performance differences of electrical submersible pump under variable speed schemes. *Int. J. Simul. Model.* **2021**, *20*, 76–86. [CrossRef]
45. Zhou, J.; Zhao, M.; Wang, C.; Gao, Z. Optimal design of diversion piers of lateral intake pumping station based on orthogonal test. *Shock. Vib.* **2021**, *2021*, 6616456. [CrossRef]
46. Zhang, L.; Wang, C.; Zhang, Y.; Xiang, W.; He, Z.; Shi, W. Numerical study of coupled flow in blocking pulsed jet impinging on a rotating wall. *J. Braz. Soc. Mech. Sci. Eng.* **2021**, *43*, 508. [CrossRef]

47. Shi, L.; Zhu, J.; Tang, F.; Wang, C. Multi-Disciplinary optimization design of axial-flow pump impellers based on the approximation model. *Energies* **2020**, *13*, 779. [CrossRef]
48. Wang, H.; Long, B.; Wang, C.; Han, C.; Li, L. Effects of the impeller blade with a slot structure on the centrifugal pump performance. *Energies* **2020**, *13*, 1628. [CrossRef]
49. Zhu, Y.; Li, G.; Wang, R.; Tang, S.; Su, H.; Cao, K. Intelligent fault diagnosis of hydraulic piston pump combining improved lenet-5 and pso hyperparameter optimization. *Appl. Acoust.* **2021**, *183*, 108336. [CrossRef]
50. Li, W. Study on Cavitation Mechanism of Engine Cooling Water Pump under Thermodynamic Effect. Master's Thesis, Jiangsu University, Zhenjiang, China, 2016.

machines

MDPI

Article

Research and Experimental Analysis of Hydraulic Cylinder Position Control Mechanism Based on Pressure Detection

Rulin Zhou [1,*], Lingyu Meng [1,2], Xiaoming Yuan [3] and Zishi Qiao [1]

1 Beijing Tianma Intelligent Control Technology Co., Ltd., Beijing 100013, China; mengly@tdmarco.com (L.M.); Qiaozs@tdmarco.com (Z.Q.)
2 School of Mechanical Electronic & Information Engineering, China University of Mining & Technology, Beijing 100083, China
3 Hebei Provincial Key Laboratory of Heavy Machinery Fluid Power Transmission and Control, Yanshan University, Qinhuangdao 066004, China; xiaomingbingbing@163.com
* Correspondence: zhourl@tdmarco.com

Abstract: This paper studies the precise position control of the hydraulic cylinder in the hydraulic support. The aim of this paper is to develop a method of hydraulic cylinder position control based on pressure and flow coupling, which takes the coupling feedback of load and flow into account, especially in the scene of cooperative control under the condition of multiple actuators and variable load. This method solves the problems of slow movement and sliding effect of hydraulic support in the traditional time-dependent hydraulic position control, as well as better realizes the intelligent and unmanned development of the fully mechanized mining face. First, based on the flow continuity equation and Newton Euler dynamic equation, the flow and stroke control model with the input and output pressure of hydraulic cylinder is established. Then, the effectiveness and correctness of the control model are verified by the comparison between the hydraulic system simulation software, AMESim, and the experiment. Finally, a test system is built. When the system pressure is large than 10 MPa, the error between the data determined by the fitting algorithm and the actual detection data is within 5%, which verifies the effectiveness of the theory and simulation model.

Keywords: hydraulic support; position control; pressure detection; flow integration; switch control

Citation: Zhou, R.; Meng, L.; Yuan, X.; Qiao, Z. Research and Experimental Analysis of Hydraulic Cylinder Position Control Mechanism Based on Pressure Detection. *Machines* **2022**, *10*, 1. https://doi.org/10.3390/machines10010001

Academic Editors: Chuan Wang, Li Cheng, Qiaorui Si and Bo Hu

Received: 12 November 2021
Accepted: 18 December 2021
Published: 21 December 2021

1. Introduction

Electro-hydraulic position servo system is a system that precisely controls the output displacement of the actuator and has been widely used in aerospace, metallurgy, engineering machinery, and other fields [1–3]. Based on the difference measured by the displacement sensor, and the target displacement to realize the position accuracy of the actuator, is the mainstream position control method to control the high-precision and high response hydraulic valve [4–7]. This control method is commonly used in variable displacement control of pump station, fault diagnosis [8,9], servo valve [10,11], and hydraulic cylinder control [12].

At present, in some scenarios requiring high precision control at home and abroad, the precise position control of hydraulic cylinder is mainly controlled by proportional valve or servo valve [13–15]. Lan Li presented a novel nonlinear model and high-precision lifting motion control method of a hydraulic manipulator driven by a proportional valve [16]. Luyue Yin proposed a novel finite-time output feedback controller with parameter adaptation for electro-hydraulic servo systems [17]. However, because of low energy efficiency and high cost of these control valves, the application scenarios are limited [18]. In recent years, many scholars have carried out new explorations in order to reduce the cost of position control system valves and improve the application scope. For example, Emeibo has created an integrated digital hydraulic actuator, which integrates the feedback and control of valve and cylinder into a whole to realize the precise control of displacement [19], and Zhou Chuanghui proposed a high-precision idea of hydraulic cylinder output position through a two-stage displacement closed-loop feedback control with ordinary electro-hydraulic control directional valve and hydraulic throttling damping as the core components [20]. Jin liyang of Zhejiang University proposed an precise control system of hydraulic cylinder based on a electro-hydraulic control directional valve, mainly from the perspective of control algorithm and model identification [21], and Jinbo et al. proposed an electro-hydraulic position control system using a large flow switch valve and small flow proportional valve in parallel to replace large flow servo valve [22]. Jiang Haiyu studied the electric and hydraulic position control system of a valve controlled cylinder based on a high-speed switch valve [23]. Zhao Ruihao proposed a novel two-position three-way electro-hydraulic proportional directional flow valve for hydraulic roof support [24]. Shi Jianpeng proposed a velocity and position combined control strategy based on mode switching control the boom and arm of hydraulic excavator [25]. Ahsan Saeedzadeh proposed a digital hydraulic circuit using a fast-switching on/off valve instead of servo valves to control the position of a hydraulic actuator [26].

In the coal industry, the hydraulic system of a fully mechanized mining face mainly uses a high flow switch hydraulic valve with 5% concentration emulsion as medium. Considering the restrictions of coal safety regulations on power consumption, reliability, cost of coal safety regulations, and the pollution level of the hydraulic system being low, the traditional electrohydraulic position control system cannot be applied to the hydraulic system of a fully mechanized mining face [27,28]. Generally, the length of a fully mechanized coal mining face is 300 m, arranged by more than 150 hydraulic supports in line, and each hydraulic support is composed of more than 10 valve-controlled cylinder units. The hydraulic system of a fully mechanized mining face has at least one main inlet and outlet, and it is equipped with pressure sensors. Therefore, the hydraulic system of a fully mechanized mining face is a complex hydraulic cylinder cluster control system. To follow up with shearer, the hydraulic support needs to carry out relevant action and attitude adjustment. Due to load difference, the action of different hydraulic cylinder units causes the mismatch of system flow and pressure. At the same time, due to the harsh working environment, the valve controlled hydraulic cylinder is mainly open-loop control and time control. Therefore, when the system pressure and load change, it is easy to cause the uncontrollable stroke of the hydraulic cylinder, which affects the attitude control of hydraulic support and ultimately affects the mining efficiency.

The straightness control of hydraulic support in a fully mechanized mining face is a key technology for the stable operation of an intelligent and unmanned mining face. With the introduction of foreign LASC (Longwall Automation Steering Committee) technology, the precise detection of straightness of a fully mechanized mining face has been realized, and the position error is at cm level [29,30], However, the precise control of the advancing hydraulic cylinder is mainly controlled by a large flow switch valve based on position detection. Although the logic cartridge valve with a large and small flow switch was adopted later, there is still room for improvement in the control accuracy and structure [31]. At the same time, other hydraulic cylinders of the hydraulic support do not have displacement sensors, and most of the control is mainly through the switch time control of the solenoid valve, which causes low system efficiency; on the other hand, it is easy to lose the position of the hydraulic supports.

As one of the key technologies for the normal, reliable, and efficient operation of the intelligent working face, the precise control of the hydraulic cylinder is not only the guarantee of the straightness control of the working face but also the basis for the efficient operation of the hydraulic support [32]. However, due to the special environment of the working face, only the advancing cylinders have displacement sensors, the rod-less chamber of the leg and the main liquid inlet pipeline have pressure sensors, the hydraulic valve is mainly a large diameter and large flow switch structure, and the hydraulic support has high quality large inertia, and leading to the precise control of the hydraulic cylinder in the hydraulic support has always been a difficult problem. Because of the most reliable sensor of the hydraulic system of the fully mechanized mining face is the pressure sensor, besides it being widely used and less costly, it is the most reliable as the system input in the actual operation process. The main contribution of this paper is combining with the matching relationship between system pressure and flow. Our paper proposes an electric-hydraulic position control system based on pressure detection, which can greatly improve the efficiency of movement and sliding of hydraulic support compared with the traditional time-dependent hydraulic position control and better realize the intelligent and unmanned development of the fully mechanized mining face. With the ordinary electric-hydraulic control directional valve as the control element, the displacement of the hydraulic cylinder is analyzed and predicted through real-time pressure monitoring, so as to achieve the precise determination of multi-stage hydraulic cylinder. The system has low cost and moderate control accuracy. It is especially suitable for an application scenario without displacement sensors but still needs location control. It has certain engineering application value.

2. Mechanism Analysis of Hydraulic Support System

Because of the hydraulic support function principle of each sub-hydraulic system is similar, the same equivalent valve controlled cylinder model is analyzed. The principle of the hydraulic system is shown in Figure 1. When the hydraulic cylinder extends, the left directional valve 2 is energized. The liquid enters the rod-less chamber through filter1, left directional valve 2, and check valve 3. The liquid in the rod chamber flows back to the tank through right directional valve 2 and return circuit breaker 5. After the hydraulic cylinder is extended in place, the left directional valve 2 is de-energized and is returned to its original position. The check valve 3 locks the rod-less chamber, and the hydraulic cylinder position is maintained. When hydraulic cylinder retracts, the right directional valve 2 is energized. The liquid enters the rod chamber through filter1 and right directional valve 2. High pressure opens check valve 3, and liquid in the rod-less chamber flows back to the tank through check valve 3, left directional valve 2, and return circuit breaker 5.

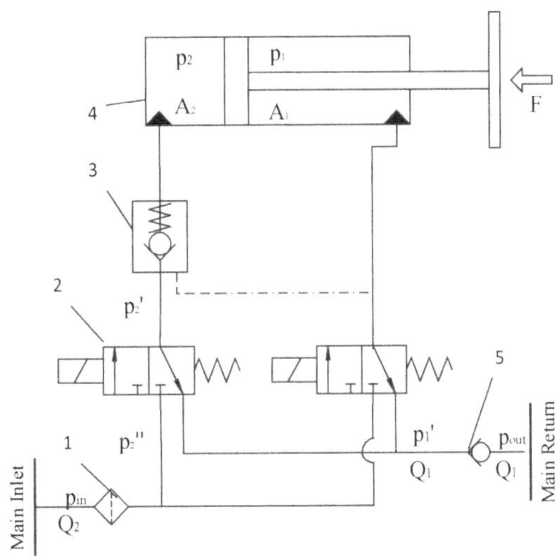

Figure 1. Schematic diagram of hydraulic system of hydraulic cylinder unit. 1. Filter. 2. Directional valve. 3. Liquid control check valve. 4. Hydraulic cylinder. 5. Return circuit breaker.

The hydraulic valves used in the hydraulic system of a fully mechanized mining face are a fixed opening switch valve spool structure, including a one-way valve and directional valve. The specific parameters can be fitted by flow resistance experiment and simulated by fixed throttling.

The equation of displacement change of the hydraulic cylinder at steady state is:

$$L = \int_{t_1}^{t_2} \frac{Q}{A} dt. \tag{1}$$

The force balance equation of rod and rod-less chambers of the hydraulic cylinder is:

$$p_2 A_2 - p_1 A_1 - F = ma. \tag{2}$$

The valve port flow equation is:

$$Q = C_V A \sqrt{\frac{2}{\rho}\Delta p} = k_p \sqrt{\Delta p}, \tag{3}$$

where Q is the total output/input flow; L is the displacement of the hydraulic cylinder; A is the corresponding area of the hydraulic cylinder chamber w/wo rod; t_1, t_2 are the start and end time; p_2, p_1 are the main inlet and return pressure, respectively; F is external joint force, and the direction is related to the direction of friction and load; a is the accelerating rate of the system; Δp is pressure difference; C_V is the flow coefficient; k_p is the comprehensive flow coefficient.

2.1. Hydraulic Cylinder Extension Model

Powered on the left working position of the solenoid valve, the following formula can be derived:

$$Q_2/Q_1 = A_2/A_1\,, \tag{4}$$

$$p_{in} - p_2'' = \Delta p_2''\,, \tag{5}$$

$$p_2'' - p_2' = \Delta p_2'\,, \tag{6}$$

$$p_2' - p_2 = \Delta p_2\,, \tag{7}$$

$$Q_2 = k_2'' \sqrt{\Delta p_2''} = k_2' \sqrt{\Delta p_2'} = k_2 \sqrt{\Delta p_2}. \tag{8}$$

We combine the above formula to obtain:

$$Q_2 = k_2^x \sqrt{\Delta p_2^x}, \tag{9}$$

where Q_2 and Q_1 are unit input and output flow, respectively; A_1 and A_2 are the area of the hydraulic cylinder chamber w/wo rod, respectively; p_{in} and p_{out} are the main inlet and return pressure, respectively; Q is the total input flow; k_2'' , k_2', k_2 are the comprehensive flow coefficient of filter, the comprehensive flow coefficient of reversing valve, and the comprehensive flow coefficient of check valve, respectively; k_2^x is the comprehensive flow coefficient from the rod chamber to the hydraulic valve at the main return when the hydraulic cylinder is extended (Equation (11)); $\Delta p_2''$, $\Delta p_2'$, Δp_2 are the pressure difference of filter, the inlet pressure difference of directional valve, and the inlet pressure difference of check valve, respectively; Δp_2^x is the pressure drop from the rod chamber to the main return when the hydraulic cylinder is extended (Equation (10)).

$$\Delta p_2^x = p_{in} - p_2, \tag{10}$$

$$k_2^x = \sqrt{\frac{1}{(\frac{1}{k_2''})^2 + (\frac{1}{k_2'})^2 + (\frac{1}{k_2})^2}}. \tag{11}$$

Similarly, the other flow equation of the hydraulic cylinder can be obtained as:

$$Q_1 = k_1^x \sqrt{\Delta p_1^x}, \tag{12}$$

where Δp_1^x is the pressure drop from the main inlet to the rod-less chamber when the hydraulic cylinder is extended. k_1^x is the comprehensive measurement coefficient from the rod-less chamber to the hydraulic valve on the main inlet side when the hydraulic cylinder is extended.

$$\Delta p_1^x = p_1 - p_{out}, \tag{13}$$

$$k_1^x = \sqrt{\frac{1}{(\frac{1}{k_1'})^2 + (\frac{1}{k_1})^2}}. \tag{14}$$

System flow equations can be obtained by solving Equations (2), (4), (9) and (12):

$$Q_2 = \sqrt{(p_{in}A_2 - p_{out}A_1 - F)/[\frac{A_1^3}{(k_1^x)^2(A_2^2)} + \frac{A_2}{(k_2^x)^2}]} = k_s \cdot f_s(p_{in}, p_{out}), \tag{15}$$

where

$$f_s(p_{in}, p_{out}) = \sqrt{(p_{in}A_2 - p_{out}A_1 - F)}, \tag{16}$$

$$k_s = \sqrt{1/[\frac{A_1^3}{(k_1^x)^2(A_2^2)} + \frac{A_2}{(k_2^x)^2}]}. \tag{17}$$

The hydraulic cylinder displacement formula is shown as follows:

$$L = \int_{t_1}^{t_2} \frac{Q_2}{A_2}\, dt. \tag{18}$$

According to the pipeline layout, the pipeline flow resistance mainly includes the friction head loss and local flow resistance loss. The general Reynolds number is much larger than 2300 for hydraulic systems of a fully mechanized mining face, so it is checked according to the turbulent state [33].

The friction head loss equation is:

$$p_f = \frac{\rho}{2} \cdot \lambda \cdot \frac{l}{d} \cdot v^2 = \frac{8\rho\lambda l q^2}{\pi^2 \cdot d_f^5}, \tag{19}$$

Where, according to Ariituri formula:

$$\lambda = 0.11 \cdot \left(\frac{\Delta}{d} + \frac{68}{\mathrm{Re}}\right)^{0.25}. \tag{20}$$

The local flow resistance loss equation is:

$$p_\xi = \frac{\rho}{2} \cdot \xi \cdot v^2 = \frac{8\rho\xi q^2}{\pi^2 \cdot d_\xi^4}, \tag{21}$$

where ρ is the emulsion density, equal to 998 kg/m^3; l is the pipe length; d_f is the pipe diameter; λ is the friction coefficient; ξ is the local resistance coefficient; Δ is surface roughness, which equals to 0.05 mm; d_ξ is the inner diameter of the pipe joint; v is the fluid velocity.

According to the above analysis, for the hydraulic system of the hydraulic cylinder unit, during the steady-state extension of the hydraulic cylinder, the hydraulic control check valve, directional valve, and pipeline joints can all be regarded as fixed damping structures. The steady-state input flow of the system can be predicted only by detecting the inlet and outlet pressure p_{in} and p_{out} of the system. Finally, the position and displacement of the hydraulic cylinder can be estimated through the integration of the flow.

2.2. Hydraulic Cylinder Retracted Model

Powered on the right working position of solenoid valve, the steady state flow continuity equation can be obtained as:

$$Q_2/Q_1 = A_1/A_2\,, \tag{22}$$

$$Q_2 = k_2^{x_2}\sqrt{p_{in} - p_1}\,, \tag{23}$$

where $k_2^{x_2}$ is the comprehensive flow coefficient from the rod chamber to the hydraulic valve at the main inlet when the hydraulic cylinder retracts.

$$k_2^{x_2} = \sqrt{\frac{1}{(\frac{1}{k_2''})^2 + \left(\frac{1}{k_1}\right)^2}}\,. \tag{24}$$

Similarly, the other flow equation of the hydraulic cylinder can be obtained as:

$$Q_1 = k_1^{x_1}\sqrt{p_2 - p_{out}}\,, \tag{25}$$

where $k_1^{x_1}$ is the comprehensive flow coefficient from the rod-less chamber to the hydraulic valve at the main return when the hydraulic cylinder retracts.

$$k_1^{x_1} = \sqrt{\frac{1}{(\frac{1}{k_1'})^2 + (\frac{1}{k_2})^2 + (\frac{1}{k_2'})^2}}\,. \tag{26}$$

System flow equations can be obtained by solving Equations (2), (22), (23) and (25):

$$Q_2 = \sqrt{(p_{in}A_1 - p_{out}A_2 + F)/[\frac{A_1}{(k_2^{x_2})^2} + \frac{A_3^3}{(k_1^{x_1})^2 \cdot A_1^2}]} = k_h \cdot f_h(p_{in},\, p_{out})\,, \tag{27}$$

where

$$f_h(p_{in},\, p_{out}) = \sqrt{p_{in}A_1 - p_{out}A_2 + F}, \tag{28}$$

$$k_h = \sqrt{1/[\frac{A_1}{(k_2^{x_2})^2} + \frac{A_3^3}{(k_1^{x_1})^2 (A_1^2)}]}\,. \tag{29}$$

The above analysis is based on the condition that the inlet and outlet pressures are known and detectable, the flow required for the action of a single hydraulic cylinder can be calculated, and the stroke displacement of the hydraulic cylinder can be calculated through flow integration. Similarly, when multiple hydraulic cylinder units apply in the fully mechanized mining face, the accumulated flow is the input flow of all units.

$$Q_Z = \sum_1^M Q_2^i\,. \tag{30}$$

3. Simulation and Experiment

3.1. Valve Control Cylinder Simulation Model

The basic working principle of the hydraulic support is to extend and retract the hydraulic cylinder through the hydraulic pump. It mainly includes three basic circuits: leg control circuit, advancing cylinder control circuit, and balance cylinder control circuit. The advancing cylinder control circuit is one of the important circuits of the hydraulic system of hydraulic support. Establishing the hydraulic simulation model of the advancing control circuit, and studying the key factors affecting the pulling and pushing of hydraulic support from the mechanism, is of great significance to improve the advancing control precision. Figure 2 shows the advancing control circuit, including advancing cylinder, hydraulic control check valve, control signal, liquid supply circuit, etc. Table 1 is the simulation parameters of the hydraulic system of the advancing control circuit.

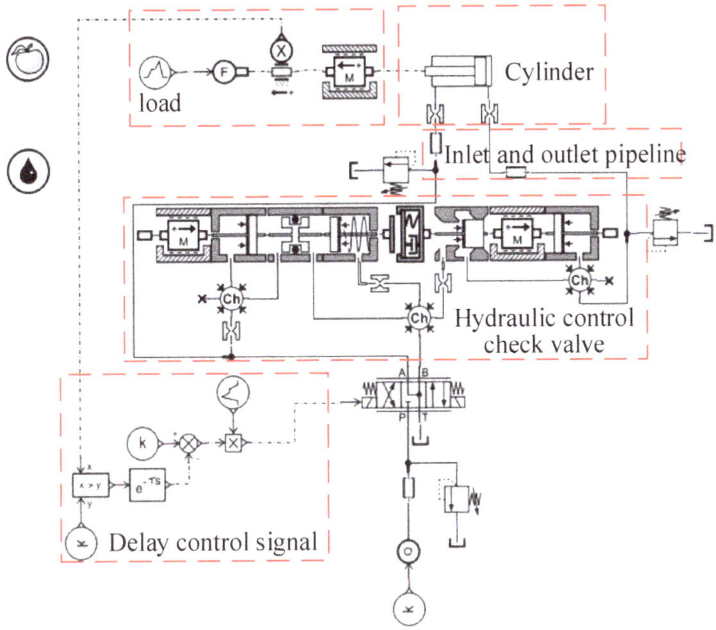

Figure 2. Hydraulic schematic diagram of advancing control circuit.

Table 1. AMESim simulation parameters of the hydraulic system of control circuit advancing.

Symbol	The Parameter Name	Magnitude
Q	Main System Flow/(L/min)	400
ρ	The emulsion density/(kg/m^3)	998
E	Bulk modulus of elasticity of emulsion/MPa	2010
p_{set}	Set pressure of system safety valve/MPa	31.5
D	Piston diameter of advancing hydraulic cylinder/mm	180
d	Diameter of piston rod of advancing hydraulic cylinder/mm	120
L	Stroke of advancing hydraulic cylinder/mm	900
d_r	Inlet diameter of advancing hydraulic cylinder/mm	11.2
d_c	Outlet diameter of advancing hydraulic cylinder/mm	11.2
m	Load mass/kg	500
$F1$	Coulomb friction of advancing hydraulic cylinder/N	7000
$F2$	Static friction of advancing hydraulic cylinder/N	10,000
h_f	Flow resistance of electro-hydraulic control directional valve/(L min^{-1}/MPa)	250/5
d_f	Valve spool diameter of hydraulic control check valve/mm	33
d_{spool}	Diameter of control piston of hydraulic control check valve/mm	13
d_{gan}	Ejector rod diameter of hydraulic control check valve/mm	7

3.2. Valve Control Cylinder Test System

In order to verify the relationship between hydraulic system parameters and hydraulic cylinder position under different working conditions, such as pressure, flow, etc., an advancing hydraulic cylinder test platform is built. The test system takes the hydraulic system of Yujialiang 52,301 working face in Shendong as a reference. The experiment adopts 400 L/min emulsion pump as liquid supply and includes hydraulic components, such as hydraulic control check valve, electro-hydraulic control directional valve (hereinafter referred to as "directional valve"), hydraulic cylinder, safety valve, return circuit breaker, and cut-off valve in the system. The experimental schematic diagram is shown in Figure 3. The P port of the directional valve is connected to the emulsion pump, one branch of the directional valve port C1 is connected to the rod chamber of the hydraulic cylinder, the other branch is connected to the control port of the hydraulic control check valve, the directional valve port C2 is connected to the hydraulic control check valve, and the locking cavity of the hydraulic control check valve is connected to the rod-less chamber of the hydraulic cylinder.

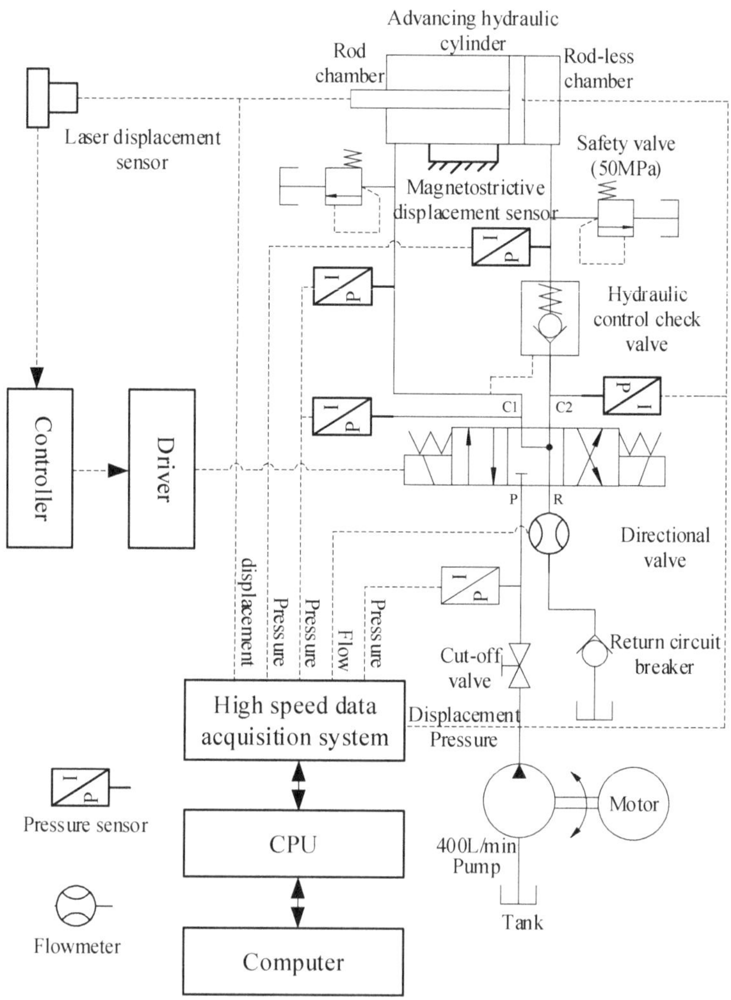

Figure 3. Hydraulic schematic diagram of the advancing test system.

Then, 40 MPa pressure sensors are connected at the inlet of the directional valve, the outlet of the directional valve port C1, the outlet of the directional valve port C2, the rod chamber of the cylinder, and the rod-less chamber of the cylinder to detect the pressure changes when the advancing hydraulic cylinder reaches to different positions. A flowmeter is set at the return of the directional valve to measure the flow of the system in real time, a 1.5 m range laser displacement sensor is set in the front of the advancing hydraulic cylinder to measure the displacement of the rod and carry out closed loop control of the rod position, and a magnetostrictive displacement sensor is installed inside the advancing hydraulic cylinder to compare the collected data with the laser displacement sensor. The data acquisition frequency of the test system is 2 kHz, i.e., a group of pressure, flow, and displacement data are collected every 0.5 ms. The parameters of test equipment are shown in Table 2, and the connection diagram of experimental equipment is shown in Figure 4.

Table 2. Parameters of experimental equipment.

Equipment	Parameter	Model
Emulsion pump	Nominal flow 400 L/min, nominal pressure 31.5 MPa	BRW400/37.5
Pressure sensor	Measurement range 0–40 MPa, measurement accuracy 0.5%	KYB18G10M1PXCX-II
Return circuit breaker	Nominal flow 400 L/min, nominal pressure 16 MPa	FD400/16
Directional valve	Nominal flow 400 L/min, nominal pressure 31.5 MPa	FHS400/31.5
Hydraulic control check valve	Nominal flow 400 L/min, nominal pressure 50 MP a	FDYA400/31.5
Magnetostrictive displacement sensor	Measurement range 900 mm, precision 0.1%	TMDXC(GUC1200/900/L).00
Laser displacement sensor	Measurement 1600 mm, precision ± 1 mm	DAN-10-150
Advancing hydraulic cylinder	Cylinder/rod diameter 180/120 mm, stroke 900 mm	TMQTC(100/70*440)SL
Flowmeter	Nominal flow 400 L/min, precision 0.5%	FD-Q 50C

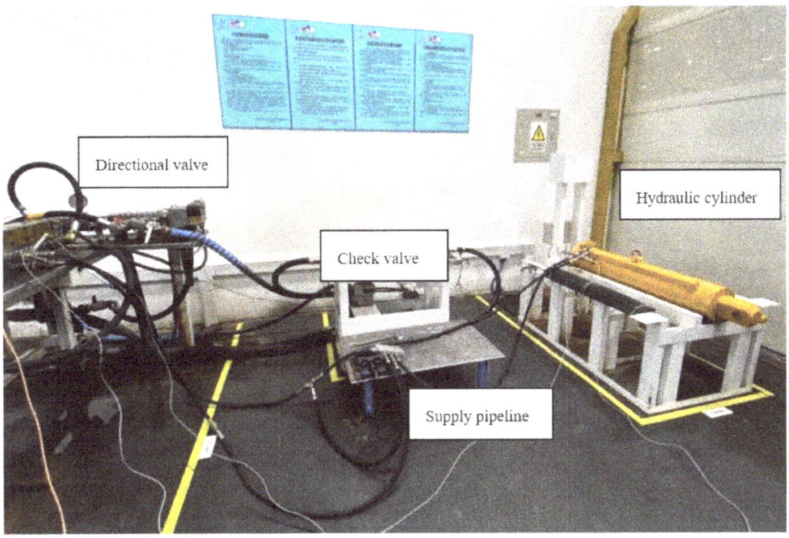

Figure 4. Connection diagram of the hydraulic system of advancing hydraulic cylinder.

3.3. Experimental Method

By adjusting the overflow valve of the emulsion pump to change the system pressure, the pressure and flow data at different positions are collected under the system pressure of 5, 10, 15, 20, 25, and 30 MPa, respectively, so as to find out the relationship between flow and pressure. In order to verify the data patterns of pressure, flow, and displacement under different pipeline specifications, the data under DN10 and DN20 diameter are collected, respectively. Figure 5 shows the data under mean filtering method of the full stroke of the advancing cylinder when the system pressure is 5 MPa with DN20 pipeline size. The analysis shows that, when the hydraulic cylinder extends and retracts, the system flow fluctuates greatly due to the short time, so as to the correlation between flow and displacement of the cylinder is weak. This is mainly due to the delay of data acquisition by the flow sensor. The flow sensor cannot reflect the transient values, such as displacement and pressure sensors. It can only measure the average flow by adjusting the response time. If the response time is short, the measurement error will be large. In this experiment, the response time sets to 1 s, i.e., the sensor outputs the data by calculating the average flow value within 1 s. The small delay leads to a certain deviation of the flow data in the initial stage and then tends to be stable. In order to verify the accuracy of the flow data after stabilization, calculate the actual flow according to Formula (31). It can be seen from Figure 5 that, when the hydraulic cylinder is retracted under system pressure 5 MPa, the calculated flow is 72.32 L/min, while the flow curve in Figure 5 shows a value 75 L/min. The relative error between the actual flow and the sensor detecting is within 5%.

$$Q = A \cdot L \times 60,000\,, \tag{31}$$

where: L—hydraulic cylinder displacement, m;
t—hydraulic cylinder action time, s;
A—hydraulic cylinder high-pressure side action area, m^2.

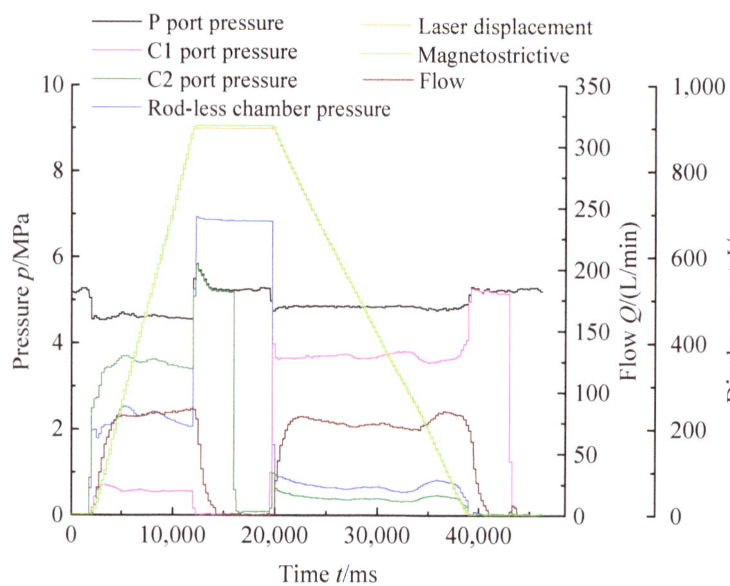

Figure 5. Full stroke mean filtering data of DN20 supply pipeline under 5 MPa system pressure.

4. Result and Discussion

4.1. Simulation Verification

In order to verify the effectiveness of the simulation model, a hydraulic system detection platform is set up. The comparison diagrams of displacement, pressure and flow between test and simulation of cyclic action of the hydraulic cylinder are shown in Figure 6.

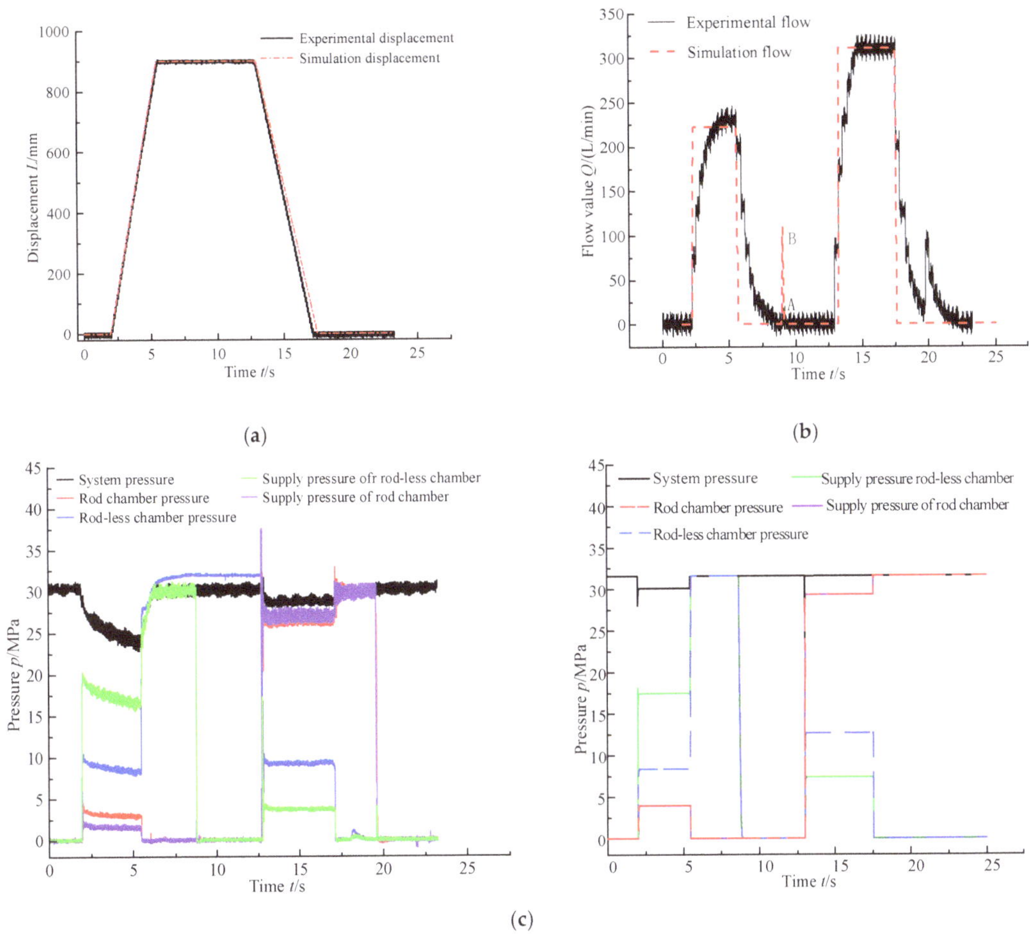

Figure 6. Comparison of the test and simulation data under 31.5 MPa system pressure: (**a**) Comparison diagram of experiment and simulation displacement; (**b**) comparison diagram of experiment and simulation outlet flow of electro-hydraulic control directional valve; (**c**) comparison diagram of experiment and simulation pressure.

The pressure of the hydraulic system is 31.5 MPa, and the liquid supply flow of the system is 400 L/min. The hydraulic cylinder extends and retracts without load, as shown in Figure 6a.

The test displacement and simulation curve of the hydraulic cylinder have good consistency. As is shown in Figure 6b, the pressure data of nodes of different components are consistent with the dynamic characteristics of simulation data. The main difference is caused by nonlinear factors (friction and elastic modulus of the system). When time $t = 9$ s, the cylinder extends to the maximum position. The directional valve is located in the middle position, and connects with the system return port. The hydraulic check valve locks cylinder rod-less chamber. At this time, the high-pressure liquid in the pipeline between the directional valve and the check valve flows to the system return port through the directional valve, causing the instantaneous change of flow, corresponding to point A of the test curve and point B of the simulation curve.

Figure 6c shows the pressure test curve and simulation curve of the extension and retraction of the cylinder. It can be seen from the experimental curve that the cylinder starts to extend at time equal to 2 s. Due to the static friction is much larger than the dynamic friction at the initial stage of start-up, the high-pressure fluid enters the rod-less chamber instantaneously, and other factors, the pressure in the rod-less chamber has a slow and slight drop at the initial stage. However, the simulation curve has the same static friction value as the dynamic friction value, resulting in stable pressure at the moment of start-up.

In conclusion, the AMESim simulation model of the advancing circuit hydraulic system is consistent with the actual system.

4.2. Simulation Result Analysis

4.2.1. Relationship between Pressure Drop of Different Components and System Flow

When the hydraulic connecting pipes of the system are DN10 and DN 20, respectively, and the flow at the liquid supply port of the directional valve increase from 0 to 500 L/min linearly within 25 s, we obtain the relationship of the square root of the pressure drop and the flow rate between different components when the hydraulic cylinder extends, as shown in Figure 7a,b.

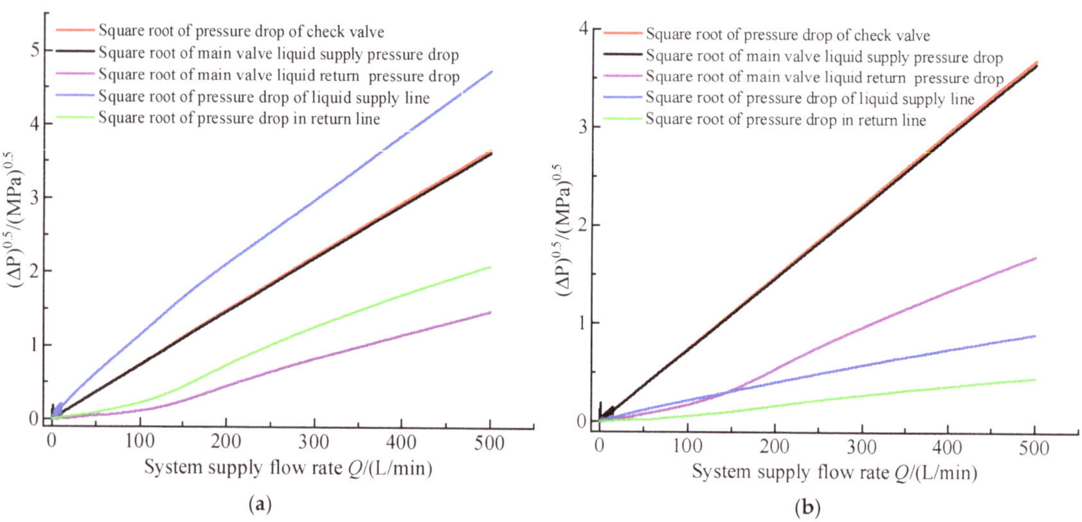

Figure 7. Relationship between liquid supply flow and pressure drop. (**a**) Relationship between liquid supply flow of DN10 pipeline and square root of pressure drop; (**b**) relationship between liquid supply flow of DN20 pipeline and square root of pressure drop.

The following conclusions are drawn by comparing Figure 7a,b: (1) with the increase of system liquid supply flow, the system flow has a linear relationship with the square root of pressure drop of directional valve and check valve, and a nearly linear relationship with pipeline; (2) the small diameter of DN10 pipeline leads to large pressure drop of inlet and outlet pipelines; and (3) different pipe diameters will also affect the pressure drop of other components of the system.

4.2.2. Analysis of Pressure Drop Caused by Different Diameter of Liquid Supply Pipeline

In order to study the relationship between the square root of system pressure and flow under different supply pipeline diameters, the supply pipeline is changed from DN10 to DN40 for batch operations to analyze the influence of the diameter of the supply pipeline on the system parameters. The simulation results are shown in Figure 8.

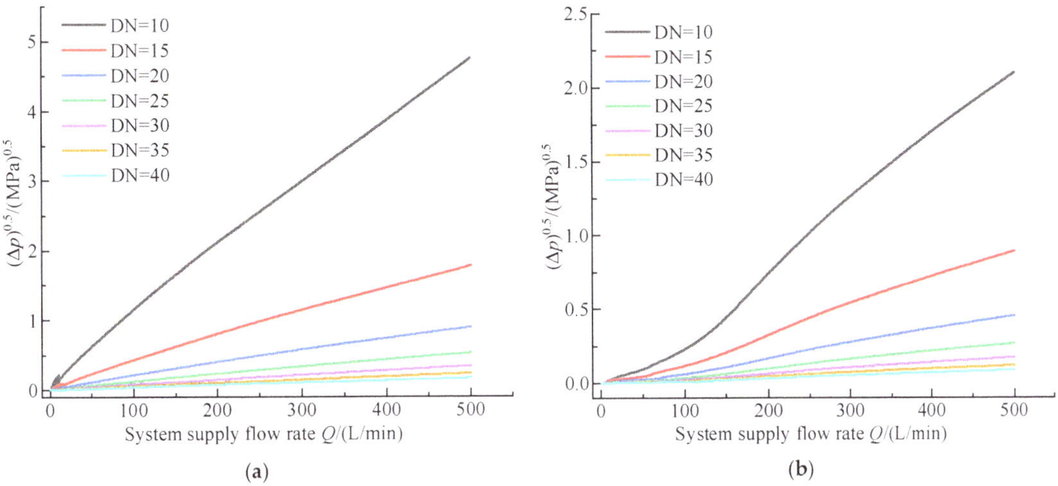

Figure 8. Relationship between supply flow and pressure drop of pipelines with different diameters: (**a**) Relationship between square root of pressure drop in inlet pipeline and supply flow; (**b**) relationship between square root of pressure drop in outlet pipeline and supply flow.

As can be seen from Figure 8a, when the system flow is less than 500 L/min, the pressure drop of the inlet pipeline decreases continuously with the increase of the diameter of the supply pipeline, and the square root of the pressure drop of the inlet pipeline is nearly linear with the supply flow. When the diameter of pipeline is above DN25, the pressure drop of the pipeline is low and has little impact in the system.

Since the outlet pipeline is directly connected with the main return, the pressure is low, so that the elastic modulus E changes greatly. It can be seen from Figure 8b that the system flow shows the approximately linear relationship with the square root of the pressure drop of the outlet pipeline.

4.2.3. Simulation Summary

The simulation shows that the flow of the hydraulic system has a linear relationship with the square root of the pressure drop of hydraulic components, pipelines, and joints, and its characteristics meet with the throttle flow equation of the valve port, so it can be equivalent to a fixed damping hole, which is also consistent with the formula results derived in the paper. The experimental platform and test data will be described in detail below.

4.3. Experimental Result Analysis

4.3.1. Relationship between Flow and Pressure Drop of Different Component

Analyzing the relationship between flow and pressure drop when the hydraulic cylinder extends and retracts, and we obtain the data between flow and pressure drop under different pump pressures, as shown in Tables 3 to 6.

Table 3. Hydraulic cylinder retraction data under DN10 pipeline diameter.

System Pressure p (MPa)	System Flow Q (L/min)	Pressure Drop Δp (MPa)		
		Δp (Directional Valve)	Δp (Pipeline)	Δp (Check Valve)
6	21.57	1.71	0.14	0.61
10.5	41.50	1.87	0.50	2.25
17	64.20	0.78	1.18	5.31
21.4	72.99	1.23	1.52	6.84
26.8	83.81	1.35	2.01	9.00
31	92.48	0.63	2.50	10.93

Table 4. Hydraulic cylinder extension data under DN10 pipeline diameter.

System Pressure p (MPa)	System Flow Q (L/min)	Pressure Drop Δp (MPa)		
		Δp (Directional Valve)	Δp (Pipeline)	Δp (Check Valve)
6	20.62	1.72	0.41	1.91
10.5	35.08	1.75	1.15	5.01
17	49.94	1.52	2.29	9.75
21.4	56.65	1.97	2.94	12.40
26.8	64.84	2.60	3.83	16.07
31	70.23	3.08	4.48	18.72

Table 5. Hydraulic cylinder retraction data under DN20 pipeline diameter.

System Pressure p (MPa)	System Flow Q (L/min)	Pressure Drop Δp (MPa)		
		Δp (Directional Valve)	Δp (Pipeline)	Δp (Check Valve)
5.2	71.22	1.15	0.15	0.28
11.5	179.74	0.91	0.97	1.79
16	222.15	0.90	1.48	2.73
22	266.95	1.30	2.14	3.95
25.8	291.09	1.51	2.54	4.67
30.6	316.37	1.77	3.00	5.48

Table 6. Hydraulic cylinder extension data under DN20 pipeline diameter.

System Pressure p (MPa)	System Flow Q (L/min)	Pressure Drop Δp (MPa)		
		Δp (Directional Valve)	Δp (Pipeline)	Δp (Check Valve)
5.2	77.61	1.08	0.18	1.26
11.5	134.60	3.09	0.54	3.36
16	159.10	4.28	0.76	4.67
22	187.64	5.95	1.05	6.44
25.8	202.95	6.91	1.23	7.58
30.6	216.28	7.83	1.40	8.60

According to the data in the table, we draw the relationship curve between flow and differential pressure. Figure 9 shows the statistical diagram of the extended and retracted flow and the square root of the differential pressure of the hydraulic cylinder under DN10 pipeline. Figure 10 shows the statistical diagram of the extended and retracted flow and the square root of the differential pressure of the hydraulic cylinder under DN20 pipeline. The analysis shows that, under different pipeline diameters, the system flow is linear with the square root of pipeline, check valve, and directional valve flow resistance. When the pipeline diameter is DN10, the system flow is low due to excessive flow resistance under the same system pressure, and the maximum flow is less than 100 L/min. When the pipeline diameter is DN20, the system flow is large due to small flow resistance under the same system pressure, and the maximum flow is greater than 300 L/min.

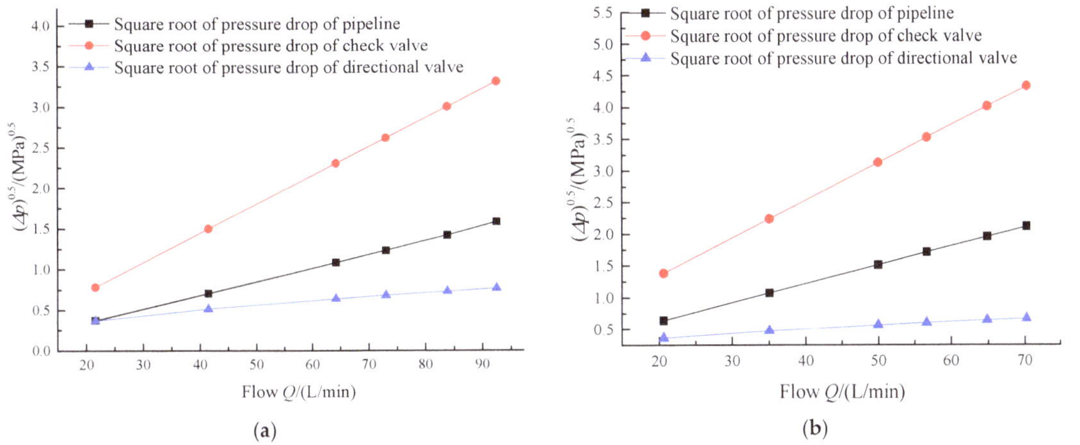

Figure 9. The relationship between flow and square root of differential pressure at all levels when the hydraulic cylinder is operated under DN10 pipeline diameter: (**a**) retracted; (**b**) extended.

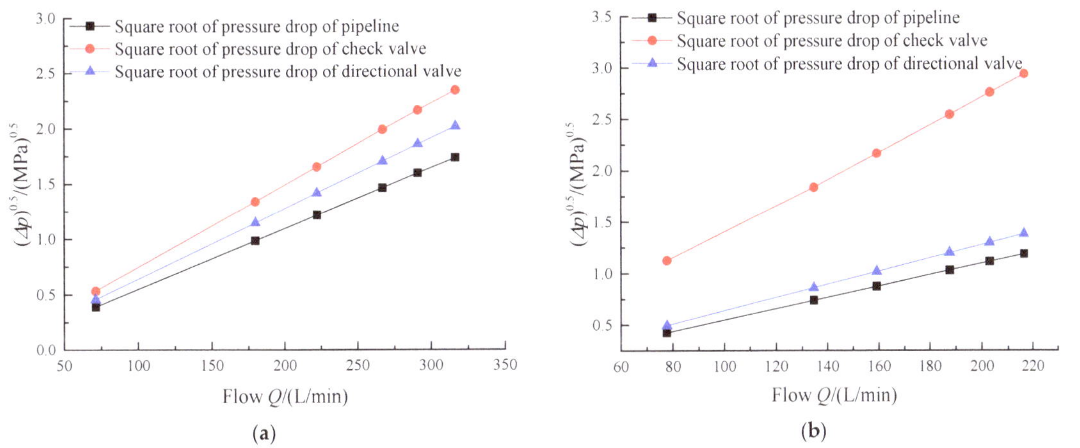

Figure 10. The relationship between flow and square root of differential pressure at all levels when hydraulic cylinder is operated under DN20 pipeline diameter: (**a**) retracted; (**b**) extended.

4.3.2. Relationship between Flow and Inlet and Outlet Pressure

In order to verify the relationship between system flow and inlet and outlet pressure under different pipeline conditions, the working state parameter tables under different working conditions are summarized, and the system parameters are shown in Table 7. According to formulas (16) and (28), the system flow is taken as the horizontal ordinate, and the output function $f(p_{in}, p_{out})$ (includes $f_s(p_{in}, p_{out})$ and $f_h(p_{in}, p_{out})$) is taken as the vertical ordinate based on the input under different working states. The experimental results are shown in Figures 11 and 12.

Table 7. System parameters table under different liquid supply conditions.

	Hydraulic Cylinder Extends				Hydraulic Cylinder Retracts			
	Flow Q (L/min)	Inlet Pressure p_{in} (MPa)	Return Pressure p_{out} (MPa)	$f_s(p_{in}, p_{out})$ $(N)^{0.5}$	Flow Q (L/min)	Inlet Pressure p_{in} (MPa)	Return Pressure p_{out} (MPa)	$f_h(p_{in}, p_{out})$ $(N)^{0.5}$
DN10 pipeline	20.62	6	0.62	0.121	21.57	6	0.62	0.073
	35.08	10.5	0.74	0.162	41.5	10.5	0.53	0.103
	49.94	17	0.82	0.207	64.2	17	0.53	0.135
	56.65	21.4	0.89	0.233	72.99	21.4	0.54	0.152
	64.84	26.8	0.83	0.261	83.81	26.8	0.57	0.171
	70.23	31	0.91	0.281	92.48	31	0.65	0.184
DN20 pipeline	77.61	5.2	0.75	0.112	71.22	5.2	0.37	0.071
	134.60	11.5	0.65	0.170	179.74	11.5	1.18	0.101
	159.10	16	0.91	0.201	222.15	16	1.84	0.117
	187.64	22	1.22	0.235	266.95	22	2.63	0.136
	202.95	25.8	1.46	0.255	291.09	25.8	3.19	0.146
	216.28	30.6	1.67	0.277	316.37	30.6	3.67	0.160

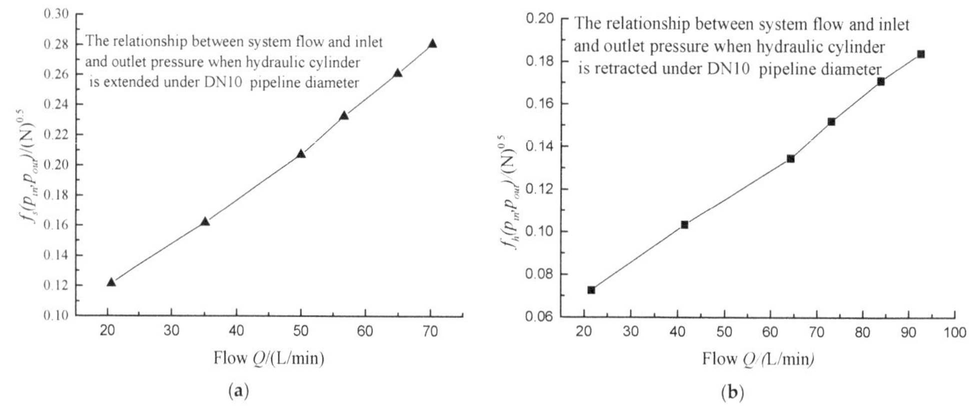

Figure 11. Relationship between flow and inlet and outlet pressure of hydraulic cylinder under DN10 diameter: (**a**) extended; (**b**) retracted.

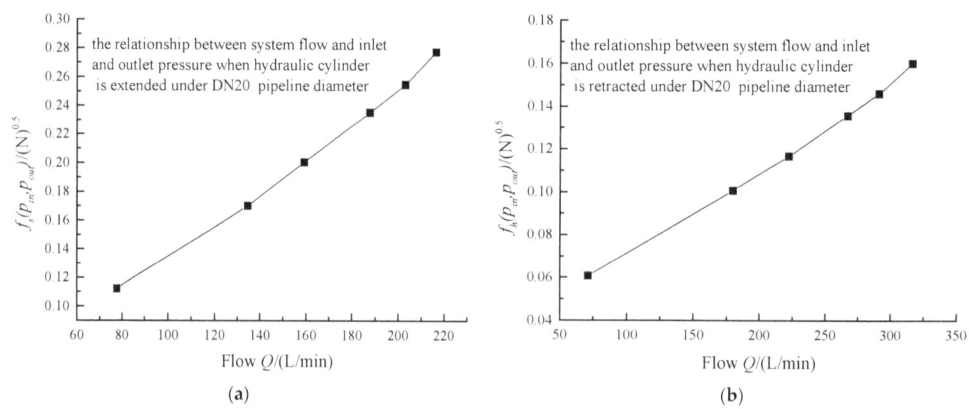

Figure 12. Relationship between flow and inlet and outlet pressure of hydraulic cylinder under DN20 diameter: (**a**) extended; (**b**) retracted.

By analyzing Figures 11 and 12, the system flow under different pipeline conditions is basically linear, with the function $f(p_{in}, p_{out})$ based on the inlet and outlet pressure as the variable. The linearity error of some low pressure areas and high pressure areas is mainly due to the change of elastic modulus in high and low pressure areas, friction, and other nonlinear loads, but the load has a small proportion in the system calibration or is equivalent to a linear variable, so the influence on the final calculation result can be ignored.

4.4. Fitting Verification

Next, assign values to Equations (15) and (27) according to the system structural parameters to calculate the fitting flow of different pipeline layouts and working states, statistically analyze the fitted flow and the actual flow, and calculate the relative error according to Equation (32). The statistical results are shown in Table 8, and the fitting relationship diagram is shown in Figure 13.

$$\delta = \frac{|Q_n - Q_s|}{Q_s} \times 100\%, \tag{32}$$

where Q_n—fitting flow; Q_s—measured flow.

Table 8. Comparison table of fitting flow and measured flow under different pipeline diameters.

	System Pressure (MPa)	5	10	15	20	25	30
Hydraulic cylinder extends under DN10 pipeline	Actual flow (L/min)	20.62	35.08	49.94	56.65	64.84	70.23
	Fitting flow (L/min)	23.16	39.04	49.89	56.15	62.90	67.72
	Relative error (%)	12.32	11.29	0.11	0.88	2.99	3.57
Hydraulic cylinder retracts under DN10 pipeline	Actual flow (L/min)	21.57	41.5	64.2	72.99	83.81	92.48
	Fitting flow (L/min)	25.894	47.234	64.53	72.656	81.738	87.952
	Relative error (%)	20.05	13.82	0.51	0.46	2.47	4.90
Hydraulic cylinder extends under DN20 pipeline	Actual flow (L/min)	77.61	134.60	159.10	187.64	202.95	216.28
	Fitting flow (L/min)	88.70	134.64	159.19	186.12	201.96	219.38
	Relative error (%)	14.29	0.03	0.06	0.81	0.49	1.44
Hydraulic cylinder retracts under DN20 pipeline	Actual flow (L/min)	71.22	179.74	222.15	266.95	291.09	316.37
	Fitting flow (L/min)	79.83	191.80	222.18	258.26	277.25	303.84
	Relative error (%)	12.09	6.71	0.01	3.25	4.75	3.96

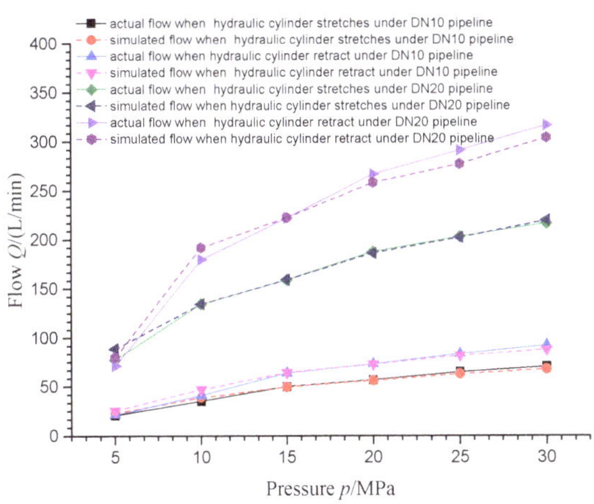

Figure 13. Comparison diagram of actual flow and fitting flow under different pipeline parameters and working conditions.

As can be seen in Figure 13 and Table 8, the coincidence degree of actual flow and fitting flow under different pipeline parameters and working conditions is good. Under the conditions of DN10 and DN20 pipelines, when the pressure is above 10 MPa, the system flow coincidence error is within 5%, and, under the conditions of 5–10 MPa, the system flow coincidence error is within 13%, which is mainly caused by the nonlinearity and large variation of the elastic model in the low pressure region. The elastic modulus is equivalent to a constant when system pressure is above 10 MPa.

5. Conclusions

The pressure and flow coupling-based precise position control was proposed for the hydraulic cylinder in the hydraulic support. The flow and stroke control model was established based on the flow continuity equation and Newton Euler dynamic equation. The hydraulic system simulation software AMESim proved the effectiveness and correctness of the control model. In addition, our proposed test system demonstrated that, when the system pressure was larger than 10 MPa, the error between the data determined by the fitting algorithm and the actual detection data was within 5%. In the future, in order to improve the accuracy and promote the application of position control based on pressure detection, we will verify the transient effects of other factors, such as load, volume, and delay on cylinder position control, as well as propose a transient control model based on volume compression formula, so as to finally realize the unification of transient model and steady-state model.

Author Contributions: Conceptualization, R.Z.; methodology, R.Z.; software, Z.Q.; validation, L.M.; formal analysis, X.Y.; investigation, L.M. and Z.Q.; resources, R.Z.; data curation, R.Z. and L.M.; writing—original draft preparation, R.Z.; writing—review and editing, L.M. and Z.Q.; supervision, R.Z.; project administration, X.Y.; funding acquisition, R.Z. All authors have read and agreed to the published version of the manuscript.

Funding: This research was funded by China Coal Technology & Engineering Group, grant number 2020-TD-MS009.

Institutional Review Board Statement: Not applicable.

Informed Consent Statement: Not applicable.

Data Availability Statement: Not applicable.

Acknowledgments: We acknowledge the funding support from CCTEG (project account code: 2020-TD-MS009) for this research and support of the Laboratory of Beijing Tianma Intelligent Control Technology Co., Ltd.

Conflicts of Interest: The authors declare no conflict of interest.

References

1. Feng, H.; Ma, W.; Yin, C.; Cao, D. Trajectory control of electro-hydraulic position servo system using improved PSO-PID controller. *Autom. Constr.* **2021**, *127*, 103722. [CrossRef]
2. Adnan, R.; Rahiman, M.H.F.; Samad, A.M. Model identification and controller design for real-time control of hydraulic cylinder. In Proceedings of the 2010 6th International Colloquium on Signal Processing & Its Applications (CSPA), Malacca, Malaysia, 21–23 May 2010.
3. Liu, H.; Wang, C.; Guo, X.; Zhao, B.; Huo, P. Active disturbance rejection control method for position servo system based on electro-hydraulic load sensing. *J. Beijing Univ. Aeronaut. Astronaut.* **2020**, *46*, 2131–2139.
4. Shi, L.; Zhu, J.; Tang, F.; Wang, C. Multi-Disciplinary optimization design of axial-flow pump impellers based on the approximation model. *Energies* **2020**, *13*, 779. [CrossRef]
5. Zhu, Y.; Li, G.; Wang, R.; Tang, S.; Su, H.; Cao, K. Intelligent fault diagnosis of hydraulic piston pump combining improved lenet-5 and PSO hyperparameter optimization. *Appl. Acoust.* **2021**, *183*, 108336. [CrossRef]
6. Tang, S.; Zhu, Y.; Yuan, S. An improved convolutional neural network with an adaptable learning rate towards multi-signal fault diagnosis of hydraulic piston pump. *Adv. Eng. Inform.* **2021**, *50*, 101406. [CrossRef]
7. Wang, H.; Long, B.; Wang, C.; Han, C.; Li, L. Effects of the impeller blade with a slot structure on the centrifugal pump performance. *Energies* **2020**, *13*, 1628. [CrossRef]

8. Wang, H.; Hu, Q.; Yang, Y.; Wang, C. Performance differences of electrical submersible pump under variable speed schemes. *Int. J. Simul. Model.* **2021**, *20*, 76–78. [CrossRef]
9. Zhu, Y.; Li, G.; Wang, R.; Tang, S.; Su, H.; Cao, K. Intelligent fault diagnosis of hydraulic piston pump based on wavelet analysis and improved AlexNet. *Sensors* **2021**, *21*, 549. [CrossRef] [PubMed]
10. Zhou, J.; Zhao, M.; Wang, C.; Gao, Z. Optimal design of diversion piers of lateral intake pumping station based on orthogonal test. *Shock Vib.* **2021**, *2021*, 6616456. [CrossRef]
11. Zhang, L.; Wang, C.; Zhang, Y.; Xiang, W.; He, Z.; Shi, W. Numerical study of coupled flow in blocking pulsed jet impinging on a rotating wall. *J. Braz. Soc. Mech. Sci. Eng.* **2021**, *43*, 508. [CrossRef]
12. Wang, H.; Qian, Z.; Zhang, D.; Wang, T.; Wang, C. Numerical study of the normal impinging water jet at different impinging height, based on Wray-Agarwal turbulence model. *Energies* **2020**, *13*, 1744. [CrossRef]
13. Zhu, P.; Zou, S.; Xie, Y. Simulation research on electro-hydraulic proportional position control system of hydraulic excavator. *Mech. Eng. Autom.* **2019**, *5*, 104–106.
14. Gao, Q.; Zhu, Y.; Luo, Z.; Wang, R.; Song, Y.; Chen, X. Hardware in the Loop Simulation for Position Control Hydraulic Cylinder Using High Speed on/of Valve. *Trans. Beijing Inst. Technol.* **2019**, *39*, 1091–1096.
15. Sun, C.; Wu, L.; Xi, G. Design and Application of Electro-Hydraulic Servo Fuzzy PID Position Control System. *Mach. Des. Manuf.* **2016**, *6*, 155–157, 162. [CrossRef]
16. Li, L.; Lin, Z.; Jiang, Y.; Yu, C.; Yao, J. Valve Deadzone/Backlash Compensation for Lifting Motion Control of Hydraulic Manipulators. *Machines* **2021**, *9*, 57. [CrossRef]
17. Yin, L.; Deng, W.; Yang, X.; Yao, J. Finite-Time Output Feedback Control for Electro-Hydraulic Servo Systems with Parameter Adaptation. *Machines* **2021**, *9*, 214. [CrossRef]
18. Ali, H.H.; Fales, R.C. A review of flow control methods. *Int. J. Dyn. Control* **2021**, *9*, 1847–1854. [CrossRef]
19. Yang, H.; Guo, Y.; Zhang, H.; Wang, X.; Lin, B.; Dang, Y. Development of the hydraulic cylinder position control system based on electromagnetic directional valve. *Mach. Des. Manuf. Eng.* **2020**, *49*, 101–104.
20. Zhou, C.; Wen, G.; Qing, X. Study of electro-hydraulic position control system using solenoid directional valve. *Eng. J. Wuhan Univ.* **2017**, *50*, 760–765.
21. Jin, L. Study on the Precise Position Control of Hydraulic Cylinder Using Solenoid Operated Directional Control Valve. Ph.D. Thesis, Zhejiang University, Hangzhou, China, 2019.
22. Jin, B.; Zhu, S.; Zhang, G.; Lin, J. Study on parallel conference control system. *J. Mech. Eng.* **2000**, *36*, 72–75+82. [CrossRef]
23. Jiang, H. Research of Electro-Hydraulic Postion Control System Based on High Speed On-Of Valve. Master's Thesis, Northeastern University, Shenyang, China, 2011.
24. Zhao, R.; Liao, Y.; Lian, Z.; Li, R.; Guo, Y. Research on the performance of a novel electro-hydraulic proportional directional valve with position-feedback groove. *Proc. Inst. Mech. Eng. Part E J. Process Mech. Eng.* **2021**, *235*, 1930–1944. [CrossRef]
25. Shi, J.; Quan, L.; Zhang, X.; Xiong, X. Electro-hydraulic velocity and position control based on independent metering valve control in mobile construction equipment. *Autom. Constr.* **2018**, *94*, 73–84. [CrossRef]
26. Saeedzadeh, A.; Rezaei, S.M.; Zareinejad, M. Energy-efficient position control of an actuator in a digital hydraulic system using on/off valve. In Proceedings of the International Conference on Robotics and Mechatronics, Tehran, Iran, 26–28 October 2016.
27. Zhou, R.; Li, S.; Wei, W.; Zhang, L.; Wang, W. Study on coupling mechanism of pressure and flow in following hydraulic system of mining face. *Coal Sci. Technol.* **2020**, *48*, 129–136.
28. Fu, X.; Wang, R.; Zhao, Y. Investigation of hydraulic system simulation and fluid feeding technology with steady pressure of working face supports. *J. China Coal Soc.* **2018**, *43*, 1471–1478.
29. Li, S. Measurement & control and localisation for fully-mechanized working face alignment based on inertial naviga tion. *Coal Sci. Technol.* **2019**, *47*, 169–174.
30. Song, Z. Application of LASC technology in mine's automatic mining of fully mechanized mining face. *Shenhua Sci. Technol.* **2018**, *16*, 26–29.
31. Wang, F. Research off precise pushing control scheme for hydraulic support and its application. *Ind. Mine Autom.* **2017**, *43*, 6–9.
32. Li, S.; Li, S.; Zhang, S.; Wang, F. Key technology and application of intelligent perception and intelligent control in fully mechanized mining face. *Coal Sci. Technol.* **2021**, *49*, 28–39.
33. Che, P.; Wu, Y.; Zhao, Y.; Zhao, Y. Pressure Loss Analysis of hydraulic support different feed liquid manner. *Hydraul. Pneum. Seals* **2013**, *33*, 69–72.

machines

Article

Effect of Nozzle Outlet Shape on Cavitation Behavior of Submerged High-Pressure Jet

Gaowei Wang [1], Yongfei Yang [1,*], Chuan Wang [2,3], Weidong Shi [1,*], Wei Li [4] and Bo Pan [5]

[1] College of Mechanical Engineering, Nantong University, Nantong 226019, China; wgwntu@outlook.com
[2] College of Mechanical and Electrical Engineering, Hainan Vocational University of Science and Technology, Haikou 571126, China; wangchuan198710@126.com
[3] College of Hydraulic Science and Engineering, Yangzhou University, Yangzhou 225009, China
[4] Research Center of Fluid Machinery Engineering and Technology, Jiangsu University, Zhenjiang 212013, China; lwjiangda@ujs.edu.cn
[5] Shandong Xinchuan Mine Electrical and Mechanical Equipment Co., Ltd., Jining 272000, China; panbo205@outlook.com
* Correspondence: yyf2020@ntu.edu.cn (Y.Y.); wdshi@ujs.edu.cn (W.S.)

Abstract: A submerged high-pressure water jet is usually accompanied by severe cavitation phenomenon. An organ pipe nozzle can greatly improve the cavitation performance of the jet, making use of the self-excited oscillation of the flow. In order to study the effect of organ pipe nozzles of different nozzle outlet shapes on cavitation behavior of submerged high-pressure jet, in this paper we build a high-pressure cavitation jet experiment system and carried out a high-speed photography experiment to study cavitation cloud characteristics of a high-pressure submerged jet. Two organ pipe nozzles with and without a whistle were compared. The dynamic characteristics of the cavitation cloud was extracted through the POD method, it was found that the result effectively reflect the dynamic characteristics of the cavitation jet. The reconstruction coefficients of mode-1 obtained by the POD can better reflect the periodic time-frequency characteristics of cavitation development. The effect of the nozzle outlet shape on the cavitation behavior of organ pipe nozzle was analyzed based on unsteady numerical simulation, and it was found that the jet generated by the nozzle with a divergent whistle had a larger vorticity in the shear layer near the outlet. Further, stronger small-scale vortex and much severe cavitation occurred from the nozzle with a divergent whistle.

Keywords: high-pressure jet; high-speed photography; cavitation; nozzle; POD

Citation: Wang, G.; Yang, Y.; Wang, C.; Shi, W.; Li, W.; Pan, B. Effect of Nozzle Outlet Shape on Cavitation Behavior of Submerged High-Pressure Jet. *Machines* **2022**, *10*, 4. https://doi.org/10.3390/machines10010004

Academic Editor: Davide Astolfi

Received: 8 November 2021
Accepted: 16 December 2021
Published: 21 December 2021

1. Introduction

Cavitation is a phase transformation phenomenon, in which the liquid pressure is reduced to the vapor pressure and the liquid phase is transformed into the gas phase, which is a harmful and unavoidable phenomenon for hydraulic machinery [1,2]. When bubbles collapse, it generates a shock wave and a micro-jet. If they act on the surface of the turbine or propeller blade, cavitation damage inevitably occurs. Therefore, cavitation elimination is an important field of fluid dynamics research [3–5]. However, when the high-energy impact load generated by the cavitation collapse is properly utilized, the impact generated by the cavitation effect can be used to improve the efficiency of high-pressure water jet cleaning, drilling and rock crushing, as well as the surface shot peening process of metal materials, which can effectively improve the fatigue strength of metal materials [6,7]. The submerged cavitating water jet is generated by the fluid movement of the cavitating cloud. The cavitation cloud in the jet process goes through different stages as a fluid movement, such as growth, shedding and collapse. Due to the destruction of the solid surface, the cavitation cloud destroys the local energy, further enhancing the cleaning and cutting effect of the jet near the target surface [8–11].

In order to master the flow characteristics of cavitation jet and the impact mechanism during cavitation formation and collapse, and improve the impact effect of a high-pressure cavitation jet, many researchers have studied the velocity field, pressure distribution and cavitation shape of a cavitation jet. Yang Y et al. [12–14] captured the unsteady flow characteristics of a high-pressure cavitation jet at three different angles through high-speed photography experiments, and extracted its growth, shedding and collapse information from high-speed images. Ryuta Watanabe et al. [15,16] applied the POD image processing technology to the high-speed image analysis of a cavitation jet, and analyzed the position of the cavitation bubble collapse through an algorithm. Peng C et al. [17] analyzed the time-frequency distribution of cavitation clouds by using the orthogonal decomposition (POD) method, and concluded that the intensity of the cavitation erosion was determined by the bubble concentration and rupture strength on the sample surface. Shridhar et al. [18] used PIV to test the cavitation bubble in the near flow field of a jet. The results showed that cavitation first appeared in the vortex ring. The formation probability of cavitation bubble was predicted by the distribution law, intensity and strain of the cavitation gas core. Nakano et al. [19] observed the evolution of cavitation near the initial submerged jet nozzle by photography and found that in the early stage, most of the cavitation bubbles were in the initial vortex and connected with each other into a vortex ring. The cavitation in the cavitation ring grew rapidly near the nozzle and decreased gradually. Paul McGinn et al. [20] studied the role of the pressure wave in the flow development and found that the dynamics of the pressure wave determined the dynamic characteristics of the bubble cloud under the condition of "choking". Ruolong Ma et al. [21] proposed a model that provides a prediction of the resonator pressure fluctuations based on the thickness of the approach boundary layer, the free stream speed and the acoustic properties of the resonator.

The nozzle is the core part of a cavitation jet and plays an important role in the cavitation performance of the jet. The jet generated by the self-excited oscillating nozzle has large structure vortex ring and a high-intensity pressure oscillation, which greatly improves the energy utilization rate of the water jet. An organ pipe nozzle is the most common oscillating nozzle [22–25]. The basic principle of cavitation is that when the jet flows through the contracted section, a feedback pressure pulsation is generated in the resonant cavity. When the feedback pressure and the inlet fluid generate standing waves, an acoustic harmonic resonance is formed, so as to generate a whirlpool vortex ring in the continuous jet flow at the nozzle exit, making the jet flow become intermittent vortex circulation [26]. Keiichi Sato et al. [27] carried out a high-speed photographic study on the cavitation jet of a contraction expansion nozzle, and also obtained the low-frequency and high-frequency signals corresponding to the pressure fluctuation of the piston pump and the shedding of the cavitation group through the statistics of the change in cavity length over time. Peng G et al. [28–30] used acrylic resin to make a transparent contraction nozzle, took a high-speed video of the cavitation distribution inside the nozzle and at the exit, and extracted the gray value of the image to obtain high-frequency and low-frequency signals respectively. By comparison, it was found that the low-frequency signal was consistent with the pulsation of the piston pump, while the high-frequency signal corresponded to the shedding frequency of the cavitation mass. Li G et al. [31] optimized the cavity of the high-pressure nozzle, and proposed a method to improve the cavity and impact performance by interaction between the fluid flow and the nozzle cavity, which improved the gravel efficiency of the jet. Jiang Y et al. [32–34] applied a CFD method with an SST turbulence model to model the sonic hydrogen jet at supersonic crossflow, the obtained results indicated that the existence of the divergent ramp decreased the jet interactions and consequently this improved the mixing rate downstream of the jet.

Although the research on water jet nozzle is relatively mature, there are few studies on the influence of an organ pipe nozzle with different nozzle outlet shape on the cavitation behavior of a submerged high-pressure jet. Therefore, a visualized experimental system of a high-pressure water cavitation jet was built in this paper. The unsteady characteristics of the cavitation clouds of an organ pipe nozzle with different outlet shapes were studied based on high-speed photography. The dynamic characteristics of the cavitation bubble cloud were extracted by the proper orthogonal decomposition method, and the influence of nozzle outlet shapes on the cavitation performance of the organ pipe was analyzed based on an unsteady numerical simulation.

2. Experimental Methods

2.1. High-Pressure Cavitation Jet Experimental System

Figure 1 shows the high-pressure cavitation jet experiment system, which was mentioned in our former works [12]. An Italian AR high-pressure plunger pump was used to provide pressure for the jet flow. The maximum working pressure of the selected plunger pump was 50 MPa, the rated speed was 1450 r/min, and its flow rate was 15 L/min. The experiment used clear water as the medium, and the experiment temperature was 25 °C. The plunger pump was powered by a frequency conversion motor and controlled by a frequency converter to control the speed of the pump. The pressure in the pipeline upstream of the nozzle was controlled by the change of speed.

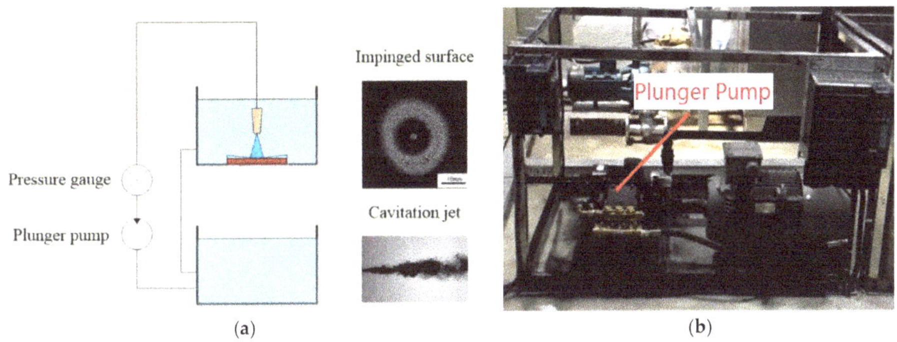

Figure 1. Test platform for high-pressure submerged jet. (**a**) Cavitation jet system. (**b**) Physical photographs of the test stand.

In the experiment and numerical study of this paper, organ pipe nozzles with different outlet shapes are mainly involved. The main geometric parameters of organ pipe nozzles include: the resonant cavity length L_1, throat tube length L_2, resonant cavity diameter D and throat tube diameter d, as well as the outlet shape. The geometry of the nozzle is shown in Figure 2. The design parameters are shown in Table 1.

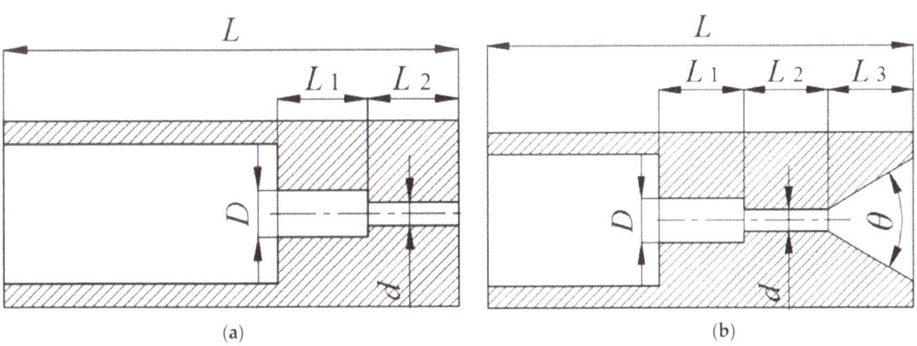

Figure 2. Geometry structure of the nozzles used in experiment. (**a**) Organ pipe nozzle (without whistle). (**b**) Organ pipe nozzle (divergent whistle).

Table 1. The design parameters of the nozzles.

L_1 (mm)	L_2 (mm)	d (mm)	L_3 (mm)	θ (°)
5	4	1	4	60

2.2. Experimental Methods of High-Speed Photography

The Ispeed high-speed camera produced by Olympus, Japan, was used to photograph the cavitation cloud of a submerged jet. The high-speed camera system is mainly composed of a high-speed camera, LED light source, mobile display panel, power supply and memory card. During the shooting process, the lens was directed at the axis of the nozzle, and the LED light source was placed on the other side of the nozzle to illuminate the lens. Frosted glass was used between the light source and the nozzle to prevent the lens from over-exposing. After adjusting the position of the camera and nozzle, we first placed a ruler with a graduated scale in the plane perpendicular to the lens axis and across the nozzle axis, and recorded it for calibrating the size of the jet cavitation. The maximum acquisition frequency of the high-speed camera was 150,000 Hz, and the highest resolution of the a picture was 1280 × 1024 pixels. The acquisition frequency set in the experiment was 20,000 Hz, and the acquisition time was 1 s for each working condition. After the shooting, the corresponding photos were exported by Ispeed high-speed camera's matching software and read into MATLAB for image processing.

2.3. POD of Cavitation Cloud Images

The proper orthogonal decomposition (POD) was first proposed by D.D. Kasambi [35], and then was widely used in the analysis of multivariable and nonlinear phenomena. The POD is a statistical technique for obtaining time-independent orthogonal basis functions (POD modes) for a field function $u(x,t)$ that can optimally reduce a given field function $u(x,t)$. For the eigenfunction φ, the average projection of $u(x,t)$ on ψ should be the maximum, and its expression is:

$$\max_{\varphi} \frac{\langle |(u,\varphi)|^2 \rangle}{||\varphi||^2} \tag{1}$$

The problem is transformed into the maximum solution of $\langle |(u,\varphi)|^2 \rangle$ under the constraint of $||\varphi||^2 = 1$ by using the variation method:

$$J[\varphi] = \langle ||(u,\varphi)||^2 \rangle - \lambda(||\varphi||^2 - 1) \tag{2}$$

For all variables $\varphi + \delta\psi$ should satisfy:

$$\frac{d}{d\delta} J\,[\varphi + \delta\psi]|_{\delta=0} = 0 \tag{3}$$

This condition can be simplified into the integral form equation:

$$\int_0^1 \langle u(x)u*(x') \rangle \varphi(x')dx' = \lambda\varphi(x) \tag{4}$$

In order to obtain the optimal basis of the eigenfunction from Equation (4), the average autocorrelation function is adopted:

$$R(x,x') = \langle u(x)u*(x') \rangle \tag{5}$$

Since the average autocorrelation function is non-negative, the sequence of eigenvalues is as follows:

$$\lambda_j \geq \lambda_{j+1} \geq 0 \tag{6}$$

Finally, the field quantity in $u(x,t)$ can be reconstructed from the decomposition mode based on the characteristic function φ_j:

$$u(x) = \sum_{j=1}^{\infty} a_j \varphi_j(x) \tag{7}$$

a_j is the reconstruction coefficient, when the flow field analyzed is a velocity field, the eigenvalue represents twice the average turbulent kinetic energy in each mode, and the first-order mode represents the most violent structure in the flow. In this paper, POD method was used to decompose the modulus of the high-speed photography image of a submerged high pressure water cavitation jet, and the evolution law of cavitation cloud in the jet flow field was analyzed. For the high-speed photographic image of a cavitation jet, $u(x)$ represents the gray value of the cavitation cloud image, while λ_j represents the weight of the corresponding modes in the reconstruction process.

3. Numerical Calculation Method

3.1. Governing Equation

3.1.1. Multiphase Flow Model

A cavitation jet belongs to a gas-liquid two-phase flow, and Euler's model or Lagrange's model can be selected according to needs when calculating its flow field. In this paper, the mixture model was used to calculate the mixture phase flow field, and its governing equation is as follows [36,37]:

$$\frac{\partial}{\partial t}(\rho_m) + \nabla \cdot \left(\rho_m \vec{v}_m\right) = 0 \tag{8}$$

$$\begin{gathered} \frac{\partial}{\partial t}\left(\rho_m \vec{v}_m\right) + \nabla \cdot \left(\rho_m \vec{v}_m \vec{v}_m\right) = -\nabla p + \nabla \cdot \left[\mu_m \left(\nabla \vec{v}_m + \vec{v}_m^{\,T}\right)\right] \\ + \rho_m \vec{g} + \vec{F} + \nabla \cdot \left(\sum_{k=1}^{n} \alpha_k \rho_k \vec{v}_{dr,k} \vec{v}_{dr,k}\right) \end{gathered} \tag{9}$$

where $\vec{v}_m$ is the average mass velocity, ρ_m is the density of the mixed phase:

$$\rho_m = \sum_{k=1}^{n} \alpha_k \rho_k \tag{10}$$

μ_m is the viscosity of the mixed phase, which is defined as follows:

$$\mu_m = \sum_{k=1}^{n} \alpha_k \mu_k \tag{11}$$

where n is the number of phases, $\vec{F}$ is the volume force, and $\vec{v}_{\mathrm{dr},k}$ represents the slip velocity of sub-phase k.

3.1.2. Cavitation Model

When the mixture model is used to calculate the cavitation multiphase flow, the gas phase volume fraction transport equation is expressed as follows:

$$\frac{\partial}{\partial t}(\alpha \rho_v) + \nabla \cdot \left(\alpha \rho_v \vec{v}_v\right) = R_e - R_c \tag{12}$$

where R_e and R_c are evaporation and condensation rates, and their values can be calculated according to the cavitation model.

Currently, the commonly used cavitation model is mainly derived from the Rayleigh-Plesset equation. The cavitation model used in this paper is the Zwart-Gerber-Belamri model, and the mass transfer expression of the model is as follows [38]:

While $P \leq P_v$,

$$R_e = F_{\text{vap}} \frac{3\alpha_{\text{nuc}}(1-\alpha_v)\rho_v}{R_B} \sqrt{\frac{2}{3} \frac{P_v - P}{\rho_l}} \tag{13}$$

While $P > P_v$,

$$R_c = F_{\text{cond}} \frac{3\alpha_v \rho_v}{R_B} \sqrt{\frac{2}{3} \frac{P - P_v}{\rho_l}} \tag{14}$$

R_B is the cavity radius, α_{nuc} is the volume fraction of the gas core in the liquid, F_{vap} is the evaporation coefficient, and F_{cond} is the condensation coefficient. The default parameters of Fluent are set as $R_B = 10^{-6}$ m, $\alpha_{\text{nuc}} = 5 \times 10^{-4}$, $F_{\text{vap}} = 50$, $F_{\text{cond}} = 0.01$. Considering the impact of turbulence on cavitation, the pressure threshold of cavitation occurrence is expressed as follows:

$$P_v = P_{\text{sat}} + \frac{1}{2} c \rho_l k_l \tag{15}$$

where ρ_l and k_l represent liquid density and liquid turbulent kinetic energy respectively, and coefficient c adopts the recommended value of Fluent 0.39.

3.1.3. Turbulence Model

The expression of turbulent kinetic energy dissipation term of DES model used in this paper is [39]:

$$Y_k = \rho \beta^* k \omega F_{\text{DES}} \tag{16}$$

$$F_{\text{DES}} = \tan\text{h}\left[(C_{\text{d}1} r_d)^{C_{\text{d}2}}\right] \tag{17}$$

$$r_d = \frac{\nu_t + \nu}{k^2 y^2 \sqrt{0.5\left(S^2 + \Omega^2\right)}} \tag{18}$$

where, $C_{\text{d}1} = 20$, $C_{\text{d}2} = 3$, S is the strain tensor, Ω is the vorticity tensor, and $k = 0.41$.

3.2. Meshing

Figure 3 shows the calculation domain of the submerged jet. The water flow must be fully developed in the pipe before reaching the nozzle, so the pipe at the nozzle inlet extends 260 mm upstream. According to the requirements of the literature, the diameter of the calculation domain should be greater than 100 times the diameter of the nozzle throat. In this paper, a cylinder with a diameter of 200 mm and a length of 500 mm was used as the submerged water calculation domain, and the size of the calculation domain met the above requirements.

The calculation area was divided into structured grids by ANSYS ICEM, and the shear layer near the jet core area and the nozzle outlet was densified. The calculation domain grid is shown in Figure 4. In order to ensure the calculation effect and reduce the amount of calculation, the distance between the grid node and the nozzle outlet was determined as the center to the downstream section and a smooth transition boundary gradually increased. The total number of grids in the final calculation area exceeded 19 million, and the grid size at the shear layer at the nozzle outlet was controlled at about 0.01 mm.

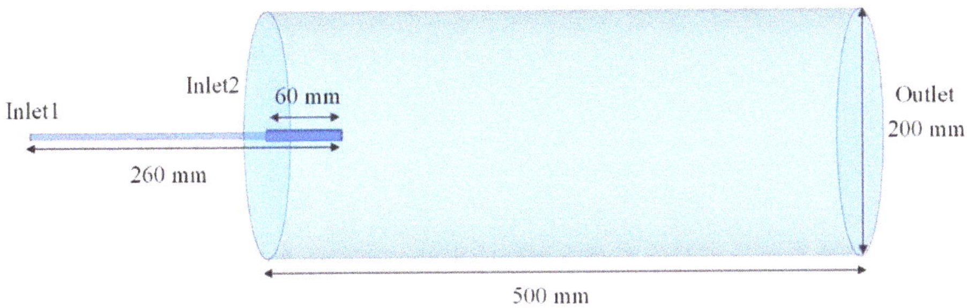

Figure 3. Calculation domain of the submerged jet.

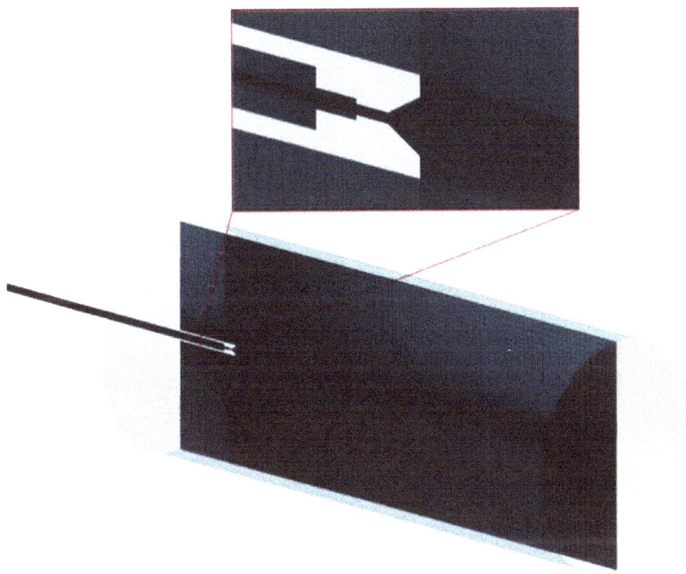

Figure 4. Mesh of calculation domain.

4. Results and Discussion

4.1. Analysis of POD Results

In order to test the effect of the POD on extracting the characteristics of the cavitating jet, a group of images of the cavitation flow field of a high-pressure water submerged jet were selected for analysis. The nozzle used in the experiment was an organ pipe nozzle with a divergent whistle, and the experimental pressure was 18 MPa. Figure 5 shows the cavitation development image within 1200 μs, and the field of view is 48 mm from the nozzle outlet to the downstream section. It can be seen from the image that the development of the cavitation cloud is periodic. At 0 μs, within a certain distance from the nozzle outlet to the downstream section, the cavitation cloud is basically continuously distributed, and the width of the cavitation cloud at the nozzle outlet is small. In the process of downstream movement, it gradually widens under the action of the flow field diffusion. There is a fracture in the cavitation cloud at $x = 30$ mm, which is the shedding phenomenon of the cavitation jet. After the shedding, the cavitation cloud in the downstream section begins to collapse, while the cavitation cloud in the upstream section continues to grow until the next shedding occurs. At 200 μs, the cavitation cloud expands near the nozzle outlet, and the growth process of the next development cycle begins. At 600 μs, the width of the cavitation cloud at the nozzle outlet reaches the peak, and then decreases and disappears completely at 950 μs. In this process, the growth and collapse process of the downstream cavitation cloud is consistent with the development trend near the nozzle outlet, and also reaches the maximum size at 600 μs. During the experiment, the shooting frequency of high-speed photography was 20,000 fps, so it can be concluded that the development cycle of the cavitation group growth, shedding and collapse of the nozzle under a pressure of 18 MPa is about 1000 Hz, but the stability of the cycle cannot be determined according to the image of a single cycle.

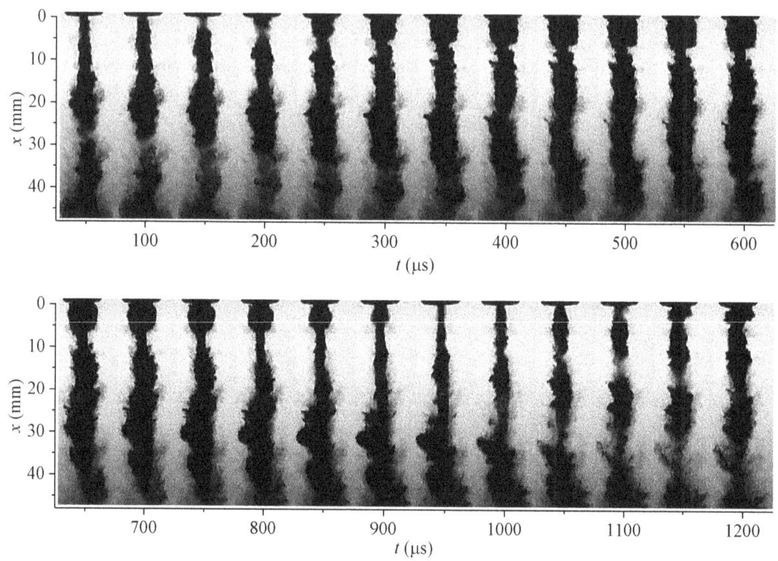

Figure 5. Cavitation cloud image of organ pipe nozzle with divergent whistle.

A total of 5000 high-speed photographic images of a continuous cavitation jet were selected for the POD. Figure 6 shows the zeroth- to seventh-order modes of the POD in the jet cavitation region, where the M0 corresponding to φ_1 represents the average change of the gray level of the cavitation images at each position in the analyzed time period. It can be seen from the M1 image that the state corresponding to the first-order mode is the growth stage of the cavitation. At this time, spherical cavitation groups are generated at the exit of the nozzle, while the cavitation from the outlet of the nozzle to the downstream 20 mm segment is in the growth stage. M2 and M1 have the same frequency and the two modes appear alternately. Combined with the M2 modal diagram, it can be seen that under this mode, the cavitation cloud falls off and the nozzle outlet segment begins to collapse. The subsequent modes of M3 have a relatively low percentage of energy, which corresponds to the expansion and contraction of the boundary layer and the fluctuation caused by the turbulence in the evolution cycle of the cavitation cluster.

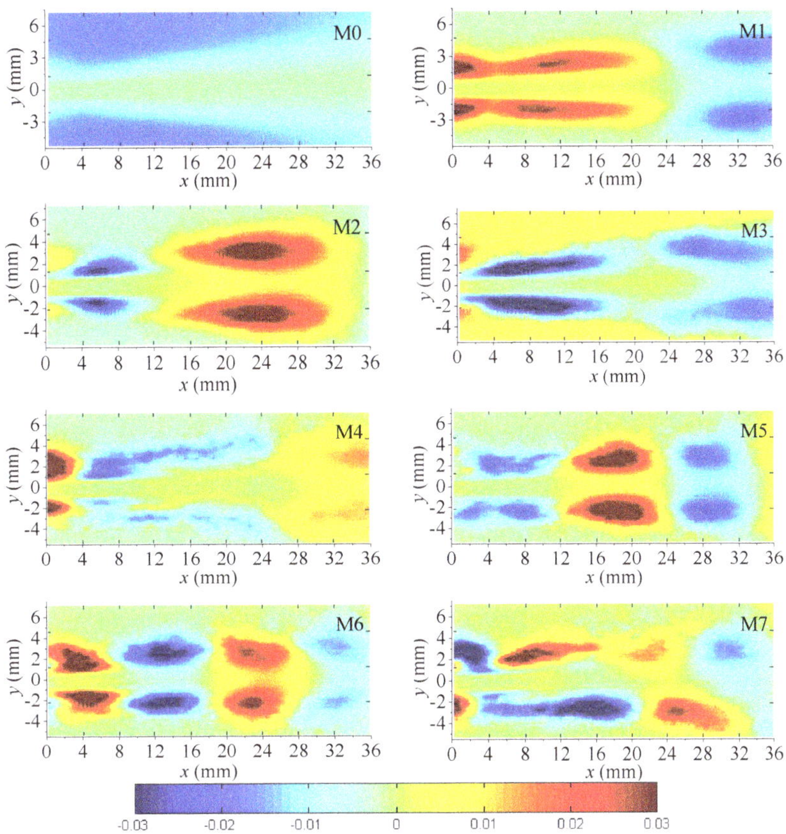

Figure 6. Spatial distribution of mode_0 to mode_7 of cavitation bubbles.

Figure 7 shows the time-domain and frequent-domain curves of the first-order to fourth-order modes. It can be seen from the figure that the time-domain waveforms of the first-order and second-order modes are consistent, with relatively stable periodic characteristics. Comparing with the corresponding frequent-domain diagram, it can be found that the frequency domain distribution of this mode is relatively concentrated at 1154 Hz, with an obvious peak value. According to Figure 6, the evolution period of the jet cavitation development, shedding and collapse is about 1000 μs under this condition, which is consistent with the first-order modal period. It can be seen that the reconstruction coefficient corresponding to the first-order mode obtained by the POD can better reflect the periodic time-frequency characteristics of cavitation development.

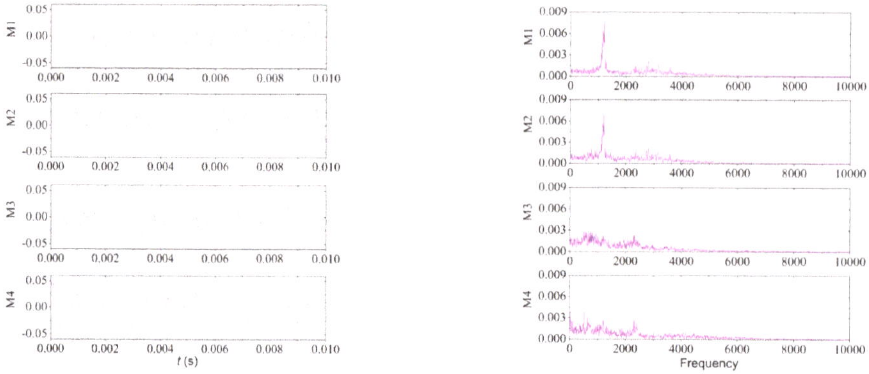

(a) (b)

Figure 7. Time coefficients of the modes in time and frequency domains. (**a**) The first four modulus time coefficients. (**b**) Frequency-domain plots of time coefficients.

4.2. Cavitation Cloud Characteristics

According to existing studies, the exit shape of the cavitation nozzle has a great influence on its cavitation performance. In this paper, a comparative study was conducted on organ pipe nozzles without a whistle and with divergent whistle nozzles. High-speed photography was taken for the two nozzles under a pressure of 20 Mpa, and their cavitation cloud morphology is shown in Figure 8. For the nozzle without a whistle, there is a long and thin columnar cavity near the outlet, which expands and contracts periodically with the passage of time, indicating that the growth of the cavitation has a fast and slow alternating process, and the period and intensity of the characteristics of the expansion region are related to the self-excited oscillation characteristics of the organ nozzle. When the outlet segment shrinks to the minimum size, shedding occurs at about 25 mm downstream of the nozzle, and then the downstream cavitation group gradually diffuses into discrete cavitation groups and collapses and disappears, with the main collapse location between 60 mm and 80 mm downstream of the nozzle outlet. The vortex shedding and merging phenomenon at the outlet of the organ tube are enhanced when the divergent whistle is added to the organ pipe nozzle, and the vortex shedding phenomenon is aggravated. The development period of the cavitation cloud is slightly longer than that of the nozzle without a whistle, but the length of the cavity expansion segment is far longer, and the boundary of the cavity is always surrounded by discrete cavitation. At this time, the falling cavitation cloud has not completely collapsed 120 mm downstream of the nozzle. It can be seen that the divergent nozzle structure has a significant effect on enhancing the cavitation performance of the high-pressure water submerged jet nozzle.

According to the above analysis, the time-domain characteristics of the reconstruction coefficients corresponding to the first-order modes correspond to the evolution process of periodic growth, shedding and collapse of the cavitation cloud. Therefore, the shedding period of jet cavities can be reflected by the time-frequency characteristics of the first-order reconstruction coefficients. Figure 9 shows the time and frequency-domain plots of the cavitation cloud by nozzles with different outlet shapes. According to the distribution in the time domain, the waveform of the organ pipe nozzle without a whistle has a certain periodicity, but the amplitude and periodicity of fluctuation are not strong, while the time domain waveform of the cavitation generated by the organ pipe nozzle with a whistle presents a regular periodic fluctuation. According to the frequency domain diagram, in the case without a whistle, the frequency waveform distribution is scattered, the main frequency is about 1000 Hz, and there are multiple peaks. In the case with a divergent whistle, the frequency distribution is concentrated between 1000 Hz and 1250 Hz, and the dominant frequency position is obvious, located at 1143 Hz. It can be seen that the whistle of a cavitation nozzle has a significant influence on the growth cycle and spatial distribution of the cavitation. The optimization of this parameter is of great significance to the design of cavitation nozzles.

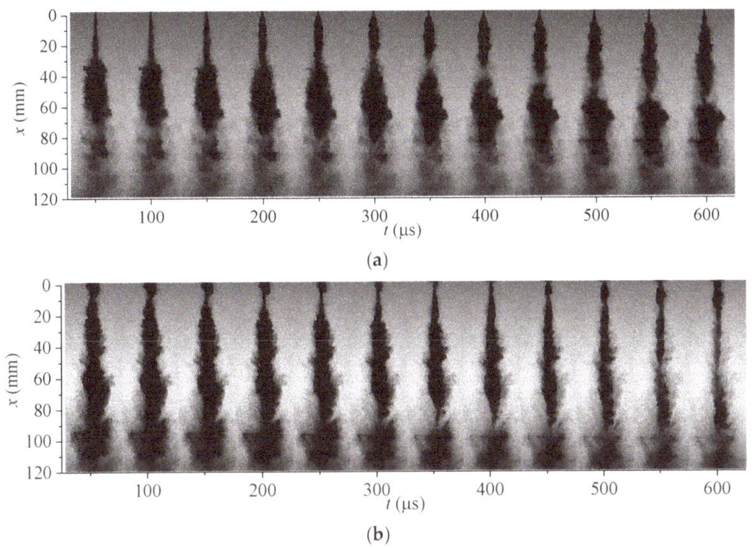

Figure 8. Cavitation cloud morphology of two nozzles. (**a**) Organ pipe nozzle without whistle. (**b**) Organ pipe nozzle with divergent whistle.

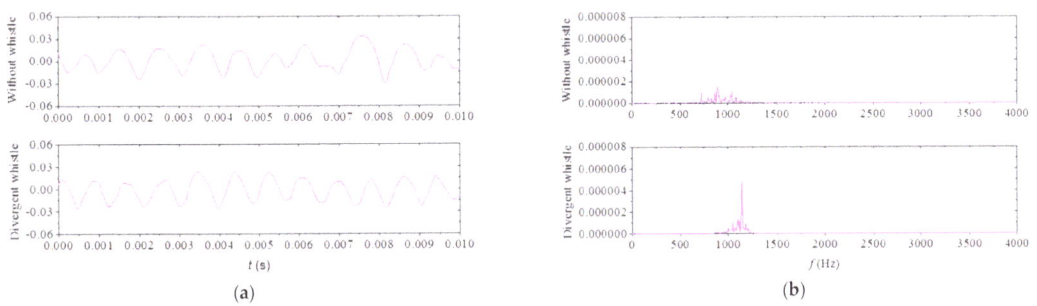

Figure 9. Time- and frequency-domain plots of cavitation cloud by nozzles with different outlet shapes. (**a**) Time-domain plot. (**b**) Frequency-domain plot.

Figures 10 and 11 respectively show the time-domain and frequency-domain distribution of the first-order modes of the two nozzle cavitation jet images under different pressures. The main frequency position at different pressures is shown in Table 2. According to the time-domain waveform, it's period becomes longer and the periodicity is more obvious with the increase of pressure. As can be seen from the frequency domain, when the pressure is 2 MPa, the main frequency of the nozzle without a whistle and that of the divergent whistle nozzle are not particularly obvious, and the waveform distribution is scattered. When the pressure rises to 6 MPa, the main frequency of the divergent whistle nozzle becomes more pronounced than that of the nozzle without a whistle, with the main frequency position at around 1800 Hz. When the pressure rises to 10 MPa, the amplitude of the two nozzles is greatest and the frequency domain peak becomes concentrated. When the pressure is further increased, the frequency domain peaks of the two nozzles are more concentrated, while the nozzles without a whistle have multiple peaks, the frequency distribution of the divergent whistle nozzle is basically around the main frequency, and the main frequency position is around 1100 Hz, indicating that the development of the cavity has a significant periodic process of growth, shedding and collapse.

Table 2. The main frequency position at different pressures.

Pressure (MPa)	2	6	10	14	18	20
Frequency without whistle (Hz)	4000	2250	1500	1250	1100	1000
Frequency with divergent whistle (Hz)	2600	1800	1400	1300	1200	1100

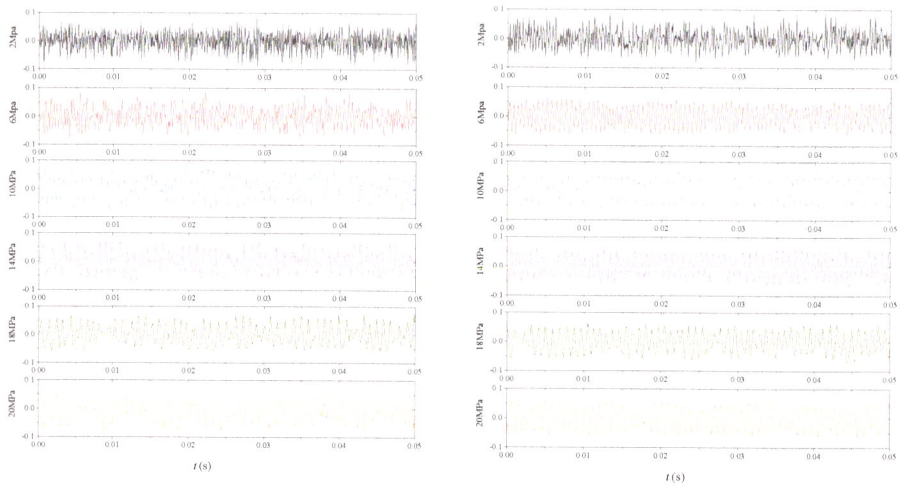

(a) (b)

Figure 10. Time-domain distribution for mode_1 of the cavitating jet images. (**a**) Cavitating jet by nozzle without whistle. (**b**) Cavitating jet by nozzle with divergent whistle.

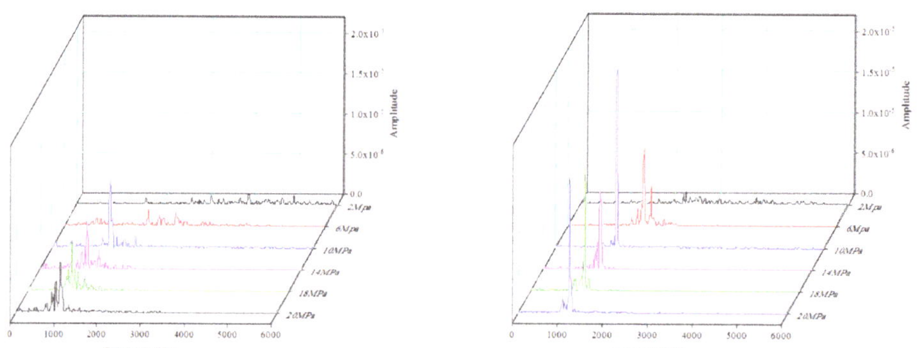

(a) (b)

Figure 11. Frequency-domain distribution for mode_1 of the cavitating jet images. (**a**) Cavitating jet by nozzle without whistle. (**b**) Cavitating jet by nozzle with divergent whistle.

4.3. Numerical Calculation of Vorticity Field

In order to directly reflect the influence of different outlet shapes on the jet flow field. Figure 12 shows the three-dimensional shape of vorticity near the outlet of two nozzles. It can be seen from the comparison that the jet vorticity of two nozzles is significantly different. Near the nozzle outlet, the vorticity iso-surface of the nozzle without a whistle is a relatively smooth cylindrical structure and there are relatively weak fluctuations on the surface, corresponding to the transition process of the outflow. The divergent whistle nozzle jet has a complete turbulent structure at the outlet. The vortex scale near the outlet is small, and the small-scale vortex continuously entrains the surrounding fluid and merges with it to form a larger-scale vortex in the process of moving downstream. On the whole, by comparing the vorticity of the two nozzle jets, it can be seen that the vorticity of the nozzle jet without a whistle is a large-scale vortex, which mainly occurs after the instability of the jet core area, while the vorticity of the divergent nozzle includes a variety of forms and scales, in which the small-scale vortex is formed by shearing and shedding in the whistle. Larger-scale vortices are generated in the entrainment process of mainstream and ambient fluids.

In order to further reveal the interaction between the vortex and the cavitation in the cavitation jet, the vorticity transport equation was introduced for the analysis, and its expression is as follows:

$$\frac{D\vec{\omega}}{Dt} = \left(\vec{\omega}\cdot\nabla\right)\vec{v} - \vec{\omega}\left(\nabla\cdot\vec{v}\right) + \frac{\nabla\rho_m \times \nabla p}{\rho_m^2} + (v_m + v_t)\nabla^2\vec{\omega} \tag{19}$$

where, the left side of the equation represents the change rate of the vorticity tensor with time, and the right side of the equation is composed of four terms in total, among which the first term represents the vortex extension caused by the velocity gradient; the second term is the vortex expansion caused by volume expansion and contraction, representing the effect of the compressibility of the fluid on vorticity; the third term is the baroclinic term caused by pressure and density gradients; the fourth term represents the vorticity change caused by viscous dissipation.

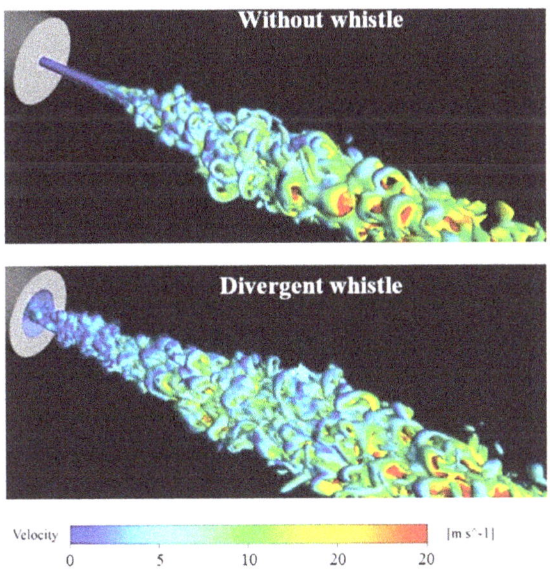

Figure 12. Vorticity iso-surface of the jet with different nozzles.

Figure 13a shows the volume fraction distribution of the jet cavitation of two kinds of organ pipe nozzles. By observing the cavitation morphology at the nozzle outlet, it can be found that the organ pipe nozzle without a whistle has a smooth tubular cavitation layer near the throat outlet, in which the cavitation begins to diffuse at the position $x = 5$ mm downstream of the throat outlet, and then the diameter of the cavitation cloud increases rapidly and presents a vortex structure. The cavitation generated by the nozzle with a divergent nozzle begins to diffuse at the outlet of the throat tube. Although the diffusion at the outlet is slow, the thickness of the cavitation layer is obviously larger than that of the nozzle without whistle. The shape and position of the cavitation at the throat of each nozzle are almost the same, which are caused by the separation vortex formed by the sudden change of the size of the resonant cavity and the throat inlet, and the cavitation volume fraction gradually decreases from the throat inlet to the downstream section. The part that does not completely disappear when reaching the outlet of the throat forms the tubular vacuole layer described above.

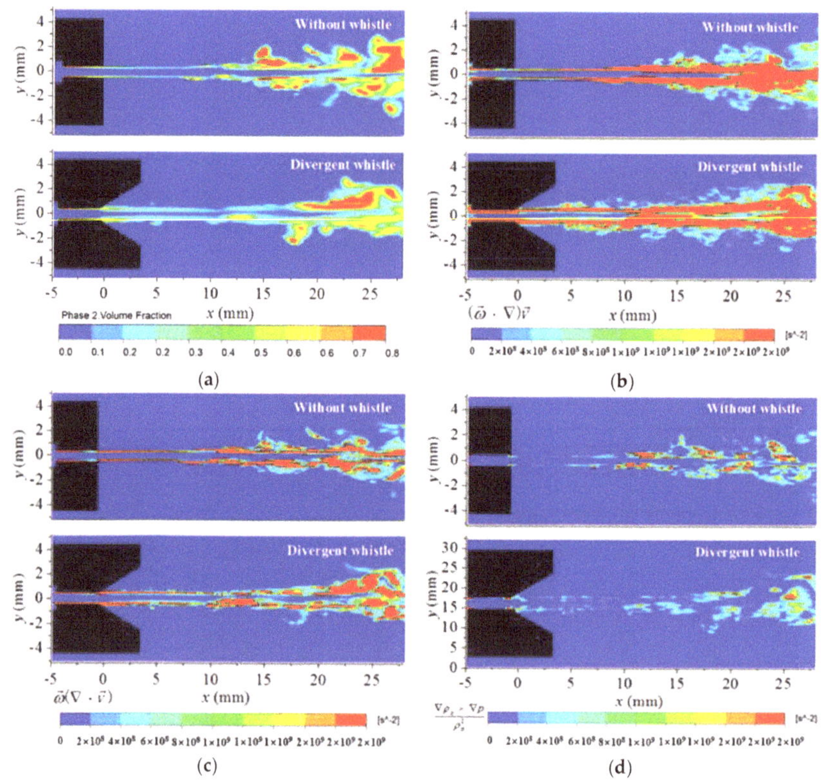

Figure 13. Different turns of vorticity transport equations of different nozzles. (**a**) Cavity volume fraction. (**b**) Vorticity extension term. (**c**) Vorticity expansion term. (**d**) Vorticity baroclinic term.

According to the distribution of items in the vorticity transport equation, the region with a higher distribution of baroclinic terms corresponds to the region with a higher volume fraction of cavitation, and the expansion term of vorticity corresponds to the shedding of cavitation, while the main generation of vorticity in the cavitation jet is reflected in the extension term of vorticity. It can be found from Figure 13b that the area with a high vortex extension term mainly starts at the nozzle outlet about 10 mm downstream. This is because although the jet velocity at the nozzle outlet is high and there is a large velocity gradient in the shear layer, the thickness of the shear layer here is thin and the diameter of the core jet is small. On the whole, the vorticity extension term is not directly related to the spatial distribution of the volume fraction, but the size of the extension term reflects the speed of the local vortex growth, while for the cavitation jet, the vortex distribution is closely related to the development of cavitation. By comparing the distribution of the vorticity extension term values of the two nozzles near the nozzle outlet, it can be found that the nozzle without a whistle has a relatively smooth tubular distribution at the outlet, and the vorticity increment at this position mainly comes from the inner wall boundary layer of the nozzle. The increment of vorticity at the nozzle outlet of the nozzle with a divergent whistle is significantly higher than that of the nozzle without a whistle. In addition to the vorticity transmitted by convection inside the nozzle, the vorticity also increases at the periphery of the shear layer, indicating that the divergent whistle can promote the generation of vorticity near the nozzle outlet.

5. Conclusions

In this paper, a visualized experimental system of a high-pressure water cavitation jet was built. The unsteady characteristics of cavitation clouds of an organ pipe nozzle with different outlet shapes were studied based on high-speed photography. The dynamic characteristics of the cavitation cloud were extracted by the proper orthogonal decomposition method (POD), and the influence of nozzle outlet shapes on the cavitation performance of the organ pipe was analyzed based on an unsteady numerical simulation. From the analysis of the results, the following conclusions can be drawn:

(1) The proper orthogonal decomposition method (POD) can effectively reflect the dynamic characteristics of the cavitation jet. The reconstruction coefficients of mode-1 obtained by the POD can better reflect the periodic time-frequency characteristics of cavitation development.
(2) The shape of the nozzle outlet has a great influence on the dynamic characteristics of cavitation. When the nozzle outlet has a divergent segment, the periodic oscillation characteristics are more obvious, the frequency domain distribution is more concentrated, and the cavitation near the nozzle outlet is more intense.
(3) The jet generated by the nozzle with a divergent whistle has a larger vorticity in the shear layer near the outlet. Further, a stronger small-scale vortex and more severe cavitation occurs from the nozzle with a divergent whistle.
(4) The region with a higher distribution of baroclinic terms corresponds to the region with a higher volume fraction of cavitation. The expansion term of vorticity corresponds to the shedding of cavitation, while the main generation of vorticity in the cavitation jet is reflected in the extension term of vorticity.

More experimental and numerical studies will be required in the future, such as using particle image velocimetry (PIV) to visualize the flow field characteristics of the jet and reveal the interference of the vortex with the jet, or study how flares affect the oscillation mechanism and improve the nozzle design theory.

Author Contributions: Conceptualization, G.W., Y.Y. and W.S.; Formal analysis, G.W., Y.Y. and C.W.; Investigation, B.P.; Data curation, W.L.; Writing—original draft preparation, G.W., Y.Y. and C.W.; Writing—review and editing, W.S. and W.L.; Visualization, B.P.; Supervision, Y.Y., W.S. and W.L.; Project administration, W.S. and W.L. All authors have read and agreed to the published version of the manuscript.

Funding: This research was funded by the National Key Research and Development Project of China (no. 2019YFB 2005300), National High-Tech Ship Scientific Research Project of China (no. MIIT [2019] 360), National Natural Science Foundation of China (no. 51979138), National Natural Science Foundation of China (No. 51979240), Jiangsu Natural Science Re-search Project (no. 19KJB470029), Jiangsu Water Conservancy Science and Technology Project (no. 2019038), and Nantong Science and Technology Project (no. JC2019155).

Institutional Review Board Statement: Not applicable.

Informed Consent Statement: Not applicable.

Data Availability Statement: Data on the analysis and reporting results during the study can be obtained by contacting the authors.

Conflicts of Interest: The authors declare no conflict of interest.

References

1. Soyama, H. Cavitating Jet: A Review. *Appl. Sci.* **2020**, *10*, 7280. [CrossRef]
2. Soyama, H. Cavitation Peening: A Review. *Metals* **2020**, *10*, 270. [CrossRef]
3. Wang, H.L.; Hu, Q.X.; Yang, Y.; Wang, C. Performance Differences of Electrical Submersible Pump under Variable Speed Schemes. *Int. J. Simul. Model.* **2021**, *20*, 76–86. [CrossRef]
4. Zhou, J.; Zhao, M.; Wang, C.; Gao, Z. Optimal Design of Diversion Piers of Lateral Intake Pumping Station Based on Orthogonal Test. *Shock. Vib.* **2021**, *2021*, 6616456. [CrossRef]
5. Tang, S.; Zhu, Y.; Yuan, S. An improved convolutional neural network with an adaptable learning rate towards multi-signal fault diagnosis of hydraulic piston pump. *Adv. Eng. Inform.* **2021**, *50*, 101406. [CrossRef]
6. Zhu, Y.; Li, G.; Wang, R.; Tang, S.; Su, H.; Cao, K. Intelligent fault diagnosis of hydraulic piston pump combining improved LeNet-5 and PSO hyperparameter optimization. *Appl. Acoust.* **2021**, *183*, 108336. [CrossRef]
7. Shi, L.; Zhu, J.; Tang, F.; Wang, C. Multi-Disciplinary Optimization Design of Axial-Flow Pump Impellers Based on the Approximation Model. *Energies* **2020**, *13*, 779. [CrossRef]
8. Wang, H.; Long, B.; Wang, C.; Han, C.; Li, L. Effects of the Impeller Blade with a Slot Structure on the Centrifugal Pump Performance. *Energies* **2020**, *13*, 1628. [CrossRef]
9. Zhang, L.; Wang, C.; Zhang, Y.; Xiang, W.; He, Z.; Shi, W. Numerical study of coupled flow in blocking pulsed jet impinging on a rotating wall. *J. Braz. Soc. Mech. Sci. Eng.* **2021**, *43*, 508. [CrossRef]
10. Wang, H.; Qian, Z.; Zhang, D.; Wang, T.; Wang, C. Numerical Study of the Normal Impinging Water Jet at Different Impinging Height, Based on Wray–Agarwal Turbulence Model. *Energies* **2020**, *13*, 1744. [CrossRef]
11. Ljiri, M.; Shimonishi, D.; Nakagawa, D.; Yoshimura, T. New water jet cavitation technology to increase number and size of cavitation bubbles and its effect on pure Al surface. *Int. J. Lightwght Mater. Manuf.* **2018**, *1*, 12–20.
12. Yang, Y.; Li, W.; Shi, W.; Wang, C.; Zhang, W. Experimental Study on Submerged High-Pressure Jet and Parameter Optimization for Cavitation Peening. *Mechanika* **2020**, *26*, 346–353. [CrossRef]
13. Yang, Y.; Li, W.; Shi, W.; Zhou, L.; Zhang, W. Experimental Study on the Unsteady Characteristics and the Impact Performance of a High-Pressure Submerged Cavitation Jet. *Shock. Vib.* **2020**, *2020*, 1701843. [CrossRef]
14. Yang, Y.; Li, W.; Shi, W.; Zhang, W.; AEl-Emam, M. Numerical Investigation of a High-Pressure Submerged Jet Using a Cavitation Model Considering Effects of Shear Stress. *Processes* **2019**, *7*, 541. [CrossRef]
15. Watanabe, R.; Kikuchi, T.; Yamagata, T.; Fujisawa, N. Shadowgraph Imaging of Cavitating Jet. *J. Flow Control. Meas. Vis.* **2015**, *3*, 106–110. [CrossRef]
16. Watanabe, R.; Yanagisawa, K.; Yamagata, T.; Fujisawa, N. Simultaneous shadowgraph imaging and acceleration pulse measurement of cavitating jet. *Wear* **2016**, *358*, 72–79. [CrossRef]
17. Peng, C.; Tian, S.; Li, G. Joint experiments of cavitation jet: High-speed visualization and erosion test. *Ocean Eng.* **2018**, *149*, 1–13. [CrossRef]
18. Shridhar, G.J.; Katz, O.K. Near-field flow structure and cavitation inception in jets. In Proceedings of the First Symposium on Turbulence and Shear Flow Phenomena, San Francisco, CA, USA, 18–23 July 1999.
19. Nakano, K.; Hayakawa, M.; Fujikawa, S. Observations of Cavitation Bubbles in a Starting Submerged Water Jet. *Trans. Jpn. Soc. Mech. Eng. Ser. B* **2003**, *69*, 360–367. [CrossRef]
20. McGinn, P.; Pearce, D.; Hardalupas, Y.; Taylor, A.; Vogiatzaki, K. Cavitation Bubble Cloud Break-Off Mechanisms at Micro-Channels. *Fluids* **2021**, *6*, 215. [CrossRef]
21. Ma, R.; Slaboch, P.E.; Morris, S.C. Fluid mechanics of the flow-excited Helmholtz resonator. *J. Fluid Mech.* **2009**, *623*, 1–26. [CrossRef]
22. Hussain, A.; Hasan, M. The 'whistler-nozzle' phenomenon. *J. Fluid Mech.* **1983**, *134*, 431–458. [CrossRef]
23. Varieras, D.; Brancher, P.; Giovannini, A. Self-sustained oscillations of a confined impinging jet. *Flow Turbul. Combust.* **2006**, *78*, 1–15. [CrossRef]
24. Favrel, A.; Junior JG, P.; Landry, C.; Müller, A.; Yamaishi, K.; Avellan, F. Dynamic modal analysis during reduced scale model tests of hydraulic turbines for hydro-acoustic characterization of cavitation flows. *Mech. Syst. Signal Process.* **2018**, *117*, 81–96. [CrossRef]
25. Fujisawa, N.; Fujita, Y.; Yanagisawa, K.; Fujisawa, K.; Yamagata, T. Simultaneous observation of cavitation collapse and shock wave formation in cavitating jet. *Exp. Therm. Fluid Sci.* **2018**, *94*, 159–167. [CrossRef]
26. Peng, K.; Tian, S.; Li, G.; Huang, Z.; Yang, R.; Guo, Z. Bubble dynamics characteristics and influencing factors on the cavitation collapse intensity for self-resonating cavitating jets. *Pet. Explor. Dev.* **2018**, *45*, 155–162. [CrossRef]
27. Sato, K.; Taguchi, Y.; Hayashi, S. High Speed Observation of Periodic Cavity Behavior in a Convergent-Divergent Nozzle for Cavitating Water Jet. *J. Flow Control. Meas. Vis.* **2013**, *1*, 102–107. [CrossRef]

28. Peng, G.; Wakui, A.; Oguma, Y.; Shimizu, S.; Ji, H. Periodic Behavior of Cavitation Cloud Shedding in Submerged Water Jets Issuing from a Sheathed Pipe Nozzle. *J. Flow Control. Meas. Vis.* **2018**, *6*, 15–26. [CrossRef]
29. Peng, G.; Yang, C.; Oguma, Y.; Shimizu, S. Numerical analysis of cavitation cloud shedding in a submerged water jet. *J. Hydrodyn.* **2016**, *28*, 986–993. [CrossRef]
30. Peng, G.; Okada, K.; Yang, C.; Oguma, Y.; Shimizu, S. Numerical Simulation of Unsteady Cavitation in a High-speed Water Jet. *Int. J. Fluid Mach. Syst.* **2016**, *9*, 66–74. [CrossRef]
31. Li, G.S.; Shen, Z.H.; Zhou, C.S.; Zhang, D.B.; Liao, H.L. Advances in Investigation and Application of Self-resonating Cavitating Water Jet. *Eng. Sci.* **2005**, *7*, 27–32.
32. Jiang, Y.; Abu-Hamdeh, N.H.; Bantan, R.A.; Li, Z. Influence of upstream angled ramp on fuel mixing of hydrogen jet at supersonic cross flow. *Aerosp. Sci. Technol.* **2021**, *119*, 107099. [CrossRef]
33. Jiang, Y.; Poozesh, A.; Marashi, S.M.; Moradi, R.; Gerdroodbary, M.B.; Shafee, A.; Li, Z.; Babazadeh, H. Effect of cavity back height on mixing efficiency of hydrogen multi-jets at supersonic combustion chamber. *Int. J. Hydrogen Energy* **2020**, *45*, 27828–27836. [CrossRef]
34. Jiang, Y.; Gerdroodbary, M.B.; Sheikholeslami, M.; Babazadeh, H.; Shafee, A.; Moradi, R.; Li, Z. Effect of free stream angle on mixing performance of hydrogen multi-jets in supersonic combustion chamber. *Int. J. Hydrogen Energy* **2020**, *45*, 25426–25437. [CrossRef]
35. Kosambi, D.D. Statistics in Function Space. *J. Indian Math. Soc.* **2016**, 115–123. [CrossRef]
36. Brennen, C.E. *Fundamentals of Multiphase Flow*; Cambridge University Press: Cambridge, UK, 2005.
37. Mikko, M.; Veikko, T.; Sirpa, K. *On the Mixture Model for Multiphase Flow*; VTT Publications: Espoo, Finland, 1996; p. 288.
38. Zwart, P.J.; Gerber, A.G.; Belamri, T. A Two-Phase Flow Model for Predicting Cavitation Dynamics. In Proceedings of the Fifth International Conference on Multiphase Flow, Yokohama, Japan, 30 May–3 June 2004.
39. Deck, S. Recent improvements in the Zonal Detached Eddy Simulation (ZDES) formulation. *Theor. Comput. Fluid Dyn.* **2012**, *26*, 523–550. [CrossRef]

machines

MDPI

Article

Numerical and Experimental Study of Hydraulic Performance and Wear Characteristics of a Slurry Pump

Guangjie Peng [1], Long Tian [1], Hao Chang [1,*], Shiming Hong [1], Daoxing Ye [2] and Baojian You [3]

1 Research Center of Fluid Machinery Engineering and Technology, Jiangsu University, Zhenjiang 212013, China; pgj@ujs.edu.cn (G.P.); 2222011050@stmail.ujs.edu.cn (L.T.); 2221911014@stmail.ujs.edu.cn (S.H.)
2 Key Laboratory of Fluid Machinery and Engineering, Xihua University, Chengdu 610039, China; dxye@mail.xhu.edu.cn
3 Shanghai Kaiquan Pump (Group) Co., Ltd., Shanghai 201804, China; youbaojian@kaiquan.com.cn
* Correspondence: changhao@ujs.edu.cn

Abstract: The slurry pump is widely used in ore mining, metal smelting, petrochemical, and other industries, mainly to transport fluid media containing large solid particles. Importantly, it is easy to damage the impeller of a slurry pump in the operation process, which greatly affects the performance of the pump. In this paper, a 25 MZ slurry pump was selected as the research object, and the Euler–Euler multiphase flow model was employed to analyze the internal flow characteristics of the slurry pump under the conditions of clear water and solid–liquid two-phase flow. Additionally, the flow characteristics of each part under different flow conditions were studied, and the effects of different particle volume concentrations, particle sizes, and pump speeds on the impeller's wear characteristics and hydraulic performance were analyzed. In order to verify the reliability and accuracy of the numerical simulation results, clean water and solid–liquid two-phase flow wear tests of the slurry pump were carried out, and the results showed that a high solid volume fraction and solid–phase slip velocity were generated at the junction of the blade leading edge and the rear cover plate, thus leading to easier wear of the blade. Therefore, enhancing the strength of the junction between the blade leading edge and the rear cover plate is beneficial for improving service life and should be considered in the design of slurry pumps.

Keywords: slurry pump; solid–liquid flow; hydraulic performance; wear characteristic

Citation: Peng, G.; Tian, L.; Chang, H.; Hong, S.; Ye, D.; You, B. Numerical and Experimental Study of Hydraulic Performance and Wear Characteristics of a Slurry Pump. *Machines* **2021**, *9*, 373. https://doi.org/10.3390/machines9120373

Academic Editor: Antonio J. Marques Cardoso

Received: 10 November 2021
Accepted: 17 December 2021
Published: 20 December 2021

1. Introduction

As an important piece of energy conversion equipment, the slurry pump has been widely used in mining, electric power, metallurgy, coal, and other industries. Both the complexity of the solid–liquid two-phase flow in the slurry pump and the limitations of two-phase flow field measurement technology have made it very difficult to obtain flow field information of an entire flow channel. Therefore, capturing the flow field characteristics of slurry pumps is the fundamental objective of their optimized design and wear investigation. In recent years, there has been much research on the motion of the solid–liquid two-phase flow. Wang et al. [1] used a CFD-DEM coupling algorithm to study the spatial distribution and motion characteristics of particles with different sizes, and the effects of particle diameter on the intensity and scale of the vortex in a guide vane were investigated; it was concluded that the existence of particles had a limited effect on the hydraulic performance of the pump in low-concentration fluids. In a numerical study of the internal flow field of the solid–liquid slurry pump, Shi et al. [2] analyzed the wear mechanism of the volute wall of a slurry pump and proposed an effective wear equation for estimating the wear strength and wear area of the slurry pump volute. Peng et al. [3] analyzed slurry flow under different particle concentrations and volume flows by using the Euler–Euler method, and the effects of particle concentration on flow resistance, reflux, and wall wear were assessed.

Li et al. [4] simulated the solid–liquid two-phase flow in a centrifugal pump by using the computational fluid dynamics discrete element coupling method. Their results showed that the increase in the wear rate was related to the increase in particle mass concentration. Tan et al. [5] utilized high-speed photography to track the movement of solid–liquid two-phase flow particles in a double-blade slurry pump, and the effects of particle diameter and density on the collision characteristics were analyzed. The results showed that with the increase in particle size, the average time through the pump first decreased and then increased. Shen et al. [6] designed a new type of longitudinal groove structure for the volute to improve the internal flow field. Wang et al. [7] combined numerical simulations and experiments to study the wear characteristics of an open impeller, and they concluded that the most serious wear area of the impeller was the middle of the pressure surface. Zhao et al. [8] simulated turbulence with a homogeneous balance model, and the relationships between concentration and external performance, velocity, pressure, and turbulent kinetic energy distribution were evaluated. Wang et al. [9] investigated the relationships between particle size, volume concentration, solid-phase volume, and inlet pressure fraction by using a mixed multiphase flow model. Wang et al. [10] combined experiments and calculation models to study the effects of different coating thicknesses on the operating characteristics of centrifugal pumps, and they reported on the effect of coating thickness on performance and pressure fluctuations. Gu et al. [11] employed the Euler model to build a non-smooth surface on an impeller blade for numerical simulation, and they found that the non-smooth surface area was better able to reduce drag and affect the performance of the pump. Yan et al. [12] carried out a full-channel numerical simulation of solid–liquid two-phase flow by using the McLaury wear model, and the distribution region and variation trend of impeller and volute wall wear caused by the change of clearance were assessed. Wang et al. [13] studied the flow characteristics and wear performance of the pump at different rotational speeds with numerical simulation and calculated the wear rate of the blade surface at different rotational speeds. Tang et al. [14] studied the solid–liquid flow in a pump and compared the contact force and collision characteristics of particles with different shapes.

Most of the abovementioned studies considered the effects of the shape and size of particles on the wear characteristics of pumps, but there has been little research on the investigation of slurry pumps under different flow conditions and pump speeds. Therefore, the authors of this paper used numerical simulations and experiments to analyze the internal flow characteristics of a slurry pump under the conditions of clear water and two-phase flow. The flow characteristics of each part under different flow conditions were studied, and the effects of different particle volume concentrations, particle sizes, and pump speed on the impeller's wear and hydraulic performance were analyzed. Additionally, in order to verify the reliability and accuracy of the numerical simulation results, clean water and solid–liquid two-phase flow wear tests of the slurry pump were carried out to provide reference for the optimization of the design of the slurry pump.

2. Analysis Model

In this paper, a 25 MZ slurry pump is selected as the research object. Its main geometric parameters are shown in Table 1, where the design flow conditions are represented by Q_{BEP} (Best Efficiency Point, BEP).

Table 1. Main parameters of 25 MZ slurry pump.

Design Flow Q_{BEP} (m^3/h)	Head H (m)	Speed n (r/min)	Blade Number	Impeller Outer Diameter D_2 (mm)	Blade Outlet Width b_2 (mm)	Inlet Diameter D_j (mm)
26	11	1500	4	178	30	75

2.1. 3D Modeling

In order to study the internal flow field of the slurry pump, the fluid domain was simplified into four parts: inlet, impeller, volute, and outlet. Based on the three-dimensional modeling software UG, the solid modeling of each fluid domain was built, and the three-dimensional model of the main components is shown in Figure 1.

(**a**) Impeller (**b**) Volute

Figure 1. (**a**,**b**) Three-dimensional models of the flow domain.

2.2. Grids

In this paper, ICEM was employed to mesh the fluid domains. Considering that the quality of the grid has a great influence on the convergence and reliability of calculation results, the calculation domain was divided into a hexahedral structured grid to better improve the accuracy of the numerical simulation. The structural grids of the main components are shown in Figure 2.

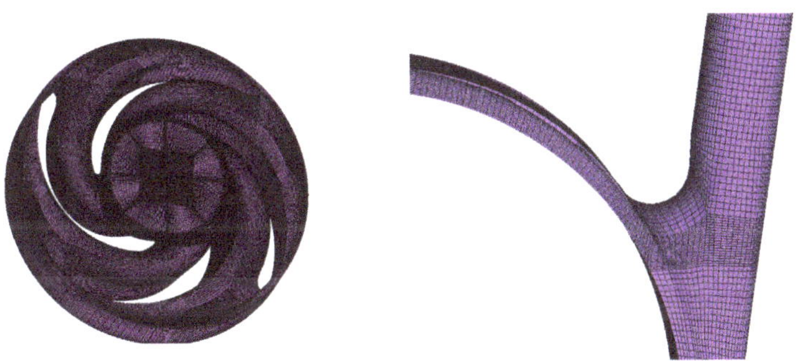

(**a**) Impeller (**b**) Volute

Figure 2. (**a**,**b**) Mesh generation of the flow domains.

Meanwhile, the greater the number and the higher the quality of grids that are generated, the more accurate the calculation results that can be obtained. However, too many grids consume many computing resources. Therefore, to ensure the accuracy of numerical simulation results and the rational use of computer resources, the grid independence of the calculation domain was conducted [15–20]. For this section, the calculation of slurry pump models with different grids was carried out, and a reasonable number of grids were obtained by comparing the fluctuation of head and efficiency. As shown in Figure 3, with the increase in grid number, the fluctuation value of the head efficiency gradually decreased. Additionally, when the grid number was 2.7 million, the fluctuation value of the external characteristic was less than 1.0%, which met the requirements of numerical simulations. Therefore, a grid number of 2.7 million was selected for the investigation.

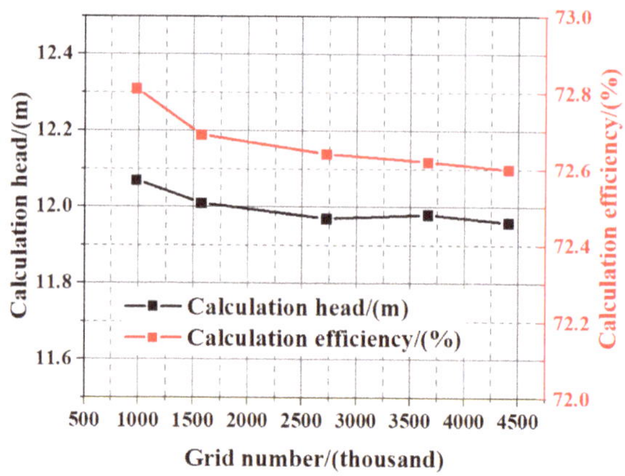

Figure 3. The grid independence.

ANSYS CFX was employed the calculation of the water, which was selected as the fluid medium, and the medium temperature was 298 K. Total pressure inflow (1 atm) and mass flow outflow were adopted, and the standard k-ε turbulence model was selected as the turbulence model. Meanwhile, the fluid domain of the impeller was set as the rotating part; the inlet section, the outlet section, and the volute were set as the stationary parts; the frozen rotor method was adopted for the interface; and the convergence accuracy was set to 1.0×10^{-5}. In this paper, the continuity equation of incompressible fluid and the Navier–Stokes equation were employed to describe the three-dimensional turbulence in a centrifugal pump.

The continuity equation is as follows:

$$\frac{\partial(\rho u_i)}{\partial x_i} = 0$$

The Navier–Stokes equation for incompressible fluid is as follows:

$$\frac{\partial(\rho u_i)}{\partial t} + \frac{\partial(\rho u_i u_j)}{\partial x_j} = -\frac{\partial p}{\partial x_j} + \rho F_i + 2\frac{\partial(\mu S_{ij})}{\partial x_j}$$

In the formula, ρ: fluid density; t: time; F_i: the volume force component of a fluid per unit mass; P: pressure; u_i and u_j (i, j = 1, 2, 3): the velocity component of the fluid; x_i and x_j (i, j = 1, 2, 3): coordinate components; μ: molecular viscosity coefficient; and S_{ij}: the fluid deformation rate tensor.

$$S_{ij} = \frac{1}{2}\left(\frac{\partial u_i}{\partial x_j} + \frac{\partial u_j}{\partial x_i}\right)$$

For the calculation of solid–liquid two-phase flow, the heterogeneous flow model of the particle in the Euler–Euler method was adopted. The flow conditions of 0.6 Q_{BEP}, 1.0 Q_{BEP}, and 1.5 Q_{BEP} were selected to carry out the calculations, and particle volume concentrations of 5%, 10%, 15%, 20%, and 25% were chosen. Meanwhile, the density of solid particles was set to 2900 kg/m^3, and the particle diameters were 0.15, 0.3, and 0.6 mm. The Gidaspow model was selected for the drag force, the Favre-averaged drag force model was selected for the turbulent dissipative force, and the dissipation coefficient was 1.0. In this paper, the effects of lift, virtual mass force, and wall lubrication force on the flow field were not considered. The inlet of the two-phase flow adopted the mass flow inlet, and the outlet adopted the average static pressure outlet. Considering the liquid without the slip velocity on the wall, the liquid phase adopted the non-slip wall. However, the wear effect of particles on the wall was considered, the solid phase adopted a free slip wall, and the SIMPLE algorithm was used to solve the pressure–velocity coupling.

Due to the different densities of solid particles and water, the density of the transport medium of the slurry pump changed after the solid particles were added to the clear water, and the calculation formula is as follows:

$$\rho' = C_V(\rho_1 - \rho_2) + \rho_2 \quad (1)$$

In the formula, ρ' is the density of solid–liquid two-phase flow, ρ_1 is the density of solid particles, ρ_2 is the density of water, and C_V is the volume concentration of particles. The calculation results of fluid density in the centrifugal pump under different solid phase volume concentrations are shown in Table 2.

Table 2. Density under different particle volume concentrations.

C_V/%	0	5	10	15	20	25	30	50
ρ' (kg/m^3)	1000	1095	1190	1285	1380	1475	1575	1950

3. Results and Discussion

3.1. Flow Field Analysis of Inlet

Figures 4 and 5 show the streamline distribution of the inlet section when using clean water and when using a medium containing 15% solid particles with a diameter of 0.6 mm under 0.6 Q_{BEP}, 1.0 Q_{BEP}, and 1.5 Q_{BEP}. It can be seen from Figure 4 that the fluid in the inlet section had a slight pre-rotation, and the pre-rotation direction was consistent with the rotation direction of the impeller under 0.6 Q_{BEP}. Under the effect of the rotation impeller, the fluid direction changed from axial flow to radial flow. Meanwhile, with the increase in the flow rate, the velocity at the impeller inlet gradually increased, and the ability of the impeller to rotate the fluid near the inlet was weakened. When the solid particles were added to the fluid, the kinematic viscosity and flow resistance of the medium were increased, and the ability of the pre-rotation was weaker than that of clean water, as shown in Figure 5. A slight pre-rotation could be observed at the impeller inlet under 1.0 Q_{BEP}, and due to the solid particles being carried in the fluid, a certain degree of wear was generated at the leading edge of the blade. Furthermore, the ability of the pre-rotation continuously decreased with the increase in flow rate. When the flow rate reached 1.5 Q_{BEP}, the fluid movement at the inlet section was relatively stable.

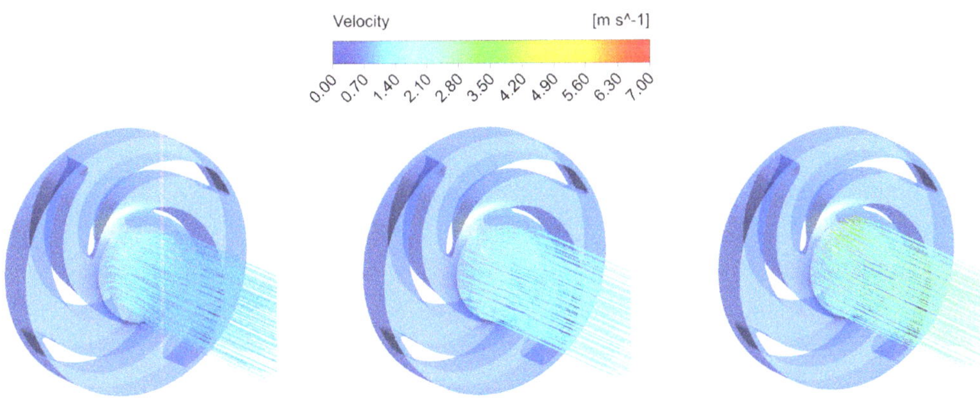

(**a**) 0.6 Q_{BEP} (**b**) 1.0 Q_{BEP} (**c**) 1.5 Q_{BEP}

Figure 4. (**a–c**) Flow distribution in the inlet section under clean water.

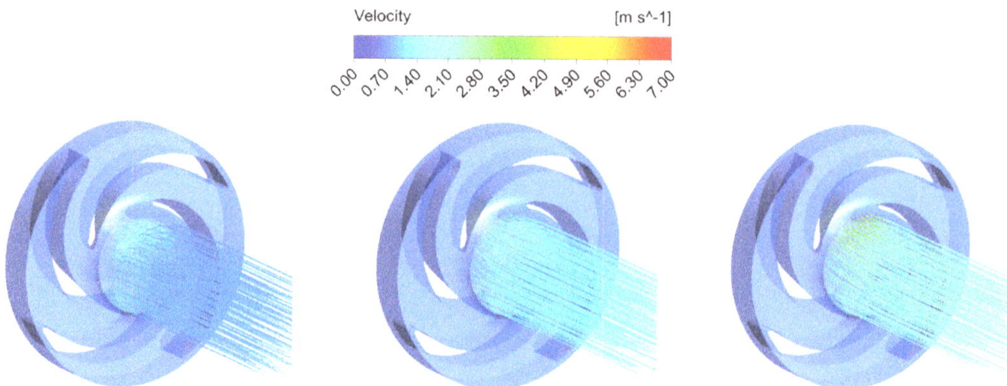

(**a**) 0.6 Q_{BEP} (**b**) 1.0 Q_{BEP} (**c**) 1.5 Q_{BEP}

Figure 5. (**a**–**c**) Flow distribution in the inlet section under two-phase flow.

3.2. Flow Field Analysis of Impeller

The solid volume fraction and slip velocity had obvious influence on the wall wear, and the collision frequency of particles against the wall were enhanced with the increase in solid volume fraction. Additionally, the increase in solid particle slip velocity improved the impaction of a single particle on the wall. However, according to a previous investigation [3], solid-phase slip velocity has a great influence on wall wear. Figure 6 shows the solid volume fraction distribution on the impeller wall under different flow conditions; the solid volume concentration was 15% and the particle size was 0.6 mm. It can be seen from the figure that the solid volume fraction of the blade pressure surface gradually moved from the leading edge to the trailing edge with the increase in flow rate. Under the 0.6 Q_{BEP} condition, the velocity of the liquid and the centrifugal force on the solid particles was low, which made the solid particles accumulate at the impeller inlet. When the flow rate increased to 1.5 Q_{BEP}, the solid particles moved to the rear cover plate under the effect of inertia force, which resulted in a large solid volume fraction at the interface between the impeller and the rear cover plate.

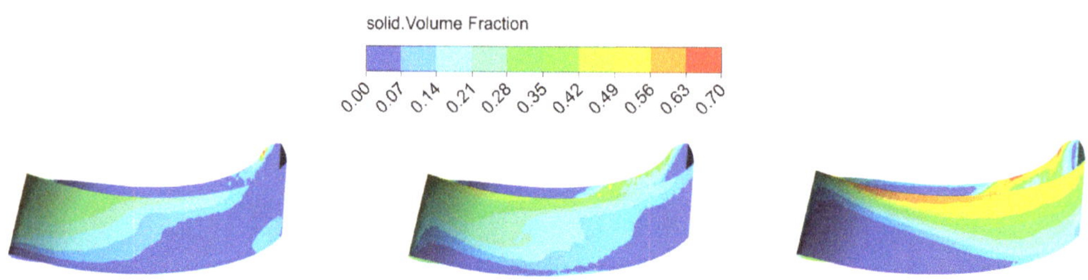

(**a**) 0.6 Q_{BEP} (**b**) 1.0 Q_{BEP} (**c**) 1.5 Q_{BEP}

Figure 6. (**a**–**c**) Solid volume fraction distribution of the blade.

3.3. Flow Field Analysis of Volute

Figures 7 and 8 present the solid volume fraction distributions of the volute wall under the flow conditions of 0.6 Q_{BEP}, 1.0 Q_{BEP}, and 1.5 Q_{BEP}; the solid volume concentration was 15% and the particle diameter was 0.6 mm. Under the 0.6 Q_{BEP} condition, the larger solid volume fraction was mainly distributed between the sixth and eighth sections, and the solid volume fraction at the tongue was lower.

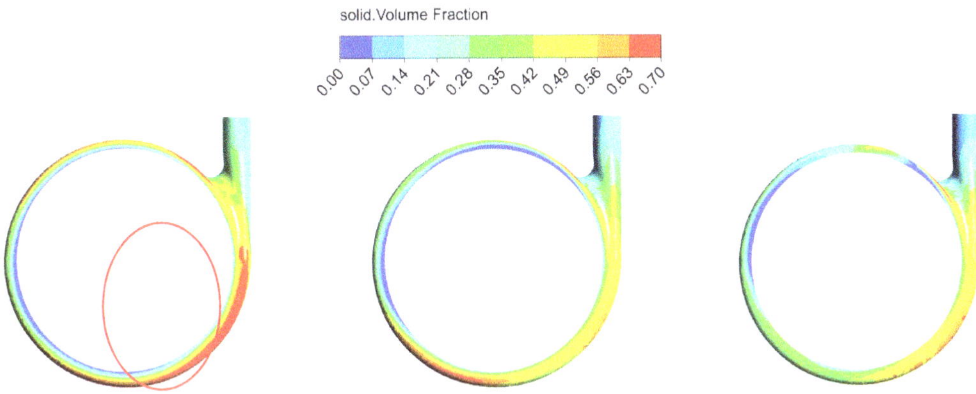

(**a**) 0.6 Q_{BEP} (**b**) 1.0 Q_{BEP} (**c**) 1.5 Q_{BEP}

Figure 7. (**a**–**c**) Solid volume fraction distribution on the rear cover.

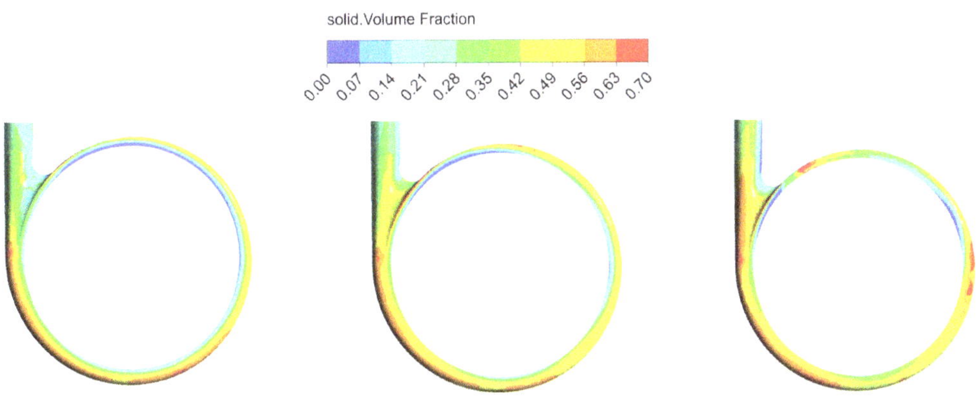

(**a**) 0.6 Q_{BEP} (**b**) 1.0 Q_{BEP} (**c**) 1.5 Q_{BEP}

Figure 8. (**a**–**c**) Solid volume fraction distribution on the front cover.

With the increase in flow rate, the region with a large solid volume fraction on the volute wall gradually moved from the front cover to the rear cover. When the fluid entered the volute from the impeller under the effect of centrifugal force, the volume fraction of solid particles near the rear cover plate was larger, resulting in the wear on the rear cover plate being higher than that on the front cover.

3.4. Effect of Particle Parameters on Wear Characteristic

In this section, the effects of particle volume concentration and particle diameter on impeller wear are discussed, and the effects of the solid volume fraction and solid slip velocity on the blade are analyzed. Abscissa indicates the relative distance in the direction of the blade streamline (streamwise); a relative distance of the blade streamline of 0 represents the leading edge of the blade, a relative distance of the streamline of 1 indicates the trailing edge of the blade, and an ordinate represents the solid phase volume fraction (solid volume fraction) and the solid phase slip velocity (solid velocity).

3.4.1. Effect of Particle Volume Concentration on Wear

Figure 9 shows the hydraulic performance curves of different concentrations under optimal flow conditions. Here, 0 indicates that the medium is clear water. It can be seen from the figure that with the increase in concentration, the head gradually and continuously decreased. With the increase in the concentration, the viscosity of the solid–liquid two-phase flow increased and the friction force and energy loss between the internal fluids gradually increased, which resulted in a decrease in the head. At the same time, with the increase in solid particle volume concentration, the enhancement of the centrifugal pump impeller torque led to the increase in motor power consumption and the decrease in centrifugal pump efficiency. When the concentration was 5%, the head decreased by 0.33% and the efficiency decreased by 1.60%. When the concentration was 25%, the head decreased by 8.9% and the efficiency decreased by 8.09%. Therefore, the head and efficiency of the pump decreased with the increases in particle volume concentration.

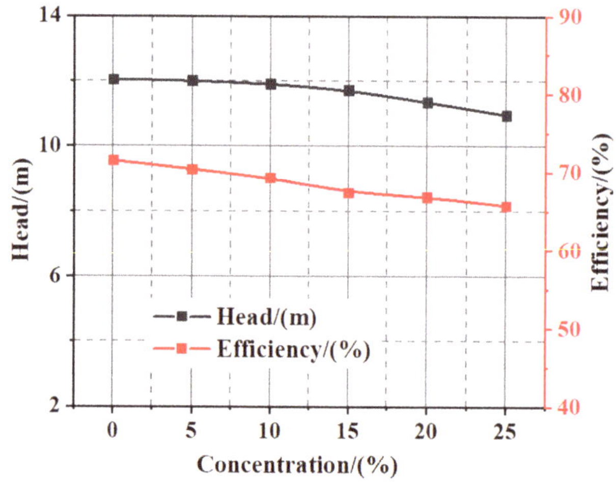

Figure 9. External characteristic curves of different concentrations.

Figure 10 shows the solid phase slip velocity distribution of the impeller when the particle size was 0.6 mm and the particle volume concentrations were 5%, 15%, and 25%. It was found that the solid phase slip velocity at the leading edge of the pressure surface was the smallest—when the particles entered the impeller, the particles flowed from the axial direction to the radial direction under the effect of centrifugal force, which decreased the slip velocity. With the increase in concentration, the slip velocity of solid particles on the pressure surface gradually increased. There was no great change in the slip velocity on the suction surface, though a sudden change in the slip velocity of the solid particles was found at the trailing edge of the blade which resulted from the flow separation near the trailing edge of the suction surface. Additionally, the slip velocity of solid particles on the pressure surface was generally larger than that on the suction surface.

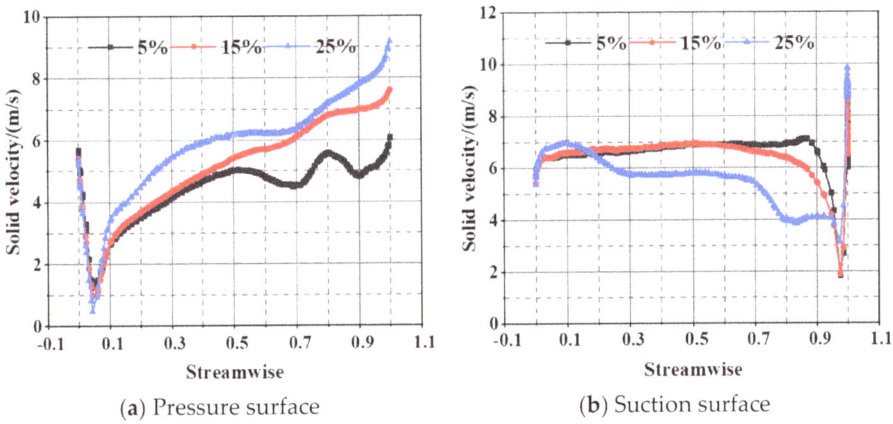

Figure 10. (**a**,**b**) Solid velocity distribution of the blade.

Figure 11 shows the solid volume fraction distribution of the impeller when the particle size was 0.6 mm and the particle volume concentrations were 5%, 15%, and 25%. It can be seen from the figure that the solid volume on the blade pressure surface gradually decreased from the leading edge to the trailing edge, with less and less accumulation of solid particles due to the work of the blade. Meanwhile, there were almost no solid particles on the suction surface of the blade, so the solid volume fraction on the pressure surface of the blade was larger than the suction surface. With the increase in the concentration, the solid particles and solid volume fraction gradually increased.

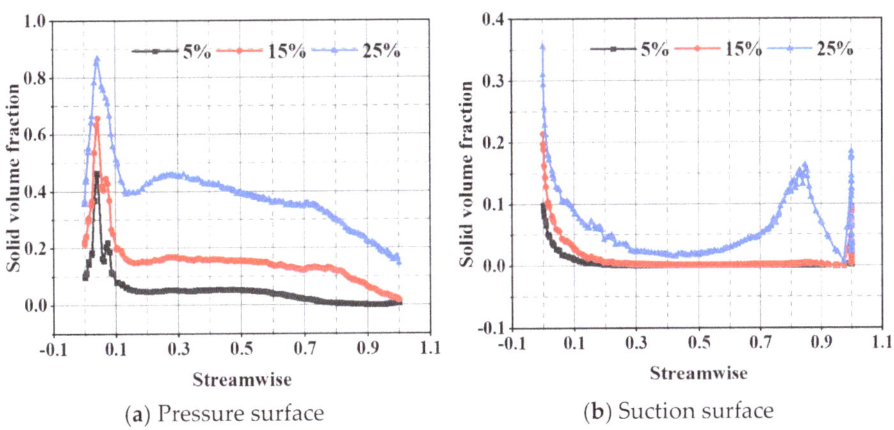

Figure 11. (**a**,**b**) Solid volume fraction distribution of blade.

3.4.2. Effect of Particle Size on Wear Characteristic

Figure 12 shows the comparison of the hydraulic performance curves of different particle diameters under design flow conditions, where a particle size of 0 mm indicates that the used medium was clear water. With the increase in the particle diameter, the energy driving the solid particles was greater and the head gradually decreased. At the same time, with the increase in solid particle diameter, the friction loss between particles and impeller increased, which led to the increases in impeller torque and motor power consumption. Therefore, as the particle diameter increased from 0.15 to 0.6 mm, the pump head decreased from 12 to 11.42 m and the efficiency decreased from 71.25% to 67.35%.

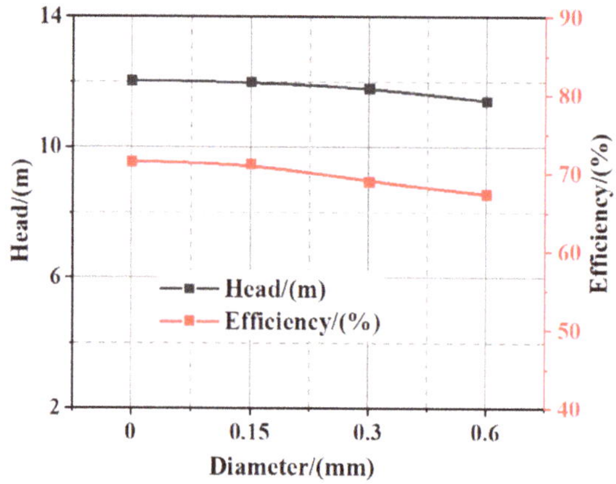

Figure 12. Characteristic curve under different particle sizes.

Figure 13 shows the distribution of the solid-phase slip velocity of the impeller when the particle volume concentration was 15% and the particle diameters were 0.15, 0.3, and 0.6 mm. It can be seen from the figure that the solid slip velocity of the pressure surface gradually increased from the leading edge to the trailing edge. Additionally, with the increase in the particle diameter, the solid slip velocity at the trailing edge of the pressure surface gradually increased. This phenomenon is attributed to the increasing radial velocity and circumferential velocity of solid particles after they entered the impeller, which led to an increase in solid-phase slip velocity. Therefore, the solid phase slip velocity on the blade surface increased with the increase in solid particle diameter.

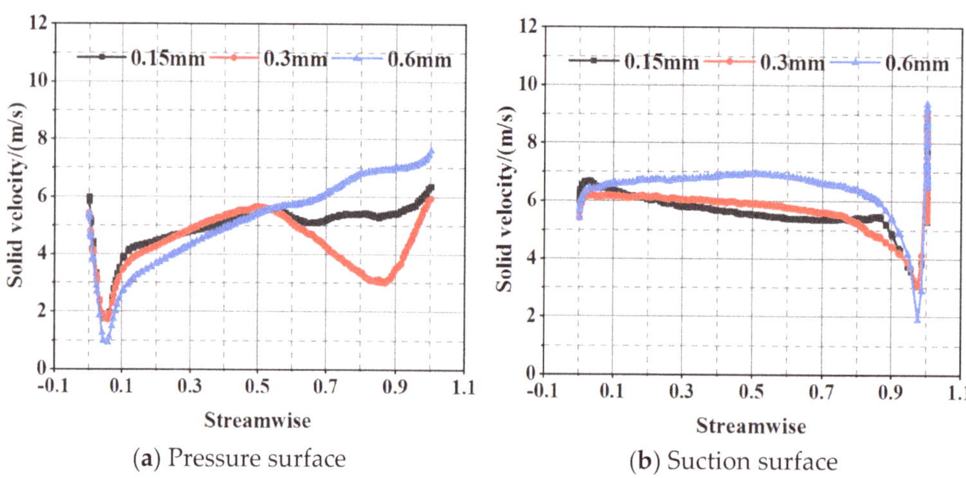

Figure 13. (**a**,**b**) Solid velocity distribution of the blade.

Figure 14 shows the solid volume fraction distribution of the impeller when the particle volume concentration was 15% and the particle diameters were 0.15, 0.3, and 0.6 mm. It was found that with the increase in solid particle diameter, the region of the larger solid volume fraction moved from the suction surface to the pressure surface. When the solid particles entered the impeller, the effects of the particle's inertia and the extrusion of the blade resulted in a larger solid volume fraction at the leading edge. At the same time, with the increase in the particle diameter, the influence of the fluid centrifugal force on the solid particles was enhanced, which made the solid particles move from the suction surface to the pressure surface. Moreover, the maximum solid volume fraction was generated at the leading edge of the blade on the pressure surface, while the maximum solid volume fraction was generated at the trailing edge of the suction surface.

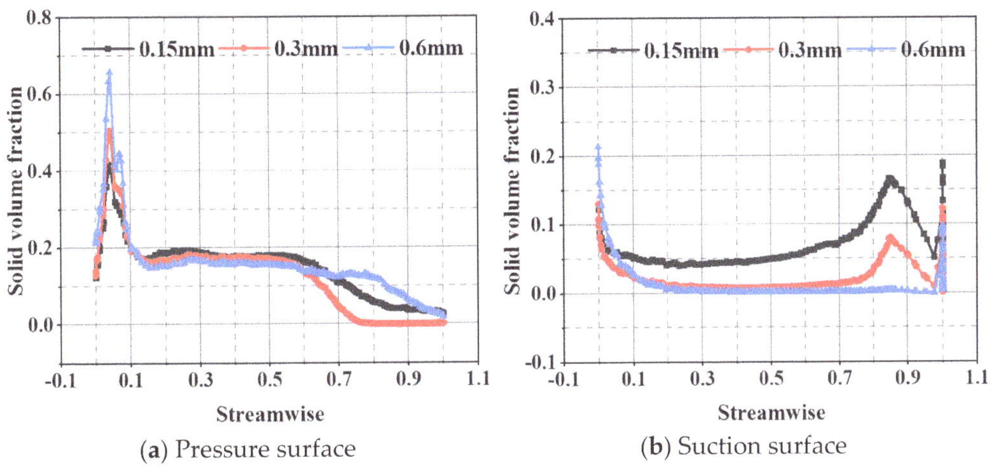

Figure 14. (**a**,**b**) Solid volume fraction distribution of blade.

3.5. Effect of Pump Speed on External Performance

In the previous analysis of solid–liquid two-phase flow, it can be seen that increases in solid volume concentration led to decreases in the head and efficiency of the slurry pump. With the increase in the solid volume concentration, the viscosity between the solid–liquid two-phase flow increased and the friction force and energy loss between the internal fluids was increased, which resulted in the decline of the head. More importantly, the slurry pump could not reach the corresponding working condition, which may be another reasons for the wear of the impeller. Therefore, in this paper, the influence of speed on hydraulic performance was studied. Under the rated flow condition, solid–liquid two-phase flow simulation calculations were carried out at the rotational speeds of 1480, 1490, 1500, 1510, 1520, and 1530 r/min and media with solid-phase volume concentrations of 0, 5%, 10%, 15%, 20%, and 25%. Then, the head value data obtained by the simulation were plotted into a three-dimensional scatter diagram, as shown in Figure 15.

According to the head scatter diagram, the relationship between the rotational speed and solid volume concentration on the head was evaluated. The head reached its minimum value at C_V = 25% and n = 1480 r/min, and it reached its maximum value at C_V = 0% and n = 1530 r/min. At the same solid volume concentration, the head of the pump increased with the increase in rotational speed. At the same speed, the head of the pump decreased with the increase in solid volume concentration.

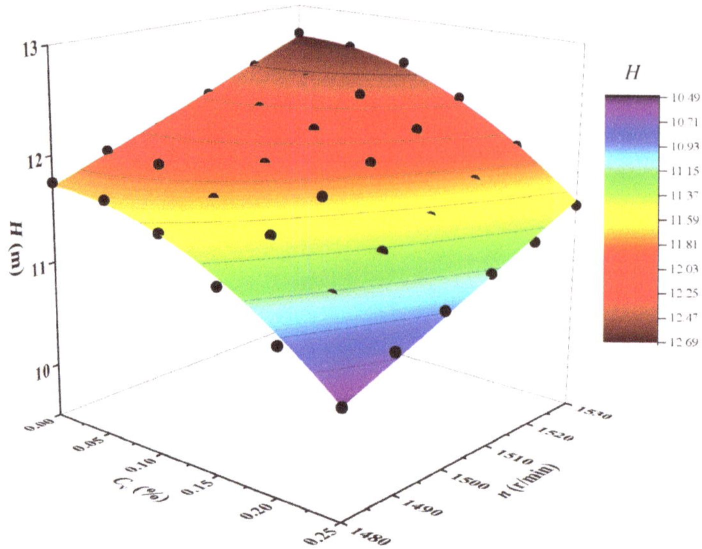

Figure 15. Three-dimensional fitting surface of different solid volume concentrations and speeds.

In order to better analyze the influence of rotational speed and solid volume concentration on the head, a three-dimensional fitting surface was built, and the relationship between the rotational speed, solid volume concentration, and the head was evaluated. In the figure, the head value of the thickened equipotential line is 12 m.

At the same speed, with the increase in solid volume concentration, the head first slowly decreased and then rapidly decreased. The surface fitting diagram showed that the fitting effect of $C_V = 10\%$ was the best and the head value of $C_V = 20\%$ was generally lower than the fitting value. On the whole, the fitting effect was able to meet the investigation requirements. From low rotational speed and high solid volume concentration to high rotational speed and low solid volume concentration, the distribution of head contours was uniform and gradually increased.

A three-dimensional fitting surface diagram was projected onto an XY plane, and solid volume concentration, rotational speed, and head diagram values were obtained, as shown in Figure 16. It can be seen from the figure that under the condition of equal head, the pump speed slowly improved with the increase in solid volume concentration. In the figure, the head value of the thickened equipotential line is 12 m. While the solid phase volume concentration increased, the pump head value could reach the corresponding required value via adjustments of the rotational speed of the pump, which ensured the pump operates at the rated flow condition and reduced the fluctuation of the shaft power. For other operating conditions or head values, the head equipotential curve could be obtained with the same method, and then the corresponding operating conditions could be reached by adjusting the pump speed. Therefore, the head first slowly decreased and then rapidly decreased with the increase in solid volume concentration, and it continuously increased with the increases in rotational speed.

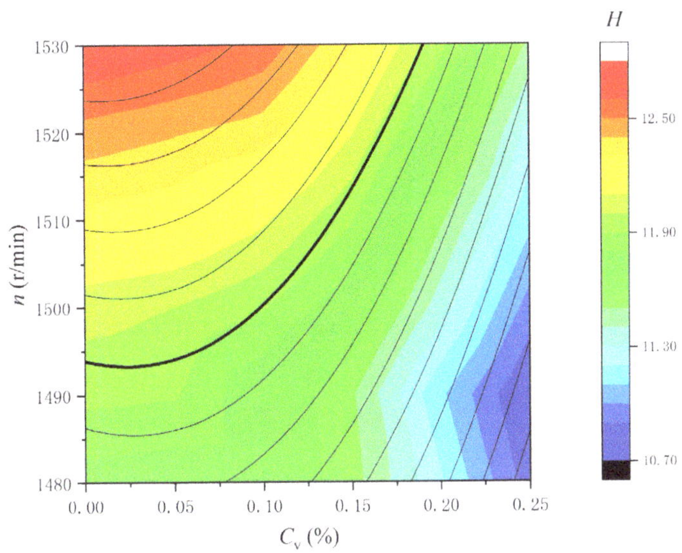

Figure 16. Equipotential head diagram of different solid concentrations and speeds.

4. Experimental Investigation

4.1. Hydraulic Performance Test

The test of the slurry pump was divided into two parts: the clear water performance test and wear performance test. The clear water performance test mainly investigated the external characteristic performance of the pump by measuring its flow rate, pressure, rotational speed, and torque, which were compared with the numerical simulation results. Meanwhile, the wear test was carried out on the slurry pump wear test bench, and the amount of wear was calculated by the impeller weight-loss method.

In order to compare the hydraulic performance between the experiment and simulation, the head and efficiency were selected as evaluation indexes and the clear water hydraulic performance under different flow conditions was assessed, as shown in Figure 17. The expressions of the head and efficiency are, respectively, as follows:

$$H = \frac{p_2 - p_1}{\rho g} + \frac{v_2^2 - v_1^2}{2g} + (z_2 - z_1) \tag{2}$$

$$\eta = \frac{\rho g Q H}{P_s} \times 100\% \tag{3}$$

In the equations, p_1, v_1, and z_1 are the pump inlet parameters; p_2, v_2, and z_2 are the pump outlet parameters; η is the pump efficiency; and P_s is the output power of the motor. During the test, a turbine flow meter (accuracy $\pm$ 0.1%) was used to measure the flow rate, and a pressure sensor (accuracy $\pm$ 0.1%) was used to measure the inlet and outlet pressure of the pump. A torque sensor (accuracy $\pm$ 0.5%) and a speed sensor (accuracy $\pm$ 0.1%) were used to measure torque and speed, respectively.

Figure 17. Hydraulic performance test bench.

Figure 18 shows the hydraulic performance results of the simulation and experiment. The leakage loss, disk friction loss, and friction loss between the bearing and sealing device were not taken into account in the numerical simulation, which resulted in the numerical simulation results being larger than the experimental data and the trend of the simulation head and efficiency curve being the same as that of the test. The maximum deviation of the head under 8 m^3/h was 4.32%, which was due to the instability of the internal flow field of the pump operation except for optimal flow conditions. However, when the flow rate was 26 m^3/h, the minimum deviation between the simulation head and the test head was 2.52% and the efficiency deviation was less than 2.56%. The calculation deviation in the full flow range was less than 5%. Therefore, the calculation model could accurately predict the performance of the pump.

4.2. Wear Performance Test

The wear test bench was mainly composed of the slurry pump, pressure gauge, pipeline, valve, motor, and test system. The test system included strong electric parts, such as the step-down starting cabinet, system distribution cabinet, and test conversion protection cabinet, and weak current parts, such as intelligent display instrument, signal conversion device, and various signal sensors. Figure 19 shows the wear test bench.

In the wear test, the wear condition of the impeller was studied by means of equal flow control and impeller weight loss measurements. The inlet and outlet of the pump were equipped with pressure gauges. The wear test was divided into the following two stages.

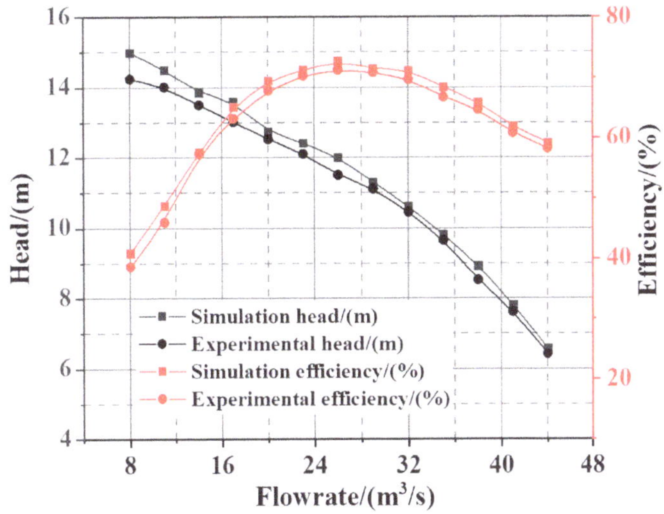

Figure 18. Comparison of external characteristics under the clean water condition.

Figure 19. Wear test bench.

1. The first stage

The total duration of the first-stage test was 24 h. The test was carried out under the condition of equal flow rate, and the running speed of the pump was 1500 r/min. Meanwhile, the solid particles were brown corundum, the density was 2900 kg/m^3, the particle size was 0.6 mm, and the solid volume was 15%. The mass of the impeller before the test was 2070 g. When the pump was running for 4 h, it was found that there was obvious wear at the intersection between the suction surface and the front cover plate, and slight wear was generated at the entrance of the blade. When the pump was running for 24 h, the weight of the impeller was 1960 g and the weight loss rate was 5.31%. Meanwhile, it was found that the blade was seriously worn at the entrance, and about 1–2 mm oval pits were created on the rear cover plate near the suction surface of the hub. Additionally, the wear of the blade suction surface and the front cover plate joint was more serious; there was an approximately 3 mm long groove on the blade, as shown in Figures 20 and 21.

Figure 20. Photograph of impeller wear in the first stage (at the inlet).

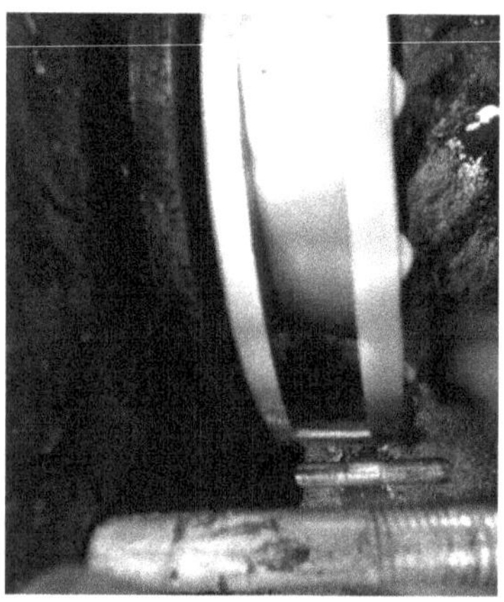

Figure 21. Photograph of impeller wear in the first stage (direction of the outlet side).

2. The second stage

The total duration of the second stage was 24 h. The test was carried out under the condition of equal flow rate, the solid particles were brown corundum, the density was 2900 kg/m^3, the particle size was 0.6 mm, the solid volume concentration increased to 25%, and the weight of the impeller before the test was 1960 g. When the pump was running for 24 h, the mass of the impeller was 1560 g and the weight loss rate was 20.41%. After the second-stage wear test, there was serious wear at the junction of the blade pressure surface and the front cover plate that extended from the blade leading edge to the trailing edge. At the same time, serious wear was generated at the junction of the impeller outlet and the rear cover plate, as shown in Figures 22 and 23.

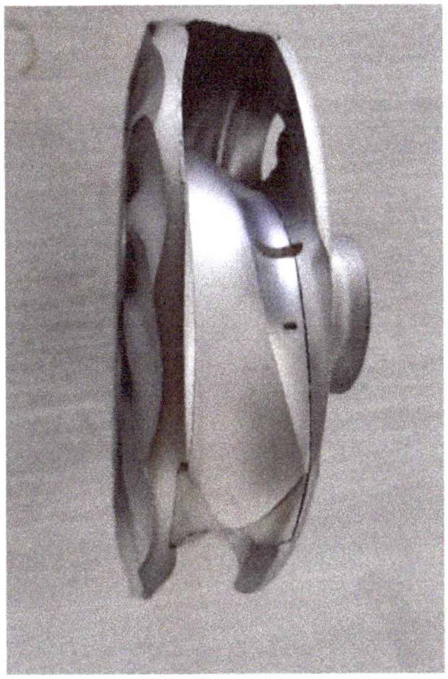

Figure 22. Photograph of the second-stage impeller wear.

Figure 23. Photograph of the second-stage impeller wear.

4.3. Analysis of Wear Distribution Characteristic

In this section, to analyze the wear distribution characteristic under the design flow condition, the calculation and experimental results with a solid volume concentration of 25% are discussed. Figure 24 shows a contrast diagram of the inlet wear of the impeller. The calculation results show that the high solid volume fraction was generated at the junction of the blade inlet and the rear cover plate. The experiment results showed that slight wear was created at the entrance of the front cover, and the rear cover was worn out. From the previous analysis of blade wear, it could be seen that the high solid-phase slip velocity was generated at the entrance of the front cover plate and the solid phase slip velocity at the leading edge of the blade was larger, which resulted in the rear cover plate being easier to wear out. Figure 25 shows the contrast diagram of impeller outlet wear. The numerical simulation results showed that the solid phase slip velocity at the junction between the blade and the rear cover plate was larger and the slip velocity gradually declined from the trailing edge to the leading edge. At the same time, the numerical simulation results were in good agreement with the experimental results. Therefore, enhancing the strength of the junction between the blade leading edge and the rear cover plate was beneficial for improving the service life of the slurry pump.

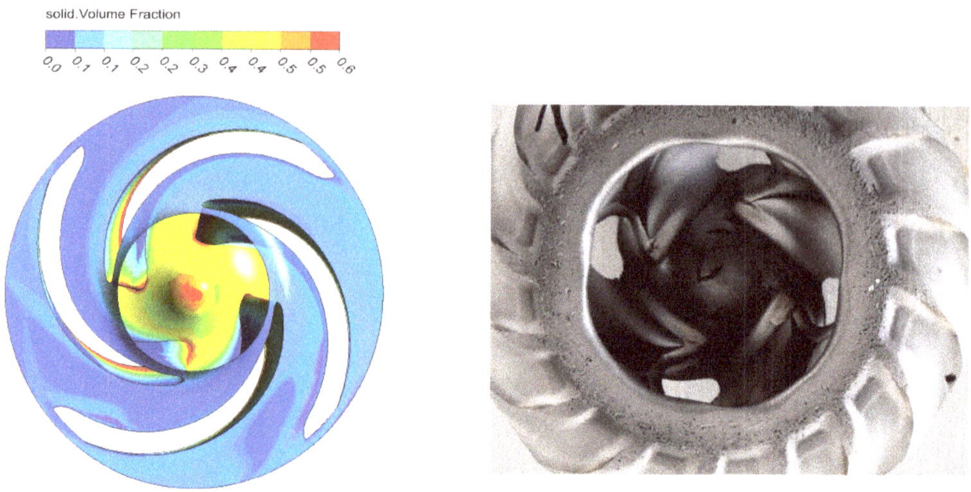

(**a**) Simulation results (**b**) Test picture

Figure 24. (**a**,**b**) Comparison of impeller inlet wear.

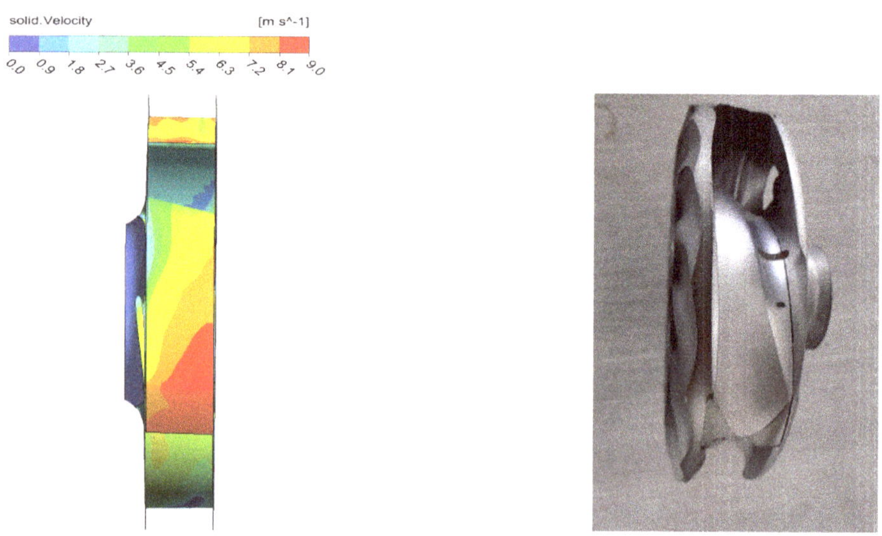

(**a**) Simulation results (**b**) Test picture

Figure 25. (**a**,**b**) Comparison of impeller outlet wear.

5. Conclusions

In this paper, a 25 MZ slurry pump was selected as the research object, and the Euler–Euler multiphase flow model was employed to analyze the internal flow characteristic of the slurry pump under the conditions of clear water and solid–liquid two-phase flow. The following conclusions were obtained:

(1) Pre-rotation was generated at the inlet section, and the pre-rotation direction was consistent with the rotation direction of the impeller under part-load conditions. Meanwhile, the ability of the pre-rotation continuously decreased with the increase in flow rate. When the flow rate reached 1.5 Q_{BEP}, the flow at the inlet section was relatively stable. Furthermore, the solid volume fraction of the blade pressure surface gradually moved from the leading edge to the trailing edge with the increase in flow rate. When the flow rate increased to 1.5 Q_{BEP}, the solid particles moved to the rear cover plate under the effect of inertia force, which resulted in a large solid volume fraction at the interface between the impeller and the rear cover plate.
(2) With the increase in the particle volume concentration, the viscosity of the solid–liquid two-phase flow increased and the friction force and energy loss between the internal fluids gradually increased, which resulted in a decrease in the head. When the concentration was 5%, the head decreased by 0.33% and the efficiency decreased by 1.60%. When the concentration was 25%, the head decreased by 8.9% and the efficiency decreased by 8.09%. Therefore, the head and efficiency of the pump decreased with increases in particle volume concentration.
(3) When the particle diameter gradually increased, the energy driving the solid particles was greater and the head gradually decreased. At the same time, with the increase in solid particle diameter, the friction loss between particles and impeller increased, which led to the increases in impeller torque and motor power consumption. Therefore, when the particle diameter increased from 0.15 to 0.6 mm, the pump head decreased from 12 to 11.42 m and the efficiency decreased from 71.25% to 67.35%.
(4) Finally, clean water and solid–liquid two-phase flow wear tests of the slurry pump were carried out. The calculation deviation in the full flow range was less than 5%, so the calculation model could accurately predict the performance of the pump. By comparing the wear distribution of numerical and experiment, it was found that high solid volume fraction and solid-phase slip velocity were generated at the junction of the blade leading edge and the rear cover plate, which made the latter easier to wear out. Therefore, enhancing the strength of the junction between the blade leading edge and the rear cover plate would be beneficial for improving the service life of a slurry pump.

In this paper, in order to simplify the analysis model, a uniformly spherical particle was employed in the calculation process. However, particles in a slurry pump have a lot of different shapes during actual operation. Therefore, the influence of particle shapes on the hydraulic performance and wear characteristics of slurry pumps will be investigated in the future.

Author Contributions: This was a joint work, and the authors worked according to their expertise and capability. G.P. and L.T.: investigation, analysis, writing, and revision; H.C.: for methodology and revision; S.H. and D.Y.: validation and revision; B.Y.: data analysis. All authors have read and agreed to the published version of the manuscript.

Funding: The authors gratefully acknowledge the support from Open Research Fund Program of State key Laboratory of Hydro-science and Engineering grant number: sklhse-2020-E-01; Open Research Subject of Key Laboratory of Fluid Machinery and Engineering (Xihua University) grant number LTJX2021-003; Open Research Subject of Hubei Key Laboratory of Hydroelectric Machinery De-sign and Maintenance (China Three Gorges University) grant number 2020KJX07; the 69th batch of general funding from the China Postdoctoral Science Foundation grant number: 2021M691298; Natural Science Research Project of Jiangsu Province Colleges and Universities: 21KJB570004, Priority Academic Program Development of Jiangsu Higher Education Institutions (PAPD).

Institutional Review Board Statement: Not applicable.

Informed Consent Statement: Not applicable.

Data Availability Statement: Not applicable.

Conflicts of Interest: The authors declare no conflict of interest.

References

1. Wang, R.; Guan, Y.; Jin, X.; Tang, Z.; Zhu, Z.; Su, X. Impact of Particle Sizes on Flow Characteristics of Slurry Pump for Deep-Sea Mining. *Shock. Vib.* **2021**, *2021*, 1–13. [CrossRef]
2. Shi, B.; Pan, J.; Wu, L.; Zhang, X.; Qiu, Y.; Zhang, Y. A Prediction Method of Wear for Volute Casing of a Centrifugal Slurry Pump. *Geofluid* **2020**, *2020*, 1–12. [CrossRef]
3. Peng, G.; Huang, X.; Zhou, L.; Zhou, G.; Zhou, H. Solid-liquid two-phase flow and wear analysis in a large-scale centrifugal slurry pump. *Eng. Fail. Anal.* **2020**, *114*, 104602. [CrossRef]
4. Li, Y.; Zeng, X.; Lv, W.; He, Z. Centrifugal pump wear for solid–liquid two-phase flows based on computational fluid dynamics–discrete element method coupling. *Adv. Mech. Eng.* **2020**, *12*, 1–17. [CrossRef]
5. Tan, M.; Lian, Y.; Liu, H.; Wu, X.; Ding, R. Visualizing test on the pass-through and collision characteristics of coarse particles in a double blade pump. *Int. J. Nav. Archit. Ocean. Eng.* **2018**, *10*, 1–8. [CrossRef]
6. Shen, Z.; Chu, W.; Yan, S.; Chen, X.; Guo, Z.; Zhong, Y. Study of the performance and internal flow in centrifugal pump with grooved volute casing. *Mod. Phys. Lett. B* **2020**, *34*, 2050268. [CrossRef]
7. Wang, Y.; Han, C.; Zhou, Y.; Lin, Z.; Ma, J.; Li, X.; Zhang, W. Wear Characteristics of Dense Fine Particles Solid-Liquid Two-Phase Fluid Centrifugal Pump with Open Impellers. *Shock. Vib.* **2021**, *2021*, 1–13. [CrossRef]
8. Zhao, W.; Zhao, G. Numerical investigation on the transient characteristics of sediment-laden two-phase flow in a centrifugal pump. *J. Mech. Sci. Technol.* **2018**, *32*, 167–176. [CrossRef]
9. Wang, Y.; Chen, B.; Zhou, Y.; Ma, J.; Zhang, X.; Zhu, Z.; Li, X. Numerical Simulation of Fine Particle Solid-Liquid Two-Phase Flow in a Centrifugal Pump. *Shock. Vib.* **2021**, *2021*, 1–10. [CrossRef]
10. Wang, Y.; Chen, J.; Xie, L.; Liu, H.-L.; Luo, K.-K. Combined experimental and computational investigation of the effect of coating on operation characteristics of solid–liquid two-phase flow centrifugal pump. *Mod. Phys. Lett. B* **2021**, *35*, 2150062. [CrossRef]
11. Gu, Y.; Liu, N.; Mou, J.; Zhou, P.; Qian, H.; Dai, D. Study on solid–liquid two-phase flow characteristics of centrifugal pump impeller with non-smooth surface. *Adv. Mech. Eng.* **2019**, *11*, 1–12. [CrossRef]
12. Yan, C.; Liu, J.; Zheng, S.; Huang, B.; Dai, J. Study on the effects of the wear-rings clearance on the solid-liquid two-phase flow characteristics of centrifugal pumps. *Symmetry* **2020**, *12*, 2003. [CrossRef]
13. Wang, P.; Zhu, X.; Li, Y. Analysis of Flow and Wear Characteristics of Solid–Liquid Two-Phase Flow in Rotating Flow Channel. *Processes* **2020**, *8*, 1512. [CrossRef]
14. Tang, C.; Kim, Y.J. CFD-DEM simulation for the distribution and motion feature of solid particles in single-channel pump. *Energies* **2020**, *13*, 4988. [CrossRef]
15. Wang, H.L.; Hu, Q.X.; Yang, Y.; Wang, C. Performance differences of electrical submersible pump under variable speed schemes. *Int. J. Simul. Model.* **2021**, *20*, 76–86. [CrossRef]
16. Zhou, J.; Zhao, M.; Wang, C.; Gao, Z. Optimal design of diversion piers of lateral intake pumping station based on orthogonal test. *Shock. Vib.* **2021**, *2021*, 1–9. [CrossRef]
17. Shi, L.; Zhu, J.; Tang, F.; Wang, C. Multi-disciplinary optimization design of axial-flow pump impellers based on the approximation model. *Energies* **2020**, *13*, 779. [CrossRef]
18. Wang, H.; Long, B.; Wang, C.; Han, C.; Li, L. Effects of the impeller blade with a slot structure on the centrifugal pump performance. *Energies* **2020**, *13*, 1628. [CrossRef]
19. Wang, C.; Chen, X.; Qiu, N.; Zhu, Y.; Shi, W. Numerical and experimental study on the pressure fluctuation, vibration, and noise of multistage pump with radial diffuser. *J. Braz. Soc. Mech. Sci. Eng.* **2018**, *40*, 481. [CrossRef]
20. Zhang, L.; Wang, C.; Zhang, Y.; Xiang, W.; He, Z.; Shi, W. Numerical study of coupled flow in blocking pulsed jet impinging on a rotating wall. *J. Braz. Soc. Mech. Sci. Eng.* **2021**, *43*, 508. [CrossRef]

machines

Article

Optimization Design of Energy-Saving Mixed Flow Pump Based on MIGA-RBF Algorithm

Rong Lu [1], Jianping Yuan [1], Guangjuan Wei [2], Yong Zhang [2], Xiaohui Lei [3] and Qiaorui Si [1,*]

1 Research Center of Fluid Machinery Engineering and Technology, Jiangsu University, Zhenjiang 212013, China; 13913446553@163.com (R.L.); yh@ujs.edu.cn (J.Y.)
2 Jiangsu Xugong Construction Machinery Institute Co., Ltd., Xuzhou 221004, China; wgj0903@163.com (G.W.); xcmgad@163.com (Y.Z.)
3 State Key Laboratory of Simulation and Regulation of Water Cycle in River Basin, China Institute of Water Resources and Hydropower Research, Beijing 100038, China; LXH@iwhr.com
* Correspondence: siqiaorui@ujs.edu.cn

Abstract: Mixed flow pumps driven by hydraulic motors have been widely used in drainage in recent years, especially in emergency pump trucks. Limited by the power of the truck engine, its operating efficiency is one of the key factors affecting the rescue task. In this study, an automated optimization platform was developed to improve the operating efficiency of the mixed flow pump. A three-dimensional hydraulic design, meshing, and computational fluid dynamics (CFD) were executed repeatedly by the main program. The objective function is to maximize hydraulic efficiency under design conditions. Both meridional shape and blade profiles of the impeller and diffuser were optimized at the same time. Based on the CFD results obtained by Optimal Latin Hypercube (OLH) sampling, surrogate models of the head and hydraulic efficiency were built using the Radial Basis Function (RBF) neural network. Finally, the optimal solution was obtained by the Multi-Island Genetic Algorithm (MIGA). The local energy loss was further compared with the baseline scheme using the entropy generation method. Through the regression analysis, it was found that the blade angles have the most significant influence on pump efficiency. The CFD results show that the hydraulic efficiency under design conditions increased by 5.1%. After optimization, the incidence loss and flow separation inside the pump are obviously improved. Additionally, the overall turbulent eddy dissipation and entropy generation were significantly reduced. The experimental results validate that the maximum pump efficiency increased by 4.3%. The optimization platform proposed in this study will facilitate the development of intelligent optimization of pumps.

Keywords: mixed flow pump; optimization platform; surrogate model; entropy generation; MIGA

Citation: Lu, R.; Yuan, J.; Wei, G.; Zhang, Y.; Lei, X.; Si, Q. Optimization Design of Energy-Saving Mixed Flow Pump Based on MIGA-RBF Algorithm. *Machines* **2021**, 9, 365. https://doi.org/10.3390/machines9120365

Academic Editor: Ahmed Abu-Siada

Received: 2 December 2021
Accepted: 15 December 2021
Published: 17 December 2021

1. Introduction

With the emergence of global extremes in recent years, the frequency of disasters such as droughts and urban flooding has suddenly increased, which greatly affect security and the economy. Because of its flexibility, the emergency pump truck has obvious advantages in dealing with urban flooding [1]. To minimize economic losses, the main requirements of dealing with urban flood are rapid drainage and long-distance transportation. Because of its characteristics of large flow and high head, mixed flow pumps have been widely used in emergency situations. The mobile pump truck is a highly integrated drainage equipment, with all power coming from the engine. Limited by the vehicle output power, improving the pump efficiency and reducing the operating energy consumption can provide protection for emergency work. Therefore, a mixed flow pump with high efficiency is required for the design of the emergency pump truck.

The impeller and diffuser are the key components of energy conversion in pumps. Much of the research mainly focuses on the meridional flow passage shape and blade profile. Based on computational fluid dynamics (CFD) technology, Kim et al. [2] performed optimization of the meridional shape of a mixed flow pump impeller to improve its suction performance. Hao et al. [3] further investigated the effect of the hub and shroud radius ratio on the hydraulic efficiency of a mixed flow pump by numerical simulation. Ji et al. [4] analyzed the effect of the blade thickness on the rotating stall of a mixed flow pump using the entropy generation method. Ikuta et al. [5] found that the forward skew blade angle has an obvious effect on positive slope characteristics of the mixed flow pumps. The positive slope region can be moved to a smaller flow rate by increasing the skew blade angle. Most vaned mixed flow pumps are equipped with an unshrouded impeller. The tip clearance between the impeller and casing may cause adverse flow phenomena such as leakage, cavitation, and so on. Li et al. [6] studied the influence of tip clearance on the rotating stall in a mixed flow pump using CFD. By investigating the effect of rotational speed on the tip leakage vortex, Han et al. [7] found that with the increase of rotational speed, the leakage flow and oscillating frequency of the tip leakage vortex will also increase. To inhibit the leakage and improve the energy performance in the unshrouded centrifugal pumps, Wang et al. [8] proposed a T-shaped blade. The CFD results show that the leakage and flow loss of the T-shaped blade is decreased. Zhu et al. [9] studied dynamics performance of the centrifugal pumps with different diffuser vane heights and found that the half vane diffuser could increase the flow uniformity and reduce the pressure pulsation intensity. The effect of the divergence angle of the diffuser on the performance of a centrifugal pump was studied by Khoeini et al. [10]; the results show that the diffuser parameters have a remarkable influence on the head and efficiency. Wang et al. [11] performed the optimization of the vaned diffuser in a centrifugal pump to improve the pump efficiency. Kim et al. [12] presented an optimization process based on a radial basis neural network model to optimize four design variables of a diffuser in a mixed flow pump, and the optimization increased in efficiency by 9.75% at the design point.

Although much of the research involves the improvement of the impeller and diffuser, the blade profile of the mixed flow pump is space-distorted and too many geometric parameters make it difficult to be fully optimized. The inverse design method (IDM) is a technique to design the blade profile by distribution of blade loading. Compared with the traditional design method, fewer design parameters are required for IDM [13]. Wang et al. [14] performed the optimization of the mixed flow impeller using IDM. The effect of different vortex distributions of the blade exit on the hydraulic performance were investigated using CFD. Lu et al. [15] proposed a modified inverse design method for the optimization of the runner blade of the mixed flow pump. The IDM is also suitable for the design of the axial impeller and diffuser [16]. Although the IDM has certain advantages, it only involves the design of the blade profile with no consideration for the meridional passage shape.

Based on the above literature, the application of CFD is almost indispensable in pump optimization. In recent years, CFD technology combined with computer aided optimization methods have been widely used in the design and optimization of fluid machinery [17]. Design of experiment (DOE) and surrogate models are the most popular auxiliary methods. The precision of the surrogate model is one of the key factors for the success of optimization. Wang et al. [18] tested the accuracy of different surrogate models in centrifugal pump optimization and the results showed that the prediction accuracy of the radial basis neural network is better than other models. Si et al. [19] implemented the multi-condition optimization of an electronic pump using DOE. The influence of each parameter on the head and efficiency was estimated, and the number of design parameters were diminished by analysis of variance (AOV). Xu et al. [20] conducted the multi-parameter optimization of a mixed flow pump based on the orthogonal experimental method and RBF neural network, while the meridional parameters of impeller were not included. Pei et al. [21] proposed a modified particle swarm algorithm to accelerate the speed of optimization, and an artificial neural network was further applied to build the mathematical model. Zhu et al. [22] applied the global dynamic criterion algorithm to the optimization of a vaned mixed flow pump, and the parallel running was realized to shorten the time consumption. Huang et al. [23] developed a modified non-dominated sorting genetic algorithm II (NSGA-II) coupled with a dynamic crowding distance (DCD) method, which contributed to the search for the pareto-front.

The similarity of the studies above is that most research only focuses on the optimization of a single hydraulic component, ignoring the interaction between the rotor and stator. Further research is needed to optimize the matching of the impeller and diffuser. In this study, a shrouded impeller was proposed to suppress the complex tip leakage flow. The MIGA-RBF algorithm combined with CFD technology was introduced to improve the hydraulic efficiency of the mixed flow pump. An automatic optimization platform integrating 3D design, meshing, and numerical simulation was built. Variables involving the meridional shape and blade profile of both the impeller and diffuser were optimized to fully consider the rotor–stator interaction. The flow regime and local energy loss were analyzed in detail. The paper is organized as follows: the relevant research status is described in Section 1; the pump information and numerical theories are briefly introduced in Section 2; the concrete optimization methods are illustrated in Section 3; the detailed results are present in Section 4; and finally, the conclusions are provided in Section 5.

2. Requirements Description

2.1. Information on Mixed Flow Pump

The pump mainly consists of an inlet, a shrouded mixed impeller, a vaned diffuser, a hydraulic motor, and an outlet. Figure 1 shows the structure of the studied pump. Table 1 shows the design and partial geometric parameters of the baseline scheme designed by the traditional one-dimensional theory.

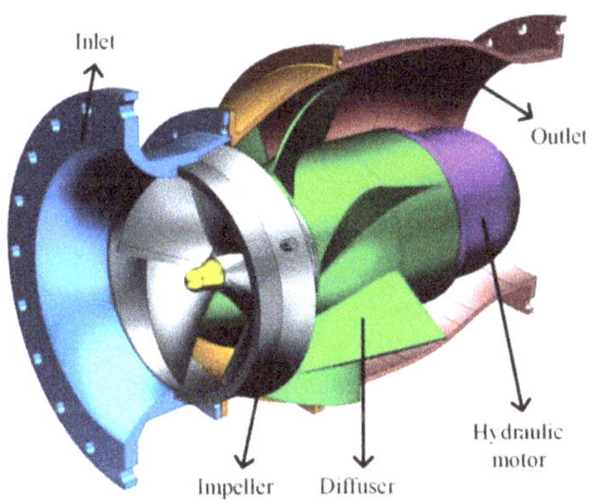

Figure 1. Structure of Mixed Flow Pump.

Table 1. Design specifications of the mixed flow pump.

Parameters	Symbols	Values
Rate flow	Q	3500 m^3/h
Rotational speed	n	1500 r/min
Head	H	16 m
Impeller suction diameter	D_s	380 mm
Impeller outlet diameter	D_2	340 mm
Impeller outlet width	b_2	125 mm
Diffuser inlet width	b_3	125 mm
Diffuser outlet angle	α_4	90°

2.2. Numerical Method

The external characteristic and inner flow regime were numerically investigated by ANSYS CFX. The governing equations listed below are discretized using the finite volume approximation.

$$\frac{\partial \overline{u_j}}{\partial x_j} = 0 \tag{1}$$

$$\frac{\partial \overline{u_i}}{\partial t} + \overline{u_j}\frac{\partial \overline{u_i}}{\partial x_j} = \overline{f_i} - \frac{1}{\rho}\frac{\partial \overline{p}}{\partial x_i} + \nu\frac{\partial^2 \overline{u_i}}{\partial x_j \partial x_j} - \frac{\partial \overline{u_i' u_j'}}{\partial x_j} \tag{2}$$

The whole computational domains contain four parts: impeller, diffuser, inlet, and discharge pipes. The inlet and outlet sections were extended more than five times the pipe diameter to consider the fully developed turbulent flow. The k–ω shear stress transport model (SST k–ω) with an automatic wall function was used as a turbulence closure model. The total pressure and mass flow were applied to the inlet and outlet boundaries, respectively. For steady simulation, the frozen rotor strategy was adopted to deal with the interface between the rotor and stator. Using the steady state result as the initial file, the transient simulation was conducted with the transient rotor–stator interface mode. The timestep for the transient case was 3.33×10^{-4}, which corresponds to 3° of the impeller rotation. The root mean square (RMS) residuals for both the steady and transient cases were selected as 10^{-4}.

Figure 2 shows the grid system in this study. Due to the advantages in the number of grids, calculation accuracy, and convergence, hexahedral grids with high quality were used for all domains. The grids in the impeller and diffuser were generated by Turbo-Grid. For inlet and outlet pipes, ICEM with O-Block strategy was adopted to discretize the domains. To treat the high velocity gradient, all near wall surfaces were refined with prism layers. The expansion ratio of near wall grids is 1.2. The first layer nodes distance was controlled to ensure the dimensionless distance, $y^+ < 50$ [24], which could meet the need of the grid for the SST k–ω turbulence model.

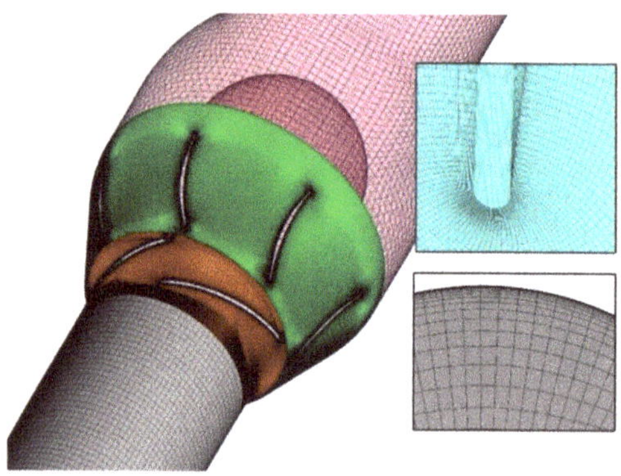

Figure 2. Hexahedral mesh of computational domains.

A grid independent check (GIC) was conducted to make sure that the simulations in the optimization process are free from errors caused by the grid number. The results of the GIC are shown in Table 2, and the grid refinement factor is approximately 1.3. When the grid number increases from 7.63 million to 9.92 million, the relative error of head is 0.24%. Finally, the grid number of 7.63 million was used for subsequent optimizations and simulations.

The comparison of the head and efficiency curves is shown in Figure 3. The CFD curves are obtained by steady simulations. The tested performance curves were acquired from an experimental study presented in Section 4.4. As the mechanical and volumetric efficiency were not considered in simulations, the results obtained by CFD are generally higher than the experimental values. The maximum relative errors of the head and hydraulic efficiency are lower than 5%. The absolute predicted deviations for the head and hydraulic efficiency at the design condition are 0.55 m and 3.83%, respectively. Thus, the numerical accuracy is suitable for the following optimization study.

Table 2. Grid independence check

Grid Number (Million)	*H* (m)	Relative Error (%)
3.47	14.87	-
4.52	15.75	5.91%
5.87	16.15	2.54%
7.63	16.39	1.49%
9.92	16.43	0.24%

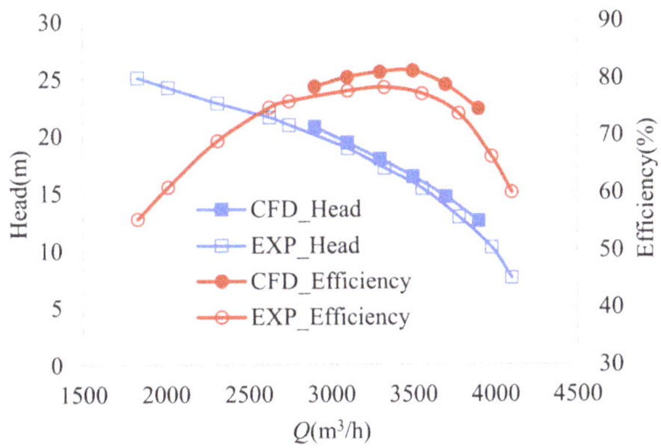

Figure 3. Comparison of performance curves.

2.3. Entropy Generation Theory

Entropy is one of the physical qualities that characterize the state of matter in thermodynamics. The nature of entropy indicates the chaos inside the system. The entropy generation theory is proposed based on the second law of thermodynamics, which effectively explains the flow direction and loss of energy. Flow losses in fluid machinery are very complex as total pressure loss cannot visualize and locate the maximum flow loss in pumps. To explain the influence of the optimization variable on hydraulic performance, the details of flow loss in the pump are revealed in depth. A visualization method of flow loss based on the entropy generation theory was proposed. The transfer equation of entropy for incompressible fluid can be described as [25]:

$$\rho\left(\frac{\partial s}{\partial t}+u\frac{\partial s}{\partial x}+v\frac{\partial s}{\partial y}+w\frac{\partial s}{\partial z}\right)=-\mathrm{div}\left(\frac{\vec{q}}{T}\right)+\frac{\Phi}{T}+\frac{\Phi_\Theta}{T^2} \quad (3)$$

where s is the specific entropy, T is the thermodynamic temperature, and u, v, and w are the Cartesian velocity components, $\vec{q}$ is the heat flux density vector, $\frac{\Phi}{T}$ and $\frac{\Phi_\Theta}{T^2}$ represent the entropy generation rate caused by dissipation and heat transfer, respectively.

According to the Reynolds averaged Navier Stokes (RANS) approach for turbulent flows, prior to time-averaging the equation, all quantities are split into time-mean and fluctuating parts; thus, the time-averaged governing equation then reads [25]:

$$\rho\left(\frac{\partial \bar{s}}{\partial t}+\bar{u}\frac{\partial \bar{s}}{\partial x}+\bar{v}\frac{\partial \bar{s}}{\partial y}+\bar{w}\frac{\partial \bar{s}}{\partial z}\right)=\overline{\mathrm{div}\left(\frac{\vec{q}}{T}\right)}-\rho\left(\frac{\partial\overline{u's'}}{\partial x}+\frac{\partial\overline{v's'}}{\partial y}+\frac{\partial\overline{w's'}}{\partial z}\right)+\left(\overline{\frac{\Phi}{T}}\right)+\left(\overline{\frac{\Phi_\Theta}{T^2}}\right) \quad (4)$$

In this research, the heat transfer is neglected; hence, only the entropy generation by dissipation ($\frac{\Phi}{T}$) is considered. The time average format of entropy generation can be expressed as [26]:

$$\overline{\left(\frac{\Phi}{T}\right)} = S = S_{\overline{D}} + S_{D'} \tag{5}$$

$$S_{\overline{D}} = \frac{\mu}{T} \cdot \{2\left[\left(\frac{\partial \overline{u}}{\partial x}\right)^2 + \left(\frac{\partial \overline{v}}{\partial y}\right)^2 + \left(\frac{\partial \overline{w}}{\partial z}\right)^2\right] + \left(\frac{\partial \overline{u}}{\partial y} + \frac{\partial \overline{v}}{\partial x}\right)^2 + \left(\frac{\partial \overline{u}}{\partial z} + \frac{\partial \overline{w}}{\partial x}\right)^2 + \left(\frac{\partial \overline{v}}{\partial z} + \frac{\partial \overline{w}}{\partial y}\right)^2\} \tag{6}$$

$$S_{D'} = \frac{\mu}{T} \cdot \{2\left[\overline{\left(\frac{\partial u'}{\partial x}\right)^2} + \overline{\left(\frac{\partial v'}{\partial y}\right)^2} + \overline{\left(\frac{\partial w'}{\partial z}\right)^2}\right] + \overline{\left(\frac{\partial u'}{\partial y} + \frac{\partial v'}{\partial x}\right)^2} + \overline{\left(\frac{\partial u'}{\partial z} + \frac{\partial w'}{\partial x}\right)^2} + \overline{\left(\frac{\partial v'}{\partial z} + \frac{\partial w'}{\partial y}\right)^2}\} \tag{7}$$

where $\overline{u}$, $\overline{v}$, and $\overline{w}$ represent the time-averaged velocity components, and u', v', and w' are the velocity fluctuation components.

The first term, $S_{\overline{D}}$, which includes the average velocity gradient, can be interpreted as the entropy generation dissipated in the average flow field. The second term,$S_{D'}$, which contains the gradients of the fluctuating velocities, cannot be obtained directly; thus, it is often called indirect or turbulent dissipation. Herwig et al. [27] found that there is a close relationship between this term and turbulent eddy dissipation. Thus, the $S_{D'}$ is defined as:

$$S_{D'} = \frac{\rho\overline{\varepsilon}}{T} \tag{8}$$

Because of the viscosity, there is a large velocity gradient near the wall. The time average variables are obviously affected by the effect of the wall surface, and it is hard to solve them accurately. Therefore, Hou et al. [28] proposed a new method to calculate the wall entropy production rate:

$$S_{\text{wall}} = \frac{\boldsymbol{\tau}_w \boldsymbol{V}}{T} \tag{9}$$

where $\boldsymbol{\tau}_w$ represents the shear force on the wall, and $\boldsymbol{V}$ is the average velocity vector of the first layer of the grid near the wall surface.

By the volume and surface integration, the entropy generation power of each term can be calculated [29]:

$$P_{ave} = \iiint S_{\overline{D}} dV \tag{10}$$

$$P_{flu} = \iiint S_{D'} dV \tag{11}$$

$$P_{wall} = \iint S_{wall} dS \tag{12}$$

The total volume entropy generation power (P_v) is defined as the sum of Equations (10) and (11) [29]:

$$P_v = P_{ave} + P_{flu} \tag{13}$$

All the statistical variables in Equations (6) and (8) were arithmetically averaged in the last revolution of the impeller.

3. Optimizing Method

To reduce the human factor and shorten the optimization time, a highly integrated optimization platform was established. Figure 4 shows the procedure of this optimization. The entire optimization process mainly consists of two stages. The main task of the first stage is to establish accurate surrogate models. In this process, the DOE method combined with the CFD method was used for sampling in m-dimensional space. Regression analysis was introduced to test the accuracy of the surrogate models. If the accuracy is lower than the threshold value, DOE will be repeatedly performed until satisfactory results are obtained. Secondly, the optimal solution was obtained by solving the surrogate models with MIGA. The optimization platform was established by the integration of CFturbo, Turbo-Grid, ICEM, and CFX. Disk Operating System (DOS) commands and script files were used to run the software in the background. The DOE program was used to drive the platform.

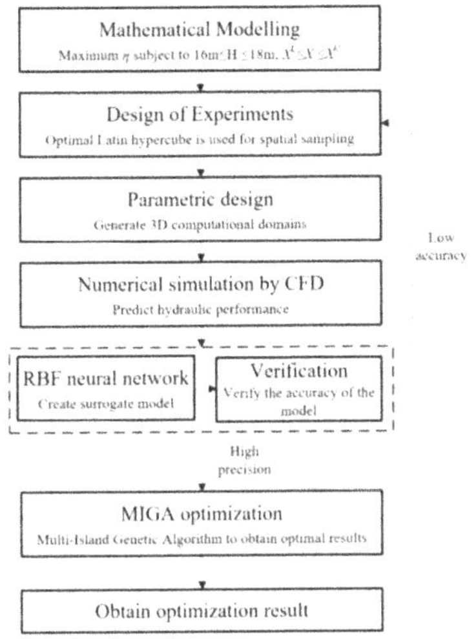

Figure 4. Procedure of the optimization design.

3.1. Mathematical Model

The main purpose of this optimization is to improve the hydraulic efficiency, which belongs to the single objective optimization. The mathematical model of this problem is as follow:

maximize

$$\eta = F(\mathbf{X}_m) \tag{14}$$

subject to

$$16\text{m} \leq H \leq 18\text{m} \tag{15}$$

$$\mathbf{X}_m^L \leq \mathbf{X}_m \leq \mathbf{X}_m^U \tag{16}$$

where η is the hydraulic efficiency under the design flow rate, $\mathbf{X} = [x_1, x_2, \ldots, x_m]$ is the vector of the m design variables. Both the meridional and blade shape of the impeller and diffuser were optimized in this study. The definition of the geometric parameters is shown in Figure 5. The meridional shape of the flow passage in optimized schemes was formed by multipoint Bezier curves, while the benchmark design adopts arcs and straight lines. To reduce the number of optimization variables appropriately, some parameters are reasonably constrained: β_1, β_2, φ_b, α_3, and φ_d vary linearly from hub to shroud, b_3 is strictly equal to b_2, while α_4 is consistent from hub to shroud. Finally, 14 were selected for the optimization. Table 3 shows the ranges of the optimization variables.

3.2. Design of Experiment

The Latin hypercube experimental design is an efficient experimental design method with the advantages of effective space filling and the ability to fit second-order or more nonlinear relationships. The optimal Latin hypercube (OLH) improves the uniformity of the Latin hypercube and makes all sampling points more evenly distributed in the design space [30]. To establish more accurate surrogate models, the OLH was used to perform the space sampling. The number of samples is related to the accuracy of the surrogate model. An accurate model requires enough sampling points; however, too many sampling points will consume a lot of time. For tradeoffs between model accuracy and sampling time, the regression analysis was performed using the coefficient of determination R^2 after DOE.

The closer R^2 is to 1, the more accurate the model will be. Usually R^2 is greater than 0.9 [31]. To ensure the accuracy of the surrogate model, if the value of R^2 is below the threshold value (0.96), additional sampling points will be added. Within each modeling process, 50 sampling points were added. R^2 is defined as follows [31]:

$$R^2 = 1 - \frac{\sum_{i=1}^{n} (y_i - \overline{y})^2}{\sum_{i=1}^{n} (\hat{y} - y_i)^2} \tag{17}$$

where n is the number of samples, $\overline{y}$ represents the average response, $\hat{y}$ is the predicted value, and y_i is the actual value.

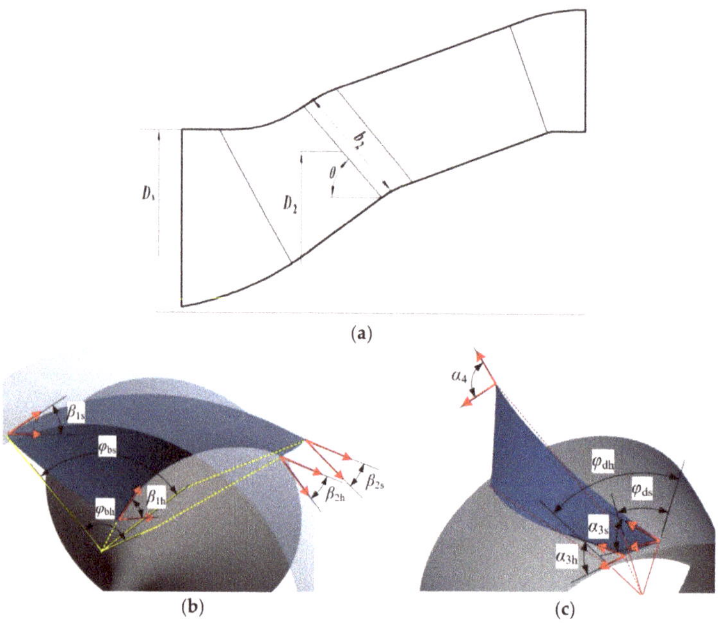

Figure 5. Schematic diagram of geometric parameters. (**a**) Definition of meridional parameters. (**b**) Definition of impeller blade angles. (**c**) Definition of diffuser vane angles.

Table 3. Ranges of optimization parameters.

Optimization Parameters	Lower Bound	Baseline	Upper Bound
Impeller suction diameter D_s (mm)	360	380	390
Impeller outlet diameter D_2 (mm)	330	340	350
Impeller outlet width b_2 (mm)	120	125	130
Inlet angle of blade hub β_{1h} (deg)	35	37	50
Inlet angle of blade shroud β_{1s} (deg)	10	15	25
Outlet angle of blade hub β_{2h} (deg)	20	23	35
Outlet angle of blade shroud β_{2s} (deg)	30	35	45
Wrap angle of blade hub φ_{bh} (deg)	90	102	110
Wrap angle of blade shroud φ_{bs} (deg)	85	91	95
Inlet angle of vane hub α_{3h} (deg)	35	41	50
Inlet angle of vane shroud α_{3s} (deg)	45	52	60
Vane outlet angle α_4 (deg)	85	90	92
Wrap angle of vane hub φ_{dh} (deg)	30	34	40
Wrap angle of vane shroud φ_{ds} (deg)	15	19	25

3.3. RBF Neural Network

The artificial neural network (ANN) is a kind of bionic computing system, which has good, nonlinear fitting, learning and updating abilities; hence, it is widely used in machine learning, optimization design, and other fields. The radial basis function (RBF) is one kind of a three layer forward neural network. Figure 6 shows the structure of the RBF [32]. It consists of three layers: input layer, hidden layer, and output layer. In the RBF neural network, the input vector is directly mapped to the hidden layer through the function, and there is no need to adjust the connection weights.

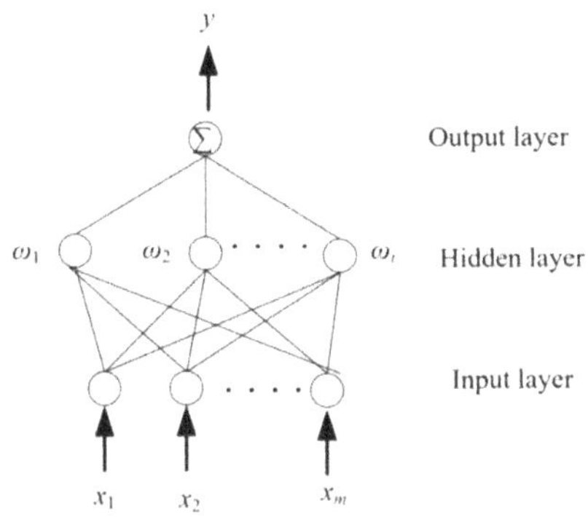

Figure 6. Principle of RBF neural network.

The independent variable of the RBF is the Euclidean distance between the test point and the sample point. The output of hidden layer is [32]:

$$R(x_m) = \exp\left(-\frac{||x_m - c_i||}{2\sigma_i^2}\right) \tag{18}$$

where c_i is the center vector of the Gaussian function, and σ_i represents the width of the ith Gaussian function.

The output layer responds to the action of the input mode, and there is a linear mapping from the output $R(x_m)$ of the hidden layer to the output layer y [32]:

$$y = \sum \omega_i \exp\left(-\frac{||x_m - c_i||}{2\sigma_i^2}\right) \tag{19}$$

where ω_i is the weight between the hidden layer and the output layer.

3.4. MIGA Algorithm

The genetic algorithm (GA) is a very classical algorithm widely used in multidisciplinary optimization. It was first proposed in 1971 based on the rule of „survival of the fittest" in Darwin's evolution theory [33]. The algorithm imitates the genetic reproduction mechanism of organisms; regards the solution space as a population with a certain principle used to encode individuals in the population; and then genetic operations on the encoded individuals was performed. The main search steps of GA include: selection, crossover, mutation, and so on. The offspring or mutated individuals replace the old population using the elitism or diversity replacement strategy and form as the new population in the next generation. The optimal solution will be obtained from the new population through iteration, which can effectively solve the problems of large-scale combination optimization or discontinuous search space. GA is one of the most prominent stochastic optimization algorithms.

After years of development, there are many types of genetic algorithms. Among them, parallel distributed genetic algorithms (PDGAs) are the most popular ones. Further, Miki et al. [34] made improvements on the PDGAs, and they divided the solution space into many parts called "islands". When performing the optimization, some individuals are selected on each "island" to conduct optimization according to the principle of the GA, and then migrate to other "islands" for the same operation at certain intervals. This is the so called MIGA. Compared with the traditional GA, the migration operation was added in MIGA. The biggest advantage of MIGA is that it is good at global search and can avoid falling into the local optimal solution. Figure 7 shows the structure of MIGA. The algorithm settings are shown in Table 4. A total of 4000 iterations were executed.

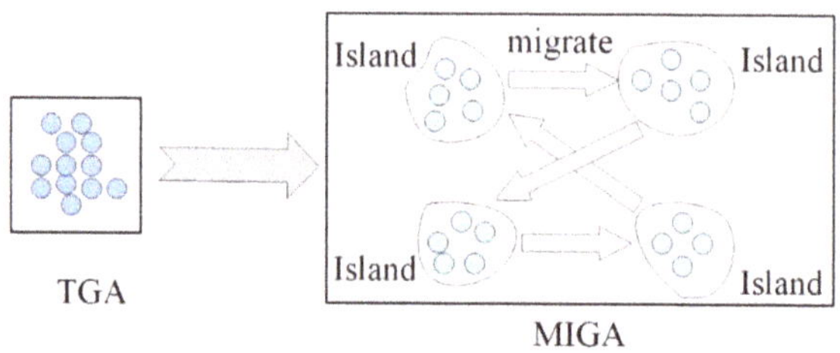

Figure 7. Structure of MIGA.

Table 4. Parameters adopted in MIGA.

Parameters	Value
Number of generations	10
Sub-population size	20
Number of islands	20
Rate of crossover	0.9
Rate of migration	0.01
Interval of migration	5
Rate of mutation	0.01

4. Results and Discussions

4.1. Regression Analysis

After repeating the experimental design five times, the accuracy of the model reached the requirements. Therefore, 250 sampling points were used to establish the RBF model. Figure 8 shows the results of the DOE. The objective variables show great fluctuations in the solution space. The maximum difference in head exceeds 3.5 m, the fluctuation amplitude of efficiency is larger than 3.5%. Therefore, the value ranges of variables are suitable for optimization.

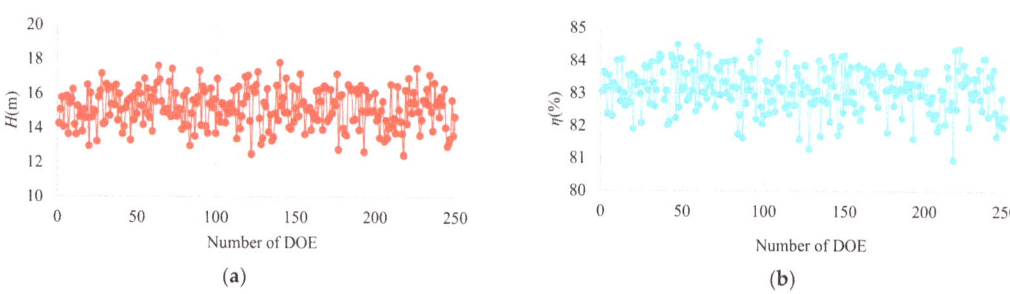

Figure 8. Results of DOE. (**a**) Head. (**b**) Efficiency.

The accuracy of the surrogate model was evaluated by regression analysis. Figure 9 shows the R^2 of both the head and efficiency, and they are all larger than 0.98. Therefore, the accuracy of these two models is sufficient for further optimization.

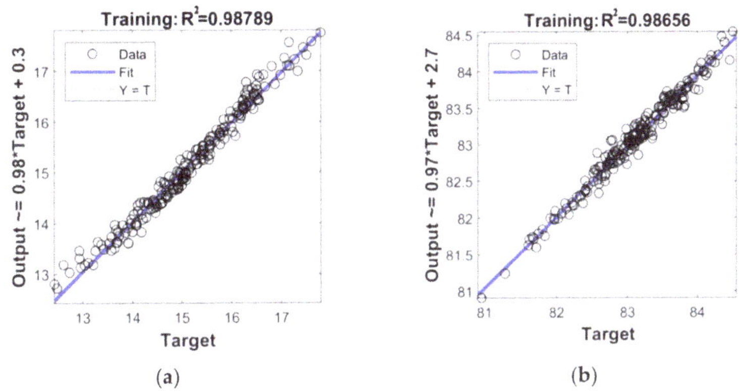

Figure 9. Regression analysis of surrogate models. (**a**) R^2 of head. (**b**) R^2 of efficiency.

4.2. Sensitivity Analysis

Sensitivity analysis is a method to study the influence of input parameters on output in a system. The sensitivity coefficients help the designer decide which parameters can be ignored in the product optimization. Correlation analysis is a linear analysis method based on the Pearson and Spearman correlation. The correlation coefficient, r, between two variables can be calculated as follows [35]:

$$r_{XY} = \frac{\Sigma(X - \overline{X})(Y - \overline{Y})}{\sqrt{\Sigma(X - \overline{X})^2}\sqrt{\Sigma(Y - \overline{Y})^2}} \tag{20}$$

It can be seen that r ranges from -1 to 1, and when $r > 0$, the two variables are positively correlated; otherwise, they are negatively correlated. It is generally believed that when the absolute value of the correlation coefficient is greater than 0.4, there is a significant correlation between the two variables. Correlation analysis is an effective tool to evaluate the influence of variables on the target and helps designers reduce the number of optimization variables.

Figure 10 shows the effects of the optimization variables on the pump head and efficiency. The blue bars indicate the positive effects and the red bars present the negative effects. The results show that b_2 and D_2 have a significant positive effect on the design head. This means that increasing the impeller outlet width and outlet diameter can significantly increase the design head of the mixed flow pump. In addition, the blade outlet angles also have an obvious positive effect on the design head, while φ_{dh}, φ_{ds}, and α_{3s} can be ignored with very small correlation coefficients. For the pump efficiency, the impeller blade shape plays an important role, especially the inlet and outlet angles. The blade outlet hub angle, β_{2h}, has the greatest positive effect on the pump efficiency. Reducing the blade inlet angle within this constraint is beneficial to improve the pump efficiency. However, efficiency is less sensitive to other parameters.

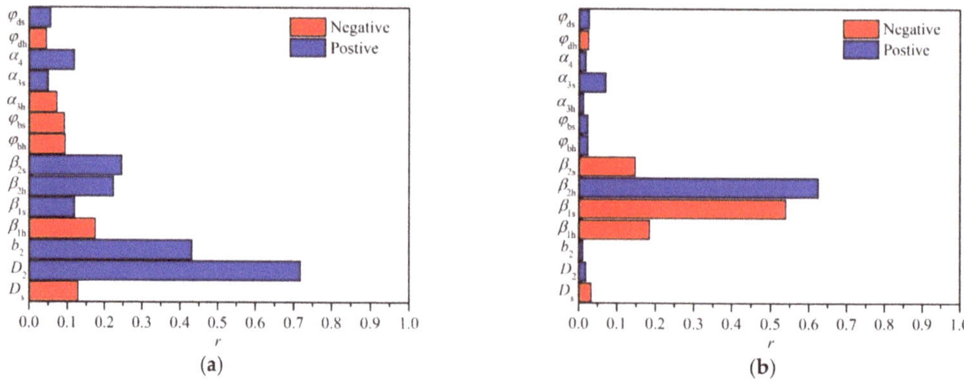

Figure 10. Correlation analysis for objectives. (**a**) Effect on head. (**b**) Effect on efficiency.

4.3. Optimization Results

By solving the RBF surrogate models, the best solution was obtained. Table 5 shows the comparison of geometric parameters between the initial and optimized schemes. The 3D geometry comparison of the mixed impeller and vaned diffuser between the initial design and the optimal scheme is shown in Figure 11. The results predicted by the surrogate models are compared with CFD in Table 6. The relative errors of the head and efficiency between the predicted values and the CFD results are 0.82% and 0.33%, respectively. Compared with the initial scheme, the efficiency under the design condition increased by 5.1%, while the deviation of head is within 0.5 m.

Table 5. Comparison of geometric parameters.

Parameters	Initial	Optimized
D_s (mm)	380	385.7
D_2 (mm)	340	341.1
b_2 (mm)	125	129.4
β_{1h} (°)	37	35.7
β_{1s} (°)	15	10.1
β_{2h} (°)	23	34.8
β_{2s} (°)	35	30.4
φ_{bh} (°)	102	109.5
φ_{bs} (°)	91	94.8
α_{3h} (°)	41	42.9
α_{3s} (°)	52	58.1
α_4 (°)	90	90.2
φ_{dh} (°)	34	30.7
φ_{ds} (°)	19	24.7

Table 6. Comparison of performance under design condition.

Case	Method	Head (m)	Efficiency (%)
Initial	CFD	16.39	80.26
Optimized	RBF	17.02	85.63
	CFD	16.88	85.36

(a) (b)

Figure 11. Geometric comparison of impeller and diffuser. (**a**) Initial scheme. (**b**) Optimized scheme.

Figure 12 shows the comparison of the velocity streamline in the impeller and guide blade runner in different spanwise. The incident angle of the blade inlet shows that there is a large inflow impact near the blade leading edge (LE) of the initial scheme, especially at span = 20%. When the flow angle is less than the blade inlet angle, flow separation occurs on the suction side (SS), increasing the velocity and flow loss near the LE. With the decrease of the blade inlet angle in the optimized scheme, the inflow direction almost fits the blade profile, the flow separation on the SS was effectively suppressed, and the velocity distribution is more uniform. In addition, the flow regime in the diffuser has also been improved. For the initial scheme, there are large-scale, low-speed regions and separation vortexes near the hub, causing serious blockage near the diffuser outlet. With the increase of spanwise, the separation vortexes move towards the diffuser outlet, and their scale decreases gradually. With the increase of the vane wrap angle, the separation phenomenon in the optimized scheme is greatly improved. Although some separation vortexes still remain at 20% spanwise, their numbers and sizes are significantly reduced. The streamlines at 80% spanwise completely align with the blade profile, and the flow separation phenomenon disappears.

According to the entropy generation theory, turbulent eddy dissipation (TED) is one of the important factors causing flow loss in the pump. Figure 13 shows the distribution of average TED at different spanwise. The TED at span = 20% and span = 50% in Figure 13a,b indicate that the flow separation on the blade SS is prone to cause dissipation loss. With the increase of spanwise, the effect of the rotor–stator interaction increased, the TED peak on the blade SS moves towards the blade TE, and the TED in the diffuser increased. Figure 13d–f illustrate that the TED on the blade surface is significantly reduced in the optimized scheme. In addition, the TED in the diffuser at span = 50% and span = 80% also decreased to a certain extent.

Figure 14 shows the comparison of volume entropy generation. The value of entropy generation reflects the magnitude of flow loss. Consistent with the results reflected in Figures 12 and 13, the peak value of entropy generation in the impeller was observed near the LE on the SS in the initial scheme, and the entropy generation fades away along the streamwise. Entropy generation in the diffuser is mainly distributed in the low-speed regions; thus the flow separation diffuser is the main cause of the diffusion loss. After optimization, entropy generation on the blade surfaces is almost eliminated, and the impeller efficiency is improved. In the diffuser, the entropy generation near the hub was still obvious in the optimized scheme, while the flow loss near the shroud is reduced, which contributes to the improvement of the energy recovery rate of the diffuser.

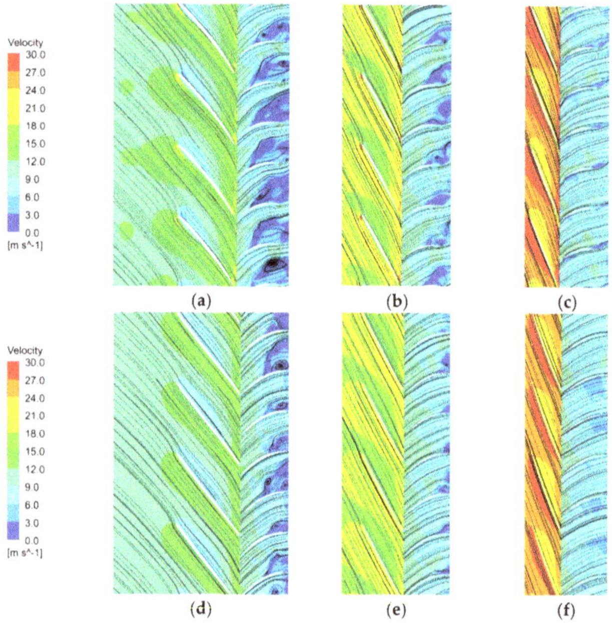

Figure 12. Velocity and streamline distribution in different spanwise. (**a**) Initial scheme at span = 20%. (**b**) Initial scheme at span = 50%. (**c**) Initial scheme at span = 80%. (**d**) Optimized scheme at span = 20%. (**e**) Optimized scheme at span = 50%. (**f**) Optimized scheme at span = 80%.

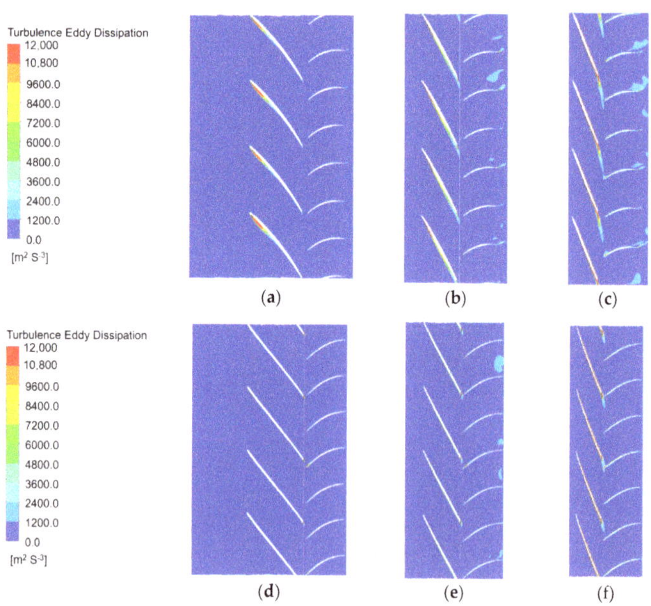

Figure 13. Turbulence eddy dissipation in different spanwise. (**a**) Initial scheme at span = 20%. (**b**) Initial scheme at span = 50%. (**c**) Initial scheme at span = 80%. (**d**) Optimized scheme at span = 20%. (**e**) Optimized scheme at span = 50%. (**f**) Optimized scheme at span = 80%.

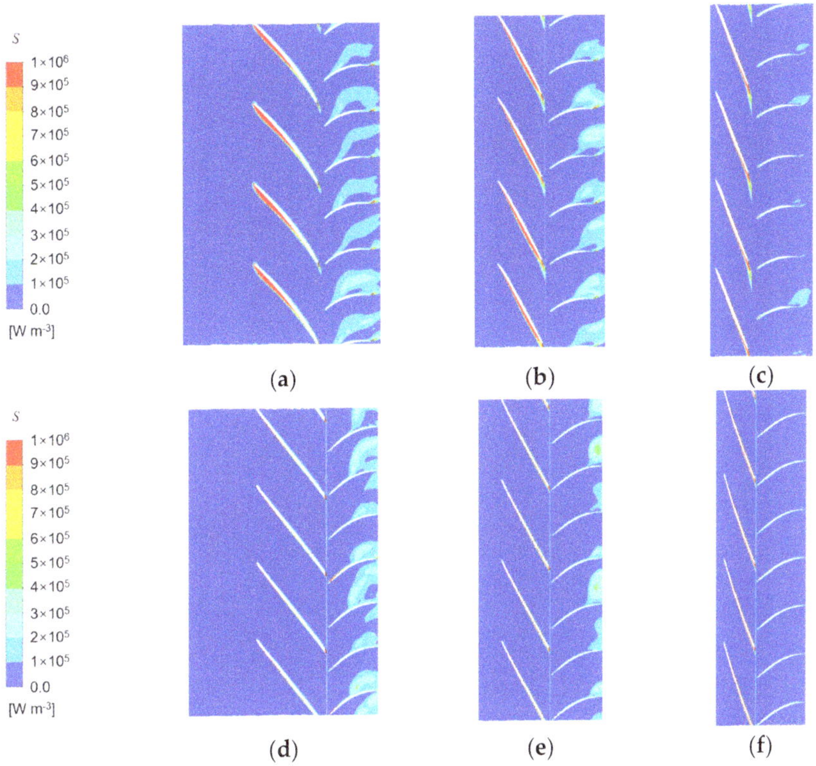

Figure 14. Comparison of volume entropy production power in different spanwise. (**a**) Initial scheme at span = 20%. (**b**) Initial scheme at span = 50%. (**c**) Initial scheme at span = 80%. (**d**) Optimized scheme at span = 20%. (**e**) Optimized scheme at span = 50%. (**f**) Optimized scheme at span = 80%.

By volume integral of entropy generation S, the total volume entropy generation power P_v of impeller and diffuser were obtained. Similar to total pressure drop, P_v represents the magnitude of flow loss in a domain. Figure 15 compares P_v in the impeller and diffuser, the flow loss in the impeller is greater than that in the diffuser. The optimization results show that P_v in the impeller and diffuser decreased by 33.3% and 19.0%, respectively. The decrease of entropy generation in the impeller is the main reason for the efficiency improvement.

Figure 16 shows the blade loading on different spanwise along the streamwise. The blade loading increases along the spanwise. The blade shape near the shroud has the greatest influence on the impeller performance. The difference is mainly found near the LE and TE. A pressure jump is observed near the TE because of the cut off. Owing to the serious flow separation near the LE of the initial scheme, the relative larger pressure difference appears in front of the blade (streamwise: 0~0.2). Additionally, the decrease of blade loading initiated before 80% streamwise. For the optimized scheme, the blade loading before 20% streamwise is greatly improved. In addition, the minimum differential pressure is closer to the TE, and the blade loading is more uniformly distributed along the streamwise.

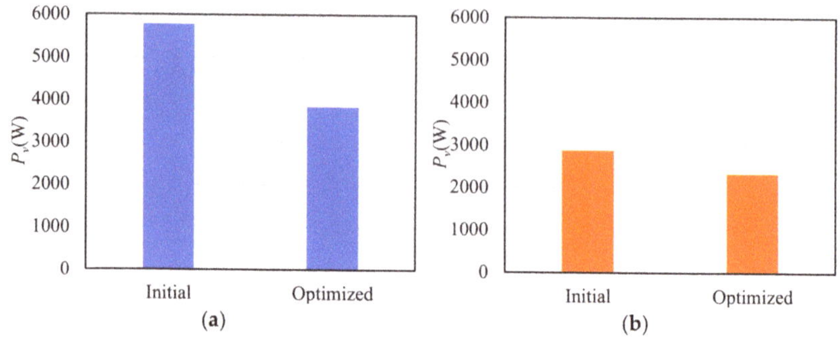

Figure 15. Comparison of the total volume entropy generation power. (**a**) P_v of impeller. (**b**) P_v of diffuser.

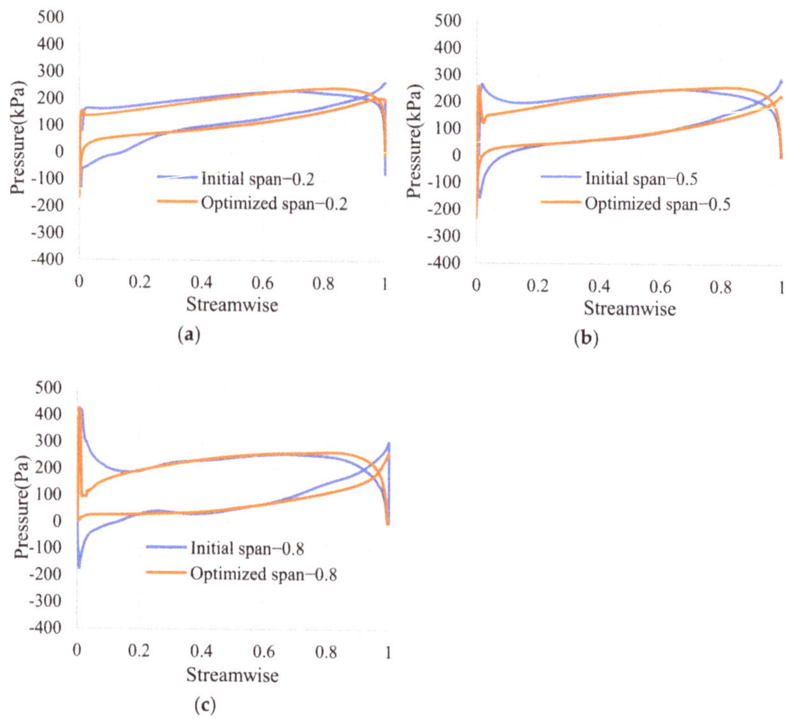

Figure 16. Distribution of blade loading. (**a**) Span = 20%. (**b**) Span = 50%. (**c**) Span = 80%.

4.4. Experimental Verification

To verify the optimization results, the hydraulic components were manufactured, and performance tests were carried out on the opening test rig. Figure 17a presents the test site of the mixed pump, and it mainly includes the tested pump, motor, pipes and so on. The pump was installed horizontally, and an elbow pipe was used to connect the pump and outlet pipe. The pump head was measured by a differential pressure transducer whose accuracy is better than 0.1%. The volume flow was measured by the electromagnetic flowmeter with an uncertainty of 0.2%. The input power of the pump was calculated according to the motor efficiency curve. The error of shaft power is less than

0.14%. The systematic uncertainty of the test rig is 0.26%, which meets the level 1 accuracy requirements specified in ISO9906-2012.

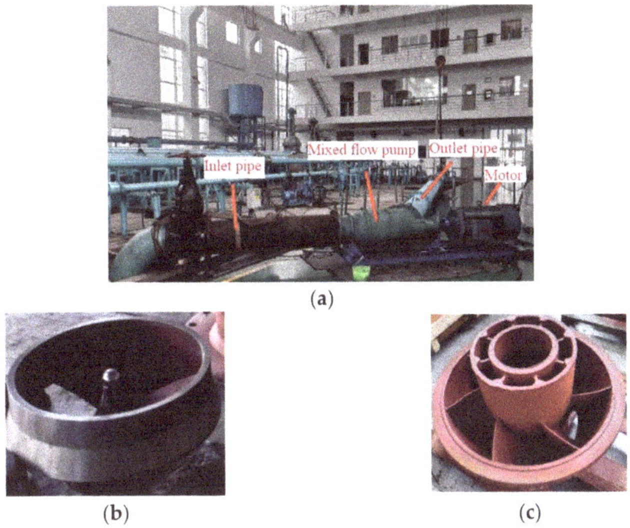

Figure 17. Experimental measurement. (**a**) Test rig. (**b**) Impeller. (**c**) Diffuser.

The performance curves of the initial and optimized schemes are compared in Figure 18. After optimization, the head and pump efficiency increased under the full flow condition. The maximum pump efficiency increased by 4.7%, and the best efficiency point shifts to large flow. At the same time, the optimization scheme alleviates the problem of the rapid drop of the head and efficiency curves under large flow conditions and broadens the high efficient operation range.

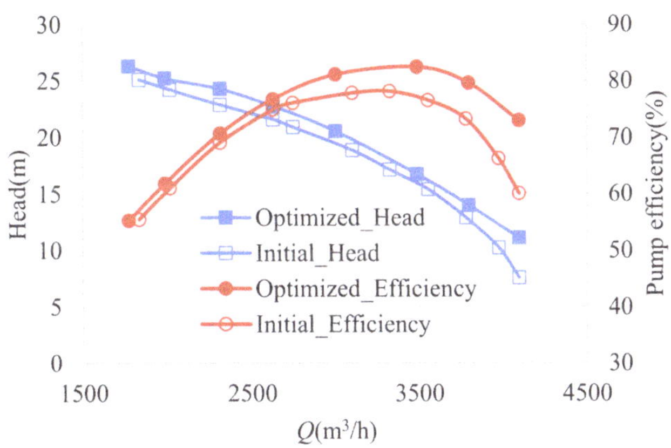

Figure 18. Comparison of experimental performance curves between initial and optimized schemes.

5. Conclusions

A vaned mixed flow pump was optimized by MIGA. An intelligent optimization platform integrating DOE, mesh generation, numerical simulation, and RBF neural network were established to shorten the optimization time. The best solution was obtained by solving the RBF model. This research has reached the following conclusions:

1. The RBF neural network has good accuracy in terms of the complex nonlinear relationships. The R^2 value of both the head and efficiency under the design condition is better than 0.98. The relative error between the predicted values and the CFD results is less than 1%.
2. The impeller outlet width, b_2, and its outlet diameter, D_2, have a significant positive effect on the design head, while the efficiency is more sensitive to the blade angles.
3. The efficiency of the optimized scheme predicted by CFD increased by 5.1%. Experimental results show that the maximum pump efficiency increased by 4.7%, and the high efficient operation range is significantly broadened.
4. The entropy generation can effectively visualize the flow loss distribution caused by turbulent dissipation and flow separation. Compared with the initial scheme, the volume entropy generation powers in the impeller and diffuser are effectively reduced, and the blade loading is more uniformly distributed along the streamwise.
5. The optimization platform proposed in this study is universal and can be applied to conduct optimization of other fluid machinery.

Author Contributions: Data curation, R.L.; validation, J.Y. and Q.S.; formal analysis, Q.S.; funding acquisition, G.W. and Y.Z.; investigation, J.Y.; resources, X.L.; supervision, J.Y.; writing—original draft preparation, R.L.; writing—review and editing, Q.S. All authors have read and agreed to the published version of the manuscript.

Funding: This research was funded by the National Key Research and Development Program of China (2020YFC1512403) and the National Natural Science Foundation of China (51976079).

Institutional Review Board Statement: Not applicable.

Informed Consent Statement: Not applicable.

Data Availability Statement: Not applicable.

Conflicts of Interest: The authors declare no conflict of interest.

Nomenclature

$u\ v\ w$	Cartesian velocity components
$x\ y\ z$	Coordinate components
t	Time
f	Body force
p	Pressure
μ	Dynamic viscosity
P	Entropy generation power
T	Temperature
D_2	Impeller outlet diameter
D_s	Impeller suction diameter
b	Width
α	Diffuser van angle
β	Impeller blade angle
φ	Wrap angle
Q	Flow rate
n	Rotational speed
m	Sample number
V	Volume
Ω	Area
H	Head
s	Entropy
$S_{\overline{D}}$	Viscous dissipation
$S_{D'}$	Turbulent dissipation
$\vec{q}$	Heat flux density vector
θ	Volute tongue angle

α	Circumferential angle
ρ	Density
ν	Kinematic viscosity
ε	Turbulent Eddy Dissipation
Subscripts	
1	Impeller inlet
2	Impeller outlet
3	Diffuser inlet
4	Diffuser outlet
i	Free index
j	Dummy index
h	Hub
s	Shroud
b	Blade
d	Diffuser
Superscripts	
$\overline{}$	Time-averaged value
$'$	Fluctuating component

References

1. Bozorgasareh, H.; Kh Alesi, J.; Jafari, M.; Gazori, H.A. Performance improvement of mixed-flow centrifugal pumps with new impeller shrouds: Numerical and experimental investigations. *Renew. Energ.* **2021**, *163*, 635–648. [CrossRef]
2. Kim, S.; Kim, Y.I.; Kim, J.H.; Choi, Y.S. Design optimization for mixed-flow pump impeller by improved suction performance and efficiency with variables of specific speeds. *J. Mech. Sci. Technol.* **2020**, *34*, 2377–2389. [CrossRef]
3. Hao, B.; Cao, S.L.; Tan, L.; Zhu, B.S. Effects of meridional flow passage shape on hydraulic performance of mixed-flow pump impellers. *Chin. J. Mech. Eng.* **2013**, *7*, 469–475.
4. Ji, L.L.; Li, W.; Shi, W.D.; Tian, F.; Agarwal, R. Effect of blade thickness on rotating stall of mixed-flow pump using entropy generation analysis. *Energy* **2021**, *236*, 121381. [CrossRef]
5. Ikuta, A.; Nitta, N.; Miyagawa, K.; Shinozuka, Y.; Tomimatsu, S. Influence of forward skew blade angle on positive slope characteristics of mixed flow pumps. *J. Phys. Conf. Ser.* **2021**, *1909*, 012076. [CrossRef]
6. Li, W.; Ji, L.L.; Li, E.D.; Zhou, L.; Agarwal, R.K. Effect of tip clearance on rotating stall in a mixed-flow pump. *J. Turbomach.* **2021**, *143*, 091013. [CrossRef]
7. Han, Y.D.; Tan, L. Influence of rotating speed on tip leakage vortex in a mixed flow pump as turbine at pump mode. *Renew. Energ.* **2021**, *162*, 144–150. [CrossRef]
8. Wang, L.K.; Lu, J.L.; Liao, W.L.; Guo, P.C.; Feng, J.J.; Luo, X.Q.; Wang, W. Numerical investigation of the effect of t-shaped blade on the energy performance improvement of a semi-open centrifugal pump. *J. Hydrodyn.* **2021**, *33*, 736–746. [CrossRef]
9. Zhu, X.Y.; Li, G.J.; Jiang, W.; Fu, L. Experimental and numerical investigation on application of half vane diffusers for centrifugal pump. *Int. Commun. Heat Mass Transf.* **2016**, *79*, 114–127. [CrossRef]
10. Khoeini, D.; Shirani, E.; Joghataei, M. Improvement of centrifugal pump performance by using different impeller diffuser angles with and without vanes. *J. Mech.* **2019**, *35*, 577–589. [CrossRef]
11. Wang, W.J.; Yuan, S.Q.; Pei, J.; Zhang, J.F. Optimization of the diffuser in a centrifugal pump by combining response surface method with multi-island genetic algorithm. *Proc. Inst. Mech. Eng. Part E J. Process Mech. Eng.* **2017**, *231*, 191–201. [CrossRef]
12. Kim, J.H.; Kim, K.Y. Analysis and optimization of a vaned diffuser in a mixed flow pump to improve hydrodynamic performance. *J. Fluids Eng.* **2012**, *134*, 071104. [CrossRef]
13. Wang, M.C.; Li, Y.J.; Yuan, J.P.; Osman, F.K. Influence of Spanwise Distribution of Impeller Exit Circulation on Optimization Results of Mixed Flow Pump. *Appl. Sci* **2021**, *11*, 507. [CrossRef]
14. Wang, M.C.; Li, Y.J.; Yuan, J.P.; Osman, F.K. Matching optimization of a mixed flow pump impeller and diffuser based on the inverse design method. *Processes* **2021**, *9*, 260. [CrossRef]
15. Lu, Y.M.; Wang, X.F.; Wang, W.; Zhou, F.M. Application of the modified inverse design method in the optimization of the runner blade of a mixed-flow pump. *Chin. J. Mech. Eng.* **2018**, *31*, 127–143. [CrossRef]
16. Bonaiuti, D.; Zangeneh, M.; Aartojarvi, R.; Eriksson, J. Parametrical Design of a Waterjet Pump by Means of Inverse Design, CFD Calculations and Experimental Analyses. *J. Fluids Eng.* **2010**, *132*, 201–215. [CrossRef]
17. Ye, W.X.; Geng, C.; Ikuta, A.; Hachinota, S.; Miyagawa, K.; Luo, X.W. Investigation on the impeller-diffuser interaction on the unstable flow in a mixed-flow pump using a modified partially averaged Navier-Stokes model. *Ocean Eng.* **2021**, *238*, 109756. [CrossRef]
18. Wang, W.J.; Pei, J.; Yuan, S.Q.; Zhang, J.F.; Yuan, J.P.; Xu, C.Z. Application of different surrogate models on the optimization of centrifugal pump. *J. Mech. Sci. Technol.* **2016**, *30*, 567–574. [CrossRef]
19. Si, Q.R.; Lu, R.; Shen, C.H.; Xia, S.J.; Yuan, J.P. An intelligent CFD-based optimization system for fluid machinery: Automotive electronic pump case application. *Appl. Sci.* **2020**, *10*, 366. [CrossRef]

20. Wu, X.F.; Tian, X.; Tan, M.G.; Liu, H.L. Multi-parameter optimization and analysis on performance of a mixed flow pump. *J. Appl. Fluid Mech.* **2020**, *13*, 199–209. [CrossRef]
21. Pei, J.; Wang, W.J.; Osman, M.K.; Gan, X.C. Multiparameter optimization for the nonlinear performance improvement of centrifugal pumps using a multilayer neural network. *J. Mech. Sci. Technol.* **2019**, *33*, 2681–2691. [CrossRef]
22. Zhu, D.; Tao, R.; Xiao, R.F.; Yang, W.; Liu, W.C.; Wang, D.J. Optimization design of hydraulic performance in vaned mixed-flow pump. *Proc. Inst. Mech. Eng. Part A J. Pow. Energ.* **2019**, *234*, 934–946. [CrossRef]
23. Huang, R.F.; Luo, X.W.; Ji, B.; Wang, P.; Yu, A.; Zhai, Z.H.; Zhou, J.J. Multi-objective optimization of a mixed-flow pump impeller using modified NSGA-II algorithm. *Sci. China Technol. Sci.* **2015**, *58*, 2122–2130. [CrossRef]
24. Zhang, F.; Appiah, D.; Hong, F.; Zhang, J.F.; Yuan, S.Q.; Adu-Poku, K.A.; Wei, X.Y. Energy loss evaluation in a side channel pump under different wrapping angles using entropy production method. *Int. Commun. Heat Mass Transf.* **2020**, *113*, 104526. [CrossRef]
25. Lu, R.; Yuan, J.P.; Wang, L.Y.; Fu, Y.X.; Hong, F.; Wang, W.J. Effect of volute tongue angle on the performance and flow unsteadiness of an automotive electronic cooling pump. *Proc. Inst. Mech. Eng. Part A J. Pow. Energ* **2020**, *235*, 227–241. [CrossRef]
26. Gu, Y.D.; Pei, J.; Yuan, S.Q.; Wang, W.J.; Zhang, F.; Wang, P.; Appiah, D.; Liu, Y. Clocking effect of vaned diffuser on hydraulic performance of high-power pump by using the numerical flow loss visualization method. *Energy* **2019**, *170*, 986–997. [CrossRef]
27. Herwig, H.; Schmandt, B. How to determine losses in a flow field: A paradigm shift towards the second law analysis. *Entropy* **2014**, *16*, 2959–2989. [CrossRef]
28. Hou, H.C.; Zhang, Y.X.; Li, Z.L.; Jiang, T.; Zhang, J.Y.; Xu, C. Numerical analysis of entropy production on a LNG cryogenic submerged pump. *J. Nat. Gas Sci. Eng.* **2016**, *36*, 87–96. [CrossRef]
29. Lin, T.; Li, X.J.; Zhu, Z.C.; Xie, J.; Li, Y.; Yang, H. Application of enstrophy dissipation to analyze energy loss in a centrifugal pump as turbine. *Renew. Energy* **2021**, *163*, 41–55. [CrossRef]
30. Lu, R.; Yuan, J.P.; Li, Y.J.; Jiang, H.Y. Automatic optimization of axial flow pump based on radial basis functions neural network and CFD. *JDIME* **2017**, *35*, 481–487.
31. Moriasi, D.N.; Arnold, J.G.; Van Liew, M.W.; Bingner, R.L.; Harmel, R.D.; Veith, T.L. Model evaluation guidelines for systematic quantification of accuracy in watershed simulations. *ASABE* **2007**, *50*, 885–900. [CrossRef]
32. Li, T.S.; Duan, S.K.; Liu, J.; Wang, L.D. An improved design of RBF neural network control algorithm based on spintronic memristor crossbar array. *Neural Comput. Appl.* **2018**, *30*, 1939–1946. [CrossRef]
33. Sinha, A.; Malo, P.; Deb, K. Evolutionary algorithm for bilevel optimization using approximations of the lower level optimal solution mapping. *Eur. J. Oper. Res.* **2017**, *257*, 395–411. [CrossRef]
34. Miki, M. A parallel genetic algorithm with distributed environment scheme. In Proceedings of the IEEE International Conference on Systems, Man and Cybernetics, Tokyo, Japan, 12–15 October 1999; IEEE: Piscataway, NJ, USA, 1999; pp. 695–700.
35. Li, Z.P.; Gao, X.J.; Lu, D.G. Correlation analysis and statistical assessment of early hydration characteristics and compressive strength for multi-composite cement paste. *Constr. Build. Mater.* **2021**, *310*, 125260. [CrossRef]

machines

MDPI

Article

Influence of Blade Type on the Flow Structure of a Vortex Pump for Solid-Liquid Two-Phase Flow

Hui Quan *, Yanan Li, Lei Kang, Xinyang Yu, Kai Song and Yongkang Wu

College of Energy and Power Engineering, Lanzhou University of Technology, Lanzhou 730050, China; Yananli0906@163.com (Y.L.); 18394666944@163.com (L.K.); 15145566732@163.com (X.Y.); sk19961116@163.com (K.S.); wforce01@163.com (Y.W.)

* Correspondence: quanh2010@163.com

Abstract: Vortex pumps have good non-clogging performance owing to their impellers being retracted into retraction cavities, but they are much less efficient than ordinary centrifugal pumps. In this paper, numerical simulations were performed on a model of the 150WX200-20 vortex pump for four different blade types, and the influence of blade structure on pump performance was determined. The simulations revealed the existence of axial vortices in the flow passage between the blades in the impeller region. The geometric characteristics of these axial vortices were more regular in two-phase solid-liquid flow than single-phase liquid flow. The presence of the solid phase reduced the vortex strength compared with the single-phase flow and suppressed the increase in size of the secondary circulation vortex. It was found, however, that the blade shape had a greater influence on the circulating flow than the presence of the solid phase. The flow state of the medium flowing out of the impeller domain had a direct effect on the circulating flow with this effect being related to the law governing the flow of the medium in the flow channel between the blades. It was found that the performance of a front-bent blade was the best and that of a curved blade the worst. This influence of blade type on the internal flow structure was used to further explain the relationship between the internal flow structure and the external characteristics of the vortex pump, the understanding of which is crucial for blade selection and hydraulic optimization.

Keywords: vortex pump; blade type; numerical simulation; internal flow structure; circulating flow

Citation: Quan, H.; Li, Y.; Kang, L.; Yu, X.; Song, K.; Wu, Y. Influence of Blade Type on the Flow Structure of a Vortex Pump for Solid-Liquid Two-Phase Flow. *Machines* **2021**, *9*, 353. https://doi.org/10.3390/machines9120353

Academic Editor: Antonio J. Marques Cardoso

Received: 25 October 2021
Accepted: 13 November 2021
Published: 15 December 2021

1. Introduction

The operational stability of centrifugal pumps in liquid transportation is important because of their strong flow energy and pulsation [1,2]. Vortex pumps have a fundamentally different structure than conventional centrifugal pumps. As shown in Figure 1, in a vortex pump, the impeller retracts into a retraction cavity, and there is a large space between the impeller and the inlet section, called the bladeless cavity. The fluid flows through the bladeless cavity, enters the impeller, and flows out of the impeller region under the action of the impeller. This outflowing fluid squeezes the circulating flow in the bladeless cavity to achieve fluid pumping. This structure endows the vortex pump with the ability to handle fluids with larger particles and longer fibrous materials. It also makes it more stable than ordinary centrifugal pumps during operation. There are fewer failures and, therefore, less maintenance time is required, and, to a certain extent, there are cost savings [3,4]. As excellent non-clogging pumps, vortex pumps are widely used in agriculture, the chemical industry, and municipal services [5,6]. The through flow and circulating flow of a vortex pump are shown in Figure 1b. As can be seen, most of the solid particles conveyed by the fluid are energized in the circulating flow and flow out of the pump through the bladeless cavity and the diffusion section. The existence of the circulating flow means that the impeller has less contact with the solid phase than in a centrifugal pump, which gives a vortex pump better non-clogging performance and reduces wear on the impeller, thus greatly extending the service life of the pump [7,8].

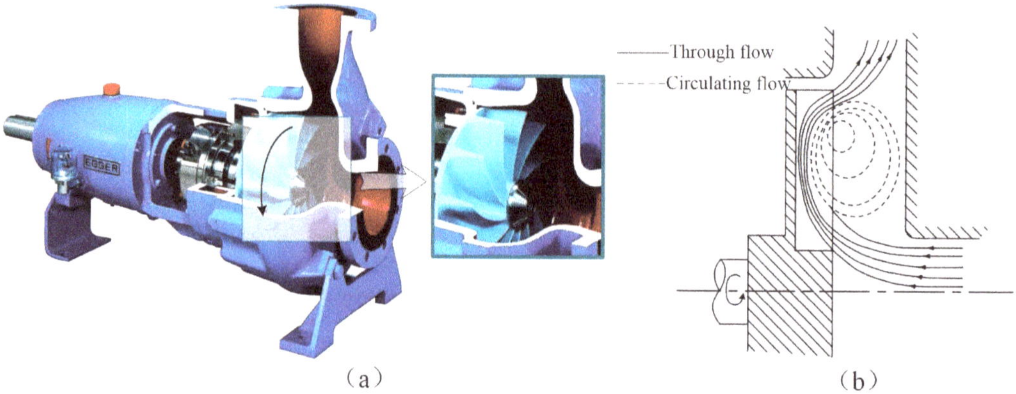

Figure 1. Schematic diagram of vortex pump and flow structure: (**a**) vortex pump structure diagram; (**b**) schematic diagram of flow structure of vortex pump.

Owing to the special structure of a vortex pump, the circulating flow within it induces the formation of a secondary vortex, which can lead to a sudden drop in the performance of the pump and affect its stability. This is mainly manifested in reduced flow performance and unit efficiency, increased water flow pulsation, induced cavitation, and blockage of the suction inlet by objects floating on the water surface [9,10]. In severe cases, there may even be damage to the pump body. Therefore, investigation of the internal flow structure of vortex pumps and development of ways to improve their low efficiency are urgent tasks in engineering practice. To study the solid-liquid two-phase pumping characteristics of a vortex pump, Steinmann et al. [11], Ye et al. [12], and Mihalić et al. [13] performed numerical simulations in which the resistance energy consumption in the pump was divided into two parts, namely, mechanical and flow loss, and they also presented empirical formulas for the mechanical efficiency (η_m) and flow efficiency (η_f) . Gerlach et al. [14] focused mainly on the effects of the solid-phase particles on the performance of vortex pumps. Chen [15] studied the flow properties inside a vortex pump, performing measurements of blade surface pressure and flow field velocity, and proposed a new flow model. Gao et al. [16] performed coupled discrete element model/computational fluid dynamics (DEM–CFD) simulations of the flow characteristics of solid particles in a vortex pump and showed that the particles followed three typical trajectories.

Under two-phase flow conditions, the solid and liquid phases move at different flow rates in the pump body owing to the difference in mass force, which is also a major factor affecting the performance of the pump. The design of multiphase vortex pumps is mostly based on a combination of clean water pump theory and practical experience [17,18]. At the same time, it is known that different blade types alter the flow patterns of solid-liquid two-phase flow in a vortex pump, which complicates the design process. Therefore, both the internal flow field structure of a vortex pump under two-phase flow conditions and the effects of different types of blades must be taken into account when laying the theoretical foundation for optimal pump design [19,20].

2. Vortex Pump Model and Numerical Method

2.1. Design of the Model Pump and Geometric Parameters

The 150WX-200-20 horizontal vortex pump was selected to establish the model. Its performance parameters were as follows: design flow $Q_d = 200\ m^3/h$, head $H_e = 20$ m, vortex pump rated speed $n = 1450$ rpm, pump efficiency $\eta = 50\%$, and shaft power $P = 26.34$ kW. The hydraulic geometric parameters are listed in Table 1.

Table 1. Vortex pump hydraulic geometric parameters.

Impeller Hydraulic Geometric Parameters		Volute Hydraulic Parameters	
Impeller outer diameter, D_2/mm	246	Volute width, L/mm	70
Impeller width, b/mm	60	The bladeless cavity base circle, D_3/mm	290
Number of blades, Z	10	Clearance between impeller outer diameter and shell, e/mm	20
Blade thickness, b_2/mm	8	The bladeless cavity throat area, F_{thr}/cm^2	110

The blades of the vortex pump were arranged in a radial array along its axis, and the volute took the form of a semi-helical pressurized water chamber. The two-dimensional hydraulic design of the impeller and volute are shown in Figure 2a,b, respectively.

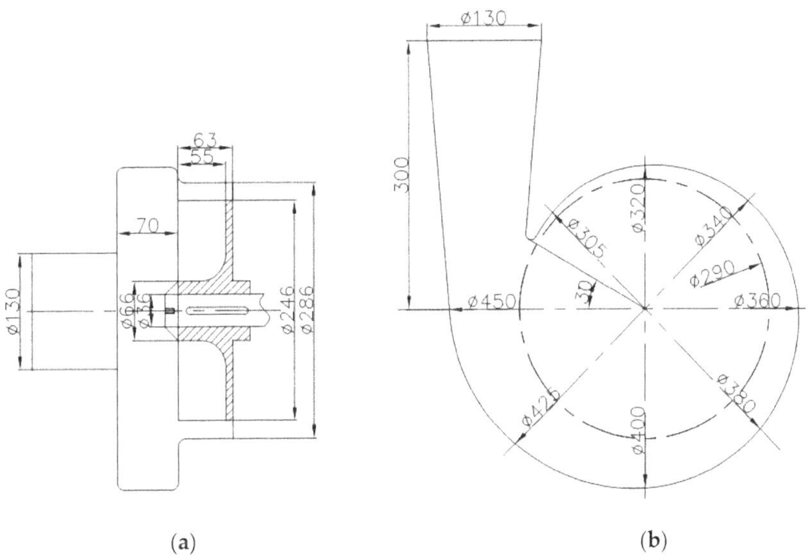

Figure 2. Hydraulic design drawing of vortex pump: (**a**) vortex pump structure design diagram; (**b**) volute hydraulic design.

2.2. Blade Design

At present, three main types of blades are used in vortex pump impellers: straight, folded, and curved. The blade type is directly related to the performance of the pump [21,22]. In this study, four semi-open blade structures were adopted, as shown in Figure 3b, including a straight blade (R30), two folded blades with a bend at one-quarter to one-third of the full length of the blade (1/4R30L30 and 1/4L30R30), and a curved blade with a wrap angle of 60° (Curl60). A schematic representation of the blade angles is shown in Figure 3a. The oblique angle from the inlet of the blade is denoted by α and the second oblique angle of inclination is denoted by β. For both angles, a prefix "L" indicates that the angle is in the same direction as that of rotation, and a prefix "R" indicates that it is in the reverse direction to that of rotation.

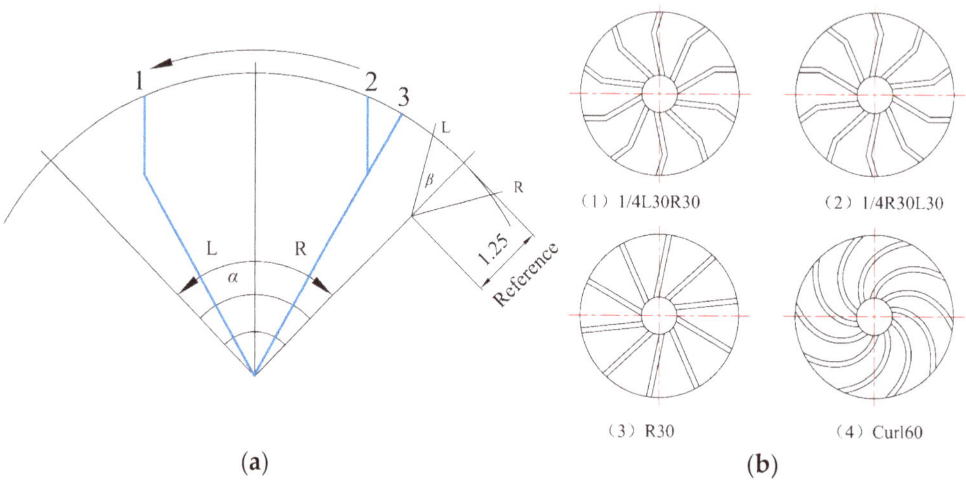

Figure 3. Blade shape design: (**a**) schematic diagram of blade angle; (**b**) four blade types.

2.3. Model Construction and Meshing

The three-dimensional model of the vortex pump was established in Creo software. The inlet section was extended, and the entire flow field was divided into two parts, namely, the volute domain and the impeller domain as shown in Figure 4. ICEM software was used to mesh the three-dimensional model and check the mesh independence.

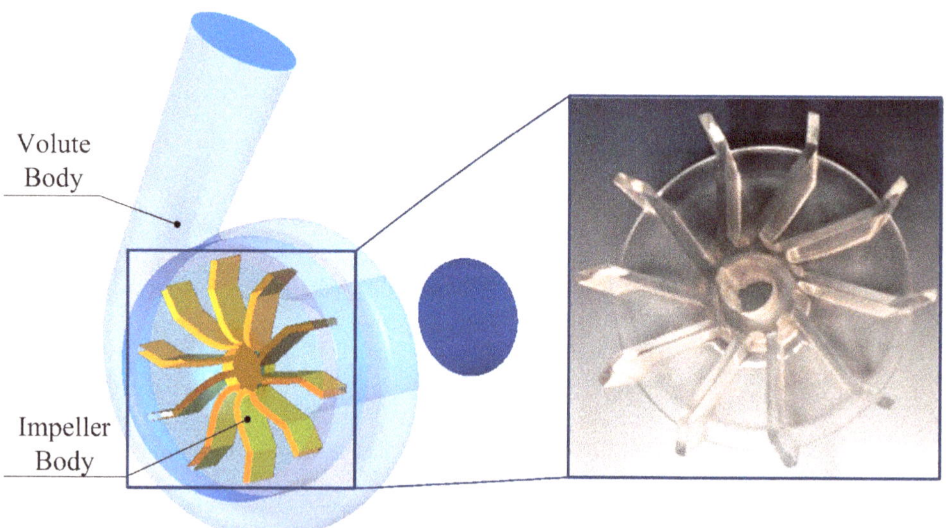

Figure 4. Area division diagram of the vortex pump.

The six sets of grids for a vortex pump with a 1/4R30L30 blade were established in a clean water medium. Numerical simulations were performed to verify the grid independence under the design conditions. The change in the head with the number of grids is shown in Figure 5a. When the number of grids is >1.2 × 10^6, the head changes only slightly, and the influence of the number of grids on the numerical calculation results is small and can be ignored. After comprehensive consideration, a mesh combination method of 1.02×10^6 in the volute area and 5.40×10^5 in the impeller area was adopted. The y+ value of the boundary layer grid was approximately 35. The meshing effect obtained by applying a structured grid is shown in Figure 5b.

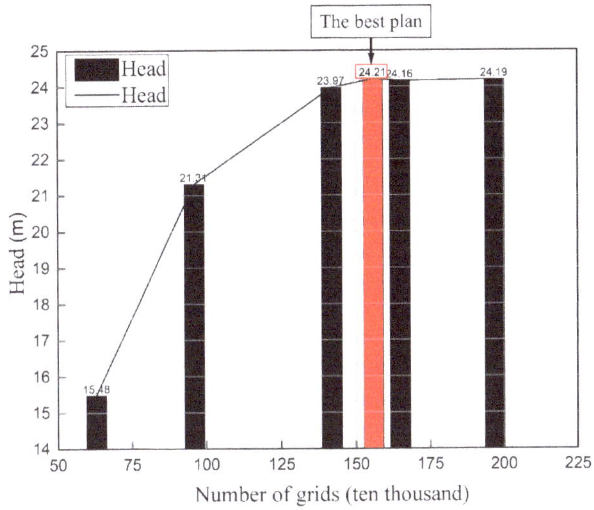

(**a**)

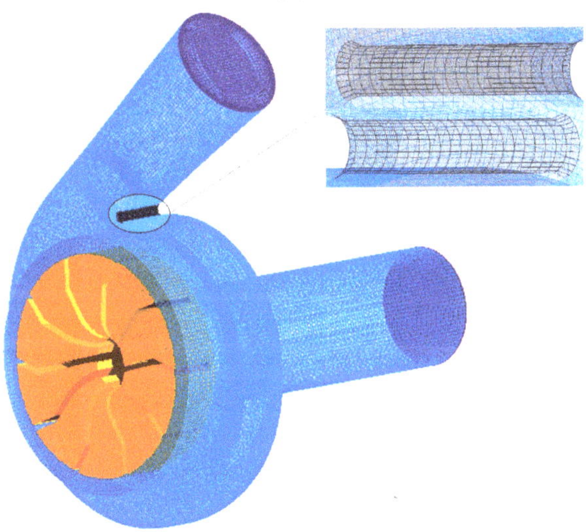

(**b**)

Figure 5. Schematic diagram of grid independence check and computational domain grid: (**a**) grid independence check; (**b**) schematic diagram of the computational domain grid.

Euler two-fluid model was used in the numerical method. The boundary conditions of inlet and outlet were velocity inlet and outflow, respectively. The inlet speed is 1.6 m/s.

A static nonslip wall condition was adopted for the volute and a moving nonslip wall condition for the impeller. Steady-state calculation was adopted for the numerical simulation of the vortex pump. Using the renormalization group (RNG) k-ε turbulence model, the SIMPLEC algorithm, and a second-order upwind discrete format provided a residual accuracy of 10^{-6}. When the residual curve reached the preset accuracy and the outlet pressure was stable, the flow field calculation result was regarded as convergent.

3. Calculation Method

3.1. Determination of Characterization Parameters

3.1.1. Section Selection

The model origin O was defined as the intersection of the symmetry plane of the bladeless cavity and the axis. The positive directions of the x-, y-, and z-axes were, respectively, the direction from the origin to the fourth section of the pressurized water chamber, the direction from the origin along the axis in the direction toward the impeller domain, and the direction of gravitational acceleration as shown in Figure 6. After the coordinate system was established, the bladeless cavity and the impeller domain were divided into equal parts, and these were divided into eight representative research sections.

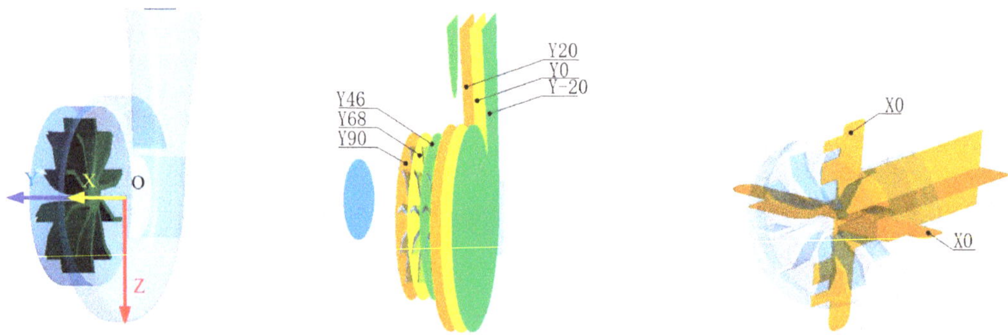

Figure 6. Cross-section selection scheme.

The definitions and nomenclature of the planes and sections are shown in Figure 6. For example, we named the plane $\{x = 0\}$, i.e., the YOZ plane, the $X0$ plane, and the $\{y = -20\}$ plane the Y-20 plane to obtain the eight sections.

It should be pointed out that the structure of the flow field in a vortex pump is complex, and only representative cross-sections could be selected for study. To display the streamlines, pressure, volume fraction distribution, and other information about each phase in each cross-section more clearly and intuitively, parts of the CFD output images were cropped to allow key points to be highlighted as shown in the $X0$ and $Z0$ sections in Figure 6.

3.1.2. Geometric Characteristics of Circulating Flow

In a solid-liquid two-phase flow, the solid phase moves along with the circulating fluid flow and, thus, when studying the circulating flow in the bladeless cavity, only the flow law for the liquid phase needs to be considered. There is a primary circulating flow I and two secondary circulating flows II and III in the pump. The flow structures in each quarter-section of the bladeless cavity are almost the same; therefore, the circulating flow can be simplified to the quarter-axis section, and there are no more than three large-scale vortex structures in each section. The vortex structures at the corresponding positions in each section are connected to form a vortex belt or circulating flow as shown in Figure 7. The geometric characteristics of the vortex structure of the circulating flow on four representative quarter cross-sections were considered to study the circulating flow.

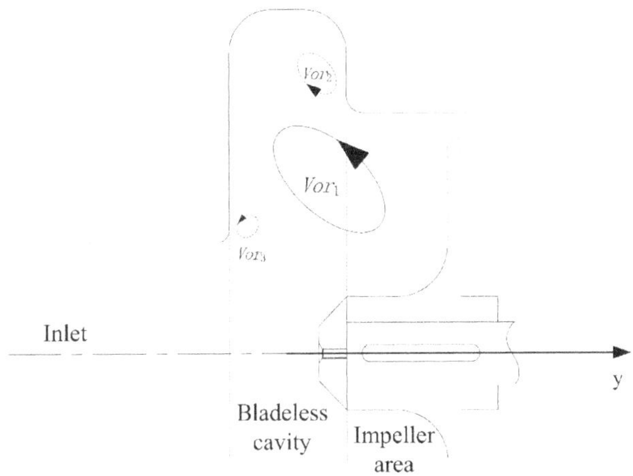

Figure 7. Schematic diagram of the circulating flow distribution.

After analysis, a circulating flow appeared in the form of a vortex structure on the X0 and Z0 sections. As shown in Figure 7, the main flow vortex is denoted by Vor1, and the secondary flow vortices are denoted by Vor2 and Vor3. In Figure 8, the red, green, and blue arcs represent the circulating flow structures formed by the Vor1, Vor2, and Vor3 vortex structures, respectively, in the bladeless cavity. When the impeller rotates, the fluid moves along the track "T→L→B→R", where T, L, B, and R represent the four quarter cross-sections. The vortex structure characteristics on the T and B cross-sections were obtained from the post-processing results of the X0 cross-section, and the L and R cross-sections were derived from the Z0 cross-section.

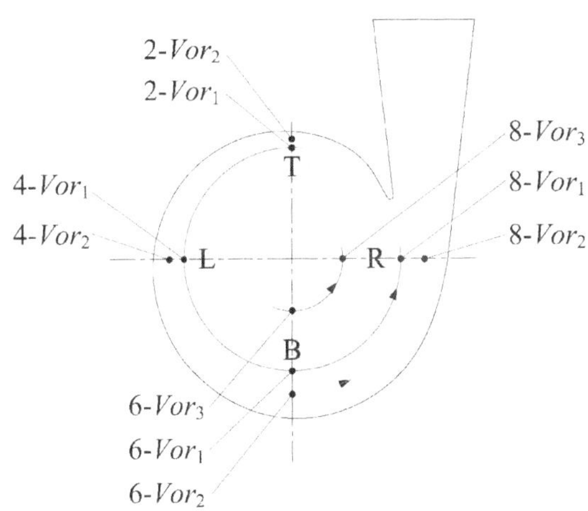

Figure 8. Existence of a vortex structure on each section in the vortex pump.

In the two-phase flow field in the vortex pump, there was not only the overall disordered flow field but also the local regular structure coexisting with it for a long time including connected turbulent fluid mass with instantaneous phase-correlated vorticity over its spatial extent. The interaction of the vortex structures between sections in the bladeless cavity may even lead to the fracture and reconnection of a vortex strip. In this paper, the performance of vortex pumps with different blade profiles were analyzed by discriminating the two-dimensional vortex structures on each section [23,24].

3.2. Description of Geometric and Physical Parameters of the Vortex Structure

(1) The surrounding fluid flows around the vortex core in an approximately circular motion. Taking the numerical simulation results of a solid volume fraction C_V = 10%, particle diameter D = 4 mm, and solid particle density ρ = 2250 kg/m^3 as an example, the geometric definition was carried out. As can be seen from the vortex intensity diagram in Figure 9a, there are two vortices rotating around the pump axis in circulating flows I and II in the bladeless cavity. From the liquid-phase vector diagram for the *X*0 section in Figure 9b, it can be observed that in the dashed area, the direction of the medium velocity vector is almost perpendicular to the section. At the same time, it can be seen from the streamline diagram for this section in Figure 9c that the surrounding fluid flows around a certain position along the axial section. From these results, it is possible to obtain a rough estimate of the position of the vortex core of the vortex structure on the inner shaft section of the pump. One method for estimating the extent of the vortex is based on the center of curvature [25]. In this method, by taking a large number of sample points on a given streamline, the center of curvature of the corresponding circle can be roughly determined. When the streamline, is approximately or exactly a circle, the center of curvature will be located in a relatively small area or at exactly the correct point, respectively. The position of the vortex core is then taken to be in this area or at this point.

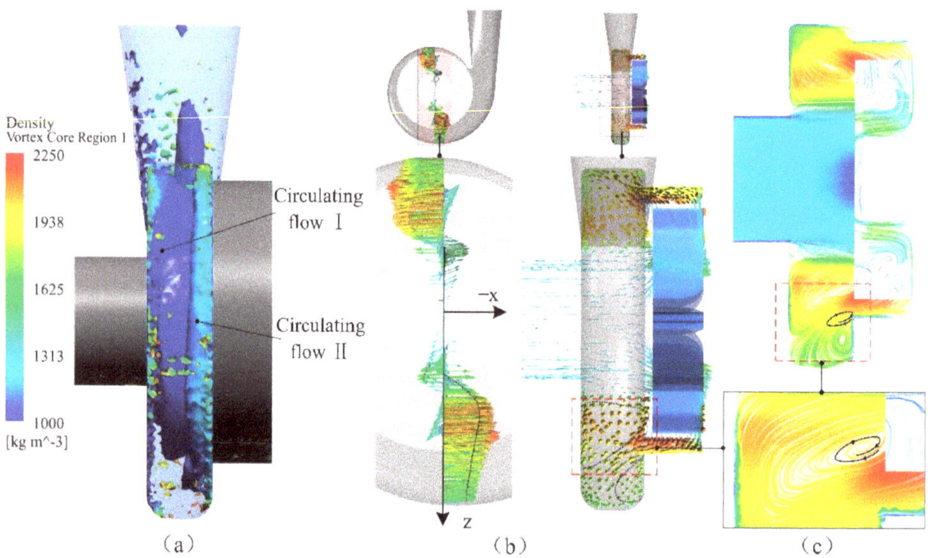

Figure 9. Schematic diagram of vortex discrimination: (**a**) cortex intensity diagram; (**b**) liquid—phase vector diagram for the X0 section; (**c**) streamline diagram.

The geometric and physical parameters of the circulating flow are thus quantified, and the streamlines near the vortex core are approximated as ellipses. For convenience of analysis, as shown in Figure 10a, the point C is regarded as the vortex core position of the flow field where the streamline is located on the section. The streamline of the *XO* section is placed in the coordinate system, where the abscissa is *y*, the ordinate is *z*, and the distance between the vortex core and the two axes is marked in the coordinate system. The distance between the vortex core and the center surface of the bladeless cavity is set as *n*, and the distance between the vortex core and the axis is set as *m*. Because the change in position of the circulating flow in each quarter section in the bladeless cavity is far larger than the error produced by the measurement, this change is obvious, and thus this method can be considered to be valid.

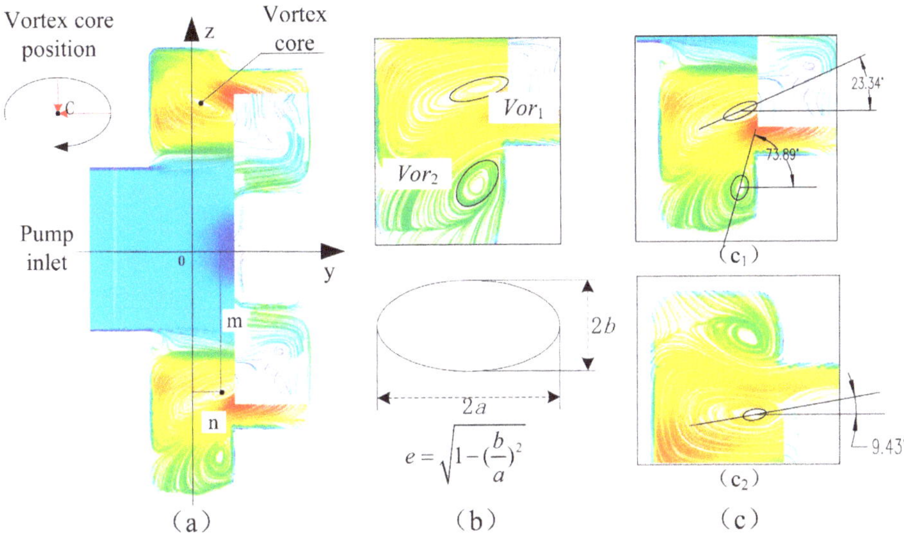

Figure 10. Schematic diagram of the position, shape, and angle of the vortex structure: (**a**) definition of the vortex core position; (**b**) vortex core shape and eccentricity calculation; (**c_1**) the positive angle of the vortex core; (**c_2**) the negative angle of the vortex core.

(2) The streamlines near the vortex core are close to elliptical, as can be seen in Figure 10b. Therefore, the eccentricity e ($0 < e < 1$), which represents the "flatness" of the ellipse, was adopted as a shape coefficient to analyze the variation of the vortex scale of the circulating flow. As shown in Figure 10b, the eccentricity can be calculated from the measured length of the semi-major and semi-minor axes of the ellipse, a and b, respectively. The closer e is to 0, the more nearly circular the shape of the vortex structure; the closer e is to 1, the flatter the vortex structure.

(3) Extending the major axis of the ellipse, as shown in Figure 10c, we took the angle between this axis and the positive direction of the y-axis as the position angle of the vortex core. The deviation between the positive y-axis and the extension of the major axis is shown in Figure 10(c_1); the direction of the deviation is positive when it points to the impeller domain. The direction of the deviation is negative if it does not point to the impeller domain as shown in Figure 10(c_2).

4. Results and Analysis

For a vortex pump, the medium flows from the inlet end through the bladeless cavity and then enters the impeller region. The direction and velocity of the outflowing medium that has been accelerated in the impeller region have direct effects on the performance of the pump. The flow of the medium in the impeller region is determined by the blade type. There are many types of vortex pump blades. This study considered only straight folded and curved blade types, although both forward and backward folded blades were compared. The effects of the blade shape on the performance of the vortex pump were analyzed.

4.1. Influence of the Blade Shape on the External Characteristics Performance of the Pump

To analyze the internal flow characteristics of the vortex pump and verify the reliability of the numerical simulation, an experimental platform was built for the 150WX-200-20 vortex pump, and an experiment to examine the external characteristics of the pump was conducted. The principle of the experimental platform is shown in Figure 11. The platform used 1/4R30L30 blades. During the experiment, a high-speed camera was used to photograph the internal flow structure in the pump, and the trajectories of the bubbles generated during pump operation were used to trace the fluid trajectory. The internal flow structures of the particles are shown in Figure 12. It can be seen from Figure 12a that the fluid flow from the inlet was initially uniform and parallel, but the fluid trajectory changed after the action of the impeller. It can be seen from Figure 12b that the main forms of fluid motion in the pump were circulating flow and through flow. Pure water medium was used in both experimental and numerical simulations. The external performance curve obtained from the experiments is shown in Figure 13. When the flow changed from $0.2Q_d$ (0.2 times design flow rate) to $1.4Q_d$, the average errors between the test head H_t and the numerical simulation head H_c and between the test efficiency η_t and the numerical simulation efficiency η_c were 2.27% and 5.26%, respectively. The test results thus demonstrate that the numerical method adopted is reliable for studies of the vortex pump. Therefore, in the subsequent investigations of vortex pump performance, numerical simulations were used instead of experiments, thereby saving time and resources.

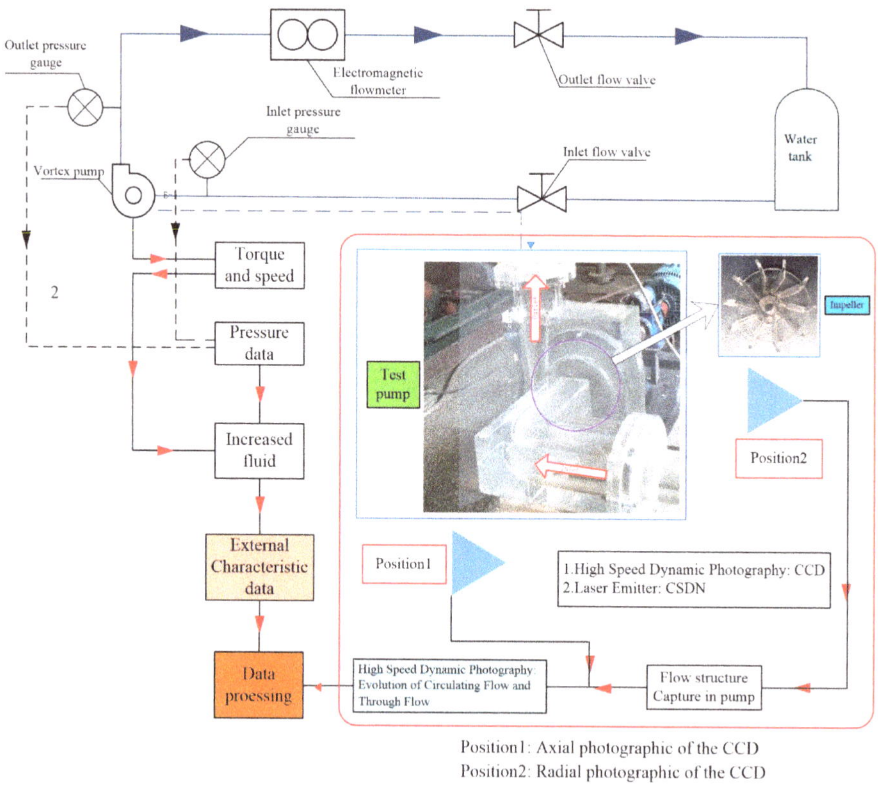

Figure 11. Schematic diagram of the experimental platform.

For numerically simulated solid-liquid two-phase flow for the four types of blades, the following solid-phase parameters were adopted: solid particle density ρ = 2250 kg/m^3, particle size = 4 mm, and solid-phase volume fraction C_v = 10%. Numerical simulations of the flow under five working conditions ranging from $0.6Q_d$ to $1.4Q_d$ were performed.

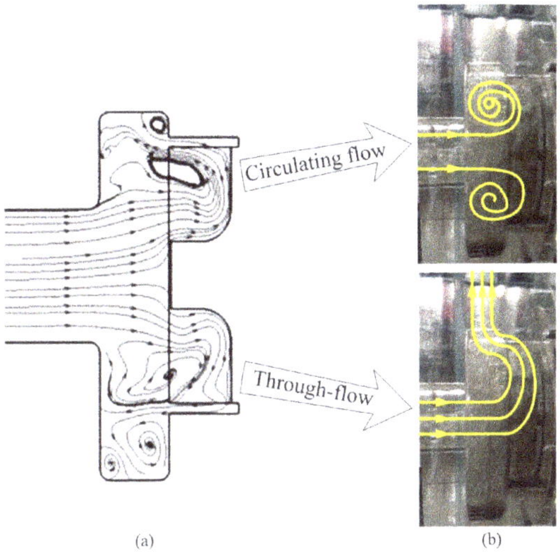

Figure 12. Fluid motion state diagram in the vortex pump: (**a**) streamline of the numerical simulation; (**b**) the internal flow structure in the pump.

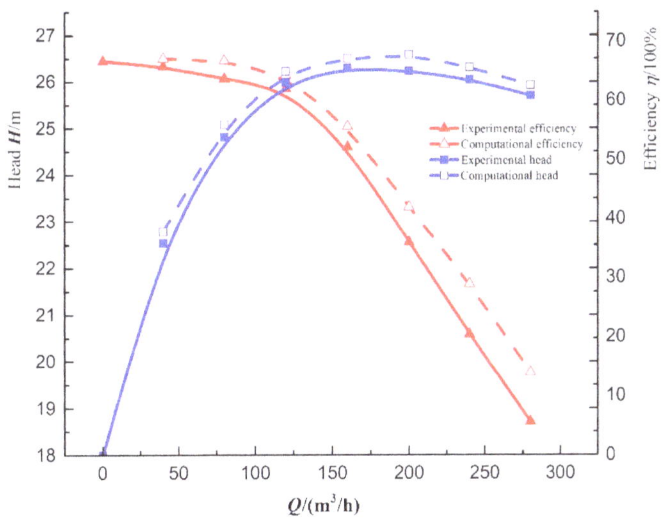

Figure 13. Hydraulic performance error verification.

Figure 14 shows a comparison of performance of the vortex pump with the same solid-liquid two-phase flow for the four different impeller blade types: 1/4R30L30, R30, 1/4L30R30, and Curl60. The four types of blades had different high-efficiency points. Under the rated conditions, the performance of the front-bent 1/4R30L30 blade was far superior to those of the other blade types. The straight R30 blade had the next highest efficiency, followed by the rear-bent 1/4L30R30 blade and finally the curved Curl60 blade. It can be seen that the efficiency of the curved blade peaked at $0.8Q_d$ and then dropped dramatically with further increases in the flow rate, and its maximum head was lower than that of the other blades. It can be noted that when the direction of deflection of the impeller blade was opposite to the direction of its angular velocity, the performance of the vortex pump was generally better.

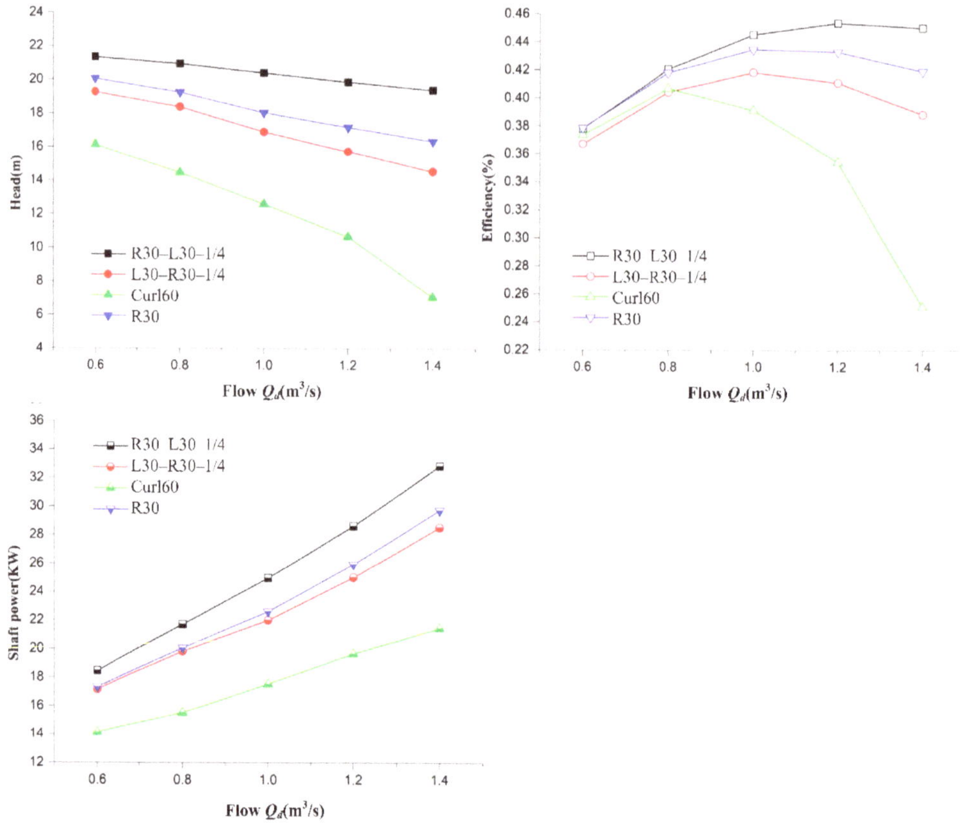

Figure 14. Performance curves of vortex pumps under different blade types.

4.2. Effect of Blade Shape on Circulating Flow in the Bladeless Cavity

4.2.1. Effect of Blade Shape on Vortex Core Position

As shown in Figure 15, different blade shapes had a greater impact on the position of the vortex core of the 2-, 4-, 6-, and 8-Vor_1 vortex structures in the vortex pump. At the same time, the position of the vortex core changed with an increasing flow rate, and there was no obvious 8-Vor_1 structure after $1.2Q_d$. The vortex cores for blade types 1/4R30L30 and R30 moved in the same way. In the y direction, with increasing flow rate, both 2-Vor_1 and 4-Vor_1 increased, whereas both 6- and 8-Vor_1 decreased. In the z direction, with an increasing flow rate, 2-, 4-, 6-, and 8-Vor_1 for 1/4R30L30 all decreased, 8-Vor_1 for R30 increased slightly, and 2-, 4-, and 6-Vor_1 showed the same trend as 1/4R30L30. Before and after the operating point, the vortex core position tended to change for both 1/4L30R30 and Curl60. For 1/4L30R30, the z coordinates of 4- and 8-Vor_1 first decreased and then increased at $1.0Q_d$ and $0.8Q_d$, respectively. The 4-, 6-, and 8-Vor_1 structures for Curl60 appeared at the best operating point of $0.8Q_d$, where the z coordinate first dropped and then rose.

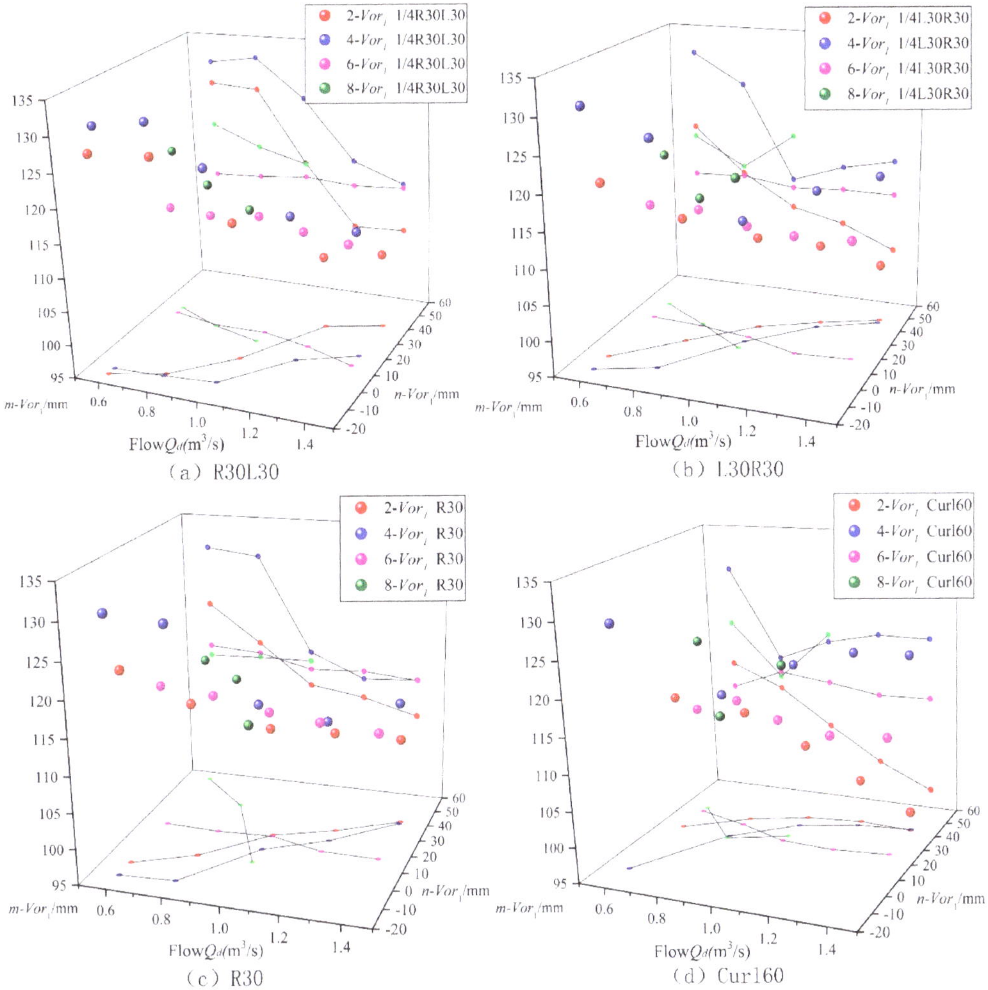

Figure 15. The position change of the vortex core in the bladeless cavity with different blade shapes: (**a**) the position change of the vortex core in 1/4R30L30; (**b**) the position change of the vortex core in 1/4L30R30; (**c**) the position change of the vortex core in R30; (**d**) the position change of the vortex core in Curl60.

The circulating flow distance is represented by the position of the vortex core in the section of the bladeless cavity. When the vortex core was closer to the bladeless cavity, the flow was disrupted by the force of the impeller rotation. After the vortex was reduced in size, it absorbed energy in the bladeless cavity and thereby reduced the performance of the vortex pump. The vortex pump attained its highest working condition when all vortex cores in all sections left the impeller region or were at a large distance from it. The position of the vortex core directly affected the head and efficiency of the pump. At the point of highest efficiency, the positions of the four vortex structures (for which there were three or four vortex cores) were all inside the bladeless cavity (n-Vor_1 < 35 mm).

4.2.2. Influence of Blade Shape on Vortex Shape

By performing numerical simulations for the four blade types, we determined the values of the eccentricities e-2-Vor_1, e-4-Vor_1, e-6-Vor_1, and e-8-Vor_1 for each blade type under the five working conditions (from $0.6Q_d$ to $1.4Q_d$). The variations of these eccentricities with the flow rate are shown in Figure 16.

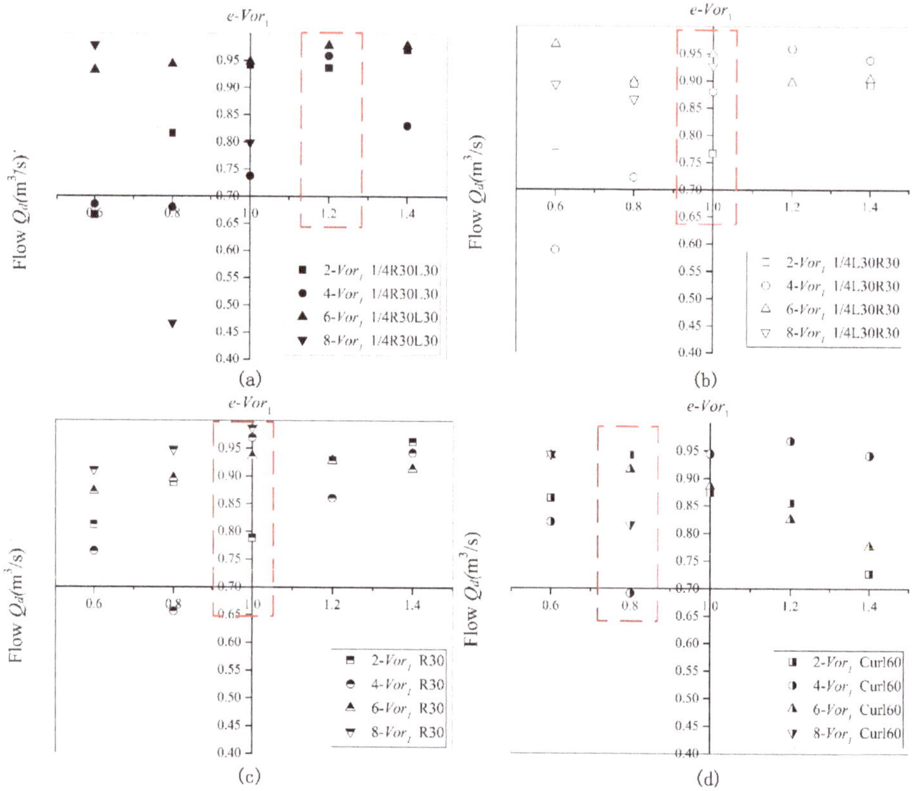

Figure 16. Variations of the shape coefficient of the vortex structure with different blade shapes: (**a**) the eccentricity of 1/4R30L30; (**b**) the eccentricity of 1/4L30R30; (**c**) the eccentricity of R30; (**d**) the eccentricity of Curl60.

It can be observed from Figure 16 that 4-Vor_1 of 1/4R30L30 and Curl60 and 2-Vor_1 of R30 and 1/4L30R30 all had a smaller eccentricity at the high-efficiency point. The greater the eccentricity of 2-Vor_1 at the high-efficiency point, the better the performance of the vortex pump. At the high-efficiency point, 2-Vor_1 was not only squeezed by the through flow but was also affected by the free vortex at the separation tongue. A higher value of eccentricity means that the circulating flow is less disturbed, which is more conducive to transport of the medium. As the cross-sectional area of the bladeless cavity increased, the shape coefficient of the circulating flow also changed during the development process. When the vortex core was close to the interface between the bladeless cavity and the impeller domain, the vortex structure became slightly rounded. When the vortex core entered the impeller, the vortex structure after the impeller domain became flat. When the shape of the vortex structure in the bladeless cavity was stable within a certain range (i.e., the difference between the maximum and minimum eccentricities was 0.2–0.25, and the difference between the maximum eccentricity and the next highest value was less than 0.05), the efficiency of the vortex pump was highest.

4.2.3. Influence of the Blade Shape on the Vortex Core Position Angle

As shown in Figure 17, regardless of the blade type at the highest operating point, the values of the vortex core position angle θ-Vor_1 were relatively close to each other. The variation of θ-Vor_1 was basically the same for the different blade types. With increasing flow, θ-2-, θ-4-, and θ-6-Vor_1 all decreased, and θ-8-Vor_1, except for the curved blade Curl60, all increased. At the optimal operating point, the greater the fluctuation range of θ-Vor_1, the better the performance of the vortex pump with the corresponding blade.

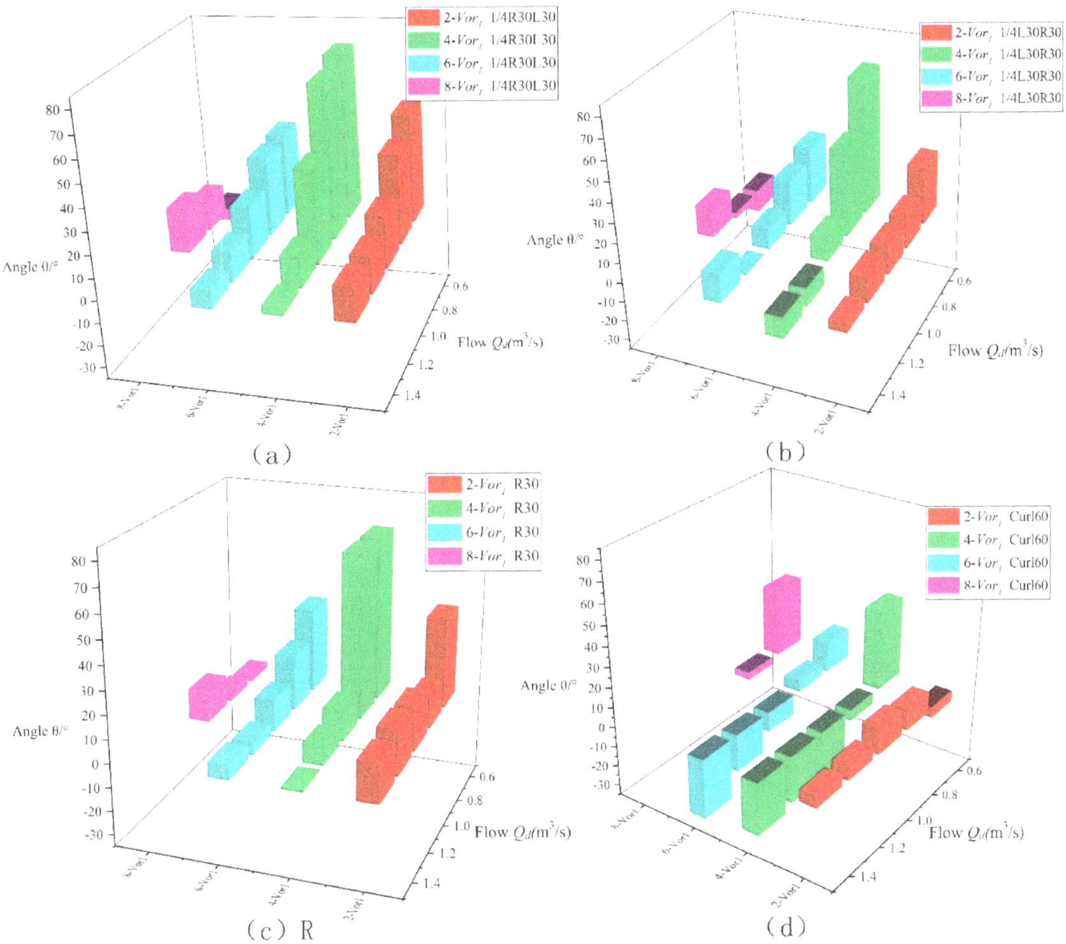

Figure 17. Changes in the angle θ between the vortex structure and the axial direction of different blade shape: (**a**) changes of the angle θ in 1/4R30L30; (**b**) changes of the angle θ in 1/4L30R30; (**c**) changes of the angle θ in R30; (**d**) changes of the angle θ in Curl60.

The front bent blade 1/4R30L30 was closest in behavior to the straight blade R30. The θ-Vor_1 values of these two blade types exhibited similar variations under most working conditions, although the range over which R30 decreased fluctuated greatly. This indicates that with the straight blade, there was poor control over the circulating flow. By contrast, the second angle of the bent blade 1/4R30L30 had a beneficial impact on the performance of the vortex pump and allowed for better control of the circulating flow. In contrast to the R30 blade, the other three blade types exhibited negative θ-Vor_1 values, especially the curved blade Curl60 and the back-bent blade 1/4L30R30.

As shown in Figure 18, when θ-Vor_1 is negative, the flow direction of the medium pumped by the impeller was parallel to the axis. In this case, the angle between through flow-Vor_1, through flow-Vor_2, and through flow and the positive y-axis were, respectively,

θ_1, θ_2, and θ, and were all approximately 0°. Part of the through flow hit the left sidewall of the bladeless cavity and part of it merged into Vor_2, and this consumed considerable energy, which degraded the performance of the pump. The direction in which the fluid flowed out after being accelerated by the impeller was directly related to the efficiency. When the angle of the vortex scale extension was larger, the efficiency of the pump with the corresponding blade type was higher.

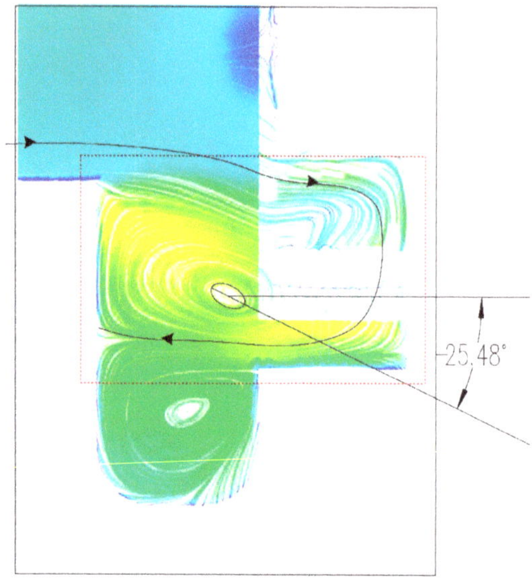

Figure 18. The flow of the vortex structure and through flow when θ-Vor_1 was negative.

5. Conclusions

(1) The four types of blades can be arranged in order of decreasing performance as follows: 1/4R30L30, R30, 1/4L30R30, and Curl60. The flow of the medium in the impeller area had a direct effect on the variations in flow structure in the bladeless cavity, thereby affecting the external characteristics of the pump;

(2) The vortex core position of the vortex structure in each section of the bladeless cavity directly affected the performance of the vortex pump. The greater the circulating flow range, the poorer the performance. The pump reached its highest working condition when and only when all the vortex cores of the vortex structure in all sections had left the impeller domain and were far from it. The farther a vortex core was from the y-axis, the less likely it was to approach the impeller, and the higher the efficiency of the pump;

(3) For the different blade shapes, the values of the eccentricities e-2-Vor_1, e-4-Vor_1, e-6-Vor_1, and e-8-Vor_1 at the cross-section of the bladeless cavity were related to the performance of the pump. Because of the squeezing of the through flow, the greater the eccentricity of 2-Vor_1 at the high-efficiency point, the better the performance;

(4) Under different blade types, the flow velocity and vector direction of the medium leaving the impeller after the impeller accelerated under the same operating conditions were different, resulting in different circulating flow structures in the bladeless cavity. When the vortex core representing the position of circulating flow at the high efficiency point had not entered the impeller domain but was close enough, the larger the position angle of the vortex core and the higher the efficiency of the vortex pump of this blade type.

Author Contributions: Data curation and writing—original draft, H.Q. and Y.L.; supervision, L.K. and X.Y.; writing—review and editing, K.S. and Y.W. All authors have read and agreed to the published version of the manuscript.

Funding: This work was partially supported by the National Natural Science Foundation of China (NSFC) (51969014, 51609113), the China Postdoctoral Science Foundation (2018M633651XB), the Natural Science Foundation of Gansu (20JR5RA456), the Outstanding Young Talents Funding Scheme of Gansu province (20JR10RA204), the Hong Liu Outstanding Young Talents Funding Scheme of Lanzhou University of Technology.

Institutional Review Board Statement: Not applicable.

Informed Consent Statement: Not applicable.

Data Availability Statement: The data presented in this study are available on request from the corresponding author.

Conflicts of Interest: The authors declare no conflict of interest.

References

1. Lu, Z.; Tao, R.; Jin, F.; Li, P.; Xiao, R.; Liu, W. The Temporal-Spatial Features of Pressure Pulsation in the Diffusers of a Large-Scale Vaned-Voluted Centrifugal Pump. *Machines* **2021**, *9*, 266. [CrossRef]
2. Wang, H.; Long, B.; Wang, C.; Han, C.; Li, L. Effects of the impeller blade with a slot structure on the centrifugal pump performance. *Energies* **2020**, *13*, 1628. [CrossRef]
3. Gao, X.; Shi, W.; Zhao, R.; Zhao, T.; Wang, H.; Mucka, P. Optimization Design and Internal Flow Field Study of Open-Design Vortex Pump. *Shock Vib.* **2021**, *2021*, 1–11. [CrossRef]
4. Quan, H.; Chai, Y.; Li, R.; Guo, J. Numerical simulation and experiment for study on internal flow pattern of vortex pump. *Eng. Comput.* **2019**, *36*, 1579–1596. [CrossRef]
5. Gerlach, A.; Thamsen, P.; Wulff, S.; Jacobsen, C. Design Parameters of Vortex Pumps: A Meta-Analysis of Experimental Studies. *Energies* **2017**, *10*, 58. [CrossRef]
6. Ju, Y.-P.; Liu, S.; Zhang, C.-H. Effect of blade shape on hydraulic performance and vortex structure of vortex pumps. *J. Hydrodyn.* **2018**, *30*, 499–506. [CrossRef]
7. Li, W.; Zhang, Y. The Vortex Pump under Highly Viscous Liquid Flow Conditions. *Arab. J. Sci. Eng.* **2018**, *43*, 4739–4761. [CrossRef]
8. Zhou, J.; Zhao, M.; Wang, C.; Gao, Z. Optimal design of diversion piers of lateral intake pumping station based on orthogonal test. *Shock Vib.* **2021**, 6616456. [CrossRef]
9. Li, W. Vortex Pump as Turbine—A Type Turbine for Energy Generation or Recovery Based on Computational Fluid Dynamics Prediction. *J. Fluids Eng.* **2019**, *141*, 15. [CrossRef]
10. Wang, H.; Hu, Q.; Yang, Y.; Wang, C. Performance differences of electrical submersible pump under variable speed schemes. *Int. J. Simul. Model.* **2021**, *20*, 76–86. [CrossRef]
11. Steinmann, A.; Wurm, H.; Otto, A. Numerical and experimental investigations of the unsteady cavitating flow in a vortex pump. *J. Hydrodyn. Ser. B* **2010**, *22* (Suppl. 1), 324–329. [CrossRef]
12. Ye, D.; Li, H.; Ma, Q.; Han, Q.; Sun, X. Numerical Investigation of Performance Improvement and Erosion Characteristics of Vortex Pump Using Particle Model. *Shock Vib.* **2020**, *2020*, 1–10. [CrossRef]
13. Mihalić, T.; Guzović, Z.; Predin, A. CFD flow analysis in the centrifugal vortex pump. *Int. J. Numer. Methods Heat Fluid Flow* **2014**, *24*, 545–562. [CrossRef]
14. Gerlach, A.; Preuss, E.; Thamsen, P.U.; Lykholt-Ustrup, F. Numerical simulations of the internal flow pattern of a vortex pump compared to the Hamel-Oseen vortex. *J. Mech. Sci. Technol.* **2017**, *31*, 1711–1719. [CrossRef]
15. Chen, H.-X. Research on turbulent flow within the vortex pump. *J. Hydrodyn.* **2004**, *16*, 701–707.
16. Gao, X.; Shi, W.; Shi, Y.; Chang, H.; Zhao, T. DEM-CFD Simulation and Experiments on the Flow Characteristics of Particles in Vortex Pumps. *Water* **2020**, *12*, 2444. [CrossRef]
17. Zhang, L.; Wang, C.; Zhang, Y.; Xiang, W.; He, Z.; Shi, W. Numerical study of coupled flow in blocking pulsed jet impinging on a rotating wall. *J. Braz. Soc. Mech. Sci. Eng.* **2021**, *43*, 508. [CrossRef]
18. Shi, L.; Zhu, J.; Tang, F.; Wang, C. Multi-Disciplinary optimization design of axial-flow pump impellers based on the approximation model. *Energies* **2020**, *13*, 779. [CrossRef]
19. Lucius, A.; Brenner, G. Numerical Simulation and Evaluation of Velocity Fluctuations During Rotating Stall of a Centrifugal Pump. *J. Fluids Eng.* **2011**, *133*, 8. [CrossRef]
20. Quan, H.; Guo, Y.; Li, R.N.; Su, Q.M.; Chai, Y. Optimization design and experimental study of vortex pump based on orthogonal test [Review]. *Sci. Prog.* **2020**, *103*, 20. [CrossRef] [PubMed]
21. Quan, H.; Cheng, J.; Guo, Y.; Kang, L.; Peng, G. Influence of Screw Centrifugal Inducer on Internal Flow Structure of Vortex Pump. *J. Fluids Eng.* **2020**, *142*, 9. [CrossRef]
22. Wang, H.; Qian, Z.; Zhang, D.; Wang, T.; Wang, C. Numerical study of the normal impinging water jet at different impinging height, based on Wray–Agarwal turbulence model. *Energies* **2020**, *13*, 1744. [CrossRef]
23. Jeong, J.; Hussain, F. On the identification of a vortex. *J. Fluid Mech.* **1995**, *285*, 69–94. [CrossRef]

24. Mariotti, A.; Galletti, C.; Salvetti, M.V.; Brunazzi, E.J.I.; Research, E.C. Unsteady Flow Regimes in a T-Shaped Micromixer: Mixing and Characteristic Frequencies. *Ind. Eng. Chem. Res.* **2019**, *58*, 29. [CrossRef]
25. Xu, H.; Cai, X.-S.; Liu, C. Liutex (vortex) core definition and automatic identification for turbulence vortex structures. *J. Hydrodyn.* **2019**, *31*, 857–863. [CrossRef]

machines

Article

Optimization of Sweep and Blade Lean for Diffuser to Suppress Hub Corner Vortex in Multistage Pump

Chao Ning [1], Puyu Cao [1,*], Xuran Gong [1] and Rui Zhu [2]

[1] National Research Center of Pumps, Jiangsu University, Zhenjiang 212013, China; 2111811011@stmail.ujs.edu.cn (C.N.); 2221811011@stmail.ujs.edu.cn (X.G.)
[2] Department of Fluid Machinery and Engineering, Xi'an Jiaotong University, Xi'an 710049, China; ruizhu@stu.xjtu.cn
* Correspondence: mafatu1988@ujs.edu.cn

Abstract: The bowl diffuser is the main flow component in multistage submersible pumps; however, secondary flow fields can easily induce a separation vortex in the hub corner region of the bowl diffuser during normal operation. To explore the flow mechanism of the hub corner separation vortex and develop a method for suppressing hub corner separation vortices, the lean and sweep of the diffuser blade were optimized using computational fluid dynamics (CFD) simulations and central composite design. Diffuser efficiency, static pressure recovery coefficient, and non-uniformity were selected as the optimization objectives. Details of the internal flow were revealed and the collaborative response relationships between blade lean/sweep parameter equations and optimization objectives were established. The optimization results show that a greater pressure difference between the pressure surface and suction surface (PS–SS) at the inlet can offset transverse secondary flow, whereas a lower PS–SS pressure difference will cause a drop in low-energy fluid in the diffuser mid-section. The blade's lean scheme suppresses the hub corner separation vortex, leading to an increase in pressure recovery and diffuser efficiency. Moreover, optimizing the sweep scheme can reduce the shroud–hub pressure difference at the inlet to offset spanwise secondary flow and enhance the hub–shroud pressure difference at the outlet, thus driving low-energy fluid further downstream. The sweep scheme suppresses the hub corner vortex, with a resulting drop in non-uniformity of 13.1%. Therefore, optimization of the diffuser blade's lean and sweep can result in less low-energy fluid or drive it further away from hub, thereby suppressing the hub corner vortex and improving hydraulic performance. The outcomes of this work are relevant to the advanced design of bowl diffusers for multistage submersible pumps.

Keywords: multistage submersible pump; bowl diffuser; parametric design; secondary flow

Citation: Ning, C.; Cao, P.; Gong, X.; Zhu, R. Optimization of Sweep and Blade Lean for Diffuser to Suppress Hub Corner Vortex in Multistage Pump. *Machines* **2021**, *9*, 316. https://doi.org/10.3390/machines9120316

Academic Editor: Davide Astolfi

Received: 27 October 2021
Accepted: 25 November 2021
Published: 26 November 2021

1. Introduction

The multi-stage submersible pump is widely used in various fields, mainly owing to its strong adaptability and easy pressurization [1]. However, complex and changeable working conditions present strict requirements for operational stability. The impeller and bowl diffuser are the main components in each stage of the pump. The head can be adjusted by changing the stage of the pump to meet the requirements of different applications. However, multi-stage submersible pumps with a bowl diffuser typically suffer from low efficiency and high operating costs. The bowl diffuser flow channel is curved and the fluid has a large impact on the inlet. Moreover, pump losses are dominated by the separation vortex in the hub corner of the diffuser. Therefore, optimization of the bowl diffuser is crucial to improving the single-stage head and overall working performance of the multistage submersible pump.

To date, numerous studies on the optimization of impellers and diffusers in pumps have been published [2,3]. To determine the algebraic relationship between the structural parameters of pumps and optimization objectives, data can be more efficiently and accurately analyzed using computational fluid dynamics (CFD) [4–6]. To obtain the optimal solution, CFD simulations are often combined with optimization design methods, such as the response surface method, neural network simulations, and orthogonal experiments [7–9]. Tong et al. [10] used numerical simulations and the Latin hypercube sampling method to construct functional relationships among independent variables and optimization objectives. Then, the second-generation genetic algorithm was used to solve the multi-objective optimization problem for a centrifugal pump. Stel et al. [11] studied the influence of pump stage on the performance of a multi-stage submersible pump using a CFD method based on the finite volume approach and investigated the transient flow characteristics in the pump under different flow rates. Heo et al. [12] compared three approximate models based on the response surface function (RSF), Kriging response surface, and a neural network for finding Pareto-optimal solutions which are set in the independent variable domain of the centrifugal pump.

The approximate model design method has been widely used in pump design [13,14]. Previous approaches could be applied to the optimization of other structural parameters, such as blade inlet and outlet angle, blade number, and blade thickness. Nonetheless, optimization techniques to achieve an optimal diffuser design that maximizes performance and stability of the overall stage are still lacking. Recently, design optimization strategies have been widely applied in the field of pneumatic fluid machinery, including the design of blade lean and sweep. A brief review is presented herein.

Rosic et al. [15] analyzed the influence of the stationary blade stacking combination on turbine performance. Razavi et al. [16] designed transonic rotor blades with different degrees of sweep and tilt. The blades were optimized using a neural network-based multi-objective optimization method, with efficiency, operating range, and stage pressure ratio as the target variables. He et al. [17,18] studied the influence of blade sweep design on transonic impeller performance through numerical simulations. The results showed that a forward-swept shroud design reduces the forward load, impact strength, and leakage vortex. A back-swept hub design suppresses the blade front load and the separation of secondary flow, thereby reducing losses near the hub. Bagshaw et al. [19] designed specially shaped cascade end walls with reverse load tilting, which can effectively inhibit the development of secondary flow in cascades. Goto et al. [20,21] used the color oil film flow display technology to capture large-scale separation vortices in the suction surface corner region of the diffuser and showed that the flow separation vortex is a source of hydraulic losses in the diffuser. Scillito et al. [22] used the large-eddy simulation method to confirm that axial compressor losses are dominated by the three-dimensional flow region near the diffuser end wall, two-dimensional laminar flow separation, and the diffuser outlet wake. The influence of inlet turbulence on sources of loss was further investigated.

In summary, blade lean and sweep design are widely used in compressor and turbine blades [23,24]; however, these approaches are rarely applied to water pumps, in particular, the bowl diffuser. The present study aimed to address these limitations by applying the design ideas and methods used for pneumatic machinery to the bowl diffuser of a multistage submersible pump. This research provides a scientific basis for follow-up research on diffuser design optimization.

The remainder of this paper is organized as follows. Section 2 describes the numerical model and simulation setup. Details of the experimental detection method are presented in Section 3. In Section 4, optimized designs of the blade lean and sweep of the bawl diffuser are presented based on the concept of parametric equations. In Section 5, the influence of various design schemes on the hydraulic characteristics of the diffuser are discussed and the collaborative response relationship between the parameter equation and the hydrodynamic performance of the diffuser is established. Section 6 discusses the effect of various optimization strategies on the hub corner separation vortex in the diffuser. Finally, the main conclusions of this work are summarized in Section 7.

2. Numerical Model and Simulation Setup

2.1. Computational Domain

In this paper, a Q80-20 multi-stage submersible pump with a bowl diffuser was selected as the research object. The basic parameters of the main flow passage parts of the multi-stage submersible pump are as follows: design flow rate, Q = 80 $m^3 \cdot h^{-1}$; single-stage head, H = 18 m; rotating speed, n = 2850 r/min. The main structural parameters of the impeller and the bowl diffuser are presented in Tables 1 and 2. Geometric models of the impeller and bowl diffuser of the multistage submersible pump are shown in Figure 1.

Table 1. The main structural parameters of the impeller.

Parameters	Value
Blade number of impeller Z	7
Diameter of inlet d_1 (mm)	97
Diameter of outlet d_2 (mm)	134
Width of blade outlet b_2 (mm)	20
Inlet angle of hub β_1 (°)	28
Outlet angle of hub β_2 (°)	34.2

Table 2. The main structural parameters of the bowl diffuser.

Parameters	Value
Blade number of diffuser Zd	8
Diameter of outlet d4 (mm)	95
Axial length e (mm)	152
Inlet angle of hub β_3 (°)	13.7
Outlet angle of hub β_4 (°)	90
Wrap angle of hub φ_{Hub} (°)	83
Wrap angle of shroud φ_{Shroud} (°)	60

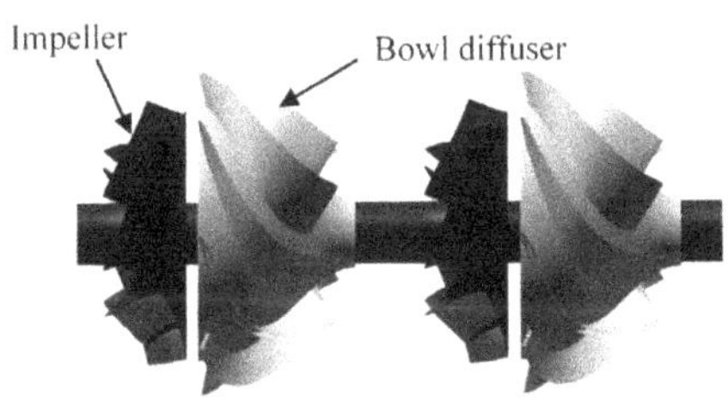

Figure 1. Geometrical model of the impeller and bowl diffuser.

According to Shi et al. [25], the internal flow characteristics in the second stage of the pump are basically the same as those in the later stages. Therefore, the following tests were based on data obtained from the second stage of the pump. To consider all stages, a large number of grid elements must be generated, which dramatically increases the calculation time. To balance computation time and numerical accuracy, Zhou et al. [26] demonstrated that two stages can be used to represent the whole pump system; therefore, the two-stage pump model was selected for the present work. The three-dimensional (3D) pump modeling software CFturbo was used to model the whole flow field of the pump, as shown in Figure 2. The calculation domain is mainly comprised of the inlet pipe, impeller, bowl diffuser, and outlet pipe. Each stage of the impeller and bowl diffuser constitutes a pressurization unit, and there are two pressurization units in total. To ensure fully developed fluid flow and improve the flow field calculation accuracy, the inlet pipe, outlet pipe, and outlet of the bowl diffuser were extended appropriately.

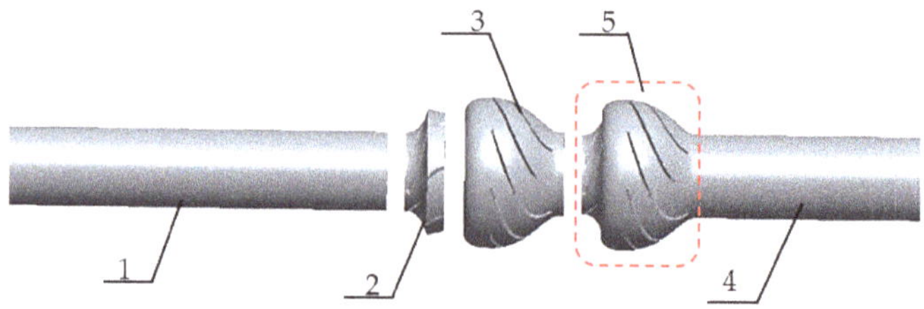

Figure 2. Multistage submersible pump calculation domain.

2.2. Mesh Generation

The ANSYS ICEM CFD software package was used to generate an unstructured mesh as the calculation domain. Key regions of the mesh were locally refined. Six grids of various sizes were selected to verify the grid independence of the calculation domain and ensure a grid quality greater than 0.3. As seen in Table 3, the simulation results become stable as the total number of grid elements increases. When the total number of elements is 6.8 million or higher, further changes in the calculated head and efficiency are very small, suggesting that the number of grid elements no longer has an effect on the calculation results. To balance computation time and solution accuracy, the total number of grid elements was selected as approximately 6.8 million. The generated mesh is shown in Figure 3.

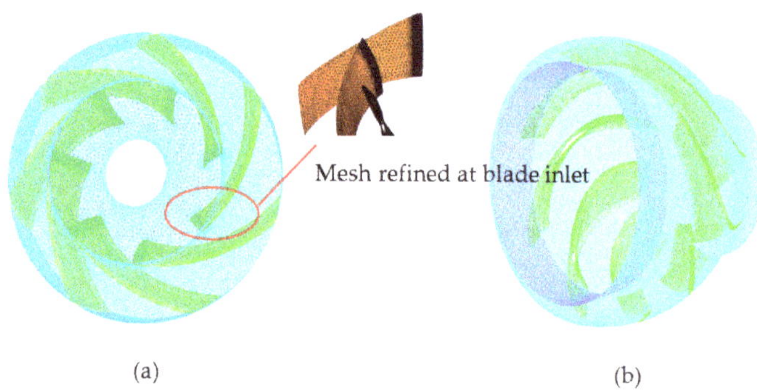

Figure 3. Mesh of impeller and diffuser for multistage submersible pump. (**a**) Mesh of impeller. (**b**) Mesh of diffuser.

Table 3. Influence of the grid number on accuracy.

Parameter	N ($\times 10^4 \times 10^4$)	H/m	η/%
value	302	33.20	95.04
	410	33.04	95.00
	553	32.86	94.91
	680	32.74	94.88
	837	32.69	94.88
	988	32.70	94.88

2.3. Turbulent Model

The standard k-ε model is based on turbulent kinetic energy (k) transport and turbulent energy dissipation rate (ε) transport and offers good robustness and economy in predicting the flow characteristics of most flow fields reasonably and accurately. However, the standard k-ε model is prone to errors when calculating flow over a complex curved wall [27].

To account for the high-speed rotation domain and large variation in curvature of the wall in the calculation domain of the multi-stage submersible pump, the RNG k-ε model proposed by Yakhot et al. [28] was selected, which is suitable for flows with separation [29]. Compared with the standard k-ε model, the RNG k-ε model contains an additional time average strain rate (Eij) in the reaction mainstream equation of ε, which can improve the accuracy for swirl flow and more reasonably deal with flow near the wall [30,31]. The two transport equations can be expressed, as follows:

The turbulent kinetic energy k transport equation:

$$\frac{\partial(\rho k u_i)}{\partial x_i} + \frac{\partial(\rho k)}{\partial t} = \frac{\partial}{\partial x_j}[(\mu + \frac{\mu_t}{\sigma_k})\frac{\partial k}{\partial x_j}] + P_k - \rho \quad (1)$$

The turbulent energy dissipation rate ε equation is:

$$\frac{\partial(\rho \varepsilon u_i)}{\partial x_i} + \frac{\partial(\rho \varepsilon)}{\partial t} = \frac{\partial}{\partial x_j}[(\mu + \frac{\mu_t}{\sigma_\varepsilon})\frac{\partial \varepsilon}{\partial x_j}] + \frac{\varepsilon}{k}(c_1^* P_k - c_2 \rho \varepsilon) \quad (2)$$

where k is turbulent kinetic energy, m^2/s^2; ε is turbulent energy dissipation rate, m^2/s^3; Pk is the pressure generating term caused by the velocity gradient; μt is the turbulent viscosity.

2.4. Simulation Setup

ANSYS CFX 17.1 was used to calculate the steady three-dimensional whole flow field of the model pump under the design conditions. The fluid in the pump was set as incompressible water. The RNG k-ε model was selected for the simulation calculations to satisfy the solution accuracy requirement. The boundary conditions were set, as follows: the inlet boundary condition was set as pressure inlet and the static pressure as 0 Pa; The outlet boundary condition was set to the mass flow outlet condition. The adiabatic nonslip solid wall boundary condition was adopted at the wall and the near wall area was treated as a scalable wall function.

The steady numerical calculation was carried out across the whole calculation domain of the multistage submersible pump. The impeller part was considered the rotating domain and the bowl diffuser part was considered the static domain. The interfaces between the rotating section and the static section were set as the dynamic and static interfaces, and the frozen rotor model was used to handle them. The General Grid Interface (GGI) was used as the grid connection method for dealing with the static interface. The root mean square (RMS) value for convergence accuracy was set to 5×10^{-5}.

3. Experimental Pump Characteristics

To verify the simulation method, the head obtained using CFD was compared with experimental head values in the flow range of 0.8Qd–1.1Qd. The test-bed, shown in Figure 4, is composed of a flow control device, a data acquisition device, and a data processing device [32,33]. The flow rate was adjusted by the valve and measured by the electromagnetic flowmeter.

The comparison of head obtained by experiment and simulation was shown in Figure 5. The average error between the numerical simulation results and the experimental results was less than 5%, and the relative error of the head under the design conditions was 2.7%. Errors between the simulation results and experimental results were within the allowable range. The results indicate that the simulation calculation can accurately predict the performance of the multistage submersible pump under the design conditions.

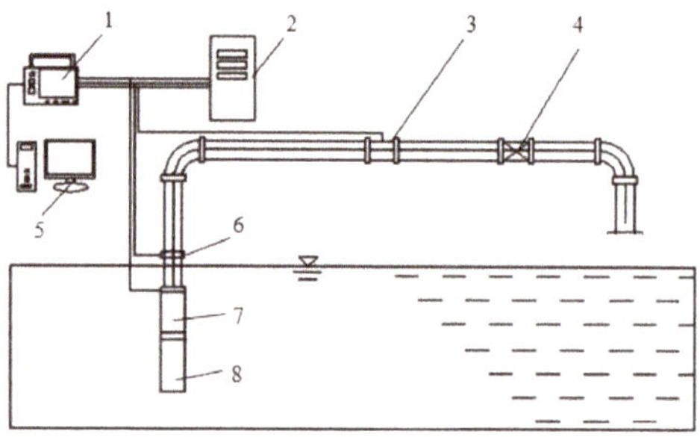

Figure 4. Testing apparatus [32]. 1. Data acquisition instrument. 2. Frequency inverter. 3. Electromagnetic flowmeter. 4. The valve. 5. Computer. 6. Pressure sensor. 7. Pump. 8. Motor.

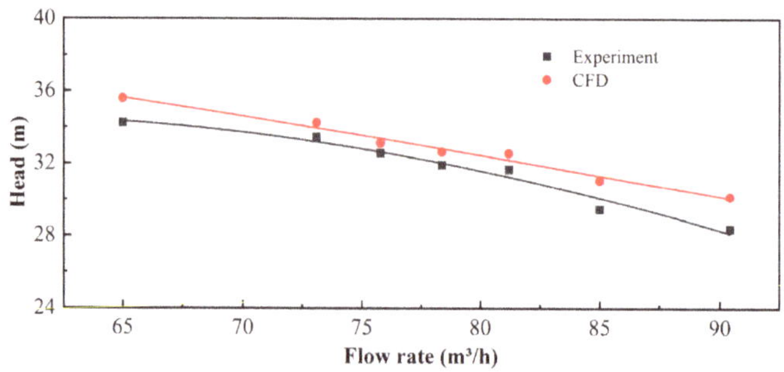

Figure 5. Comparison of head obtained by experiment and simulation.

4. Optimization Schemes

4.1. Design of Blade Lean

The coordinate system of the blade inlet profile was established, as shown in Figure 4. The origin of the coordinate system is defined as the intersection between the blade inlet edge and the hub. The positive direction of the ordinate axis is from the hub to the shroud of the diffuser, expressed by the spanwise coefficient Sp_1. The circumferential direction is the positive direction of the abscissa, represented by $f(Sp_1)$. The starting angle of any point at the inlet edge of the blade refers to the angle between the line connecting the point to the center point and the axial plane with a starting angle of zero, indicated by θ in Figure 6.

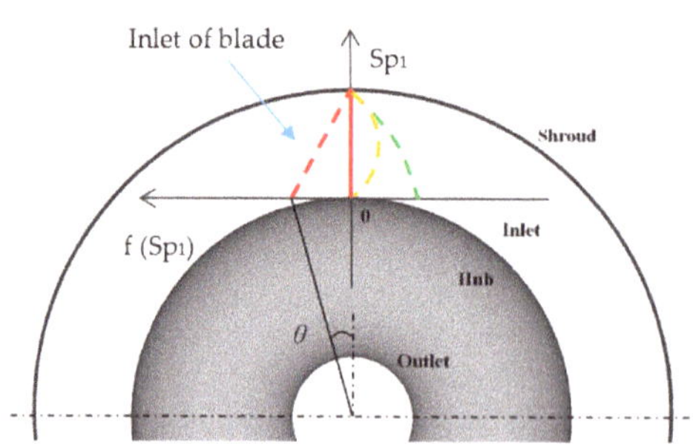

Figure 6. Blade lean coordinate system and blade leading edge angle of diffuser (from inlet view).

It is assumed that the parabolic equation governing the blade inlet profile is:

$$f(Sp_1) = a_0 + a_1 Sp_1 + a_2 Sp_1^2 \tag{3}$$

where spanwise coefficient $Sp_1 \in [0,1]$, 0 for the hub and 1 for the shroud; a_0, a_1, and a_2 are the parameters to be optimized. The difference between the starting angle of the hub and the shroud is referred to as the starting angle difference $\Delta\theta$, defined as

$$\Delta\theta = \theta_{Hub} - \theta_{Shroud} \quad (4)$$

where θ_{Hub} is the starting angle of the hub at the blade inlet, $\theta_{Hub} = a_0$; θ_{Shroud} is the starting angle of the shroud at the blade inlet, $\theta_{Shroud} = a_0 + a_1 + a_2$, $\Delta\theta = a_1 + a_2$. The factors and the levels used in the central composite designs are presented in Table 4.

Table 4. Factors and levels for the central composite design.

Levels	Factors		
	a_0	a_1	a_2
−1.682	−12.07	−17.73	8.93
−1	−9	−15	12
0	−4.5	−11	16.5
1	0	−7	21
1.682	3.07	−4.27	24.06

4.2. Design of Sweep

Figure 5 shows the sweep coordinate system in the meridian plane of the diffuser. The positive direction of the longitudinal axis is defined as the direction from the hub to the shroud, expressed by the spanwise coefficient Sp_2. The axis is the positive direction of the abscissa, represented by $f\ (Sp_2)$. The origin is the intersection between the hub and the blade outlet edge. It is assumed that the parabolic equation governing the blade inlet profile is

$$f(Sp_2) = b_0 + b_1 Sp_2 + b_2 Sp_2^2 \quad (5)$$

where Sp_2 is the spanwise coefficient; $Sp_1 \in [0,1]$, 0 for hub and 1 for shroud.

As shown in Figure 7, b_0 is located at the edge of the outlet on the hub, b_1 is located at the edge of the outlet on the shroud, and b_2 is located at the edge of the inlet on the hub, selected as the independent factors. The values are presented in Table 5. The sweep angle β is the angle between the new outlet edge of the blade and the original outlet edge of the blade. When the position of the blade outlet edge on the hub remains unchanged and the position of the shroud moves in the positive direction along the abscissa, β is negative; when moving in the negative direction along the abscissa, β is positive. Here, the positive and negative signs indicate direction only.

Table 5. Factors and levels for the central composite design.

Levels	Factors		
	b_0	b_1	b_2
−1.682	−3.05	10.61	−55.36
−1	−1	13	−54
0	2	16.5	−52
1	5	20	−50
1.682	7.05	22.39	−48.64

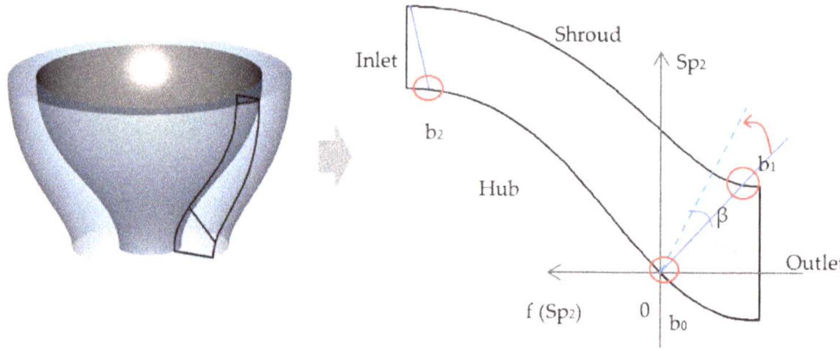

Figure 7. Sweep coordinate system and blade sweep angle of diffuser sweep design (from meridional view).

4.3. Analysis Parameters

Diffuser efficiency η, static pressure recovery coefficient Cp, and non-uniformity ζ_i were selected to evaluate the hydrodynamic performance of the diffuser before and after optimization. The diffuser efficiency is $\eta = P_{t4}/P_{t3}$. P_{t3} is the total pressure at the inlet of the diffuser and P_{t4} is the total pressure at the outlet of the diffuser. Static pressure recovery coefficient is $Cp = (P_{s4} - P_{s3})/P_{s3}$, P_{s3} is the static pressure at the inlet of the diffuser and P_{s4} is the static pressure at the outlet of the diffuser. Cp indicates the potential for converting kinetic energy into static pressure energy as fluid flows through the diffuser. An increase in Cp indicates enhanced static pressure recovery ability.

The non-uniformity ζ_i is an index for quantitatively evaluating flow uniformity at the outlet of the diffuser. The efficiency and operating stability of the pump are inversely affected by flow uniformity in the diffuser and impeller. As flow uniformity increases at the outlet of the diffuser, ζ_i decreases; conversely, as the flow becomes less uniform, ζ_i increases. The formula for calculating the non-uniformity ζ_i is

$$\zeta_i = \frac{1}{Q}\int_{Ai} \sqrt{(V_z - V_{F,av,i})^2}dA \tag{6}$$

where Q is the design flow rate; V_z is the local axial velocity in the flow section, m/s. Here, the section is the outlet surface of the diffuser and $V_{F,av,i}$ is the average velocity at the outlet surface of the diffuser, m/s.

5. Results

5.1. Response Surface of the Blade Lean Optimized Diffuser

Factor a_0 represents the starting angle of the blade on the hub surface, a_1 is related to the position of the axis, and a_2 affects the opening size of the parabola. The influence of pairs of factors on the response value was analyzed by fixing any one of the three factors a_0, a_1, and a_2 to zero. Figure 8a,b show that the opening of the response surface is downward, and the trend is consistent when a_2 is at a medium level or a_1 is at a medium level. The radian of the curve increases when the starting angle on the hub surface is $-4.5°$ (medium a_0), the symmetrical axis moves to hub (high a_1), and the opening of parabola decreases (high a_2), which improves the diffuser efficiency. Figure 8c shows that the interaction between factors a_1 and a_2 is significant when a_0 is medium. The interaction between a_1 and a_2 results in optimal diffuser efficiency when both a_1 and a_2 are high.

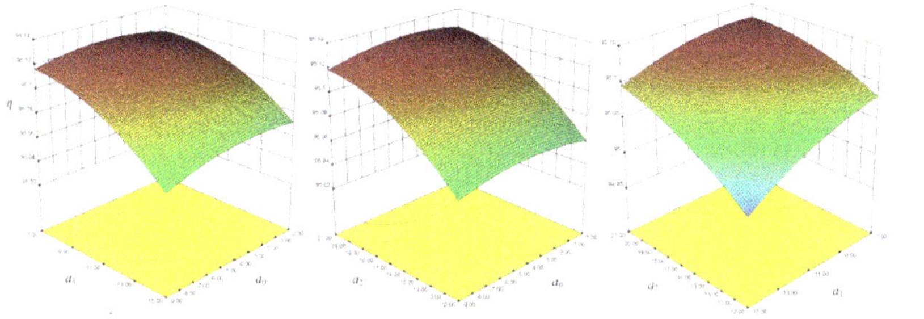

(a) (b) (c)

Figure 8. Response surface of diffuser efficiency in blade lean optimization (**a**) $a_2 = 0$. (**b**) $a_1 = 0$. (**c**) $a_0 = 0$.

Figure 9a shows that the static pressure recovery coefficient reaches the optimal value when a_0 is above the medium level and a_1 is at the medium level on the low side. A medium–high level of a_0 indicates that the starting angle of the hub at the blade inlet $\theta \in (0°, 4.5°)$. A medium–low level of a_1 indicates that the symmetrical axis of the curve deviates from the shroud, therefore, movement of the symmetrical axial can improve the conversion rate of kinetic energy to static pressure in the diffuser. Figure 9b shows that the static pressure recovery coefficient is optimal when a_0 is medium level and a_2 is medium–low level. A medium level of a_0 results in a starting angle of the inlet edge on the hub of 4.5° and when a_2 is medium–low, the curve of the inlet edge changes gently as the static pressure recovery coefficient increases. Figure 9c shows that the static pressure recovery coefficient reaches the optimal value when a_1 is at a medium–low level and a_2 is at a medium level. A medium–low level of a_1 indicates that the symmetry axis of the curve is inclined towards the shroud and a medium level of a_2 indicates that the opening size and bending size of the curve are moderate. Therefore, the static pressure recovery coefficient of the diffuser will decrease if the inlet edge bends excessively or too gently. Importantly, the static pressure recovery coefficient is improved when the inlet profile is moderately bent with an axially symmetrical offset to the shroud.

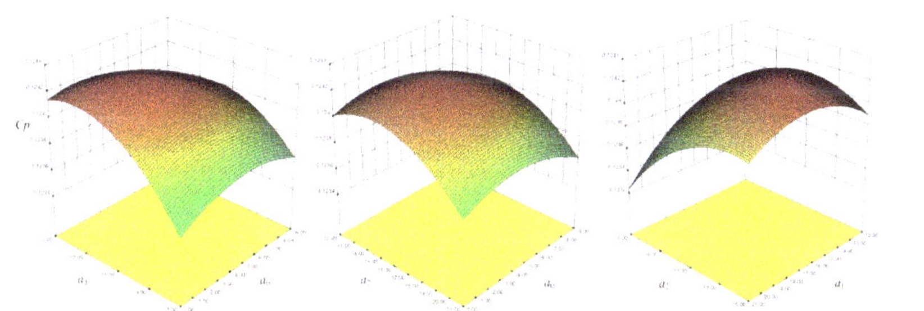

(a) (b) (c)

Figure 9. Response surface of pressure recovery coefficient in blade lean optimization (**a**) $a_2 = 0$. (**b**) $a_1 = 0$. (**c**) $a_0 = 0$.

Figure 10 shows the response surface with non-uniformity. Figure 10a shows that low levels of a_0 and a_1 minimize non-uniformity. When a_0 is low, the starting angle of the inlet edge of the diffuser on the hub is 9°. When a_1 is low, the axis of the curve moves from the hub side to the center of the spanwise end, the internal flow uniformity of the diffuser improves, and non-uniformity decreases. Figure 10b shows that when a_0 and a_2 are both at low levels, non-uniformity is lowest. When a_1 is at a medium level, the contour lines of the diffuser non-uniformity are evenly distributed, and the variation of the response surface is relatively gentle. The non-uniformity decreases with decreasing a_2 but is less affected by a_0. Figure 10c shows that both a_2 and a_1 have the lowest non-uniformity at low levels. When the starting angle on the hub of the inlet side is large, the symmetrical axis of the inlet profile is at 1/2 of the spanwise direction and the range of starting angles for each flow surface in the spanwise direction increases. Thus, the internal flow characteristics and flow uniformity in the diffuser can be improved.

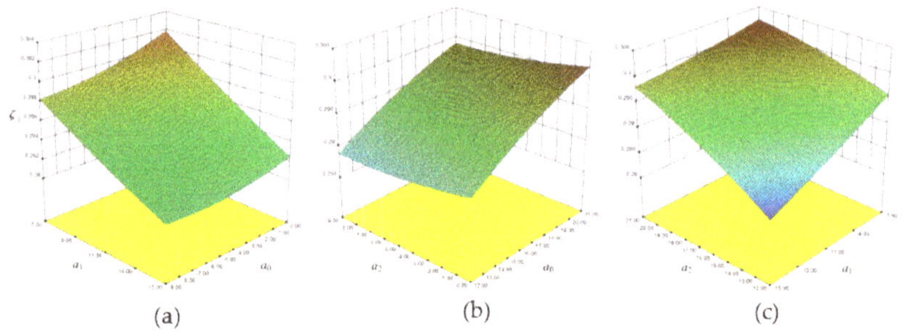

Figure 10. Response surface of non-uniformity in blade lean optimization. (**a**) $a_2 = 0$. (**b**) $a_1 = 0$. (**c**) $a_0 = 0$.

The factor a_2 has the largest influence on the diffuser efficiency, followed by a_1, and the influence of a_0 is the smallest. The diffuser efficiency increases with the increase of a_1 and a_2, increasing first and then decreasing with the increase of a_0. The influence of factor a_1 on the static pressure recovery coefficient of diffuser is the largest, followed by a_2, and a_0 is the smallest. The static pressure recovery coefficient increases first and then decreases with the decrease of the three factors. Moreover, a_1 and a_2 have significant effects on the diffuser non-uniformity, while a_0 has little effect on the non-uniformity. When the three factors are at a low level, the non-uniformity is the lowest, and the outlet uniformity of the diffuser is the best. The optimal blade lean scheme was obtained by considering the actual operating conditions of the multi-stage submersible pump and the effects of a_0, a_1, and a_2 on diffuser efficiency, static pressure recovery coefficient, and non-uniformity, as shown in Table 6. The results show that when a_0 is at the medium level of −4.4, a_1 has the low level of −12 and a_2 has the medium level of 15.16 both the diffuser efficiency and the static pressure recovery coefficient improve.

Table 6. Optimal solution of blade lean scheme.

	a_0	a_1	a_2	η (%)	Cp	ζ_i
Blade lean optimized diffuser	−4.4	−12	15.16	95.1	0.1242	0.292
Original diffuser	5	−5	0	94.8	0.1210	0.321

Structural changes to the inlet edge of the blade lean optimized diffuser and the original diffuser are shown in Figure 11. The inlet edge profile equation and spanwise distribution of the initial angle are illustrated in Figure 12. The starting angle of the blade lean optimized diffuser has a curved distribution, and the starting angle of the original diffuser has a linear distribution. After optimization, the governing equation of the blade inlet profile is $f\ (Sp_1) = -4.4-12Sp + 15.16Sp^2$, with axis $Sp = 0.4$. The starting angle of the hub is −4.4° and the starting angle of the shroud is −1.24°. The efficiency of the optimized diffuser is 0.32% higher than that of the original diffuser, and the static pressure recovery coefficient is 2.64% higher. However, the non-uniformity is 9% lower.

Figure 13 shows the static pressure and streamlines in the diffuser outlet section. Static pressure at the outlet of the blade lean optimized diffuser increases significantly, and the static pressure recovery coefficient is 2.64% higher than that of the original diffuser. This is because the blade inlet profile of the blade lean optimized diffuser changes from a straight line to a curve, the structure of the leading edge is more in line with the fluid flow trend, making it difficult to flow off, as shown in Figure 13b. Figure 14 shows the circumferential velocity distribution along the spanwise wall at the inlet edge of the diffuser. The fluid velocity at the shroud of the blade lean optimized diffuser is significantly lower than that of the original diffuser. The blade lean scheme improves the static pressure conversion capacity of the diffuser and effectively reduces the circumferential velocity component of the fluid and velocity difference at the leading edge of the diffuser. Therefore, the local hydraulic loss of the diffuser is reduced and the diffuser efficiency and static pressure recovery coefficient are improved.

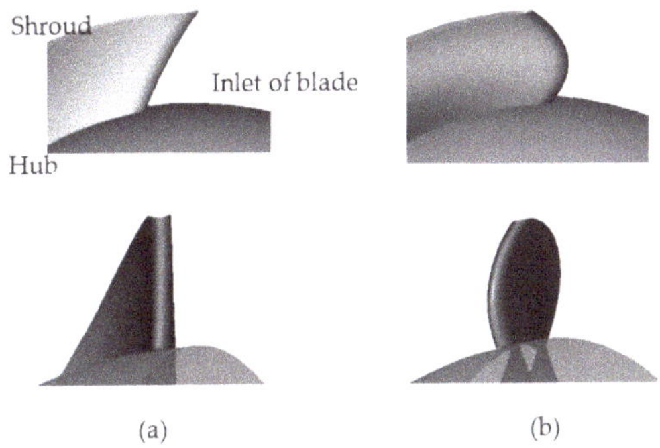

Figure 11. Blade inlet edge profile of diffuser in circumferential view (first row) and inlet view (second row). (**a**) Original diffuser. (**b**) Blade lean optimized diffuser.

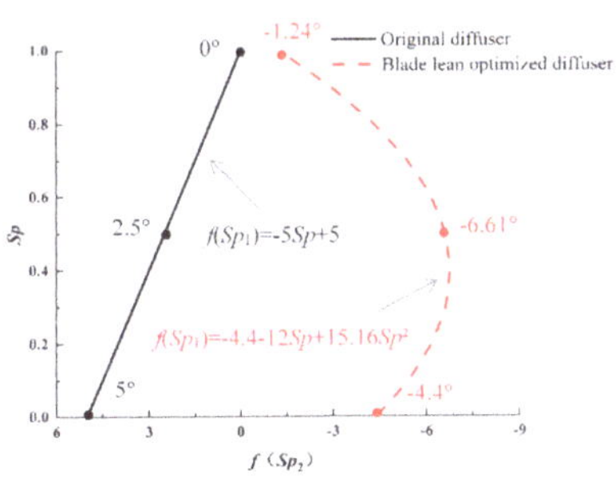

Figure 12. Profiles of inlet edge and spanwise distribution of stacking angle.

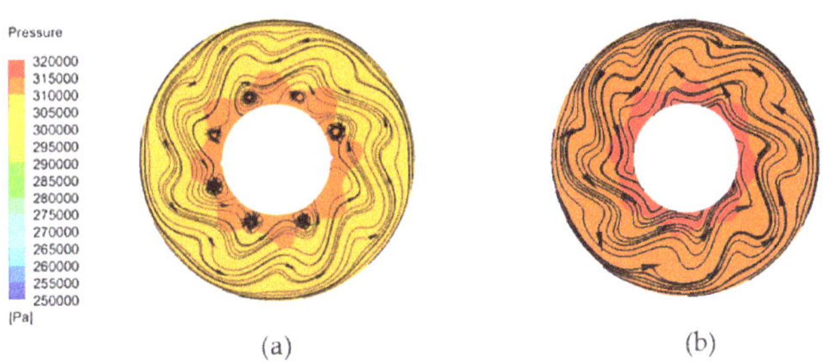

Figure 13. Comparison of static pressure and streamlines at diffuser outlet surface. (**a**) Original diffuser and (**b**) Blade lean optimized diffuser.

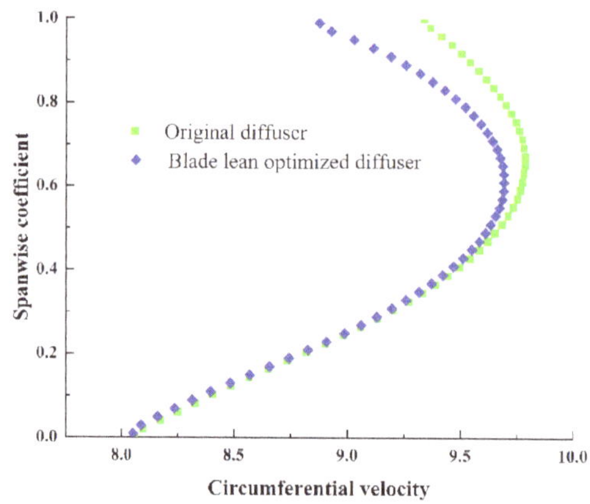

Figure 14. Comparison of pitch-averaged circumferential velocity distribution at leading edge of diffuser blade.

5.2. Response Surface of the Sweep Optimized Diffuser

As described in Section 4.2, b_0 is located at the outlet edge on the hub, b_1 is located at the outlet edge on the shroud, and b_2 is located at the inlet edge on the hub. One factor, b_0, b_1, or b_2, was fixed to zero and the influence of the other pair of factors on the response was analyzed. Figure 15a shows the influence of interaction between b_0 and b_1 on the diffuser efficiency when b_2 is at a medium level. High levels of b_0 and b_1 resulted in the optimal diffuser efficiency. Figure 15b shows the effect of interaction between b_0 and b_2 on the diffuser efficiency when b_1 is at a medium level. Factor b_2 has a large impact on diffuser efficiency, which increases when the outlet edge on the hub moves to the outlet of the diffuser (medium–high b_2). Figure 15c shows that the interaction between b_1 and b_2 has a significant influence on diffuser efficiency when b_0 is at a medium level. A medium–high level of b_2 and a high level of b_1 resulted in the highest diffuser efficiency.

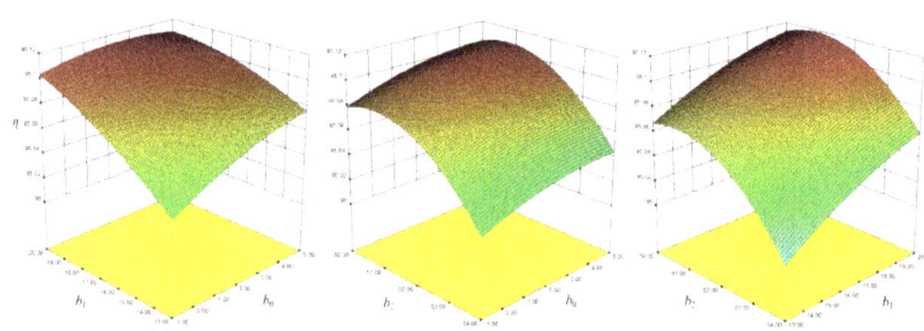

Figure 15. Response surface of diffuser efficiency in sweep optimization. (**a**) $b_2 = 0$. (**b**) $b_1 = 0$. (**c**) $b_0 = 0$.

Figure 16a shows that the static pressure recovery coefficient reaches the highest value when b_0 is low and b_1 is high. When the blade outlet on the hub moves towards the inlet of the diffuser (low b_0) and the blade outlet on the shroud side moves towards the outlet of the diffuser (high b_1), the static pressure recovery capacity of the diffuser improved. Figure 16b shows that the static pressure recovery coefficient is optimal when b_0 is low and b_2 is medium. Figure 16c shows that the static pressure recovery coefficient reaches the highest value when b_1 is high and b_2 is medium. The blade outlet edge on the hub moves towards the inlet of the diffuser (low b_0) and the blade outlet edge on the shroud moves towards the outlet of diffuser (high b_1), whereas the position of the blade inlet edge does not change at medium b_2 and the static pressure recovery coefficient can be increased.

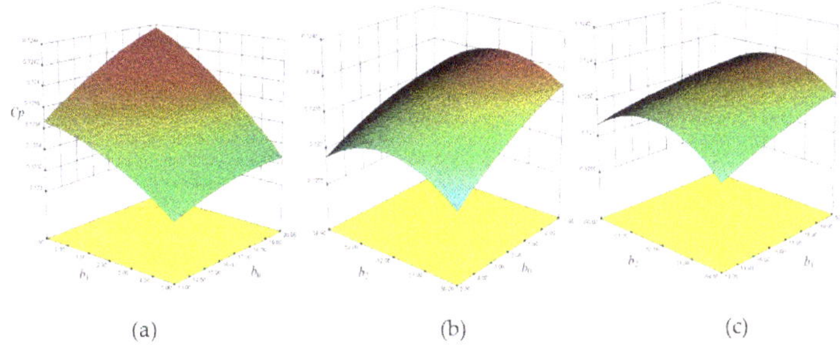

Figure 16. Response surface of pressure recovery coefficient in sweep optimization. (**a**) $b_2 = 0$. (**b**) $b_1 = 0$. (**c**) $b_0 = 0$.

Figure 17a shows that non-uniformity decreases when high b_0 interacts with high b_1. The outlet edge of the diffuser moves from the shroud to the hub (high b_0 and b_1), which can improve the flow uniformity in the diffuser. Figure 17b shows that non-uniformity is relatively low when b_0 is high. Factor b_2 has no obvious effect on non-uniformity and provided b_0 moves to a higher level, flow non-uniformity in the diffuser will decrease. A high level of b_0 is the key to optimizing flow uniformity in the diffuser but does not improve the static pressure recovery coefficient. Figure 17c shows that flow non-uniformity in the diffuser is low when b_1 is high. The change in direction of flow on the inlet side of the diffuser hub has very little effect on non-uniformity in the diffuser, while moving the outlet side of the diffuser towards the outlet direction (high b_0 and b_1) improves the flow uniformity.

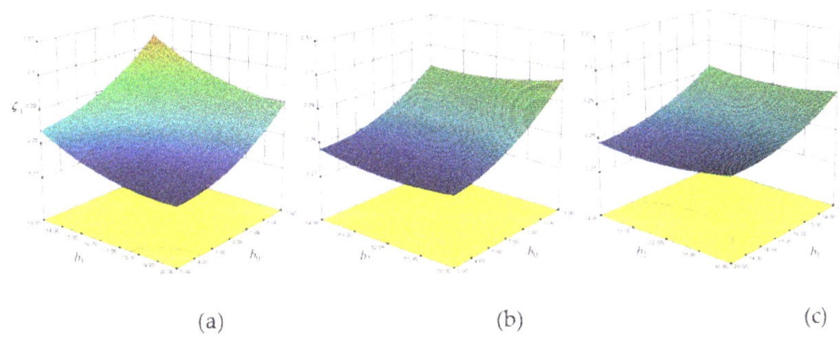

Figure 17. Response surface of non-uniform in sweep optimization. (**a**) $b_2 = 0$. (**b**) $b_1 = 0$. (**c**) $b_0 = 0$.

According to the measured data and response surface analysis, the factor b_2 has the largest influence on the diffuser efficiency, followed by b_1, and the influence of b_0 is the smallest. The diffuser efficiency exhibits an upward trend with increasing b_1 and b_0 and an inverted-U trend with increasing b_2. The influence of factor b_0 on the static pressure recovery coefficient of diffuser is the largest, followed by b_2, and b_1 is the smallest. The static pressure recovery coefficient increases with decreasing b_0, increases slowly with increasing b_1, and first increases and then decreases with increasing b_2. At the same time, non-uniformity decreases with increasing b_0 and b_1, whereas changes in b_2 have very little effect on flow uniformity. The optimal solution was obtained by considering the influence of b_0, b_1, and b_2 on diffuser efficiency, static pressure recovery coefficient, and non-uniformity, as shown in Table 7. The results show that when b_0 has the medium–low level of 1.5, b_1 has the high level of 20, and b_2 has the medium level of –52, the diffuser efficiency and static pressure recovery coefficient can be improved.

Table 7. Optimal solution of sweep scheme.

	b_0	b_1	b_2	η (%)	Cp	ζ_i
Sweep optimized diffuser	1.5	20	−52	95.21	0.1245	0.279
Original diffuser	0	14	−52.4	94.8	0.1210	0.321

Figure 18 shows the inlet and outlet profile positions of the original diffuser and the sweep optimized diffuser on the meridian plane. In the original pump, $b_0 = 0$, $b_1 = 14$, and $b_2 = -52.4$. In the sweep scheme, the values are $b_0 = 1.5$, $b_1 = 20$, and $b_2 = -52$. The position of the shroud surface at the blade outlet (b_1) changes greatly. When the shroud edge of the blade outlet moves towards the outlet of the diffuser, the influence of b_1 on the diffuser can be clearly observed.

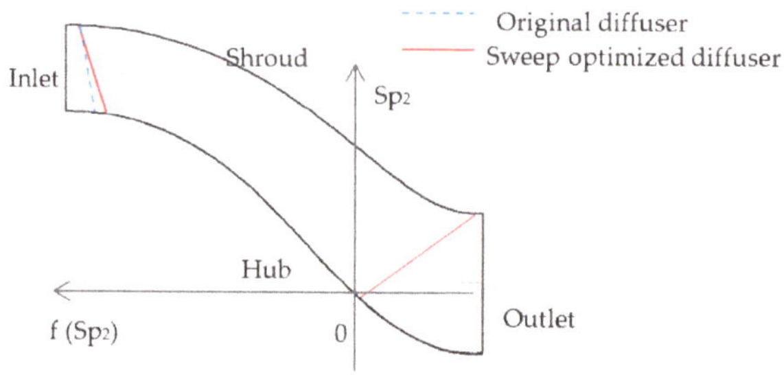

Figure 18. Comparison of sweep optimized meridional plane and the original of diffuser.

The optimized blade trailing edge extends the diffuser outlet on the shroud side, as shown in Figure 18. The increased blade trailing edge can better drain the fluid to the diffuser outlet, whereas there is sufficient time for velocity exchange with the high-speed main body when the fluid flows through the extended blade trailing edge (Figure 19b), thereby reducing the velocity gradient and improving flow uniformity at the outlet of the diffuser. In Figure 19a, low-energy fluid in the original diffuser does not fully mix with the high-speed mainstream and large differences in velocity lead to increased hydraulic losses in the diffuser, which affects the outlet flow uniformity. Therefore, the flow inside the sweep optimized diffuser is more reasonable. Compared with the original diffuser, flow non-uniformity is reduced by 13.1%.

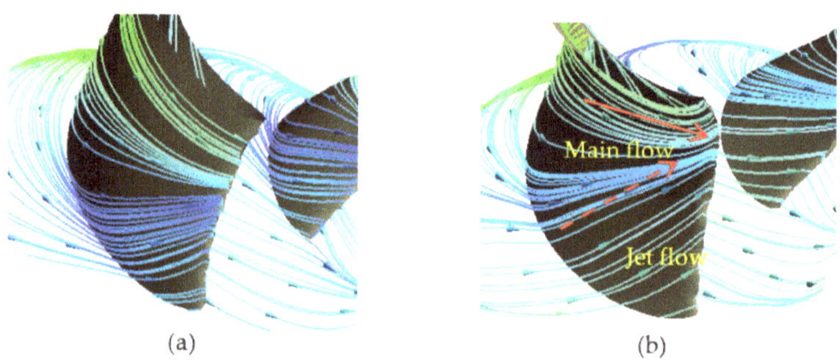

Figure 19. Comparison of the streamlines of blade outlet edge. (**a**) Original diffuser. (**b**) Sweep optimized diffuser.

6. Discussion

The origin of the hub corner separation vortex can be explained by reference to Figure 20. The vortex originates from two positions: the inlet of the diffuser, represented by

A, and the central of the diffuser, represented by B. In region A1, the curvature difference between the meridian shroud and the hub surface leads to the development of spanwise differential pressure, driving secondary flow from the shroud to the hub, which scours the inlet edge, thus forming a spanwise pressure difference, represented by region L. As the flow develops, swirling flow occurs at the hub in region A2 and rushes against the transverse pressure gradient towards the PS. In the center of the diffuser (region B1), the working fluid must flow against the reverse pressure gradient, resulting in an increase in low-momentum fluid on the hub. Then, the increased amount of low-momentum fluid is deflected towards the SS under the transverse differential pressure at the hub in region B2. Finally, a large hub corner separation vortex appears in the corner of the blade, accounting for about 1/3 of the area of the hub. In a previous study [20], the spanwise secondary flow from shroud to hub at the trailing edge of the diffuser was shown to scour low-energy fluid, thereby inhibiting the hub corner vortex. However, our results suggest that low-energy fluid accumulated on the suction surface cannot overcome the spanwise pressure difference between the shroud and the hub and can only reverse the flow along the hub surface, forming a secondary flow angle to suppress the vortex in region B3.

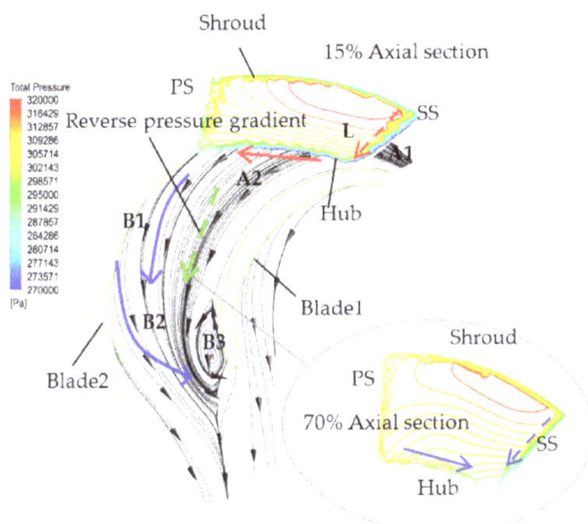

Figure 20. Onset of hub corner separation depicted by limiting streamlines and total pressure.

To explain the improvement in diffuser efficiency and diffuser uniformity, the spanwise/transverse pressure difference in the diffuser were quantified, as shown in Figures 21 and 22. At the inlet of the diffuser, the spanwise pressure difference in the blade lean optimized diffuser is significantly higher, resulting in enhanced transverse secondary flow (Figure 21). The spanwise pressure difference in the sweep optimized diffuser is significantly lower than in the blade lean optimized diffuser; therefore, the transverse pressure difference caused by spanwise secondary flow at the inlet of the diffuser is weakened. Moreover, both the transverse pressure differences at the inlet of the sweep optimized diffuser and blade lean optimized diffuser increase (Figure 22), which offsets the transverse pressure difference and suppresses the hub corner separation vortex.

At the outlet of the diffuser, the spanwise pressure difference at the trailing edge in the blade lean optimized diffuser is lower than in the original diffuser, as shown in Figure 19, which is not conductive to improving the diffuser performance. However, the spanwise pressure difference at the trailing edge in the sweep optimized diffuser is highest; therefore, low-energy fluid in the corner area becomes caught up in the main flow and is dragged toward the outlet of the diffuser, thus improving flow uniformity in the diffuser. When the low-energy fluid at the trailing edge of the diffuser is restrained and the difference between the mainstream velocity is weakened, mixing losses at the diffuser outlet are reduced and the hydraulic efficiency of the diffuser is improved.

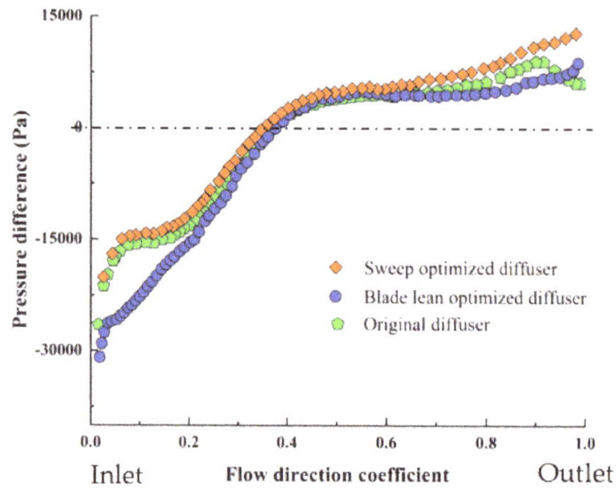

Figure 21. Comparison of spanwise pressure difference from hub to shroud of suction surface.

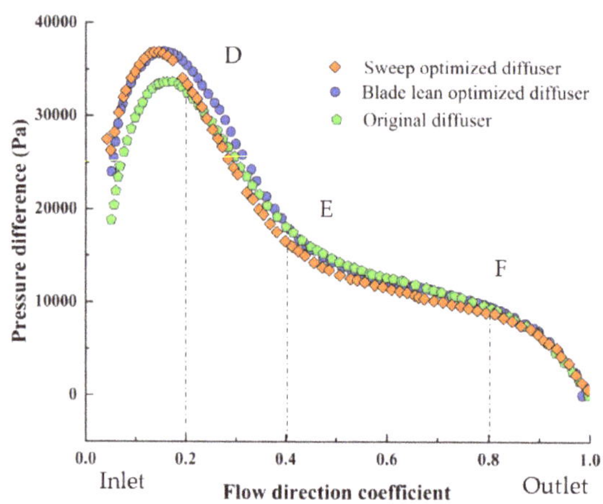

Figure 22. Comparison of transverse pressure difference from PS to SS on hub surface.

The reduced low-energy fluid in the center of the diffuser and the spanwise differential pressure drainage at the trailing edge of the diffuser lead to more uniform outflow. In Figure 21, the spanwise pressure difference in the center of the diffuser changes from negative to positive, and the direction is reversed. Therefore, the low-energy fluid originally returns from the hub to the low-pressure area in the center, which suppresses the corner separation vortex and improves the outlet flow uniformity in the diffuser. In Figure 22, the transverse pressure difference in the sweep optimized diffuser is less than in the blade lean optimized diffuser in the DF stage; however, the reverse pressure gradient in the DE stage is greater than in the blade lean optimized diffuser. The amount of low-energy fluid in the center of the diffuser increases and the reduced transverse pressure difference in the center cannot reduce the amount of low-energy fluid in the suction surface corner. However, the transverse pressure difference between the PS and SS in the EF stage decreases; therefore, the reverse pressure gradient in the flow direction decreases and the amount of low-energy fluid in the corner decreases.

Figure 23 compares the static pressure on the shroud of the diffuser under different scenarios. The static pressure on the shroud of the blade lean optimized diffuser is higher than that of the original diffuser, and the increase in static pressure on the PS is more obvious. Therefore, the blade lean optimized diffuser can convert more kinetic energy in the fluid into static pressure energy and the static pressure recovery coefficient of the diffuser can be improved. The static pressure on the shroud of the sweep optimized diffuser does not always increase; however, it is higher than those of other diffusers at the stage of 10–65% flow direction and decreases at the outlet section. The reason for this is that the flow direction at the outlet of the sweep optimized diffuser changes greatly and the distance between the working fluid flowing through the diffuser and the trailing edge of the blade increases. Thus, the static pressure decreases and the static pressure recovery coefficient of the diffuser does not increase significantly.

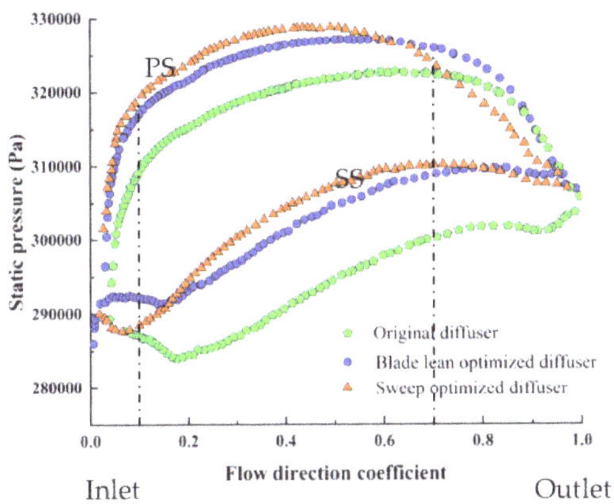

Figure 23. Static pressure comparison around shroud surface of diffuser.

7. Conclusions

Aimed at controlling the hub corner separation vortex and corresponding high hydraulic losses in the bowl diffuser of a multistage submersible pump, the diffuser was optimized in terms of blade lean and sweep. The blade structure can be controlled digitally and accurately using parametric equations. CFD simulations and central combination tests were used to optimize different blade lean and sweep schemes. The internal flow field and static pressure distribution in the diffuser were thoroughly investigated.

At the hub corner of the diffuser, the low-energy fluid induced by spanwise and transverse pressure at the inlet of the diffuser blade overcomes the streamwise pressure gradient and accumulates in the mid-section of the suction surface. However, the accumulated low-energy fluid cannot overcome the spanwise pressure difference between the shroud and the hub, the flow direction is reversed toward the inlet, and the hub corner vortex forms. Optimization of the blade lean and sweep can improve the transverse pressure in the diffuser. In particular, the hub–shroud spanwise pressure can be enhanced in order to drive the low-energy fluid towards the center of the blade, rather than reversing it. This suppresses the hub corner vortex and improves the hydraulic performance of the diffuser.

In the blade lean scheme, the transverse pressure difference at the inlet of the diffuser increases, which inhibits the formation of the hub corner separation vortex. However, the spanwise pressure difference at the inlet of the suction surface increases, and the spanwise pressure difference at the trailing edge decreases, which does not improve the performance of the diffuser. Moreover, since the structure of the leading edge of the blade changes and the fluid flows close to the blade after entering the diffuser, flow separation is difficult to induce. The fluid kinetic energy and static pressure energy are fully converted, the static pressure recovery coefficient increases by 2.64% compared with the original diffuser, and the diffuser efficiency increases by 0.32%. Optimization of blade lean can improve the static pressure recovery coefficient and diffuser efficiency but does not improve flow uniformity in the diffuser.

The sweep scheme can reduce the spanwise pressure difference and transverse pressure difference at the inlet of the diffuser, increase the spanwise pressure difference at the outlet of the diffuser, inhibit the formation of the hub corner separation vortex, and improve the flow uniformity of the diffuser. However, the increase in the reverse pressure gradient in the flow direction of the diffuser limits the improvement in diffuser efficiency. Sweep design can effectively reduce flow non-uniformity in the diffuser; however, the influence of the static pressure recovery coefficient is small.

Author Contributions: Conceptualization, P.C. and C.N.; Formal analysis, C.N. and X.G.; Methodology, C.N. and R.Z.; Project administration, P.C.; Software, R.Z.; Writing—original draft, C.N.; Writing—review & editing, X.G. All authors have read and agreed to the published version of the manuscript.

Funding: This research was funded by the Natural Science Foundation of Jiangsu Province: BK 20190847, China Postdoctoral Science Foundation: 2019M661744, the National Natural Science Foundation of China: 51879120, and the Priority Academic Program Development of Jiangsu Higher Education Institutions (PAPD).

Institutional Review Board Statement: Not applicable.

Informed Consent Statement: Not applicable.

Data Availability Statement: The data that support the findings of this study are available from the corresponding author upon reasonable request.

Acknowledgments: A huge thanks is due to the editor and reviewers for their valuable comments to improve the quality of this paper.

Conflicts of Interest: The authors declare no conflict of interest.

Abbreviations

Cp	Static pressure recovery coefficient
CCD	Center composite design
H	Head (m)
PS	Pressure surface of blade
SS	Suction surface of blade
k	Turbulent kinetic energy (m^2/s^2)
P_{in}	Total pressure of pump inlet (Pa)
P_{out}	Total pressure of pump outlet (Pa)
P_{t3}	Total pressure of diffuser inlet (Pa)
P_{t4}	Total pressure of diffuser outlet (Pa)
P_{s3}	Static pressure of diffuser inlet (Pa)
P_{s4}	Static pressure of diffuser outlet (Pa)
ρ	Density of fluid (kg/m^3)
μ_t	Turbulent viscosity ($kg/m{\cdot}s$)
ε	Turbulent energy dissipation (m^2/s^3)
ϕ_{Hub}	Wrap angle of hub (°)
ϕ_{Shroud}	Wrap angle of shroud (°)
θ	Starting angle (°)
$\Delta\theta$	Starting angle difference (°)
θ_{Hub}	Starting angle of hub flow surface (°)
θ_{Shroud}	Starting angle of shroud flow surface (°)
β	Sweep angle of diffuser (°)
η	Efficiency of diffuser (%)
ζ_i	Non-uniformity

References

1. Wang, H.L.; Hu, Q.X.; Yang, Y.; Wang, C. Performance Differences of Electrical Submersible Pump under Variable Speed Schemes. *Int. J. Simul. Model.* **2021**, *20*, 76–86. [CrossRef]
2. Zhou, J.; Zhao, M.; Wang, C.; Gao, Z. Optimal Design of Diversion Piers of Lateral Intake Pumping Station Based on Orthogonal Test. *Shock. Vib.* **2021**, *2021*, 6616456. [CrossRef]
3. Shi, L.; Zhu, J.; Tang, F.; Wang, C. Multi-Disciplinary Optimization Design of Axial-Flow Pump Impellers Based on the Approximation Model. *Energies* **2020**, *13*, 779. [CrossRef]
4. Zhang, L.; Wang, C.; Zhang, Y.; Xiang, W.; He, Z.; Shi, W. Numerical study of coupled flow in blocking pulsed jet impinging on a rotating wall. *J. Braz. Soc. Mech. Sci. Eng.* **2021**, *43*, 508. [CrossRef]
5. Wang, H.; Long, B.; Wang, C.; Han, C.; Li, L. Effects of the Impeller Blade with a Slot Structure on the Centrifugal Pump Performance. *Energies* **2020**, *13*, 1628. [CrossRef]
6. Wang, H.; Qian, Z.; Zhang, D.; Wang, T.; Wang, C. Numerical Study of the Normal Impinging Water Jet at Different Impinging Height, Based on Wray–Agarwal Turbulence Model. *Energies* **2020**, *13*, 1744. [CrossRef]
7. Bellary, S.A.I.; Husain, A.; Samad, A. Effectiveness of meta-models for multi-objective optimization of centrifugal impeller. *J. Mech. Sci. Technol.* **2014**, *28*, 4947–4957. [CrossRef]
8. Bellary, S.A.I.; Samad, A.; Couckuyt, I.; Dhaene, T. A comparative study of kriging variants for the optimization of a turbomachinery system. *Eng. Comput.* **2016**, *32*, 49–59. [CrossRef]
9. Huang, K.L.; Yuan, J.P.; Si, Q.R.; Lin, G. Numerical simulation of pressure pulsation in multistage centrifugal pump under multi-operation condition. *J. Drain. Irrig. Mach. Eng.* **2019**, *37*, 387–392.
10. Tong, S.G.; Zhao, H.; Liu, H.Q.; Tong, Z.M.; Yue, Y.U.; Tang, N. Optimization calculation method for efficiency of multistage split case centrifugal pump. *J. Zhejiang Univ. Eng. Sci.* **2019**, *53*, 988–996.
11. Stel, H.; Sirino, T.; Ponce, F.; Chiva, S.; Morales, R. Numerical investigation of the flow in a multistage electric submersible pump. *J. Pet. Sci. Eng.* **2015**, *136*, 41–54. [CrossRef]
12. Heo, M.W.; Ma, S.-B.; Shim, H.S.; Kim, K.Y. High-efficiency design optimization of a centrifugal pump. *J. Mech. Sci. Technol.* **2016**, *30*, 3917–3927. [CrossRef]
13. Murugesan, C.; Rudramoorthy, R. Numerical And Experimental Study Of Single stage And Multistage Centrifugal Mixed Flow Submersible Borewell Pumps. *Int. J. Fluid Mach. Syst.* **2016**, *9*, 107–118. [CrossRef]
14. Kim, J.-H.; Cho, B.-M.; Kim, S.; Lee, Y.-K.; Choi, Y.-S. Steady and Unsteady Flow Characteristics of a Multi-stage Centrifugal Pump under Design and Off-design Conditions. *Int. J. Fluid Mach. Syst.* **2019**, *12*, 64–70. [CrossRef]
15. Rosic, B.; Xu, L. Blade Lean and Shroud Leakage Flows in Low Aspect Ratio Turbines. *J. Turbomach.* **2011**, *134*, 031003. [CrossRef]
16. Razavi, S.R.; Sammak, S.; Boroomand, M. Multidisciplinary Design and Optimizations of Swept and Leaned Transonic Rotor. *J. Eng. Gas Turbines Power* **2017**, *139*, 122601. [CrossRef]
17. He, X.; Zheng, X. Mechanisms of Sweep on the Performance of Transonic Centrifugal Compressor Impellers. *Appl. Sci.* **2017**, *7*, 1081. [CrossRef]
18. He, X.; Zheng, X. Performance improvement of transonic centrifugal compressors by optimization of complex three-dimensional features. *Proc. Inst. Mech. Eng. Part G J. Aerosp. Eng.* **2016**, *231*, 2723–2738. [CrossRef]
19. Bagshaw, D.A.; Ingram, G.; Gregory-Smith, D.G.; Stokes, M.R.; Harvey, N.W. The design of three-dimensional turbine blades combined with profiled endwalls. *Proc. Inst. Mech. Eng. Part A J. Power Energy* **2008**, *222*, 93–102. [CrossRef]
20. Goto, A.; Zangeneh, M. Hydrodynamic Design of Pump Diffuser Using Inverse Design Method and CFD. *J. Fluids Eng.* **2002**, *124*, 319–328. [CrossRef]
21. Goto, A.; Nohmi, M.; Sakurai, T.; Sogawa, Y. Hydrodynamic Design System for Pumps Based on 3-D CAD, CFD, and Inverse Design Method. *J. Fluids Eng.* **2002**, *124*, 329–335. [CrossRef]
22. Scillitoe, A.; Tucker, P.G.; Adami, P. Numerical Investigation of Three-Dimensional Separation in an Axial Flow Compressor: The Influence of Freestream Turbulence Intensity and Endwall Boundary Layer State. *J. Turbomach.* **2016**, *139*, 021011. [CrossRef]
23. Şahin, F.C. Experimental investigation on flow improvement in compressor cascades. *Int. J. Energy Res.* **2016**, *41*, 526–539. [CrossRef]
24. Ananthakrishnan, K.; Govardhan, M. Influence of fillet shapes on secondary flow field in a transonic axial flow turbine stage. *Aerosp. Sci. Technol.* **2018**, *82–83*, 425–437. [CrossRef]
25. Shi, W.; Zhou, L.; Lu, W.; Xu, L.; Li, W. Numerical Simulation and Experimental Study of Different Stages Deep-Well Centrifugal Pump. *J. Comput. Theor. Nanosci.* **2013**, *10*, 2897–2901. [CrossRef]
26. Zhou, L.; Shi, W.; Li, W.; Agarwal, R. Numerical and Experimental Study of Axial Force and Hydraulic Performance in a Deep-Well Centrifugal Pump With Different Impeller Rear Shroud Radius. *J. Fluids Eng.* **2013**, *135*, 104501. [CrossRef]
27. Yuan, S.Q.; He, W.T.; Si, Q.R.; Yuan, J.P.; Zhang, H.Y.; Cui, Q.L. Numerical simulation on gas-liquid two-phase flow in centrifugal pump based on MUSIG model. *J. Drain. Irrig. Mach. Eng.* **2021**, *39*, 1–7.
28. Yakhot, V.; Orszag, S.A. Renormalization group analysis of turbulence. I. Basic theory. *J. Sci. Comput.* **1986**, *1*, 3–51. [CrossRef]
29. Lam, S.H. On the RNG theory of turbulence. *Phys. Fluids A Fluid Dyn.* **1992**, *4*, 1007–1017. [CrossRef]
30. Mompean, G. Numerical simulation of a turbulent flow near a right-angled corner using the Speziale non-linear model with RNG K–ε equations. *Comput. Fluids* **1998**, *27*, 847–859. [CrossRef]

31. Cao, P.; Zhu, R.; Yin, G. Spike-type disturbances due to inlet distortion in a centrifugal pump. *Renew. Energy* **2021**, *165*, 288–300. [CrossRef]
32. Zhou, L.; Wang, W.; Hang, J.; Shi, W.; Yan, H.; Zhu, Y. Numerical Investigation of a High-Speed Electrical Submersible Pump with Different End Clearances. *Water* **2020**, *12*, 1116. [CrossRef]
33. Zhou, L.; Bai, L.; Shi, W.; Li, W.; Wang, C.; Ye, D. Numerical analysis and performance experiment of electric submersible pump with different diffuser vanes number. *J. Braz. Soc. Mech. Sci. Eng.* **2018**, *40*, 89. [CrossRef]

machines

Article

Investigation on the Transient Characteristics of Self-Priming Pumps with Different Hub Radii

Hao Chang [1], Shiming Hong [1], Chuan Wang [2,3], Guangjie Peng [1,*], Fengyi Fan [1] and Daoxing Ye [4]

1 Research Center of Fluid Machinery Engineering and Technology, Jiangsu University, Zhenjiang 212013, China; changhao@ujs.edu.cn (H.C.); 2221911014@stmail.ujs.edu.cn (S.H.); ffyfwt@163.com (F.F.)
2 College of Mechanical and Electrical Engineering, Hainan Vocational University of Science and Technology, Haikou 571126, China; wangchuan198710@126.com
3 College of Hydraulic Science and Engineering, Yangzhou University, Yangzhou 225009, China
4 Key Laboratory of Fluid Machinery and Engineering, Xihua University, Chengdu 610039, China; dxingye@163.com
* Correspondence: pgj@ujs.edu.cn

Abstract: Self-priming pumps, important fluid equipment, are widely used in the disaster relief and emergency fields. Meanwhile, the impeller is the only rotational unit of the self-priming pump, which plays an essential part in the power capability of the pump. In this paper, impellers with different hub radii are proposed; by comparing the internal flow characteristics, blade surface load, pressure pulsation characteristics, and radial force distribution of each scheme, the relationship between transient characteristics and hub radius is obtained. The results present that the impeller with a large hub radius can not only weaken the pressure pulsation, blade surface load, and radial force distribution, but also improve the ability of the blade to work on the internal flow field. Finally, the relevant hydraulic experiment is conducted, with the difference between the experiment and calculation below 3%, which ensures the accuracy of the calculation results.

Keywords: transient characteristics; self-priming pump; numerical simulation; hub radius

Citation: Chang, H.; Hong, S.; Wang, C.; Peng, G.; Fan, F.; Ye, D. Investigation on the Transient Characteristics of Self-Priming Pumps with Different Hub Radii. *Machines* **2021**, 9, 311. https://doi.org/10.3390/machines9120311

Academic Editor: Antonio J. Marques Cardoso

Received: 26 October 2021
Accepted: 16 November 2021
Published: 25 November 2021

1. Introduction

The impeller is the only rotational unit of the self-priming pump, which plays an essential part in the power capability of the pump. More and more scholars pay attention to the performance improvement of the centrifugal pump by optimizing the geometric parameters of the impeller [1–9]. Jaiswal et al. [10] optimized the input power of a centrifugal pump based on the multi-objective genetic algorithm; by modifying the impeller blade exit angle, the flow separation at the blade trailing edge is improved effectively, and the efficiency improvement of the optimal model is greater than 10%. Qian et al. [11] selected blade exit angle, outlet diameter, and blade wrap angle of the secondary impeller as optimization parameters, as well as axial force and hydraulic performance, as the response objectives to improve the performance of the multistage pump. Meanwhile, the multiple regression model is established, which can effectively reflect the relationship between the performance parameters with geometric dimensioning. Chen et al. [12] designed six impellers with different blade inlet angles based on ANSYS Bladegen and NX software, and the calculations are carried out to analyze the influence of blade inlet angle on the performance of the single blade pump. Lin et al. [13] found that obvious pressure pulsation will be generated by the high-speed rotational impeller. However, the intensity of the rotor-stator can be eliminated by employing the impeller with a bionic sinusoidal tubercle trailing edge, and the vortex generated at the trailing edge can be effectively suppressed.

Posa et al. [14] adopted the Large Eddy Simulations to investigate the effect of the diffuser inlet angle on the pressure fluctuation; the results found that the decrease in the diffuser inlet angle will lead to separation on the pressure side under the design flow rate. Thakkar et al. [15] used the response surface methodology to study the influence of blade outlet angle, blade wrap angle, and blade outlet width on the head and efficiency of a sanitary centrifugal pump. Twenty-five optimized schemes are generated based on the Latin hypercube sampling method, and the head and efficiency are increased 9.154% and 10.15%, respectively. Shigemitsu et al. [16] analyzed the effect of blade number on the performance of a centrifugal pump for low viscous fluid by experiment and calculation; the results present that the head and shaft power can be improved by increasing the blade number. Shi et al. [17] analyzed the energy conversion ability of the multiphase pump impeller by adjusting the blade wrap angle; with the increase in wrap angle, the energy conversion presents a trend of decreasing first and then increasing. Wang et al. [18] proposed 50 sets of impellers based on the Latin Hypercube Sampling method, and the Pareto-based genetic algorithm was employed to optimize the impeller geometry parameters of pump-as-turbines. Zhang et al. [19] investigated the energy loss of the side channel pump with different wrap angles based on the entropy loss production method; the results present that the highest entropy losses are generated in the impeller and the side channel passages with the smaller wrap angle. Peng et al. [20] selected the blade outlet angles with 16 degrees, 20 degrees, 24 degrees, 28 degrees, and 32 degrees to analyze the internal flow characteristic of the centrifugal pump, and the relationship between the blade outlet angle and the cavitation performance is obtained. However, most investigations focus on the impeller optimization by adjusting the blade wrap angle, blade number, inlet angle, and outlet angle; a few analyses about hub radius optimization are carried out.

Therefore, in this paper, a self-priming pump with different hub radii is systematically investigated by numerical simulations and experiments, while the internal flow characteristics, blade surface load, pressure pulsation characteristics, and radial force distribution are analyzed. Finally, the relevant hydraulic experiment is conducted, and the difference between the experiment and calculation is below 3%, which ensures the accuracy of the calculation results.

2. Numerical Model and Grids

In this paper, a self-priming centrifugal pump is selected as the analysis object; the design flow rate is 500 m^3/h, the rotation speed is 2200 r/m, and the design head is 45 m. Meanwhile, the calculation model consists of the inlet section, outlet section, impeller, volute, and self-priming chamber, as shown in Figure 1. Considering the sufficient evolution of inflow and outflow, the length of the inlet and outlet section extends to five times the diameter of the impeller inlet and the volute outlet, respectively. At the same time, the calculation domain is modeled by UG software, and structured grids are employed on all models by ANSYS ICEM, while local encryption is employed on the leading edge of the blade and tongue of the volute.

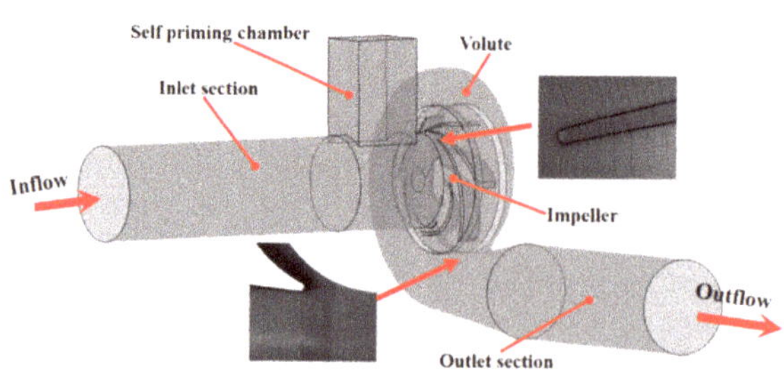

Figure 1. The calculation model.

Moreover, grid sensitive analysis is carried out to improve the investigation accuracy; by comparing the head and efficiency of the grid number at 5×10^6, 4×10^6, 3×10^6, 2×10^6, and 1×10^6, it can be concluded that the head and efficiency gradually increase with the rise of the grid number. However, when the grid number is more than 2×10^6, the deviation of the head and efficiency of the different grid schemes is less than 1%, as shown in Figure 2. Therefore, considering the consumption of calculation resources and time, the grid number of 2×10^6 is selected for further investigation.

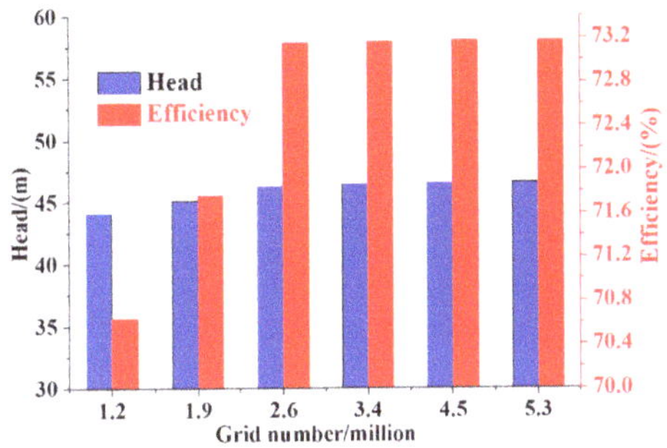

Figure 2. Grid sensitive analysis.

3. Analysis Scheme

In this paper, the geometric parameters, such as impeller inlet diameter, outlet diameter, outlet width, inlet angle, outlet angle, and shroud radius, remain the same, while the hub radius is 1.5, 2, 2.5, and 3 times the shroud radius, and the corresponding schemes is 1, 2, 3, and 4, respectively. Furthermore, the SST k–ω turbulence and water (at normal temperature) are employed, while the calculation model of the impeller is set as a rotational part; the calculation model of the inlet, outlet, impeller, volute, and the self-priming chamber is set as the stationary part; and the interface between the rotational and stationary part is set as a frozen rotor interface. Meanwhile, the boundary condition is set as pressure inlet and mass flow outlet. According to the previous investigation [21], the time step is set as 4.54×10^{-4} s; RMS residual is set as 0.00001. The transient simulation is carried out to analyze the influence of the hub radius on the performance of the self-priming centrifugal pump. Schemes of impellers with different meridians are shown in Figure 3.

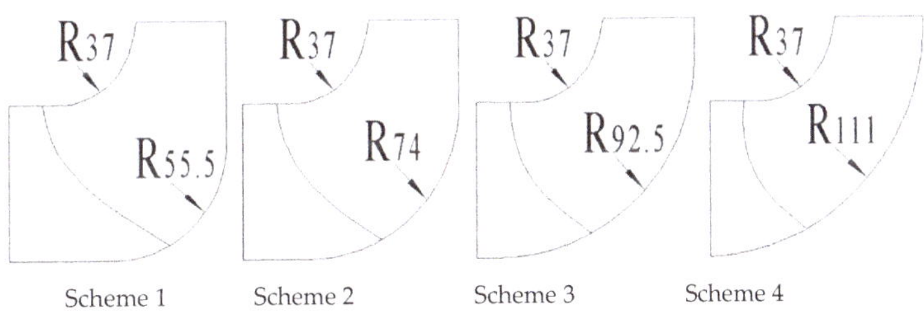

Figure 3. The schemes of impellers with different meridians.

4. Analysis of Internal Flow Characteristics

Figure 4 presents the velocity distribution of the inner circumference of the four schemes under part-load flow conditions. It found that the velocity distribution of the impeller inlet on the hub (Span 0) is relatively disordered, and the different degrees of low-speed clusters are generated on the pressure surface. Meanwhile, with an increase in the hub radius, the area of the low-speed zone gradually decreases. As the circumferential surface moves towards the shroud, the area of the low-speed clusters decreases further, and the formation location shifts from the leading edge of the pressure surface to the trailing edge. However, the low-speed clusters are transferred from the pressure surface to the suction surface on Span 0.5, and the area of the low-speed clusters with a smaller hub radius was still significantly large. When the circumferential surface moves from Span 0.5 to Span 1, the low-speed clusters gradually shift from the trailing edge to the leading edge, and the area continuously increases. What is more, the streamline contour is pushed onto the pressure distribution diagrams to further investigate the flow characteristics with different meridian profiles, as shown in Figure 5. The four schemes have similar pressure distribution on the same circumferential surface, but there are large differences in the streamline contour.

As shown in Figure 5a, an obvious vortex is generated at the trailing edge of the pressure surface in Scheme 1; by comparing the velocity distribution, it can be found that the vortex is mainly generated in the low-speed clusters. At the same time, under the effect of rotor-stator interaction between the impeller and volute tongue, obvious vortices are formed at the trailing edge of the blade close to the volute tongue. However, with the increase in hub radius, the strength of the vortex decreases continuously. As the circumferential surface moves forward to the shroud, the vortex of each scheme is reduced, and the flow channels affected by the vortices in Scheme 1 are also significantly declined. Furthermore, the backflow at the trailing edge of the suction surface gradually shifts to the leading edge. At the same time, the intensity of the vortex gradually weakens with the increase in the curvature of the hub.

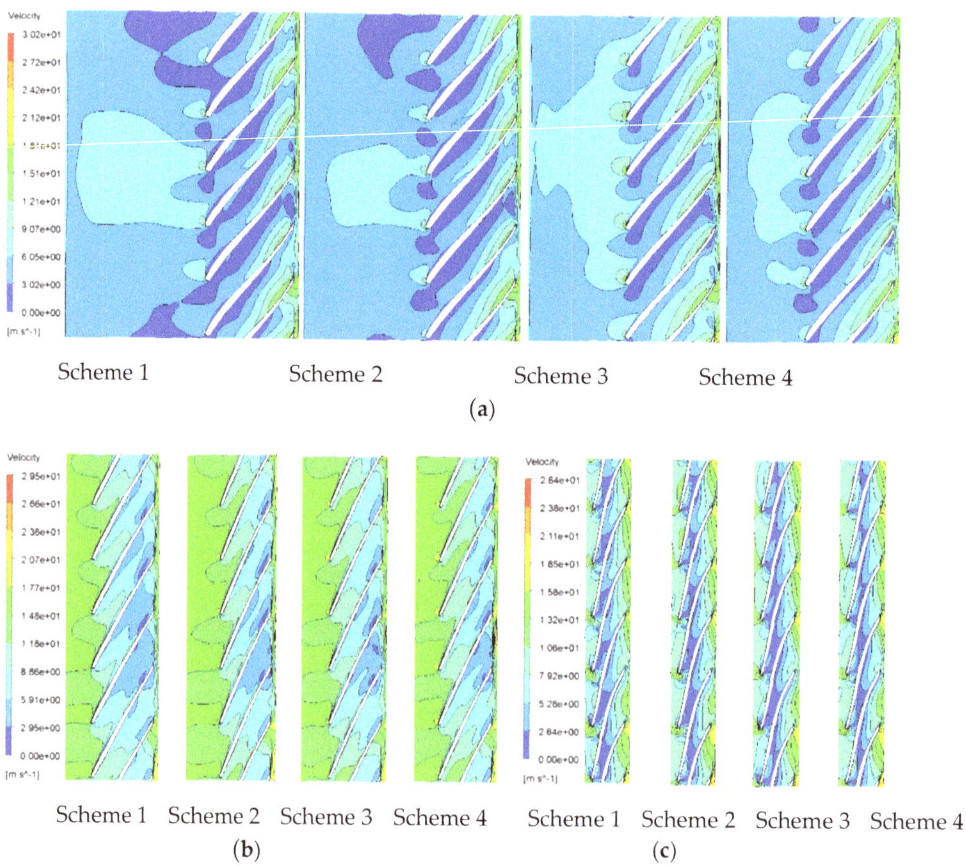

Figure 4. The circumferential velocity distribution under 300m^3/h. (**a**) Velocity distribution of thecircumferential surface of Span 0. (**b**) Velocity distribution of Span 0.5. (**c**) Velocity distribution of Span 1.

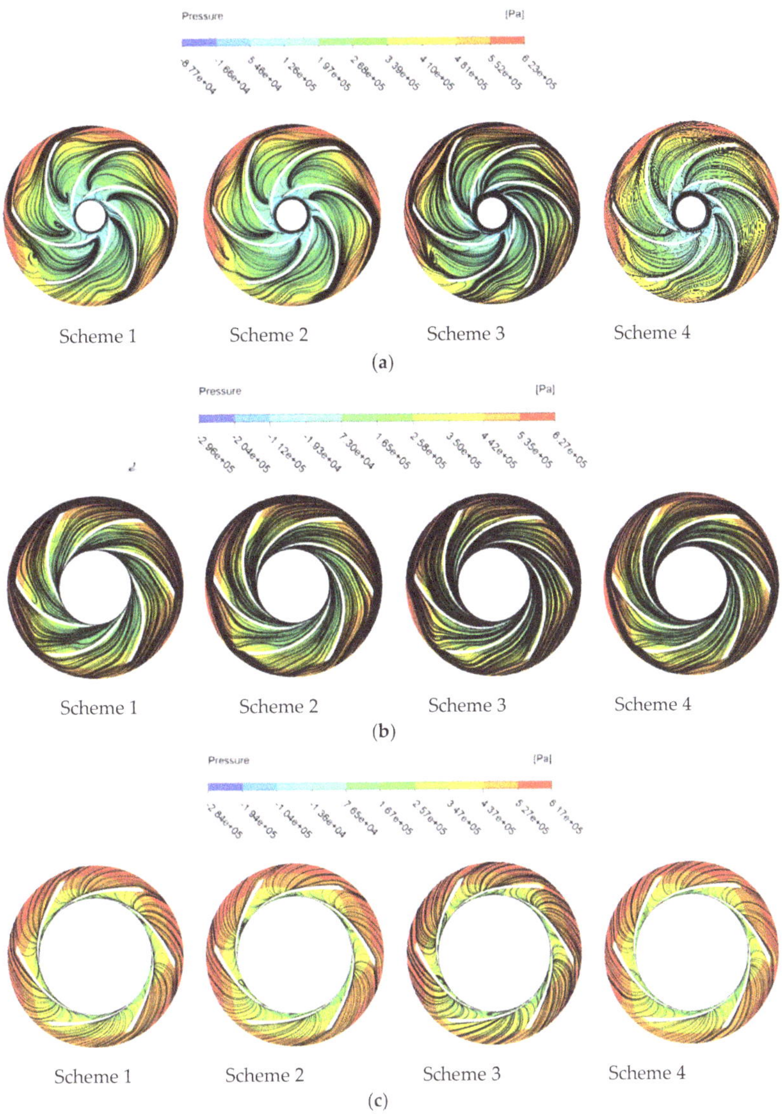

Figure 5. The circumferential pressure distribution and streamline under 300 m^3/h. (**a**) Pressure distribution of the circumferential surface of Span 0. (**b**) Pressure distribution of the circumferential surface of Span 0.5. (**c**) Pressure distribution of the circumferential surface of Span 1.

Figure 6 presents the velocity distribution of the impellers under optimal flow conditions; the area of low-speed clusters significantly decreases compared with part-load flow conditions. Furthermore, low-speed clusters can be found on the pressure and suction surface of Scheme 1, and the area on the pressure surface is significantly larger than the suction surface. When the radius of the hub continues to increase, the area of the low-speed clusters on the pressure surface of Schemes 3 and 4 continuously reduces, which is mainly concentrated in the middle of the blade. When the circumferential surface moves from the hub to the shroud, the low-speed area is gradually declined and transferred to the middle of the suction surface. Furthermore, the low-speed area of the scheme with a large hub radius is still larger than the small one.

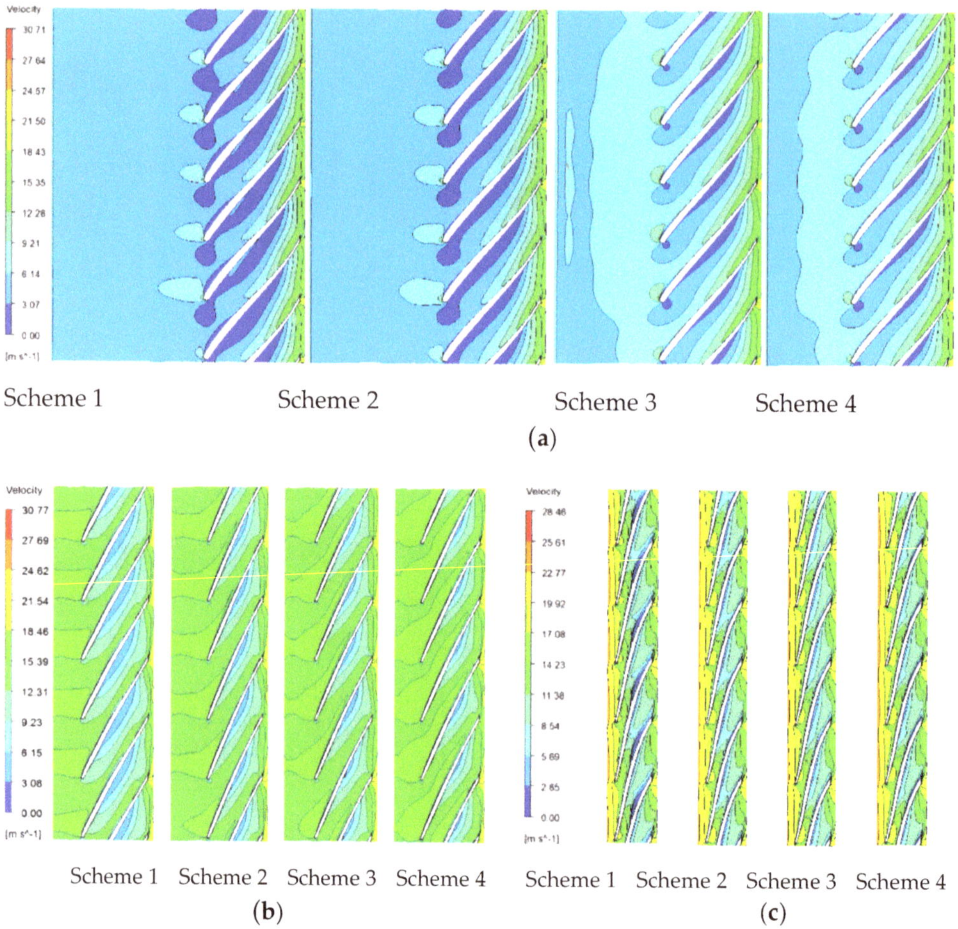

Figure 6. The circumferential velocity distribution under 500 m^3/h. (**a**) Velocity distribution of the circumferential surface of Span 0. (**b**) Velocity distribution of Span 0.5. (**c**) Velocity distribution of Span 1.

Figure 7 shows the pressure and streamline distribution under optimal flow conditions. It can be found that the pressure distribution of each scheme is uniform on different circumferential surfaces, which has a similar variation rule. However, with the increase in hub radius, the area of high pressure at the trailing edge gradually decreases. By comparing the streamline diagrams of the internal flow field, obvious vortices are formed at the leading edge of the pressure surface under Span 0. The vortex is constantly weakened and eliminated with the rising of the hub radius. Furthermore, as the circumferential surfaces moves towards the shroud, only a few vortices are generated at the leading edge, and the intensity of the vortices is significantly weakened. What is more, no vortices are found in the impeller flow channels; the streamlines are smooth under Span 0.5, but only a small amount of vortices are formed on the leading edge of the suction surface under Span 1.

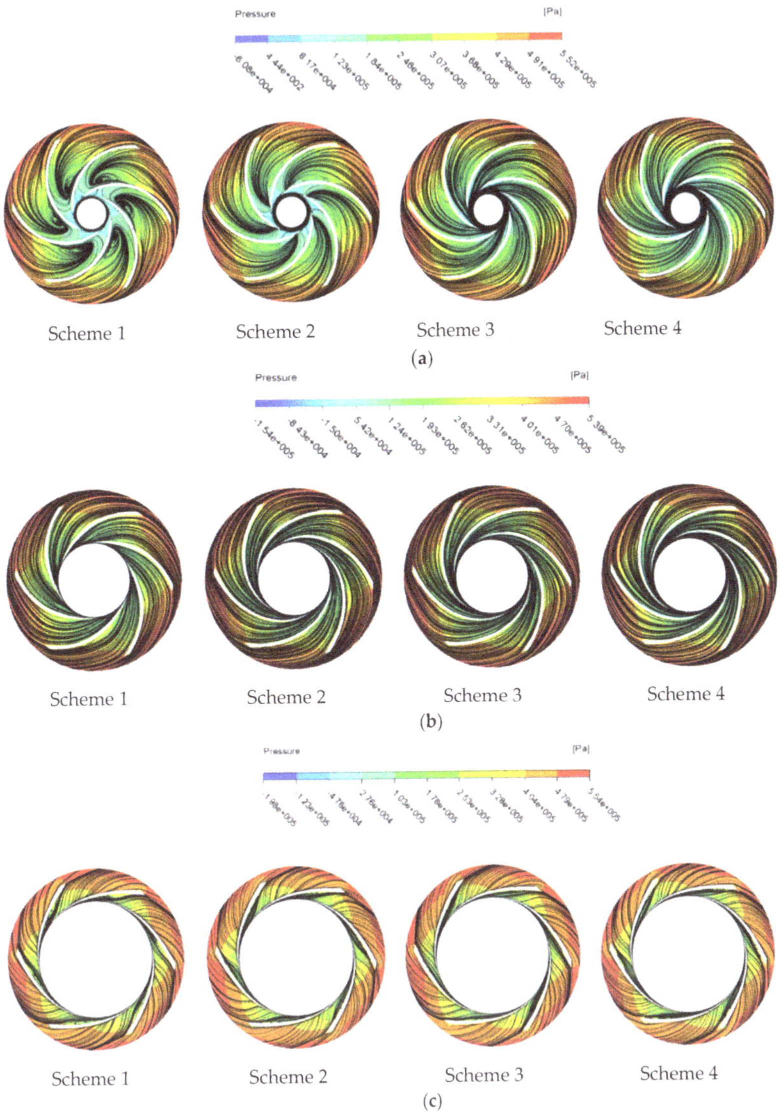

Figure 7. The circumferential pressure distribution and streamlines under 500 m^3/h. (**a**) Pressure distribution of the circumferential surface of Span 0. (**b**) Pressure distribution of the circumferential surface of Span 0.5. (**c**) Pressure distribution of the circumferential surface of Span 1.

As shown in Figure 8, according to the velocity distribution of each scheme under overload flow conditions, it can be seen that the variation trend of the internal flow field under optimal flow conditions is similar to the part-load condition; also, under optimal flow conditions, a large area of the low-speed zone is generated at the leading edge with a smaller hub radius scheme, while the area of the low-speed zone continues to decrease as the hub radius increases. At the same time, the low-speed zone in Scheme 1 extends from the leading edge to the trailing edge. However, as the hub radius continues to rise, the area of the low-speed zone continues to shrink and concentrates in the middle of the blade pressure surface. As the circumference moved from the hub to the shroud, the low-speed zone on the blade pressure surface decreased continuously until it completely disappeared. What is more, the low-speed region does not appear on the suction surface, which is different from the velocity distribution under part-load conditions and design-flow conditions.

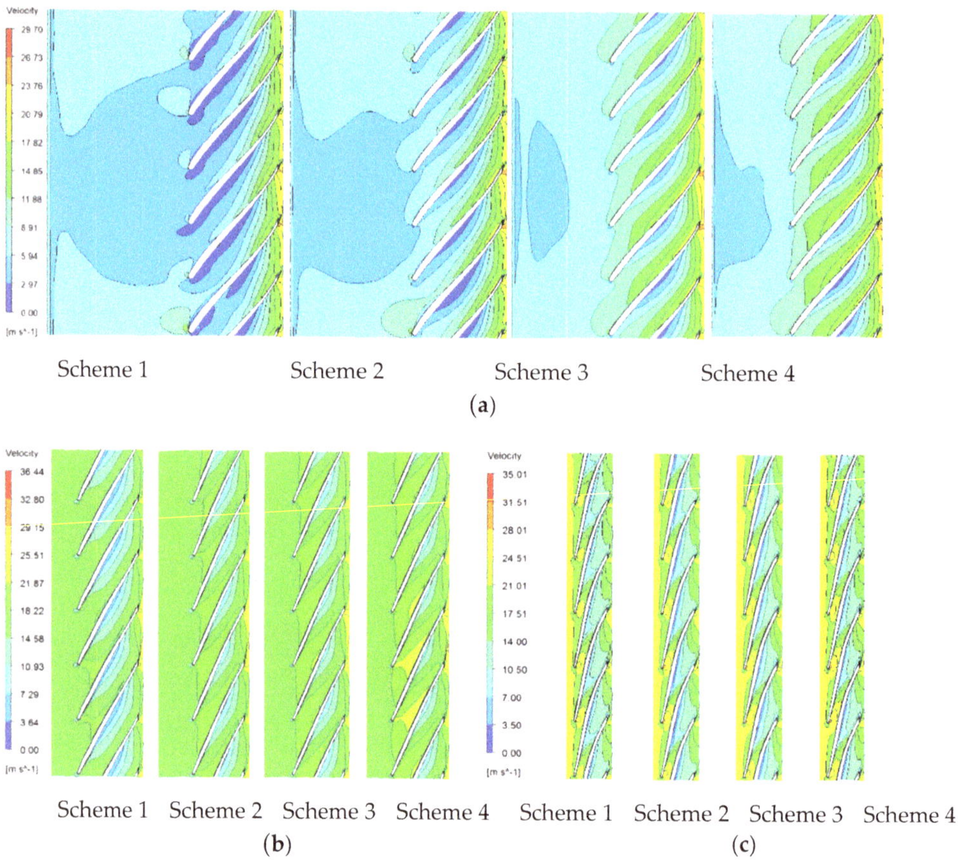

Figure 8. The circumferential velocity distribution under 700 m^3/h. (**a**) Velocity distribution of the circumferential surface of Span 0. (**b**) Velocity distribution of Span 0.5. (**c**) Velocity distribution of Span 1.

Figure 9 presents the pressure and streamline distribution of different schemes under overload conditions. Compared with part-load conditions and overload conditions, except for the slight vortices found on Span 0 in Schemes 1 and 2, the internal streamlines of the other schemes are all smooth. When the circumferential surface moved from the shroud to the hub, no obvious vortices were observed. At the same time, the area of the high-pressure area gradually decreases with the rising of the hub radius. The main reason for this phenomenon is that the curvature of the hub is small, which causes a large flow area in the impeller. When the water flows axially into the impeller along the inlet pipe, the radial flow area increases, which results in serious diffusion, and the velocity distribution is very disorderly. The backflow is formed at the hub, and the vortex appears in the impeller. When the radius of the hub increases continuously, the bending degree of the whole meridian channel increases and flow conditions in the impeller improve significantly, which makes it more in line with the optimal flow rule. Therefore, the reasonable design of the back cover of the impeller is of great significance to restrain the backflow and improve internal flow conditions.

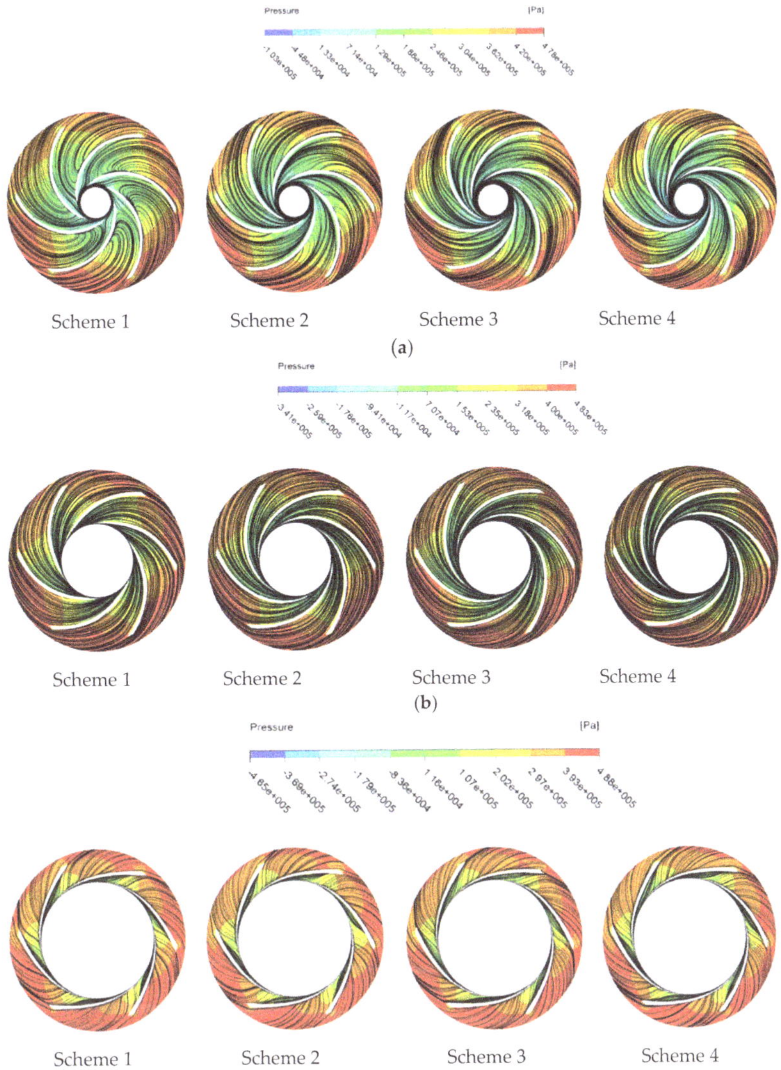

Figure 9. The circumferential pressure distribution and streamlines under 700 m^3/h. (**a**) Pressure distribution of the circumferential surface of Span 0. (**b**) Pressure distribution of the circumferential surface of Span 0.5. (**c**) Pressure distribution of the circumferential surface of Span 1.

5. Analysis of Blade Surface Load Characteristics

The blade surface load can not only effectively explain the stress condition of the blade in operation, but also reflect the energy conversion ability of the blade. Therefore, the analysis of blade surface load is beneficial to improve the safe and stable operation of the blade. Figure 10 compares the blade surface load distributions at the different flow conditions (300 m^3/h, 500 m^3/h, and 700 m^3/h) on Span 0.5, where the abscissa represents the relative position along the blade streamline direction. When the relative streamline position is 0, it represents the blade leading edge; when the relative streamline position is 1, it represents the blade trailing edge. The vertical coordinate is the blade surface pressure value, monitored at the corresponding streamline position. What is more, it can be seen from the figure that blade surface load is mainly composed of two curves, which represent the pressure on the pressure surface and the suction surface, respectively, and the difference between the two curves is the blade surface load. By comparing the blade surface load distribution under part-load conditions, optimal flow conditions, and overload conditions, it can be found that the blade load distribution has a similar variation trend. The pressure on the pressure surface and the suction surface shows an upward trend as a whole. However, the curve of the pressure surface first drops sharply near the leading edge of the blade and then rises at a constant speed. Meanwhile, there is a position where the pressure on the pressure surface is equal to the suction surface, so the blade surface load is zero, namely, the first zero load point. This phenomenon is attributed to the collision between the incoming flow and the leading edge of the blade, since the velocity direction of the liquid changes suddenly under the effect of the blade, which causes dramatic variation in the pressure. At the same time, by comparing the different meridian schemes, it is found that the first zero load point gradually moves away from the leading edge when the hub radius continues to increase.

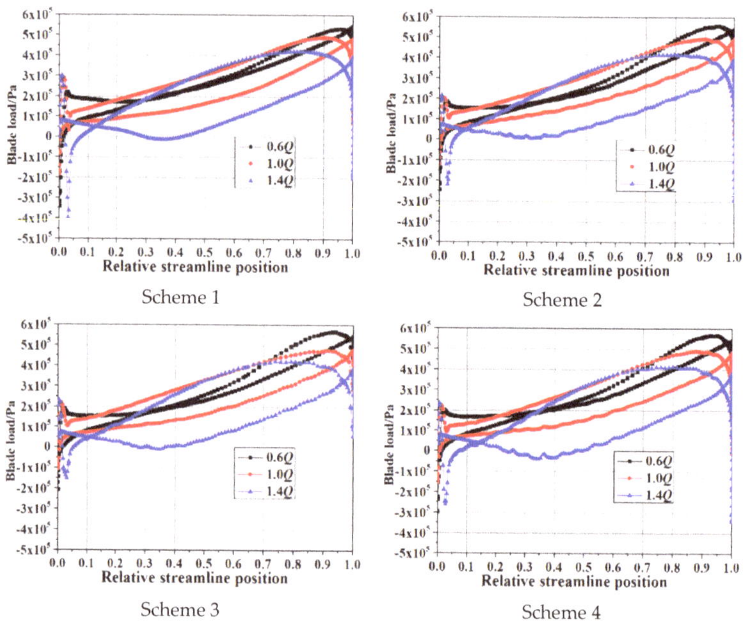

Figure 10. The blade loading under different flow conditions.

After the relative streamline position passes the first zero load point, the pressure on both the pressure surface and the suction surface increases continuously, but the growth rate on the pressure surface is significantly higher than the suction surface, which results in the load on the blade surface gradually increasing. Then, the growth rate on the pressure surface gradually decreases, while the pressure on the suction surface still keeps an upward trend, making the load on the blade surface decline gradually. Therefore, when the pressure on the pressure face reaches the maximum, the blade load reaches the maximum; namely, the energy conversion capacity of the blade is the strongest. By comparing the maximum load points under the different flow conditions, it can be found that, with the continuous increase in flow, the position of the maximum load point is further away from the trailing edge. Then, the pressure on the pressure surface continued to drop, and the descending rate gradually accelerated. The pressure on the suction surface rises uniformly, which intersects with the pressure curve of the pressure surface near the trailing edge; that is, the pressures on the pressure surface and on the suction surface are equal for the second time. The formation of the second zero load point is attributed to the influence of the stator-rotor interaction between tongue and blade. By comparing the relative positions of the second zero load point under the different flow conditions, it can be seen that with the increase in flow rate, the second zero load point is constantly away from the trailing edge of the blade.

Therefore, by comparing the blade surface loads under the different flow conditions, it can be found that the blade load increases first and then decreases along with the streamline position, while the blade surface load continues to enhance as the flow increases. At the same time, as the hub radius increases, the blade surface load continuously increases, which effectively improves the ability of the blade to work on the internal flow field.

6. Analysis of Pressure Pulsation Characteristics

6.1. Distribution of Pressure Fluctuation Monitoring Points

In order to ensure the stable and efficient operation of the self-priming centrifugal pump, the pressure fluctuation characteristics caused by pressure fluctuation and mechanical vibration must be fully considered. The pressure pulsation can not only destroy sealing parts and fixed support parts, but also induce noise. In particular, vibration will aggravate the fatigue and failure of the self-priming centrifugal pump, shortening the service life of the whole unit significantly. Therefore, the pressure pulsation can not only directly describe the relationship between the pressure in the flow channel with time, but also indirectly reflect the instability of the self-priming centrifugal pump. Therefore, in this section, by setting up pressure pulsation monitoring points at the trailing edge of the impeller flow channel, the variation of pressure pulsation with time by numerical simulation can be analyzed. Since the swept angle by a single impeller is 60°, the trailing edge of the blade is taken as the starting edge, and the trailing edge of the secondary blade is taken as the ending edge. When the blade rotates 10°, pressure pulsation monitoring points are set on the circumferential surfaces of Span 0, Span 0.5, and Span 1, respectively. Among them, the pressure pulsation monitoring points on the circumference of Span 0 are C1, C2, C3, C4, and C5; the pressure pulsation monitoring points on the circumference of Span 0.5 are B1, B2, B3, B4, and B5; and the pressure pulsation monitoring points on the circumference of Span 1 are A1, A2, A3, A4, and A5, respectively. The pressure pulsation monitoring points are all distributed on the calculation domain of the rotating impeller, as shown in Figure 11.

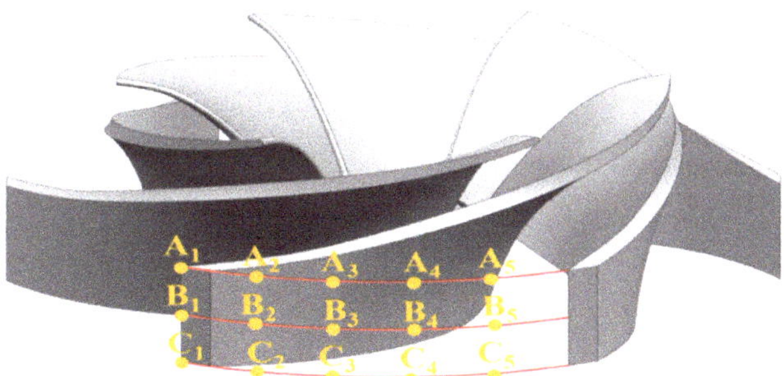

Figure 11. The position of pressure pulsation monitoring points.

6.2. Analysis of Pressure Pulsation

In order to improve the accuracy of the analysis, the calculated pressure pulsation data are processed into a dimensionless format, and the pressure pulsation distribution at the trailing edge of the different schemes under the three flow conditions are obtained, respectively, as shown in Figures 12–15. Among them, the dimensionless pressure pulsation data can be calculated by the following formula [22]:

$$\overline{p}(node) = \frac{1}{N}\sum_{j=0}^{N-1} p(node, t_0 + j\Delta t) \tag{1}$$

$$\widetilde{p}(node, t) = p(node, t) - \overline{p}(node) \tag{2}$$

$$C_p = \frac{\sqrt{\frac{1}{N}\sum_{j=0}^{N-1} \widetilde{p}^2(node, t + \Delta t)}}{\frac{1}{2}\rho u_2^2} \tag{3}$$

where $\overline{p}$ and $\widetilde{p}$ represent the time-averaged and periodic pressure, respectively, N and Δt are the total time step and time step, respectively, u_2 represents the peripheral velocity of the impeller outlet, Cp is the dimensionless pressure pulsation coefficient, and ρ represents the liquid density.

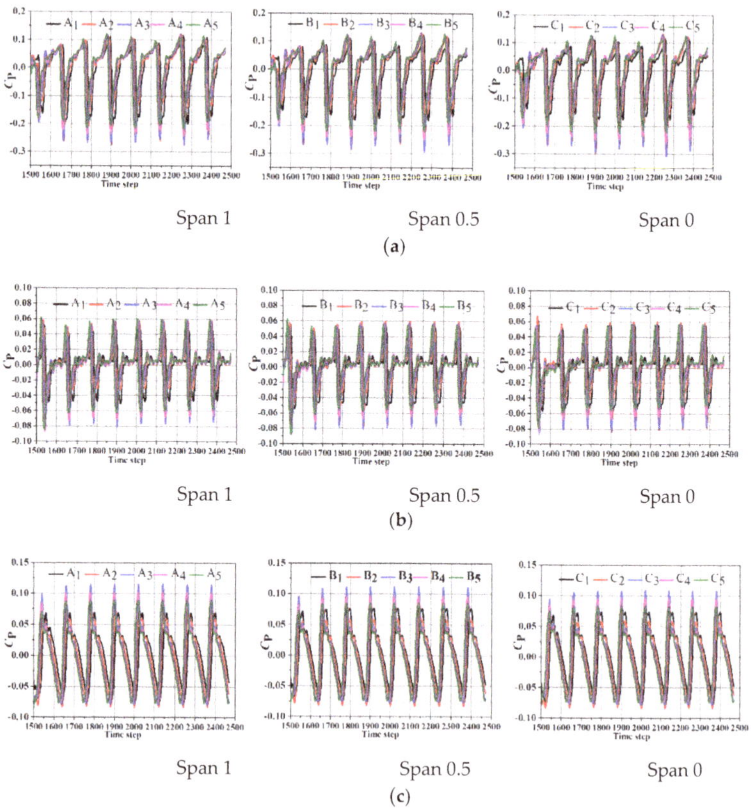

Figure 12. The pressure fluctuations of Scheme 1 under different flow conditions. (**a**) The pressure fluctuations of Scheme 1 under 300 m^3/h. (**b**) The pressure fluctuations of Scheme 1 under 500 m^3/h. (**c**) The pressure fluctuations of Scheme 1 under 700 m^3/h.

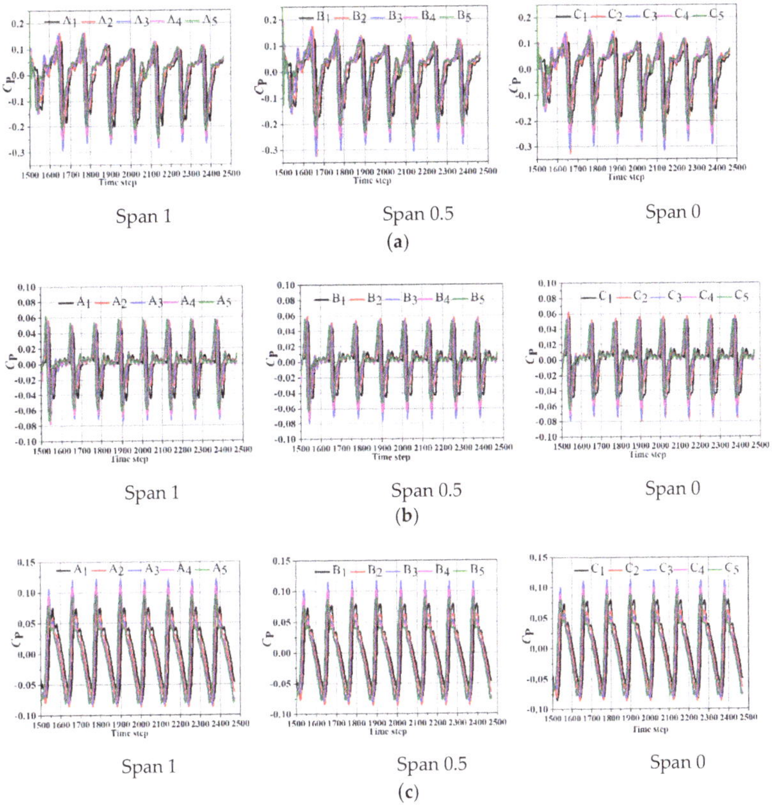

Figure 13. The pressure fluctuations of Scheme 2 under different flow conditions. (**a**) The pressure fluctuations of Scheme 2 under 300 m^3/h. (**b**) The pressure fluctuations of Scheme 2 under 500 m^3/h. (**c**) The pressure fluctuations of Scheme 2 under 700 m^3/h.

The pressure pulsation distribution on Span 1, Span 0.5, and Span 0 under the three flow conditions of Scheme 1 are shown in Figure 12. It can be found that each scheme presents periodic peaks and troughs under the different flow conditions. Since monitoring points are set in the rotation domain of the impeller, only one stator-rotor interaction is generated in one revolution; thus, each monitoring point only has one trough in a calculation period. Meanwhile, the pressure pulsation amplitude of each group of monitoring points is quite different; the pressure pulsation amplitude of each monitoring point is the largest and the negative amplitude is greater than the positive amplitude under part-load conditions; when the flow is increased to optimal flow conditions, the pressure pulsation amplitude is reduced to the minimum, and the negative amplitude is greater than the positive amplitude. However, as the flow increased to overload conditions, the pressure pulsation amplitude increases instead, but it is still lower than part-load conditions, and the positive amplitude is greater than the negative amplitude.

By comparing the pressure pulsation amplitudes on different circumferential surfaces, it can be seen that as the monitoring point moves from the shroud to the hub, the pressure pulsation amplitude increases continuously. At the same time, due to the different locations of the monitoring points, there is a significant phase difference in the pressure pulsation peaks. The No. 5 monitoring point is the first one to sweep the tongue. Therefore, the first wave peak is generated at the No. 5 monitoring point, and the other wave peaks all appear behind the No. 5 monitoring point. Furthermore, it can be found that the order of the pressure pulsation amplitude is: 3 > 2 > 4 > 5 > 1, so the positions where the pressure pulsation amplitude changes drastically are mainly concentrated in the middle of the impeller outlet. When the monitoring point is far from the A3, B3, and C3 positions, the pressure pulsation amplitude is smaller.

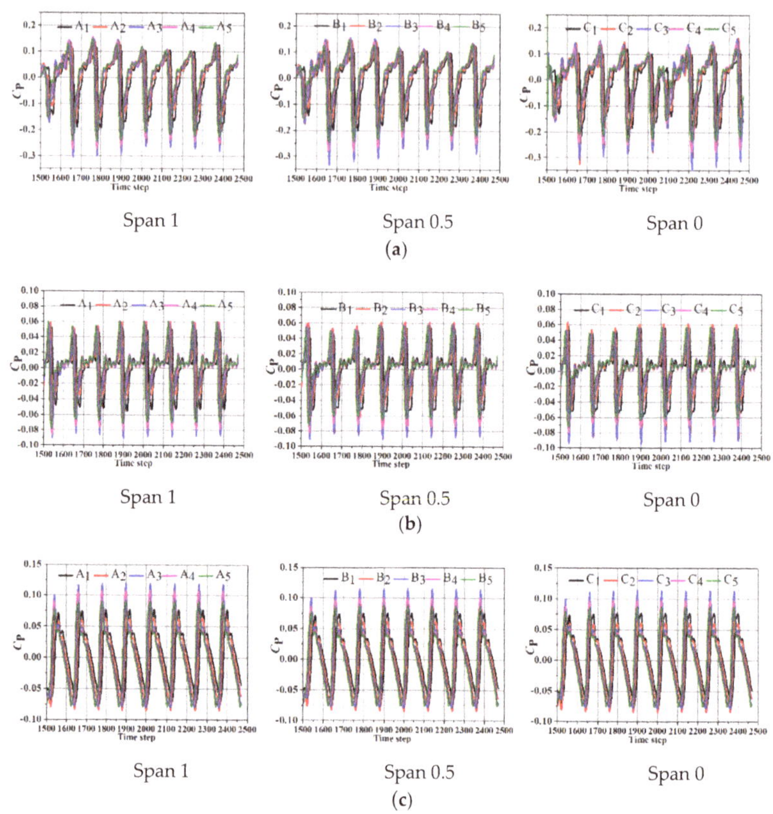

Figure 14. The pressure fluctuations of Scheme 3 under different flow conditions. (**a**) The pressure fluctuations of Scheme 3 under 300 m^3/h. (**b**) The pressure fluctuations of Scheme 3 under 500 m^3/h. (**c**) The pressure fluctuations of Scheme 3 under 700 m^3/h.

Figures 13–15 are the pressure pulsation distributions of Scheme 2, Scheme 3, and Scheme 4, respectively. Through comparison, it can be found that each scheme has a similar variation trend. The pressure pulsation amplitude of the monitoring point is the largest, and the negative amplitude is greater than the positive amplitude under part-load conditions. However, the pressure pulsation amplitude is reduced to the minimum, and the negative amplitude is greater than the positive amplitude under optimal flow conditions. When the flow continues to increase to overload conditions, the amplitude increases instead and the positive amplitude is greater than the negative amplitude. The main reason for this phenomenon is the velocity of the water flowing from the impeller continues to increase, which collides with the water in the volute. Due to the large difference in velocity and direction, a great impact loss is formed in the volute, which results in severe pressure pulsation. After the impact, the velocity of the water flowing out of the impeller is reduced to the same as the flow field of the volute; part of the energy is dissipated during the collision, and the remaining kinetic energy is transferred to the water in the volute, which makes the pressure of the volute continue to rise. When pump operation is under overload conditions, the velocity of water flowing out of the impeller is faster; meanwhile, the velocity of water in the volute increases significantly, so that the velocity difference with the impeller is smaller. Although collisions are generated under overload conditions, the collision intensity and energy dissipation are significantly lower than under part-load conditions. Therefore, the obvious pressure pulsation will be generated under part-load and overload conditions.

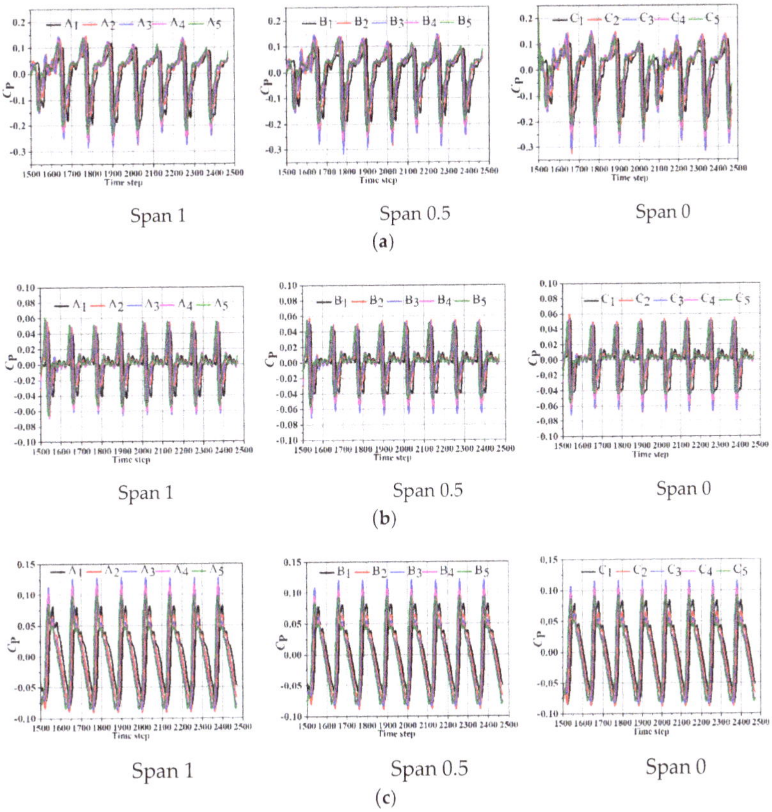

Figure 15. The pressure fluctuations of Scheme 4 under different flow conditions. (**a**) The pressure fluctuations of Scheme 4 under 300 m^3/h. (**b**) The pressure fluctuations of Scheme 4 under 500 m^3/h. (**c**) The pressure fluctuations of Scheme 4 under 700 m^3/h.

At the same time, as the hub radius continues to increase, the pressure pulsation amplitudes of the corresponding monitoring points all show a downward trend, indicating that the impeller with a large hub radius is beneficial to weaken the pressure pulsation. In addition, it is further verified that the internal flow field of Scheme 4 is relatively smooth in the analysis of internal flow characteristics, which has an important effect on improving the running stability of the self-priming centrifugal pump.

7. Analysis of Radial Force Characteristics

Since the geometric parameters of the self-priming centrifugal pump are designed based on the internal flow under optimal flow conditions, which results in the velocity and pressure distribution being symmetrical and uniform. However, when the self-priming centrifugal pump operation deviates from optimal flow conditions, the symmetry distribution of pressure and velocity in the flow field is destroyed, and the pressure push on the impeller is uneven. Therefore, the feedback on the impeller is the generation of radial force, and the deflection is generated on the pump under alternating stress, which reduces the service life of the pump shaft and causes damage to related parts such as bearings and seals.

The radial force on the impeller is mainly concentrated on the blade surface, the shroud, and the hub, which can be decomposed into two components perpendicular to the axial direction. In this paper, the impeller of the self-priming centrifugal pump rotates around the Z-axis, so the radial force on the impeller can be decomposed into the component force F_X in the X-axis direction and the component force F_Y in the Y-axis. In the process of numerical simulation, the resultant radial force can be calculated by detecting F_X and F_Y, and the relevant calculation relationship can be expressed as [23]:

$$F_R = \sqrt{(F_X^2 + F_Y^2)} \tag{4}$$

$$\theta_R = \arctan(F_Y/F_X) \tag{5}$$

where F_R represents the resultant force of radial forces and θ_R represents the angle between the component forces.

In order to improve the accuracy of radial force analysis, dimensionless treatments are carried out on the radial force to obtain the radial force coefficient C_F. The calculation formula can be expressed as:

$$C_F = \frac{2F_R}{\rho u_2^2 D_2 b_2} \tag{6}$$

where C_F represents the dimensionless radial force coefficient, ρ represents the liquid density, u_2 represents the peripheral velocity of the impeller outlet, D_2 represents the outer diameter of the impeller, and b_2 represents the width of the impeller outlet.

Figure 16 presents the polar coordinate distribution of radial force. It can be seen from the figure that the direction and magnitude of the radial force of each scheme have a similar variation trend. With the rotation of the impeller, six wave peaks and troughs appear in the radial force distribution, which is equal to the number of impeller blades, and the overall distribution is symmetrical. In addition, the wave peaks and troughs of each scheme are generated at the same angles. However, by comparing the radial force under the different flow conditions, it can be found that the radial force is largest under part-load conditions. When the flow rate increases to optimal flow conditions, the radial force decreases to the minimum. When the flow rate continues to increase to overload conditions, the radial force increases instead, but the magnitude is still lower than the radial force under part-load conditions. This phenomenon is similar to the analysis results of pressure pulsation in the previous investigation.

This phenomenon is attributed to that self-priming centrifugal pump operating under part-load conditions; the velocity and direction in the volute flow field are inconsistent with the impeller, therefore, the collision is generated between the impeller and the volute, resulting in uneven velocity distribution at the impeller outlet and the flow in the impeller is in relative disorder, thereby causing a relatively large radial force under part-load conditions. When the flow rate increases to design-flow conditions, the velocity direction is consistent with the design direction, the internal flow field is relatively smooth, the collision loss caused by the water flowing into the volute is the lowest, and the velocity distribution of the impeller outlet is uniform, so the radial force formed on the impeller is relatively low. When the self-priming centrifugal pump runs under overload conditions, the velocity of water flowing out of the impeller is faster. At this time, the water velocity in the volute improves significantly with the increase in the flow rate. However, the velocity difference with the water in the impeller is small. Although there will still be a collision with energy loss, the collision intensity and energy dissipation are significantly lower than under part-load conditions. Furthermore, an uneven flow state is formed at the exit of the impeller, thereby forming a certain degree of radial force. It shows that the magnitude and direction of the radial force are directly affected by the velocity distribution at the impeller exit. Comparing the radial force of the different hub radii schemes, it can be found that the radial force of Scheme 4 is symmetrically distributed as a whole, and the magnitude is lower than other schemes. The phenomenon is mainly caused because the internal flow channel of the impeller and the volute are better matched by employing a larger hub radius, plus the internal flow field is smoother, which results in a more uniform and stable flow state, and the radial force generated upon the impeller is relatively low.

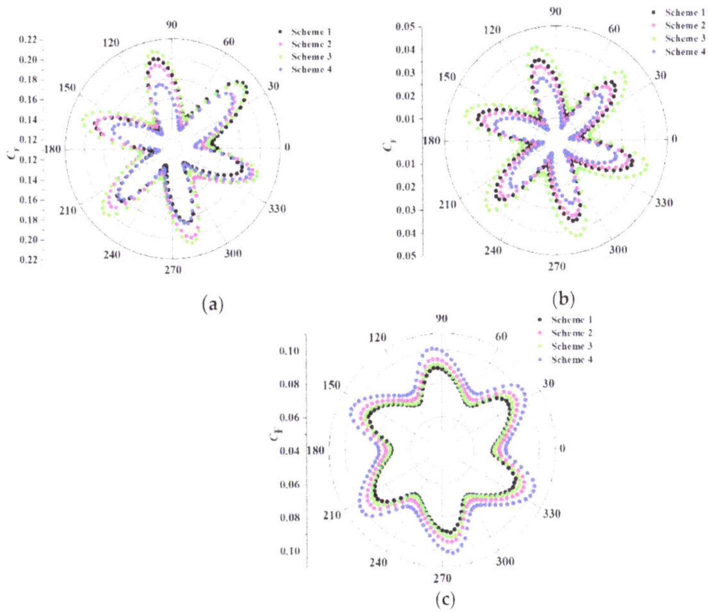

Figure 16. Radial force distribution under different flow conditions. (**a**) Radial force distribution under 300 m^3/h. (**b**) Radial force distribution under 500 m^3/h. (**c**) Radial force distribution under 700 m^3/h.

7.1. Analysis of Hydraulic Performance

To compare the hydraulic performance of each scheme, head and efficiency are selected as indicators for analysis. The head represents the height that the water can be raised by a self-priming centrifugal pump, and the efficiency represents the energy conversion rate of the self-priming centrifugal pump; the expression of head and efficiency can be expressed as:

$$H = \frac{P_O - P_I}{\rho g} + \frac{V_O^2 - V_I^2}{2g} + (Z_O - Z_I) \tag{7}$$

$$\eta = \frac{60gHQ}{\pi M n} \tag{8}$$

where $P_O - P_I$, ${V_O}^2 - {V_I}^2$, $Z_O - Z_I$ represent the pressure, velocity, and potential energy difference after operating the self-priming centrifugal pump, respectively; M represents the torque, n is the rotation speed of the impeller, and Q represents the flow rate. To reduce the errors generated in the analysis process, the dimensionless treatment on head and efficiency is carried out in this paper. The dimensionless calculation formula can be expressed as:

$$\Phi = \frac{60Q}{\pi^2 D_2 b_2 n} \tag{9}$$

$$\Psi = \frac{7200Hg}{\pi^2 D_2^2 n^2} \tag{10}$$

Figures 17 and 18 reveal the relationship between the head and flow rate as well as the relationship between the efficiency and flow rate of each scheme, respectively. It can be seen that the head of Scheme 4, with a larger hub radius, is lower than other schemes. Through the analysis of the internal flow characteristics in the previous section, it is concluded that more turbulent vortices are generated in the internal flow field of Scheme 1 with a smaller hub radius; the impeller flow passage is blocked by the vortex structure, which significantly reduces the effective flow area; and the actual flow of the self-priming centrifugal pump is lower than the internal flow channel without a vortex. Considering that the head shows a decreasing trend as the flow increases, therefore, the head of Scheme 1 is higher. At the same time, a large amount of energy dissipation is caused by the formation of vortices in the internal flow field of Scheme 1, which results in the efficiency of Scheme 1 being lower. When the hub radius continues to increase, the curvature of the entire meridian flow channel rises, the flow state in the impeller is continuously improved, and the vortex is gradually eliminated, making it more in line with the ideal internal flow rule. Therefore, the increase in hub radius is beneficial for improving the internal flow law and increasing the energy-conversion capability of the self-priming centrifugal pump.

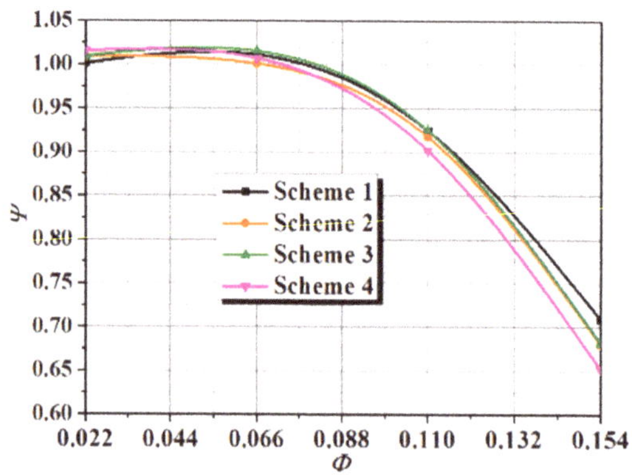

Figure 17. Head of each scheme under different flow conditions.

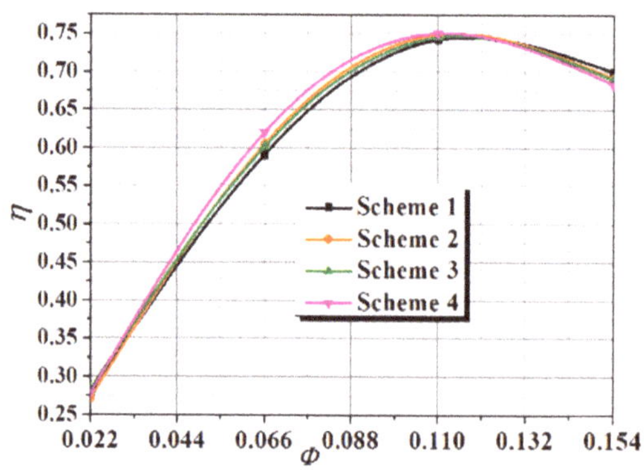

Figure 18. Efficiency of each scheme under different flow conditions.

7.2. Test Measurement Standards and Device Accuracy

To further verify the accuracy of the calculation results, the impeller of Scheme 4 is manufactured and the relevant hydraulic experiment is conducted; the impeller model of Scheme 4 is shown in Figure 19a. During the test, pressure sensors were used to monitor the inlet and outlet pressures, and the measurement errors of the inlet and outlet pressure sensors were both ±0.5%. At the same time, the measurement error of the turbine flowmeter is ±1.0%, while the measurement error of the torque sensor and the speed sensor are ±0.5% and ±1.0%, respectively. The measurement error of the test is ±1.22%, and the test bench is shown in Figure 19b.

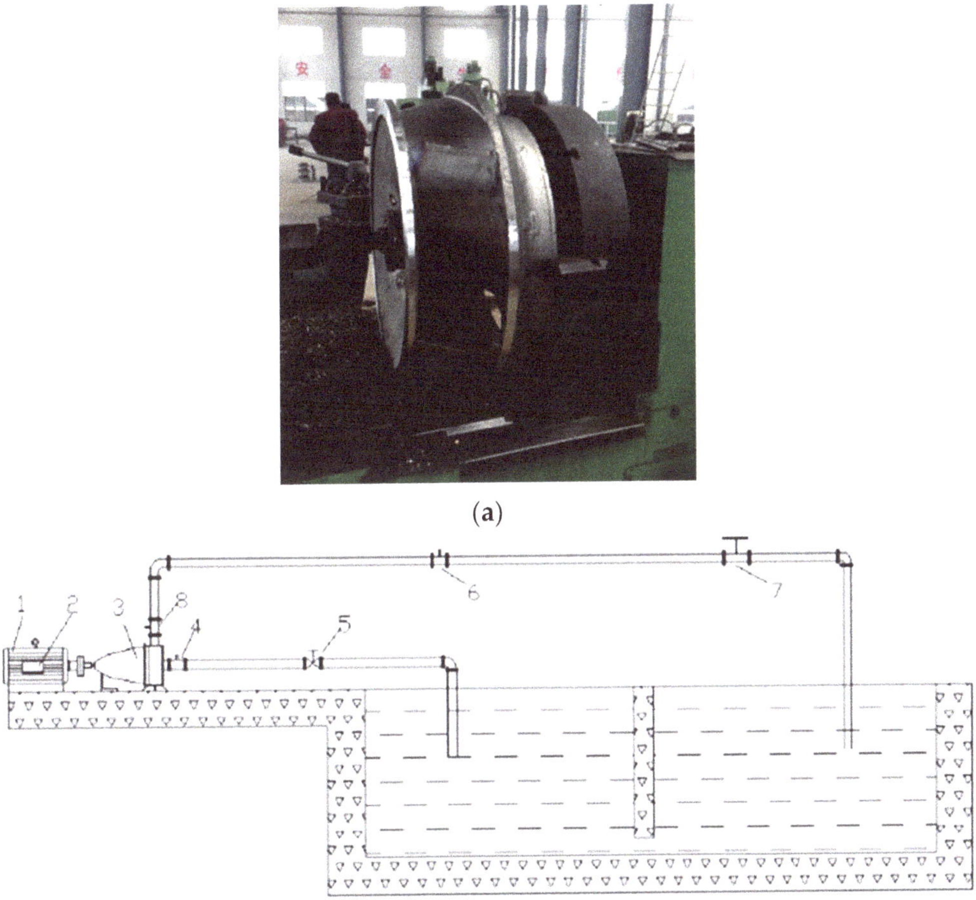

(a)

(b)

Figure 19. The test bench. (**a**) The test bench for Scheme 4. (**b**) Test bench: 1. Motor; 2. Torque sensor; 3. Self-priming centrifugal pump; 4. Inlet pressure sensor; 5. Inlet valve; 6. Turbine flowmeter; 7. Outlet valve; 8. Outlet pressure sensor.

The comparison results of the test and numerical calculations are shown in Figure 20; the test results have the same variation trend with the calculation results. Considering the complexity of the calculation structure, the front pump cavities, rear pump cavities, seals, and bearings are not included in the calculation domain, which leads to the calculation results of the self-priming centrifugal pump possessing a slight deviation. The result of the test is slightly lower than the calculation, but the difference is below 3%, which is within the allowable range of the error in the analysis process. Therefore, it is concluded that the model and setting of the calculation are reliable.

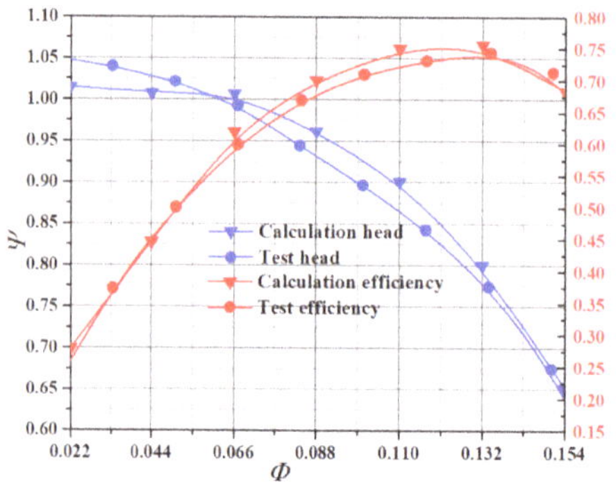

Figure 20. The comparison results of the test and numerical calculations.

8. Conclusions

In this paper, a self-priming pump with different hub radii was systematically investigated by numerical simulations and experiments; the internal flow characteristics, blade surface load, pressure pulsation characteristics, and radial force distribution were analyzed. The following conclusions can be drawn.

According to the analysis of blade surface load characteristics, the results present that the pressure on the pressure surface and the suction surface show an upward trend as a whole. Furthermore, it can be found that the blade load increases first and then decreases along with the streamline position, while the blade surface load continues to enhance as the flow increases. At the same time, as the hub radius increases, the blade surface load continuously increases, which effectively improves the ability of the blade to work on the internal flow field.

The velocity of the water flowing from the impeller continues to increase, which collides with the water in the volute. In addition, due to the large differences in velocity and direction, great impact loss is formed in the volute, which results in severe pressure pulsation. Therefore, the pressure pulsation is reduced to the minimum under optimal flow conditions. In addition, as the hub radius continues to increase, the pressure pulsation amplitudes of the corresponding monitoring points all show a downward trend, indicating that the impeller with a large hub radius is beneficial to weaken the pressure pulsation.

By comparing the radial force of the different hub radii schemes, it can be found that six wave peaks and troughs are generated in the radial force distribution, which is equal to the number of impeller blades, and the overall distribution is symmetrical. In addition, the wave peaks and troughs of each scheme are generated at the same angles. Considering this, the internal flow channel of the impeller and the volute are better matched by employing a larger hub radius, and the internal flow field is smoother. Thus, the radial force generated upon the impeller is relatively low, the radial force of Scheme 4 is symmetrically distributed as a whole, and the magnitude is lower than other schemes. Finally, the relevant hydraulic experiment is conducted, and the difference between the experiment and calculation is below 3%, which ensures the accuracy of the calculation results.

Author Contributions: Conceptualization, H.C.; methodology, S.H.; validation, C.W.; investigation, G.P.; writing—review and editing, F.F. and D.Y. All authors have read and agreed to the published version of the manuscript.

Funding: This research was funded by Open Research Fund Program of State Key Laboratory of Hydro-science and Engineering grant number:sklhse-2020-E-01; Open Research Subject of Key Laboratory of Fluid Machinery and Engineering (Xihua University) grant number LTJX2021-003; Open Research Subject of Hubei Key Laboratory of Hydroelectric Machinery Design & Maintenance (China Three Gorges University) grant number 2020KJX07; the 69th batch of general funding from the China Postdoctoral Science Foundation grant number: 2021M691298; Natural Science Research

Project of Jiangsu Province Colleges and Universities: 21KJB570004, Priority Academic Program Development of Jiangsu Higher Education Institutions (PAPD).

Institutional Review Board Statement: Not applicable.

Informed Consent Statement: Not applicable.

Data Availability Statement: Not applicable.

Conflicts of Interest: The authors declare no conflict of interest.

References

1. Wang, H.; Hu, Q.; Yang, Y.; Wang, C. Performance differences of electrical submersible pump under variable speed schemes. *Int. J. Simul. Model.* **2021**, *20*, 76–86. [CrossRef]
2. Zhou, J.; Zhao, M.; Wang, C.; Gao, Z. Optimal design of diversion piers of lateral intake pumping station based on orthogonal test. *Shock. Vib.* **2021**, *2021*, 6616456. [CrossRef]
3. Wang, H.; Long, B.; Wang, C.; Han, C.; Li, L. Effects of the impeller blade with a slot structure on the centrifugal pump performance. *Energies* **2020**, *13*, 1628. [CrossRef]
4. Zhang, L.; Wang, C.; Zhang, Y.; Xiang, W.; He, Z.; Shi, W. Numerical study of coupled flow in blocking pulsed jet impinging on a rotating wall. *J. Braz. Soc. Mech. Sci. Eng.* **2021**, *43*, 508. [CrossRef]
5. Shi, L.; Zhu, J.; Tang, F.; Wang, C. Multi-Disciplinary optimization design of axial-flow pump impellers based on the approximation model. *Energies* **2020**, *13*, 779. [CrossRef]
6. Wang, H.; Qian, Z.; Zhang, D.; Wang, T.; Wang, C. Numerical study of the normal impinging water jet at different impinging height, based on Wray–Agarwal turbulence model. *Energies* **2020**, *13*, 1744. [CrossRef]
7. Tang, S.; Zhu, Y.; Yuan, S. An improved convolutional neural network with an adaptable learning rate towards multi-signal fault diagnosis of hydraulic piston pump. *Adv. Eng. Inform.* **2021**, *50*, 101406. [CrossRef]
8. Zhu, Y.; Li, G.; Wang, R.; Tang, S.; Su, H.; Cao, K. Intelligent fault diagnosis of hydraulic piston pump combining improved lenet-5 and pso hyperparameter optimization. *Appl. Acoust.* **2021**, *183*, 108336. [CrossRef]
9. Deng, Q.; Pei, J.; Wang, W.; Lin, B.; Zhang, C.; Zhao, J. Energy Loss and Radial Force Variation Caused by Impeller Trimming in a Double-Suction Centrifugal Pump. *Entropy* **2021**, *23*, 1228. [CrossRef]
10. Jaiswal, A.K.; Siddique, M.H.; Paul, A.R.; Samad, A. Surrogate-based design optimization of a centrifugal pump impeller. *Eng. Optim.* **2021**, 1–18. [CrossRef]
11. Qian, C.; Luo, X.; Yang, C.; Wang, B. Multistage pump axial force control and hydraulic performance optimization based on response surface methodology. *J. Braz. Soc. Mech. Sci. Eng.* **2021**, *43*, 136. [CrossRef]
12. Chen, J.; Shi, W.; Zhang, D. Influence of blade inlet angle on the performance of a single blade centrifugal pump. *Eng. Appl. Comput. Fluid Mech.* **2021**, *15*, 462–475. [CrossRef]
13. Lin, Y.; Li, X.; Li, B.; Jia, X.; Zhu, Z. Influence of Impeller Sinusoidal Tubercle Trailing-Edge on Pressure Pulsation in a Centrifugal Pump at Nominal Flow Rate. *J. Fluids Eng.* **2021**, *143*, 091205. [CrossRef]
14. Posa, A. LES study on the influence of the diffuser inlet angle of a centrifugal pump on pressure fluctuations. *Int. J. Heat Fluid Flow* **2021**, *89*, 108804. [CrossRef]
15. Thakkar, S.; Vala, H.; Patel, V.K.; Patel, R. Performance improvement of the sanitary centrifugal pump through an integrated approach based on response surface methodology, multi-objective optimization and CFD. *J. Braz. Soc. Mech. Sci. Eng.* **2021**, *43*, 24. [CrossRef]
16. Shigemitsu, T.; Nakaishi, E.; Maeda, M.; Araki, Y. Influence of Blade Number on Performance and Internal Flow Condition of Centrifugal Pump for Low Viscous Fluid Food. *Int. J. Fluid Mach. Syst.* **2021**, *14*, 132–141. [CrossRef]
17. Shi, G.; Liu, Z.; Xiao, Y.; Wang, Z.; Luo, Y.; Luo, K. Energy conversion characteristics of multiphase pump impeller analyzed based on blade load spectra. *Renew. Energy* **2020**, *157*, 9–23. [CrossRef]
18. Wang, L.; Asomani, S.N.; Yuan, J.; Appiah, D. Geometrical Optimization of Pump-As-Turbine (PAT) Impellers for Enhancing Energy Efficiency with 1-D Theory. *Energies* **2020**, *13*, 4120. [CrossRef]
19. Zhang, F.; Appiah, D.; Hong, F.; Zhang, J.; Yuan, S.; Adu-Poku, K.A.; Wei, X. Energy loss evaluation in a side channel pump under different wrapping angles using entropy production method. *Int. Commun. Heat Mass Transf.* **2020**, *113*, 104526. [CrossRef]
20. Peng, G.; Chen, Q.; Zhou, L.; Pan, B.; Zhu, Y. Effect of Blade Outlet Angle on the Flow Field and Preventing Overload in a Centrifugal Pump. *Micromachines* **2020**, *11*, 811. [CrossRef]
21. Chang, H.; Agarwal, R.K.; Li, W.; Zhou, L.; Shi, W. Numerical and experimental study of a vortex structure and energy loss in a novel self-priming pump. *Processes* **2019**, *7*, 701. [CrossRef]
22. Chang, H.; Li, W.; Shi, W.; Liu, J. Effect of blade profile with different thickness distribution on the pressure characteristics of novel self-priming pump. *J. Braz. Soc. Mech. Sci. Eng.* **2018**, *40*, 518. [CrossRef]
23. Nishi, Y.; Fukutomi, J.; Fujiwara, R. Radial thrust of single-blade centrifugal pump. *Int. J. Fluid Mach. Syst.* **2011**, *4*, 387–395. [CrossRef]

machines

Article

Experimental Investigation of Unsteady Pressure Pulsation in New Type Dishwasher Pump with Special Double-Tongue Volute

Yilei Zhu [1], Jinfeng Zhang [1], Yalin Li [1,2,*], Ping Huang [1], Hui Xu [2] and Feng Zheng [2]

[1] National Research Center of Pumps, Jiangsu University, Zhenjiang 212013, China; 2211811038@stmail.ujs.edu.cn (Y.Z.); zhangjinfeng@ujs.edu.cn (J.Z.); huangping@ujs.edu.cn (P.H.)
[2] Ningbo FOTILE Kitchen Ware Co., Ltd., Ningbo 315336, China; xuh@fotile.com (H.X.); zhengfeng@fotile.com (F.Z.)
* Correspondence: ylli@ujs.edu.cn

Abstract: A pressure pulsation experiment of a dishwasher pump with a passive rotation double-tongue volute was carried out and compared with the pressure pulsation of a single-tongue volute and a static double-tongue volute. The pressure pulsation of the three volute models was compared and analyzed from two aspects of different impeller speeds and different monitoring points. The frequency domain and time–frequency domain of pressure pulsation were obtained by a Fourier transform and short-time Fourier transform, respectively. The results showed that the average pressure of each monitoring point on the rotating double-tongue volute was the smallest and that on the single-tongue volute was the largest. When the impeller rotates at 3000 rpm, there were eight peaks and valleys in the pressure pulsation time domain curve of the single-tongue volute, while the double-tongue volute was twice that of the single-tongue volute. Under different impeller speeds, the changing trends of pressure pulsation time and frequency domain curves of static and rotating double-tongue volutes at monitoring point p1 are basically the same. Therefore, a volute reference scheme with passive rotation speed is proposed in this study, which can effectively improve the flow pattern and reduce pressure inside the dishwasher pump, and also provide a new idea for rotor–rotor interference to guide the innovation of dishwashers.

Keywords: dishwasher pump; pressure pulsation; rotor–rotor interaction; impeller speeds; double-tongue volute; experiment

Citation: Zhu, Y.; Zhang, J.; Li, Y.; Huang, P.; Xu, H.; Zheng, F. Experimental Investigation of Unsteady Pressure Pulsation in New Type Dishwasher Pump with Special Double-Tongue Volute. *Machines* **2021**, 9, 288. https://doi.org/10.3390/machines9110288

Academic Editor: Dan Zhang

Received: 25 October 2021
Accepted: 11 November 2021
Published: 14 November 2021

1. Introduction

Today, dishwashers are used by more and more families. Compared with manual dishwashing, a dishwasher can not only shorten the cleaning time but also save water and labor, making dishwashing simple. Therefore, domestic and foreign companies have launched various models of dishwasher to meet the market, for example, the DWA5—1513 dishwasher from TOSHIBA, the SC73M12TI dishwasher from SIMENS, the NP-8LZK5RX dishwasher from Panasonic, and the RX600 dishwasher from Midea, etc. What these dishwashers have in common is that they perform the cleaning function through the passive rotation of the built-in spray arm. The dishwasher pump system is an important part of the dishwasher system, which can complete the water transport and spraying function. This paper is based on FOTILE company dishwasher pump system research. The dishwasher of this company is accepted by many people because it can achieve open cleaning without pipelines. The pipeline-free cleaning replaces the pipeline with a double-tongue volute flow channel that can be passively rotated. The symmetrical design of the volute is conducive to its rotation, and the jet hole is set on the volute spray arm to realize the function of water spraying. This avoids the problem of fouling and is more convenient for daily cleaning and maintenance. Of course, the nozzle structure, injection angle, and jet velocity will affect the cleaning effect [1,2]. However, the pressure pulsation during the operation of the dishwasher pump will cause the vibration of the dishwasher, which will reduce the stability of the system and reduce the service life [3–5]. Dishwashers are household appliances accompanied by pressure pulsation noise, which seriously affects the user experience.

An important part of the pump system is the volute structure and parameters (volute tongue angle, cross-section shapes, volute differ shape, etc.). It will affect the pump performance and internal flow [6,7]. In recent years, scholars have studied many volute structures and volute parameter optimization to improve pump performance and reduce unstable flow. Jin et al. [8] studied the influence of the gap between the volute tongue and the impeller on the pressure pulsation and the performance of the double suction pump. Results suggested that as the gap increased, the value of the head and the pressure pulsation were reduced, and the efficiency was increased. Alemi et al. [9] considered the parameters (cutwater gap, tongue shape, and volute tongue angle) of a centrifugal pump. The research found that the best volute tongue angle is 5°, which lets the radial force be lower 40% than others at the design point. Patil et al. [10] investigated four volute tongue clearances of 6%, 8%, 10%, and 12.5% of impeller diameter in the centrifugal blower and found that the volute tongues' clearance decreases from 12.5% to 6% of impeller diameter, and the total pressure and efficiency increase by 19.52% and 21.90%, respectively, in a full discharge condition. Shim et al. [11] carried out multi-objective optimization for the double-volute centrifugal pump with a diaphragm, and the pressure pulsation of the optimized model was significantly reduced. Yang et al. [12] studied four parameters of the volute of centrifugal pump. It was found that the volute designed with a round cross-section shape and the volute spiral development areas designed according to the constant swirl can achieve high efficiency. A large volute throat area could obtain smooth pump performance, and the optimal radial gap between the impeller and the volute tongue is 15 mm.

Scholars use a numerical simulation and experimental test as the main methods to study the pressure pulsation in the pump. Usually, test methods are used to validate the accuracy of simulations [13–16]. Li et al. [17] completed numerical analysis on the reduction of pressure pulsation for a double-section centrifugal pump and proposed that a combination of multi-blades, a larger radial gap, and staggering arrangements can reduce the pressure pulsation inside the pump. Zhang et al. [18] explored the pressure pulsation in a nuclear reactor coolant pump. It was confirmed that pressure pulsation acting on the impeller's blade is mainly dominated by the impeller rotating frequency, the vane passing frequency, and the double blade-passing frequency, and the pressure pulsation acting on the blade's pressure surface is more intense than on the suction surface. Tan et al. [19] completed numerical analysis of the blade rotational angle on the pressure pulsation of a mixed-flow pump and proved that the main frequency of pressure pulsation in the impeller was dominated by shaft frequency or blade frequency. Meng et al. [20] found that the peak efficiency of the runner with splitter blades in an ultra-high head turbine is higher than that of the impeller without splitter blades. Xia et al. [21] explained that the rotating stall of the impeller leads to rotor–dynamic instabilities and causes strong vibrations when the pump-turbine is in pump mode. Wu et al. [22] investigated the pressure pulsation in a centrifugal pump by experiment. The results showed that an unstable separation vortex in the impeller flow passages causes the frequency to be located in the range from 100 to 145 Hz. Shi et al. [23] studied the pressure pulsation characteristics in a full tubular pump under different flow conditions and found that the dominant frequency of the pressure pulsation decreases as the flow rate increases. Zhang et al. [24] found that the rotor–stator interface-induced component of the vaneless area pressure pulsation can be effectively decreased by enlarging the vaneless area. Song et al. [25] performed an experimental investigation of pressure pulsation in an axial flow pump. The results showed that the floor-attached vortex can induce a low-frequency pulsation of 2.12 Hz, which let the amplitudes of pressure pulsation be larger than those without the floor-attached vortex.

At present, the research on the pressure pulsation in the pump is mostly based on the static condition of the volute, and the research on the pressure pulsation under the passive rotation of the volute is still rare. Therefore, in order to reveal the new 'rotor–rotor' interference mechanism between the high-speed rotation of the impeller and the low-speed passive rotation of the volute, this paper studies the pressure pulsation characteristics of the double-tongue volute under passive rotation at different impeller speeds through an innovative rotating pressure test device. In addition, the pressure pulsation characteristics of the single-tongue volute and the double-tongue volute at rest are compared and studied, which provides a reference for the design and optimization of the open pump.

2. Open Test Rig Setup for Dishwasher Pump

As shown in Figure 1, the open test setup of the dishwasher pump was set at the National Research Center of Pumps, Jiangsu University. The test rig was inside a transparent water tank in order to observe the working state of the dishwasher more conveniently, and the tank is 400 × 440 × 400 mm. The shaft of the motor at the bottom of the water tank connects the impeller to control the operation of the dishwasher pump.

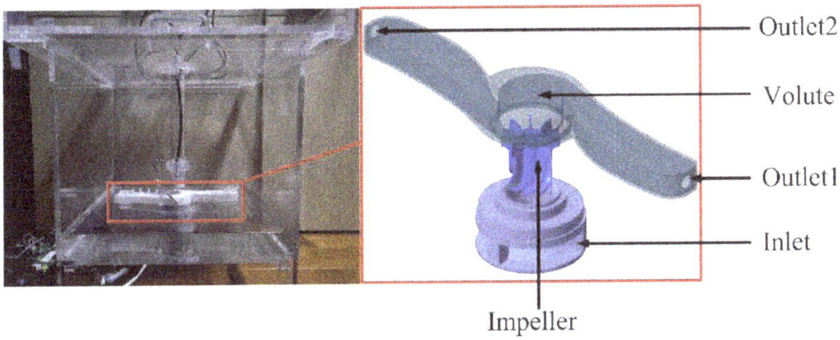

Figure 1. Open test rig of the dishwasher pump.

The dishwasher pump consists of a compound impeller with eight blades and a spray arm of the volute type. The bottom end of the compound impeller is a forward-curved axial flow blade, the top is a radial centrifugal blade, and there is no obvious front shroud and back shroud. The volute had two symmetrical flow channels that helped the spray arm rotate. In particular, an unsealed connection between the impeller and the volute is also designed to enhance the cleaning effect of the dishwasher by passively rotating the spray arm. The specific design parameters of the dishwasher pump are shown in Figure 2 and listed in Table 1. The flow rate Q and head H under design point were 55 L/min and 2 m, respectively. The impeller speed n was designed as 3000 rpm.

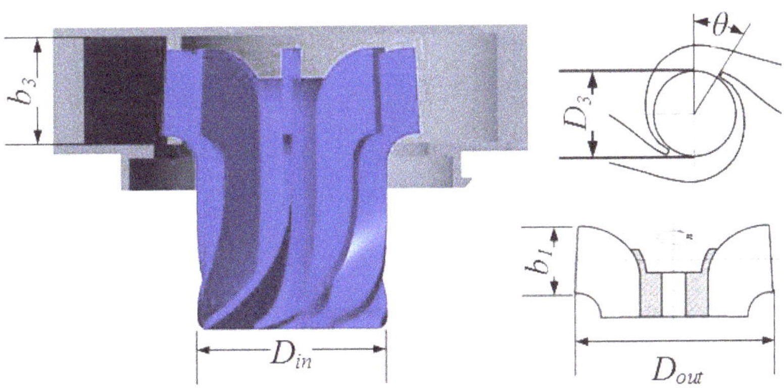

Figure 2. Dimension labeling of dishwasher pump.

Table 1. Essential parameters of the dishwasher pump.

	Parameter	Value
Impeller	Impeller inlet diameter D_{in} (mm)	32
	Impeller outlet width b_1 (mm)	14.15
	Impeller outlet diameter D_{out} (mm)	43.3
	Number of blades Z	8
Volute	Volute inlet width b_3 (mm)	17.8
	Volute base circle diameter D_3 (mm)	46
	Volute tongue angle θ (°)	34

3. Experimental Tests

Figure 3 shows a schematic diagram of data acquisition and the transient pressure test setting of the dishwasher pump. The whole test system consists of two parts. The first system was to change the speed of the impeller. By adjusting the speed control button on the panel, the panel will feedback a number. One hundred times, the number is the rotating speed of the impeller, and the unit is rpm. The second system was to measure the transient pressures of the dishwasher pump. The pressure pulsation sensors (SCYG314) produced by Senno Sci-tech Co., Ltd. were supported by a 15 V DC power source. The accuracy of each pressure sensor was 0.2%, and the range of measurement was 0–20 kpa. The sensors were connected to the data acquisition unit by a converter, and the computer accepted the converted current signal by a data line. The test software of pressure pulsation is smart and integrates functions that can change the sampling time and frequency. In order to ensure the accuracy of the test, there was zero calibration of the sensor before the formal test. The sampling frequency fs was set at 10,000 Hz, and the data were acquired for 2 s when the dishwasher pump reached a stable condition. The sampling frequency is much larger than blade frequency and satisfies Shannon's sampling law [26–29]. In addition, in order to solve the problem of the pressure pulsation test in the process of volute rotation, the rotating pressure pulsation test device was designed. The top of the device is bolted to the cover of the water tank, and the sensor probe extending from the bottom of the device is placed in a preset monitoring hole on the volute. The part of the device near the sensor probe can be rotated by the electricity slip ring, which can not only ensure the normal electricity consumption of the sensor but also effectively avoid the wire winding problem in the rotation test process. The whole device was waterproofed to ensure the normal operation of the sensor.

In order to verify that the test repeatability could be reliable, repeated multiple measurements for the pressure pulsation of monitoring point p1 were carried out. According to the obtained data, the time domain curves of the pressure pulsation of the three tests are drawn, as shown in Figure 4a, and they are periodic distribution in the time of the impeller rotating for two cycles, and the curve trend is basically the same. It is clearly shown that the maximum deviation of average pressure is 1.9% from Figure 4b.

The positions of pressure pulsation monitoring points of the volute of the dishwasher pump are shown in Figure 5. The pressure monitoring point p1 is on the side of the volute, and it is positive in the impeller outlet direction. The pressure monitoring points p2, p3, p4, p5, and p6 were distributed in the top of the volute, and in turn, it extends from the tongue position of the volute to the outlet of the volute.

To investigate the influence of the impeller speeds on the pressure pulsation in the dishwasher pump, five different speeds of the impeller were modified by the impeller speed control system, which was previously mentioned in Figure 3. As shown in Figure 6, in order to further study the influence of the volute form and the passive rotation of the volute on the pressure pulsation, the pressure pulsation tests were carried out on the single-tongue volute and the double-tongue volute at rest and compared with the double-tongue volute rotation model. In order to make the comparative test more scientific and rigorous, each pressure pulsation monitoring point position of three different volute models was consistent during the test.

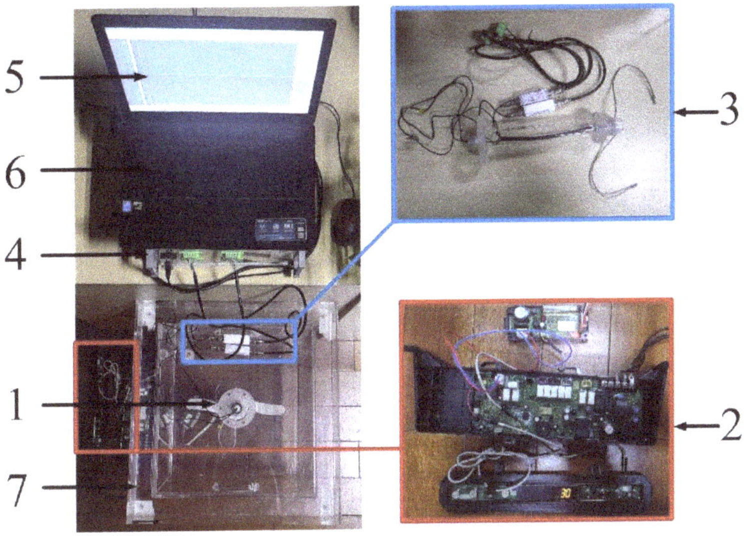

Figure 3. Dishwasher pump test system. 1: dishwasher pump, 2: impeller speed regulator, 3: rotating pressure pulsation test device, 4: data-acquisition system, 5: software interface, 6: personal computer, and 7: water tank.

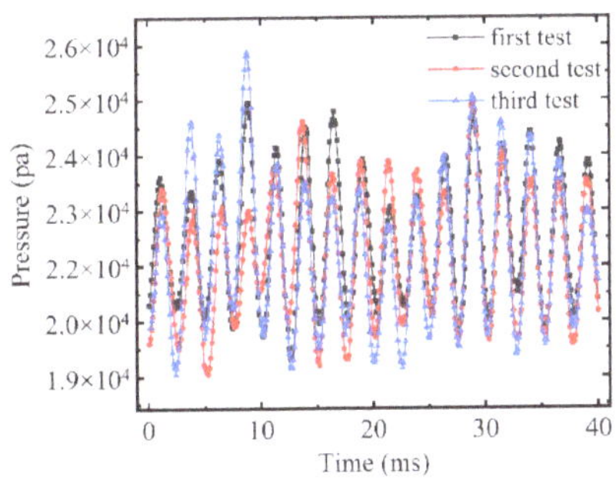

(**a**)

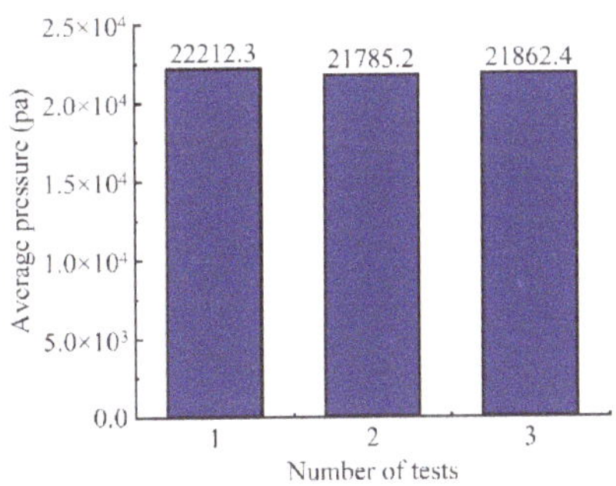

(**b**)

Figure 4. Three pressure pulsation tests. (**a**) Pressure pulsation curves at three tests; (**b**) comparison of average pressure at three tests.

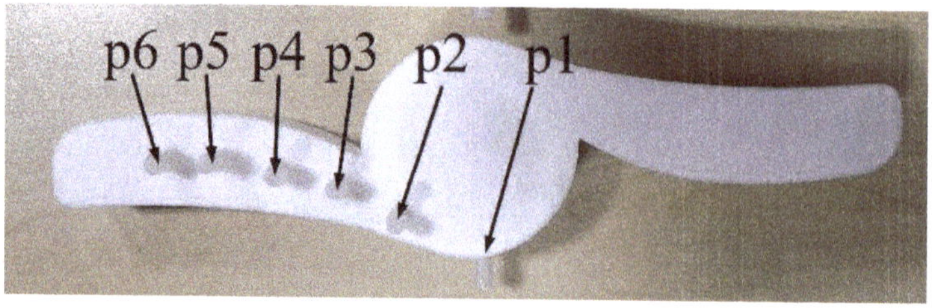

Figure 5. Pressure pulsation monitoring points positions.

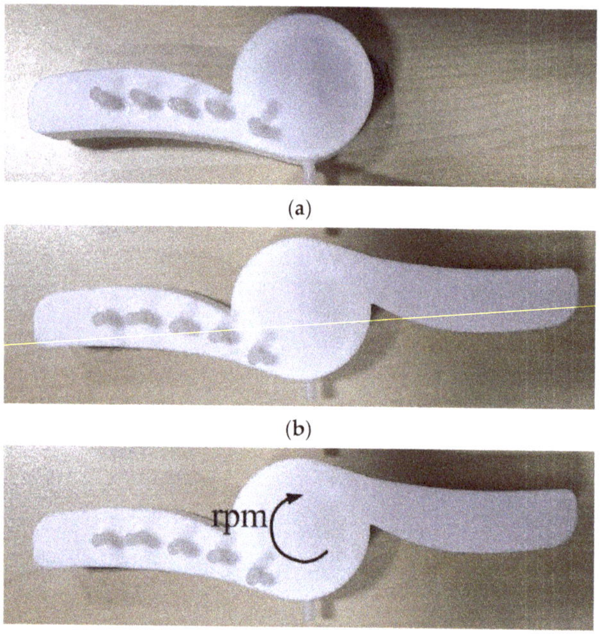

(**a**)

(**b**)

(**c**)

Figure 6. Tested volutes. (**a**) single-tongue volute; (**b**) double-tongue volute (static); (**c**) double-tongue volute (rotate).

4. Result and Discussion

To facilitate the normalization of pressure pulsation data, the time–frequency analysis method was introduced to describe the pressure pulsation. The method of time–frequency can not only give the frequency of the pressure pulsation but also show the information about the frequency domain representation changes with time. This method is using a short-time Fourier transform (STFT) to transform the time domain signal of pressure pulsation [30–32]. The source signal is divided into several small signal segments by the window function. Firstly, the signal of each segment is converted to one-dimensional by Fourier transform, and then the two-dimensional time–frequency diagram is obtained by the translation of the window function.

$$S_x(t, f) = \int_t [x(t) \times w(t - t')] \times e^{-j2\pi f t} dt \tag{1}$$

where $x(t)$ is the source signal, and $w(t - t')$ is the window function. The window function is Hanning.

4.1. Pressure Pulsation Analysis with Different Volutes

4.1.1. Time–Frequency Domain Analysis of Pressure Pulsation

Figure 7 shows the time–frequency domain of pressure pulsation for three volutes at six different monitoring positions. It shows the dynamic pressure pulsation is unsteady and dependent on time. The main frequency of pressure pulsation of the single-tongue volute is 1 times blade frequency, while the main frequency of pressure pulsation of the double-tongue volute is 2 times blade frequency, and the secondary frequency is 1 times blade frequency, indicating that the static and dynamic interference is the root cause of pressure pulsation. The amplitude pulsation of pressure pulsation of the double-tongue volute at 1000 Hz is also strong at monitoring points p1, p2, and p6. Compared with the stationary double-tongue volute, the amplitude of the double-blade frequency of the rotating double-tongue volute will reach a maximum between 400 and 1200 ms, which is higher than the amplitude of the main frequency at other times (Figure 7i,l). This is probably caused by the new dynamic interference between the volute and the rotating impeller, which is also used as a rotating component.

4.1.2. Comparative Analysis of Pressure Fluctuation Average Pressure

Figure 8 shows a comparison of the average pressure at different pressure pulsation monitoring points when the dishwasher pump with three different types of volute. It can be seen that the average pressure of the single-tongue volute is the largest at the monitoring point p4, while that of the double-tongue volute reaches the maximum at the monitoring point p3. In three different volute models, the average pressure of each monitoring point is the largest at the same impeller speed due to the minimum flow passage of the single-tongue volute. Similarly, in the process of passive rotation, the double-tongue volute can effectively alleviate the squeezing effect of water flow on the volute, making the average pressure minimum. The variation trend of average pressure of three volute models from p1 to p3 at the monitoring point is the same. However, the change of the rotating double-tongue volute from p3 to p6 at the monitoring points is more smooth, and the maximum deviation of the average pressure is 1.4%. The maximum deviations of average pressure of the single-tongue volute and static double-tongue volute are 4.5% and 4.9%, respectively. This is because when the double-tongue volute rotates, the flow pattern in the volute channel is improved, and the pressure distribution is uniform.

4.1.3. Dominant Frequency Amplitudes of the Pressure Pulsation

The dominant frequency amplitudes of the pressure pulsation for three volutes at different measuring points are shown in Figure 9. On the double-tongue volute, the amplitude of dominant frequency of the pressure pulsation increases first and then decreases from p1 to p6 at the monitoring point and reaches the maximum at the monitoring point p2. The amplitude of dominant frequency of pressure pulsation at different monitoring points of the single-tongue volute has no obvious regularity, and the amplitude of p1 reaches the maximum at the monitoring point. This is because the experiment is carried out at the impeller speed of 3000 rpm, which is much higher than the design speed of the single-tongue volute. The leakage vortex at the outlet of the impeller disturbs the flow field around the monitoring point p1, changes the stable structure of the flow field at the outlet of the impeller, and leads to severe pressure pulsation in the flow field. The larger impeller speed increases the flow velocity in the vortex chamber and deteriorates the flow pattern, which is the reason for the irregular amplitude of the dominant frequency of the pressure pulsation in the single-tongue volute.

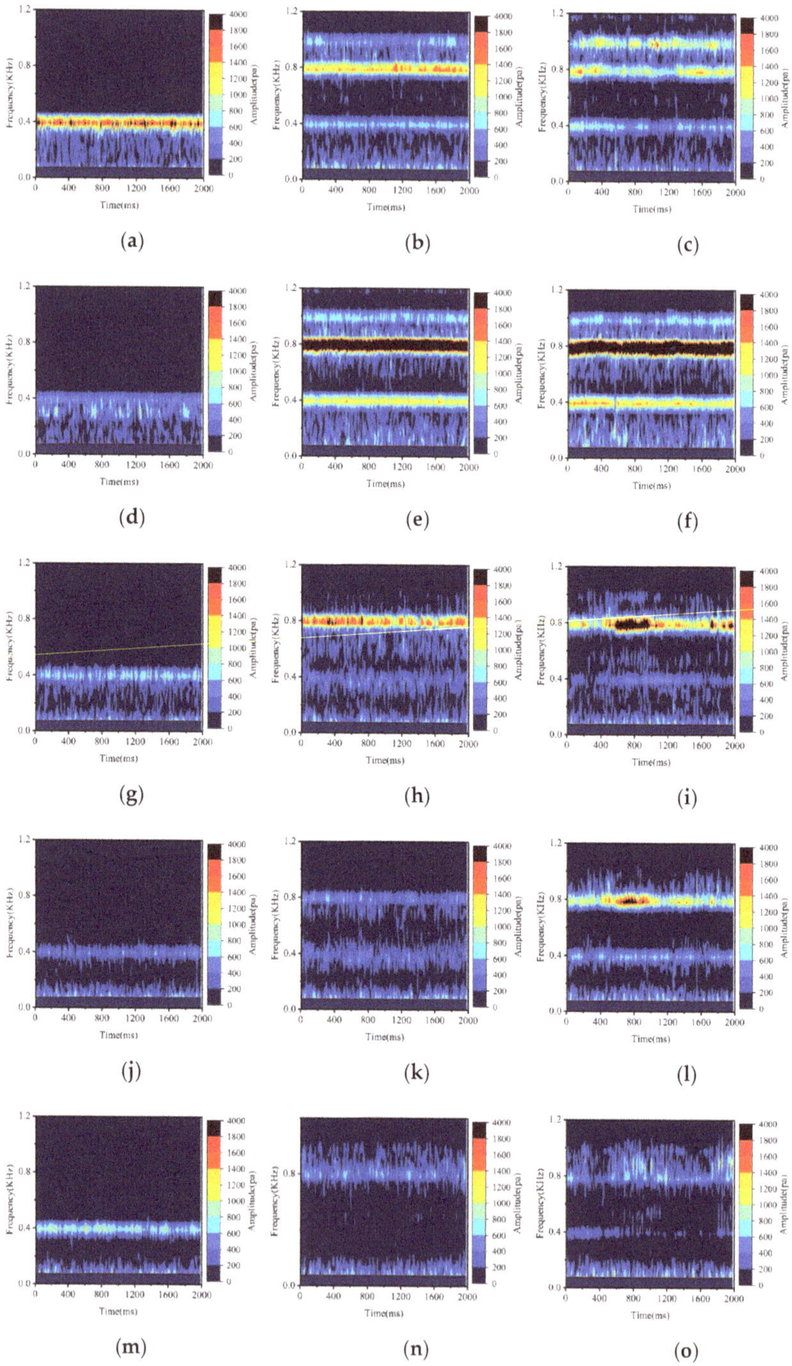

Figure 7. *Cont.*

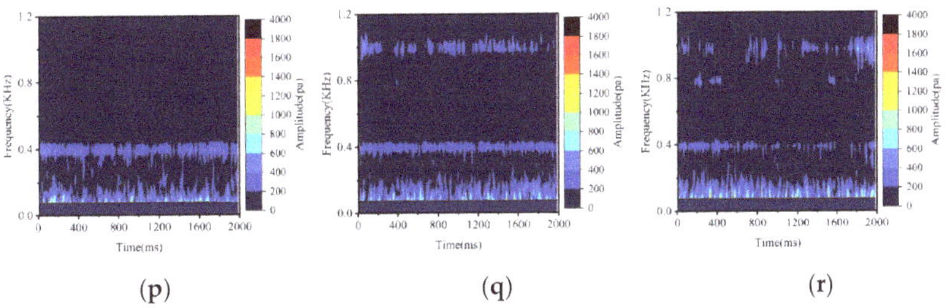

Figure 7. Time–frequency domain of pressure pulsation for three volutes at different positions. (**a**) p1 at single-tongue volute; (**b**) p1 at double-tongue volute (static); (**c**) p1 at double-tongue volute (rotate); (**d**) p2 at single-tongue volute; (**e**) p2 at double-tongue volute (static); (**f**) p2 at double-tongue volute (rotate); (**g**) p3 at single-tongue volute; (**h**) p3 at double-tongue volute (static); (**i**) p3 at double-tongue volute (rotate); (**j**) p4 at single-tongue volute; (**k**) p4 at double-tongue volute (static); (**l**) p4 at double-tongue volute (rotate); (**m**) p5 at single-tongue volute; (**n**) p5 at double-tongue volute (static); (**o**) p5 at double-tongue volute (rotate); (**p**) p6 at single-tongue volute; (**q**) p6 at double-tongue volute (static); (**r**) p6 at double-tongue volute (rotate).

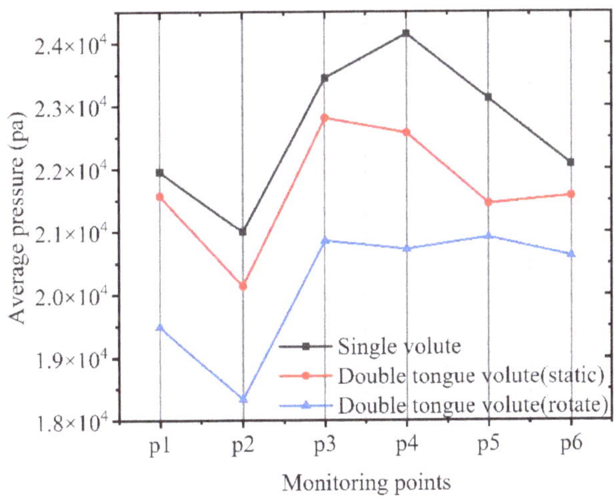

Figure 8. The variation trend of average pressure with different volutes.

4.2. Pressure Pulsation Analysis at Different Impeller Speeds

4.2.1. Time–Domain Analysis of Pressure Fluctuation

Pressure pulsation appears when the rotating impeller blades sweep through the tongue of the volute. Apparently, the number of blades skimming over the volute tongue and the structure of the volute play an important role in the change of pressure pulsation. Figure 10 shows the time domain variation of pressure pulsation at p1 for three volutes at different impeller speeds. With the decrease of impeller speeds, the pressure peak value and the amplitudes of pressure pulsation decrease. When the impeller rotates at 3000, 2500, and 2000 rpm, the variations of pressure pulsation with different volutes all have certain periodicity. When the impeller rotates at 1500 and 1000 rpm, the variations of pressure pulsation with different volutes all have certain periodicity, the pressure fluctuates violently, and the pressure distribution is uneven. This is because, at low rotational speeds, the pressure and flow rate of the blade acting on the flow will decrease, which keeps air in the pump chamber. The coupling of water and air makes the flow field in the dishwasher pump more complex, and the flow pattern becomes unstable, which lets the time domain curve of pressure pulsation be uneven.

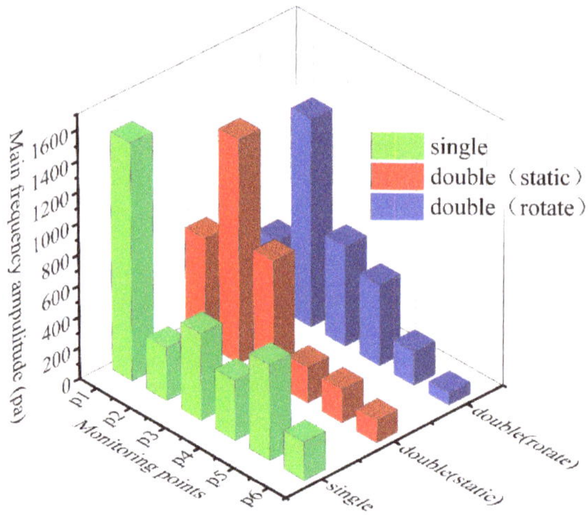

Figure 9. Dominant frequency amplitudes of pressure pulsation for three volutes at different measuring points.

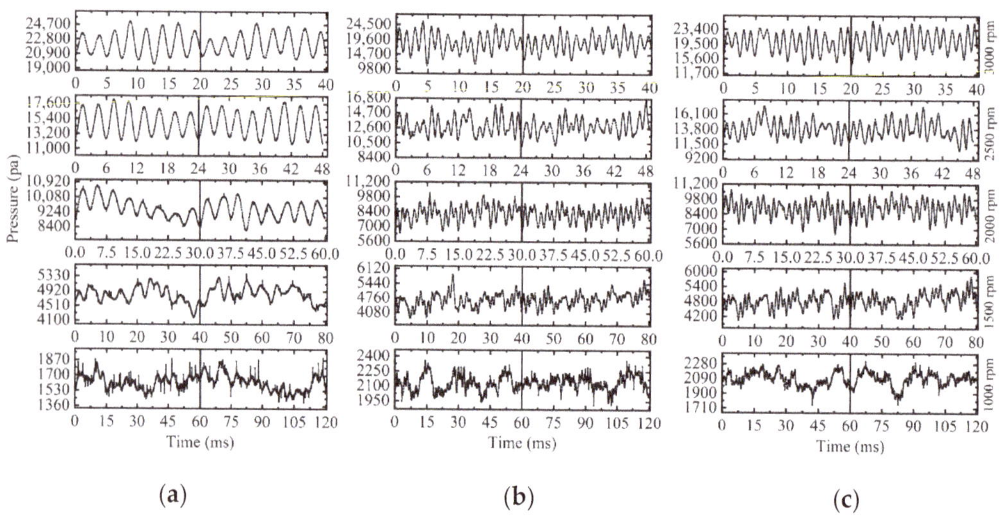

Figure 10. Time domain of pressure pulsation for three volutes at different impeller speeds. (**a**) Single-tongue volute; (**b**) double-tongue volute (static); (**c**) double-tongue volute (rotate).

As shown in Figure 10a, the curves of pressure pulsation have eight peaks and troughs when the impeller rotates at 3000, 2500, and 2000 rpm, and the dishwasher pump has a single-tongue volute. The number of cycles corresponds to the number of blades [33–35]. In addition, from Figure 10b,c, it is apparent that the curves of pressure pulsation have 16 peaks and troughs when the pump has a double-tongue volute and the impeller is 3000 and 2500 rpm. It is the same as the number of volute tongues multiplied by the number of blades. It is explained that the rotor–stator interference between the impeller and volute is the fundamental cause of pressure pulsation. Compared with the pump double-tongue volute under static and rotating conditions, the amplitudes of pressure pulsation of the pump with static volute are significantly higher than those of the pump with a rotating volute at different impeller speeds.

4.2.2. Frequency–Domain Analysis of Pressure Fluctuation

The shaft frequency (f_n) and blade frequency (f_{BPF}) of the pump are determined by the rotational speeds of the impeller (n), and blade frequency depends on the axial frequency and blade number. Equation (2) lists the relationship between the f_n (kHz), f_{BPF} (kHz) and n (rpm) [36–38].

$$\begin{aligned} f_n &= 1/\tfrac{60}{n} \\ f_{BPF} &= 8f_n \end{aligned} \quad (2)$$

The calculated parameters of the pump at different impeller speeds are given in Table 2.

Table 2. Shaft frequency and blades frequency at different impeller speeds.

Variable	Frequency/kHz				
	3000 rpm	2500 rpm	2000 rpm	1500 rpm	1000 rpm
f_n	0.05	0.042	0.033	0.025	0.017
f_{BPF}	0.4	0.333	0.267	0.2	0.133

To obtain frequency domain curves of pressure pulsation at different impeller speeds, pressure pulsation data of monitoring point p1 were transformed by FFT [39–41]. Table 3 shows the dominant frequency of pressure pulsation obtained at different impeller speeds. It can be seen that the dominant frequencies at each impeller speed are distributed in the shaft frequency and multiple blade frequency. For the pump with a single-tongue volute, the dominant frequency of pressure pulsation is blade frequency when the impeller rotates at 3000, 2500, 2000, and 1500 rpm. When the impeller speed is 1000 rpm, the dominant frequency turns to shaft frequency, and the dominant frequency decreases with the decrease of impeller speed. For the pump with the double-tongue volute, the dominant frequency is concentrated at twice the blade frequency when the impeller rotates at 3000 and 2500 rpm. When the impeller speeds are 2000, 1500, and 1000 rpm, the dominant frequency is double-blade frequency, triple-blade frequency, and shaft frequency, respectively.

Table 3. Dominant frequency of pressure pulsation at different impeller speeds.

Volute Condition	Frequency/kHz				
	3000 rpm	2500 rpm	2000 rpm	1500 rpm	1000 rpm
Single-tongue volute	0.396	0.339	0.265	0.024	0.016
Double-tongue volute (static)	0.791	0.662	0.793	0.199	0.016
Double-tongue volute (rotate)	0.784	0.658	0.793	0.198	0.016

Figure 11 shows the frequency domain variation of pressure pulsation at p1 for three volutes at different impellers speeds. By comparing (a) with (b) in Figure 11, it is clear that the dominant frequency of the double-tongue volute at different impeller speeds is higher than that of the single-tongue volute. In addition, the amplitude range of the double-tongue volute is obviously larger than that of the single-tongue volute. This is because, compared with the single-tongue volute, the double-tongue volute has two symmetrical chambers, and the flow pattern becomes more complex, which causes the pressure pulsation of the flow field more violent. As shown in Figure 11b,c, we can find that the frequency domain variation of the rotating double-tongue volute under the same impeller speeds has little difference to that of the static double-tongue volute.

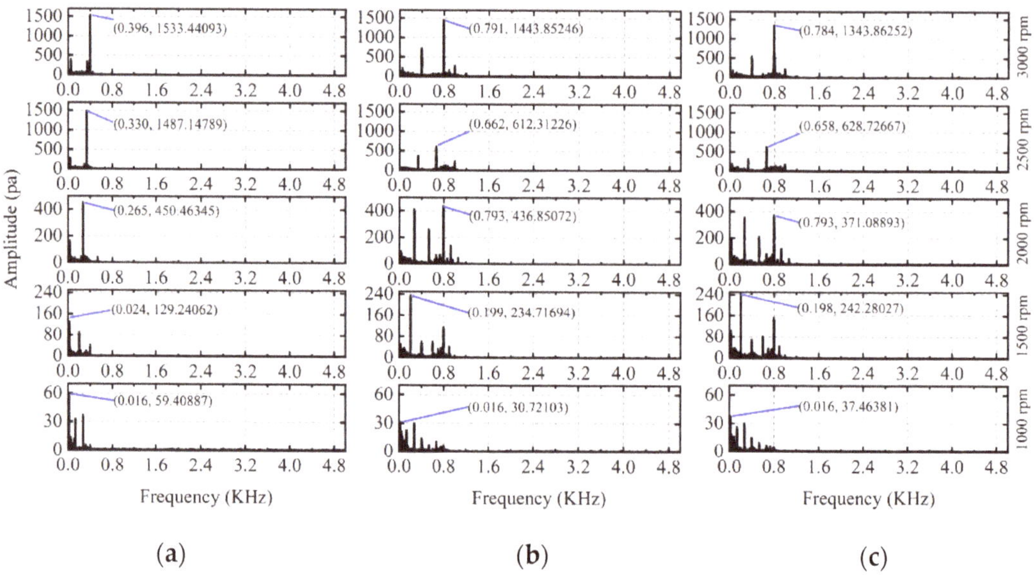

Figure 11. Frequency domain of pressure pulsation for three volutes at different impeller speeds. (**a**) Single-tongue volute; (**b**) double-tongue volute (static); (**c**) double-tongue volute (rotate).

4.2.3. Dominant Frequency Amplitudes of the Pressure Pulsation

The dominant frequency amplitudes of the pressure pulsation are carried out to evaluate the pressure pulsation test to explore the influence of the impeller speeds on the pressure pulsation in the dishwasher pump. Figure 12 shows the dominant frequency amplitudes of the pressure pulsation at different speeds in the pump with three different volutes' conditions. The amplitude of dominant frequency of pressure pulsation of three different volutes decreases with the decrease of impeller speeds. This is because reducing the speed of the impeller makes the pressure of the impeller on the flow decrease and the energy of the pressure pulsation decrease. Compared with the double-tongue volute, when the impeller speed is greater than 2500 rpm, the amplitude of the dominant frequency of the pressure pulsation is not obvious. When the impeller speed is 3000 rpm, the amplitude of the single-tongue volute increases by only 3%, and the amplitude of the double-tongue volute increases by 57% and 53%, respectively, when it is static and rotating. This is because the design flow rate of the double-tongue volute is higher than that of the single-tongue volute at the same impeller speed, and the dishwasher uses an open pump. When the impeller speed is greater than the critical speed, the water will leak out from the impeller outlet and the volute connection. The dishwasher uses an open pump. When the impeller speed is greater than the critical speed, the water flow will leak out from the impeller outlet and the volute connection, which makes the pressure in the pump not change greatly.

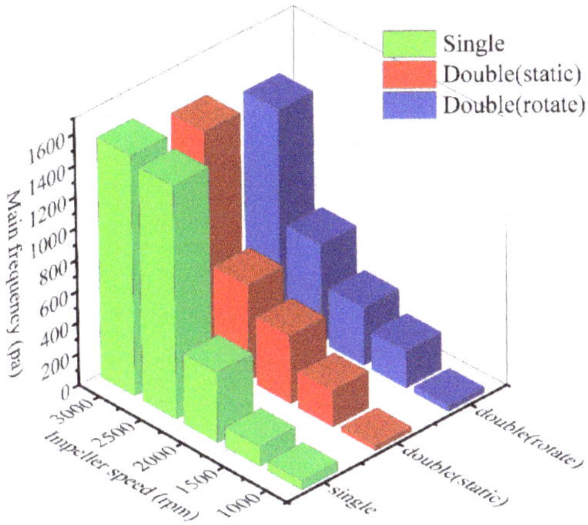

Figure 12. Dominant frequency amplitudes of pressure pulsation for three volutes at different impeller speeds.

5. Conclusions

Experimental tests were carried out to analyze the effects of the single-tongue volute, the static double-tongue volute, and the rotating double-tongue volute on the pressure pulsation characteristics of the dishwasher pump. The pressure pulsation characteristics caused by rotor–stator interference and rotor–rotor interference of the dishwasher pump were analyzed by time domain analysis, frequency domain analysis, and time–frequency domain analysis.

(1) Compared with the single-tongue volute and the static double-tongue volute, the average pressure of each monitoring point in the dishwasher pump with a rotating double-tongue volute is the smallest, and from the pressure monitoring point from p3 to p6, the average pressure changes more gently, and the maximum deviation value is only 1.4%. The amplitude of the main frequency of the pressure pulsation of the double-tongue volute increases first and then decreases from the monitoring point p1 to p6 and reaches the maximum at the monitoring point p2. However, the amplitude of the main frequency of the single-tongue volute changes irregularly, and the maximum amplitude appears at the monitoring point p1. Therefore, for the dishwasher, the volute with the passive rotation has the best results, which can not only reduce the pressure and reduce vibration noise but also perform all-around cleaning, improving dishwasher efficiency.

(2) When the impeller rotates at 3000 rpm, the number of peaks and valleys of the pressure pulsation time domain curve of the single-tongue volute is eight, while that of the double-tongue volute is two. The main frequency of the single-tongue volute is 0.396 kHz, which is concentrated near the blade frequency. The main frequencies of the static double-tongue volute and the rotating double-tongue volute are 0.791 kHz and 0.786, respectively, which are concentrated near the double-blade frequency.

(3) Under the high impeller speeds of 2000, 2500, and 3000 rpm, the pressure pulsation time domain curves present periodicity, and the main frequencies are given priority with the blade frequency and integer times of the blade frequency. At the low impeller speeds of 1000 and 1500 rpm, the pressure pulsation becomes disordered, the periodicity of the time-domain curves disappears, and the main frequency is mainly axial frequency.

Author Contributions: Writing—draft preparation, Y.Z.; writing—review and editing, Y.L.; methodology, Y.L.; software, J.Z.; conceptualization, J.Z. and Y.L.; formal analysis, J.Z. and Y.L.; funding acquisition, J.Z. and Y.L.; project administration, H.X. and F.Z.; supervision, P.H., H.X. and F.Z. All authors have read and agreed to the published version of the manuscript.

Funding: This work is supported by the project of National Natural Science Foundation of China (No. 51809120), the Natural Science Foundation of Jiangsu Province (No. BK20180871), the Project Funded by China Postdoctoral Science Foundation (No. 2018M640462), the Natural Science Foundation of the Jiangsu Higher Education Institutions of China (No. 18KJB470005), the Key Research and Development Plan Project of Jiangsu Province (No. BE2019009), and A Project Funded by the Priority Academic Program Development of Jiangsu Higher Education Institutions (PAPD).

Institutional Review Board Statement: Not applicable.

Informed Consent Statement: Not applicable.

Data Availability Statement: Not applicable.

Conflicts of Interest: The authors declare no conflict of interest.

References

1. Zhang, L.; Wang, C.; Zhang, Y.C. Numerical study of coupled flow in blocking pulsed jet on a rotating wall. *J. Braz. Soc. Mech. Sci.* **2021**, *43*, 508. [CrossRef]
2. Wang, H.L.; Qian, Z.D.; Zhang, D. Numerical study of the normal impinging water jet at different impinging height, based on Wray–Agarwal turbulence model. *Energies* **2020**, *13*, 1744. [CrossRef]
3. Li, D.Y.; Wang, H.J.; Qin, Y.L. Numerical simulation of hysteresis characteristic in the hump region of a pump- turbine model. *Renew. Energy* **2018**, *115*, 438–447. [CrossRef]
4. Li, D.Y.; Wang, H.J.; Qin, Y.L. Mechanism of high amplitude low frequency fluctuations in a pump-turbine in pump mode. *J. Mech. Sci. Technol.* **2018**, *126*, 668–680. [CrossRef]
5. Xia, L.S.; Cheng, Y.G.; Cai, F. Pressure pulsation characteristics of a model pump-turbine operating in the S-shaped region: CFD simulations. *Int. J. Fluid Mach. Syst.* **2017**, *10*, 287–295. [CrossRef]
6. Tang, S.N.; Zhu, Y.; Yuan, S.Q. An improved convolutional neural network with an adaptable learning rate towards multi-signal fault diagnosis of hydraulic piston pump. *Adv. Eng. Inform.* **2021**, *50*, 101406. [CrossRef]
7. Zhu, Y.; Li, G.P.; Wang, R. Intelligent fault diagnosis of hydraulic piston pump combining improved LeNet-5 and PSO hyperparameter optimization. *Appl. Acoust.* **2021**, *183*, 108336. [CrossRef]
8. Jin, J.M.; Jung, K.J.; Kim, Y.H. Effects of gap between impeller and volute tongue on a pressure fluctuation in double-suction pump. *J. Mech. Sci. Technol.* **2020**, *34*, 4879–4903. [CrossRef]
9. Alemi, H.; Nourbakhsh, S.A.; Raisee, M. Effect of the volute tongue profile on the performance of a low specific speed centrifugal pump. *Proc. Inst. Mech. Eng. Part A J. Power Energy* **2015**, *229*, 210–220. [CrossRef]
10. Patil, S.R.; Chavab, S.T.; Jadhav, N.S. Effect of volute tongue clearance variation on performance of centrifugal blower by numerical and experimental analysis. *Mater. Today* **2018**, *5*, 3883–3894. [CrossRef]
11. Shim, H.S.; Afzal, A.; Kim, K.Y. Three-objective optimization of a centrifugal pump with double volute to minimize radial thrust at off-design conditions. *Proc. Inst. Mech. Eng. Part A J. Power Energy* **2016**, *230*, 598–615. [CrossRef]
12. Yang, S.H.; Kong, F.Y.; Chen, B. Research on pump volute design method using CFD. *Int. J. Rotating Mach.* **2011**, *2011*, 137860. [CrossRef]
13. Wang, H.L.; Long, B.; Wang, C. Effects of the impeller blade with a slot structure on the centrifugal pump performance. *Energies* **2020**, *13*, 1628. [CrossRef]
14. Zhou, J.R.; Zhao, M.M.; Wang, C. Optimal design of diversion piers of lateral intake pumping station based on orthogonal test. *Shock Vib.* **2021**, *2021*, 6616456. [CrossRef]
15. Wang, H.L.; Hu, Q.X.; Yang, Y. Performance differences of electrical submersible pump under variable speed schemes. *Int. J. Simul. Model.* **2021**, *20*, 76–86. [CrossRef]
16. Shi, L.J.; Zhu, J.; Tang, F.P. Multi-Disciplinary optimization design of axial-flow pump impellers based on the approximation model. *Energies* **2020**, *13*, 779. [CrossRef]
17. Li, Q.Q.; Li, S.Y.; Wu, P. Investigation on reduction of pressure fluctuation for a double-suction centrifugal pump. *Chin. J. Mech. Eng.* **2021**, *34*, 192–209. [CrossRef]
18. Zhang, J.F.; Appiah, D.; Zhang, F. Experimental and numerical investigations on pressure pulsation in a pump mode operation of a pump as turbine. *Energy Sci. Eng.* **2019**, *7*, 1264–1279. [CrossRef]
19. Tan, L.; Yu, Z.Y.; Xu, Y. Role of blade rotational angle on energy performance and pressure fluctuation of a mixed-flow pump. *Proc. Inst. Mech. Eng. Part A J. Power Energy* **2017**, *231*, 227–238.
20. Meng, L.; Zhang, S.P.; Zhou, L.J. Study on the pressure pulsation inside runner with splitter blades in ultra-high head turbine. *IOP Conf. Ser. EES* **2014**, *22*, 032012. [CrossRef]

21. Xia, L.S.; Cheng, Y.G.; Zhang, X.X. Numerical analysis of rotating stall instabilities of a pump- turbine in pump mode. *IOP Conf. Ser. EES* **2014**, *22*, 032020. [CrossRef]
22. Wu, D.H.; Ren, Y.; Mou, J.G. Investigation of pressure pulsations and flow instabilities in a centrifugal pump at part-load conditions. *Int. J. Fluid Mach. Syst.* **2017**, *10*, 355–362. [CrossRef]
23. Shi, L.J.; Yuan, Y.; Jiao, H.F. Numerical investigation and experiment on pressure pulsation characteristics in a full tubular pump. *Renew. Energy* **2021**, *163*, 987–1000. [CrossRef]
24. Zhang, F.; Lowys, P.Y.; Houdeline, J.B. Pump-turbine rotor-stator interaction induced vibration: Problem resolution and experience. *IOP Conf. Ser. EES* **2021**, *774*, 012124. [CrossRef]
25. Song, X.J.; Liu, C. Experimental investigation of pressure pulsation induced by the floor-attached vortex in an axial flow pump. *Adv. Mech. Eng.* **2019**, *11*, 1–13. [CrossRef]
26. Liu, Y.P.; Li, X.J.; Li, B.W. Influence of impeller sinusoidal tubercle trailing-edge on pressure pulsation in a centrifugal pump at nominal flow rate. *J. Fluids Eng.* **2021**, *143*, 091205.
27. Duan, X.H.; Tang, F.P.; Duan, W.Y.; Zhou, W.; Shi, L. Experimental investigation on the correlation of pressure pulsation and vibration of axial flow pump. *Adv. Mech. Eng.* **2019**, *11*, 1–11. [CrossRef]
28. Ni, D.; Zhang, N.; Gao, B. Dynamic measurements on unsteady pressure pulsations and flow distributions in a nuclear reactor coolant pump. *Energy* **2020**, *198*, 117305. [CrossRef]
29. Zhang, N.; Yang, M.G.; Gao, B. Experimental and numerical analysis of unsteady pressure pulsation in a centrifugal pump with slope volute. *J. Mech. Sci. Technol.* **2015**, *29*, 4231–4238. [CrossRef]
30. Wang, W.J.; Pei, J.; Yuan, S.Q. Experimental investigation on clocking effect of vaned diffuser on performance characteristics and pressure pulsations in a centrifugal pump. *Energy* **2018**, *90*, 286–298. [CrossRef]
31. Allen, J.B. Short term spectral analysis, synthesis, and modification by discrete Fourier transform. *IEEE Trans. Acoust. Speech Signal Process.* **1977**, *25*, 235–238. [CrossRef]
32. Sun, T.Z.; Zhang, X.S.; Zhang, J.Y. Experimental study on the unsteady natural cloud cavities: Influence of cavitation number on cavity evolution and pressure pulsations. *J. Mar. Sci. Eng.* **2021**, *9*, 1-13. [CrossRef]
33. Wang, C.; Zhang, Y.X.; Li, Z.W. Pressure fluctuation–vortex interaction in an ultra-low specific-speed centrifugal pump. *J. Low Freq. Noise Vib. Act. Control.* **2019**, *38*, 527–543. [CrossRef]
34. Bai, L.; Zhou, L.; Han, C.; Zhu, Y.; Shi, W. Numerical study of pressure fluctuation and unsteady flow in a centrifugal pump. *Processes* **2019**, *7*, 354. [CrossRef]
35. Bosioc, A.I.; Tanasa, C.; Muntean, S. Unsteady pressure measurements and numerical investigation of the jet control method in a conical differ with swirling flow. *IOP Conf. Ser. EES* **2010**, *12*, 012017. [CrossRef]
36. Zhou, Q.; Zhao, X.W.; Pei, L. Investigation on pressure pulsation and modal behavior of the impeller in a nuclear reactor coolant pump. *Energy Sci. Eng.* **2021**, *9*, 1440–1449. [CrossRef]
37. Yang, J.; Liu, J.; Liu, X.H. Numerical study of pressure pulsation of centrifugal pumps with the compressible mode. *J. Therm. Sci.* **2019**, *28*, 106–114. [CrossRef]
38. Gao, Y.S.; Cheng, J.; Huang, J.H. Simulation analysis and experiment of variable-displacement asymmetric axial piston pump. *Appl. Sci.* **2017**, *7*, 328. [CrossRef]
39. Yang, Y.; Zhou, L.; Shi, W.D. Interstage difference of pressure pulsation in a three-stage electrical submersible pump. *J. Petrol. Sci. Eng.* **2021**, *196*, 107683. [CrossRef]
40. Zhou, Q.; Li, H.K.; Pei, L. Research on non-uniform pressure pulsation of the diffuser in a nuclear reactor coolant pump. *Nucl. Eng. Technol.* **2020**, *53*, 1020–1028. [CrossRef]
41. Muntean, S.; Tanasa, C.; Bosioc, A.I. Investigation of the plunging pressure pulsation in a swirling flow with precessing vortex rope in a straight diffuser. *IOP Conf. Ser. EES* **2016**, *49*, 082010. [CrossRef]

MDPI
St. Alban-Anlage 66
4052 Basel
Switzerland
Tel. +41 61 683 77 34
Fax +41 61 302 89 18
www.mdpi.com

Machines Editorial Office
E-mail: machines@mdpi.com
www.mdpi.com/journal/machines

www.ingramcontent.com/pod-product-compliance
Lightning Source LLC
Chambersburg PA
CBHW061459220926
51850CB00004B/129